T0190276

Magnetic Measurement Techniques for Materials Characterization

Victorino Franco • Brad Dodrill

Editors

Magnetic Measurement Techniques for Materials Characterization

 Springer

Editors
Victorino Franco 🆔
Department of Condensed Matter Physics
University of Seville
Seville, Spain

Brad Dodrill 🆔
Lake Shore Cryotronics
Westerville, OH, USA

ISBN 978-3-030-70445-2   ISBN 978-3-030-70443-8   (eBook)
https://doi.org/10.1007/978-3-030-70443-8

This Springer imprint is published by the registered company Springer Nature Switzerland AG
The registered company address is: Gewerbestrasse 11, 6330 Cham, Switzerland

Preface

Magnetic Materials pervade every aspect of our daily lives, being used in a wide variety of important commercial and technological applications. These range from mobility applications, as an integral part of the power train of personal mobility devices (PMD), to anti-counterfeiting measures implemented in bank notes, passing through the storage of information in multiple formats. The sustainability of our energy-hungry way of life depends on the discovery and use of efficient and environmentally friendly energy conversion devices. Many of these energy conversion processes rely on magnetic materials and their improvement would have a significant influence on society. From a materials perspective, permanent magnetic materials are used in electrical motors, hybrid vehicles, etc. Soft magnetic materials are used for high frequency power electronics, power conditioning, and grid integration systems. Magnetic thin films and multilayers are used in high density recording media. Magneto-caloric effect materials have application in magnetic refrigeration systems. Nanomagnetic materials have applications in spintronics, medical diagnostics, and targeted drug delivery.

The discovery of new materials for any specific application goes through different stages that involve formulation, synthesis, and characterization. And while characterization techniques in quality control laboratories can be performed in a routine way, the use of the same characterization methods in the materials discovery process requires a thorough understanding of the experimental techniques used for their characterization. When the materials properties are not previously known, it is necessary to evaluate if the newly found features are intrinsic to the material, if they are related to the specific characteristics of the sample under study, or if they are the consequence of some peculiarities of the experimental technique being employed. The use of advanced and costly equipment as a "black box," rather extended nowadays among those researchers who start on a new topic, can lead to blunt artifacts in the measurements or erroneous interpretation of the data. While the scientific literature has valuable research articles on the different techniques, it is rare in the literature to find a single source where the most relevant techniques are presented and compared, showing their advantages and limitations.

This book aims to fill this gap and will discuss the most commonly used techniques for characterizing magnetic material properties and will present examples of their application to several relevant magnetic materials.

The target audience of the book is rather broad, including graduate students starting their research on magnetic materials, more senior researchers who want to know more details about the fundamentals of the techniques that they are using or they plan to use, and those who are changing their focus of research into other types of materials and would like to know which are the appropriate techniques for the characterization of the properties that they are interested in.

Magnetism is probably the only field of research in which two systems of units unofficially coexist in the scientific literature. The reluctance to change is not only due to tradition, but also because even the equations of magnetism change in the different systems of units. The recent redefinition of the International System of Units (SI) in 2019 had some fundamental effects on the magnetic constants. Therefore, Part I of the book starts with a chapter on the units for magnetic quantities. We hope that the coexistence of different systems of units in magnetism will soon fade out, with all of us adhering to SI. There are some research subtopics in magnetism where the centimeter-gram-second (CGS) system of units still prevails and, therefore, you will still find such legacy units in some of the chapters of this book.

The rest of the book is structured into five parts. The first four, consisting of 15 chapters, focus on the description of the measurement techniques, showing the underlying principles that make them operate in the way they do and possible limitations for their use. We have grouped the techniques by the principle of measurement. Magnetization is most frequently measured by using inductive- or force-based methods, which are described in Part II. These methods provide average information of the magnetic response of a material. In current technology, it is necessary to have more local information and the ability to visualize the different domains in which magnetic moments arrange; this is dealt by imaging techniques that use visible light, X-ray photons or electrons. These techniques are presented in Part III. There are additional techniques that are for more specific applications, or so general that they do not fit in the previous more restricted approaches. These are grouped in Part IV. Field sensing is necessary in any magnetic technique, but it is also possible to use it to infer the magnetization of a sample by detecting the field produced by it. In this part, we have also included two chapters on neutron scattering targeted to different audiences: the first is a description of the possibilities of neutron techniques oriented to those who are new to the technique, and the second is a thorough description of the fundamentals accompanied by practical examples. While most of the previous techniques are quasistatic or involve relatively low frequencies, technology is continuously evolving towards higher frequencies in the MHz to GHz range. Characterization of materials in these higher frequency ranges yield fundamental information about magnetization dynamics and are described in Part V.

Part VI of the book, consisting of 10 chapters, presents a set of current examples in which the magnetic properties of the materials are relevant, the peculiarities associated with measuring that kind of material or property are discussed, and illustrative results are presented. First-order reversal curves (FORC) has become a go-to technique for the characterization of hysteresis of any kind. In magnetism, it is used in fields as different as geomagnetism and nanomagnetism. Efficient energy conversion using magnetic materials is related to soft magnetic materials, permanent magnets, and magneto-caloric materials. In this latter case, it is shown that magneto-caloric characterization is, by itself, a tool to study phase transitions even for materials that will never be used in a magnetic refrigerator. The study of thermomagnetic hysteresis by FORC is also included in this chapter. Magnetic actuation or sensing by magnetostrictive materials, new trends in information storage technology using heat-assisted magnetic recording (HAMR), and the biological and medical applications of magnetic nanoparticles close the spectrum of emerging applications.

The idea for this book was developed by the Editors in October 2018 while Victorino Franco was a visiting scientist at Lake Shore Cryotronics. In December of 2018, Springer Nature authorized the Editors to develop the book. The Editors then proposed a table of contents, identified and invited authors to contribute chapter content. The broad distribution of scientific topics of the chapters in the book is also matched by a geographical distribution of the sources. Authors of the different chapters are all specialists in their respective fields and are from 9 countries: the United States, Canada, Austria, Germany, Romania, Spain, Switzerland, the United Kingdom, and Japan. Magnetics research nowadays is focused both on fundamental science and technological applications, thus authors from Academia, Industry and National Laboratories have contributed (each sector contributing 16, 6 and 4 chapters, respectively). This distribution is also evident in the editors of the book, with one being from Academia and the other from Industry.

Every project is always subject to contingencies and, in the case of this book, it was given the name of the novel Coronavirus COVID-19. This severely impacted the schedules of the contributing authors in 2020. All authors had to implement remote (virtual) learning curriculum, or adapt to remote working conditions with limited access to labs, which caused delays in research programs, etc. Despite the COVID-19 situation, authors never waivered in their enthusiasm for the book, and in their commitment to contribute high-quality content.

We would like to thank Springer Nature for providing us with the opportunity to develop this book. We are also deeply indebted to all authors for their excellent contributions, and their support of this project. Each of them had their public or private funding sources, listed at the end of each chapter. The whole collection is too long to be mentioned here. But we want to explicitly acknowledge all those funding agencies, foundations, and companies that financially support advancements in science that ultimately improve our society. At the same time, we also acknowledge the enthusiastic support of Lake Shore Cryotronics and the University of Seville that enabled the successful completion of this endeavor. We hope this book proves to be a valuable resource for researchers, engineers, and students working in the field of magnetism and, equally importantly, to those who are new to this rich field of research.

Seville, Spain   Victorino Franco

Westerville, OH, USA   Brad Dodrill

Contents

About the Editors

Victorino Franco is a professor in the Condensed Matter Physics Department of the University of Seville, Spain. His main research interests cover magnetic materials for energy applications, including soft magnetic and magnetocaloric materials. He has published more than 200 peer-reviewed technical articles on these topics. He received the Young Scientist Award from the Royal Physical Society of Spain in 2000 and was named IEEE Magnetic Society Distinguished Lecturer in 2019. He has served as chair of the Spain Chapter of the IEEE Magnetics Society, chair of the Magnetic Materials Committee of the Minerals, Metals & Materials Society (TMS), and member of the Steering Committee of the European School of Magnetism. He has been Editor and Publications Chair of several Magnetism and Magnetic Materials (MMM) conferences and the General Chair of the 2022 Joint MMM-INTERMAG Conference.

Brad Dodrill graduated from the Ohio State University in 1982 with a BSc degree in Physics. He completed 2 years of graduate studies in Physics and Electrical Engineering and took a position with Lake Shore Cryotronics in 1984 as a Research Scientist. He was promoted to Systems Applications Engineer (1989–1996) and served a key role in the development of a commercial AC susceptometer/DC magnetometer and a vibrating sample magnetometer. He later served as an Applications Scientist and Product Manager (1996–2002), and in 2002 he was appointed VP of Sales. In 2018 he transitioned to his current position as VP of Applications and Senior Scientist. He has 49 paper publications to his credit and holds 3 U.S. patents. He has lectured at numerous universities and given invited and contributed talks at technical conferences in the U.S., Europe, Asia, and Mexico on vibrating sample magnetometry, AC susceptometry, alternating gradient magnetometry, and on the use of the first-order reversal curve (FORC) measurement and analysis technique for the characterization of magnetic materials.

Part I
Units in Magnetism

Units for Magnetic Quantities

Ronald B. Goldfarb

Abstract The centimeter-gram-second (CGS) system of units was adopted by the pioneers of electromagnetism in the nineteenth century. By the early twentieth century, two limitations of the CGS system became apparent: its inability to gracefully incorporate the electrical units common in engineering and inconvenient factors of 4π in electromagnetic equations. Giovanni Giorgi was most responsible for the development of the rationalized meter-kilogram-second-ampere system, which evolved into the International System of Units (SI). In 2019, the SI was redefined in terms of seven defining constants of nature, which set the value of the elementary charge. A direct consequence is that the value of the magnetic constant, the permeability of vacuum, is no longer fixed in the SI. Some conversions from CGS electromagnetic units to SI units in an updated conversion table thus involve the redefined permeability of vacuum, whereas other conversions require only powers of 10 and factors of 4π. The effect on magnetism and magnetic measurements is more philosophical than practical.

Keywords Magnetism · Magnetism history · Magnetic units · Electromagnetic units · International System of Units · Giorgi system · Permeability of vacuum · Magnetic constant · Conversion table · Units of measure · Magnetic quantities · International Bureau of Weights and Measures

1 The Centimeter-Gram-Second System of Units

In 1873, the same year that James Clerk Maxwell published the first edition of *A Treatise on Electricity and Magnetism,* the Committee for the Selection and Nomenclature of Dynamical and Electrical Units, under the leadership of William Thomson (later known as Lord Kelvin), presented its first report at the 43rd

R. B. Goldfarb (✉)
National Institute of Standards and Technology, Boulder, CO, USA
e-mail: ron.goldfarb@nist.gov

© Springer Nature Switzerland AG 2021
V. Franco, B. Dodrill (eds.), *Magnetic Measurement Techniques for Materials Characterization*, https://doi.org/10.1007/978-3-030-70443-8_1

meeting of the British Association for the Advancement of Science. It formally recommended the adoption of the centimeter-gram-second (CGS) system of units [1].

The following year, noting that "students usually find peculiar difficulty in questions relating to units," the Committee commissioned a book to explain the new CGS system and give examples of its application to physical measurements [2]. The book, authored by the Committee's secretary, Joseph David Everett, contained an appendix that reproduced the Committee's first report to the British Association [3].

However, the appendix omitted the dissent for the record by Committee member George Johnstone Stoney, who objected that "the centimetre was recommended as the unit of length against my earnest remonstrance," stating that "it is far too small." Stoney predicted that "the metre must in the end be accepted as the standard unit of length" [1] (the British spelling "metre" is used in the original). Indeed, the Committee's recommendation reversed the decision of its predecessor, the British Association's Committee for Standards of Electrical Resistance, which had adopted the meter-gram-second (MGS) system [4, 5]. But by 1873, the CGS system was preferred over the MGS system because it had the advantage "of making the value of the density of water practically equal to unity" [1].

The CGS system is an "absolute" system, that is, one based on the fundamental mechanical units of length L, mass M, and time T. Thus, the quantities in the electrostatic (ESU) and electromagnetic (EMU) subsystems of CGS all resolve to whole or fractional powers of centimeters, grams, and seconds. For example, the dimensions for magnetic moment in EMU are $L^{5/2} M^{1/2} T^{-1}$, with units $cm^{5/2} \cdot g^{1/2} \cdot s^{-1}$. Although magnetic moment has no named unit in EMU (recourse is often made to writing "emu" as a pseudo-unit), the units for magnetic moment correspond to those for the ratio of ergs per gauss: $cm^2 \cdot g \cdot s^{-2} / cm^{-1/2} \cdot g^{1/2} \cdot s^{-1}$. (The name "erg" was recommended as the unit for work and energy by the British Association in 1873. The name "gauss" was assigned, initially, to magnetic field strength by the International Electrical Congress in 1900 and, later, to magnetic flux density by the International Electrotechnical Commission in 1930.)

It was the intent of the British Association's Committee for the Selection and Nomenclature of Dynamical and Electrical Units that "one definite selection of three fundamental units be made once for all" so "that there will be no subsequent necessity for amending it" [1].

It was not to be.

2   The Rationalized Meter-Kilogram-Second-Ampere System

One of Oliver Heaviside's many accomplishments was the reformulation of Maxwell's cartesian equations in compact vector calculus notation. He believed that the factor of 4π in electromagnetic equations was simply an illogical convention, and he made a strong case for rationalization of the CGS system, that is, removal of the irrational number 4π in most equations, including those of Maxwell [6].

Giovanni Giorgi viewed rationalization as an optional but convenient adjunct to a four-dimensional, meter-kilogram-second (MKS) system, in which the fourth, electromagnetic unit was initially not specified [7]. Giorgi respectfully submitted preprints of his papers to Heaviside, who was 21 years his senior and quite famous. Heaviside was skeptical, as evidenced by his notations on Giorgi's correspondence, currently in the archives of the International Electrotechnical Commission [8]. In Giorgi's typewritten letter of 11 March 1902 to Heaviside, he outlined the differences between their two systems: "My object was in fact not only to get rid of the 4π, but to bring the practical electrical units into agreement with a set of mechanical units of reasonable size, and then to have a system which is absolute and practical at the same time." Years later, Giorgi extended the classical definition of an absolute system of units by noting the equivalence of mechanical and electrical energy and thus applied the coveted "absolute" adjective to his four-dimensional MKS system [9].

The meaning of the permeability of vacuum μ_0 was central to Giorgi's system [10]. He noted, "In my system, [μ_0] is not a numeric, nor do I assume any special value for it; it is a physical quantity, having dimensions, and to be measured by experiment" [11]. Thus, he regarded both μ_0 and the permittivity of vacuum ε_0 as subject to experimental refinement, with $\mu_0 \approx 1.256 \times 10^{-6}$ henries per meter and $\varepsilon_0 \approx 8.842 \times 10^{-12}$ farads per meter, and both subject to the condition that $(\mu_0 \, \varepsilon_0)^{-\frac{1}{2}}$ is equal to the speed of light $c \approx 3 \times 10^8$ m/s. He noted that his four-dimensional system "is neither electrostatic nor electromagnetic, because neither the electric nor the magnetic constant of free ether is assumed as a fundamental unit" [12].

Opposition to the full adoption of Giorgi's system was led by Richard Glazebrook, a former student and intellectual heir of Maxwell, who served as the chair of the Symbols, Units, and Nomenclature (SUN) Commission of the International Union of Pure and Applied Physics. The SUN Commission accepted the three-dimensional MKS as a parallel system but with μ_0 as just a fixed scaling factor with respect to the CGS system [10].

3   The International System of Units

Eventually, in 1954, the 10th General Conference on Weights and Measures (CGPM) approved the ampere as the fourth base unit, thereby formalizing the "MKSA" practical system of units. In 1960, the 11th CGPM adopted the name *Système International d'Unitès*, with the abbreviation "SI," for the practical system of units. In the SI, the "definition of the ampere was based on the force between two current carrying conductors and had the effect of fixing the value of the vacuum magnetic permeability μ_0 (also known as the magnetic constant) to be exactly $4\pi \times 10^{-7}$ H·m^{-1} $= 4\pi \times 10^{-7}$ N·A^{-2}" [13].

On 16 November 2018, in Versailles, France, the 26th CGPM adopted the most significant change in units of measure since 1954. It went into effect on

20 May 2019, World Metrology Day. The revised SI fixed the values of formerly measurable constants: the Planck constant, h; the elementary charge, e; the Boltzmann constant, k; and the Avogadro constant, N_A, thereby, individually or in combination, redefining the units kilogram, ampere, kelvin, and mole. The cesium 133 hyperfine transition frequency, $\Delta\nu_{Cs}$; the luminous efficacy of radiation of frequency 540×10^{12} Hz, K_{cd}; and the speed of light in vacuum, c, had already been fixed by the CGPM in 1967, 1979, and 1983, respectively, which defined the units second, candela, and meter [13].

The motivation for the use of defining constants is explained carefully in the 9th edition of the *SI Brochure* [[13], pp. 125–126]:

> Historically, SI units have been presented in terms of a set of—most recently seven—base units. All other units, described as derived units, are constructed as products of powers of the base units.
>
> Different types of definitions for the base units have been used: specific properties of artefacts such as the mass of the international prototype for the unit kilogram; a specific physical state such as the triple point of water for the unit kelvin; idealized experimental prescriptions as in the case of the ampere and the candela; or constants of nature such as the speed of light for the definition of the unit metre.
>
> To be of any practical use, these units not only have to be defined, but they also have to be realized physically for dissemination. In the case of an artefact, the definition and the realization are equivalent—a path that was pursued by advanced ancient civilizations. Although this is simple and clear, artefacts involve the risk of loss, damage or change. The other types of unit definitions are increasingly abstract or idealized. Here, the realizations are separated conceptually from the definitions so that the units can, as a matter of principle, be realized independently at any place and at any time. In addition, new and superior realizations may be introduced as science and technologies develop, without the need to redefine the unit. These advantages—most obviously seen with the history of the definition of the metre from artefacts through an atomic reference transition to the fixed numerical value of the speed of light—led to the decision to define all units by using defining constants.
>
> The choice of the base units was never unique, but grew historically and became familiar to users of the SI. This description in terms of base and derived units is maintained in the present definition of the SI, but has been reformulated as a consequence of adoption of the defining constants.

Instead of the definition of the ampere fixing the value of μ_0, the 2019 revision of the SI defines the ampere in terms of the fixed value of e. As a result, the value of μ_0 must be determined experimentally. Similarly, the permittivity of vacuum $\varepsilon_0 = 1/(\mu_0 c^2)$ must be determined experimentally (as it was before c was fixed in 1983). The product $\mu_0 \varepsilon_0 = 1/c^2$ remains exact. The experimental value of μ_0 is now based on that of the dimensionless fine-structure constant α, the coupling constant of the electromagnetic force: $\mu_0 = 2h\alpha/ce^2$, where h is the newly fixed Planck constant, c is the fixed speed of light in vacuum, and e is the newly fixed elementary charge (equal to the absolute value of the electron charge). The relative standard uncertainties in μ_0, ε_0, and α are identical [14].

It was reasonable to fix the value of e instead of μ_0 because, by the 1990s, the realization of the ampere was by Ohm's law, the Josephson effect for voltage, and the quantum Hall effect for resistance (both in terms of the 1990 recommended values of e and h [15]), not by the force on currents in parallel wires. A definition of the ampere and the kilogram in terms of fixed values of e and h, respectively, brought the practical quantum electrical standards into exact agreement with the SI [13].

4 Conversion Factors

Conversion tables are helpful for magnetics researchers who want to compare data appearing in published articles. The need will diminish with time as the SI becomes universal for instruction in electromagnetism. Magnetics researchers who currently measure in SI units and analyze using SI equations do not have to worry about conversion factors, but even they occasionally need to refer to published data in EMU.

Units of measure have been examined and reexamined vigorously. The monograph by Silsbee is noteworthy for its completeness [16]. The appendixes in the textbooks by Jackson [17] and Coey [18] are good resources. Few articles deal specifically with units for magnetic properties. Bennett et al. published a conversion guide especially for magnetics in which they pointed out, to the surprise of many, that "emu" is not actually a unit [19]. During an evening panel discussion on magnetic units at the 1994 Joint Magnetism and Magnetic Materials—International Magnetics Conference, different perspectives were advanced by seven practitioners [20], some of whom recapitulated their recent articles or prefaced their future articles on the subject [21, 22, 23].

In the MKSA system and the SI of 1960, μ_0 served both as a conversion factor and as a means for rationalization with respect to EMU. Thus, the 2019 revision of the SI, which made μ_0 an experimental constant, has consequences for magnetics. A conversion guide for magnetic quantities from EMU to SI may now distinguish between conversions based on an experimental determination of μ_0 and conversions based on rationalization of EMU. As first noted by Davis, conversion factors to CGS systems, such as EMU, which made use of the exact relation $\{\mu_0/4\pi\} \equiv 10^{-7}$, are no longer exactly correct after the SI revision of 2019 [24] (The curly brackets mean that one removes the units associated with the quantity within.)

Table 1 is a conversion guide from EMU to SI that reflects the redefinition of the SI. Conversion factors formerly based on the fixed permeability of vacuum $\{\mu_0\} \equiv 4\pi \times 10^{-7}$ are here replaced explicitly by the symbol $\{\mu_0\}$. However, factors based only on the conversion of centimeters to meters, grams to kilograms, and rationalization of EMU retain the factor of 4π; for example, the sum of the three axial demagnetizing factors of an ellipsoid is 4π in EMU and unity in the SI.

Magnetism in the SI is concordant with the Sommerfeld constitutive relation $B = \mu_0(H + M)$ for magnetic flux density B, magnetic field strength H, and magnetization M. However, magnetic polarization J and magnetic dipole moment j, derived from the Kennelly convention, $B = \mu_0 H + J$, are also recognized. In both conventions, B and H have units different from each other.

Table 1 Conversion of units for magnetic quantities. In the right column, $\{\mu_0\}$ refers to the numerical value of μ_0, the recommended value of which may change slightly over time. Factors of 4π originate from the conversion of unrationalized EMU to rationalized SI units. In the absence of units, a dimensionless quantity is labeled with its associated system of units (EMU or SI). The arrows ($\rightarrow$) indicate correspondence, not equality. From [10], after [25]

SI Symbol	SI Quantity	Conversion from EMU and Gaussian Units to SI Units [a]
Φ	Magnetic flux	$1\ \text{Mx} = 1\ \text{G·cm}^2 \rightarrow 10^{-8}\ \text{Wb} = 10^{-8}\ \text{V·s}$
B	Magnetic flux density, magnetic induction	$1\ \text{G} \rightarrow 10^{-4}\ \text{T} = 10^{-4}\ \text{Wb/m}^2$
μ	Permeability [b]	$1\ (\text{EMU}) \rightarrow \{\mu_0\}\ \text{H/m} = \{\mu_0\}\ \text{N/A}^2 = \{\mu_0\}\ \text{Wb/(A·m)}$
H	Magnetic field strength, magnetizing force	$1\ \text{Oe} \rightarrow 10^{-4}/\{\mu_0\}\ \text{A/m}$
m	Magnetic moment	$1\ \text{erg/G} = 1\ \text{emu} \rightarrow 10^{-3}\ \text{A·m}^2 = 10^{-3}\ \text{J/T}$
j	Magnetic dipole moment	$1\ \text{erg/G} = 1\ \text{emu} \rightarrow 10^{-3}\{\mu_0\}\ \text{Wb·m}$
M	Magnetization, volume magnetization	$1\ \text{erg/(G·cm}^3) = 1\ \text{emu/cm}^3 \rightarrow 10^3\ \text{A/m}$
		$1\ \text{G} \rightarrow 10^{-4}/\{\mu_0\}\ \text{A/m}$
J, I	Magnetic polarization, intensity of magnetization	$1\ \text{G} \rightarrow 10^{-4}\ \text{T} = 10^{-4}\ \text{Wb/m}^2$
σ	Specific magnetization, mass magnetization	$1\ \text{erg/(G·g)} = 1\ \text{emu/g} \rightarrow 1\ \text{A·m}^2/\text{kg}$
χ	Susceptibility, volume susceptibility	$1\ (\text{EMU}) \rightarrow 4\pi\ (\text{SI})$
$\chi_\rho,\ \chi_m$	Specific susceptibility, mass susceptibility	$1\ \text{cm}^3/\text{g} \rightarrow 4\pi \times 10^{-3}\ \text{m}^3/\text{kg}$
w, W	Energy product, volume energy density [c]	$1\ \text{erg/cm}^3 \rightarrow 10^{-1}\ \text{J/m}^3$
N, D	Demagnetizing factor	$1\ (\text{EMU}) \rightarrow (4\pi)^{-1}\ (\text{SI})$

[a] EMU are the same as Gaussian units for magnetostatics: Mx = maxwell, G = gauss, Oe = oersted. SI: Wb = weber, T = tesla, H = henry, N = newton, J = joule.
[b] In the SI, relative permeability $\mu_r = \mu/\mu_0 = 1 + \chi$. In EMU, permeability $\mu = 1 + 4\pi\chi$. Relative permeability μ_r in the SI corresponds to permeability μ in EMU.
[c] In the SI, $w\ [\text{J/m}^3] = B\ [\text{T}] \cdot H\ [\text{A/m}] = \mu_0\ [\text{Wb/(A·m)}] \cdot M\ [\text{A/m}] \cdot H\ [\text{A/m}]$. In EMU, $w\ [\text{erg/cm}^3] = (4\pi)^{-1} B\ [\text{G}] \cdot H\ [\text{Oe}] = M\ [\text{erg/(G·cm}^3)] \cdot H\ [\text{Oe}]$.

In EMU, $B = H + 4\pi M$, where B and H have the same units with different names, gauss (G) and oersted (Oe). As has been noted, "the magnetization, when written as $4\pi M$, is also in gausses and may be thought of as a field arising from the magnetic moment. When magnetization is expressed simply as M (the magnetic moment m per unit volume), its units are erg·G^{-1}·cm^{-3}. In terms of base units, erg = cm^2·g·s^{-2} and G = $\text{cm}^{-1/2}$·$\text{g}^{1/2}$·s^{-1}; therefore, erg·G^{-1}·cm^{-3}, the units for M, are dimensionally but not numerically equivalent to G" [21].

In the table, dimensionless quantities are labeled with their associated system of units (EMU or SI) to distinguish them. In magnetic materials with permeability μ, $B = \mu H$, where μ is dimensionless in EMU. The conversion of dimensionless volume susceptibility χ from EMU to SI is based on the correspondence between $\mu = 1 + 4\pi\chi$ in EMU and relative permeability $\mu_r = \mu/\mu_0 = 1 + \chi$ in SI; that is, $4\pi\chi$ (EMU) corresponds to χ (SI); $\{\mu_0\}$ is not involved. This also follows from the definition $\chi = M/H$, in both EMU and SI, and $4\pi\chi$ (EMU) having units of gausses per oersted (dimensionless). The conversion of specific (mass) susceptibility follows from that of volume susceptibility.

The SI redefinition of the ampere implies that the EMU abampere (the prefix "ab" means "absolute") does not convert exactly to 10 amperes, as was similarly footnoted by Quincey and Brown in relation to the abcoulomb and coulomb [26]. This affects the conversion of magnetic field strength H from oersteds (the named unit for gilberts per centimeter, which corresponds to $(4\pi)^{-1}$ abamperes per centimeter) to amperes per meter by requiring the use of $\{\mu_0\}$. Alternatively, the conversion factor of $10^{-4}/\{\mu_0\}$ in the table may be considered to arise from the equivalence of oersted and gauss in EMU, the conversion of gauss to tesla, and the relationship $B = \mu_0 H$ in vacuum. The same factor is used in the table for the conversion of magnetization M, when formulated as $4\pi M$ in gausses, to amperes per meter.

Conversions based on transformations from gausses to teslas and ergs to joules do not involve $\{\mu_0\}$. For example, magnetization in gausses converts to magnetic polarization in teslas without involvement of $\{\mu_0\}$. However, magnetic moment, when expressed in EMU as ergs per gauss (or "emu"), converts to magnetic dipole moment in weber meters with a required factor of $\{\mu_0\}$.

5   Epilogue

While the accepted value of $\{\mu_0\}$ will change slightly over time with changes in the experimental fine-structure constant α, $\{\mu_0\}$ is currently equal to $1.256\,637\,0621 \times 10^{-6} \pm 0.000\,000\,0019 \times 10^{-6}$, based on the latest quadrennial adjustment to the fundamental physical constants by the International Science Council's Committee on Data [27]. That is, the value of $\{\mu_0\}$ is equal to $4\pi \times 10^{-7}$ to nine significant figures. Thus, the distinction between $\{\mu_0\}$ and $4\pi \times 10^{-7}$ is largely philosophical and hardly practical; their difference is much smaller than the total uncertainty in any magnetic measurement.

In the revised SI, it is compelling to regard B as the primary magnetic field vector, μ_0 as an experimental constant, and H as an arithmetically derived auxiliary vector [10]. For displays of measurement data, the symbol B_0 could be used for applied magnetic field in units of teslas, much as $\mu_0 H$ is sometimes used, where B_0 is distinguished from the flux density B in magnetic materials. Magnetic volume susceptibility χ should remain defined as M/H (dimensionless), not M/B_0, because M/H is embedded historically in EMU, the MKSA system, and the SI.

Acknowledgments Valuable comments by Richard Davis (Bureau International des Poids et Mesures) demonstrated deep insight into the International System of Units. Montserrat Rivas (Universidad de Oviedo), Barry Taylor (National Institute of Standards and Technology), Michael Coey (Trinity College Dublin), Brad Dodrill (Lake Shore Cryotronics), and Victorino Franco (Universidad de Sevilla) offered constructive suggestions on the manuscript.

References

1. W. Thomson, G.C. Foster, J. Clerk Maxwell, G.J. Stoney, F. Jenkin, C.W. Siemens, F.J. Bramwell, J.D. Everett, First report of the Committee for the Selection and Nomenclature of Dynamical and Electrical Units, *Report of the Forty-Third Meeting of the British Association for the Advancement of Science*, pp. 222–225 (1873). https://www.biodiversitylibrary.org/item/94452
2. W. Thomson, G.C. Foster, J. Clerk Maxwell, G.J. Stoney, F. Jenkin, C.W. Siemens, F.J. Bramwell, W.G. Adams, B. Stewart, J.D. Everett, Second report of the Committee for the

Selection and Nomenclature of Dynamical and Electrical Units," *Report of the Forty-Fourth Meeting of the British Association for the Advancement of Science,* p. 255 (1874). https://www.biodiversitylibrary.org/item/94451

3. J.D. Everett, *Illustrations of the Centimetre-Gramme-Second (C.G.S.) System of Units* (Taylor and Francis, London, 1875). https://archive.org/details/b22295641

4. A. Williamson, C. Wheatstone, W. Thomson, W.H. Miller, A. Matthiessen, F. Jenkin, C. Bright, J. Clerk Maxwell, C.W. Siemens, B. Stewart, J.P. Joule, C.F. Varley, Report of the Committee on Standards of Electrical Resistance, *Report of the Thirty-Fourth Meeting of the British Association for the Advancement of Science,* pp. 345–367 (1864). https://www.biodiversitylibrary.org/item/93072

5. A.E. Kennelly, I.E.C. adopts MKS system of units. Trans. Am. Inst. Electr. Eng. **54**, 1373–1384 (1935). https://doi.org/10.1109/T-AIEE.1935.5056934

6. O. Heaviside, *Electromagnetic Theory, Vol. 1, Ch. 2* (The Electrician, London, 1893), pp. 116–127. https://archive.org/details/electromagnetict01heav

7. G. Giorgi, Unità razionali di elettromagnetismo, Atti dell' Associazione Elettrotecnica Italiana, **5**, 402–418 (1901). Republished as: "Sul sistema di unità di misure elettromagnetiche," Il Nuovo Cimento, **4**, 11–30 (1902), https://doi.org/10.1007/BF02709460. English translation in: C. Egidi, Ed. *Giovanni Giorgi and His Contribution to Electrical Metrology,* (Politecnico di Torino, Turin, Italy, 1990), pp. 135–150.

8. IEC, *Giovanni Giorgi's Correspondence with Oliver Heaviside* (Int. Electrotech. Comm., Geneva, 1902). https://www.iec.ch/history/historical-documents

9. G. Giorgi, *Memorandum on the M.K.S. System of Practical Units,* (International Electrotechnical Commission, London, 1934). Republished: IEEE Magn. Lett., **9**, 1205106 (2018). https://doi.org/10.1109/LMAG.2018.2859658

10. R.B. Goldfarb, Electromagnetic units, the Giorgi system, and the revised International System of Units. IEEE Magn. Lett. **9**, 1205905 (2018). https://doi.org/10.1109/LMAG.2018.2868654

11. G. Giorgi, Rational electromagnetic units, in *Electrical World and Engineer,* vol. 40, (McGraw-Hill, New York, 1902), pp. 368–370; editorial comment on p. 355. https://books.google.com/books?id=e15NAAAAYAAJ

12. G. Giorgi, Proposals concerning electrical and physical units. J. Inst. Electr. Eng. **34**, 181–185 (1905). https://doi.org/10.1049/jiee-1.1905.0013

13. BIPM, *The International System of Units*, 9th edn. (Bureau Int. des Poids et Mesures, Sèvres, France, 2019). https://www.bipm.org/utils/common/pdf/si-brochure/SI-Brochure-9.pdf

14. R.B. Goldfarb, The permeability of vacuum and the revised International System of Units (editorial). IEEE Magn. Lett. **8**, 1110003 (2017). https://doi.org/10.1109/LMAG.2017.2777782

15. N. Fletcher, G. Rietveld, J. Olthoff, I. Budovsky, M. Milton, Electrical units in the new SI: Saying goodbye to the 1990 values. NCSLI Measure J. Meas. Sci. **9**, 30–35 (2014). https://doi.org/10.1080/19315775.2014.11721692

16. F.B. Silsbee, Systems of electrical units. J. Res. Natl. Bur. Stand. **66C**, 137–183 (1962). https://doi.org/10.6028/jres.066C.014. Republished: Nat. Bur. Stand. Monograph 56 (1963). https://nvlpubs.nist.gov/nistpubs/Legacy/MONO/nbsmonograph56.pdf

17. J.D. Jackson, *Classical Electrodynamics*, 3rd edn. (Wiley, New York, 1998), pp. 775–784. https://www.wiley.com/en-us/Classical+Electrodynamics%2C+3rd+Edition-p-9780471309321

18. J.M.D. Coey, *Magnetism and Magnetic Materials* (Cambridge University Press, Cambridge, 2010), pp. 590–594 and inside back cover. https://doi.org/10.1017/CBO9780511845000

19. L.H. Bennett, C.H. Page, L.J. Swartzendruber, Comments on units in magnetism. J. Res. Natl. Bur. Stand. **83**, 9–12 (1978). https://doi.org/10.6028/jres.083.002. Republished from: Proc. 1975 Magn. Magn. Mater. Conf., Philadelphia, Penn., *AIP Conf. Proc.* **29**, xix–xxi (1976). New York: Am. Inst. Phys.

20. C.D. Graham, S. Chikazumi, R. Street, A. Arrott, J.M.D. Coey, M.R.J. Gibbs, R.B. Goldfarb, Panel discussion on units in magnetism. IEEE Magn. Soc. Newsl. **31**(4), 5–8 (1994) https://www.ieeemagnetics.org/files/newsletters/IEEEMS-N-31-4-Oct-1994.pdf. Republished: *Magnetic and Electrical Separation,* **6**, 105–116 (1995)

21. R.B. Goldfarb, Magnetic units and material specification, in *Concise Encyclopedia of Magnetic and Superconducting Materials*, ed. by J.E. Evetts, 1st edn., (Pergamon Press, Oxford, 1992), pp. 253–258. https://www.worldcat.org/title/concise-encyclopedia-of-magnetic-superconducting-materials/oclc/924972388

22. A.S. Arrott, "Magnetism in SI and Gaussian units," Ch. 1.2, in *Ultrathin Magnetic Structures I*, ed. by J.A.C. Bland, B. Heinrich, (Springer-Verlag, Berlin & Heidelberg, 1994), pp. 7–19. https://www.springer.com/gp/book/9783540219552

23. J. Crangle, M. Gibbs, Units and unity in magnetism: A call for consistency. Phys. World **7**, 31–32 (1994). https://doi.org/10.1088/2058-7058/7/11/29

24. R.S. Davis, Determining the value of the fine-structure constant from a current balance: Getting acquainted with some upcoming changes to the SI. Am. J. Phys. **85**, 364–368 (2017). https://doi.org/10.1119/1.4976701

25. R.B. Goldfarb, F.R. Fickett, *Units for Magnetic Properties, National Bureau of Standards Special Publication 696* (U.S. Gov. Printing Office, Washington, D.C., 1985). https://archive.org/details/unitsformagnetic696gold

26. P. Quincey, R.J.C. Brown, A clearer approach for defining unit systems. Metrologia **54**, 454–460. https://doi.org/10.1088/1681-7575/aa7160

27. CODATA, Committee on Data of the International Science Council, Task Group on Fundamental Physical Constants (2018). https://physics.nist.gov/cuu/Constants/index.html, https://codata.org/initiatives/strategic-programme/fundamental-physical-constants

Part II
Inductive and Force-Based Techniques for Measuring Bulk Magnetic Properties

Vibrating Sample Magnetometry

Brad Dodrill and Jeffrey R. Lindemuth

Abstract A magnetometer is an instrument to measure the magnitude and direction of a magnetic field. The most commonly used magnetometric technique to characterize magnetic materials is vibrating sample magnetometry (VSM). VSMs can measure the magnetic properties of magnetically soft (low coercivity) and hard (high coercivity) materials in many forms: solids, powders, single crystals, thin films, or liquids. They can be used to perform measurements from low to high magnetic fields employing electromagnets, Halbach rotating permanent magnet arrays, or high-field superconducting magnets. They can be used to perform measurements from very low to very high temperatures with integrated cryostats or furnaces, respectively. And, they possess a dynamic range extending from 10^{-8} emu (10^{-11} Am^2) to above 10^3 emu (1 Am^2), enabling them to measure materials that are both weakly magnetic (ultrathin films, nanoscale structures, etc.) and strongly magnetic (permanent magnets). In this chapter, we will discuss the VSM measurement technique and its implementation in an electromagnet. We will also discuss relevant extensions of the technique that provide variable temperature capability, a vector VSM for magnetic anisotropy studies, and implementation of data acquisition algorithms for first-order reversal curve (FORC) measurements for characterizing magnetic interactions and coercivity distributions in magnetic materials. We will present typical measurement results over a range of experimental conditions for various materials to demonstrate the VSM capability for magnetic materials characterization.

Keywords Magnetometry · Hysteresis · Vibrating sample magnetometer, VSM · Vector VSM, magnetic anisotropy · Variable temperature measurements · First-order reversal curve · FORC

B. Dodrill (✉) · J. R. Lindemuth
Lake Shore Cryotronics, Inc., Westerville, OH, USA
e-mail: brad.dodrill@lakeshore.com

© Springer Nature Switzerland AG 2021
V. Franco, B. Dodrill (eds.), *Magnetic Measurement Techniques for Materials Characterization*, https://doi.org/10.1007/978-3-030-70443-8_2

1 Magnetic Measurement Techniques

Magnetometry techniques can be classified into several broad classes: inductive-based, for example, vibrating sample magnetometry (VSM) and superconducting quantum interference device (SQUID) [1] magnetometry; force-based, for example, alternating gradient magnetometry (AGM) [2] and Faraday [3] and Gouy balance [4]; and optically based, for example, nitrogen vacancy defects in diamond [5].

Some magnetic materials such as nanowires, nanoparticles, thin films, etc., typically possess weak magnetic signatures, owing to the small amount of magnetic material that is present. Thus, one of the most important considerations in determining which type of magnetometer is best suited to specific materials is its sensitivity, as this determines the smallest magnetic moment that may be measured with acceptable signal-to-noise ratio. Measurement speed, i.e., the time required to measure a hysteresis loop, is also important because it determines sample throughput, and it is particularly important for First Order Reversal Curve (FORC) measurements because a typical series of FORCs can contain thousands to tens of thousands of data points. The final consideration is the temperature and field range over which measurements are to be performed, and this is dictated largely by the magnetic materials that are being studied.

Commercial VSM systems provide measurements to field strengths of ~34 kOe (3.4 T) using conventional electromagnets [6, 7], as well as systems employing superconducting magnets to produce fields to 160 kOe (16 T) [8, 9]. In an electromagnet-based VSM, the magnetic field can be swept at up to 10 kOe/s (1 T/s), and a typical hysteresis loop measurement can take as little as a few seconds to a few minutes, and a typical series of FORCs takes minutes to hours. When used with superconducting magnets, higher field strengths are possible, which are necessary to saturate some magnetic materials such as rare-earth permanent magnets; however, the measurement speed is inherently slower due to the speed at which the magnetic field can be varied using superconducting magnets due to their large inductance. Field sweep rates are typically limited to 200 Oe/s (20 mT/s), and thus a typical hysteresis loop measurement can take tens of minutes or more, and a typical series of FORCs can take a day or longer. Magnetometers employing superconducting magnets are more costly to operate since they require liquid helium. Cryogen-free systems employing closed-cycle refrigerators, or liquefiers that recover helium in liquid helium-based systems, are available, but these represent an expensive capital equipment investment. The noise floor of commercially available VSMs is in the 10^{-7}–10^{-8} emu (10^{-10}–10^{-11} Am^2) range.

The most common measurement used to characterize a material's magnetic properties is measurement of the major hysteresis or $M(H)$ loop as illustrated in Fig. 1. The main parameters extracted from the hysteresis loop that are used to characterize the properties of magnetic materials include the saturation magnetization M_s (the magnetization at maximum applied field), the remanence M_r (the magnetization at zero applied field after applying a saturating field), and the coercivity H_c (the field required to demagnetize the sample).

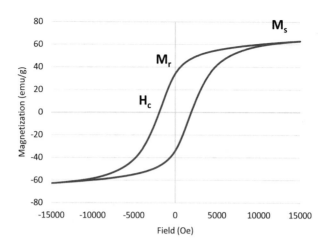

Fig. 1 A typical hysteresis loop showing the extracted parameters M_s, M_r, and H_c

More complex magnetization curves covering states with field and magnetization values located inside the major hysteresis loop, such as minor hysteresis loops and FORCs, can give additional information that can be used for characterization of magnetic interactions [10].

All VSM results presented in this chapter were recorded using a Lake Shore Cryotronics Model 8600 electromagnet-based VSM.

2 Electromagnet-Based VSM

The vibrating sample magnetometer was originally developed by Simon Foner [11] of MIT's Lincoln Laboratory. Foner patented the VSM technology [12] and sold an exclusive license to Princeton Applied Research Corporation (PARC) to develop and market the VSM. The early VSMs were called Foner magnetometers.

In an electromagnet-based VSM, a magnetic material is vibrated within a uniform magnetic field H generated by an electromagnet, inducing an electric current in suitably placed sensing coils. The resulting voltage induced in the sensing coils is proportional to the magnetic moment of the sample. Variable temperature measurements can be performed from <4.2 to 1273 K using integrated cryostats and furnaces, respectively.

Figure 2 shows a schematic representation of an electromagnet-based VSM. A variable magnetic field in the *x* direction is produced by an electromagnet energized by an appropriate bipolar power supply. Four-coil transverse detection or sensing coils [13] are mounted on the pole faces of the magnet, two on each face. The coils are balanced so as to produce zero signal (voltage) in the absence of a sample. A Hall probe, which is connected to a gaussmeter, is also mounted on the electromagnet

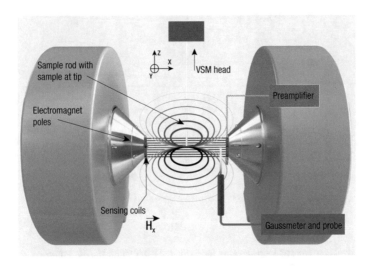

Fig. 2 Schematic representation of a VSM. The red and black contours represent the dipole magnetic field of a magnetized sample

pole face for closed-loop control of the magnetic field. A sample of any form (solid, powder, thin film, etc.) is placed in a suitable non-magnetic sample holder which is attached to the end of the VSM sample rod, which is in turn attached to the VSM head. The sample is vibrated in the z direction within the sensing coils, and the resulting induced voltage is passed through a preamplifier and then to a narrow bandwidth lock-in amplifier (LIA). The LIA reference is phased locked to the head drive vibration frequency.

The voltage induced in the VSM sensing coils is given by:

$$V_{\text{emf}} = mAfS \tag{1}$$

where:

m = magnetic moment.
A = amplitude of vibration.
f = frequency of vibration.
S = sensitivity function of the VSM sensing coils.

S is determined by calibrating the VSM with a magnetic calibrant [14], i.e., a material with known magnetization at a specified applied field H.

A VSM's sensitivity depends on a number of factors:

- Electronic sensitivity.
- Noise rejection through signal conditioning.
- Amplitude and frequency of mechanical drive.
- Thermal noise of sensing coils.

- Optimized design and coupling (proximity) of sensing coils to the sample under test.
- Vibration isolation of the mechanical head assembly from the electromagnet and VSM sensing coils.
- Minimization of environmental mechanical and electrical noise sources, which can deleteriously effect VSM sensitivity.

It is clear from Eq. (1) that increasing A, f, or S will improve moment sensitivity; however, there are practical limitations to each. Frequencies of less than ~100 Hz are typically used so as to minimize eddy current generation in magnetic materials that are electrically conductive, and it is also important to avoid frequencies that are close to the line frequency and its higher-order harmonics. The vibration amplitude should be sufficiently small to ensure that the sample is not subjected to inhomogeneous magnetic fields arising from the field source. S may be increased by optimizing the design of the sensing coils (i.e., number of windings, coil geometry, etc.) and by increasing the coupling between the sense coils and the sample under test (i.e., minimize gap spacing). When the sensing coils and sample are very close together, finite sample size effects [15] can lead to errors in the measured sample magnetization. These errors can be mitigated by using a calibrant that is geometrically identical to the sample. At first glance, it would seem that all that one needs to do to increase S is to maximize the number of windings in the coils; however, this increases the resistance of the coils, which in turn increases their thermal noise which negatively impacts their signal-to-noise ratio. Finally, increasing signal averaging of the LIA also improves signal-to-noise ratio.

VSM Sensitivity Examples: Figure 3 shows typical noise measurement results at 100 ms/point (top) and 10 s/point (bottom) averaging. Note that the vertical axis is expressed in nemu $= 10^{-9}$ emu (10^{-12} Am^2). The RMS noise values are noted in the figure caption.

Figure 4 shows typical low moment measurement results for a CoPt bit-patterned (bit size <100 nm) magnetic media (thin film) sample with saturation moment m_{sat} < 20 μemu (20 × 10^{-9} Am^2). The hysteresis loop was recorded to ±5 kOe (0.5 T) in 25 Oe (2.5 mT) steps at 100 ms/point averaging. The total loop measurement time was 1 min 25 s. Figure 5 shows results for a synthetic antiferromagnetic thin film with m_{sat} < 2 μemu (2 × 10^{-9} Am^2). The hysteresis loop was recorded to ±500 Oe (50 mT) in 2.5 Oe (0.25 mT) steps at 2 s/point averaging. The total loop measurement time was 28 min.

3 VSM Components and Extensions

Aside from the electromagnet and associated bipolar power supply, the principal VSM components include the sensing coils, vibration head, control and measurement electronics, and data acquisition software. In this section, we will discuss the coils, head, and electronics.

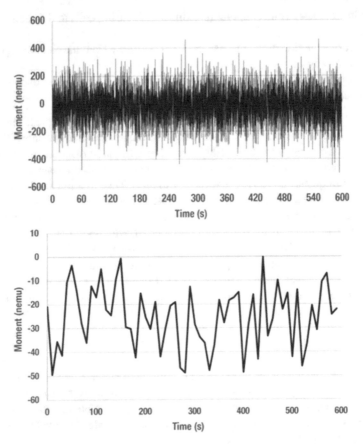

Fig. 3 (**a, b**) Noise at 100 ms/point and 10 s/point (**a**, top) and 10 sec/point (**b**, bottom) averaging. The observed noise is 119.5 nemu and 13 nemu RMS, respectively

Sensing Coils: Various transverse detection or sensing coil configurations have been proposed, but the most commonly employed in an electromagnet VSM is the four-coil configuration proposed by Mallinson [16] shown in Fig. 6. Practical detection coil arrangements utilize an even number of coils to minimize sensitivity to sample position. The coils are typically balanced (i.e., geometrically identical with precisely the same number of windings). A perfectly balanced coil set produces zero signal in the absence of a sample. A nonzero signal is a consequence of not having a perfectly balanced coil set, but such offsets are usually small and can be removed with appropriate nulling electronics. Finding the appropriate balance between the number of windings and the resistance of the coils is important to maximize their sensitivity without creating thermal noise. The size (or diameter in the case of circular coils) of the coils should be larger than the sample under test and at the same time sufficiently small to ensure they are in a homogeneous magnetic field. In the configuration shown in Fig. 6, the axes of the coils are parallel

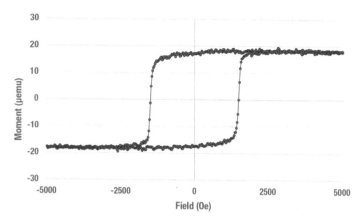

Fig. 4 Hysteresis loop for a 20 μemu CoPt nanomagnet array

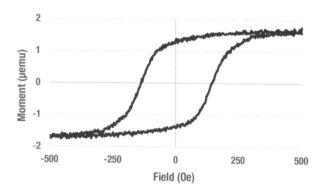

Fig. 5 Hysteresis loop for a < 2 μemu synthetic antiferromagnetic thin film

to the applied field (*x*) and transverse to the sample vibration (*z*). A magnetic sample is properly positioned within the sensing coils by adjusting the *xyz* position of the sample via micrometers attached to the vibrating head to maximize the signals in the *z* and *y* directions and minimize the signal in the *x* direction.

Vibrating Head: The vibrating head must provide a vibration of constant frequency and amplitude as a function of time. If either drifts, then the voltage induced in the sensing coils drifts, which produces an apparent drift in a samples magnetization. Frequencies should not be close to the line frequency or its higher-order harmonics and should be less than 100 Hz to minimize eddy currents in electrically conductive materials. The drive amplitude should be sufficiently small to ensure that a sample is not subjected to an inhomogeneous magnetic field, and it should be less than the sensing coil diameter. The head should provide a stable reference signal for lock-in detection of the signal induced in the sensing coils. And, finally, the head should be either passively or dynamically decoupled from the

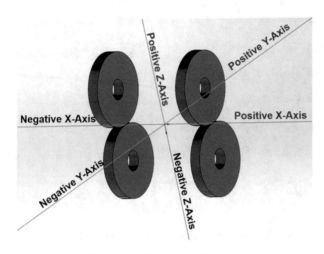

Fig. 6 Four-coil Mallinson configuration with *xyz* coordinates

electromagnet and sensing coils to minimize induced coherent noise in the coils due to vibration coupling.

Home-built VSMs commonly use a simple loud speaker [17–20] or stepper motor [21] as the head drive. While these are simple to implement and use inexpensive commercially available parts, they also have their limitations. When using loudspeakers, the drive amplitude tends to be small at typical operating frequencies, which limits VSM sensitivity, and they can't typically drive more massive samples. More sophisticated electromechanical assemblies incorporating dynamically decoupled drives for vibration isolation have been developed [6]. A cross-sectional schematic of an electromechanical VSM head is shown in Fig. 7.

VSM Electronics: A VSM requires gaussmeter and Hall sensor for closed-loop field control, electronics to drive the VSM head, a lock-in amplifier (LIA) to provide the head with an AC reference signal and to measure the induced voltage in the sensing coils, and a temperature controller if the VSM is equipped with variable temperature apparatus. Home-built VSMs commonly use a sine wave generator fed into a power amplifier to drive the head, a commercial gaussmeter or multimeter interfaced to the Hall sensor, and a commercial LIA. Data acquisition and control software is commonly written in LabVIEW or Python. A schematic of the electronics is illustrated in Fig. 8.

VSM Variable Temperature Measurements: VSMs can be configured with crysotats or furnaces for low- and high-temperature measurements, respectively. Such fixtures may be custom built or procured from commercial sources. Wet cryostats may be used with either liquid helium or nitrogen for measurements to 4.2 and 77.3 K, respectively. These can be continuous flow or bath cryostats. A liquid helium bath cryostat may be pumped to achieve temperatures to <2 K. Closed-cycle refrigerators (CCR) providing cryogen-free operation can be adapted to VSMs as well; however, special care must be taken to decouple the CCR cold-head vibrations

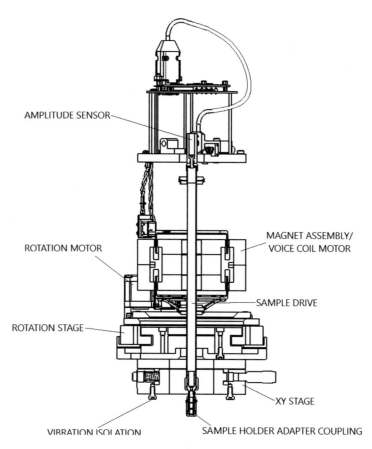

Fig. 7 Cross-sectional view of an electromechanical VSM head [6]

from the VSM or its sensitivity will degrade. High-temperature furnaces providing temperatures to as high as 1273 K (1000 °C) can be similarly adapted to VSMs. Figure 9 shows cross-sectional views of a continuous flow cryostat that can operate with either liquid helium or nitrogen (left) and a high-temperature furnace assembly (right).

Variable Temperature Measurement Examples: Figure 10 shows low-temperature (120–300 K) field-cooled (FC) and zero-field-cooled (ZFC) magnetization curves for Fe_3O_4 core (10 nm)/CoO shell (3 nm) nanoparticles (NPs). The peak in the ZFC curve at each applied field corresponds to the blocking temperature T_B. Figure 11 shows hysteresis loops for a high-temperature $YBa_2Cu_3O_{7-x}$ (YBCO) superconductor at $T = 10, 20, 30, 40, 50, 60, 70$, and 80 K. Figure 12 shows a high-temperature magnetization curve for a NiFe alloy sample recorded with an applied field of 5 kOe (0.5 T) for temperatures ranging from 323 K to 1123 K (50–850 °C). The lower- and higher-temperature transitions correspond to the Curie temperatures (T_c) of Ni and Fe, respectively.

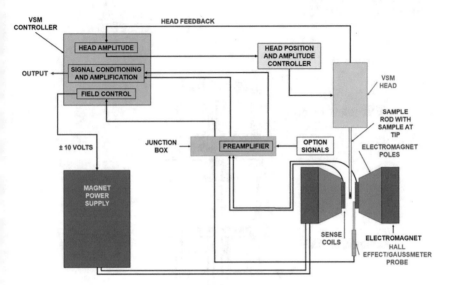

Fig. 8 Schematic of VSM electronics

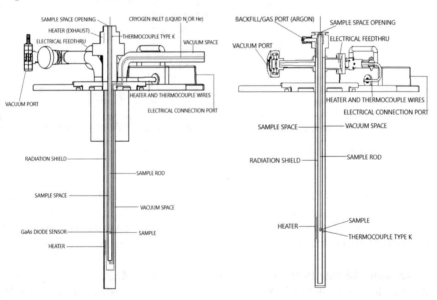

Fig. 9 Cross-sectional views of a typical VSM cryostat (left) and furnace (right) [6]

Vector VSM: A vector VSM (VVSM), or biaxial VSM [22–26], provides measurement of the angular dependence of the vector components of the total magnetization. The VVSM can be used to determine anisotropy constants from torque curves. The usual method for anisotropy determinations is torque magnetometry, which directly measures the macroscopic torque exerted by an applied field on the

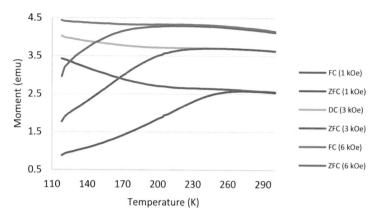

Fig. 10 Low-temperature zero-field-cooled (ZFC)/field-cooled (FC) magnetization curves for Fe_3O_4 core (10 nm)/CoO shell (3 nm) nanoparticles (NPs)

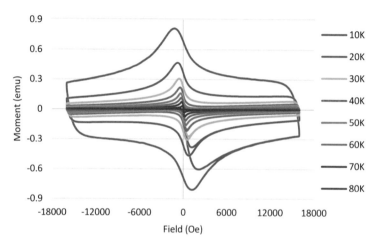

Fig. 11 $M(H)$ at $T = 10, 20, 30, 40, 50, 60, 70$, and 80 K for a YBCO high-temperature superconductor

magnetization of a sample that is not precisely aligned with the field. Although torque magnetometers enjoy the advantage of high sensitivity and accuracy, they are unsuited to other magnetic measurements of interest, e.g., major and minor hysteresis loops, remanence curves, first-order reversal curves (FORC), etc. A normal or single-axis VSM measures only M_{II}, the component of the magnetization parallel (or longitudinal) to the applied field, while the VVSM measures both M_{II} and $M_\perp$, the magnetization component perpendicular (or transverse) to the applied field. $M_\perp$ is directly related to the macroscopic torque, and hence the VVSM provides information that is essentially identical to that provided by a torque magnetometer.

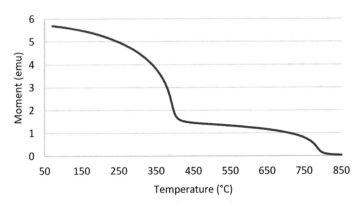

Fig. 12 High-temperature magnetization curve for a NiFe alloy sample

Fig. 13 Schematic top view
of the VVSM

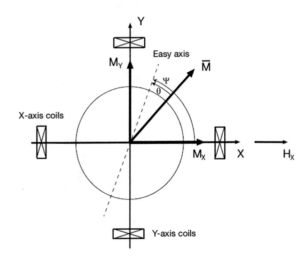

Measurement Methodology: A schematic top view illustration of the VVSM is shown in Fig. 13 with different angles defining the directions of magnetization and applied field. One pair of sensing coils is parallel to the applied field and senses the magnetization component longitudinal to the field $M_{||} = M_x$. A second set of coils is mounted at right angles to the applied field and senses the magnetization component transverse to the field $M_\perp = M_y$. Hence, the VVSM may be used to measure anisotropy in the xy plane. Sample rotation, ψ, in the xy plane is achieved by rotating the sample about the z axis via a computer-controlled motor attached to the VSM head. For measuring in-plane (IP) anisotropy, a sample is mounted to a bottom mount sample holder so that the applied field is parallel to the sample plane. For measuring out-of-plane (OOP) anisotropy, a sample is mounted to a side mount sample holder so that the orientation of the applied field can be varied from parallel

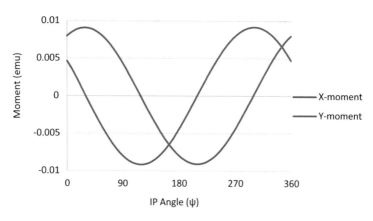

Fig. 14 M_x and M_y as a function of IP angle from $0°$ to $360°$ for a magnetic tape at $H = 0$

(IP) to perpendicular (OOP) with respect to the sample plane. The easy and hard axis of magnetization of either IP or OOP samples may be determined by measuring M_x and M_y as a function of angle at remanence ($H = 0$). Figure 14 shows M_x and M_y as a function of angle from $0°$ to $360°$ for a magnetic tape with IP anisotropy. At remanence, $M_y = 0$ and $M_x = $ maximum when the easy axis is aligned with the x-axis coils (i.e., parallel to the applied field direction). Alternatively, $M_y = $ maximum and $M_x = 0$ when the easy axis is aligned with the y-axis coils (i.e., perpendicular to the applied field direction).

Theory: Anisotropy Constants from Vector Magnetization Data

The energy density of the system is given by the sum of anisotropy and magnetic potential energy densities:

$$E_T = E_a + E_p$$

In the case of uniaxial symmetry, we have:

$$E_a = K_o + K_1\sin^2\theta + K_2\sin^4\theta + \ldots..$$

where θ is the angle between **M** and the easy direction of magnetization of the sample, and K_o is a constant and independent of angle. The potential energy is given by $-\mu_o$ **M•H**; hence:

$$E_p = -\mu_o MH \cos(\psi - \theta)$$

where ψ is the angle between **H** and the easy direction of magnetization (see Fig. 13). If $K_2 << K_1$:

$$E_T \approx K_o + K_1 \sin^2\theta - \mu_o\, MH\cos(\psi-\theta)$$

At equilibrium, the total energy is minimized, which requires that $dE_T/d\theta = 0$; hence:

$$K_1 \sin(2\theta) = \mu_o\, MH\sin(\psi-\theta).$$

The expression on the left is the torque $\tau = -dE_a/d\theta$ exerted on the magnetization M due to the intrinsic anisotropy of the material, and the expression on the right is the torque exerted on M by the applied magnetic field. Since $M\sin(\psi - \theta) = M_\perp$, the transverse magnetization:

$$K_1 \sin(2\theta) = \mu_o\, M_\perp H.$$

When $M_\perp$ is a maximum, $\sin(2\theta) = 1$. Hence:

$$K_1 = \mu_o\, M_\perp^{\,max} H.$$

Measurement of $M_\perp$ as a function of angle ψ therefore allows determination of the anisotropy constant K_1 and torque τ.

An alternative method of extracting a value for K_1 and also a value for K_2 is by fitting the torque curve to a Fourier series. All data in the torque curve is then used to calculate the anisotropy, not just the peak value.

Keeping terms in the energy to order $\sin^4\theta$, the torque is:

$$\tau(\theta) \approx (K_1 + K_2)\sin(2\theta) - 2K_2\sin(4\theta). \tag{2}$$

K_1 and K_2 can both be extracted by least squares fitting the torque curve to this function. If the magnitudes of the Fourier components for 2θ and 4θ are $\tau_{2\theta}$ and $\tau_{4\theta}$, respectively, then $K_2 = -\tau_{4\theta}/2$, $K_1 = \tau_{2\theta} - K_2$. If we only keep terms to order $\sin^2\theta$ in the energy, then $K_1 = \tau_{2\theta}$.

Vector VSM examples: Figures 15 and 16 show the major hysteresis loops, $M_x(H)$ and $M_y(H)$, respectively, as a function of IP angle for a uniaxially anisotropic magnetic tape. The loops were measured to applied fields of ±16 kOe (1.6 T) as a function of IP angle (ψ) from $0°$ ($H \parallel$ easy axis) to $90°$ ($H \parallel$ hard axis) in $15°$ increments.

$M_y(\psi)$ for IP angles ranging from $0° \le \psi \le 360°$ in $2.5°$ increments at an applied field of 25 kOe (2.5 T) was measured, and the torque curve in dyne-cm derived from the Fourier analysis of the $M_y(\psi)$ results is shown in Fig. 17. Least squares fitting the torque curve to Eq. (2) yields anisotropy constants of $K_1 = 5.1 \times 10^{-2}$ dyne-cm and $K_2 = -6.53 \times 10^{-4}$ dyne-cm.

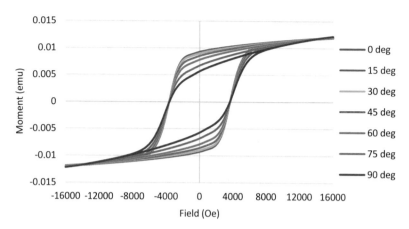

Fig. 15 $M_x(H)$ versus IP angle for a uniaxially anisotropic magnetic tape

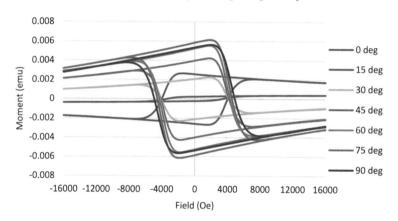

Fig. 16 $M_y(H)$ versus IP angle for a uniaxially anisotropic magnetic tape

4 First-Order Reversal Curves (FORC)

While the most common measurement used to characterize a material's magnetic properties is measurement of the major hysteresis or $M(H)$ loop as illustrated in Fig. 1, more complex magnetization curves covering states with field and magnetization values located inside the major hysteresis loop, such as minor hysteresis loops and first-order reversal curves (FORCs), can give additional information that can be used for characterization of magnetic interactions.

FORC is relevant to any hysteretic magnetic material that is comprised of fine micron- or nano-scale particles, grains, etc. It has been extensively used by earth and planetary scientists studying the magnetic properties of natural samples (rocks, soils, sediments, etc.) because FORC can distinguish between single-domain (SD), multi-domain (MD), and pseudo single-domain (PSD) behavior and because it

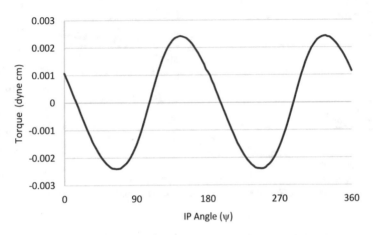

Fig. 17 Torque curve (dyne-cm) as a function of IP angle for a uniaxially anisotropic magnetic tape at $H = 25$ kOe (2.5 T)

can discriminate between different magnetic mineral species [27, 28]. FORC has proven to be useful in better understanding the nature of magnetization reversal and interactions in magnetic nanowires [29–33], nanomagnet arrays [34–37], thin-film magnetic recording media [38–40], exchange bias magnetic nanodot arrays [41] and thin-film magnetic multilayers [42, 43], nanostructured permanent magnet materials [44, 45], soft magnetic bilayers [46], and magnetocaloric effect (MCE) materials [47]. It has also been used to differentiate between phases in multiphase magnetic materials because it is very difficult to unravel the complex magnetic signatures of such materials from a hysteresis loop measurement alone [48, 49].

A FORC is measured by saturating a sample in a field H_{sat}, decreasing the field to a reversal field H_a, then measuring moment versus field H_b as the field is swept back to H_{sat}. This process is repeated for many values of H_a, yielding a series of FORCs as shown in Fig. 18 for a ferrite permanent magnet sample. The FORC distribution $\rho(H_a, H_b)$ is the mixed second derivative:

$$\rho(H_a, H_b) = -(1/2)\,\partial^2 M(H_a, H_b)/\partial H_a \partial H_b$$

A FORC diagram is a 2D or 3D contour plot of $\rho(H_a, H_b)$. It is common to change the coordinates from (H_a, H_b) to:

$$H_c = (H_b - H_a)/2, \quad H_u = (H_b + H_a)/2$$

H_u represents the distribution of interaction or reversal fields, and H_c represents the distribution of switching or coercive fields of the hysterons. The 2D FORC diagram for the ferrite permanent magnet sample is shown in Fig. 19.

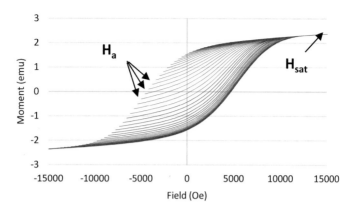

Fig. 18 Measured first-order reversal curves for a permanent ferrite magnet

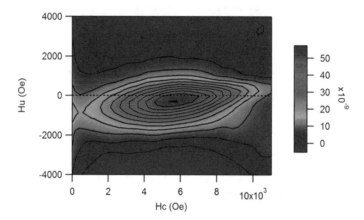

Fig. 19 FORC diagram for a permanent ferrite magnet

A FORC diagram not only provides information regarding the distribution of interaction and switching fields but also serves as a "fingerprint" that gives insight into the domain state and nature of interactions occurring in magnetic materials. In a FORC diagram, entirely closed contours are usually associated with SD behavior, while open contours that diverge toward the H_u axis are associated with MD, and open and closed contours together are associated with PSD. The peak in the FORC distribution is usually centered at a switching field H_c that correlates with the coercivity as determined from a hysteresis loop measurement. If the peak in the FORC distribution is centered at an interaction field $H_u = 0$, this means that interactions between particles, grains, etc. are weak. Conversely, if the peak is shifted toward positive H_u they are strong. Multiple peaks in a FORC diagram mean there are multiple magnetic phases in a material. And the very shape of the FORC distribution provides insight into the nature of interactions (dipolar, exchange) that are occurring in a magnetic material.

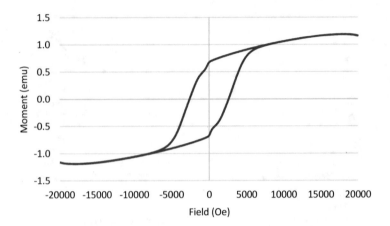

Fig. 20 Hysteresis $M(H)$ loop for BaFe$_{12}$O$_{19}$ nanoparticles

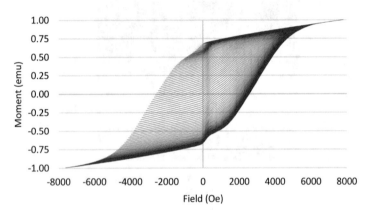

Fig. 21 FORCs for BaFe$_{12}$O$_{19}$ nanoparticles

There are a number of open-source FORC analysis software packages such as FORCinel [50] and VARIFORC [51]. In the results that follow, FORCinel was used to calculate the FORC distributions and plot the FORC diagrams.

FORC Examples: To demonstrate the utility of FORC measurements and analysis for differentiating magnetic phases, FORC data were acquired at $T = 300$ K on a sample consisting of nanometer-sized (~60 nm) barium hexaferrite BaFe$_{12}$O$_{19}$ exchange-coupled nanocomposite.

Figures 20 and 21 show the measured hysteresis $M(H)$ loop and FORCs, respectively. Figure 22 shows the resultant FORC diagram. There is a subtle "kink" in the $M(H)$ loop (Fig. 21) at low fields suggesting the presence of a low- and high-coercivity phase. The FORC diagram (Fig. 23) shows two peaks corresponding to the low- and high-coercivity components, and the region between the two peaks is related to the exchange coupling between the two phases [52].

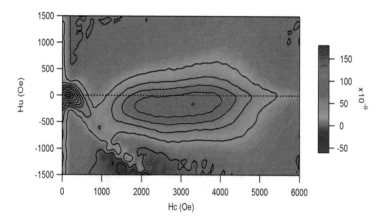

Fig. 22 FORC diagram for $BaFe_{12}O_{19}$ nanoparticles, the low- and high-coercivity phases are clearly differentiated

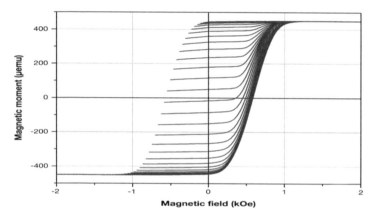

Fig. 23 FORCs for an array of nickel nanowires

Figure 23 shows a series of FORCs measured at $T = 300$ K for a periodic array of Ni nanowires with a mean diameter of 70 nm and interwire spacing of 250 nm. Figure 24 shows the FORC diagram and shows the distribution of coercive and interaction fields resulting from coupling between adjacent nanowires. The long tail centered at $H_u = 0$ has been theoretically shown to be due to interwire interactions occurring between higher coercivity wires within the array [53].

A typical series of FORCs can contain thousands to tens of thousands of data points, making the measurement very time-consuming if the measurement speed of the magnetometer is slow or if long signal averages are required to improve signal-to-noise ratio for low moment samples. Figure 25 shows 100 FORCs (8818 data points) recorded at 500 ms/point averaging in 1 h and 20 min for a CoPt bit-patterned magnetic media (thin-film) sample with saturation moment $m_{sat} < 100$ μemu

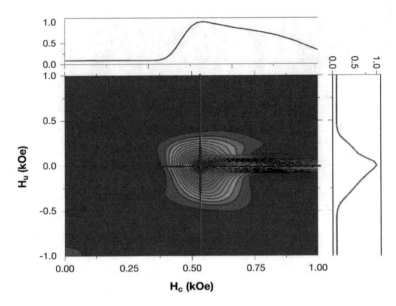

Fig. 24 FORC diagram for an array of nickel nanowires. The upper and right curves correspond to the profile of the distribution along the horizontal and vertical marked lines, respectively

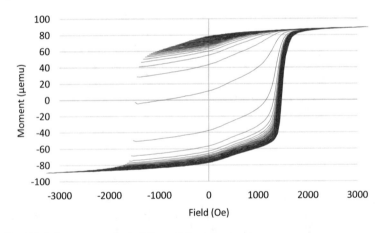

Fig. 25 FORCs for an array of sub-100 nm CoPt nanomagnets

$(1 \times 10^{-7} \, \text{Am}^2)$. This is a fraction of the time that would be required if using a superconducting magnet-based magnetometer. Figure 26 shows the resultant FORC diagram. Note that the peak in the FORC distribution is shifted toward negative interaction fields (H_u) and that the distribution has a "boomerang" shape. These features are usually associated with exchange interactions [54] suggesting that exchange interactions are occurring between adjacent CoPt nanomagnets within the array.

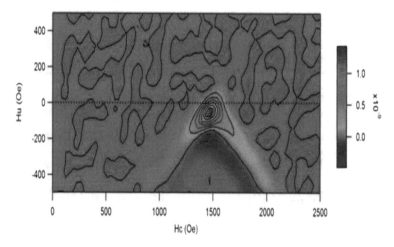

Fig. 26 FORC diagram for an array of sub-100 nm CoPt nanomagnets

5 Summary

Vibrating sample magnetometry (VSM) is the most common technique employed to characterize the magnetic properties of all manner of magnetic materials in any form as a function of magnetic field and temperature. In this chapter, we have described the principles of operation and its implementation in an electromagnet and discussed extensions for both low- and high-temperature measurements, vector measurements for torque curve and anisotropy constant determinations, and FORC measurements and analysis for characterization of magnetic interactions and coercivity distributions. We have also presented typical measurement results demonstrating the sensitivity, speed, and versatility of the VSM for magnetic materials characterization.

References

1. R.K. Dumas, T. Hogan, Recent advances in SQUID magnetometry, in Magnetic Measurement Techniques for Materials Characterization. Springer Nature (2021)
2. B.C. Dodrill, H.S. Reichard, Alternating gradient magnetometry, in Magnetic Measurement Techniques for Materials Characterization. Springer Nature (2021)
3. B.L. Morris, A. Wold, Faraday balance for measuring magnetic susceptibility. Rev. Sci. Instrum. **39**, 1968 (1937)
4. A. Sella, *Gouy's Tube* (Royal Society of Chemistry, 2010)
5. L. Rondin, J.P. Tetienne, T. Hingant, J.F. Roch, P. Maletinsky, V. Jacques, Magnetometry with nitrogen-vacancy defects in diamond. Rep. Prog. Phys. **77**, 056503 (2014)
6. Lake Shore Cryotronics, USA, www.lakeshore.com
7. Microsense, USA, www.microsense.net

8. Quantum Design, USA, www.qdusa.com
9. Cryogenic Limited, UK, www.cryogenic.co.uk
10. I.D. Mayergoyz, *Mathematical Models of Hysteresis and their Applications*, 2nd edn. (Academic Press, 2003)
11. S. Foner, Versatile and Sensitive Vibrating Sample Magnetometer. Rev. Sci. Instrum. **30**, 548 (1959)
12. US Patent # 2,946,848
13. A. Zieba, S. Foner, Detection coil, sensitivity function, and geometry effects for vibrating sample magnetometer. Rev. Sci. Instrum. **53**, 1344 (1982)
14. National Institute of Standards & Technology (NIST) standard reference materials (SRM) 772a (Ni) and 2853 (YIG)
15. J. Lindemuth, J. Krause, B. Dodrill, Finite sample size effects on the calibration of vibrating sample magnetometer. IEEE Trans. Magn. **37**, 4 (2001)
16. J. Mallinson, Magnetometry coils and reciprocity. J. Appl. Phys. **37**, 2514 (1966)
17. A. Niazi, P. Poddar, A.K. Rastogi, A precision, low-cost vibrating sample magnetometer. Curr. Sci. **79**, 1 (2000)
18. R.M. El-Alaily, M.K. El-Nimr, S.A. Saafan, M.M. Kamel, T.M. Meaz, S.T. Assar, Construction and calibration of a low-cost and fully automated vibrating sample magnetometer. J. Magn. Magn. Mater. **386**, 25 (2015)
19. D. Jordan, D. Chavez, D. Laura, L.M. Hilario, E. Moenteblanco, A. Gutarra, L. Felix, Detection of magnetic moment in thin films with a home-made vibrating sample magnetometer. J. Magn. Magn. Mater. **456**, 56 (2018)
20. V. Dominguez, A. Quesada, J.C. Mainuez, L. Moreno, M. Lere, J. Spottorno, F. Giacomone, J.F. Ferandez, A. Hernando, M.A. Garcia, A simple vibrating sample magnetometer for macroscopic samples. Rev. Sci. Instrum. **89**, 034707 (2018)
21. V.I. Nizhankovskii, L.B. Lugansky, Vibrating sample magnetometer with a step motor. Meas. Sci. Technol., 1533 (2006)
22. J.P.C. Bernards, G.J.P. van Engelen, H.A.J. Cramer, An improved detection coil systems for a biaxial vibrating sample magnetometer. J. Magn. Magn. Mater. **123**, 141 (1993)
23. E.O. Samel, T. Bolhuis, J.C. Lodder, An alternative approach to vector vibrating sample magnetometer detection coil setup. Rev. Sci. Instrum. **69**, 9 (1998)
24. B.C. Dodrill, Model 8600 Vector Vibrating Sample Magnetometer, Lake Shore Cryotronics Application Note (2020)
25. S.U. Jen, J.Y. Lee, Method of easy axis determination of uniaxial magnetic films by vector vibrating sample magnetometer. J. Magn. Magn. Mater. **271**, 237 (2004)
26. P. Stamenov, J.M.D. Coey, Vector vibrating sample magnetometer with a permanent magnet flux source. J. Appl. Phys. **99**, 08D912 (2006)
27. C.R. Pike, A.P. Roberts, K.L. Verosub, Characterizing interactions in fine magnetic particle systems using first order reversal curves. J. Appl. Phys. **85**, 6660 (1999)
28. A.P. Roberts, C.R. Pike, K.L. Verosub, First-order reversal curve diagrams: A new tool for characterizing the magnetic properties of natural samples. J. Geophys. Res. **105**, 461 (2000)
29. A. Rotaru, J. Lim, D. Lenormand, A. Diaconu, J. Wiley, P. Postolache, A. Stancu, L. Spinu, Interactions and reversal field memory in complex magnetic nanowire arrays. Phys. Rev. B **84**(13) (2011) 134431
30. O. Trusca, D. Cimpoesu, J. Lim, X. Zhang, J. Wiley, A. Diaconu, I. Dumitru, A. Stancu, L. Spinu, Interaction effects in Ni nanowire arrays. IEEE Trans. Magn. **44**(11), 2730 (2008)
31. A. Arefpour, M. Almasi-Kashi, A. Ramazani, E. Golafshan, The investigation of perpendicular anisotropy of ternary alloy magnetic nanowire arrays using first order reversal curves. J. Alloys Comp. **583**, 340 (2014)
32. B. C. Dodrill, L. Spinu, First-Order-Reversal-Curve Analysis of Nanoscale Magnetic Materials, Technical Proceedings of the 2014 NSTI Nanotechnology Conference and Exposition, CRC Press (2014)
33. A. Sharma, M. DiVito, D. Shore, A. Block, K. Pollock, P. Solheid, J. Feinberg, J. Modiano, C. Lam, A. Hubel, B. Stadler, Alignment of collagen matrices using magnetic nanowires and

magnetic barcode readout using first order reversal curves (FORC). J. Magn. Magn. Mater. **459**, 176 (2018)

34. F. Beron, L. Carignan, D. Menard, A. Yelon, in *Extracting Individual Properties from Global Behavior: First Order Reversal Curve Method Applied to Magnetic Nanowire Arrays, Electrodeposited Nanowires and their Applications*, ed. by N. Lupu, (INTECH, Croatia, 2010), p. 228

35. R. Dumas, C. Li, I. Roshchin, I. Schuller, K. Liu, Magnetic fingerprints of sub-100 nm Fe Nanodots. Phys. Rev. B **75**, 134405 (2007)

36. D. Gilbert, G. Zimanyi, R. Dumas, M. Winklhofer, A. Gomez, N. Eibagi, J. Vincent, K. Liu, Quantitative decoding of interactions in tunable nanomagnet arrays. Sci. Rep. **4**, 4204 (2014)

37. J. Graffe, M. Weigand, C. Stahl, N. Trager, M. Kopp, G. Schutz, E. Goering, Combined first order reversal curve and X-ray microscopy investigation of magnetization reversal mechanisms in hexagonal Antidot arrays. Phys. Rev. B **93**, 014406 (2016)

38. B. Valcu, D. Gilbert, K. Liu, Fingerprinting Inhomegeneities in recording media using first order reversal curves. IEEE Trans. Magn. **47**, 2988 (2011)

39. A. Stancu, E. Macsim, Interaction field distribution in longitudinal and perpendicular structured particulate media. IEEE Trans. Magn. **42**(10), 3162 (2006)

40. M. Winklhofer, R.K. Dumas, K. Liu, Identifying reversible and irreversible magnetization changes in prototype patterned media using first- and second-order reversal curves. J. Appl. Phys. **103**, 07C518 (2008)

41. R. Dumas, C. Li, L. Roshchin, I. Schuller, K. Liu, Deconvoluting reversal modes in exchange biased nanodots. Phys. Rev. B **144410** (2012)

42. R. Gallardo, S. Khanai, J. Vargas, L. Spinu, C. Ross, C. Garcia, Angular dependent FORC and FMR of exchange biased NiFe multilayer films. J. Phys. D. Appl. Phys. **50**, 075002 (2017)

43. N. Siadou, M. Androutsopoulos, I. Panagiotopoulos, L. Stoleriu, A. Stancu, T. Bakas, V. Alexandrakis, Magnetization reversal in [Ni/Pt](6)/Pt(x)/[Co/Pt](6) multilayers. J. Magn. Magn. Mater. **323**(12), 2011 (1671)

44. T. Schrefl, T. Shoji, M. Winklhofer, H. Oezeit, M. Yano, G. Zimanyi, First order reversal curve studies of permanent magnets. J. Appl. Phys. **111**, 07A728 (2012)

45. M. Pan, P. Shang, H. Ge, N. Yu, Q. Wu, First order reversal curve analysis of exchange coupled SmCo/NdFeB nanocomposite alloys. J. Magn. Magn. Mater. **361**, 219 (2014)

46. M. Rivas, J. Garcia, I. Skorvanek, J. Marcin, P. Svec, P. Gorria, Magnetostatic interaction in soft magnetic bilayer ribbons unambiguously identified by first order reversal curve analysis. Appl. Phys. Lett. **107**, 132403 (2015)

47. V. Franco, F. Beron, K. Pirota, M. Knobel, M. Willard, Characterization of magnetic interactions of multiphase magnetocaloric materials using first order reversal curve analysis. J. Appl. Phys. **117**, 17C124 (2015)

48. B.C. Dodrill, First-Order-Reversal-Curve Analysis of Nanocomposite Permanent Magnets, Technical Proceedings of the 2015 TechConnect World Innovation Conference and Expo, CRC Press (2015)

49. C. Carvallo, A.R. Muxworthy, D.J. Dunlop, First-order-reversal-curve (FORC) diagrams of magnetic mixtures: Micromagnetic models and measurements. Phys. Earth Planet. Inter. **154**, 308 (2006)

50. R.J. Harrison, J.M. Feinberg, FORCinel: An improved algorithm for calculating first-order reversal curve distributions using locally weighted regression smoothing. Geochem. Geophys. Geosyst. **9**, 11 (2008)

51. R. Egli, VARIFORC: An optimized protocol for calculating non-regular first-order reversal curve (FORC) diagrams. Glob. Planet. Chang. **203**, 110 (2013)

52. Y. Cao, M. Ahmadzadeh, K. Xe, B. Dodrill, J. McCloy, Multiphase magnetic systems: Measurement and simulation. J. Appl. Phys. **123**(2), 023902 (2018)

53. C. Dubrota, A. Stancu, What does a first order reversal curve really mean: A case study: Array of ferromagnetic nanowires. J. Appl. Phys. **113**, 043928 (2013)

54. D. Roy, P.S.A. Kumar, Exchange spring behaviour in $SrFe_{12}O_{19}$-$CoFe_2O_4$ nanocomposites. AIP Adv. **5** (2015)

Recent Advances in SQUID Magnetometry

Randy K. Dumas and Tom Hogan

Abstract This chapter aims to provide a contemporary overview of Superconducting quantum interference device (SQUID)-based magnetometry. As there are many existing, and well-written, resources devoted to SQUID-based magnetometry (Clarke and Braginski. The SQUID Handbook Vol. 1: Fundamentals and Technology of SQUIDs and SQUID Sytems, New York: Wiley, 2004, Clarke and Braginski. The SQUID Handbook, Vol. 2: Applications of SQUIDs and SQUID Systems, New York: Wiley, 2004, Fagaly, Rev. Sci. Instrum 77:101101, 2006), the goal of this chapter is to highlight some of the most recent innovations rather than to repeat much of what has already been published. For example, advances in sample transport, magnet technology, and SQUID detection modes have dramatically improved measurement sensitivity and throughput over the past decade. A discussion on techniques to improve measurement accuracy is also provided near the end of this chapter.

Keywords Magnetometry · Superconducting quantum interference device · SQUID · DC SQUID · Josephson junction · Superconducting magnet · Cryogenics · Gradiometer · Flux-locked loop · Persistent switch · SQUID-VSM

1 Introduction to SQUID Magnetometry

A magnetometer's primary function is to measure the extrinsic material quantity of *magnetic dipole moment* as a function of an externally varied parameter, typically applied magnetic field or temperature, or in some cases time. As discussed at length in other chapters in this book, there are many different types of magnetometers that rely on a variety of physical mechanisms (e.g., induction, force, optical, etc.) to measure the magnetic moment with varying degrees of precision, accuracy, and

R. K. Dumas (✉) · T. Hogan
Quantum Design Inc., San Diego, CA, USA
e-mail: dumas@qdusa.com

© Springer Nature Switzerland AG 2021
V. Franco, B. Dodrill (eds.), *Magnetic Measurement Techniques for Materials Characterization*, https://doi.org/10.1007/978-3-030-70443-8_3

ease of use. As the SQUID, which relies on superconductivity and the Josephson effect, intrinsically requires cryogenic temperatures to function, minimal additional infrastructure is needed to then provide a variable temperature environment (1.8–400 K typical) and the cooling necessary for a superconducting magnet, which allows for measurements in high (7 T typical) fields. In sum, the SQUID-based magnetometer therefore provides the experimenter the ability to study a wide variety of materials over a large environmental parameter space with extremely high precision (10^{-8} emu typical).

2   Primary Components of a SQUID-Based Magnetometer

Sample Transport

As with any induction-based magnetometer, it is necessary to move the sample through (or at least in close proximity to) a gradiometer or pickup coil. A gradiometer in this context is a specifically designed coil of wire which is sensitive to changes in the magnetic flux due to the moving magnetic sample. The second-order gradiometer typically used in SQUID-based magnetometers will be described in more detail in the following section. To enable multiple measurement modes (e.g., DC extraction, SQUID-VSM, AC susceptibility), it is critical not only to reliably translate the sample over a long (~6 cm) distance at a well-defined speed but also to vibrate the sample at a given frequency (~10 Hz) and amplitude (0.1–8 mm) with high precision.

Second-Order Gradiometer

The fundamental principle of operation in an induction-based magnetometer is to move a sample of some unknown moment through, or near, a set of pickup coils; Faraday's law then relates the associated change in magnetic flux with the induced electric fields which cause charges to flow in a conductor. If the details of the coil construction are known, and the motion is precisely controlled, the task then becomes to quantify the results of this induction, from which the moment can be determined. For a gradiometer comprised of wire in the normal state, it would be sufficient to simply measure the voltage across the terminals of the sets of wire turns as the sample is moved in some characteristic way through the coils. The magnitude of the observed voltage is equivalent to the time derivative of the flux through the gradiometer, and the moment can readily be calculated.[1] The dissipative nature

---

[1]This is, in fact, the underlying principle of operation for VSM instruments which do not employ SQUID detection.

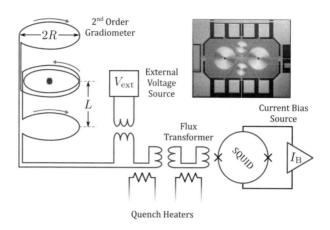

Fig. 1 Superconducting elements, including the SQUID, flux transformer, and gradiometer coils, are shown in red. Normal wiring is shown in black with external electronics indicated by green. The gradiometer geometry parameters R and L are labeled; gray arrows indicate the directionality of the coil windings. A small blue circle shows the position of a vibrating sample during a VSM measurement. The inset image depicts an example of a SQUID chip configuration

of normal (i.e., non-superconducting) wire and sensitivity limits of commercially available electronics, however, set a lower bound on the size of moments which can be resolved.

For the most precise inductive magnetometers, the gradiometer coil is instead made from a closed loop of *superconducting* wire. Here, measuring the induced current is not straightforward. A voltmeter is not helpful since a superconductor cannot support a voltage drop across itself, and no real ammeter would have sufficiently low input impedance (when compared with the true zero resistance of a superconducting state) not to perturb the system. Instead, the method used to couple to the gradiometer is to wire it in a series with another coil to make a superconducting flux transformer as shown schematically in Fig. 1.

Of the many potential gradiometer designs, a balanced second-order gradiometer configuration is typically chosen with the geometry shown in Fig. 1: the upper coil is wound clockwise, the center coil is then comprised of twice as many counterclockwise wound coils, and the bottom coil is again wound clockwise. This configuration is chosen to reduce noise in the detection circuit caused by uncontrolled fluctuations in the magnetic field and rejects external magnetic fields as well as linear field gradients. Further, such an arrangement also minimizes the influence of drifts in the (potentially large) applied magnetic field as the nearly balanced design ensures that any flux change in the center coil is cancelled by the flux change in the top and bottom coils. Any small difference in the enclosed area between the sets of turns of the gradiometer will lead to a small imbalance and as the external field is ramped, allowing persistent currents to build up in the detection coil. To eliminate these, it is crucial to include a mechanism allowing for a small portion of the detection coil to be temporarily quenched (i.e., heated above the superconducting transition temperature) to dissipate any such trapped currents.

For the utmost sensitivity, it would be ideal for the gradiometer to be as small as possible in order to better couple to the sample (i.e., not have magnetic field flux lines return *within* the pickup loop). Conversely, to minimize effects related to sample geometry and radial offsets within the gradiometer, it is desirable to ensure that the detection coils are relatively large compared to the sample size. If the sample occupies a sufficiently small volume within the gradiometer, it can be easily modeled as a point-dipole source. There is therefore a trade-off that needs to be optimized, and realistically, experimental samples will always fill a nonzero volume within the gradiometer and in turn impact the measurement accuracy. Unless the sample is the exact same size and shape as the reference sample used to calibrate the system, the reported moment value will have a systematic offset from the true moment. There are several techniques to then calculate additional scale factors to account for different sample sizes and shapes and improve measurement accuracy, which are discussed later.

It is a property of a closed superconducting circuit that the total flux enclosed by its constituent elements (i.e., the various sets of wire turns) remains constant. This results in a "flux linkage" between the larger turns of the gradiometer itself and the smaller ones which couple it to the detection circuit. Any change in flux resulting from translating a magnetic moment through the primary coils induces a screening current which causes a flux change of the same magnitude to be produced at the secondary coils. Since the conserved quantity is flux, a proper choice of the loop area and number of turns of the secondary coils can produce significantly larger magnetic fields than those of the sample which generated the flux in the primary pickup coil.

In this way, the flux generated by the sample can be "transported" to a different spatial location to couple into a circuit which can measure it directly. As before, if the details of the gradiometer construction are known, and the sample (for now) can reasonably be approximated as a point dipole, then an analytic solution for the flux as a function of position can readily be determined from the application of basic magnetostatics. A measurement of that flux, as the sample is translated according to a known scheme, is thus an indirect measurement of the moment itself, which is the goal. One of the most sensitive techniques used to measure small changes in magnetic flux in modern magnetometers makes use of a superconducting quantum interference device, or a SQUID.

The DC SQUID

SQUID construction in principal is relatively simple—a device called a Josephson junction is placed in line with a superconducting loop of wire. Historically, commercial SQUID magnetometers utilized radio frequency (RF) SQUIDs, which employ a single junction. The direct-current (DC) SQUID requires two nearly identical Josephson junctions within the superconducting loop, which can prove to be challenging to fabricate. However, advances in film deposition, lithography, and device fabrication have made producing DC SQUIDs at the commercial scale feasible. While slightly more involved to fabricate, the DC SQUID has several advantages over RF SQUIDS. For example, DC SQUIDs do not require RF signals to function, thus simplifying the electronics, and they typically offer a much larger dynamic range. The functionality of a DC SQUID is built up subsequently in the following sections.

Properties of SC Loops

First, then properties of a simple, singular loop of superconducting wire are examined. It can be shown that any supercurrent flowing in this loop is subject to constraints which derive from the London equations. In the case of a multiply-connected region (such as a loop), a quantity referred to as the "fluxoid" is both quantized and conserved [4, 5] and can take a nonzero value. For a loop with self-inductance L subject to an applied magnetic flux Φ_a and passing a supercurrent I_S, the fluxoid Φ' can be represented as [6]:

$$\Phi' = \Phi_a + LI_s + \oint \Lambda \, J_s \cdot dl$$

where J_s is the supercurrent density and Λ is a phenomenological quantity related to the density of superconducting carriers and their effective charge and mass. The conservation condition simply means that the value of the fluxoid quantity remains constant in time:

$$\frac{\partial}{\partial t} \Phi' = 0$$

$$\therefore \Phi' = C$$

while the quantization condition is represented as:

$$\Phi' = C = n\Phi_0$$

where Φ_0 is the fundamental flux quantum.[2] The expression can be further simplified if the line integral of the current density is evaluated along a contour

---

[2]This is sometimes referred to a "fluxon" and has a value of $\Phi_0 = 2.067 \cdot 10^{-15}$ Wb.

chosen to be within the bulk of the material. Since all the supercurrent travels in the skin, $J_S = 0$ in the bulk. This leaves a more basic relation governing the flux inside the loop:

$$n\Phi_0 = \Phi_a + LI_S$$

Thus, after undergoing a transition to the superconducting state, the total flux penetrating a loop is always a fixed integer number of flux quanta. A few cases are discussed briefly to make this point clear:

1. A loop is cooled to the superconducting state with zero applied field, making $n = 0$. A small amount of flux $\Phi_a = 0.25\ \Phi_0$ is then applied. The induced screening current would then be $I_S = -0.25\frac{\Phi_0}{L}$, where the negative sign denotes the direction of current such that the applied flux is canceled. The screening current will scale linearly in this fashion with the applied flux indefinitely until the screening current exceeds the critical current of the superconductor and the state is quenched.
2. Alternatively, a loop is cooled to the superconducting state under a small applied field $\Phi_a = 0.75\ \Phi_0$. In this case, the smallest screening current[3] required to bring the total flux to an integer number would then be $I_S = +0.25\frac{\Phi_0}{L}$, where the positive sign denotes the direction of current which *aids* the applied flux. Here, $n = 1$. If the applied flux is increased to $\Phi_a = 1.5\ \Phi_0$, however, the expected screening current changes magnitude *and* direction to become $I_S = -0.5\frac{\Phi_0}{L}$, in order to preserve the $n = 1$ condition henceforth.

With this property of superconducting loops now articulated, the other primary, and necessary, component of a SQUID is discussed: the Josephson junction.

Josephson Junctions

These junctions occur when two superconductors are joined by a thin barrier of normal (non-superconducting) material. If this barrier is thin enough (typically defined as being smaller than the coherence length of the Cooper pairs in the superconductor), then it will exhibit some unique properties. Namely, if the current is below some critical threshold current I_C defined by the comprising materials and geometry of the junction, then there will be no observed voltage drop across the normal segment of the junction: $\Delta V = 0$. Above the critical current, the behavior rapidly recovers that of a typical Ohmic element, where the observed voltage drop

---

[3]Generally, the system will choose the state which satisfies the quantization condition and requires the least amount of screening current. For this example, while $I_S = -0.75\frac{\Phi_0}{L}$ also satisfies the condition with $n = 0$, this state requires $\Delta U = \frac{1}{2}L\left(0.75^2 - 0.25^2\right)\left(\frac{\Phi_0}{L}\right)^2 = 0.25\frac{\Phi_0^2}{L}$ more energy and so is not "preferred."

Fig. 2 A representation of the generalized current-voltage (IV) response curve of a Josephson junction is shown in blue. The dashed red line corresponds to the Ohmic resistance of the normal portion of the device once the current through the junction exceeds I_C beyond the narrow transition region

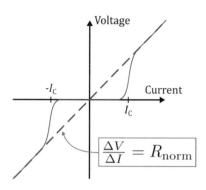

is simply proportional to the flowing current and scales with the resistance of the normal segment of the junction. This behavior is summarized in Fig. 2.

The transition region, where $I \approx I_C$, is of particular interest: here, a very small variation in current results in a large change in the measured voltage. The utility of a Josephson junction in constructing a SQUID lies largely in the steepness of the IV curve slope in this transition region, which sets the sensitivity of the device.

SQUID Functionality

A complete DC SQUID is constructed by placing two Josephson junctions (indicated schematically with an "X") in series within a superconducting loop, as shown in Fig. 1. To operate the SQUID, it is first cooled below the superconducting transition with no applied field, and then a bias current $I_B \approx 2I_C$ is applied. This bias current is tuned such that it is just sufficient to bring the Josephson junction of each separate "branch" near the transition region of their nominally identical IV curves.

Consider now the impact of adding some small amount of applied flux $\Phi_a = 0.45 \ \Phi_0$ to the SQUID. The superconducting loop wants to maintain the quantization of total flux condition for $n = 0$, so a screening current of $I_S = -0.45 \frac{\Phi_0}{L}$ is observed. If this applied flux is increased to $\Phi_a = 0.55 \ \Phi_0$, what then? From the discussion of the basic loop, one anticipates a screening current of $I_S = -0.55 \frac{\Phi_0}{L}$, but instead a current of $I_S = +0.45 \frac{\Phi_0}{L}$ is actually observed. What happened?

As mentioned prior, the system "prefers" to meet the quantization condition with as low of an energy cost as possible, so instead of canceling the increasingly larger applied flux Φ_a, the induced current changes direction to produce *additional flux* in the same direction, such that now $n\Phi_0 = \Phi_a + \Phi_{ind}$ is satisfied for $n = 1$. The difference in the case of the DC SQUID is that while the condition of flux quantization remains enforced, from the inclusion of the Josephson junctions on the circuit, the total flux in the loop is no longer *conserved*. As the threshold of $\Phi_a = n \cdot 0.5 \ \Phi_0$ is crossed, the junctions temporarily revert to the normal state, and

Fig. 3 The induced current in a DC squid is shown by the blue curve, which is periodic in the applied flux. The discontinuous jumps represent the allowance of a flux line corresponding to one flux quanta Φ_0 to enter the SQUID loop

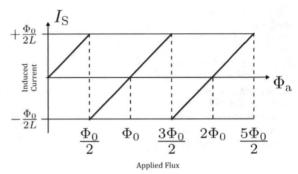

an additional flux line of value Φ_0 enters the loop and the induced current begins flowing in the opposite direction. As $\Phi_a \to n\Phi_0$, the screening current approaches zero. If the applied flux increases further, this process simply repeats itself, resulting in the periodic behavior shown in Fig. 3.

A consequence of this is that the induced current is no longer a single-valued function of the applied flux. Knowing the current relates the applied flux only insofar as its value relative to the nearest modulus $n\Phi_0$, meaning the DC SQUID by itself is not generally useful to determine the *absolute* amount of flux it is subject to. Instead, the utility lies in the ability to measure very precisely small *relative changes* in the magnetic flux coupled in.

Next, the voltage observed across one of the Josephson junctions of the DC SQUID is considered. As discussed earlier, the measured voltage is a function of the current passed by the junction. The current through each junction is simply a sum of the bias and screening currents to give $I_{JJ} = \frac{I_B}{2} \pm I_S$. Since the bias current is chosen to bring the device into the transition region in the IV curve, now small changes in the screening current I_S will register as comparatively large changes in voltage due to the steepness of the curve. This is the fundamental operating principle leveraged by SQUID detection generally, and their function is readily summarized as a high-gain current-to-voltage amplifier [3].

Flux Locking

Since the absolute amount of flux applied to the SQUID cannot be determined from the voltage measurement alone, a feedback circuit, shown in Fig. 1, is required to ascertain this value. The scheme described here constitutes an *external* feedback implementation (though the corresponding internal version can be implemented as well). The setup is as follows: the SQUID is quenched with zero applied field (imagine the sample is far away from the pickup coils of the gradiometer); this dissipates any current flowing within the loop. When it reenters the superconducting state, there are no screening currents. The SQUID is then biased to near the transition region of the IV curve of its Josephson junctions.[4]

As the sample translates toward the gradiometer, a small amount of flux is "seen" by the pickup coils. This flux is coupled to the SQUID by the flux linkage property of the primary and secondary coils of the gradiometer. Thus, the SQUID is now subject to a small amount of applied field, and in turn the resulting screening current is generated to maintain the flux quantization (assume for the moment that this is less than 0.5 Φ_0 worth of flux). The detection circuitry sees the corresponding change in the voltage across a junction compared with the baseline from the biasing current. Since the inductance of the DC SQUID is known, it can be inferred from the voltage the precise amount of additional flux that originates from the sample.

To balance this additional flux, a new component is added to the system. An additional source of flux, controlled by an external voltage source, is coupled into the gradiometer loop, as shown in Fig. 1. The feedback circuit applies a voltage V_{ext} to generate a current I_{ext} through an inductor L_{ext} to produce some flux $I_{ext} \cdot L_{ext} = \Phi_{ext}$; this current is chosen such that $\Phi_{ext} = -\Phi_a$. As a result, no additional screening current persists in the gradiometer loop, and in turn, no additional current is induced in the DC SQUID. As a result, the DC SQUID remains at the nominal center of the transition region of the IV curve. In other words, the SQUID is essentially used as a very sensitive null detector. The electronics applying the adjustment (to cancel out the applied flux) are sufficiently fast that the sample can be moved as quickly as the transport motor can carry it, and the SQUID is not reset. This condition, of keeping the flux through the DC SQUID constant, is referred to as "flux locking," and the feedback scheme as a whole is a "flux-locked loop."

Ultimately, it is the voltage applied to the external inductor which is interpreted as the "signal" from the SQUID. This voltage is directly proportional to the amount of flux from the sample which penetrates the area enclosed by the gradiometer coils. With the precise position of the sample known by way of the transport motor, a calibrated SQUID detection circuit readily allows for sample flux as a function of position, $\Phi(z)$, to be measured. The sample moment can be extracted from this data using several techniques, which are examined in the following section.

3 Detection Modes

As with any magnetometer which relies on magnetic induction, the sample must move through, or in close proximity, to the pickup coils. A traditional vibrating

---

[4]Though not denoted explicitly in the figures here, it should be mentioned that the bias current is itself modulated (typically at frequencies of a few 100 s of kHz) with some small additional amplitude. This permits lock-in detection of the associated voltage drop across the SQUID junctions, increasing the overall signal-to-noise ratio of the measurement.

sample magnetometer (VSM), which relies on conventional Cu-wire pickup coils, requires that the sample be continuously vibrated, usually at frequencies near 50 Hz. Conversely, the all-superconducting nature of the SQUID and second-order gradiometer enforces no such requirement on continuous sample motion. In fact, the sample could in theory move through the gradiometer quasi-statically. However, in the interest in increasing measurement throughput, the sample is either translated or vibrated within the gradiometer such that a typical measurement takes ~1 s. The following sections will describe two distinct detection modes where the sample is either translated (the so-called DC scan) or vibrated (the so-called SQUID-VSM mode) within the second-order gradiometer. The distinct advantages of each detection mode will then be compared.

Traditional DC Scan

The traditional DC scan relies on moving the sample through the entire length of the second-order gradiometer. The resulting SQUID response, measured via the aforementioned compensation (or feedback) voltage from the flux-locked loop within the SQUID detection electronics, is then recorded as a function of sample position, defining the so-called SQUID waveform, $V(z)$, shown in Fig. 4 (black circles). The basic functional form of the SQUID waveform can be generally expressed as:

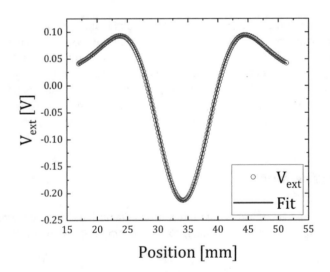

Fig. 4 SQUID signal, and corresponding fit (red line), as a function of position as small magnetic sample is translated through the second-order gradiometer

$$V(z) = O + Sz + C \left\{ 2\left[R^2 + (z - z_0)^2 \right]^{-3/2} - \left[R^2 + (L + z - z_0)^2 \right]^{-3/2} \right.$$
$$\left. - \left[R^2 + (-L + z - z_0)^2 \right]^{-3/2} \right\}$$

where O is an offset voltage, S is a slope term proportional to the linear SQUID drift, C provides the amplitude or scale of the response, R is the radius of the gradiometer, L is half the length of the gradiometer, and z_0 is the sample center position. This fit function assumes the sample can be represented as a physically small (relative to the gradiometer) magnetic dipole. By simply fitting the experimental SQUID waveform, the extracted amplitude C is then directly proportional to the magnetic moment via a system-dependent calibration factor. The sample center position, z_0, can be fixed or allowed to vary slightly to correct for longitudinal sample centering errors or a small drift in the sample position as the temperature is varied. The quality of resulting fit to the experimental data (Fig. 4, red line) is directly related to the accuracy of the resulting calculated moment. If a given waveform is poorly fit, then the resulting moment data point should be discarded. If the resulting fits are persistently poor over a large parameter space, and the instrument itself is known to be working properly and outside sources of noise have been considered, this could be indicative of:

- A sample which is physically too large and can therefore no longer be considered a point dipole.
- A sample magnetic moment which is too small and below the detection limit of the instrument.
- A large and/or inhomogeneous background of the sample holder.

These issues can often be remedied simply by changing the sample size and improving the sample holder, techniques which are discussed later.

Squid-VSM

The SQUID-VSM [7] detection mode relies on sinusoidally oscillating the sample at the center of the second-order gradiometer at an angular frequency of $\omega = 2\pi f$. For small oscillation amplitudes, A, the local voltage profile is approximately quadratic, namely, $V(z) \approx Cz^2$, where $z = A\sin(\omega t)$. Therefore, the generated voltage will pass through a minimum and maximum twice during each oscillation period, resulting in a generated voltage with twice the oscillation frequency (2ω) when compared with the physical vibration. This is shown schematically in Fig. 5. The time dependence of the resulting voltage signal can then be expressed as:

$$V(t) \approx \frac{CA^2}{2} \left[1 - \cos(2\omega t) \right]$$

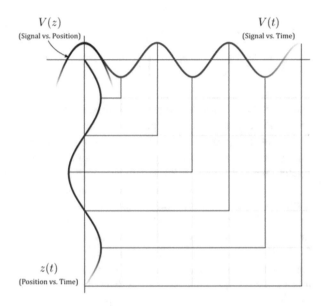

Fig. 5 A schematic depiction of the oscillating parameters in VSM mode. A portion of the response function $V(z)$ is shown in red for only the peak at the center; due to the geometry of the gradiometer,s this can be locally approximated as a quadratic function. The periodic displacement about the center of the gradiometer of the sample in time $z(t)$ with frequency f is shown in blue. Due to the symmetric nature of $V(z)$, this produces a signal $V(t)$ which oscilates in time with frequency $2f$, as shown in green. The drop lines highlight the doubling in frequency of the time-varying signal as compared with mechanical vibration which generates it

Lock-in-based detection is used to measure amplitude of the voltage response component with twice the physical oscillation frequency. Several aspects of $V(t)$ can be exploited to improve measurement precision and accuracy, for example:

- As the generated voltage signal varies at $2f$, SQUID-VSM is less susceptible to mechanical noise, which primarily presents itself at the fundamental physical oscillation frequency, f.
- The induced voltage is proportional to A^2, resulting in a large dynamic range. Small magnetic moments will benefit from using the largest possible amplitude to maximize the generated voltage. Conversely, very large magnetic moments can be still be measured by employing the lowest available range of oscillation amplitudes.
- The magnitude of the induced voltage response is independent of the oscillation frequency. Therefore, relatively slow ~10 Hz frequencies can be employed. A lower frequency minimizes both mechanical noise issues and complications from eddy current generation in the sample or sample environment.
- The relatively small oscillation amplitudes (<8 mm typical) ensure that any temperature and magnetic field gradients seen by the sample are much smaller than those experienced during a traditional DC scan (30–60 mm typical).

Strengths and Weaknesses of each Detection Mode

While each detection mode described above utilizes the output voltage generated by the SQUID detection electronics, the resulting magnetic moment is extracted in two very different ways. The DC scan mode relies on fitting the *spatial variation* of the voltage, $V(z)$, to a specific functional form that assumes a point (or at least relatively small) dipole moving through the second-order gradiometer. Very differently, the SQUID-VSM mode relies on simply measuring the amplitude of the time varying voltage, $V(t)$, by using lock-in detection as the sample is slowly oscillated at the center of the gradiometer. These differences can be leveraged to the benefit of the experimenter depending on the needs and constraints of the measurement.

For example, if the sample mounting technique results in any components not rigidly mounted to the holder, and the sample holder has a diameter that spans a significant fraction of the sample chamber diameter or is particularly massive (as is commonly the case for pressure cell sample holders), then the relatively slow and gentle motion of the sample provided by a DC scan will generally yield superior results. If a high density of data points is desired, the SQUID-VSM detection mode offers much higher acquisition rates, typically ~1 s per point. The SQUID-VSM detection mode is also preferential for measurements of very low magnetic moment samples for two primary reasons: Firstly, the SQUID-VSM mode will generally have an intrinsically lower noise floor owing to the narrow-band lock-in amplifier-based detection scheme. Secondly, as will be discussed in more detail later, if background subtraction is required to minimize the contribution from the sample holder, a simple point-by-point technique can be utilized.

4 Sample Environment

As the SQUID and certain other components of the circuitry require a 4 K cryogenic temperature environment, the same cooling infrastructure can be leveraged to generate large magnetic fields and provide a variable temperature environment for the sample.

Temperature Control and Thermometry

Temperature control is made possible by placing the sample in a sealed variable temperature insert along with an exchange gas (usually high purity helium). As placing a temperature sensor at the sample location would generate an undesirable magnetic background, thermometry must be performed away from the sample location, generally at the sample chamber wall. The helium exchange gas then provides the necessary medium to equilibrate the sample to the sample chamber temperature, where the reading is taken. It is therefore critical to allow a sample to properly thermalize with the sample chamber wall for the most accurate temperature readings. For especially massive sample holders, e.g., a pressure cell, this can take a significant length of time to achieve thermal equilibrium. It is also crucial that no oxygen contaminates the sample chamber: oxygen undergoes a variety of complex magnetic transition in the 30–60 K temperature range that are easily detected in a SQUID-based magnetometer and can unnecessarily complicate interpretation of the desired sample response. Finally, as a DC scan measurement relies on moving the sample through the second-order gradiometer, potentially travelling distances up to 60 mm, it is critical that the temperature profile within the sample chamber be as uniform as possible.

Magnetic Field Control

The 4 K cryogenic environment also enables the use of superconducting magnets, allowing the user to conduct magnetometry at fields far above what can be achieved using a conventional dipole electromagnet. The vast majority of commercial SQUID-based magnetometers utilize superconducting solenoids to producing vertical axial fields. As the SQUID detection electronics are incredibly sensitive, in addition to optimizing the uniformity of the magnetic field, great care must also be taken to ensure measurement noise, from both the boiling cryogenic bath and magnet power supply, is minimized.

The superconducting magnet cannot be simply submerged in a cryogen bath as the bubbling diamagnetic helium would strongly couple to the gradiometer, resulting in a significant amount of measurement noise. Therefore, alternate techniques must be devised to efficiently cool the magnet. In situations which do rely on submerging the magnet in the cryogenic liquid, only the outer diameter of the superconducting magnet should be exposed to the boiling cryogen. The inner diameter, and the location of the second-order gradiometer, can then be conduction cooled via a thermally conductive medium (usually copper or a related alloy), which is in direct contact with the cryogen bath. Yet another, and often preferred, technique relies on keeping the superconducting magnet out of the cryogen bath entirely, and instead relying on the cold helium vapor above the bath keeps the magnet below the critical temperature of its superconducting coils.

An additional source of noise often derives from the magnet power supply (PSU). Given the inductive nature of the solenoid magnet, any variation in the amount of electric current will necessarily correlate directly to a corresponding change in the magnetic field. There are two primary techniques to mediate this parasitic power supply noise.

The first relies on a traditional superconducting persistent switch, depicted schematically in Fig. 6(a). The persistent switch comprises a region of the superconducting path that has been thermally sunk to a nearby resistive element of a separate circuit. This small segment of the superconducting path can then be toggled

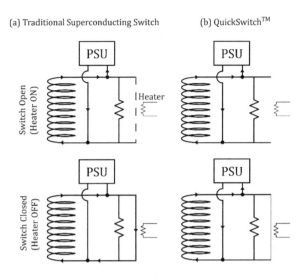

Fig. 6 Magnet charging protocols. (**a**) The traditional persistent switch allows for the magnet power supply to be discharged and is ideal for long-term measurements requiring stable fixed fields. (**b**) The QuickSwitch™, which requires the power supply to remain energized, can open and close quickly due to its small thermal mass, facilitating rapid measurements over many fields. When closed, the QuickSwitch™ shunts high-frequency noise from the magnet power supply away from the superconducting solenoid

between a normal (switch open) and superconducting (switch closed) state simply by sourcing current to the resistive element which locally heats the segment above its critical temperature. When the persistent switch is open, the magnet power supply can efficiently apply a potential difference across the "open" superconducting circuit. This applied voltage is proportional to the rate of change of the current in the solenoid. Thus, the power supply can charge the magnet to a specified field value by choosing the correct amount of voltage and length of time to apply it. After the desired current/field is achieved, the persistent switch is allowed to cool and therefore close, thus short-circuiting the magnet.

At this point, the current supplied by the magnet power supply can be ramped to zero. With the exception of small transient relaxation of the field due to flux creep, as long as the magnet windings remain superconducting, the field will, from a practical standpoint, persist indefinitely. In addition to completely eliminating any power supply noise issues, being able to reduce the current from the power supply to zero has the added benefit of significantly reducing liquid helium boil-off in the case of non-superconducting magnet leads.

As the wire that constitutes the persistent switch must also be able to carry the same current as the superconducting magnet, it has to be thick enough to support a large (several 10s of amps) electric current. It can therefore take a significant amount of time (10s of seconds) to heat and cool this segment of wire to "flip" the switch. Furthermore, before opening the switch to change the magnetic field, the voltage of the magnet power supply must be increased to sufficiently accelerate charges to match the current within the magnet, which takes additional time. If there is a significant difference between the current in the magnet and that supplied by the power supply, a magnet quench can occur in which the magnet will no longer be superconducting. While superconducting magnets have protection diodes to prevent any damage in the event of a quench, a rapid boil-off of any liquid cryogens will still occur. The traditional persistent switch technique is ideal for extended measurements, e.g., moment vs. temperature or time, which requires a limited number of fixed magnetic fields. However, measurements requiring a large number of magnetic fields, e.g., a moment vs. field hysteresis loop, could take prohibitively long as the additional overhead in heating/cooling the persistent switch and charging the power supply can easily add 30 seconds or more per measurement point. An obvious solution to this particular problem would be to simply leave the magnet power constantly energized and the switch open all the time for quick field response during measurements that require ramping. The drawback, as cited earlier however, is the resulting field noise. Thus, the user is forced to prioritize either data quality or throughput when using systems that employ a persistent switch.

A relatively recent variant of the superconducting switch described above can actually resolve both the issues of magnetic field noise and slow response when changing field. This switch purposefully utilizes a very thin superconducting wire segment with a much larger normal-state resistance, as shown in Fig. 6(b). With the switch open (switch heater on), the magnet can be charged to a given field value, as is accomplished in the traditional persistent switch described above. Just as before, little to no electric current will pass through the relatively resistive parallel path as the superconducting magnet will act as an efficient DC current shunt. However, when the switch is closed (switch heater off), the low inductance parallel path defined by the switch will efficiently shunt higher-frequency AC noise from the power supply away from the superconducting solenoid and ultimately the detection circuit. Note that with the power supply always energized, the superconducting switch does not have to carry the same high electric current as the traditional switch described prior and can therefore be made much thinner. By virtue of being thin, the thermal mass of the superconducting switch is much smaller, and the switch is thus more susceptible to rapid heat transfer and can therefore open and close much more quickly (<500 ms typical). Additionally, this arrangement relies on keeping the magnet power supply continually connected to the magnet (in the so-called "driven" mode), and therefore changes to the magnetic field can be made rapidly. As current is supplied continuously to the magnet it becomes essential to use superconducting magnet leads to minimize the heat load on the cryogens and mitigate boil-off. This QuickSwitch™ [8] technology provides a "best of both worlds" scenario in which the field generated by the superconducting magnet can be quickly changed while simultaneously shunting away parasitic noise.

5   Improving Measurement Accuracy

The commercial SQUID-based magnetometer is capable of *precisely* measuring the magnetic moment, often with better than 10^{-7} emu sensitivity. However, measurement *accuracy* is dependent on many variables including:

- Magnetic response of the sample holder (background signal).
- Sample centering—both longitudinally and radially within the gradiometer.
- Sample size and shape.
- Magnetic field accuracy.

Failure to account for these aforementioned variables can easily result in differences between the reported and true moments of 10% or more. The following sections will discuss techniques to improve measurement accuracy.

Sample Mounting Considerations: Basics

In terms of minimizing the background signal contribution of the sample holder, the user should always first use a sample holder which is as uniform as possible, particularly along its length direction (i.e., along the axis of the gradiometer). Manufacturers will often provide suitable sample holders, often made of quartz or brass, which have been explicitly designed for this purpose. Clear drinking straws have also historically been a favorite sample holder within the SQUID-based magnetometer community. Note that imperfections in the sample holder, e.g., cracks, holes, dents, etc., can potentially manifest as a detected magnetic moment. For very small moment samples, such imperfections in the sample holder become significant if these two signals are of comparable magnitude.

Once a suitably uniform holder is found, the sample needs to be adhered such that it is mechanically stable (over the intended temperature and magnetic field range) and will not break free or become loose when the sample is moved within the gradiometer. Ideally, the best-case scenario is that no adhesives are needed. For example, some samples (thin films on rigid substrates, powder-filled capsules, etc.) can be pressure-fit by hand within a drinking straw. More often than not though, some adhesive is necessary. Cryogenic varnish (GE7031 is a commonly used type) can provide a secure hold from sub-kelvin temperatures up to 400 K. Common rubber cement is also a versatile adhesive that holds well over a wide temperature range and is easily removed from many types of samples. Whatever adhesive is chosen should be free of magnetic contaminants and used sparingly.

Background Subtraction

Most often SQUID-based magnetometry is sought by the experimenter because of its inherent high sensitivity and ability to measure small ($<10^{-7}$ emu) moments. This can be, however, a blessing and a curse in that the induced signals from magnetically impure or intrinsically inhomogeneous sample holders (e.g., a pressure cell) can often compete with that of the sample of interest. In these situations, sample holder background subtraction is required. Depending on the measurement mode (traditional DC scan or SQUID-VSM) chosen, background subtraction is treated differently, as discussed below. Central to any background subtraction protocol is to ensure that the sample holder remains unchanged when measuring with and without the sample present.

Background Subtraction: Traditional DC Scan

For a traditional DC scan, the magnetic moment is calculated by fitting an entire $V(z)$ waveform generated as the sample is translated through the gradiometer, as discussed previously. Therefore, background subtraction relies on first subtracting a background waveform, $V_{BG}(z)$, from the total sample + background waveform, $V_{S+BG}(z)$. Ideally, this will then result in a waveform from the sample alone, $V_{S+BG}(z) - V_{BG}(z) = V_S(z)$, which itself must be fit to extract the magnetic moment value. Example data of $V_{S+BG}(z)$ and $V_{BG}(z)$ waveforms are seen in Fig. 7(a), showing only a subtle difference between the two. These measurements, which were performed under a 7 T field, clearly illustrate how a significant background

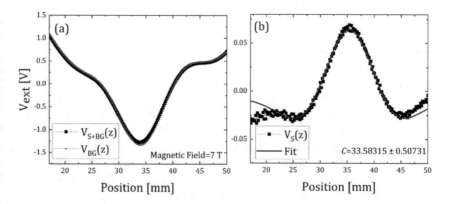

Fig. 7 Background subtraction using a DC scan measurement mode. (**a**) SQUID voltage waveforms as a function of position of the sample holder with (black squares) and without (red circles) the sample present. (**b**) By subtracting the two waveforms shown in (**a**), the SQUID voltage waveform of the sample alone can be extracted (black squares). By fitting this response (red line), the resulting moment can be calculated

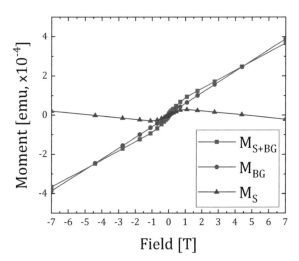

Fig. 8 Background subtraction using the SQUID-VSM measurement mode. The measured magnetic moment of the sample holder with (black squares) and without (red circles) the sample present can be simply subtracted from one another to extract the moment of the sample alone (blue triangles)

signal can completely mask the response of the sample, which has a magnetic moment of only ~20 μemu at 7 T in this case. Only after subtracting $V_{BG}(z)$ from $V_{S+BG}(z)$ can the resulting $V_S(z)$ waveform be reasonably well fit, as seen in Fig. 7(b). The scale parameter C is proportional to the magnetic moment (with the proportionality set by system-dependent calibration constants). Note that the fit is still far from ideal, likely due to the difficulty in ensuring the sample holder is completely unchanged between measuring $V_{S+BG}(z)$ and $V_{BG}(z)$ (except for the presence of the sample itself of course).

Background Subtraction: SQUID-VSM

Background subtraction is more straightforward when using the SQUID-VSM detection mode as no $V(z)$ waveform must be fit to extract the moment. A simple point-by-point subtraction of background moment, M_{BG}, from the sample + background moment, M_{S+BG}, yields the response of the sample only, $M_{S+BG} - M_{BG} = M_S$. An illustrative example is shown in Fig. 8 for a thin permalloy film on a Si substrate (mounted such that the applied field is perpendicular to the plane of the film). Such point-by-point subtraction is most easily accomplished by measuring the moment for the exact same field or temperature control points. Note, for this example, the sample holder was purposefully chosen to yield a large response (relative to the sample), and therefore M_{S+BG} and M_{BG} are of a similar magnitude. After the described subtraction operation, the resulting data appear consistent with the expected behavior for this material. More specifically, a saturation field of approximately 1 T and the weak diamagnetism of the Si substrate are clearly observed. To extract the response of the permalloy film alone, one would then subtract the linear diamagnetic response of the Si substrate.

Sample Centering and Size/Shape Effects

The accuracy of the reported moment is sensitive to how well the sample is centered longitudinally and, as is often overlooked, radially in the gradiometer. Longitudinal misalignments will result in a reported value smaller than the true moment, while radial offsets will report a moment larger than the true one. A SQUID-based magnetometer is typically calibrated against a given reference sample which provides the overall scale constants to accurately convert measured voltages into magnetic moments. For the most accurate results, an experimental sample to be measured should be of the same size/shape as the reference sample. In instances where the sample geometry unavoidably deviates from the shape of the reference, then additional corrections must be applied to ensure an accurate value.

A variety of analytical or finite element techniques [9, 10] can be employed to calculate the additional scale factors needed to correct for radial offsets and varying sample sizes and shapes when the utmost accuracy is required. A test case showing hysteresis loops of a 4 mm x 4 mm ferromagnet thin film on a Si substrate measured with the applied field parallel and perpendicular to the film plane is shown in Fig. 9(a). Not only is this test sample significantly different in size/shape compared to the reference sample used to calibrate the system, but both field orientations will require their own unique scale factor given that the sample couples differently to the gradiometer for each. The different loop shapes are expected due to the shape anisotropy of the thin magnetic film, but one would expect the measured moment to overlap at high fields, which is clearly not consistent with the uncorrected measurement. By using a sample size/shape calculator employing a

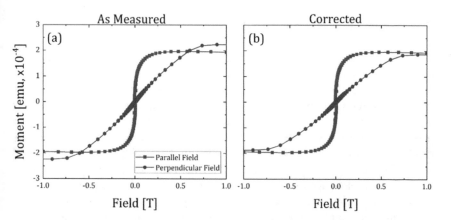

Fig. 9 Sample geometry effects and corrections. (**a**) As-measured response of a 4 mm x 4 mm thin-film sample with the applied field parallel (black squares) and perpendicular (red circles) to the film plane. Note the different saturation moments at large field magnitudes. (**b**) Additional scale factors are then applied to correct for sample geometry effects

finite element method, scale factors of 0.996 and 0.834 are calculated for the parallel and perpendicular field orientations, respectively. The corrected dataset is shown in Fig. 9(b), clearly indicating much better agreement (overlapping at high fields) of the measured moment.

Superconducting Solenoids-Remanent Field

SQUID-based magnetometers often utilize superconducting solenoids to generate fields simply for magnetizing a sample and/or exploring specific regions of a material's phase diagram. While a superconducting magnet can easily generate a field on the order of several teslas, they can present additional challenges when trying to accurately set fields of less than tens of oersteds due to subtle phenomena such as flux trapping and pinning. Further, the construction of such solenoids within a SQUID-based magnetometer leaves no clear location for an independent field sensor, e.g., a Hall sensor, and the reported magnetic field is often based only on the current through the windings as reported from the magnet power supply. If a significant remanent field exists, field errors up to several tens of oersted can be present. To minimize the magnet remanence, the field should first be set to 1 T and then reduced to zero using an oscillating approach algorithm such that the field will alternatively slew to positive and negative values with a decreasing amplitude. Yet another way to reduce the remanent field is to perform a magnet reset operation, in which a portion of the superconducting magnet is temporarily heated above its critical temperature. While these techniques are suitable if one would like to, for example, zero-field cool a sample, they are not practical in many measurement situations which rely on measuring at large and small fields within the same experiment. As an example, such remanent field errors can present significant artifacts to hysteresis loop measurements of ferromagnetic samples with low coercivities, often resulting in an "inverted" loop where the switching field along the descending (ascending) branch of the hysteresis loop occurs for positive (negative) fields, as shown in Fig. 10(a) for an Fe thin-film test sample with an expected coercivity near 20 Oe. Obviously the as-measured data are unrealistic and due solely to the remanent field effect in the superconducting solenoid.

This remanent field artifact can be mostly remedied by measuring a sample with zero coercivity (hysteresis) and a known susceptibility, e.g., a paramagnet such as Pd, as shown in Fig. 10(b). The hysteresis observed in Fig. 10(b) is due to the remanent field. However, if the magnetic susceptibility is known, the true field can be calculated by simply dividing the measured moment by the product of the susceptibly and the mass of the paramagnetic reference sample: $H_{True} = \frac{measured\ moment}{\chi_g \cdot mass}$. This true field can then simply be substituted for the reported field (Fig. 10c), providing a corrected dataset with the expected hysteretic dependence.

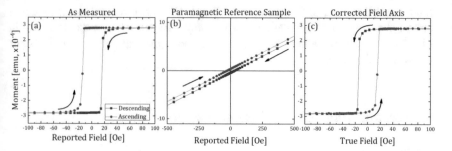

Fig. 10 Correcting for low-field remanent field errors. (**a**) As measured response of a magnetically soft Fe thin-film sample exhibiting an unphysical "inverted" hysteresis loop. (**b**) By measuring a paramagnetic sample of a known susceptibility, the true field can be calculated to (**c**) provide a corrected field axis

6 Additional Capabilities of the SQUID Magnetometer

Concentric with the superconducting solenoid, modulation and trim coils can expand the measurement capabilities of the SQUID magnetometer. The modulation coils are wound as a long solenoid that when energized uniformly modulate the field strength at the sample position by a small fraction. In contrast, the trim coils are wound in such a fashion that when energized only modify the field profile above and below the sample location. Together, the modulation and trim coils can be used to actively null (i.e., reduce the magnitude and curvature) the generally inhomogeneous remanent field left over from the superconducting solenoid. Such "ultra-low-field" capabilities are often needed for experiments that require zero-field cooling or hysteresis measurements of magnetically soft materials.

The modulation coils can also provide the necessary excitation field for AC susceptibility measurements. For this application, the trim coils can be utilized to remove a significant fraction of the frequency-dependent background signals inherent to AC susceptibility measurements which often make quantitative results difficult to obtain. Removing these background signals dramatically improves the measurement precision.

A variety of specialized sample holders also expand the measurement parameter space. Several examples include:

- **High Temperatures:** A sample holder with an integrated resistive heater and thermometer can be used to increase the maximum available temperature, typically up to 1000 K, by locally heating the sample.
- **Sub-Kelvin Temperatures:** A specialized closed-cycle He^3 insert can be integrated within the sample chamber to decrease the measurement temperature below 0.5 K.
- **Photomagnetism:** Sample holders with integrated fiber optic cables allow for magnetometry studies in the presence of electromagnetic radiation.

- **Electrical Transport:** Sample holders with integrated wiring allow for electrical resistance measurements and voltage-controlled magnetism studies of, e.g., multiferroic samples.
- **High Pressure:** Specially designed hydrostatic and diamond anvil cells allow for magnetometry measurements in excess of 1 GPa.
- **Sample Rotation:** Motorized sample holders allow for in situ rotation of anisotropic samples.

Due to their complex design, the specialized sample holders mentioned above typically have a relatively large and inhomogeneous background signature. For measurements of small magnetic signals, one must typically perform a careful background subtraction protocol, as discussed earlier in this chapter, for improved accuracy.

7 Conclusions

In summary, the SQUID magnetometer is an ideal instrument for the magnetician who requires the utmost sensitivity and versatility to efficiently characterize their samples over a wide range of temperatures and magnetic fields. The latest advances described in this chapter, most notably the SQUID-VSM detection mode, have dramatically improved measurement throughput and precision.

Acknowledgments The authors would like to thank Dr. Andreas Amann for fruitful discussions and careful reading of the manuscript.

References

1. J. Clarke, A. Braginski (eds.), *the SQUID Handbook Vol. 1: Fundamentals and Technology of SQUIDs and SQUID Sytems* (Wiley, New York, 2004)
2. J. Clarke, A. Braginski (eds.), *The SQUID Handbook, Vol. 2: Applications of SQUIDs and SQUID Systems* (Wiley, New York, 2004)
3. R. Fagaly, Superconducting quantum interference device instruments and applications. Rev. Sci. Instrum. **77**, 101101 (2006)
4. F. London, On the problem of the molecular theory of superconductivity. Phys. Rev. **74**(5), 562–573 (1948)
5. T.K. Hunt, J.E. Mercereau, Fluxoid conservation by superconducting thin film rings. Phys. Rev. **135**(4A), A944–A950 (1964)
6. T. Orlando, 6.763 applied superconductivity, in *Massachusetts Institute of Technology: MIT OpenCourseWare*, (2005)
7. D. Hurt, S. Li, A. Amann, Versatile SQUID Susceptometer with multiple measurement modes. IEEE Trans. Magn. **49**, 3541 (2013)
8. J. Diederichs, A. Amann and M. Simmonds, Superconducting QuickSwitch. United States of America Patent 8,134,434, 2012

9. P. Stamenov, J. Coey, Sample size, position, and structure effects on magnetization measurements using second-order gradiometer pickup coils. Rev. Sci. Instrum. **77**, 015106 (2006)
10. M. Hayden, V. Lambinet, S. Gomis, G. Gries, Note: Evaluation of magnetometry data acquired from elongated samples. Rev. Sci. Instrum. **88**, 056106 (2017)

AC Susceptometry

Neil R. Dilley and Michael McElfresh

Abstract AC susceptometry is a highly sensitive and versatile tool which has long been used in materials characterization for the identification of magnetic or superconducting phase transitions and in probing magnetic relaxation processes among other applications. The reader is introduced to the principle of the technique, the origins of AC magnetic responses in materials, AC susceptometer design, and considerations for good experiment protocol. The chapter concludes with research examples of AC susceptometry including its application to high-Tc superconductivity and how this showcases both the strengths and limitations of the technique.

Keywords AC susceptometry · Magnetic relaxation · Screening · Superconductor · HTS · Superparamagnetism · Spin glass · Molecular magnetism

1 Introduction

The technique of AC susceptometry has been a valuable experimental tool for investigating both static and dynamic magnetic properties of materials for almost a century [1]. The term "AC susceptometry" refers to the application of a time-varying magnetic field stimulus to a sample specimen and the detection of its dynamic magnetic response. The typical setup is essentially a transformer using the sample as the core, where a small and frequency-selectable oscillatory magnetic field is generated in the primary windings and an inductive voltage is measured across the secondary windings. The popularity of the technique owes to several factors:

N. R. Dilley (✉)
Birck Nanotechnology Center, Purdue University, West Lafayette, IN, USA
e-mail: ndilley@purdue.edu

M. McElfresh
Matsfide, Inc., Livermore, CA, USA

V. Franco, B. Dodrill (eds.), *Magnetic Measurement Techniques for Materials Characterization*, https://doi.org/10.1007/978-3-030-70443-8_4

- Ease of the sample preparation, i.e., no need for any electrical contacts to be made.
- Familiarity of the measurement principles to students of physics and electrical engineering.
- Compactness of the design which permits it to be used at cryogenic temperatures, in ovens, and in high-pressure environments.
- High sensitivity afforded by lock-in amplifiers and nulling techniques in the susceptometer, enabling small samples (e.g., single crystals) to be measured.
- Ability to study magnetic response over several decades of frequency and thus better understand the underlying mechanisms for the magnetic dynamics.
- Phase-sensitive detection of the sample's magnetic response at the drive frequency or at harmonic multiples (further benefits of using lock-in amplifiers), enabling the instrument to quantify dissipative and nonlinear magnetic responses even at small signal levels.

The research applications of AC susceptometry span various disciplines of physics, chemistry, materials science, and engineering and mostly fall into categories of magnetic relaxation, screening effects in conductors, identifying phase transitions, or as a sensitive probe of the differential magnetic response. Possible variables in the measurement include temperature, DC applied magnetic field, AC frequency, or AC applied magnetic field. A ramp of temperature during AC measurements may reveal a sharp peak in the AC susceptibility at a magnetic ordering temperature, while a ramp of frequency can identify a relaxation time constant in a superparamagnet. The screening currents in a conductor will lead to a similar peak in the frequency response of the AC susceptibility, in this case originating from the skin depth of the currents. The richness of the phenomena which produce signatures in the AC magnetic response, together with the multidimensional space of parameters that can be varied, is both an asset and a liability to the technique as researchers often must work hard to determine the origin of the features.

There are some well-written reviews of the AC susceptibility technique in the literature to which we refer the interested reader [2–5]. The present chapter offers a practical perspective which hopes to edify a casual reader, properly orient a newcomer to the field, offer advice and tips to a student performing AC measurements, and provide a refreshing overview to veterans of the field.

Section "Theory of AC Susceptibility" provides the theoretical framework for understanding AC susceptibility and describes the origin of AC magnetic response in materials; Section "AC Susceptometer Design" familiarizes the reader with the basic aspects of AC susceptometer design; Section "Working with an AC Susceptometer" provides practical guidance for successful measurements; Section "Research Examples" gives research examples from several fields which illustrate the strengths and broad applicability but also the challenges in interpreting results of AC magnetic measurements.

Theory of AC Susceptibility

We start with the equation describing magnetic fields in matter, where the magnetic field B is the sum of the material's magnetization density M and the applied field H:

$$B = \mu_0 (H + M) \tag{1}$$

The constant μ_0 is the permeability of free space. Consider that the applied field H is made of a constant H_{dc} and sinusoidal component at frequency $\omega = 2\pi f$:

$$H(t) = H_{dc} + H_{ac} \cos(\omega t) \tag{2}$$

For this chapter, we will assume that H_{dc} and H_{ac} are both directed along the z-axis, and we will consider the magnetization component along the z-axis, unless otherwise stated, so we will dispense with the vector notation and the z subscript: $H = H_z \to H$ and $M = M_z \to M$.

Since the material's magnetization is a function of the applied field, we know that $M(t)$ must be periodic with the same fundamental frequency ω. With the benefit of Fourier's theorem for any periodic function, we can express this as:

$$M(t) = M_{dc}(H_{dc}) + H_{ac} \sum_{n=1}^{\infty} \left[\chi_n' \cos(n\omega t) + \chi_n'' \sin(n\omega t) \right] \tag{3}$$

The coefficients of the Fourier expansion of χ describe components of the differential magnetic susceptibility. The $n = 1$ terms of the expansion correspond to a linear response of the material's magnetization to the applied field, and they are almost always dominant. Higher-order harmonics ($n > 1$) due to nonlinear magnetic response will be discussed later. To build some intuition about the above equation, let us set $H_{dc} = 0$, assume that $M_{dc}(0) = 0$, apply a small H_{ac}, and let the AC frequency $\omega \to 0$ which produces a change of δM to find:

$$\chi_1' = \delta M / H_{ac} \equiv \chi_{dc}(H) = \frac{dM_{dc}}{dH} \tag{4}$$

At low frequencies, the χ' term (in-phase with the drive field) is ascribed to instantaneous magnetization of the material and can be thought of as the local slope of the magnetization curve $M(H)$. χ' is referred to as the reactive or dispersive component, while χ'' is the dissipative or absorptive component. The dissipative aspect of χ'' can be seen by computing the energy loss E per unit volume per period T of AC field using an expression for energy loss in magnetic materials (see Ref. [6]) and the periodic expressions for M and H given above:

$$E = \mu_0 \oint H dM = \mu_0 \int_0^T H \frac{dM}{dt} dt = \mu_0 H_{ac}^2 \chi_1'' \tag{5}$$

Of all the terms in the integral that result from putting the time derivative of Eq. 3 into Eq. 5, only $\cos^2(\omega t)$ does not average to zero when integrated over one period. That implies that only the χ_1'' fundamental dissipative response contributes to energy loss. This dissipation can be quantified graphically by the enclosed area of the small hysteresis loop traced out over one AC cycle in the $M(H)$ plane. Magnetic hysteresis in $M(H)$ indicates that M lags in time behind the applied H due to a viscosity, and the energy expended through this viscosity is measurable as heating of the material. When the magnetization curve $M(H)$ is reversible, this loss will be zero. Note that that energy loss, and hence χ_1'', cannot be a negative value.

AC susceptibility results are reported in terms of χ' and χ'' or equivalently as the magnitude χ and phase φ which are related as follows:

$$\chi = \sqrt{\chi'^2 + \chi''^2} \tag{6}$$

$$\varphi = \tan^{-1}\left(\chi''/\chi'\right) \tag{7}$$

These quantities can be visually represented in the complex (Argand) plane where real values (χ') lie along the x-axis and imaginary values (χ'') along the y-axis, and we write the complex susceptibility $\tilde{\chi} = \chi' + i\chi''$. We see that the phase corresponds to the lag and losses mentioned above such that $\varphi = 0$ when $\chi'' = 0$. In this chapter, we will use χ when referring to the magnitude of the AC susceptibility (Eq. 6) and $\tilde{\chi}$ when explicitly dealing with the complex AC susceptibility.

The dynamic magnetization $M(t)$ arises from two distinct phenomena. Clearly, the sample's *magnetism* seen at DC fields will be reflected in χ' as discussed above, and magnetic hysteresis leads to χ''. But in the case of an electrically conducting sample, there is an additional contribution to $M(t)$: *screening*, or eddy currents which oppose the changing magnetic field in the material. Both of these effects will be discussed below.

Units and Conventions

In this chapter, we will work in SI units (amp, meter, joule, tesla) instead of CGS units (emu, cm, gauss, oersted), though many instruments report in CGS due to them formerly being favored in physics research. Magnetic susceptibility per unit volume is dimensionless in both systems, with $\chi_{SI} = 4\pi \chi_{cgs}$. In this chapter, it is assumed that χ refers to the AC (rather than the DC) volume susceptibility unless otherwise stated (see Ref. [7] for help with magnetic unit conversions).

Demagnetizing Factors

When the magnetic flux vector $\boldsymbol{B}$ exits from the surface of the pole of a uniformly magnetized bar magnet, the magnetization goes to zero at the surface because

M exists only inside the material; meanwhile, $\boldsymbol{B} = \mu_0(\boldsymbol{H} + \boldsymbol{M})$ must be continuous (from the Maxwell equation $\nabla \cdot \boldsymbol{B} = 0$). Continuity of $\boldsymbol{B}$ requires that there be a component of $\boldsymbol{H}$ existing only inside the material and in a direction opposing $\boldsymbol{M}$, which furthermore starts and stops at the poles of the magnet. This is called the demagnetizing field $\boldsymbol{H}_d$. The field inside the sample is then $\boldsymbol{H} = \boldsymbol{H}_a + \boldsymbol{H}_d = \boldsymbol{H}_a - \mathbb{N}\boldsymbol{M}$, where $\boldsymbol{H}_a$ is the externally applied field and $\mathbb{N}$ is the demagnetizing tensor which is defined only by the sample shape. For ellipsoidal samples magnetized along a principle axis, $\mathbb{N}$ becomes a scalar (N), and the magnetization is uniform in the material. In all other sample geometries, the magnetization in a DC magnetic field will be non-uniform. This is a complication that is usually put aside in analyzing volume-averaged magnetometry data but should be kept in mind. The demagnetizing factor N can assume a value between 0 and 1, with some common examples being a sphere $(N = 1/3)$, a needle magnetized along its axis $(N \approx 0)$, and a thin plate magnetized normal to its surface $(N \approx 1)$. Due to the modified internal magnetic field, the measured AC susceptibility $\widetilde{\chi}_{ext}(H_a)$ is therefore different than the material's intrinsic susceptibility $\widetilde{\chi}(H)$:

$$\widetilde{\chi}_{ext} = \frac{\widetilde{\chi}}{1 + N\widetilde{\chi}} \tag{8}$$

Due to both $\widetilde{\chi}$ and $\widetilde{\chi}_{ext}$ being complex, separation into χ' will contain both χ_{ext}' and χ_{ext}'' terms, and the same is true for χ''. We refer the reader to Ref. [2] for the explicit expressions.

The main points to note about demagnetizing effects in magnetic measurements are:

- The relative systematic error introduced into the measured susceptibility is $N\widetilde{\chi}$ which will be relevant for a material with high susceptibility and shaped to have a significant value of N. In all other cases, the correction can be dismissed.
- In strongly diamagnetic materials such as superconductors in the screened state $(\chi' = -1)$, $|\chi_{ext}| > |\chi|$ because the internal field $\gg H_a$; this leads to an apparent compression of the measured $M(H)$ curves along the field axis.
- When $\chi \gg 1$, such as in ferromagnetic materials, this implies $\chi_{ext} < \chi$, which results in an apparent expansion of the measured $M(H)$ curves along the field axis, sometimes referred to as a "shearing" effect [6].

Magnetic Relaxation

Electrons interact with an applied magnetic field through both their intrinsic spin and their orbital motion. First, let us consider the various types of electronic magnetism that exist in the absence of any type of magnetic ordering such as ferro-, anti-ferro-, or ferrimagnetism. For electrons in insulators, or for those electrons in metals which remain localized, unpaired electron spins show Curie-Weiss spin *paramagnetism*

($\chi_{dc} > 0$) and van Vleck spin paramagnetism, while paired electron spins exhibit Larmor orbital *diamagnetism* ($\chi_{dc} < 0$). In metals, additional magnetic responses come into play from the electrons in conduction bands: Pauli paramagnetism and Landau diamagnetism arise from the spin and orbital motion of these electrons, respectively. In all these cases, the response to a magnetic field is instantaneous on the experimental time scales relevant to this chapter, so there is no dissipation arising from these magnetic responses. The largest effect is generally Curie-Weiss paramagnetism (when unpaired electrons are present) which follows a $\chi_{dc} \propto 1/T$ law with a typical volume susceptibility of $\chi_{dc} \sim 10^{-2}$ even at room temperature. In contrast, the other susceptibilities are typically $|\chi_{dc}| \sim 10^{-5}$ or smaller and depend on temperature weakly (see Ref. [8] for a detailed discussion of the various magnetic responses listed above).

Dynamic effects in magnetism come into play when interactions between electron spins in paramagnetic materials give rise to magnetic ordering and the spins become "frozen" into an ordered phase. Near the ordering temperature, then, we expect the time scale for magnetic relaxation to cross through the time scale for AC magnetic measurements and, as we will show below, results in a resonant behavior in the magnetic response to an AC applied field. The simplest way to model a process of "freezing" the spins is to slow the dynamics using a damping coefficient α which retards the alignment of spins to an applied field as expressed in this first order differential equation:

$$M = \chi_{dc} H_{ac} - \alpha \frac{dM}{dt} \qquad (9)$$

A small applied field H_{ac} would instantly lead to a magnetization M according to the DC susceptibility X_{dc}, but it is slowed by the presence of damping.

If we take $H_{ac} = H_0 e^{-i\omega t}$, we can solve the above equation for the complex magnetization M and write that in terms of the magnetic susceptibility as:

$$\tilde{\chi}(\omega) = \frac{M(\omega)}{H_{ac}(\omega)} = \frac{\chi_{dc}}{1 - i\omega\tau} \qquad (10)$$

where the relaxation time constant is given by $\tau = \alpha \chi_{dc}$. That is, if the field H were stepped from $H = H_{dc}$ to $H = 0$ at time $t = 0$, the moment would relax as $M(t) = M(0)e^{-\frac{t}{\tau}}$, which is the familiar exponential decay seen, for instance, when monitoring the voltage across a capacitor in an RC circuit while it is being discharged. The complex susceptibility can be broken into the real and imaginary parts which are plotted in Fig. 1:

$$\frac{\chi'}{\chi_{dc}} = \frac{1}{1 + \omega^2 \tau^2} \qquad (11)$$

$$\frac{\chi''}{\chi_{dc}} = \frac{\omega\tau}{1 + \omega^2 \tau^2} \qquad (12)$$

Fig. 1 Frequency (ω) dependence of AC susceptibility for Debye relaxor with time constant τ

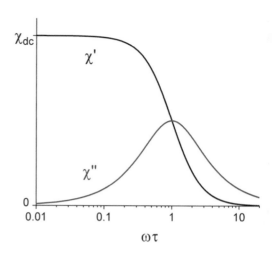

The peak in χ'' occurs at $\omega\tau = 1$ and provides a practical method for identifying a relaxation time constant in a material using variable frequency AC susceptibility. While this peak in χ'' is often referred to as a resonance, we should point out that this is not the harmonic oscillator self-resonance of spins as is observed in magnetic resonance measurements (NMR, EPR, FMR).

There are three regions of interest in Fig. 1:

1. Low-frequency $\omega\tau \ll 1$: the moments can follow the applied field in phase ($\chi' \approx \chi_{dc}$), and at low frequencies, the dissipation term $\alpha dM/dt$ from Eq. 9 is small ($\chi'' \approx 0$).
2. High-frequency $\omega\tau \gg 1$: the moments are not able to respond fast enough to the applied field due to the damping, so all magnetic response falls off (χ', $\chi'' \to 0$).
3. Intermediate-frequency $\omega\tau \approx 1$: the moments are lagging behind the applied field, so the in-phase response χ' decreases and dissipation χ'' increases. Since χ'' must eventually fall back to zero at high frequency, there results a maximum where the oscillating field is most effectively dissipating heat in the sample.

The above model is known as Debye relaxation [9] and was first formulated in the context of relaxation in dielectrics in response to AC applied electric fields. It has been very helpful in the study of systems like superparamagnets and single-molecule magnets, where spins "freeze" at low temperatures, though it is important to note that this model ignores interactions between spins. The Debye model can be extended to incorporate these interactions by considering a distribution of relaxation times instead of a single τ, as discussed in a recent review [5]. As a final example, the sharp peak in $\chi''(T)$ in magnetic phase transitions has long been used to identify the magnetic ordering temperature in zero DC applied field. Once again, this peak occurs when the AC field frequency matches the relaxation rate of the spins ($\omega\tau \approx 1$) as they go from the paramagnetic to the frozen ordered state. More examples of magnetic relaxation effects will be given in Sect. "Research Examples" below.

Screening

An electrically conducting material will attempt to screen out changes in the magnetic field in its interior, an effect known as Lenz's law which derives from the Maxwell equation $\nabla \times E = -\frac{\partial B}{\partial t}$, where E is the induced electric field which creates the eddy currents.

In a material with isotropic resistivity ρ placed in an external magnetic field H oscillating at frequency ω as described above, Maxwell's equations lead to this relation [10]:

$$\nabla^2 H = -\frac{i\omega\tilde{\mu}}{\rho} H = -\frac{2i}{\delta^2} H \tag{13}$$

The material's magnetic permeability μ enters into the equation, which is related to the complex susceptibility by $\tilde{\mu} = \mu_0 \left(1 + \tilde{\chi}\right)$. The skin depth:

$$\delta = \left[2\rho/\left(\omega\tilde{\mu}\right)\right]^{\frac{1}{2}} \tag{14}$$

describes the screening of the magnetic field: at a depth δ inside the material, the magnetic field has fallen to $1/e$ ($\sim37\%$) of its value at the surface. Magnetic susceptibility due to screening is found by solving the above differential equation is much simpler in a high symmetry geometry, such as a sphere of radius r, where the measured susceptibility is found to be [2]:

$$\chi'_{ext} = \frac{9}{4} \frac{(\delta/r)\,[\sinh{(2r/\delta)} - \sin{(2r/\delta)}]}{\cosh{(2r/\delta)} - \cos{(2r/\delta)}} - \frac{3}{2} \tag{15}$$

$$\chi''_{ext} = \frac{9}{4} \frac{(\delta/r)\,[\sinh{(2r/\delta)} + \sin{(2r/\delta)}]}{\cosh{(2r/\delta)} - \cos{(2r/\delta)}} - \frac{9}{4}\left(\delta^2/r^2\right) \tag{16}$$

Figure 2 plots the real and imaginary susceptibilities as a function of r/δ which shows a peak in χ'' when the field penetration is approximately equal to the sample size. As r/δ approaches zero (low AC frequency), both components of the screening susceptibility go to zero as would be expected in the limit of static magnetic field. At large values of r/δ (small skin depth, which occurs as $\rho \to 0$ or at high frequency), $\chi'_{ext} \to -3/2$. Since the demagnetizing factor $N = 1/3$ for a sphere, then the material's intrinsic susceptibility $\chi' \to -1$ (see Eq. 8) which corresponds to $B = 0$ inside the material, that is, a state of perfect diamagnetism.

At this point, several comments are in order:

1. This screening susceptibility is an electrical transport phenomenon, except that currents circulate in the sample around the axis of the AC field rather than in a straight line as for standard electrical transport. It is therefore a reflection of sample geometry and any sample imperfections which affect the conduction

Fig. 2 AC susceptibility of a conducting sphere of radius r as a function of AC frequency ω

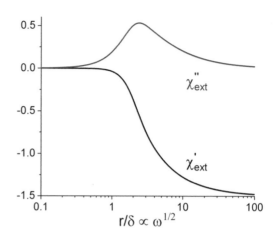

paths. In a uniform sample of known geometry, however, the screening effect can be exploited to measure the resistivity without making electrical contact to the sample which will be discussed below in Sect. "Research Examples".

2. Sample properties are probed roughly within a depth δ of the surface. If $\delta \ll r$, the interior of the sample is shielded from the applied magnetic field and thus does not participate in the measurement. The possibility of screened material must be considered when making statements about the average magnetic properties of a specimen.

3. As seen in Eq. 14 above, the skin depth is a function of the permeability $\tilde{\mu}$, so a magnetic conductor may have a much smaller skin depth than even a low resistivity metal of the same shape. This is important in the design of transformer cores working at power line frequency (50–60 Hz): eddy currents are detrimental to performance, so the core is composed of laminate steel sheets each with thickness $t \ll \delta$ (see Chap. 13 in Ref. [11]) which are insulated from each other. For higher-frequency magnetic applications, the cores are composed of (ferromagnetic) ferrites which are insulators and therefore free of screening effects.

AC Susceptometer Design

There are two elements necessary for the design of a mutual inductance AC susceptometer: (1) a drive coil to produce the alternating field H_{ac} and (2) a detection coil around the sample to inductively measure the sample's response. A susceptometer with only these elements allows for the study of susceptibility as a function of frequency and drive amplitude of H_{ac} at ambient conditions. To achieve any measure of sensitivity, the mutual inductance between the drive and detection coils should be balanced, so that an inductive voltage is generated only when the

Fig. 3 Schematic
cross-section cut of the
solenoid coils of an inductive
ac susceptometer. For each
coil, dark (light) windings are
passing into (out of) the page

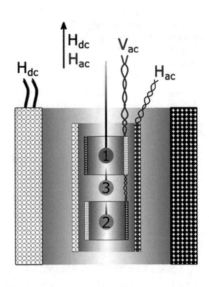

sample is placed in the detection coil. This is often strived for [11, 12] by having two
identical series detection coils (innermost red coils in Fig. 3) placed symmetrically
within the drive coil which are wound in opposite directions to balance each other's
coupling to the drive coil (blue coil in Fig. 3). In reality, it is not possible to remove
all mutual inductance by coil design, so a further sophistication is to mitigate this
by having a "trim" drive coil which is inductively coupled to the detection coils
and is driven at a selectable phase and amplitude relative to the main drive coil
to further null the imbalance. The trim coil calibration as a function of temperature
and AC frequency must be mapped out with no sample in the detection coil. Another
technique to remove imbalance effects is shown in Fig. 3: the two detection coils are
coaxial which allows the sample (green) to be positioned in either coil (position 1
or 2) by attaching to a sample rod accessible from the top of the instrument. This
provides the great advantage of separating the sample signal from the imbalance
between drive and detection coils: when placed in the other detection coil, the signal
due to the sample will be reversed, but any coil imbalance will remain the same. A
further advantage of a variable position sample system is that a trim coil can be
used at each measurement point by placing the sample between the detection coils
(position 3 in Fig. 3) during the imbalance nulling procedure [13, 14].

Assuming any oppositely wound detection coil design, the voltage in the two
detection coils (1,2) of N windings each and cross-section A is generated by time-
changing magnetic flux $\Phi = N \int \boldsymbol{B} \cdot d\boldsymbol{A}$ according to Faraday's law:

$$V_{ac}(t) = -\frac{d\left(\Phi_1 - \Phi_2\right)}{dt} = -\mu_0 N A \frac{dM}{dt}$$

$$= \mu_0 N A \omega H_{ac} \sum_{n=1}^{\infty} n \left[\chi_n'' \cos(nwt) - \chi_n' \sin(nwt)\right]$$

$$(17)$$

Note that the time derivative creates a 90-degree phase shift in the detected voltage relative to the sample's magnetization. The benefit of the oppositely wound detection coils is nulling the applied field H since it appears equally in both Φ_1 and Φ_2. The above equation assumes the sample and detection coils both have a cross-section A; however in real systems, the coils will be larger than the sample so the coupling, referred to as the filling factor, will be lower, and this will reduce the detected AC voltage.

To investigate the sample's dependence on DC magnetic field, the susceptometer is placed within an external field source (H_{dc} and outermost solenoid in Fig. 3). It is common to use a superconducting solenoid operating at liquid helium temperature for generating fields up to 10 tesla or greater. For low-temperature studies, the DC magnet remains at the liquid helium temperature, while the AC drive and detection coils are typically located in the variable temperature space in order to increase sample coupling and reduce coupling to other materials in the cryostat. In the case of SQUID-based AC susceptometry, however, the detection coils also use superconducting wire and thus must be held near liquid helium temperature. In our instrument description below, we will assume a conventional conducting susceptometer unless explicit comparison is made to SQUID-based designs.

The drive coil is connected to an AC current source with high enough compliance voltage to overcome the component of voltage due to the inductance L of the drive coil: $V = -L \cdot dI/dt$. This compliance puts an upper limit on the drive frequency, typically in the range of 10 kHz, though a cryogenic AC susceptometer design optimized for frequencies from kHz to MHz range has recently been reported [15]. The detection coil is connected to a lock-in amplifier which is phase-referenced to the drive coil current. Copper wire is chosen for drive coils (due to low resistance) and detection coils (due to low Johnson noise). Since the induced voltage is proportional to the drive frequency (Eq. 17), it places a lower limit of typically 1–10 Hz on the measurement. More detection coil windings, while increasing the flux Φ, will reduce the maximum usable frequency due to the larger coil inductance.

If the detection coils are the superconducting pickup coils for a SQUID gradiometer, however, the SQUID output voltage $V \propto \Phi$ is not directly dependent on frequency, and this allows SQUID AC susceptometry to extend down to 0.01 Hz or lower [16], limited ultimately by voltage drifts in the SQUID due generally to $1/f$ noise and primarily from SQUID drifts induced by magnetic flux relaxation ("flux creep") in the superconducting wiring of the magnet used to create the DC applied field. Ultra-low-frequency SQUID AC measurements allow for the suppression of screening signals and detection of magnetic phase transitions in situations where the sample holder contains conducting material, such as a Be-Cu high-pressure cell in which a small sample specimen is placed inside: as $\omega \to 0$, screening currents from the cell approach zero (see Fig. 2), while the magnetic response of the sample approaches χ_{dc} (see Fig. 1).

Inductive coupling should be minimized elsewhere in the system wiring, and for this reason, it is important to wind the wires leading to the drive and detection coils separately as tightly and uniformly twisted pairs. Capacitively coupled noise from the environment is reduced by electrically shielding the cable and referencing this shield to a quiet analog ground in the measurement electronics. It is important to connect the instrument to ground through only one path in order to avoid currents in the shielding due to "ground loops." The other elements of the susceptometer are ideally free of any magnetic materials or conductors within the range of the AC drive field, since these will introduce parasitic signals which will change with AC frequency, temperature, and DC magnetic field and will also change the phase and amplitude of the drive field H_{ac} as seen at the sample location. These background signals can complicate matters greatly and must be minimized.

Note that the detection of the time-dependent magnetization can also be done by methods other than magnetic induction. Other magnetic detection schemes like X-ray magnetic circular dichroism (XMCD) and magneto-optical Kerr effect (MOKE) can also be phase-detected in the presence of a small H_{ac} to obtain χ' and χ'' as has been demonstrated for monolayer magnetic films [17].

If one orients the AC drive coils and AC detection along the same axis but perpendicular to the applied DC magnetic field, then the "transverse susceptibility" χ_T can be measured. It was theoretically predicted [18] and experimentally verified using such a susceptometer design [19] that χ_T has a singularity at the anisotropy field of a ferromagnet (see Sect. "Differential Probe of Magnetism"). Note that magnetic resonance techniques (NMR, EPR, FMR) are designed to measure the transverse susceptibility and do so using a waveguide transmission line in place of a drive coil due to the high frequencies (MHz–GHz) required. Another coil-based AC susceptometer design [20] uses a high-frequency (MHz) resonant LC "tank circuit" for the detection with the sample placed in the coil of the inductor L, so that changes in sample permeability μ at the resonant frequency will change the inductance and induce a shift in the circuit's resonance frequency $= 1/\sqrt{LC}$.

As we mention resonant circuits for measuring magnetic properties, it is worth pointing out the usual AC susceptometer design depicted in Fig. 3 does not operate near resonance of the LCR circuits comprised of the inductances L, stray capacitances C, and wire resistances R in the drive or detection wiring. In fact, measures are taken to avoid exciting any resonances (often occurring in the range of 0.1–1 MHz) that will induce "ringing" and loss of sensitivity at the measurement frequency.

Working with an AC Susceptometer

After reading the preceding discussion of the many factors which place limits on the performance of even the most carefully designed AC susceptometer, the experimentalist will hopefully appreciate that these factors must be kept in mind during sample preparation, mounting, and measurement of magnetic properties using a dynamic technique like AC susceptometry. Starting with the sample itself, its geometry determines the demagnetizing factor N (see Sect. "Demagnetizing Factors"). Choosing a regular geometry like a cylinder [21] is often practical; an ellipsoidal or thin needle-shaped sample is ideal but often not practical. Recall that in non-ellipsoidal samples, the magnetization in the sample will not be homogeneous. One consequence is that the best approximation for N will depend on how the magnetic moment is measured: if the detection coil height h is much less than the height of a sample, then it averages the flux over a surface near the middle of the sample and the *fluxmetric* N_f should be used, while if the detection coil measures the entire sample volume, then the *magnetometric* N_m should be used (the latter design is depicted in Fig. 3) [22]. AC susceptometers typically measure the volume averaged magnetic moment, so we will assume $N = N_m$ in this chapter.

The susceptometer's sample holder should be chosen to be non-magnetic and uniform along the length which couples to the detection coils. It should also be free of metallic parts whose geometry allows screening currents to be generated. It is important to realize that these conditions are never perfectly met, e.g., there is a gap in the sample holder where the sample is clamped, so the goal for the experimentalist is to ensure that the sample signal dominates that from the sample holder or that the sample and holder signals can be separated in post-measurement analysis.

Phase calibration of an AC susceptometer is necessary in order to discern χ' and χ'' and can be done using a sample with a large signal and known phase such as a paramagnetic insulator ($\varphi = 0$) like Dy_2O_3 or gadolinium gallium garnet (GGG), or a superconductor at low fields ($\varphi = 180°$). Phase of the coil response is dependent on AC frequency and also changes with temperature due to coil thermal contraction and other instrument background signals mentioned above. Such a reference sample should be kept at hand, and the experimentalist must check any results against both the sample blank and this reference sample in order to have confidence in their conclusions. Since phase calibration is affected by demagnetizing effects, this correction (Sect. "Demagnetizing Factors") must be applied to both the reference and research samples.

An important experimental practice in all cases is to perform the intended measurement on a sample "blank" in which all materials are present except for the material under investigation (e.g., the blank substrate on which a superconducting film will later be deposited).

Amplitude calibration of the AC response is yet more challenging and requires measuring χ of a standard sample with known magnetic moment over the range of AC frequency, temperature, and sometimes DC magnetic field relevant to the experiment. For this reason, many research articles report χ in "arbitrary units (au)" and focus on identifying phase transitions and analyzing the phase φ of the sample's response. A susceptometer design [12] which achieves absolute amplitude and phase calibration of χ uses an *in situ* calibration scheme to temporarily imbalance the detection coils by a fixed amount, thus producing a signal which emulates a paramagnetic sample with a known moment. The sample is placed between the detection coils (position 3 in Fig. 3, as done when nulling with a trim coil) during the process so that it does not contribute to the imbalance. This scheme is done at each measurement point, and the susceptibility is reported in units of m^3 (SI) or emu/Oe (cgs). Dividing by the sample volume produces $\tilde{\chi}_{ext}$ (see Eq. 8).

The sample temperature must be monitored during the measurement and therefore should be thermally anchored to the susceptometer through a thin layer of cryogenic grease like Apiezon N or H grease (in the case of a stationary sample design for low temperatures) or coupled to the susceptometer via >0.1 torr of helium thermal exchange gas (in the case of a design which allows repositioning of the sample between detection coils; He exchange gas is effective for temperatures above 1.5 K). Accurately assessing the sample temperature requires great care in AC susceptibility measurements, especially at low temperatures, due to thermal gradients created by Joule heating from the resistance of the drive coil or eddy current heating in metal parts of the susceptometer or a conducting sample. The measurement of a sample with strong temperature dependence (say, GGG) at the temperatures of interest should show an asymptotic approach to thermal equilibrium as AC drive frequency and AC amplitude decrease. In this sense, the sample is being used as a thermometer to identify heating in the coil system.

Research Examples

Differential Probe of Magnetism

The in-phase AC susceptibility χ' produces a sensitive measure of the local slope dM_{dc}/dH at low frequencies. This inherently differential measurement reveals subtle features in the DC magnetization curve that may be lost when numerically differentiating even a very densely spaced $M_{dc}(H)$ curve. In the context of ferromagnetic materials, this principle has been used to determine the magnetic anisotropy: the tendency of the ferromagnetic polarization to align along a certain "easy axis." The anisotropy energy can be quantified as an equivalent magnetic field by applying a large enough DC magnetic field perpendicular to the easy axis (known as the "hard axis") to overcome that anisotropy. The anisotropy field H_A corresponds to the saturation in the magnetization (or DC magnetic moment "m (emu)" in Fig. 4) at high fields applied along the hard axis. It was also shown that the transverse susceptibility χ_T peaks at the anisotropy field [18], as we mentioned in Sect. "AC Susceptometer Design". However, transverse susceptometers are uncommon due to the inconvenient coil geometry. In one study, it was shown that H_A could be identified using a conventional AC susceptometer [23]. The polycrystalline samples (with randomly oriented grains) were challenging because only a fraction of the grains were magnetized along the hard axis during the measurement, while the rest of the material contributed a large broad background signal due to grains which were partially or fully aligned with the applied field. Identifying H_A within the $M_{dc}(H)$ curve is not practical in such a case. The sample's in-phase AC susceptibility χ' (shown as the dashed line $M' = \chi' \cdot H_{ac}$ in Fig. 4) reveals an inflection point at H_A, which implies that the derivative $d\chi'/dH = d^2M_{dc}/dH^2$ must have a peak there. In a finite AC excitation field H_{ac}, it can be shown that, in addition to the signal from the sample at ω which is proportional to χ', there is

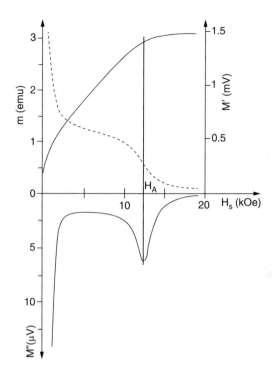

Fig. 4 Measured high-field DC magnetic moment m and AC response of a polycrystalline ferromagnet; moments M' and M'' are defined in the text. (Figure reproduced from Ref. [23] with permission from Elsevier)

a signal at 2ω which is proportional d^2M_{dc}/dH^2 [23], and in Fig. 4, we indeed see a peak in $M'' = d^2M_{dc}/dH^2$ (note the difference in notation and that this is *not* related to χ''). Said differently, d^2M_{dc}/dH^2 is the curvature of $M_{dc}(H)$ which will peak when the magnetization crosses over from a linear dependence at intermediate fields to saturation at H_A. The lock-in amplifier can easily be set to measure the 2ω signal instead of the fundamental at ω. The 2ω signal depends on the amplitude of the AC drive as $V_{2\omega} \propto H_{ac}^2$ (compared to $V_\omega \propto H_{ac}$) because nonlinearity in the $M_{dc}(H)$ curve only becomes evident at larger AC drive amplitudes. At larger amplitude of H_{ac}, the signal of the peak will be higher but broadened due to the averaging effect of the AC field, so the researcher must determine the optimal value for H_{ac} for their purposes. This example shows the power of differential magnetic measurements and AC harmonics analysis in revealing subtle features in the magnetic response which would otherwise remain hidden. The principles of using AC modulation to reject large background signals and harmonic detection to highlight nonlinear responses are important for experimentalists to keep in mind.

Determining the Skin Depth in Conducting Samples

The screening AC susceptibility of a conductor displays a characteristic frequency dependence which depends on the material resistivity and sample geometry, as

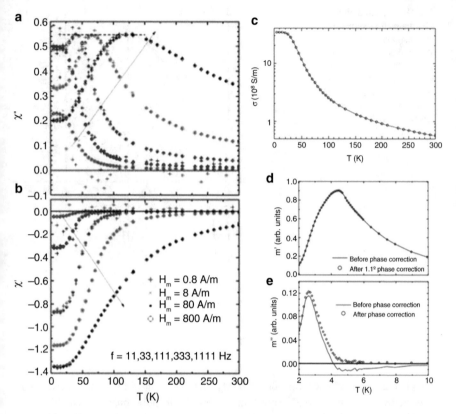

Fig. 5 Temperature and frequency dependence of AC susceptibility of a Cu cylinder, with arrows indicating direction of increasing frequency (**a** and **b**, left), electrical conductivity derived from the measurement data (**c** upper right), and example of magnetic relaxation in a molecular magnet after calibration correction based on the new Cu cylinder standard (**d** and **e**). (Figure reproduced from Ref. [21] with permission from AIP)

shown in Sect. "Screening". If a conductor has a known and regular geometry, then it is possible to measure the electrical resistivity without the need of wire contacts as is usually required [24]. The authors of Ref. [21] used this principle to establish a precisely machined copper cylinder as a calibration check of a commercial AC susceptometer which can be used over the entire temperature range of the instrument. The temperature dependence of χ' and χ'' for this sample at various AC drive frequencies from 11 Hz (red) to 1111 Hz (black) is plotted in Figs. 5a and 5b. Arrows in both panels indicate the direction of increasing frequency. The skin depth (Eq. 14), together with the knowledge that copper resistivity ρ drops at low temperature, can be used to understand all the features in the data. Using the theoretical solution for screening in the known cylindrical geometry (in this case, diameter = height = 5 mm), the temperature dependence of the conductivity ($\sigma = 1/\rho$) was also inferred (see Fig. 5c). Note that the diamagnetic screening in χ'

increases at lower temperatures due to the drop in ρ and gets close to full screening for the highest frequency and lowest temperature (full screening corresponds to $\chi' = -1.45$ due to demagnetizing effects in the cylinder). The peak in χ'' occurs when $\delta = 0.57r$ where r is the radius of the cylinder [25], and this peak occurs at higher temperature when using higher AC frequency. At low frequencies, the skin depth remains larger than the sample dimension at all temperatures, so there is no peak in χ''. The susceptibility is independent of AC drive field H_{ac} (called "H_m" in Figs. 5a and 5b) as expected, with drive amplitudes from $H_m = 0.8$ A/m to 800 A/m (0.01–10 Oe) indicated by different data symbols which overlay each other except at lowest AC drive where the precision is poor due to the low signal generated in the detection coil. After quantifying the temperature-dependent electrical conductivity of the copper sample, the authors of Ref. [21] go on to propose it as a calibration standard for AC susceptometry and check it against the standards of Pd and Dy_2O_3 used in that commercial apparatus. Because of the well-known screening properties of a copper cylinder, this allowed for a further refinement of the susceptometer at all temperatures and AC frequencies. The resulting small correction in the phase φ of the AC susceptibility at low temperatures removed a $\chi'' < 0$ artifact that is sometimes reported at low temperatures (see Fig. 5e). A negative dissipation is unphysical, but this systematic error can easily result from a small phase error when $\chi' \gg \chi''$ as is the case in the molecular magnet sample measured in Figs. 5d and 5e (see Sect. "Superparamagnetism in Molecular Magnets" for more about molecular magnetism).

Measuring Spin Relaxation in Magnetic Materials

Relaxation in a Paramagnetic Salt at Low Temperatures

One of the primary applications of AC susceptibility is possibly in its power to quantify temperature and field dependence of magnetic relaxation owing to the broad frequency range accessible to AC susceptometers. Casimir and du Pré [26] may have been the first to analyze slow-spin relaxation observed in a magnetic material. Their goal was to understand data on the AC magnetic properties of a paramagnetic salt, an insulating material with large magnetic moments which follow the $\chi_{dc} \sim 1/T$ Curie paramagnetic law down to very low temperatures $T < 1$ K. The application of a small magnetic field produces some magnetic alignment of the spins with the field, and this large decrease in magnetic entropy leads to an increase in the temperature of the spins due to thermodynamics (this *magnetocaloric effect* is exploited for magnetic refrigeration). Both these effects take place instantly on the time scale of a typical AC susceptometer. However, the heat generated in the spins must equilibrate with the rest of the crystal lattice, and at very low temperature, the spin-lattice thermal coupling is weak so that the thermal time constant for the sample under study in Ref. [26] was $\tau_{SL} \sim 0.01$ seconds. The cooling of the spins during thermal equilibration produces a large change in the magnetic susceptibility according to the $1/T$ Curie law, and this is registered easily on the susceptometer as a magnetic relaxation. The time dependence of the relaxation is found to follow the Debye relaxation model mentioned earlier where the response at low frequencies ($\omega \ll 1/\tau_{SL}$) is referred to as the isothermal susceptibility because the spins have time to thermally equilibrate with the lattice, while at high frequencies ($\omega \gg 1/\tau_{SL}$), the spin temperature fluctuates too fast to exchange any heat with the lattice, and this limit is referred to as the adiabatic susceptibility. The peak in χ'' occurs for $\omega \approx 1/\tau_{SL}$ because the spin heat is continually being dissipated to the lattice during the measurement. At low frequencies, $\chi'' \to 0$ in the DC field limit, and at high frequencies, $\chi'' \to 0$ because no heat is exchanged. Note that the adiabatic susceptibility ($\chi'(\omega \to \infty)$) is zero in the simple Debye model we have described in Sect. "Magnetic Relaxation", but in the above example, there is clearly a paramagnetic response of the spins at high frequencies, so a constant χ_S must be added to χ' in Eq. 11. Because an adiabatic magnetic response is often present in systems studied for spin relaxation phenomena, the χ_S term is included in order to fit the experimental data to the model.

Superparamagnetism in Molecular Magnets

Single-domain ferromagnetic materials at the nanoscale can fluctuate spontaneously when the energy barrier to spin reversal $E_A \ll k_B T$, where k_B is the Boltzmann constant and T is the temperature in kelvin. The barrier $E_A = KV$, where V is the particle volume and K is the ferromagnet's intrinsic magnetic anisotropy energy density which causes the spins to align along a certain easy axis. This behavior is termed *superparamagnetism* since the single-domain particle behaves as a paramagnetic macrospin and follows the Curie law at high temperature. At lower temperatures such that $k_B T < KV$, thermal spin reversal is slowed, and the magnetic relaxation time constant τ becomes observable with AC susceptometry. For particles which are immobilized and non-interacting, there will be a single time constant which follows the Arrhenius law for thermal activation over a barrier, which in the context of superparamagnets is also referred to as Néel relaxation: $\tau_N = \tau_0 e^{KV/k_B T}$. The inverse attempt frequency τ_0 is a material property and sets the lower bound for the reversal time. It is typically assumed to be on the order of 10^{-9} s but can be 2–3 orders of magnitude longer in some cases. Particle size will greatly affect the dynamics due to the exponential dependence on particle volume, so it is very helpful to have a narrow distribution of particle sizes. Remarkably, it is possible to chemically synthesize bulk crystals of identical superparamagnets such as the $Mn_{12}Ac$ cluster [27] in which the clusters of 12 Mn ions are isolated from each other by organic ligands (abbreviated as "Ac" in the chemical formula) and magnetically order at very low temperatures with a total spin $S = 10$. Fig. 6 plots the temperature dependence of AC susceptibility at various AC frequencies from 10 to 1000 Hz where χ' (data points) has a crossover from the Curie $1/T$ trend at high temperature to a low susceptibility at low temperature where spin reversal is blocked, while χ'' (lines) shows a peak which occurs when $\omega \cdot \tau_N \sim 1$. This is akin to Fig. 1 inverted along the x-axis where high temperatures correspond to short τ_N and vice versa.

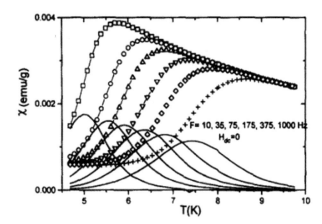

Fig. 6 Blocking behavior in molecular magnet Mn12 complex revealed in the temperature dependence of χ' (data points) and χ" (lines) at various AC frequencies. The peaks shift to higher temperature as frequency increases. (Figure reproduced from Ref. [27] with permission from Elsevier)

Plotting τ_N vs. T found at various frequencies and fitting to the Néel relaxation equation reveals the material properties KV and τ_0. The time constant in the low-temperature "blocked" state is so slow that molecular magnets display magnetic hysteresis in $M_{dc}(H)$ with high coercivity and dramatic step-like features in the field dependence which are interpreted as resonant magnetic tunneling between spin multiplet states [28]. The peak in $\chi''(T)$ identifies a "blocking temperature" T_B, but it is important to note that the value depends directly on the AC frequency as seen in Fig. 6 and in the Néel relaxation equation which shows an exponential temperature dependence of relaxation time.

Superparamagnetism in Nanoparticles

Magnetic nanoparticles coupled to drugs or other functional biological units are of great interest in medicine due to the ability to move them to a target region remotely with magnetic field gradients (DC magnetic fields pass through the body essentially unhindered). Furthermore, magnetic relaxation of the particles in AC magnetic fields can be exploited by choosing an AC frequency to match the peak in χ'' and generate localized heating, referred to as magnetic hyperthermia. AC susceptibility is a critical tool in this field which is used to characterize the particle sizes and magnetic properties in order to determine their suitability for applications. The nanoparticles in this case are in a liquid solution, unlike the above example of molecular magnets which are locked in a crystal lattice. To prevent agglomeration of the magnetic material due to dipole interactions between particles, each must have a sufficiently thick non-magnetic coating around it. In addition to Néel relaxation explained in the previous example, magnetic particles in solution are able to change

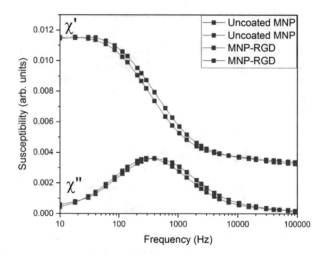

Fig. 7 Brownian relaxation of superparamagnetic nanoparticles of iron oxide. (Figure reproduced from Ref. [29] with permission from Elsevier)

the direction of the magnetic moment in an applied field by physical rotation of the whole particle. This is known as Brownian relaxation, and it follows the law: $\tau_B = (\pi \eta D_H^3)/(2k_B T)$, where η is the dynamic viscosity of the fluid, and D_H is the hydrodynamic diameter of the particle including the magnetic core and coatings. The time constant depends linearly on hydrodynamic volume, in contrast to Néel relaxation which has an exponential dependence on core volume. At small particle sizes, the time scales for these two relaxation mechanisms can overlap and complicate the determination of either the hydrodynamic size or core volume. The frequency-dependent AC properties of an aqueous solution of magnetite (Fe_3O_4) nanoparticles coated with dextran, a water-soluble polysaccharide, are shown in Fig. 7 [29]. The average D_H of the particles was found to be 116 nm, and in this regime, $\tau_N \gg \tau_B$ so that the shift in the observed peak in χ'' can be attributed to a change in D_H between the dextran-coated particles ("uncoated MNP" in Fig. 7) and those with an additional RGD tripeptide polymer. The shift to lower frequencies was found to correspond to an increase in the hydrodynamic diameter of 6 nm which was attributable to the tripeptide molecule.

Spin Glasses

While the freezing of magnetic fluctuations in superparamagnets can be explained by the magnetic properties of non-interacting particles, spin freezing can also occur at low temperatures due to *magnetic frustration*, and at the freezing temperature T_g, it will lead to a sharp peak in χ' and an onset of dissipation in χ'' below T_g. In some solids, there are interacting magnetic moments, yet there is no transition to a periodic long-range order at low temperatures (e.g., ferro-, ferri-, or anti-

Fig. 8 Spin glass freezing in AuMn (2.98 at % Mn). (Figure reproduced from Ref. [30] with permission from APS)

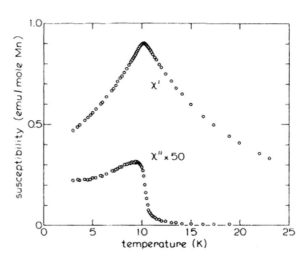

ferromagnetic). The magnetic interactions between moments vary in direction and magnitude non-periodically in the material, and this leads to a collective freezing at low temperatures in one of the many possible metastable configurations. "Frustration" refers to the inability of the ensemble of moments to find a unique spin arrangement which minimizes the energy of the system. In disordered alloys, this is known as a *spin glass*, while in crystallographically ordered but magnetically frustrated systems, it is called a *spin ice*. The most thoroughly studied spin glass systems are disordered alloys of a nonmagnetic metallic host with a few percent of a miscible magnetic element substituting at random sites, e.g., AuMn, CuMn, AgMn, AuFe, or CuCo. In Fig. 8 are temperature-dependent AC data on AuMn alloy (3% Mn), where a freezing temperature was identified by a sharp peak in χ' and also a dissipation (χ'') peak similar to the molecular magnet from Fig. 6. However, unlike the molecular magnet blocking behavior, the χ' peak in spin glasses is very weakly dependent on AC frequency. Note that the researchers took care to remove screening effects from the conducting sample by crushing the sample and mixing it with Al_2O_3 powder to electrically isolate the grains. Due to the fact that magnetic interactions are central to the spin glass state, it is understandable that the simple Debye model for non-interacting spins does not fit the observed frequency dependence of the AC response. Regarding the temperature dependence, a modification to Arrhenius activation, known as the Vogel-Fulcher law, is seen to fit spin glass systems: $\tau = \tau_0 e^{E/k_B(T-T_0)}$ where E is an activation energy and T_0 is a parameter which describes the strength of the interactions between moments and represents the temperature at which the relaxation time diverges. Note that this will be below the experimentally determined T_g due to the finite time scale of measurements.

As relaxation becomes slower than the longest time scale for conventional AC susceptometry ($\tau_{max} \approx 1$ s), researchers turn to SQUID AC susceptometry ($\tau_{max} \approx 1$ min) or relaxation using repeated measurements of the DC magnetic moment ($\tau_{max} \approx 1$ day). Any slow measurements are plagued by instrument and environmental drifts, so it is crucial to separate background and sample contributions. Methods to achieve this include measuring a "null" reference sample of similar signal size which shows no relaxation or a "blank" without any sample to rule out other instrument drifts.

Temperature-dependent DC magnetometry $M_{dc}(T)$ also provides a method of revealing the onset of magnetic irreversibility that occurs at freezing transitions (frustrated spin systems, superparamagnets) or magnetic phase transitions. The usual "ZFC/FC" method is to cool the sample in zero field well below the transition temperature of interest (ZFC), apply a small magnetic field (typically 1–10 mT), start measurements while warming through the transition until well above it, and continue measurements in the applied field upon cooling again (FC). The ZFC and FC curves will separate at an irreversibility temperature that is related to the peak in χ' and for a similar reason: irreversibility appears at a temperature where magnetic relaxation of the moments becomes much longer than the measurement time for a DC magnetic measurement (1–10 s). The ZFC/FC method is also used in the superconducting state to probe the onset of flux pinning (to be discussed in the next section).

Superconductivity

Superconductors are a certain class of conductors which undergo a phase transition at low temperatures to a state of zero electrical resistance. They have the remarkable property that under certain circumstances the magnetic field is spontaneously expelled from the interior: $B = 0$, or $\chi = -1$. However, in all cases, the interaction with a magnetic field is very different from other materials. As we have pointed out in Sect. "Theory of AC Susceptibility", we can expect signatures in the AC susceptibility arising from both the dramatic change in resistivity (producing a peak in χ'' due to screening) and the onset of strong diamagnetism ($\chi' < 0$) that occurs in the superconducting state. The situation becomes even more subtle when a magnetic field is applied at $T < T_c$ and the perfect diamagnetic state yields to the *mixed state* in which magnetic field penetrates into the superconductor in the form of quantized flux lines. This vortex matter can take the form of a periodic or disordered lattice and interacts with defects (voids, grain boundaries, inclusions, etc.) in the superconductor, which tend to pin them in place. Depending on conditions, the lattice can behave as a solid or glass (pinned strongly by defects) or a liquid (free to move under magnetic forces). The magnetism of the vortex matter under AC and DC applied fields, together with the diamagnetism of the superconducting state and the screening properties of electrically connected regions of the superconductor, richly showcases both magnetic relaxation and screening phenomena at the same time and requires careful examination in order to identify the relevant physics in a given measurement of AC susceptibility. We will describe one such superconductor system below in order to illustrate these points.

There are thousands of materials that are known to superconduct, including elements, intermetallic compounds and alloys, ceramics, and even organic com-

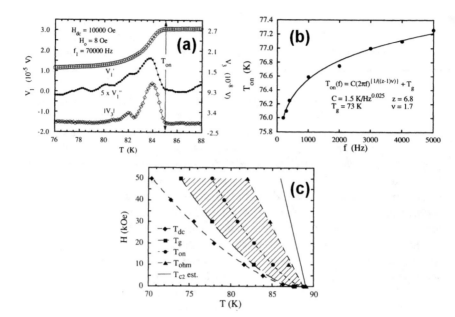

Fig. 9 (**a**) Temperature dependence the first and third harmonic AC voltages for a $YBa_2Cu_3O_7$ (YBCO) thin film; (**b**) frequency dependence of the onset of the AC response for a $YBa_2Cu_3O_7$ (YBCO) thin film; (**c**) H(T) boundaries determined using different experimental methods for a $YBa_2Cu_3O_7$ (YBCO) thin film. (Figures **a–c** reproduced from Ref. [34] with permission from APS)

pounds. The onset temperature for superconductivity (T_c) was observed to be below 30 K in all of these materials until a large new class of copper oxide-based ceramics were discovered in the 1980s which became known as high-temperature superconductors (HTS) because the T_c could be as high as 150 K (the highest T_c values, however, have been recently reported in hydrides when subjected to extremely high hydrostatic pressures, with LaH_{10} currently holding the record of $T_c \approx 250$ K at a pressure of $P = 170$ GPa [31]).

Prior to the discovery of HTS materials, it was very common to use AC susceptibility and electrical transport measurements interchangeably when determining T_c and its dependence on applied magnetic field, H. Although many understood that these measurements were actually measuring different phenomena—a voltage drop in one case and screening in the other—it is the rapid change in resistivity as T_c is approached in these low-T_c materials that results in a near coincidence of the two measurements. The resistivity is measured directly in one case, and in the other, the change in resistivity results in a change in the screening depth which produces the conditions for the behavior shown in Fig. 9a, in which the AC susceptibility loss component (χ'') goes through a maximum when the penetration depth is on the order of the sample dimension. To have this coincidence of measurements required that there was an onset of strong flux pinning at T_c.

After the discovery of HTS materials, what was initially thought to be $T_c(H)$ was observed to follow a glass-like dependence $[(1 - t)^{3/2}$ where $t = T/T_c]$, and for a time, this dependence was attributed to the combination of low coherence length and granular nature of the earliest HTS samples [32]. However, after studying high-quality single-crystal and thin-film YBCO samples, it became clear that the onset of strong flux pinning was in all cases following a glass-like dependence and a new region of the field-temperature $(H\text{-}T)$ diagram existed between the onset of superconductivity, which became identified as $T_{c2}(H)$ [33] and the boundary where flux pinning was strong enough to allow a persistence condition $(\rho = 0)$. This is referred to as the vortex glass transition $T_g(H)$ which is the same boundary initially thought to be $T_c(H)$ for HTS materials. In the HTS case, the results of the two measurements no longer coincided in applied magnetic fields. Within this region between $T_{c2}(H)$ and $T_g(H)$, the sample is superconducting and flux vortices exist, but their pinning is not strong enough to maintain a persistent state.

Shown in Fig. 9a are χ for a YBCO film with an applied field $H_{dc} = 1$ T, AC field $H_{ac} = 0.8$ mT, at $f = 70$ kHz. Both DC and AC fields are applied normal to the surface of the film. Several features are notable including the fact that both the first and third harmonic $\chi(T)$ change temperature dependence at the same temperature identified as $T_{on}(H)$ in Fig. 9a. In contrast to earlier suggestions, which identified the peak in the first harmonic as T_{c2} and the peak in the third harmonic as T_g, it can be seen in the superconducting $H\text{-}T$ phase diagram in Fig. 9c that the spread in temperature between T_{c2} and T_g at $H = 1$ T is actually much larger than the first and third harmonic peak separations and the order in temperature is actually reversed, i.e., the third harmonic peak is higher in T, not lower.

In order to better understand what phenomena are actually being measured by AC susceptometry, the results of DC magnetization and $E\text{-}J$ transport measurements (voltage-current curves normalized to the sample dimensions) are included in the $H\text{-}T$ diagram in Fig. 9c [34] for a YBCO thin film with H applied normal to the film surface. The $E\text{-}J$ transport curves were used to determine several of these boundaries that can help provide insight into the χ measurements. From the $E\text{-}J$ isotherms, the temperature at which the curvature changes from positive to negative identifies the boundary below which a persistent state exists, identified as T_g, while T_{ohm} identifies the temperature above which only dissipative behavior (linear $E\text{-}J$) is observed. The onset T_{on} of the change in χ is observed between T_g and T_{ohm} for reasons to be discussed below. The onset of superconductivity, $T_{c2}(H)$, is estimated from DC magnetization using Ginzburg-Landau theory of reversible superconductors with knowledge of the penetration depth and coherence length [33], while T_{dc} is the point where irreversibility appeared in the temperature-dependent DC magnetization ZFC/FCC measurements and was the method originally used to try to measure $T_c(H)$.

Either the temperature of the peak (T_p, as described in Sect. "Theory of AC Susceptibility") or the onset of the χ'' screening response can be used to identify the point where the resistivity condition is such that the screening skin depth, δ, is on the order of the sample size [35]. In Fig. 9c, the temperature of the onset T_{on} of both first and third harmonic χ_{ac} responses, measured at $f = 70$ kHz, is plotted as a function of H and is seen to fall well above T_g . Plotted in Fig. 9b is T_{on} at $H = 5$ T measured over a range of frequencies. Fig. 9b fits the data to the form $T_{on}(f) = T_g + C(2\pi f)^{\{1/[(z-1)\nu]\}}$ with $z = 6.8$ and $\nu = 1.7$ determined from scaling of the E-J data using the vortex glass model of Fisher [36] which allows the determination T_g by extrapolation to $f = 0$. This frequency dependence results from the existence of a metastable current (or magnetization) above T_g. Due to this frequency dependence, it is not possible to determine T_g from a temperature sweep at a single frequency; however, measuring T_{on} at a series of frequencies allows a fairly accurate determination of T_g to be made using AC susceptometry.

In a high-field superconductor with vortex pinning, the presence of a third harmonic in AC susceptibility is a direct consequence of magnetic hysteresis during the AC excitation as pointed out by Bean [37]. Within the model for flux pinning put forward in that work, the predicted third harmonic magnetization $M_3 \propto 1/J_c$, which explains qualitatively the peak in χ_3' near T_c. While the Bean model is applicable to HTS below T_g, where time-independent hysteresis loops in DC magnetization are observed, it is not completely clear that this model applies when flux pinning is weak, and DC hysteresis loops are not observable, as it is in the region between T_g and T_{c2}, yet a third harmonic is observed in that region. This is, however, consistent with the metastable vortex state and associated magnetic relaxation mentioned above [36].

We have seen that the AC susceptibility measurements of HTS were initially misunderstood as it was not appreciated that the physics of HTS was different than that of low-Tc superconductors. The combination of effects occurring in HTS could not have been understood nor distinguished without additional measurements and a relevant theory to explain the physics. In addition, the physics of HTS is such that measurement at a single frequency cannot identify the key feature (T_g) that AC susceptometry excels at identifying. The sensitivity of AC susceptometry to all of the relevant phenomena can be both a strength and a liability. The hatched region in Fig. 9c shows that we can measure crossovers that are dependent on frequency and not just phase transitions.

Summary

As has been shown, AC susceptibility is a very versatile method that can be used to study magnetic relaxation phenomena, changes in resistance of materials, and magnetic susceptibility. It can be a fairly simple method to implement and has been widely used for rapid screening of materials. This is especially true of phenomena like phase transitions where a variable like temperature or magnetic field can be swept and the location of the transition or crossover observed. The sample can be held stationary, and it usually does not require a specific geometry, thereby simplifying sample preparation and allowing for relatively rapid measurements. In addition, electrical contacts are not needed.

The fact that there are different ways that the AC susceptibility signal can be generated means that additional independent measurements are generally required to understand the physics that is producing the observed response. We saw that in the case of HTS, AC susceptibility results were initially improperly interpreted and that it took some time before the behavior could be understood. This required various other measurements to clarify the associated observations. Initial misunderstandings resulted primarily from assumptions and methods used in the past on low-T_c superconductors which were no longer valid for the new HTS materials.

AC susceptibility measurements can also be complicated by the competition between screening of a sample's interior—due to the sample's conductivity—and other phenomena to be observed, like magnetic susceptibility or relaxation. In addition, the temperature dependence of a sample's conductivity can add further complexity to a measurement.

AC susceptibility can be used to quantitatively determine physical parameters like resistivity or magnetic susceptibility. In this case, the measurement is more involved, and the system will require much more care including calibration and runs with control samples. Several commercially available systems exist that allow for quantitative measurements of physical properties, but the need for control samples and calibration checks is just as great in these systems.

References

1. L. Hartshorn, A precision method for the comparison of unequal mutual inductances at telephonic frequencies. J. Sci. Instrum. **2**, 145 (1925). https://doi.org/10.1088/0950-7671/2/5/301
2. R.B. Goldfarb, M. Lelental, C.A. Thompson, Alternating-field Susceptometry and magnetic susceptibility of superconductors, in *Magnetic Susceptibility of Superconductors and Other Spin Systems*, ed. by R. A. Hein, T. L. Francaville, D. H. Liebenberg, (Springer Science+Business Media, LLC, 1991), pp. 49–80
3. F. Gömöry, Characterization of high-temperature superconductors by AC susceptibility measurements. Supercond. Sci. Technol. **10**(8), 523–542 (1997). https://doi.org/10.1088/0953-2048/10/8/001
4. M. Bałanda, AC susceptibility studies of phase transitions and magnetic relaxation: Conventional, molecular and low-dimensional magnets. Acta Phys. Pol. A **124**(6), 964–976 (2013). https://doi.org/10.12693/APhysPolA.124.964
5. C.V. Topping, S.J. Blundell, A.C. susceptibility as a probe of low-frequency magnetic dynamics. J. Phys. Condens. Matter **31**(1) (2019). https://doi.org/10.1088/1361-648X/aaed96
6. S. Chikazumi, *Physics of Ferromagnetism*, 2nd edn. (Clarendon Press; Oxford University Press, Oxford: New York, 1997)
7. "Magnetic units." https://www.ieeemagnetics.org/index.php?option=com_content&view=article&id=118&Itemid=107.
8. N.W. Ashcroft, N.D. Mermin, *Solid State Physics, College Ed* (W.B. Saunders, Philadelphia, 1976)
9. P.J. Debye, *Polar Molecules* (Chemical Catalog Company, 1929)
10. L.D. Landau, E.M. Lifshits, L.P. Pitaevskiĭ, *Electrodynamics of Continuous Media*, 2nd edn. (Pergamon, Oxford [Oxfordshire]; New York, 1984)

11. M. Nikolo, Superconductivity: A guide to alternating current susceptibility measurements and alternating current susceptometer design. Am. J. Phys. **63**(1), 57–65 (1995). https://doi.org/10.1119/1.17770

12. AC Measurement System (ACMS). Quantum Design, Inc., 1994.

13. "AC Measurement System II (ACMS-II)." Quantum Design, Inc., 2014.

14. A. Amann, M. Nallaiyan, L. Montes, A. Wilson, S. Spagna, Fully automated AC Susceptometer for Milli-kelvin temperatures in a DynaCool PPMS. IEEE Trans. Appl. Supercond. **27**, 4 (2017). https://doi.org/10.1109/TASC.2016.2639480.

15. E. Riordan et al., Design and implementation of a low temperature, inductance based high frequency alternating current susceptometer. Rev. Sci. Instrum. **90**, 7 (2019). https://doi.org/10.1063/1.5074154

16. A.D. Hibbs et al., A SQUID-based ac susceptometer. Rev. Sci. Instrum. **65**(8), 2644–2652 (1994). https://doi.org/10.1063/1.1144664.

17. A. Aspelmeier, M. Tischer, M. Farle, M. Russo, K. Baberschke, D. Arvanitis, Ac susceptibility measurements of magnetic monolayers: MCXD, MOKE, and mutual inductance. J. Magn. Magn. Mater. **146**(3), 256–266 (1995). https://doi.org/10.1016/0304-8853(95)00025-9

18. A. Aharoni, E.M. Frei, S. Strikman, D. Treves, The reversible susceptibility tensor of the stoner-Wohlfarth model. Bull. Res. Counc. of Israel **6**, 215–238 (1957)

19. L. Pareti, G. Turilli, Detection of singularities in the reversible transverse susceptibility of an uniaxial ferromagnet. J. Appl. Phys. **61**(11), 5098–5101 (1987). https://doi.org/10.1063/1.338335

20. H. Srikanth, J. Wiggins, H. Rees, Radio-frequency impedance measurements using a tunnel-diode oscillator technique. Rev. Sci. Instrum. **70**(7), 3097–3101 (1999). https://doi.org/10.1063/1.1149892

21. D.X. Chen, V. Skumryev, Calibration of low-temperature ac susceptometers with a copper cylinder standard. Rev. Sci. Instrum. **81**, 2 (2010). https://doi.org/10.1063/1.3309779

22. D.X. Chen, J.A. Brug, R.B. Goldfarb, Demagnetizing factors for cylinders. IEEE Trans. Magn. **27**(4), 3601–3619 (1991). https://doi.org/10.1109/20.102932

23. G. Turilli, Measurement of the magnetic anisotropy field by performing the second derivative of the magnetization curve in a continuous field. J. Magn. Magn. Mater. **130**(1–3), 377–383 (1994). https://doi.org/10.1016/0304-8853(94)90697-1

24. Y.A. Kraftmakher, Measurement of electrical resistivity via the effective magnetic susceptibility. Meas. Sci. Technol. **2**(3), 253 (1991)

25. N.R. Dilley, M. Baenitz, Measurement of electrical resistivity without contacts using the ACMS option. Quantum Design (2009) [Online]. Available: https://www.qdusa.com/siteDocs/appNotes/1084-306.pdf

26. H.B.G. Casimir, F.K. du Pré, Note on the thermodynamic interpretation of paramagnetic relaxation phenomena. Physica **5**(6), 507–511 (1938). https://doi.org/10.1016/S0031-8914(38)80164-6

27. M.A. Novak, R. Sessoli, A. Caneschi, D. Gatteschi, Magnetic properties of a Mn cluster organic compound. J. Magn. Magn. Mater. **146**(1–2), 211–213 (1995). https://doi.org/10.1016/0304-8853(94)00860-4

28. S.M.J. Aubin et al., Resonant magnetization tunneling in the trigonal pyramidal Mn(IV)Mn(III)3 complex [Mn4O3Cl(O2CCH3)3(dbm)3]. J. Am. Chem. Soc. **120**(20), 4991–5004 (1998). https://doi.org/10.1021/ja974241r

29. K.K. Narayanasamy, M. Cruz-Acuña, C. Rinaldi, J. Everett, J. Dobson, N.D. Telling, Alternating current (AC) susceptibility as a particle-focused probe of coating and clustering behaviour in magnetic nanoparticle suspensions. J. Colloid Interface Sci. **532**, 536–545 (2018). https://doi.org/10.1016/j.jcis.2018.08.014

30. C.A.M. Mulder, A.J. van Duyneveldt, J.A. Mydosh, Frequency and field dependence of the ac susceptibility of the AuMn spin-glass. Phys. Rev. B **25**(1), 515–518 (1982). https://doi.org/10.1103/PhysRevB.25.515.

31. A.P. Drozdov et al., Superconductivity at 250 K in lanthanum hydride under high pressures. Nature (London) **569**(7757), 528–531 (2019). https://doi.org/10.1038/s41586-019-1201-8.

32. K.A. Müller, M. Takashige, J.G. Bednorz, Flux trapping and superconductive glass state in La2CuO4:Ba. Phys. Rev. Lett. **58**(11), 1143–1146 (1987). https://doi.org/10.1103/PhysRevLett.58.1143.

33. Z. Hao, J.R. Clem, M.W. McElfresh, L. Civale, A.P. Malozemoff, F. Holtzberg, Model for the reversible magnetization of high- type-II superconductors: Application to high-Tc superconductors. Phys. Rev. B **43**(4), 2844–2852 (1991). https://doi.org/10.1103/PhysRevB.43.2844

34. J. Deak et al., Irreversibility line in YBa2Cu3O7 thin films: Correlation of transport and magnetic behavior. Phys. Rev. B **49**(9), 6270 (1994)

35. J.R. Clem, Ac losses in type II superconductors, in *Magnetic Susceptibility of Superconductors and Other Spin Systems*, ed. by R. A. Hein, T. L. Francaville, D. H. Liebenberg, (Springer Science+Business Media, LLC, 1991), p. 177

36. M.P.A. Fisher, Vortex-glass superconductivity: A possible new phase in bulk high-Tc oxides. Phys. Rev. Lett. **62**(12), 1415–1418 (1989). https://doi.org/10.1103/PhysRevLett.62.1415

37. C.P. Bean, Magnetization of high-field superconductors. Rev. Mod. Phys. **36**(1), 31–39 (1964). https://doi.org/10.1103/RevModPhys.36.31

DC Hysteresigraphs for Hard and Soft Materials

Thomas Bapu

Abstract The DC hysteresigraphs we are discussing here are devices that utilize an inductive magnetic measurement technique combined with a closed magnetic circuit for the evaluation of the magnetic material properties of bulk samples. Two distinct variants of these machines will be discussed to cover both hard and soft magnetic material testing of larger, bulk physical samples. The DC hysteresigraph for hard magnetic materials is focused on the measurement of higher-coercivity magnetic materials such as permanent magnets. It utilizes a high-field (1–3 T) laboratory electromagnet yoke to magnetize the sample under test and create the closed magnetic circuit during the measurement. The DC hysteresigraph for soft magnetic materials utilizes a toroidal/ring-shaped specimen of the sample under test as the closed magnetic circuit itself and incorporates windings about the toroid to magnetize the sample. Once magnetized, inductive coil sensing is utilized in both devices to acquire flux density information while varying the applied field. Depending on the device, this reveals either the second quadrant demagnetization curve or the entire major magnetic hysteresis loop and associated magnetic material properties for the sample under test.

Keywords Hysteresigraph · Second quadrant demagnetization curve · Hysteresis loop · B-H loop · Permanent magnet properties · Soft magnetic properties · Ring testing

T. Bapu (✉)
Lake Shore Cryotronics, Inc., Westerville, OH, USA
e-mail: thomas.bapu@lakeshore.com

© Springer Nature Switzerland AG 2021
V. Franco, B. Dodrill (eds.), *Magnetic Measurement Techniques for Materials Characterization*, https://doi.org/10.1007/978-3-030-70443-8_5

1 Introduction to DC Hysteresigraphs for Hard Magnetic Materials

We will begin with the DC hysteresigraphs for hard magnetic materials. These devices were designed with the purpose of measuring the properties of permanent magnets. They are essential equipment in permanent magnet material manufacturer's sites as well as their customer's facilities for ongoing quality control (QC) evaluation and refinement of product. They are also found in testing labs for independent evaluation of material properties as well as laboratories focused on research of novel permanent magnet materials as they draw closer to their production form (Fig. 1).

In typical use, a saturated sample under test is placed within the air gap of a laboratory electromagnet. Inductive sensing coils read the flux density from the sample. Current from a magnet power supply applies an increasing negative magnetizing field to the sample sufficient to drive the sample through its second quadrant demagnetization curve. The curve is processed to reveal various material properties of the permanent magnet, as show in Fig. 2 below. If the sample began the measurement in a saturated magnetic state, these properties are representative of data from the major hysteresis loop.

Higher-coercivity samples require external equipment such as capacitive discharge magnetizers to saturate the sample prior to test. But with a sufficiently high-field electromagnet, some materials with lower coercivities can be fully

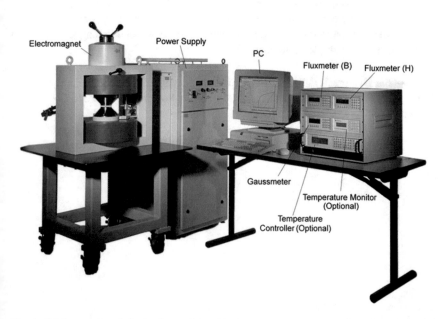

Fig. 1 DC hysteresigraph for hard materials with electromagnet, power supply, and instrumentation, courtesy of Lake Shore Cryotronics [1]

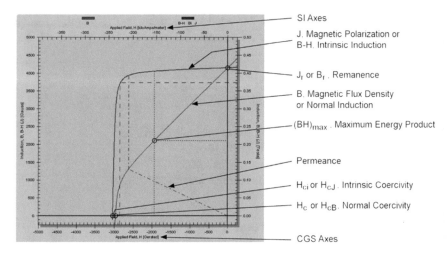

Fig. 2 Second quadrant demagnetization curve

magnetized and reverse magnetized within the magnet yoke, allowing the entire hysteresis loop to be measured. The initial magnetization curve and recoil loops may also be acquired with the device. Measurements at variable temperatures are possible with the correct supplemental equipment.

2   Why Do we Need a DC Hysteresigraph for Hard Magnetic Materials?

In the modern era, there is some overlap between this device and other measurement technologies. Simpler devices such as Helmholtz coils coupled with integrating fluxmeters or more complex equipment such as VSMs [2] and SQUIDs [3] can also be used to perform similar measurements. But those devices are not as well suited for the specific task at hand. VSMs and SQUIDs cannot be used with very large or massive samples, or if the material being measured is so strongly magnetic that it overloads the VSM or SQUID detection circuitry. Helmholtz coils only measure a single operating point on the hysteresis loop. And all of these techniques measure samples in an open magnetic circuit condition which requires corrections for sample demagnetization to accurately determine intrinsic material parameters.

   A DC hysteresigraph for hard magnetic materials is optimized to perform rapid magnetic measurements of permanent magnet materials in their larger, bulk form while operating within a closed magnetic circuit. A closed magnetic circuit is *ideally* free of the self-demagnetization fields (and associated corrections) that exist when a bulk sample is tested in an open-circuit condition.

In an open-circuit condition, the applied magnetic field strength, H_{app}, is modified by the following equation to resolve the resultant H internal field, H_{int} [4]:

$$H_{int} = H_{app} - N_d M$$

where
H_{int} = internal or effective field, seen by sample [A/m].
H_{app} = applied magnetic field strength [A/m].
N_d = sample under test demagnetization factor.
M = volume magnetization of the sample under test [A/m].
and:

$$B = \mu_0 (H + M.)$$

where
B = magnetic flux density [Tesla]
μ_0 = permeability of free space = $4\pi \times 10^{-7}$ [Henry/m]

Failure to correct this error alters the H field and distorts the M vs. H loop shapes during open-circuit testing.

Magnetic circuit closure during measurement of magnetic material properties is of primary concern when a sample's magnetic length dimension is small compared to the dimensions of its cross-section or diameter. This is a common condition for samples diced or cored from production permanent magnet geometries. A large length to cross-section or diameter ratio minimizes the impact of self-demagnetizing fields. Bulk samples, however, generally have a very poor ratio, often less than 1, so they are heavily influenced by self-demagnetization fields. Bulk permanent magnet samples are typically larger than 3 mm × 3 mm in cross-section and greater than 3 mm in magnetic length dimension, with typical samples in the 10 mm cube or cylinder range. Maximum sample dimensions are limited by the size of the pole cap and field uniformity.

Tables of self-demagnetization correction factors (N_d) do exist for simple, ideal shapes (cylinders, ellipsoids, etc.) [5, 6]. However, in practice, applying analytically determined correction factors to samples being measured does not always yield accurate corrections. And if the corrected loop shapes appear unnaturally overcorrected or undercorrected (as indicated by "shear" in the shape of the hysteresis loop as shown in Fig. 3), the user's best guess for correction factors replaces analytical data tables, making the resulting measurement exercise somewhat subjective. DC hysteresigraphs utilize a closed magnetic circuit path. This provides for accurate determination of the second quadrant of the major hysteresis loop without the need for self-demagnetization field corrections.

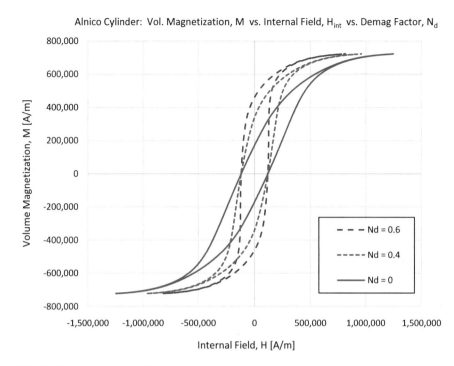

Fig. 3 Sheared and corrected hysteresis loop

3 Materials to Test

Alnico, hard ferrites, samarium cobalt (SmCo), and neodymium iron boron (NdFeB) permanent magnets are some of the materials most frequently measured by such equipment. But new classes of materials such as MnBi and nanostructured exchange-coupled soft (low coercivity)/hard (high coercivity) materials commonly referred to as exchange-spring magnets [7, 8] are now being developed and analyzed toward eventual commercialization. These developments are motivated by procurement concerns and the rising cost of rare-earth materials used today in the higher energy product (BH_{max}) materials such as SmCo and NdFeB. The goal is to develop permanent magnet materials with high energy product but with either no or less rare-earth constituents.

Some testing of soft magnetic materials is possible with this type of equipment, but it requires more care from the operator to ensure greater attention is paid to minimize air gaps in the sample to pole cap interfaces to avoid self-demagnetizing field effects. But typically, soft magnetic material specimens with coercivities below 8000 A/m are not evaluated with a large laboratory electromagnet. A return path with greater precision, such as a double Fahy-type yoke [9], is preferred. A Fahy-type yoke has provisions for the sample to be clamped to the ends of a C-shaped iron return path to minimize air gaps in the circuit. A double Fahy yoke duplicates the C-shaped structure to create a more symmetric return path and is represented in Fig. 15.

In addition to testing homogenous, solid bulk hard magnetic samples, ferrofluids may also be measured, but some interpretation of the results is required due to effects related to shear forces in the suspended media. Thin-film samples are not compatible with this measurement technique due to instrument sensitivity limitations.

4   Theory

In order to extract the magnetic material properties from bulk test samples of permanent magnet materials, we need to start with the basic relation between magnetic flux density, magnetic polarization, and magnetic field strength [10]. In SI units:

$$B = \mu_0 H + J \tag{1}$$

Magnetic polarization, J, is commonly represented on the Y axis for hysteresigraph measurements as opposed to volume magnetization, M. They are related as follows:

$$J = \mu_0 M$$

Sensing coils will inductively measure B. With the addition of an H field sensor, typically an H coil of similar sensitivity to the B coil, we have enough information to determine B, H, and J. Coaxial B and H sense coils in an electromagnet gap are illustrated in Fig. 4.

Since the sensing coils are inductive devices, an electronic fluxmeter is performing the integration of the induced voltages due to the flux linked by the B surrounding coil as the magnetic field varies. The apparent change in magnetic flux density is determined using the fluxmeter's integrated voltage divided by the area of the coil and number of turns. Here, we will follow IEC 60404–5 derivations [10]:

$$\Delta B_{ap} = B_2 - B_1 = \frac{1}{AN} \int_{t_1}^{t_2} U \, dt$$

where
B_2 = magnetic flux density at instant 2 [Tesla]
B_1 = magnetic flux density at instant 1 [Tesla]
A = cross-sectional area of sample under test [m^2]
N = number of turns on the search coil
$\int_{t_1}^{t_2} U \, dt$ = integrated induced voltage from time t$_1$ to t$_2$ [Weber's]

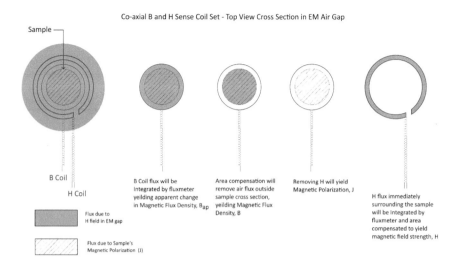

Fig. 4 Coaxial B and H sense coil in electromagnet gap

Since the B surrounding coil contains a combination of sample flux and air flux, we need to remove the air flux contribution to arrive at the change in magnetic flux density of the sample:

$$\Delta B = \Delta B_{ap} - \mu_0 \Delta H \frac{(A_t - A)}{A}$$

where
ΔH = change in magnetic field strength [A/m]
A_t = cross-section of search coil [m^2]

The magnetic field strength, H, can be obtained with the presence of another coil, similar in sensitivity to the B coil. This H sensing coil can be located next to the B coil and sample. If the H coil is similar in shape to the B coil and situated beside the B coil, the coil set is said to be using a side-by-side coil arrangement as shown in Fig. 5 below.

The side-by-side coil arrangement has the effect of measuring the H field in a localized area off to one side of the sample, and cannot sense the air flux immediately surrounding the sample. An alternative coil arrangement can remedy this. Winding an H coil with a sensing area that immediately surrounds the periphery of the B coil eliminates the localized area concern and "sees" the H field immediately surrounding the sample. Shrinking this coil's diameter so the entirety of the H coil is located very close to the B coil further improves the coil's ability to "see" the same H field as the B coil and sample. This is referred to as a coaxial coil arrangement. Side-by-side coils are useful in variable temperature coil sets as their symmetric shapes can counter some errors that occur during thermal expansion/contraction.

Fig. 5 Two common B and
H coil set configurations

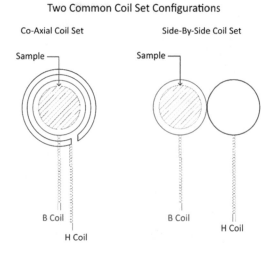

Other variations in coil set design and signal acquisition exist. In some circumstances, a Hall probe and Gaussmeter are used to acquire magnetic field strength, instead of an H coil. However, the use of an H coil and fluxmeter is preferred as it enables the use of the superior surrounding H coil technique, described earlier.

In a DC hysteresigraph, magnetic polarization, J, cannot be directly acquired from a simple coil wound around a sample under test [11]. J must be extracted from the acquired B and $\mu_0 H$ acquisition channels exploiting the fact that $J = B - \mu_0 H$. This can be performed pre- or post-fluxmeter integration. The post-fluxmeter calculation primarily involves mathematical subtraction of the integrated voltages. Pre-fluxmeter methods subtract $\mu_0 H$ from the B coil electrically to create a J-compensated coil set. The B and H coils can be balanced and wired in series opposition to generate J instead of B. The J and H signal are then sent to independent integrating fluxmeters. This results in a coil set specifically optimized for extraction of magnetic polarization, J.

For simplicity, only B and H coils are shown in the block diagrams (J is calculated post-fluxmeter).

5 Measurement Equipment/Block Diagram

Figure 6 shows a block diagram of a typical DC hysteresigraph for hard magnetic materials. The laboratory electromagnet provides the applied field for the measurement and the magnetic yoke to close the magnetic circuit path. The electromagnet should provide sufficient field uniformity within the measurement volume containing the sample under test and associated coils. The electromagnet should have an easy-to-manipulate pole adjuster and locking mechanism to fine-tune the air gap of the electromagnet to specifically fit the length of each sample

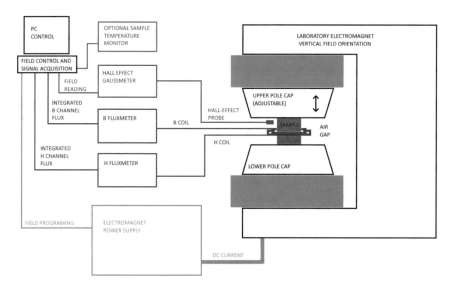

Fig. 6 Block diagram of DC hysteresigraph for hard materials

under test. Typically, a 2:1 ratio of pole cap face diameter to magnet air gap is utilized. That, along with a few other geometric constraints, ensures a 1% field uniformity in the volume where the sample and sensors reside. For convenience with sample exchange, a vertical field orientation is preferred. The lower pole of the electromagnet is generally held in a fixed location, while the upper pole height is manipulated to accommodate various sample lengths.

A magnet power supply must energize the electromagnet to sufficiently high magnetic fields. For second quadrant demagnetization curves, it should be able to drive the sample beyond the intrinsic coercivity field H_{cJ} of the sample under test. To measure an entire major hysteresis loop, the electromagnet and power supply combination must provide sufficient magnetizing field to fully saturate the sample in both forward and reverse field directions. A bipolar magnet power supply or other design with a smooth zero-current crossing is useful to avoid discontinuities in the measurements near zero field. Since the measurement incorporates a continuously varying field, a variety of methods can be utilized to control field. This can range from simple, linear programming of a voltage mode power supply to utilization of current mode supplies and more sophisticated proportional-integral-derivative (PID) field control loops.

A B,H coil set incorporating inductive sensing coils will be used to measure flux from the sample and magnetizing field during the measurement. These will be connected to electronic integrating fluxmeters. Since electronic fluxmeters drift, a fluxmeter with a microprocessor-aided integrator drift feature is preferred for minimal user interaction in managing integrator drift.

A Hall effect gaussmeter will be used in the process of calibration. It can be used for measuring magnetizing fields when calibrating sense coil area turns for the B or

H coils. They may also be used for setting magnetizing fields as part of a calibration method using physical standard reference materials for *B* coils.

High-purity nickel is often used as a material reference standard for *B* channels of hysteresigraphs. NIST does not offer reference standards in suitable shapes and sizes for DC hysteresigraphs for hard magnetic materials. Therefore, anneal state and purity should be well understood when using standards. Typically 99.995% or higher-purity nickel is used. Transfer standards comprised of well-characterized reference magnets may also be used for calibration or system verification. Compensation for the temperature coefficient of induction of the reference material is required to account for variation in temperature during instrument calibration.

A personal computer (PC) is used to acquire voltages from the fluxmeters and gaussmeter. The PC can also handle field sweep programming for the electromagnet power supply. The complexity of the field sweep is up to the user to determine. Typically, the sweep is slow enough to minimize eddy current effects. Second quadrant demagnetization curve sweep times on the order of tens of seconds to a couple minutes are typically employed to yield a reasonable compromise between going slow enough to extract DC properties and going fast enough to minimize integrator drift in the fluxmeter. An example of a typical second quadrant demagnetization curve is shown in Fig. 7.

6   Making a Measurement

Next, we will go through the process of performing a measurement:

1. Optionally saturate the sample under test in an external magnetizer (in the case of high-coercivity samples that cannot be saturated in the electromagnet).
2. Introduce sample under test.

   (a) Remove the B and H coil set from the electromagnet gap.
   (b) Reset the fluxmeter integrators to establish a zero flux reference.
   (c) Insert the sample under test in the coil set, and place the coil set into the electromagnet gap.
   (d) Fine-tune the electromagnet gap to fit the sample length.

3. Magnetize sample under test.

   (a) Fully magnetize the electromagnet in the positive field direction.
   (b) Reduce the current to zero quickly, within the limitations of the magnet power supply.

4. Smoothly increase negative magnetizing field, and continue demagnetization of the sample until intrinsic coercivity, H_{cJ}, is reached. (The B vs. H and J vs. H curves are calculated and plotted during the sweep).

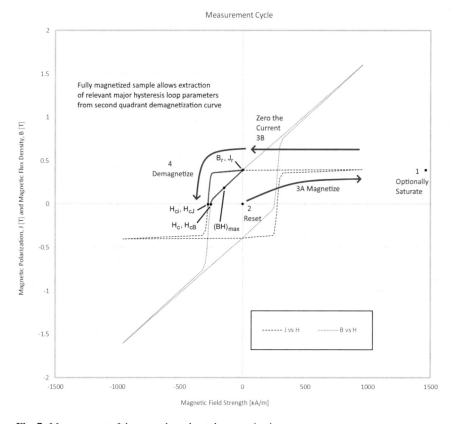

Fig. 7 Measurement of the second quadrant demagnetization curve

5. Return the field to 0.
6. Remove the coil set from the electromagnet and the sample under test from the coil set.
7. Check integrator drift by determining how far the indicated flux reading is from zero established in step 2b.

The data that was acquired reveals a second quadrant demagnetization curve. Various parameters can now be extracted from the curve, but the most common include but are not limited to:

B_r or J_r, remanence, [tesla].
H_c or H_{cB}, normal coercivity [A/m].
H_{ci} or H_{cJ}, intrinsic coercivity [A/m].
$(BH)_{max}$, maximum energy product [J/m^3].

Sample Magnetization

To determine the material properties of a sample, the sample under test must be magnetically saturated. For lower-coercivity samples, this simply involves magnetizing within the gap of the electromagnet until saturation is achieved. Typically, an applied field of three to five times the coercivity field is required. For high-coercivity magnets, this is not possible within the electromagnet gap. In those instances, a separate apparatus is used to achieve saturation. Often, a capacitive discharge pulse magnetizer is used. The magnetizer stores energy in an array of capacitor banks and then quickly discharges this energy in the form of a rapid, very high-current pulse (on the order of 100,000 Amperes) into a low turns count and low inductance air core solenoid that contains the sample under test.

Since external magnetization occurs in a separate device, an open-circuit condition will be seen by the sample prior to return to the electromagnet for measurement. It is accepted that a saturated, high-coercivity sample placed into the closed magnetic circuit of the hysteresigraph and re-magnetized beyond +1600 kA/m will regain most of the flux density lost by the prior open-circuit condition. This is an experimental reality.

Initial Curves

If an initial curve is to be performed, the sample should be demagnetized prior to the measurement. This is often performed in the electromagnet yoke. Apply a current high enough to saturate the sample, and then perform an alternating polarity decay of field in sufficiently small steps to smoothly demagnetize the sample to zero induction.

7   Alternative Measurement Sensors

Both the B,H surrounding coaxial coil arrangement and the B,H side-by-side coil arrangement utilize a non-magnetic structure (usually phenolic) to house the coil set. This phenolic structure is inserted and removed into the electromagnet air gap during each test cycle and is optimized for use with samples of a specific cross-section.

An alternative method of inductive coil sensing are embedded pole coils as illustrated in Fig. 8. It involves moving the B and H sensing coils to a position below the pole cap's surface. The B coil is located directly under the sample. The H coil is next to it but only measures air flux. The B and H coils are now wound upon an iron core identical to the pole caps of iron composition.

EMBEDDED POLE COILS

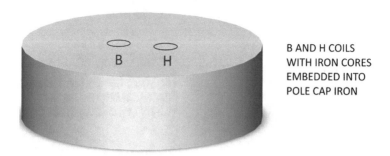

B AND H COILS
WITH IRON CORES
EMBEDDED INTO
POLE CAP IRON

Fig. 8 Embedded pole coil with *B* and *H* coils

Maxwell's equations and Gauss' theorem tell us that the normal component of flux density, *B*, on either side of a transition boundary of two materials with different permeabilities is identical [12]. Therefore, a coil embedded within the pole cap surface may be used to determine the flux density of samples placed in direct contact with it. Using the embedded coil to read the flux density of the iron in the pole cap will also tell us the flux density in the sample. As long as the sample covers the embedded coil's cross-section, the flux density of the sample can be determined. The flux from the coil is divided by the area of the embedded coil to determine the sample flux density. This method is commonly used in quality control applications for test of nonstandard or highly varied geometries. A limitation of this sensing technique is that it requires that the pole caps are operating far away from the saturation region of the pole cap iron. This technique is commonly used for the measurement of low energy product permanent magnets such as Alnico and hard ferrites. Higher-coercivity materials such as NdFeB are typically not measured using this technique since the sense coils will become nonlinear as the iron in the sense coil core approaches saturation. Air gaps between sample and pole caps will degrade this technique, as they introduce a third material for flux to move through (pole iron to air gap to sample), which complicates the two-material boundary transition condition we started with.

Embedded pole coils are often wired so that the B and H coils are in series opposition to acquire magnetic polarization, J, and a Hall probe is used as the magnetic field strength sensor.

More complex geometries can also be evaluated with embedded pole coils. Upper and lower pole caps can be fabricated with radii that match the inner radius and outer radius of motor arc segments as shown in Fig. 9. Embedded *B* and *H* coils can be shaped and constructed to follow these contours as well. With these arc-shaped embedded pole coils, an entire ferrite motor arc segment can be placed into the air gap of an electromagnet and evaluated without the destructive process of extracting

B AND H COILS WITH IRON
CORES EMBEDDED INTO
POLE CAP IRON

UPPER AND LOWER RADII
MATCH ARC SEGMENT

OPTIMIZED FOR NON-DESTRUCTIVE
TESTING OF ARC-SHAPED
FERRITE MOTOR MAGNETS

Fig. 9 Arc-shaped embedded pole coils with *B* and *H* coils

a core test specimen. Some attention must be placed on pole cap design as well as the interpretation of results, particularly when measuring arc segments with large differences in radii, creating slight non-uniformities in H fields.

8   Temperature Measurements

Magnetic materials respond to variation in temperature. As samples approach their Curie temperatures, they lose their permanent magnetic properties. Usually, flux densities and energy products tend to decrease with increasing temperatures. In certain cases, however, coercivities may actually see regions of increase with temperature [13]. And since practical magnetic materials are used in a wide variety of ambient/operating environments, knowledge of magnetic performance as a function of temperature is desirable.

Temperature monitoring of samples is preferred for magnetic materials, particularly those with high-temperature coefficients of induction and coercivity (hard ferrites, NdFeB, etc.). The lower the Curie point, the more significant the temperature effects. For this reason, even room temperature measurements should be monitored to account for variations in the measured response as a function of ambient temperature.

Ancillary equipment may be added to a DC hysteresigraph to allow for measurements above and below room temperature. Typically, a heated pole cap allows for measurements from room temperature up to +100 to 150C (+373 to 423 K). Thermal conduction can be exploited with a pole-mounted cartridge heater to raise the temperature of the sample under test. The cartridge heater or other heating element is connected to a power supply. A temperature sensor is used for feedback control of the pole cap/sample temperature. The large mass of the electromagnet can absorb the incremental added thermal energy from the heated

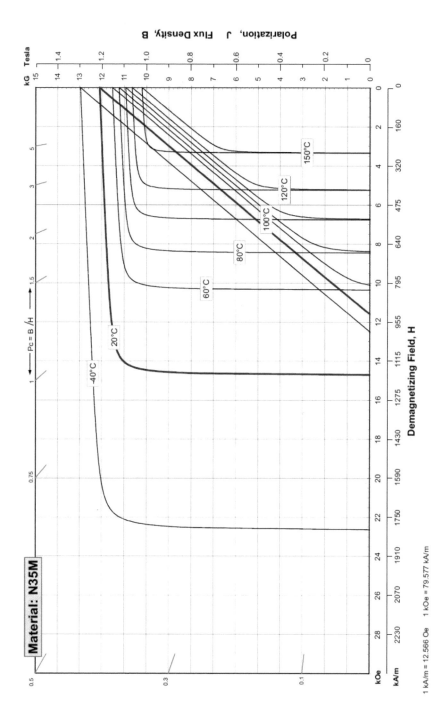

Fig. 10 Variable temperature of second quadrant demagnetization curves, courtesy of Arnold Magnetic Technologies [14]

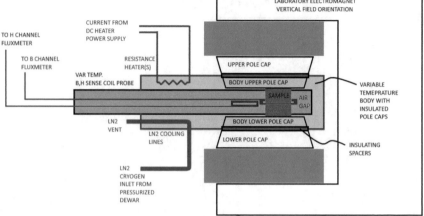

Fig. 11 DC hysteresigraph with variable temperature chamber in electromagnet gap

pole without significant negative effects. Figure 10 shows a series of second quadrant demagnetization curves at temperatures ranging from 40 to 150 °C for a permanent magnet sample.

Temperature measurements at lower temperatures require a variable temperature chamber to be fitted to the electromagnet as shown in Fig. 11. This temperature chamber encloses the sample, pole caps, sense coil sets, and variable temperature components within a nonmagnetic, yet thermally conductive enclosure (typically aluminum). The body of the temperature chamber contains the pole caps, heaters, and liquid nitrogen cooling lines. A removable variable temperature coil set contains the sample and sensing coils and temperature sensors. This entire assembly is placed between the poles of the electromagnet. Thin, thermally insulating spacers isolate the temperature chamber from the electromagnet. The thin spacers add to the reluctance of the magnetic circuit, but these losses are tolerated owing to the added utility of variable temperature measurement capability. The reluctance in the magnetic circuit reduces the flux in the circuit. It is proportional to the spacer's thickness divided by both the cross-section of the interface and permeability of the spacer.

With the additional thermal isolation from the electromagnet via the insulating spacers, additional heating power as well as cooling power can be administered to the chamber, creating a combination of high- and low-temperature chamber, with temperature capability ranging from below −150 to over +250C (123–523 K). A high/low variable temperature chamber and supplemental equipment are shown in Fig. 12.

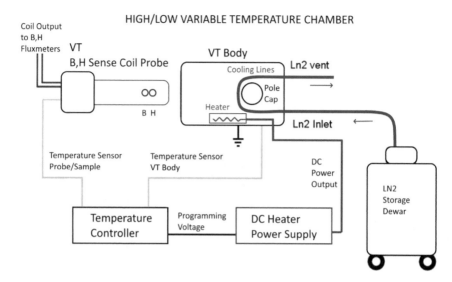

Fig. 12 Variable temperature chamber and supplemental equipment

9 Limitations of this Technique

Not Useful for Thin Films

The volume of a thin-film sample provides insufficient total flux for the typical electronic fluxmeter and sense coil combination used in a DC hysteresigraph. Therefore, a DC hysteresigraph does not have sufficient sensitivity to measure thin films.

Quality of the Closed Circuit

Establishing an ideal closed magnetic circuit is the goal, but not fully realized. There is always some short circuited flux escaping from/within an ideal closed circuit. Of primary interest is when a reluctance occurs at the interface between the pole caps and the sample under test. In operation, the air gap between the magnet pole caps is initially set slightly larger than the sample's length. Once energized, the gap between the magnet pole caps will decrease slightly due to the pull forces created by the energized electromagnet. In a perfect world, this reduction in electromagnet gap allows the sample and pole caps to touch. In reality, however, that is not always the case. Too little gap, the pull forces of the electromagnet will break more brittle samples (e.g., hard ferrites). Too much gap creates an air gap in the magnetic circuit. Various methods are utilized to optimize the air gap—spacer shims are used to define the air gap and protect the sample; pole locks, electronic pole adjusters, and gap monitors are also used to minimize this effect. These small reluctances can be estimated and compensated for, but the corrections are small and are generally not applied.

Assumes Homogeneous Materials

B surrounding coils measure the average flux density seen within the coil's effective diameter and across its height. Material properties are biased toward the sample's mid-section and will not reveal pressing density, material distribution, or other non-homogeneity issues occurring along the sample's magnetic length.

H Field Accuracy

Poor H field uniformity in the air gap of the electromagnet, particularly in the vicinity of the sample, will result in a poor representation of the applied field, *H*, particularly when using a side-by-side *B,H* coil set.

Low-Coercivity Materials

Soft materials with coercivities below 8000 A/m are not deemed suitable for this measurement technique. However, they are measured on occasion. In that case, self-demagnetization fields can easily yield shears in the shape of the B vs. H curve, and corrections for self-demagnetization factors may be required. Care should be taken to minimize magnetic reluctances by ensuring excellent sample to pole cap interfaces (highly parallel contact surfaces and minimized air gaps from the sample to pole cap face). In this case, an alternate yoke design, such as a double Fahy yoke or a toroidal specimen measured in a DC hysteresigraph for soft magnetic materials, will yield superior results.

10   Introduction to DC Hysteresigraphs for Soft Magnetic Materials

There are DC hysteresigraphs specifically designed for the measurement of bulk samples of soft magnetic materials. These larger forms of magnetically soft materials are often present in return paths of a variety of magnetic circuits. Motor stators, rotors, and flux carrying elements of solenoids, relays, etc. are just a few of the large-scale commercial applications where the performance of the magnetic circuit is critical. And optimization of these products for reduced mass and efficiency also facilitates changes in the flux carrying portion of designs. Soft magnetic materials are highly impacted by anneal state and machining and forming stresses endured in mass production. Additionally, lot-to-lot variation of procured material is not uncommon. For these reasons, DC hysteresigraphs are present in quality control labs of soft magnetic material manufacturers, their customer's sites, independent testing groups, and labs focused on development of soft materials in near-production form.

11   Why Do we Need a DC Hysteresigraph for Soft Magnetic Materials?

As with permanent magnets, soft magnetic materials are also impacted by the geometry-based self-demagnetization effects described earlier. And once again, the larger samples are not suitable for VSM or SQUID testing, and open-circuit testing brings along the associated concerns regarding the accuracy of measured hysteresis loop shapes. One form of the DC hysteresigraph solves this problem by utilizing a toroidal or ring specimen. The toroid shape creates an excellent closed magnetic circuit which eliminates the need to correct for self-demagnetization fields and therefore yields highly accurate B,H loop shapes. By exchanging a high-field electromagnet and massive power supply with a smaller, high-precision power supply allows for a superior field control at the lower magnetizing fields required for measurement of low-coercivity samples.

12   Theory and Test Circuit

Figure 13 is a block diagram of a DC hysteresigraph for soft materials. The toroid or ring-shaped test specimen will function much like a DC transformer. We can magnetize the sample under test by applying current to a primary winding randomly distributed about the sample's circumference. In this case, the ring is considered uniformly magnetized as long as it adheres to a simple geometry constraint for a thin-walled ring (the ring's inner diameter is greater than 0.82 times the ring's outer diameter) [15]. The gauge of the winding should be chosen to carry the current required to fully magnetize the sample. A precision DC power supply is required to provide current to magnetize the sample. A bipolar supply is preferred to avoid switching transients near zero field that are caused by unipolar power sources and their associated reversal hardware. An ammeter (usually in the form of a precision resistor shunt + precision voltmeter) should be installed in the magnetizing circuit to monitor magnetizing current.

SOFT MAGNETIC RING SPECIMEN TEST SETUP

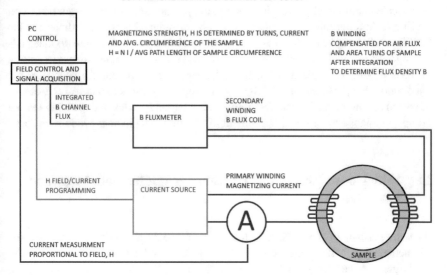

Fig. 13 Block diagram of DC hysteresigraph for soft materials

Below, we will follow ASTM A773 derivations [15].

The primary magnetizing winding generates magnetizing field, H:

$$H = \frac{NI}{l_m}$$

where

H = magnetic field strength [A/m]

N = number of turns in the primary winding

l_m = mean magnetic path length (average circumference of the toroid) [m]

The addition of a secondary winding to the ring creates the B flux sensing "coil." The resistance of the winding should be small compared to the input impedance of the fluxmeter's input stage, or a correction for loading effects should be made as it can otherwise impact the integrator circuit's accuracy. The secondary winding picks up a voltage responding to the instantaneous flux change within the sample as the applied field is varied. Utilizing a fluxmeter, we can integrate the voltage change and determine the DC voltage proportional to the change in flux. Dividing the cross-sectional area of the toroidal sample allows us to derive the observed flux density, B_{obs} [Tesla].

We must then correct for air flux in the B winding to arrive at an accurate flux density, B.

The correction constant, K_2, is defined as:

$$K_2 = \frac{a_C - A_C}{A_C}$$

where
  a_C = cross-sectional area of B coil [m^2]
  A_C = cross-sectional area of test specimen [m^2]
  And the flux density correction is:
  $B = B_{obs} - K_2\mu_0 H$

The flux reported by the fluxmeter should be acquired and coordinated with the magnetizing current reported by the ammeter. The signal acquisition from the fluxmeter, current monitoring from the ammeter, and the power supply programming are performed by a PC. Since there is no physical reference standard to conduct this experiment, we rely on accurate fluxmeter and ammeter calibration to ensure accuracy of the electronics. If the ammeter is comprised of a discrete shunt and voltmeter, those devices should also be calibrated.

13   Making a Measurement

Next, we will go through the process of performing a measurement on a low-coercivity material:

1. Demagnetize the sample (if the initial curve and associated permeabilities are of interest).
2. Zero the fluxmeter.
3. Magnetize the sample to saturation.
4. Smoothly sweep the magnetizing field to drive the sample through remanence, coercivity, and again saturation in the negative direction.
5. Repeat step 4 but toward the positive applied field direction.
6. Return the current to 0.

Re-center the B axis of the measurement by averaging the extremes of the flux density readings at +/− saturation. If this central value does not equal zero, a fault may have occurred due to poor demagnetization, integrator drift, or failure to magnetize the sample to saturation.

The data that are acquired reveal the material's major hysteresis loop as illustrated in Fig. 14. Various parameters can now be extracted from the curve, but the most common include but are not limited to:

- Permeability (initial, relative, and maximum): $\mu = B/H$ [Henry/m].
- Remanent induction: B_r [T].
- Normal coercivity: H_c [A/m].
- Saturation induction: B_{sat}[T].

Minor loops, recoils loops, and initial curves may also be generated with this instrumentation.

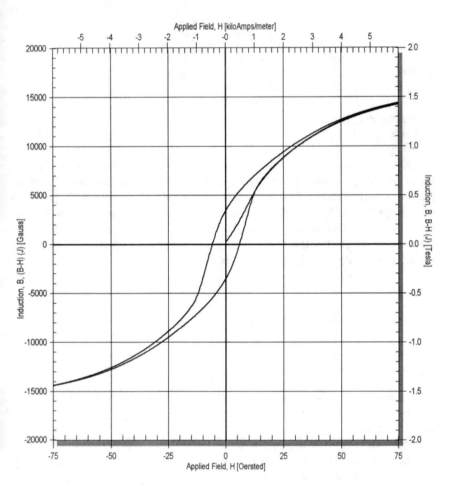

Fig. 14 Measured hysteresis loop

Magnetizing Field Control

The nature of control of the magnetizing field is somewhat dependent on the user's preferences. Some use a fixed value of dH/dt to control field. More sophisticated methods modulate current to control dB/dt. The correct methodology is really a function of what variable the user deems critical for their measurements. The measurement is considered a DC technique, so very rapid field sweep rates are discouraged during measurement acquisition. This is balanced by the need to complete the measurement quickly enough to avoid excessive fluxmeter drift.

Initial Curve

If an initial curve is to be measured, the sample should be demagnetized prior to the measurement. Demagnetization of the sample can be performed by applying a current high enough to saturate the sample and then performing an alternating polarity logarithmic (or similar) decay of current in sufficiently small steps to smoothly demagnetize the sample to zero induction.

14 Alternative Testing Circuits for Bulk Soft Magnetic Materials

A variation of this instrumentation is used for testing larger soft magnetic materials in rod or bar form. A yoke similar in concept to the DC electromagnet is used for testing rod-shaped specimen. It is essentially a smaller electromagnet that uses a clamping method to make the specimen one leg of the magnetic circuit path. A permanent magnetizing winding is also located in the circuit around the yoke's iron core. This is referred to as a Fahy yoke [9]. Another interpretation of this concept has two symmetric return paths, magnetizing coils, and cores and is referred to as a double Fahy yoke and is illustrated in Fig. 15. B sensing is performed by a surrounding B coil and fluxmeter. H sensing can be performed by H coil and fluxmeter or a Hall effect probe and gaussmeter.

Another variation of the ring specimen hysteresigraph uses an Epstein frame for testing strip steel samples as shown in Fig. 16. An Epstein frame recreates the closed circuit path of a "ring" in four linear segments. Each segment is a stack of lamination steel strips. The strips are stacked and overlap at the corners. The Epstein frame uses a pre-formed H magnetizing and B sense winding coils, both of which cover much of the segment lengths. The pre-formed windings make sample exchange far

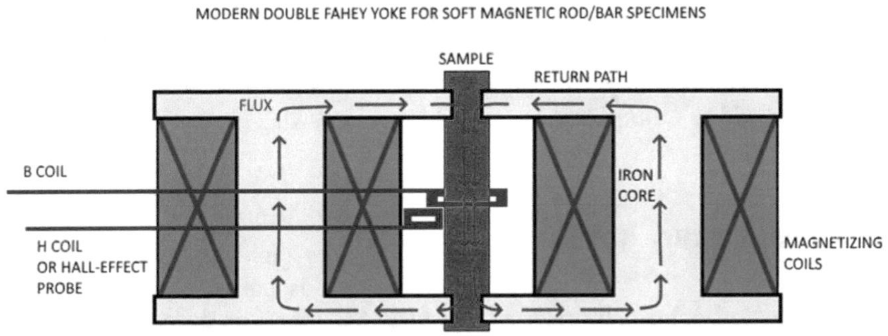

MODERN DOUBLE FAHEY YOKE FOR SOFT MAGNETIC ROD/BAR SPECIMENS

Fig. 15 Modern double Fahy yoke, used with soft magnetic rod and bar specimens

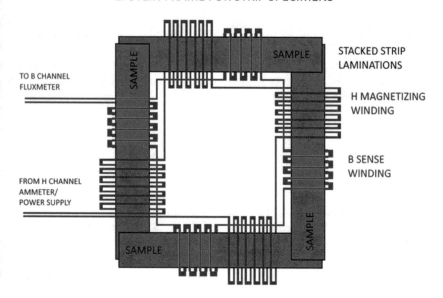

Fig. 16 Epstein frame for strip specimens to simulate a "ring" sample

easier than the winding process for ring specimens. The frame incorporates a mutual inductor setup for air flux compensation. The dimensions and turns count have become standardized in an effort to maintain measurement consistency of results across various user's Epstein frames.

Hysteresis loop AC testing is also possible in the case of ring and Epstein testing setups. The samples remain the same, but some changes are required in the test instrumentation. For AC testing, the H field is excited by an AC sine wave generated by an AC power supply/linear amplifier. An RMS or peak AC voltmeter with shunt or ammeter measures the magnetizing field strength. A mean/average response AC voltmeter replaces the fluxmeter for reading the B flux. The ring specimen or Epstein frame itself is essentially unchanged. AC core loss measurements may also be determined with a Wattmeter in the circuit.

15   Temperature Measurements

Since many technologically relevant soft magnetic materials do not exhibit large temperature coefficients of induction or coercivity, tests do not usually compensate for variations in room temperature. However, temperature should be recorded, particularly for samples that are more temperature-sensitive. Some temperature increase is expected due to the nature of driving current through the magnetizing winding and is accepted. Oil cooling of the sample under test is sometimes required for higher-power test setups.

Elevated variable temperature testing using a DC hysteresigraph for soft magnetic materials is possible. The means to achieve this is fairly straightforward and commonly available. A laboratory oven may be used as an environmental test chamber. The ring specimen is placed in the oven and allowed to equilibrate. The normal test methodology is then performed. Thermocouples may be added to monitor the sample temperature. This method's primary drawback is that changing resistance of the B winding as temperature is varied and influences the drift of the electronic fluxmeter. Once the drift characteristics of the elevated temperature testing are controlled, elevated temperature testing may be performed.

16 Limitations of this Technique

Specimen Shape

The ring-shaped specimen is not always a convenient shape. Fortunately, many motor housings, stators, etc. are close to the correct shape, though they do not always adhere to the thin-walled ring geometry constraint (ID > 0.82 OD). In this case, the properties from the test are considered to be "core" properties (with results dependent on shape of the test "core") as opposed to geometry-independent material properties. Care should be taken with respect to orientation when fashioning a ring specimen from a bulk material as the ring's orientation averages some anisotropy effects. Wire electrical discharge machining (EDM) processing is considered an acceptable method to create a ring specimen without altering its anneal state.

Need to Wind the Ring

For higher-field experiments, a large turns count may be required. Winding a toroid in this manner may be time-consuming or require specialized toroid winding equipment. Epstein frames are a practical variant of the ring specimen that eliminates this requirement, as the strip-shaped specimens are placed inside pre-formed windings.

Integrator Drift

Electronic fluxmeters will drift. This makes extended duration measurements more prone to flux reading errors. Computer-controlled drift management and correction aids in reducing these effects. Variable temperature measurements and associated changes in winding resistance also contribute to fluxmeter stability concerns.

17 Summary

DC hysteresigraphs allow for the measurement of bulk samples of new magnetic materials as they approach commercialization and ensure the quality control of materials already in mass production. In this chapter, we have discussed the basis of operation of the equipment to make room temperature DC hysteresigraphs measurements, including sensing techniques and instrumentation. We have also described the supplemental equipment required to extend that measurement capability to variable temperatures. Two basic types of machines were described: a laboratory electromagnet-based DC hysteresigraph for the measurement of magnetically hard magnetic materials and a wound ring specimen-based DC hysteresigraph for the measurement of magnetically soft materials. Both utilize inductive sensing within a closed magnetic circuit and integrating fluxmeters to process flux from the sample to quickly generate a second quadrant demagnetization curve or the entire hysteresis loop.

These test methods have their limitations. However, when rapid measurements of bulk forms of magnetic materials are required, the DC hysteresigraph is an extremely efficient tool for extracting a sample's magnetic material properties.

References

1. Lake Shore Cryotronics, USA, www.lakeshore.com
2. B.C. Dodrill, J.R. Lindemuth, *Vibrating Sample Magnetometry, in Magnetic Measurement Techniques for Materials Characterization* (Springer Nature, 2021)
3. R.K. Dumas, T. Hogan, *Recent Advances in SQUID Magnetometry, in Magnetic Measurement Techniques for Materials Characterization* (Springer Nature, 2021)
4. D. Jiles, *Introduction to Magnetism and Magnetic Materials*, vol 40 (Chapman & Hall, 1996)
5. D. Chen, J.A. Brug, R.B. Goldfarb, Demagnetization factors for cylinders. IEEE Trans. Magn. **27**(4), 3601–3619 (1991)
6. D. Chen, E. Pardo, A. Sanchez, Demagnetizing factors for rectangular prisms. IEEE Trans. Magn. **41**(6), 2077–2088 (2005). https://doi.org/10.1109/TMAG.2005.847634.
7. N.V. Rama Rao, A.M. Gabay, G.C. Hadjipanayis, Anisotropic fully dense MnBi permanent magnet with high energy product and high coercivity at elevated temperatures. J. Phys. D. Appl. Phys. **46**, 6 (2013)
8. X. Rui, J.E. Shield, Z. Sun, L. Yue, Y. Xu, D.J. Sellmyer, Z. Liu, D.J. Miller, High-energy product exchange-spring FePt/Fe cluster nanocomposite permanent magnets. J. Magn. Magn. Mater. **305**(1), 76–82 (2006)
9. R.M. Bozorth, Ferromagnetism IEEE Press, 850–851 (1993)
10. IEC 60404-5:2015, Magnetic materials—Part 5: Permanent magnet (magnetically hard) materials—Methods of measurement of magnetic properties, IEC, 6–10 (2015)

11. R.W. Kubach, A Hysteresigraph for plotting magnetization curves, AD-783 740. Air Force Materials Lab, 8–10 (1966)
12. *Maxwell equations, Encyclopedia of Mathematics, EMS Press, 2001 [1994]*
13. MMPA Standard No. 0100-00, Standard specifications for permanent magnet materials. Magnetic Materials Producers Association **10**, 15 (2001)
14. Arnold Magnetic Technologies, USA, www.arnoldmagnetics.com
15. ASTM A773-14, Standard test method for direct current magnetic properties of low coercivity magnetic materials using Hysteresigraphs, ASMT 03.07 magnetic properties. Section **19.1**, 170 (2019)

Radio-Frequency Transverse Susceptibility as a Probe to Study Magnetic Systems

Sayan Chandra and Hariharan Srikanth

Abstract The direct measurement of magnetic anisotropy in magnetic systems provides invaluable insights into the physical processes that enable researchers to engineer materials for a wide range of applications including spintronics, magnetic memory, hyperthermia treatments, and magnetic refrigeration. This chapter showcases the transverse susceptibility (TS) measurement technique using a self-resonant tunnel diode oscillator circuit as a versatile tool for the direct measurement of the effective magnetic anisotropy. We provide a brief history of the development of the theoretical description of TS followed by the evolution of experimental measurement configurations over the past two decades. The applicability of the TS technique to investigate magnetic anisotropy in different systems is highlighted. We summarize that the TS technique has proven to be a highly sensitive and reliable method that complements the conventional magnetic measurement techniques, thereby facilitating a comprehensive understanding of complex magnetic systems.

Keywords Transverse susceptibility · Tunnel diode oscillator · Magnetic anisotropy · Magnetization dynamics

1 Introduction

The study of reversible susceptibility (RS) of a material as a function of temperature and magnetic field reveals fascinating fundamental physical processes occurring in a material. Over the past several decades, reversible susceptibility in ferromagnetic, ferrimagnetic, paramagnetic, and superconducting materials has been actively investigated around the world. While reversible parallel susceptibility

S. Chandra
Department of Physics and Astronomy, Appalachian State University, Boone, NC, USA

H. Srikanth (✉)
Department of Physics, University of South Florida, Tampa, FL, USA
e-mail: sharihar@usf.edu

© Springer Nature Switzerland AG 2021
V. Franco, B. Dodrill (eds.), *Magnetic Measurement Techniques for Materials Characterization*, https://doi.org/10.1007/978-3-030-70443-8_6

(χ_p) has been studied for over a century, experimental measurements of reversible transverse susceptibility (TS) (χ_T) have gained traction in the last 20 years and have provided insightful understanding of a variety of systems, including paramagnetic susceptibility in salts, penetration depth in superconductors, superparamagnetism in nanocomposites, and deciphering magnetic phase diagrams in strongly correlated electron systems, particularly oxides and ferrites. The gain in popularity of TS measurement technique has established it as a reliable method and is now featured as an alternative probe for magnetic systems by multiple scientific equipment manufacturing companies.

2 Theoretical Background

The development of the TS measurement technique discussed in this chapter is based off the seminal theoretical work on the Stoner-Wohlfarth model by Aharoni et al. in 1957 [1]. Prior to Aharoni et al., the first theoretical calculations on TS were carried out by Gans in 1909 [2], followed by Grimes and Martin in 1954 [3] where they considered the magnetization reversal occurring due to domain wall motion. The Stoner-Wohlfarth (SW) model simplifies the magnetization reversal by considering only rotational processes in ellipsoidal ferromagnetic particles, each assumed to be a single domain, having uniaxial anisotropy energy that may be a combination of magnetocrystalline, magnetostriction, or shape anisotropy. These particles will be referred to as SW particles from here on. To simplify the calculation, all SW particles were considered identical, with random orientation and no interparticle interaction energy. The reversible susceptibility tensor χ_{ij} of the SW particle was defined as $\chi_{ij} = dM_i/dH_j$, where the diagonal terms included the longitudinal susceptibility component χ_p (along the bias magnetic field) and the two transverse susceptibility components χ_{T1} and χ_{T2} (along the orthogonal directions to the bias magnetic field).

The description of a randomly oriented SW particle of volume V, anisotropy K_1, and saturation magnetization M_S is shown in Fig. 1a. Here, the DC bias magnetic field $\vec{H}_{DC}$ is along the z-axis, the easy axis of the SW particle lies along the prolate of the ellipsoid at an angle (θ_K, φ_K), and M_S makes an angle (θ_M, φ_M) such that the total energy (E) of the particle is minimized and is in equilibrium as:

$$\frac{\partial E}{\partial \theta_M} = 0; \quad \frac{\partial^2 E}{\partial \theta_M^2} \geq 0 \text{ and } \frac{\partial E}{\partial \varphi_M} = 0; \quad \frac{\partial^2 E}{\partial \varphi_M^2} \geq 0 \tag{1}$$

$\vec{H}_{RF}$ is a small perturbation field along the x-axis that is necessary for the experimental measurement; however in the theoretical treatment, we consider $H_x = H_y = 0$. The total energy, $E = E_K + E_H$, of a SW particle can be expressed as the sum of the anisotropy energy, $E_K = -K_1\left(\vec{M_S}.\vec{\mu_K}\right)^2$, and the Zeeman energy,

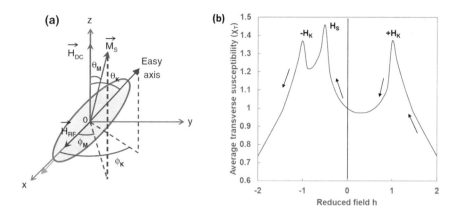

Fig. 1 (a) Diagram of a TS measurement geometry of a uniaxial magnetic particle [4], (b) typical TS unipolar curve as a function of H_{DC} for a collection of randomly oriented particles. The arrows show the direction of change in DC magnetic field

$E_H = -\vec{M_S}.\vec{H}_{DC}$, which accounts for the interaction between the magnetization and $\vec{H}_{DC}$. Here, $\vec{\mu}_K$ is the unit vector along the easy axis. By defining the reduced magnetic field, $h = \vec{H}_{DC}.\vec{M_S}/2K_1$, and $\chi_0 = \vec{M_S}^2/3K_1$, the TS for the case $\varphi_K = 0$ is given by:

$$\chi_{T1} = \frac{3}{2}\chi_0 \left(\cos^2\varphi_K \frac{\cos^2\theta_M}{h\cos\theta_M + \cos2\,(\theta_M + \theta_K)} + \sin^2\varphi_K \frac{\sin(\theta_K - \theta_M)}{h\sin\theta_K} \right) \tag{2}$$

Extending the calculation for a randomly oriented array of SW particles, the average TS can be expressed as:

$$\langle\chi_T\rangle = \frac{1}{2\pi} \int_0^{2\pi} \int_0^{\pi/2} \chi_{T1}\sin\theta_K \, d\theta_K \, d\varphi_K \tag{3}$$

$$\langle\chi_T\rangle = \frac{3}{4}\chi_0 \int_0^{\pi/2} \left(\cos^2\varphi_K \frac{\cos^2\theta_M}{h\cos\theta_M + \cos2\,(\theta_M + \theta_K)} + \sin^2\varphi_K \frac{\sin(\theta_K - \theta_M)}{h\sin\theta_K} \right) \sin\theta_K \, d\theta_K \tag{4}$$

This indicates that for the RS, tensor is diagonalized in the case of randomly oriented particles by the components χ_p and $\langle\chi_T\rangle = \langle\chi_{T1}\rangle = \langle\chi_{T2}\rangle$. The theoretical values of $\langle\chi_T\rangle$ calculated using Eq. (4) for a system of SW particles having positive uniaxial anisotropy oriented at random are plotted in Fig. 1b. As can be seen, three peaks emerge, one that corresponds to the magnetic switching field (H_S) and coincides with the reversible parallel susceptibility and two peaks located at $\pm H_K$ at the effective magnetic anisotropy field.

3 Measurement Technique

The direct measurement of the effective anisotropy field was first done by Pareti et al. in 1986 by carefully measuring the out-of-phase component in an electronic bridge circuit where an AC field $\left(\vec{H}_{RF} \right)$ of magnitude ~1 Oe at a frequency of 1–60 kHz was applied to study $BaFe_{12}O_{19}$ particles ranging in diameter from 150 to 3000 nm [5]. Using this setup, the authors were able to successfully measure the anisotropy and switching fields (Fig. 2b) as predicted in Aharoni's paper; however, the measurements were limited to room temperature. In addition, the particles studied were not "nano" by the modern definition (<100 nm) but polycrystalline and thereby deviated from the definition of a SW particle. Furthermore, the setup suffered from mechanical vibrations of the coil systems, which gave rise to spurious signals at the same frequency as the drive signal. The vibrations originated from the Lorentz interactions between the bias field H and the AC current that was supplied to the drive coil. The spurious signal diminished at a higher frequency of 50 kHz which was determined by performing measurements with and without the sample. This hints to the fact that the signal-to-noise ratio of TS measurements could be enhanced by operating at higher frequencies.

Oscillators have been considered as useful transducers for making measurements due to the great precision with which frequency can be measured [6]. They have higher sensitivity than AC bridge circuits, for example, the best commercial capacitance bridges have sensitivity $\Delta C/C \sim 10^{-8}$, whereas oscillators can detect changes $\sim \Delta C/C \sim 10^{-10}$. Furthermore, oscillators can be designed to operate at MHz frequencies; however, AC bridges are limited from DC to few kHz. Additionally, oscillators consume less power and operate at lower voltages than AC bridges. These advantages of oscillators over conventional AC bridge circuits motivated the research on tunnel diode oscillators for the measurement of magnetic materials.

Inspired by the first work on tunnel diode oscillators for low-temperature applications by O.W.G Heybey at Cornell University in 1962 and Boghosian et al. in 1966, Van Degrift showed for the first time in 1975 that tunnel diode oscillators (TDO) can be used for ultrasensitive measurements, with a precision of 0.001 ppm at low temperatures [6]. Following this, TDOs that operate in the negative differential resistance (NDR) (Fig. 3a) regime gained popularity to investigate phenomena like RF penetration depth of cuprate superconductors, borocarbide superconductos, and phase transitions in complex oxide systems [7–11]. Srikanth and co-workers at the University of New Orleans adopted this technique and extended it to investigate magnetic switching behavior and effective anisotropy in SW particles [12–20]. This proved to have several advantages over the setup presented by Pareti with significant improvements over previous instruments from other groups.

Fig. 2 (**a**) Experimental apparatus used to measure transverse susceptibility. C1 is the driving field (*h*) coil; C2 and C3 are the pickup coils; *P* indicated the electromagnet poles. The directions of both bias (*H*) and driving (*h*) fields are indicated. (**b**) TS as a function *f* magnetic field for different particle sizes [5]

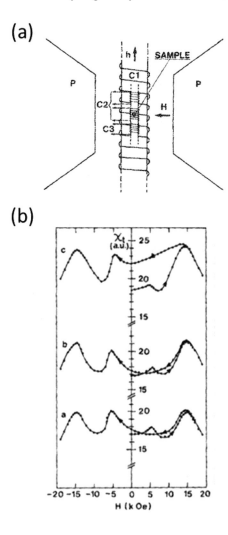

Let us understand how the self-resonant oscillations are sustained in a TDO operating at the NDR. The tunnel diode is connected to a LC tank circuit and forward biased by an external source precisely to drive it to the NDR. The LC tank circuit is maintained at a constant amplitude resonance where the power supplied by the tunnel diode sustains the continuous oscillations at a frequency determined by the expression:

$$\omega = 1/\sqrt{LC} \tag{5}$$

The negative AC resistance of the tunnel diode cancels losses of the tuned circuit. In an ideal case, we expect minimal drift and noise in the resonance frequency; however, factors like parasitic impedance can introduce high-frequency parasitic

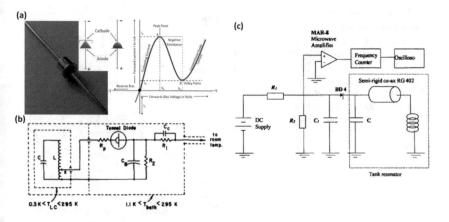

Fig. 3 (a) Tunnel diode IN3716 (left) and I-V characteristic of a tunnel diode showing the NDR. Inset shows the electronic symbol of a tunnel diode. Circuit diagram used for TDO (b) Van Degrift [6] and (c) Srikanth et al. [14]

oscillations, thereby deteriorating oscillator performance. Degrift showed that by carefully choosing a parasitic suppression resistor (R_p in Fig. 3b), one can avoid unintentional circuit oscillations arising from stray capacity of the diode. This resulted in a 40-fold improvement in TDO stability over previous studies.

For the TS measurements, the sample is placed inside a gelcap that is then snugly inserted into the inductor coil (Fig. 4f). The respective geometry of the applied magnetic fields is also shown with the condition that $|\vec{H}_{DC}| \gg |\vec{H}_{RF}|$. Following the theoretical paper by Aharoni et al., the DC magnetic field is swept from positive to negative saturation for a unipolar measurement. Any change in the inductance (ΔL) will translate to a change in the resonance frequency ($\Delta\omega$) obtained by differentiating Eq. (5):

$$\frac{\Delta\omega}{\omega} = -\frac{\Delta L}{2L} \tag{6}$$

The total inductance of the coil is the sum of the inductance from the empty coil (L_0) and contribution from the sample (L_S). The complex permeability of the sample is given by $\mu = \mu' + i\mu''$, which in terms of magnetic susceptibility (χ) can be expressed as $\mu = \mu_0(1 + \chi)$. For a coil of volume V_0, length l, and a total number of turns, N, the total inductance can be written as:

$$L = \mu_0 \left(N/l \right)^2 (V_0 + \chi V_S) \tag{7}$$

where V_S is the volume of the sample such that the fill fraction of the coil is $\eta = V_S/V_0$. A differential change in the inductance is related to change in the susceptibility (χ) as:

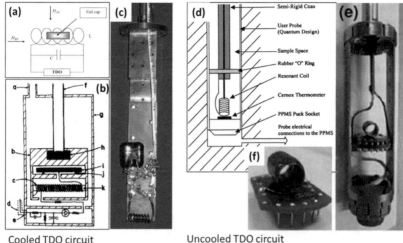

Cooled TDO circuit   Uncooled TDO circuit

bias.

Fig. 4 (**a**) Schematic of the LC tank depicting the sample inside the inductor coil and the respective RF and DC field directions. Schematic of the (**b**) TDO circuit inside a cryostat and (**c**) image of the sample holder with TDO circuit and coil [21]; (**d**) schematic of the (**d**) TDO circuit integrated to a quantum design physical property measurement system (PPMS), (**d**), (**e**) and (**f**) show images of the multifunctional probe and coil [14]

$$\Delta L = \mu_0 \left(\frac{N}{l} \right)^2 \Delta\chi V_S \tag{8}$$

Therefore, dividing Eq. 8 by Eq. 7, we get:

$$\frac{\Delta L}{L} = \frac{\Delta\chi V_S}{(V_0 + \chi V_S)} \tag{9}$$

In the case of $V_0 \ll \chi V_S$, we get:

$$\frac{\Delta L}{L} \approx \frac{\Delta\chi}{\chi} \tag{10}$$

And for the case of $V_0 \gg \chi V_S$, which is associated with a very low fill fraction:

$$\frac{\Delta L}{L} \approx \Delta\chi\eta \tag{11}$$

From Eqs. (6), (10), and (11), we get:

$$\frac{\Delta\omega}{\omega} = -\frac{\Delta L}{2L} = -\frac{\Delta\chi}{2\chi} \text{ and } \frac{\Delta\omega}{\omega} = -\frac{\Delta L}{2L} = -\frac{\Delta\chi\eta}{2} \tag{12}$$

We also know from classical electrodynamics that the complex impedance of a material with resistivity (ρ) is given by $Z = (1 + i) \frac{\rho}{\delta}$, where δ is the skin depth $\delta = \sqrt{2\rho/\mu\omega}$. It can be shown that $\Delta L \propto \Delta Z \propto \Delta\sqrt{\mu\rho}$. It is clear from the expression that a measurement of the resonance frequency shift ($\Delta\omega$) will directly correlate to change in both material resistivity and permeability. Therefore, the TDO technique enables a contactless measurement of the spin and charge dynamics in a single experiment. This makes it attractive to study strongly correlated electron systems that show strong coupling of the structural, electronic, and magnetic properties.

4 Choice of Circuit Components

The greatest slope of the negative resistance region corresponds to the minimum value for the magnitude of the negative resistance (R_n). Optimal performance of the circuit can be obtained by adjusting the DC voltage bias to the inflection point of the current-voltage curves of the tunnel diode using a voltage divider (resistances R_1 and R_2 in Fig. 3c) [6]. At the onset of self-oscillations, the AC voltage amplitude ($v_d \sim 10\ mV$) is expected to vary about the DC voltage bias.

Typical germanium diodes have peak and valley voltages of about 40 mV and 300 mV, respectively, but they widely differ in terms of the peak current which is limited by $|R_n|$. For sustained oscillations, the rf power (P) loss in the LC tank circuit should match the rf power supplied by the tunnel diode. Therefore, the tunnel diode chosen should satisfy $R_n \approx -v_d^2/P \approx -10^{-4}/P\Omega$. A large bypass filter (C_l in Fig. 3b) is chosen such that it provides isolation from thermally induced impedance changes within the coaxial cable. If the oscillation amplitude is small, a preamplifier may be used to properly trigger the frequency counter as done by Srikanth et al. High-Q chip capacitors were used. The inductor in the tank circuit was hand-wound using AWG42 insulated copper wire around a hollow ceramic tube (0.5 cm diameter) and potted with polystyrene epoxy resin. A coating of GE varnish was applied to improve thermal conductivity (Figs. 4e, f). Tapped inductors may be used to dictate the impedance of the LC tank at resonance, with the inductance values between 1 and 5 μH. Using these guidelines, the damping factor of the circuit ($\gamma = R/2L$) was regulated to satisfy the condition $\gamma \ll \omega$ at resonance, which yields well-defined oscillations with sinusoidal line form (Fig. 5b). This is also associated with high Q which, in combination with the capacitance of the coaxial cable, determines the upper limit for the resonance frequency of the TDO circuit. Furthermore, the entire TDO circuit except the inductor coil L, the DC power supply, and the measurement instruments were shielded for stability in resonance frequency. This was done by mounting the circuit on a PCB and enclosing it in an aluminum box. The ground of the circuit was connected to the aluminum chassis which was also grounded. The sinusoidal nature (Fig. 5b) of the stable oscillations was monitored over ~24-h time period to account for the stability of the resonance frequency from the central value of ~6 MHz. The drift in resonance over a short period of time (20–30 min) which is the realistic time for a TS unipolar curve

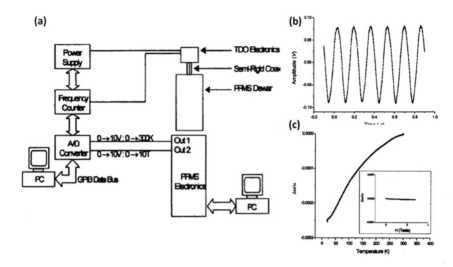

Fig. 5 (**a**) Layout of the complete TDO measurement system displaying the computer control and data acquisition stages as implemented by Srikanth et al. [14]; (**b**) snapshot of the sinusoidal output waveform from the TDO circuit with the central resonance frequency at ~6 MHz; and (**c**) temperature-dependent shift in the resonance frequency of an empty coil. Inset shows the magnetic field dependence of the coil cooled to 100 K

measurement was found to be about 2–3 Hz which is ~1 ppm. Such drift has little influence on the measurements; however, it could not be completely removed.

Two approaches were adopted for the integration of the TDO circuit into the cryostat: (a) cooling the tunnel diode inside the cryostat (Figs. 4b, c) [21] and (b) cooling the inductor coil only, while the tunnel diode remained at room temperature (Figs. 4d–f). The current-voltage characteristic curves for tunnel diodes are highly temperature-dependent near room temperature, but at temperatures below 30 K, they appear to change their properties by only a few parts per million per Kelvin. Therefore, in the case where the tunnel diode circuit is cooled inside the cryostat, one must optimize the DC bias voltage as a function of temperature for the stable operation of the TDO. This can be avoided in the second case where only the inductor coil is cooled and the TDO circuit always remains at room temperature. In either case, the circuit must be connected to the measurement instruments using a coaxial cable with minimal temperature-dependent change in impedance. Srikanth et al. used a semirigid RG 402 coaxial cable assembly with the end at room temperature terminated by a type A (SMA) connector that could be mated to the rest of the TDO circuit. The lower (cold) end of the coaxial cable held the inductor coil by connecting it to the inner and outer shielding. The coaxial cable was electrically insulated by wrapping in Teflon tape, and baffles were connected along the probe arm as heat shields. The integration of the TDO circuit was done using a customized cryogenic user probe that fitted into the bore of a commercial quantum design physical property measurement system; however, in principle, this can be designed to work with any cryostat in the presence of a magnetic field. The temperature and magnetic field specifications of the cryostat and the superconducting/electromagnet determine the measurement ranges of the TDO setup.

A temperature-dependent shift in the resonance frequency is unavoidable in the TDO circuit which arises from the decrease in resistivity of the Cu wire making up the coil and the effect of thermal contraction. However, a careful characterization of this relative shift as a function of temperature revealed a drift of ~1400 Hz as the inductor coil was cooled from room temperature to 20 K. This relative shift in resonance frequency ($\Delta\omega = \omega_{RT} - \omega(T)$) at a given temperature, T, can be expressed as $\Delta\omega/\omega_{RT}$ as shown in Fig. 5c.

5 Magnetic Systems Investigated Using Transverse Susceptibility

In this section, we discuss selected studies to highlight the versatility of the TS measurement technique to probe various magnetic phenomena.

Dipolar Interactions within Soft Ferrite Particles

With the development of elegant chemical synthesis routes in the early 2000s, researchers were able to realize the closest resemblance to SW particles. These particles typically are sub-20 nm in diameter and monodisperse that can be approximated to have uniaxial anisotropy and are termed *superparamagnetic*. A tremendous surge in the research on the synthesis and physical property characterization of superparamagnetic particles followed in the subsequent years targeted toward potential applications in cancer treatment using magnetic hyperthermia. In this context, some of the fundamental questions that needed to be answered were as follows: (a) *What role does interparticle interaction have on the magnetic anisotropy of the ensemble of superparamagnetic particles?* (b) *How does the magnetic anisotropy evolve when such particles undergo superparamagnetic to ferromagnetic transitions?* Answers to these questions provide valuable knowledge necessary to synthesize designer superparamagnetic particles for intended applications.

Poddar et al. performed one of the first comprehensive investigations in determining the dipolar interactions in soft ferrite nanoparticles and used TS to analyze the evolution of anisotropy as a function of dipolar interactions [4]. In their study, they used high-quality monodisperse 15-nm manganese zinc ferrite (MZFO) nanoparticles and varied their concentration in paraffin wax matrix which is a nonmagnetic host. By varying the concentration, the authors were able to tune the average interparticle distance, thereby controlling the dipolar interactions and magnetic anisotropy. By diluting the MZFO particle concentration in paraffin wax, the authors found a steady decrease in the blocking transition temperature (Fig.

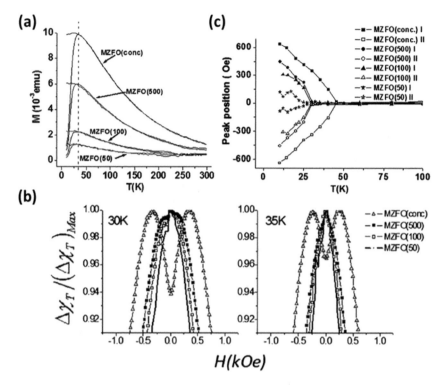

Fig. 6 (**a**) Temperature-dependent field-cooled (FC) and zero-field-cooled (ZFC) magnetization curves, (**b**) at 30 and 35 K, and (**c**) anisotropy peak locations versus temperature for different concentrations of MZFO in paraffin wax [4]

6a) and coercive field (not shown), which is a signature of the superparamagnetic to ferromagnetic transition. In their experiment, they consider the anisotropy field dispersion of the randomly oriented particles to be the same for samples of all concentrations.

Figure 6b shows the representative TS curves obtained at 30 and 35 K for different concentrations of MZFO, where one can see the enhanced anisotropy (consistent with anisotropy peaks at higher DC magnetic fields) for higher concentration of particles. A careful analysis of the anisotropy peak positions as a function of temperature (Fig. 6c) showed the effect of dipolar interaction strength on the effective anisotropy of the nanoparticles. They showed that at temperatures above the blocking transition temperatures, as the nanoparticle ensemble becomes superparamagnetic, the anisotropy peaks disappear. Interestingly, the authors found an asymmetry in the anisotropy peak height and width, which they attributed to interparticle interaction strength and concluded that the asymmetry disappears for the most concentrated sample. The different energy landscapes seen by the interacting versus non-interacting/weakly interacting particles manifested into their effective anisotropy field probed by TS.

Magnetic Anisotropy in Exchange-Biased Nanostructures

Over the years, the TS technique has been adopted to investigate anisotropy-driven magnetic phenomena occurring in variedly shaped nanostructures like nanowires, nanorods, core/shells, and hybrid nanoparticles [12, 23–28]. One such intensely studied magnetic phenomenon is the exchange bias effect that is known to develop due to anisotropy and interfacial effects between two or more materials with different magnetic states. The exchange bias effect is accompanied by the horizontal shift of the magnetization M(H) hysteresis loops along the field-cooled direction, but in some cases, there is an additional vertical shift in the M(H) loops. In this case we present here, Chandra et al. used the TS technique to investigate the origin of the asymmetry in field-cooled M(H) loops in core/shell nanoparticles [22]. Figure 7a shows TEM images of core/shell Co/CoO nanoparticles (~19 nm), which are a complex ferromagnet (Co)/ anti-ferromagnet (CoO) system and are known to exhibit the exchange bias effect. By analyzing the temperature- and field-dependent magnetization and unipolar TS curves (Fig. 7b), the authors identified the distinct freezing and blocking temperature of the core and shell. The TS measurements showed compelling evidence of the competing Zeeman and anisotropy energy that leads to the origin of anisotropy in the exchange bias loops. They quantified the competing nature of the two energies by analyzing the difference in TS peak heights of the magnetization and return curves in the bipolar TS scans (Fig. 7c). The location of the anisotropy peaks obtained from the TS curves (Fig. 7d) gives a comprehensive picture of effective anisotropy in such complex nanostructures as they undergo multiple temperature-dependent changes in their magnetic state.

Magnetic Anisotropy in Ferromagnetic Epitaxial Thin Films

The applicability of the TS measurement technique in ferromagnetic thin films was first demonstrated by Spinu et al. where they studied chromium oxide (CrO_2) films (~200 nm) that were grown epitaxially on TiO_2 substrates [17, 18].

The study was conducted to obtain direct estimation of the anisotropy along the well-defined magnetic easy [001] (*c-axis*) and hard [010] (*b-axis*) axes (Fig. 8a). The accurate determination of anisotropy field in such materials is critical to the design of spin-based devices that take advantage of its spin-polarized states and large magnetoresistance. The TS measurements showed evidence of varying anisotropy due to temperature-dependent strain arising from lattice mismatch between the film and the substrate. The sharp contrast in the anisotropy field associated with the hard and easy axes of CrO_2 films is shown in the respective TS curves obtained at room temperature. Upon saturation of the magnetization (M_S), the anisotropy constant K_1 obtained from the magnetic anisotropy field ($H_K = 2 K_1/M_S$) for the case of $\Psi = 0$ (hard axis) was 1.9×10^5 erg/cm^3. This was in excellent agreement with previously measured values of K_1 for CrO_2 using other techniques, thereby establishing TS as

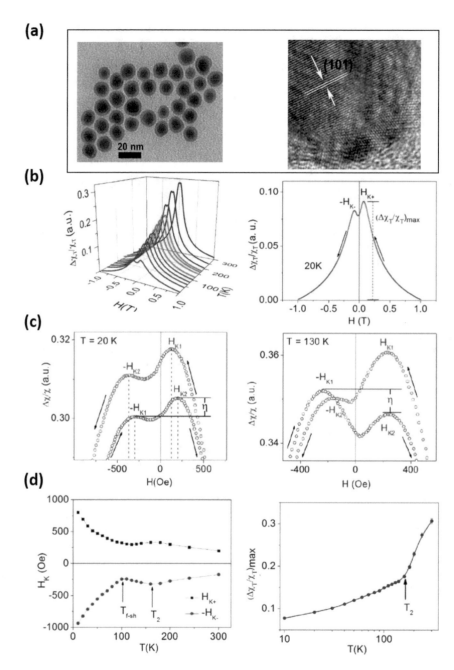

Fig. 7 (**a**) TEM and HRTEM image of Co/CoO core/shell nanoparticles, (**b**) unipolar TS curves for different temperatures (left) and representative TS curve at 20 K (right); (**c**) bipolar TS curves at 20 K (left) and 130 K (left); (**d**) temperature-dependent anisotropy field [22]

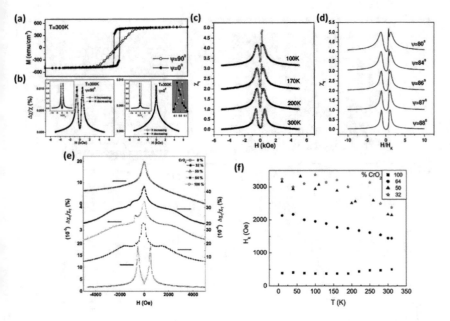

Fig. 8 (a) M(H) hysteresis loops and (b) TS curves for CrO_2 films along the magnetic hard and easy axes, TS as a function of (c) temperature and (d) angular orientation with respect to applied DC magnetic field [17, 18]; (e) TS curves of CrO_2/Cr_2O_3 bilayer films for varying concentrations of CrO_2 and (f) magnetic anisotropy field for different concentrations of CrO_2 as a function of temperature [29]

a reliable technique to probe ferromagnetic thin films. Their analysis demonstrated a property of CrO_2 that the sample behaved like a Stoner-Wohlfarth single-domain particle, i.e., coherent rotation dominated the magnetization process. Further, high sensitivity of the TS technique could detect misalignment in the angle between the applied DC field and the easy axis in the perpendicular geometry consistent with the coherent rotation model (Figs. 7c, d). This angular deviation between the applied field and the easy axis of the film was attributed to a net magnetoelastic energy term related to the in-plane strain. This caused the easy axis to deviate by a small angular orientation from the *c-axis* toward the *b-axis*. This study on ferromagnetic films demonstrated the versatility of the TS technique to provide valuable insights into the rich anisotropic properties of technologically important oxides.

A few years later, Frey et al. performed a detailed investigation of the anisotropic properties of epitaxial bilayer CrO_2/ Cr_2O_3 (ferromagnetic/antiferromagnetic) thin films [29]. This work showed TS as an excellent method to directly probe dynamic magnetization and the effective anisotropy (magnetic field direction-dependent) in exchange-biased thin films composed of a spin-polarized ferromagnet (CrO_2) and Cr_2O_3 that is known to be magnetoelectric. The TS measurements shed light on the dependence of anisotropy field to the relative thicknesses of CrO_2 and Cr_2O_3 films (see Table 1 and Figs. 7e, f) and how the interfacial coupling and the magnetoelectric effects were intrinsically correlated.

Table 1 Saturation magnetization, anisotropy field, and anisotropy constants for CrO_2/Cr_2O_3 bilayer films with different concentrations of CrO_2 obtained from room temperature and low-temperature measurements [29]

%CrO$_2$	Effective thickness (nm)	M_s (emu/cc) (RT)	H_k (Oe) TS (RT)	K_{eff} (erg/cc) (observed) (RT)	K_{eff} (erg/cc) (calculated)	H_k (Oe) (cal)	H_k (Oe) TS (LT)	M_s (emu/cc) (LT)	K_{eff} (erg/cc) (LT)
100	200	436	514	1.1×10^5	1.1×10^5	514	390	660	1.3×10^5
64	128	291	1448	2.1×10^5	4.6×10^4	314	2130	423	4.5×10^5
50	99	227	2075	2.3×10^5	6.2×10^4	548	3150	326	5.1×10^5
32	64	136	2100	1.4×10^5	7.7×10^4	1136	3230	212	3.4×10^5

Magnetocrystalline Anisotropy-Driven Phase Transition in Strongly Correlated Electron Systems

Rare-earth-based compounds like manganites and cobaltites exhibit fascinating magnetic behavior associated with magnetic and structural phase transitions and charge ordering, phenomena that are closely tied to the magnetocrystalline anisotropy in such strongly correlated electron systems [30, 31]. Several research groups across the world have studied these materials with potential applications in magnetocalorics and spintronics. Frey et al. utilized the TS measurement technique to investigate half-doped $Pr_{0.5}Sr_{0.5}CoO_3$ that is known to exhibit anomalous magnetism driven by magnetocrystalline anisotropy that results in a structural transition at ~120 K where the magnetization undergoes a ferromagnetic to ferromagnetic transition associated with the rotation of the anisotropy vector [31].

TS measurements in such a magnetic system were instrumental in identifying the nature of magnetocrystalline anisotropy that was not evident in traditional magnetometry and transport measurements. The TS measurements were able to precisely identify the anisotropy fields and switching fields as a function of temperature. From the TS plots (Fig. 9a), the authors identified a peak that was observed only upon decreasing the magnetic field after positive (or negative) saturation. They found the presence of this peak to be associated with a crossover field describing the transition between the lower field-cooled and higher field-cooled magnetic states. The temperature evolution of the anisotropy field, switching field, and the crossover field (Fig. 9b) provided qualitative evidence of the magnetocrystalline anisotropy-driven structural transition between the two ferromagnetic states. The transition was associated with a dramatic increase in the measured anisotropy. The temperature-dependent switching field intensity obtained from the TS curves shown (Fig. 9c) is from those crystallites in the sample whose hard axes were aligned with the DC magnetic field. By comparing the relative change in magnitude of the parallel susceptibility and TS switching peaks, the authors demonstrated compelling evidence of the rotation in the anisotropy vector direction at the phase transition temperature. They concluded that TS is a very useful method to obtain insightful information on the unusual magnetic properties of doped perovskites.

6   The Future of Transverse Susceptibility

Without a doubt, the TS method is indeed unique and proved to be a well-grounded technique in advancing the direct measurement of magnetic anisotropy. The original TS setup used a tunnel diode to sustain self-resonant oscillations (TDO) when biased in the negative differential resistance regime; however, tunnel diodes are rare electronic components that are not easily available. Figueroa et al. implemented the use of CMOS transistors as active devices instead of tunnel diodes to generate

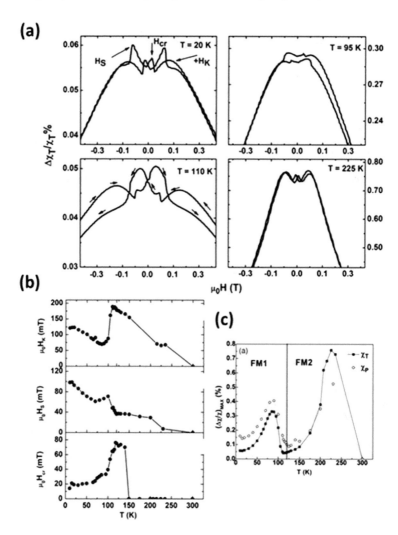

Fig. 9 (**a**) Bipolar TS scans at 20, 95, 110, and 225 K, (**b**) temperature-dependent anisotropy peaks, crossover fields and switching behavior, and (**c**) evolution of the switching field intensity with temperature obtained from TS and perpendicular susceptibility curves [31]

the self-resonant oscillations [32]. It consists of a simple inverter cell in cross-coupled topology, which compensated the energy loss in the LC tank. Among the benefits of considering CMOS transistors as the active element are (a) significantly lower parasitic losses and noise rendering the TS measurements more sensitive and (b) reproducible resonant circuits with robust performance and reduced costs. The response of the CMOS transistors can be easily modelled, thereby minimizing errors in determining the shift in resonance frequency from the inductance change. We believe the future of the TS research and development lies with the making of highly sensitive CMOS-based frequency tunable ultra-stable resonant circuits for temperature- and magnetic field-dependent investigations. That being said, the first theoretical and experimental studies on transverse susceptibility and the successful capability of the TDO-based circuits in studying complex anisotropy-dependent physical phenomena in a variety of magnetic systems are remarkable.

Acknowledgments We thank the numerous researchers (past and present) in the Functional Materials Laboratory at the University of South Florida as well as collaborators, who contributed to joint publications on transverse susceptibility studies of a variety of magnetic systems. A large portion of the research at USF has been supported over the years by the Office of Basic Energy Sciences of the US Department of Energy through Award number DE-FG02-07ER46438.

References

1. A. Aharoni, E.H. Frei, S. Shtrikman, D. Treves, Bull. Res. Counc. Isr. **6A**, 215 (1957)
2. R. Gans, Ann. Physik **29**, 301 (1909)
3. D.M. Grimes, D.W. Martin, Reversible susceptibility of Ferromagnetics. Phys. Rev. **96**, 889–896 (1954)
4. P. Poddar, M.B. Morales, N.A. Frey, S.A. Morrison, E.E. Carpenter, H. Srikanth, Transverse susceptibility study of the effect of varying dipolar interactions on anisotropy peaks in a three-dimensional assembly of soft ferrite nanoparticles. J. Appl. Phys. **104** (2008)
5. L. Pareti, G. Turilli, Detection of singularities in the reversible transverse susceptibility of an uniaxial Ferromagnet. J. Appl. Phys. **61**, 5098–5101 (1987)
6. C.T. Van Degrift, Tunnel-diode oscillator for 0.001 ppm measurements at low-temperatures. Rev. Sci. Instrum. **46**, 599–607 (1975)
7. T. Coffey, Z. Bayindir, J.F. DeCarolis, M. Bennett, G. Esper, C.C. Agosta, Measuring radio frequency properties of materials in pulsed magnetic fields with a tunnel diode oscillator. Rev. Sci. Instrum. **71**, 4600–4606 (2000)
8. H.C. Okean, *Chapter 8 Tunnel Diodes, Semiconductors and Semimetals* (Elsevier, 1971), pp. 473–624
9. R.B. Clover, W.P. Wolf, Magnetic susceptibility measurements with a tunnel diode oscillator. Rev. Sci. Instrum. **41**, 617 (1970)
10. F. Bolzoni, F. Leccabue, O. Moze, L. Pareti, M. Solzi, Magnetocrystalline anisotropy of Ni and Mn substituted Nd2fe14b compounds. J. Magn. Magn. Mater. **67**, 373–377 (1987)
11. L. Pareti, F. Bolzoni, M. Solzi, K.H.J. Buschow, Magnetocrystalline anisotropy in Nd2-Xtbxfe14b. J Less-Common Met. **132**, L5–L8 (1987)
12. H. Srikanth, E.E. Carpenter, L. Spinu, J. Wiggins, W.L. Zhou, C.J. O'Connor, Dynamic transverse susceptibility in au-Fe-au nanoparticles. Mat. Sci. Eng. A-Struct. **304**, 901–904 (2001)
13. H. Srikanth, L. Spinu, T. Kodenkandath, J.B. Wiley, J. Tallon, Vortex dynamics and magnetic anisotropy in RuSr2GdCu2O8. J. Appl. Phys. **89**, 7487–7489 (2001)
14. H. Srikanth, J. Wiggins, H. Rees, Radio-frequency impedance measurements using a tunnel-diode oscillator technique. Rev. Sci. Instrum. **70**, 3097–3101 (1999)
15. L. Spinu, C.J. O'Connor, H. Srikanth, Radio frequency probe studies of magnetic nanostructures. IEEE Trans. Magn. **37**, 2188–2193 (2001)
16. L. Spinu, H. Srikanth, E.E. Carpenter, C.J. O'Connor, Dynamic radio-frequency transverse susceptibility in magnetic nanoparticle systems. J. Appl. Phys. **87**, 5490–5492 (2000)
17. L. Spinu, H. Srikanth, A. Gupta, X.W. Li, G. Xiao, Probing magnetic anisotropy effects in epitaxial CrO$_2$ thin films. Phys. Rev. B **62**, 8931–8934 (2000)

18. L. Spinu, H. Srikanth, C.J. O'Conner, A. Gupta, X.W. Li, G. Xiao, Switching behavior and its strain dependence in epitaxial CrO2 thin films. IEEE Trans. Magn. **37**, 2596–2598 (2001)
19. L. Spinu, H. Srikanth, J.A. Wiemann, S. Li, J. Tang, C.J. O'Connor, Superparamagnetism and transverse susceptibility in magnetic nanoparticle systems. IEEE Trans. Magn. **36**, 3032–3034 (2000)
20. L. Spinu, A. Stancu, C.J. O'Connor, H. Srikanth, Effect of the second-order anisotropy constant on the transverse susceptibility of uniaxial ferromagnets. Appl. Phys. Lett. **80**, 276–278 (2002)
21. A. Hansen, A Tunnel Diode Oscillator for Magnetic Susceptibility Measurements (2013). http://ufdc.ufl.edu/AA00057594/00001
22. S. Chandra, H. Khurshid, M.H. Phan, H. Srikanth, Asymmetric hysteresis loops and its dependence on magnetic anisotropy in exchange biased co/CoO core-shell nanoparticles. Appl. Phys. Lett. **101** (2012)
23. S. Chandra, A. Biswas, S. Datta, B. Ghosh, A.K. Raychaudhuri, H. Srikanth, Inverse magnetocaloric and exchange bias effects in single crystalline $La_{0.5}Sr_{0.5}MnO_3$ nanowires. Nanotechnology, 24 (2013)
24. S. Chandra, A.I. Figueroa, B. Ghosh, A.K. Raychaudhuri, M.H. Phan, P. Mukherjee, H. Srikanth, Fabrication and magnetic response probed by RF transverse susceptibility in $La_{0.67}Ca_{0.33}MnO_3$ nanowires. Physica B **407**, 175–178 (2012)
25. S. Chandra, N.A.F. Huls, M.H. Phan, S. Srinath, M.A. Garcia, Y. Lee, C. Wang, S.H. Sun, O. Iglesias, H. Srikanth, Exchange bias effect in au-Fe_3O_4 nanocomposites. Nanotechnology **25** (2014)
26. N.F. Huls, M.H. Phan, A. Kumar, S. Mohapatra, S. Mohapatra, P. Mukherjee, H. Srikanth, Transverse susceptibility as a biosensor for detection of Au-Fe_3O_4 nanoparticle-embedded human embryonic kidney cells. Sensors-Basel **13**, 8490–8500 (2013)
27. J. Bartolome, A.I. Figueroa, L.M. Garcia, F. Bartolome, L. Ruiz, J.M. Gonzalez-Calbet, F. Petroff, C. Deranlot, F. Wilhelm, A. Rogalev, N. Brookes, Perpendicular magnetic anisotropy in Co-Pt granular multilayers. Low Temp. Phys. **38**, 835–838 (2012)
28. C.M. Bonilla, I. Calvo, J. Herrero-Albillos, A.I. Figueroa, C. Castan-Guerrero, J. Bartolome, J.A. Rodriguez-Velamazan, D. Schmitz, E. Weschke, D. Paudyal, V.K. Pecharsky, K.A. Gschneidner, F. Bartolome, L.M. Garcia, New magnetic configuration in paramagnetic phase of $HoCo_2$. J. Appl. Phys. **111** (2012)
29. N.A. Frey, S. Srinath, H. Srikanth, M. Varela, S. Pennycook, G.X. Miao, A. Gupta, Magnetic anisotropy in epitaxial CrO_2 and CrO_2/Cr_2O_3 bilayer thin films. Phys. Rev. B **74** (2006)
30. C. Leighton, D.D. Stauffer, Q. Huang, Y. Ren, S. El-Khatib, M.A. Torija, J. Wu, J.W. Lynn, L. Wang, N.A. Frey, H. Srikanth, J.E. Davies, K. Liu, J.F. Mitchell, Coupled structural/magnetocrystalline anisotropy transitions in the doped perovskite cobaltite $Pr_{1-x}Sr_xCoO_3$. Phys. Rev. B **79** (2009)
31. N.A.F. Huls, N.S. Bingham, M.H. Phan, H. Srikanth, D.D. Stauffer, C. Leighton, Transverse susceptibility as a probe of the magnetocrystalline anisotropy- driven phase transition in $Pr_{0.5}Sr_{0.5}CoO_3$. Phys. Rev. B **83** (2011)
32. A.I. Figueroa, J. Bartolome, J.M.G. del Pozo, A. Arauzo, E. Guerrero, P. Tellez, F. Bartolome, L.M. Garcia, Low temperature radio-frequency transverse susceptibility measurements using a CMOS oscillator circuit. J. Magn. Magn. Mater. **324**, 2669–2675 (2012)

Alternating Gradient Magnetometry

Brad Dodrill and Harry S. Reichard

Abstract Force methods involve determination of the apparent change in weight for a material when placed in an inhomogeneous magnetic field. The equipment required for such force methods is either an electromagnet or superconducting magnet and a microbalance for force measurements. A commercial variant of these methods is the alternating gradient magnetometer (AGM) with a noise floor ranging from 10^{-8} to 10^{-9} emu (10^{-11}–10^{-12} Am^2). Commercial AGM systems can be used for room temperature measurements at fields up to the moderate ~30 kOe (3 T), which are achievable with electromagnets.

Keywords Force magnetometry · Hysteresis · Alternating gradient magnetometer, AGM · High-sensitivity magnetic measurements

1 Introduction

The alternating gradient force magnetometer (AGFM, or simply AGM) is the direct descendant of the Faraday balance, a very early technique for characterizing magnetic materials by measuring the force exerted on a magnetic sample by an externally applied magnetic field.

More particularly, the AGM makes use of the magnetic force exerted on a magnetic dipole by a non-uniform magnetic field. The relevant relation is shown in Eq. 1:

$$\mathbf{F} = \nabla \left(\mathbf{m} \cdot \mathbf{B} \right) \tag{1}$$

where.

$\mathbf{F}$ = force (N).

B. Dodrill (✉) · H. S. Reichard
Lake Shore Cryotronics, Inc., Westerville, OH, USA
e-mail: brad.dodrill@lakeshore.com

© Springer Nature Switzerland AG 2021
V. Franco, B. Dodrill (eds.), *Magnetic Measurement Techniques for Materials Characterization*, https://doi.org/10.1007/978-3-030-70443-8_7

Fig. 1 Alternating gradient
magnetometer—basic
principle

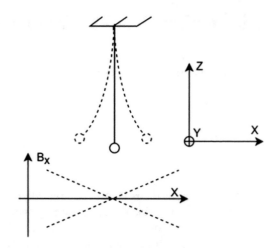

$\mathbf{m}$ = magnetic moment (Am^2).
$\mathbf{B}$ = magnetic field (T).

A practical instrument responds to a single component of the vector force, usually the component aligned with a coordinate axis. If this axis is denominated x, we have:

$$F_x = m_x \left(\frac{\partial B_x}{\partial x} \right) + m_y \left(\frac{\partial B_y}{\partial x} \right) + m_z \left(\frac{\partial B_z}{\partial x} \right) \tag{2}$$

Figure 1 depicts a typical configuration: gradient, in the X-direction, of the X-component of the gradient field B_x operates on the X-directed component of the sample magnetic moment to produce a force (also in the X-direction), corresponding to the first term of Eq. 2. The field gradient is made to reverse direction at some relatively low frequency, e.g., 500 Hz, so the resultant force is modulated accordingly. The magnetic moment of the sample can be calculated from the magnitude of the field gradient and the measured value of the alternating force.

The second and third terms of Eq. 2 represent undesired responses that, in the usual case, are minimized by sample isotropy (such that m_y and m_z are much smaller than m_x) and by symmetrical gradient coil pairing that minimizes $\frac{\partial B_y}{\partial x}$ and $\frac{\partial B_z}{\partial x}$ over the sample volume.

Figure 2 is a diagram of a practical embodiment of the alternating gradient magnetometer [1], with additional detail of sample and force sensor shown in Fig. 3.

In the apparatus of Figs. 2 and 3, the AGM probe utilizes a piezoelectric bimorph as a force sensor. Two fused silica extensions connect the bimorph to a glass carrier (typically 5 mm square, 100 μm thick), on which the sample is secured. A field-controlled electromagnet applies a uniform field, to which the AC gradient field is added, directed along the X-axis, generated by the alternating gradient coils.

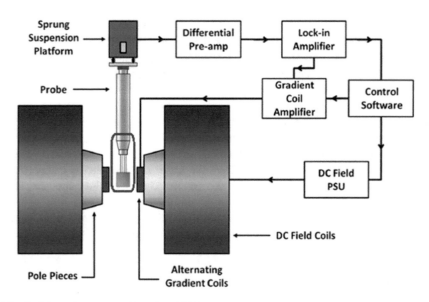

Fig. 2 Schematic representation of an AGM

Fig. 3 Sample and force sensor (detail)

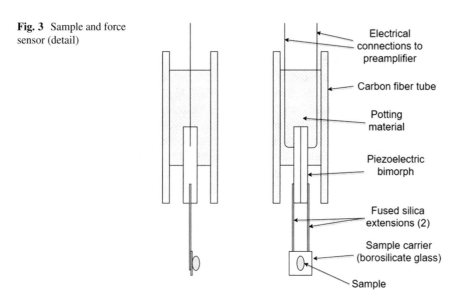

An electrical signal at the same frequency as the gradient coil excitation, and proportional to the force on the sample, is generated by the piezoelectric bimorph. This signal is amplified and detected by a lock-in amplifier (LIA). Control software serves to tune the gradient coil excitation frequency to the mechanical resonance of the sample bimorph assembly.

Various other arrangements are possible, and AGMs have been realized in many different configurations [2–11], but the gradient coils are invariably arranged so that the gradient field is zero at the geometric center (the nominal sample position). The magnitude of the gradient field to which the sample is exposed is therefore a function of sample dimensions, as well as the amplitude of the field gradient.

While the gradient field is zero at the geometric center, it is important to note that the gradient (of the gradient field) is _not_ zero at the geometric center. Ideally, the gradient is constant over the volume occupied by the sample. Gradient coils are typically connected in series opposition. In principle, a parallel-opposition arrangement could also be employed, but the series configuration is better suited to achieving a stable field gradient at the geometric center, where the sample is positioned.

All results presented in this chapter were recorded using a commercial Model 2900 AGM and Model 8600 vibrating sample magnetometer (VSM) from Lake Shore Cryotronics [12].

2 Sensitivity

The AGM is one of the most sensitive techniques for measuring the magnetic properties of materials, making it ideal for performing field-dependent measurements such as major hysteresis loops, minor hysteresis loops, first-order reversal curves (FORC), etc. on magnetically diluted or weak samples whose magnetization (or magnetic moment) is small. Owing to its high sensitivity, it can perform such measurements faster than other magnetometric techniques because less signal averaging is required to obtain adequate signal-to-noise ratio (SNR).

Using a quartz tuning fork as the piezoelectric sensor and a gradient field of 5 kOe/cm (50 T/m), Todorovic et al. [7] achieved noise levels of 1 pemu (10^{-15} Am^2). Richter et al. [13] reported noise levels of 100 pemu (10^{-13} Am^2) at a gradient field of 4.7 kOe/cm (47 T/m). Both instruments were optimized to measure microscopic samples with linear dimensions <10 μm.

Using a much smaller gradient field 150 Oe/cm (1.5 T/m), and adapted for macroscopic samples with linear dimensions of several mm, the Lake Shore Cryotronics Model 2900 [12] specified 10 nemu (10^{-11} Am^2) standard deviation at an averaging time of 1 sec/point. This is approximately ten times lower noise than is achievable using a vibrating sample (VSM) or superconducting quantum interference device (SQUID) magnetometer at the same signal averaging.

When using conventional VSM or SQUID magnetometers to acquire very high-density data sets (e.g., minor loops, FORC) for magnetically weak samples, measurement times can extend over many hours or even days. For these types of applications, the signal-to-noise advantage of ten of the AGM can translate into a hundred-fold reduction in measurement time.

The factors that impact the sensitivity of the AGM include:

- Acoustic forces (noise).

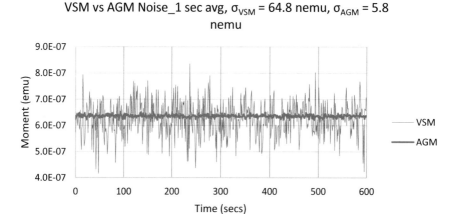

Fig. 4 VSM and AGM noise measurements at 1 s/point averaging

- Inertial forces (conducted vibration).
- Preamplifier noise.
- Magnitude of the gradient field.
- Thermal noise [7] of sample and carrier (Brownian motion).
- Choice of operating frequency.

The piezoelectric bimorph is effectively a microphone, and hence acoustic noise is the primary limitation of AGM sensitivity. Thus, to achieve optimal sensitivity, audible environmental noise sources (e.g., HVAC systems, electronics cooling fans, etc.) should be minimized as much as possible by either locating the AGM in a room where such noise sources are minimized or by enclosing the AGM in a sound-proof enclosure.

Figure 4 shows typical AGM and VSM noise measurements recorded at 1 s/point averaging. The AGM noise $\sigma = 5.8$ nemu (5.8×10^{-12} Am^2) is approximately ten times lower than the VSM noise $\sigma = 64.8$ nemu (64.8×10^{-12} Am^2).

Figure 5 shows AGM and VSM hysteresis loop measurement results for a synthetic anti-ferromagnetic thin film with saturation moment $m_{sat} < 2$ μemu (2×10^{-9} Am^2). The hysteresis loops were recorded to ± 500 Oe (50 mT) in 2.5 Oe (0.25 mT) steps. The VSM data were recorded at 2 s/point averaging (total loop measurement time $= 28$ mins), while the AGM data were recorded at 100 ms/point averaging (total loop measurement time $= 1$ min 30 s). The results show that an AGM can measure hysteresis loops (and other field-dependent data) for very low moment samples with comparable signal-to-noise ratio in less than one tenth the time required in a VSM.

Figure 6 shows an AGM hysteresis loop for a thin-film sample with saturation moment $m_{sat} \approx 0.6$ μemu (6×10^{-10} Am^2). The hysteresis loop was recorded to ± 5 kOe (0.5 T) in 40 Oe (4 mT) steps at 100 ms/point averaging, with a total loop measurement time of 56 s. Obtaining similar results with comparable signal-to-noise ratio using conventional VSM or SQUID magnetometers would require $\approx$

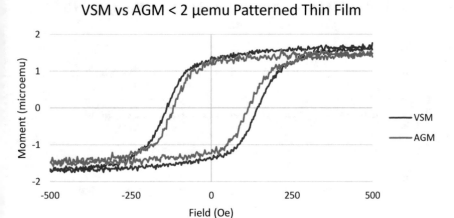

Fig. 5 VSM and AGM hysteresis loops for $a < 2$ μemu thin film. The AGM data were collected approximately 20 times faster than the VSM data

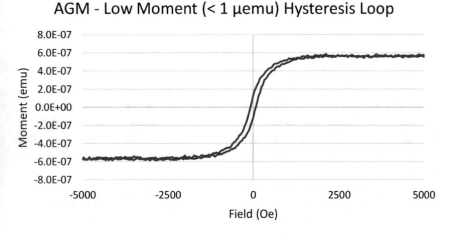

Fig. 6 A 56-s AGM hysteresis loop for $a < 1$ μemu sample

5–10 s/point signal averaging, and the measurement would take approximately 50–100 times longer.

Examples of the speed with which an AGM can acquire high-density data are illustrated in Figs. 7 and 8. Fig. 7 shows 19 ascending minor hysteresis loops for a thin-film recording media sample with saturation moment $m_{sat} = 400$ μemu (4×10^{-7} Am^2). The minor loops were recorded from an initially demagnetized state to maximum applied fields of ± 2 kOe (0.2 T) in 15.8 Oe (1.58 mT) steps at 100 ms/point averaging. The data set consists of 4844 points and the total measurement time was only 9 min 24 s.

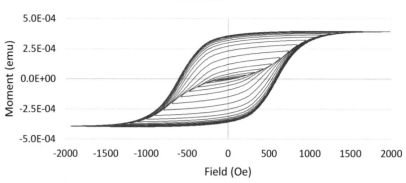

Fig. 7 Ascending minor hysteresis loops for a thin-film recording media sample

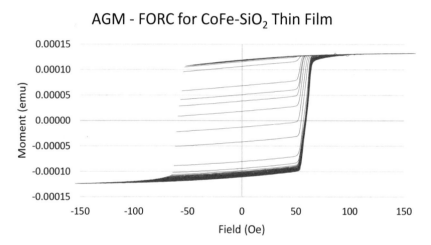

Fig. 8 FORCs (23,585 data points) for CoFe nanoparticles dispersed in a 30% volume fraction of SiO_2

Figure 8 shows 176 FORCs (23,585 data points) for a thin-film nanocomposite consisting of $\approx$ 10 nm CoFe nanoparticles dispersed in a 30% volume fraction of non-magnetic SiO_2. The saturation moment is only 130 μemu (1.3×10^{-7} Am^2). At 50 ms/point averaging, these data were recorded in only 58 min. Similar results could be obtained in an electromagnet-based VSM [12], but the measurement would take approximately 2 times longer, and similar results using a superconducting magnet-based VSM or SQUID magnetometer would require measurement times 10–20 times longer. Two factors contribute to this longer measurement time: (1) longer signal averages would be required to obtain comparable signal-to-noise ratio, and (2) while the magnetic field can be swept at rates up to 10 kOe/s (1 T/s) in an electromagnet, superconducting magnets are usually limited to $\approx$ 200 Oe/s (20 mT/s).

3 AGM Limitations

While sensitivity is the principal advantage of the AGM technique, there are significant limitations relative to other magnetometric techniques.

Environmental Effects: As noted previously, environmental factors can significantly degrade AGM sensitivity by an order of magnitude or more. Audible noise, air flow fluctuations (HVAC), ambient temperature fluctuations, electronic noise sources, and mechanical vibrations due to either nearby machinery or human traffic can all impact AGM performance. To achieve the best sensitivity, the AGM should either be located in a room where such noise sources are minimized or be enclosed within a sound-proof enclosure that also minimizes air flow and temperature fluctuations.

Sample Accommodation: While a VSM, for example, can accommodate samples of any form (solids, thin films, powders, liquids) with dimensions ranging from a few to tens of millimeters (mm) and sample masses ranging from <1 milligram (mg) to tens of grams, the AGM can only accommodate relatively small samples (typically <5 mm) that are not too massive (typically < tens of milligrams). And, while small solid samples (thin films, single crystals, etc.) can be easily affixed to the AGM probe, powder and liquid samples need to be encapsulated in a non-magnetic sample holder that in turn can be secured to the probe. Additionally, while a VSM can measure strongly magnetic samples (e.g., permanent magnets) with large magnetic moments (i.e., hundreds of emu), it is challenging to measure them in an AGM. This is because the AGM probe is suspended from the AGM head with springs; thus, the freely suspended probe to which a strongly magnetic sample is attached will be attracted to, and thus will move toward, the electromagnet poles at high applied fields. To measure such samples in the AGM, one must measure physically small, low-weight samples to minimize the moment and, therefore, the force.

AGM Probe Drift: Although the force sensing piezoelectric bimorph used in the commercial 2900 AGM is subject to long-term drift of the bimorph transducer constant owing to "aging" of the piezoelectric ceramic, this is rarely a limitation. Users of the AGM typically perform a calibration using a magnetic moment standard, such as NIST SRM2853 (yttrium iron garnet), on a daily, or weekly, basis.

Typically the operating frequency of the AGM is set to the mechanical resonant frequency of the AGM probe, consisting of piezoelectric bimorph, fused silica extensions, and sample carrier. In this situation, the instrument response is strongly influenced by ambient temperature fluctuations, which affect resonant frequency and quality factor Q. These effects can be mitigated by operating off-resonance, where their impact is greatly reduced, although operating off-resonance also slightly reduces AGM sensitivity.

Robustness: VSM systems are very robust and often serve as the "work horse" magnetometer in multi-user magnetics labs. The AGM probes, however, are very fragile and require careful and delicate handling when attaching and removing samples to prevent the fused silica extensions from breaking.

Variable Temperature Measurements: While VSMs and SQUIDs are capable of temperature-dependent measurements from low (<4.2 K) to high (1273 K) temperatures, the AGM is somewhat limited for variable temperature measurements. The materials used in the 2900 AGM probe cannot withstand high temperatures. For low-temperature operation, the piezoelectric bimorph must be removed from the variable temperature environment because its force constant is temperature-dependent, leading to a reduction in AGM sensitivity. Because the low-temperature cryostat employed in the 2900 AGM is a liquid cryogen (liquid helium or nitrogen)-based flow cryostat, the flowing gas causes instabilities/movement of the freely suspended AGM probe, leading to an additional reduction of sensitivity. For low-temperature measurements, the AGM sensitivity will typically degrade by a factor of 10–100, rendering it less sensitive than VSMs or SQUIDs.

Low-temperature operation can be improved by isolating the AGM probe from the cryostat gas flow, although this partially thermally isolates the sample from the cryostat which leads to temperature differences (errors) between the sample and cryostat. Static exchange gas-type cryostats which minimize instabilities/movement of the AGM probe can also be employed, but these tend to have slower thermal response times than gas flow cryostats, which increase the time required to conduct typical variable temperature magnetic measurements. Other techniques that address AGM operation at variable temperatures have been implemented as well [14–19].

High-Field Measurements: Faraday [20] and electromechanical [21] force magnetometers have been developed for high-field (9 T) and very low-temperature measurements (100 mK). The 2900 AGM could be adapted to high-field superconducting magnets. The electromagnet shown in Fig. 2 could be replaced by a room temperature bore, horizontal field split-coil superconducting magnet which would provide for high-field room temperature measurements to $\approx$ 60–70 kOe (6–7 T). The AGM could be similarly adapted to a cold bore horizontal field superconducting magnet for low-temperature measurements; however, the limitations discussed previously regarding low-temperature AGM operation would still apply.

Magnetically Soft (Low Coercivity) Materials: The AGM cannot be used to measure very low-coercivity (sub-mT) samples. This is a consequence of the measurement technique itself [8, 15]. The small-amplitude AC gradient field produces "distortions" in the hysteresis loops of soft magnetic materials, e.g., minor loop excursions, etc. These effects can be minimized by reducing the magnitude of the gradient field amplitude, but this in turn reduces the force, and hence the sensitivity, of the AGM.

Figure 9 shows hysteresis loops for a 3-nm-thick permalloy (NiFe) thin film with intrinsic coercivity of <1 Oe (0.1 mT) and saturation moment $m_{sat} \approx 30$ μemu (30×10^{-9} Am^2) at gradient field amplitudes of 15 Oe/cm (150 mT/m)

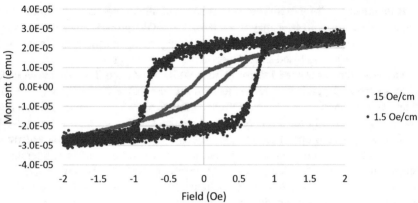

Fig. 9 AGM hysteresis loops for a 3-nm-thick NiFe thin film versus gradient field amplitude

and 1.5 Oe/cm (15 mT/m). The higher-field gradient distorts the loop leading to an erroneous coercivity determination. The lower-field amplitude correctly measures the loop but at the expense of increased moment noise because of the reduced AGM sensitivity.

4   Summary

In this chapter, we have described the principles of operation of the AGM measurement technique and its advantages and limitations relative to conventional VSM and SQUID magnetometers. We've also presented typical measurement results that demonstrate its sensitivity and resultant measurement speed for magnetically diluted or weak samples.

References

1. J. Sagar, Optimization of Heusler Alloy Thin Films for Spintronic Devices, PhD thesis, University of York (2013)
2. H. Zijlstra, A vibrating reed magnetometer for microscopic particles. Rev. Sci. Instrum. **41**, 124 (1970)
3. R.T. Lewis, A faraday type magnetometer with an adjustable field independent gradients. Rev. Sci. Instrum. **42**, 31 (1970)
4. P.J. Flanders, An alternating gradient magnetometer. J. Appl. Phys. **63**, 3940 (1988)
5. P.J. Flanders, A vertical force alternating gradient magnetometer. Res. Sci. Instrum. **61**, 839 (1990)

6. G.A. Gibson, S. Schultz, A high sensitivity alternating gradient magnetometer for quantifying magnetic force microscopy. J. Appl. Phys. **69**, 5880 (1991)
7. M. Todorovic, S. Schultz, Miniature high sensitivity quartz tuning fork alternating gradient magnetometer. Appl. Phys. Lett. **73**, 24 (1998)
8. J.W. Harrell, Effect of AC gradient field on magnetic measurements with an alternating gradient magnetometer. J. Magn. Magn. Mater. **205**, 121 (1999)
9. L. Thomas, A. Rahmani, P. Renaudin, A. Wack, High sensitivity in plane vector magnetometry using the alternating gradient force method. J. Appl. Phys. **93**, 7062 (2003)
10. M. Barbic, Sensitive measurement of reversible parallel and transverse susceptibility by alternating gradient magnetometry. Rev. Sci. Instrum. **75**, 5016 (2004)
11. M. Perez, R. Ranchal, I. Vazquez, P. Cobos, C. Aroca, Combined alternating gradient force magnetometer and Susceptometer. Rev. Sci. Instrum. **86**, 015110 (2015)
12. Lake Shore Cryotronics, USA; www.lakeshore.com
13. H.J. Richter, K.A. Hempel, J. Pfeiffer, Improvement of sensitivity of the vibrating reed magnetometer. Rev. Sci. Instrum. **59**, 1388 (1998)
14. T. Frey, W. Jantz, R. Stibal, J. Appl. Phys. **64**, 6002 (1988)
15. K. O'Grady, V.G. Lewis, D.P.E. Dickson, Alternating gradient force magnetometry: Applications and extension to low temperatures. J. Appl. Phys. **73**, 5608 (1993)
16. V.G. Lewis, *Development and Applications of an Alternating Gradient Force Magnetometer, PhD Thesis* (University of Wales, Bangor, 1995)
17. D. Krasa, K. Petersen, N. Petersen, *Variable Field Translation Balance, Encyclopedia of Geomagnetism and Paleomagnetism* (Springer-Verlag, 2007)
18. N. Petersen, K. Petersen, Comparison of the temperature dependence of irreversible and reversible magnetization of basalt samples. Phys. Earth Planet. Inter. **169**, 89 (2008)
19. The Series 300 Force Magnetometer from George Associates offered variable temperature over the range 2 K to 1150 K, but appears to be no longer available
20. T. Sakakibara, H. Mitamura, T. Tayama, H. Amitsuka, Faraday force magnetometer for high sensitivity magnetization measurements at very low temperatures and high fields. Jpn. J. Appl. Phys. **33**(1), 9A (1994)
21. T. Sakon, M. Motokawa, Electromechanical magnetization measurements at ultra low temperatures and high magnetic fields. Rev. Sci. Instrum. **71**, 3473 (2000)

Nanomechanical Torque Magnetometry

Joseph E. Losby, Vincent T. K. Sauer, and Mark R. Freeman

Abstract Over recent decades, there has been considerable effort devoted toward the miniaturization of mechanical magnetometry platforms. With the backdrop of ever-increasing detection sensitivity, torque magnetometry has provided new insight into materials with spin textures confined to nanoscale geometries. Torque sensing allows for broadband characterization of a wide range of magnetic materials, through quasistatic hysteresis to spin resonances. In this chapter, we present the basic principles of nanomechanical torque magnetometry, outline its capabilities, and offer a guide for application of the method.

Keywords Magnetometry · Nanomagnetism · Magnetic torque · Mechanical torque sensing · Hysteresis · Spin resonance · Nanofabrication · Cavity optomechanics

1 Introduction

As are the subjects of this volume, magnetic properties of materials can be probed through a plethora of means, including electromagnetic interactions (inductively detected spin resonance, Faraday and magneto-optical Kerr effects) and particle scattering (magnetoresistance, Hall effects, neutron diffraction, muon spin relaxation). The focus of this chapter is on mechanical methods and more specifically on mechanical torque [1, 2], arguably the archetypal physical probe of magnetism (see Ref [3], Ch. 1). Because magnetization is manifest through dipoles, torque is the fundamental coupling to fields; forces are an extremely useful but particular consequence of gradients in dipolar potential energies.

J. E. Losby · V. T. K. Sauer · M. R. Freeman (✉)
Department of Physics, University of Alberta, Edmonton, AB, Canada
e-mail: mark.freeman@ualberta.ca

© Springer Nature Switzerland AG 2021   151
V. Franco, B. Dodrill (eds.), *Magnetic Measurement Techniques for Materials Characterization*, https://doi.org/10.1007/978-3-030-70443-8_8

2   Torque Sensing and Magnetometry Basics

Fig. 1 illustrates the conceptual elements of a mechanical measurement of magnetic torque. Fundamentally, a magnetic torsion stiffness must exist against rotation of magnetization within the sample, in order for a nonzero torque to develop and hence for the rotation of an external applied field, ultimately, to do mechanical work that results in displacement of the specimen's physical support. In the limiting (and unachievable) case of a perfectly spherical specimen of a perfectly isotropic material, the magnetic torsion stiffness will be zero, and there will be no torque. Figure 1 indicates a scenario for a monocrystalline magnet that has been polished into a sphere and attached and centered to the circular end face of a cylindrical torsion rod and with a specific crystallographic orientation relative to the cylinder axis. Observations of magnetic torque in such a configuration probe the magnetocrystalline anisotropy intrinsic to the material.

Traditional (millennia-old) measurements of magnetic torque concentrated on the equilibrium, constant, or DC orientation of the specimen in an external field. The past century has seen the development of very sensitive mechanically resonant AC measurements, as were required famously for pioneering measurements of Einstein and de Haas. The figure also indicates an additional advantage of more recent implementations of AC mechanical measurements, in which all three orthogonal components of magnetic torque can be monitored simultaneously through separate mechanical resonances (in the case of Fig. 2, one twisting/torsional mode and two bending/cantilever modes). Other compelling features of mechanical torque measurements include their scalability across many orders of magnitude in sample size, thanks to micro- and nanofabrication, and their compatibility with other mechanisms that pump angular momentum into the magnetic system, such as spin resonances and spin currents. In an emerging direction, ultrasensitive torque measurements are being performed in experiments invoking levitation via optical, ion, or superconducting traps to eliminate solid mechanical supports [4–6].

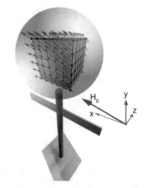

Fig. 1 Schematic representation of a mechanical torque sensor affixed to a magnetic monocrystalline sphere. Magnetic moments, located at the lattice points, are canted away from the magnetocrystalline easy axis (e.g., along the y direction) due to an applied external field H_0

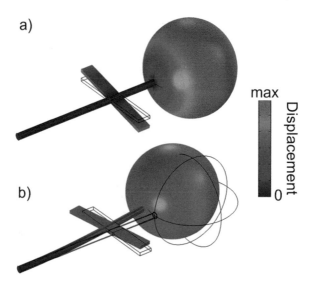

Fig. 2 Simulated mechanical profiles of the (**a**) torsional and (**b**) lateral flexing modes for the basic sensor of Fig. 1. The wire frame indicates the resting position of the structure

3 Experimental Apparatus, Sensor Design, and Fabrication

Optical Detection Schemes

The most sensitive, and generally noninvasive, methods of mechanical displacement detection have so far been through free-space or integrated optical schemes. In the former, the most straightforward implementation is an optical "beam bounce" method off of a (mirrored) armature of a torque sensor which experiences the highest amount of deflection, and the modulation of the reflected light intensity is recorded. As sensor dimensions are further miniaturized toward increased sensitivity and higher operating frequencies, sub-picometer displacement detection has been accomplished by incorporating free-space lasers with interferometers. Two common approaches are path-stabilized Michelson interferometry, where the light reflected off of the oscillating sensor is modulated with a reference beam, and Fabry-Perot interferometry, in which the optical modulation is enhanced through interaction with an etalon created by the torque sensor and a substrate (both shown schematically in Fig. 1. of [7]). The etalon gap width is ideally tuned to the appropriate optical wavelength and can be lithographically defined through removal of a sacrificial layer between the sensor and substrate or by affixing a mirror behind the mechanical sensor.

Recent advances in resonant optical cavity transduction of nanoscale mechanical resonators have granted extraordinary sensitivities, including the detection of standard quantum-limited motion. "Cavity optomechanics" approaches generally rely on the interaction of a moving object with an optical standing wave, thus affecting its path length (dispersive coupling), or by mechanical motion creating absorption or other losses in the resonant optical system (dissipative coupling) [8]. In each case, the mechanical transduction occurs through modulation of the output power from the system.

Apparatus

The key components of a measurement system for high-sensitivity mechanical displacement detection through free-space or integrated optical schemes are shown in Fig. 3. The torque sensor (A) can be housed in a vacuum chamber to primarily minimize viscous (squeeze-film) damping effects with the atmosphere (usually pressures on the order of a millitorr are sufficient), though operation in ambient conditions can be achieved. The sensors (and/or light source) are often mounted on multi-axis piezo-actuated stages with positioning resolution consistent with the scale of the sensor for focusing and alignment of the optical readout. Temperature control can be achieved by, for example, mounting the device chip on a cold finger connected to nitrogen or helium closed-cycle cryostats. Adequate vibration damping of vacuum and/or cooling lines to the chamber is necessary, and ideally all components in the setup within applied magnetic fields should be non-magnetic.

As discussed earlier, the optical cavity (B) can be designed using various methods. In- and out-coupling of light can be done directly through free-space optics [9] or via evanescent coupling to tapered optical fibers (with diameters suited for single-mode optical transmission) [10]. For integrated systems, light is coupled into and detected out of integrated waveguides through grating couplers [11] or bonded fibers.

Fig. 3 The mechanical torque sensor (A) is measured using an optical cavity (B) created by light source (C). Magnetic fields (D) are applied at the specimen and consist of either DC bias fields or AC dither fields along desired axes. The signal acquisition system (E) measures the specimen motion using the optical modulations detected at the photodetector (F)

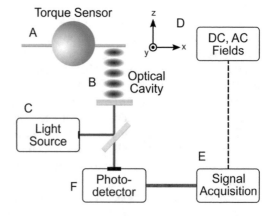

Continuous-wave (CW) lasers are often used as the light source (C) for both free-space and integrated optics implementations of torque sensing, as nanomechanical resonators operating in the MHz range are ideal for measuring equilibrium to slow magnetization processes. General requirements for the light source are operation wavelengths beyond the range of high optical absorption coefficients in the materials comprising the optomechanical system, stable power output in order to minimize mechanical frequency shifts, and low-power (micro- to milliwatt range) operation. Channels of thermal conduction in nanoscale torque sensors generally occur through clamping points which are designed to be narrow in order to decrease mechanical spring constants and increase deflection amplitudes, making higher laser fluences problematic to their operation. Tunable narrow-linewidth CW lasers with wave-length operation through optical resonances are generally implemented in cavity optomechanics setups. Pulsed ultrafast lasers have been used for stroboscopically measuring fast nanomechanical systems [12] and non-equilibrium magneto-optics. Low-measurement duty cycles are required to allow thermal relaxation between pulses, but applications in torque magnetometry can be anticipated as a future thread.

Uniform magnetic fields applied at the specimen (D) are generally classified as static (DC) with magnitudes ideally capable of saturating the magnetization and smaller (on the order of tens to hundreds of microtesla) AC fields which serve to perturb ("dither") the spin texture to generate torques. The applied field directions for both the DC and AC fields depend on the torque directions of interest. High DC fields are achieved with the use of electromagnets or permanent magnets. For low-field (1 T or lower) operation, permanent magnets offer less instrumental complexity than required for electromagnets. For field variation, the magnets are placed on linear (motorized) stepper rails and offer very precise field stability for repeated, fine hysteresis measurements. AC fields can be generated from coils placed near the specimen, with the simplest implementation being a single loop of magnet wire placed above the torque sensor (which will generally allow up to tens of MHz operation, and extended through the use of tuned LC circuits). Printed circuit boards with impedance-matched transmission lines can grant local AC fields up to several GHz and can be designed using multiple layers to generate fields along different directions [13].

Data acquisition (E) in most cases is done in the frequency domain using lock-in amplification with the demodulation and output (AC field drive) frequencies set to the mechanical resonance of the torque sensor. Operating frequencies in the MHz range immunize the measurements from $1/f$ technical noise sources, with a trade-off being that care is required at radio frequencies to eliminate direct crosstalk from drive to photoreceiver. Computer control of optical wavelength, magnetic drive frequency, and bias field strength are required for hysteresis measurements and precursor optical and mechanical characterizations. Phase-locked loop capability is important for tracking mechanical frequency shifts arising from the contribution of the field-dependent DC magnetic torsion stiffness.

Torque Sensor Design

Computer simulations play a very significant role in the present evolution of mechanical torque studies of magnetism. The principles are not complicated, but neither the mechanics of the sensors nor the magnetics of the specimens are fully amenable to analytical descriptions. Numerical simulations play a role in initial experimental design and again in interpretation of measurements made with as-fabricated structures.

Torque sensor design is guided by finite element mechanical simulations. The mechanical resonance frequencies can be predicted, usually within 20% or better using bulk elastic constants and densities, and the mechanical displacement profiles visualized. The mechanical quality factors of the resonant modes are more difficult to predict from simulation, but reasonable guesses can be made based on the literature and from experience with similar devices. $Q \sim 1000$ is a good starting estimate for operation in vacuum. This is a very low Q in absolute terms, and there is a large phase space to be exploited for increased torque sensitivity through reduction of mechanical dissipation.

Continuing from mechanical mode simulation, sensor integration with a displacement detection modality can be optimized. State-of-the-art displacement detectors using microwave or optical microcavities require separate finite element electromagnetic simulations. In practice, the capability to resolve thermomechanical noise remains a good litmus test of sensitivity. With sensitivity limited by Brownian motion, the accessible science from a given study might be independent of other parameters, such as overall measurement time (provided the signal integrations are not degraded by $1/f$ noise).

Within an initially unbound design parameter space, the torsion stiffness of the sensor introduces a trade-off between mechanical frequency and torque sensitivity: higher torsional compliance yields larger displacements and lower resonant frequencies. With the addition of constraints imposed by the fabrication process (minimum feature size, integrating the magnetic material, device robustness, and fabrication yield) and by the measurement (sample environment, temperature, pressure, fields, additional infrastructure within the sample locale for signal transduction), the same sample geometry and/or fabrication approach might not scale over orders of magnitude in resonant frequency. For example, whereas MHz-range cantilever and doubly clamped torsion devices are conveniently fabricated from silicon-on-insulator platforms, a kHz-range device of similarly thin material may necessitate patterning a "trampoline" from a silicon nitride window in a silicon frame [14, 15].

Torque Sensor Fabrication

The ongoing progress in nanofabrication methods allows for feature sizes on the order of tens of nanometers to be defined in a wide variety of materials, making it feasible to design and fabricate torque sensors with high geometric precision (which is of special importance for integrated optomechanical cavities). The fabrication of hybrid magneto-mechanical torque sensors is generally approached from the direction of affixing magnetic specimens to (non-magnetic, and often pre-fabricated) mechanical sensors or by creating the sensor completely out of magnetic material. A common theme behind each is the need to protect (or avoid completely) the magnetic material from exposure with incompatible processing used in standard lithography such as dry or wet chemical etching, which can induce pits and other geometric defects or induced chemistry and/or damage to the magnetic lattice from implantation during focused ion beam milling (creating a magnetic "dead layer" in the region exposed to ions).

Silicon-on-insulator (SOI) substrates have been widely utilized as precursors for nanomechanical resonators in both cases of free-space or integrated optical detection. The SOI architecture (engineered to minimize current leakage in integrated circuits) comprises a thin silicon "device layer" (which can range from several microns down to tens of nanometers thick) atop a silicon oxide insulating layer which is sandwiched by a bulk silicon "handle" layer below. Mechanical structures are defined in the device layer generally through electron beam or optical lithography in an appropriate photosensitive resist, which serves as a mask for subsequent etching of the silicon device layer. Sections of the oxide layer are then removed in order to release the mechanical structures. In the case of Fabry-Perot detection, the thickness of the oxide layer can be tuned such that the etalon created by the device and handle layers provide maximum constructive interference of the reflected intensity through the oscillation cycle of the mechanical resonator. In the case of integrated optics (waveguides and optomechanical devices) in silicon, the thickness of the device layer is set to around 220 nm to optimize single-mode optical confinement at telecommunication wavelengths.

Another popular starting material for nanomechanical systems are silicon nitride membranes, which are commercially available in differing stresses and thicknesses (down to 10 s of nanometers). The mechanical structures can be defined lithographically or through focused ion beam milling, though, owing to their thinness, care must be taken during the processing steps. With high optical transmittance at visible and telecommunication wavelengths, they are ideal candidates for membrane-in-the-middle optomechanical systems.

The recent emergence of single-crystal diamond optomechanics aims to take advantage of their remarkable optical, mechanical, and thermal properties [16, 17]. Diamond in the infrared and visible wavelengths has a maximum theoretical transmission of 70%, owing to its large bandgap, and allows for high optical power confinement. The optical properties combined with a high Young's modulus (about five times higher than single-crystal silicon), high thermal conductivity, and low mechanical dissipation make diamond an ideal candidate for integration with optomechanical systems.

Several methods have been and are continuing to be developed for the attachment of magnetic material to torque sensors and are dependent on the material of interest and their compatibility with fabrication processing. For micro- to nanometer-scale geometrically confined magnetic thin films, where specific shapes are to be controllably patterned, pre-fabricated sensors are coated with photoresist using procedures to protect from effects of stiction, and the shapes are exposed using optical or electron beam lithography. The developed photoresist serves as a mask for liftoff of the deposited magnetic thin film [18]. Thin films can also be deposited onto thin membranes, through predefined shadow masks (e.g., through holes in another thin membrane), and the mechanical sensor can be defined around the magnetic material with focused ion beam milling [19]. For materials not amenable to standard lithographic processes, such as those that need to be grown epitaxially on specialized surfaces (an example being synthetic ferrimagnets such as yttrium iron garnet), the focused ion beam can be used to mill micromagnetic shapes out of a bulk material, which can be "plucked and placed" onto the sensor using appropriate nanomanipulation tools [13]. Mechanical resonators fully comprised of magnetic material have also been developed [20].

4   Magnetic Properties and Parameters

Magnetic Hysteresis

Nanomechanical torque magnetometry is ideal for the measurement of equilibrium magnetic hysteresis and slow dynamics (such as those that are thermally activated) of nanoscale materials. The ability to measure the equilibrium response of single specimens, as opposed to arrays of similar structures (as required by most other methods for sufficient signal-to-noise ratio), grants access to magnetic "fingerprints" unresolvable through averaging across multiple structures. These are often on account of fabrication-induced inhomogeneities in the magnetic microstructure (such as surface and edge roughness or grain boundaries), which affect local magnetic properties (saturation magnetization, anisotropy, and exchange coupling).

The interaction of the high exchange energy present in domain walls with local energy minima created by inhomogeneities (pinning sites) manifests as the Barkhausen effect, which are observed as small, abrupt transitions (steps) in the hysteresis loop. An example illustrating the sensitivity of nanomechanical torque detection is shown through the measurement of thermally activated hopping of a single vortex core (a high exchange density zero-dimensional domain wall) between adjacent pinning sites, of which the hopping rates as a function of applied field and temperature, as shown in Fig. 4a, can be used to quantitatively determine the energy barriers. The vortex core, with a width on the order of 10 nanometers, can also be rastered across the magnetic specimen using externally applied fields to subtly probe minor hysteresis loops induced by disorder caused by the Barkhausen effect, as shown in Fig. 4b, which can be used to recreate the 2D magnetic energy landscape with resolution on the order of 10 meV (Fig. 4c).

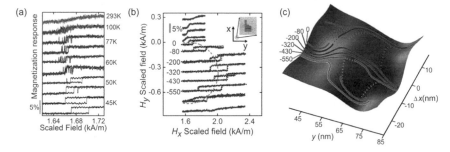

Fig. 4 Quantification of the Barkhausen effect. (**a**) Thermally activated vortex core hopping between pinning sites in a 1 μm diameter, 42-nm-thick permalloy disk as a function of temperature. At room temperature, no hysteresis is observed as the hopping rate is much higher than the measurement bandwidth. The hopping rate is decreased at room temperature and telegraph noise is observed. At lower temperatures, thermal activation is further suppressed and shown through small variation of transitions in the minor hysteresis. (**b**) Mapping minor hysteresis loops over small steps of H_x and H_y and (designated by inset values, in Am^{-1}). (**c**) The energy landscape from the minor hysteresis data in (**b**) constructed using an analytical model of vortex core pinning [21]. Minimum energy paths are solid lines, while dashed lines connect vortex core hopping regions. Reprinted with permission from [22]

Mechanical Torque Detection of Spin Resonance

Torque magnetometry has been applied primarily to magnetostatic studies where the magnetic system remains in equilibrium with the total applied field, and furthermore the magnetic torque determines instantaneously the mechanical drive torque. Beyond this, nonequilibrium spins undergoing precession and relaxation have a time-rate-of-change angular momentum that can be registered as a net mechanical torque. The pioneering demonstration of electron spin resonance detection via mechanical torques was carried out by Alzetta et al. in 1967 [23]. Their system consisted of a paramagnetic free radical solid specimen (DPPH) suspended with a torsion fiber pendulum that was RF driven at magnetic resonance with an amplitude modulation at the mechanical torsional resonance frequency. A DC field applied along the torsion axis provided the net polarization in the specimen, and the angular momentum transferred through spin-lattice relaxation registered as a resonant mechanical torque. The same group would later miniaturize their system geometry to the micrometer scale in 1996 [24].

For application to ferromagnets, the ingenious method of Alzetta et al. must be modified to enable concurrent sensing of both the equilibrium magnetization (hysteresis) and dynamics (spin resonances). The application of the DC bias field along the torsion axis in the original method results in the equilibrium magnetization being parallel to the necessary torque direction which, following the orthogonality condition in the magnetic torque expression, would not yield a mechanical deflection. A "torque-mixing" method lifts this constraint by generating a net AC mechanical torque from a driven spin resonance with the DC field

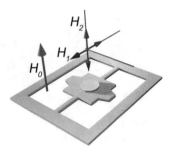

Fig. 5 Applied field configurations and sensor geometry enabling torque mixing magnetic resonance spectroscopy. The torque sensor supports a micromagnetic disk on a paddle, operating at a fundamental torsional mode (axis along the torsion rods) at frequency f_{mech}. H_0 biases the magnetization along z-direction, while H_1 (at f_1) drives the transverse magnetic resonance in-plane with the sensor. H_2 (at f_2) is applied such that $f_1 - f_2 = f_{mech}$

applied perpendicular to the torsion axis, in the usual magnetometry configuration [13]. In the geometry of Fig. 5, the static field H_0 is applied in the z-direction, polarizing the spins in a micromagnetic specimen mounted on a nanomechanical torsional resonator. An RF field at H_1 (at frequency f_1) pumps (in a classical picture) the precession of magnetic moments at f_{res}. An additional RF field H_2 (at f_2) multiplies with H_1 to generate sum and difference torque-mixing components which are proportional to the magnetic resonance amplitude and linear to H_1 and H_2. By spacing f_1 and f_2 such that their difference is the mechanical resonance frequency f_{mech}, the spin dynamics can be read out with high sensitivity. An additional RF tone (at f_3, detuned slightly off f_{mech}) applied in the z-direction (the standard torque magnetometry configuration) allows for demodulation of the equilibrium magnetization signal. Sweeping the two torque-mixing RF tones (f_1 and f_2 with constant separation f_{mech}) while monitoring the signal at f_3 allows for characterization of the spin dynamics as well as the underlying equilibrium magnetization, both of which are necessary for a comprehensive understanding of the system. The torque-mixing magnetic resonance detection scheme is broadband, as long as the mechanical resonance frequency does not far exceed the spin-lattice relaxation rate. With ever-increasing sensitivities granted by the advances in optomechanical detection, torque magnetometry has potential for becoming a standard tool for magnetic resonance spectroscopy in small, geometrically confined spin systems.

Multi-Axis Torque Magnetometry

Geometrically confined micromagnetic structures investigated to date have predominantly been thin-film based. In standard torque magnetometry measurements of thin specimens, the magnetic torque is parallel to the mechanical rotation axis, generated

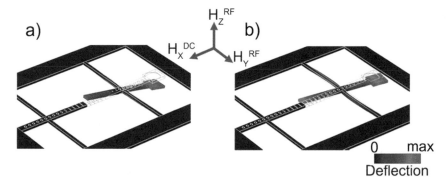

Fig. 6 Two-axis torque sensor geometry. (**a**) The fundamental torsion mode, τ_y, is driven via torques directed in the y-direction. (**b**) Out-of-plane magnetic torques are transduced through the τ_z mode

as a cross-product of the net magnetic moment and perpendicular applied field. This torque is dominated by the in-plane net moment as shape anisotropy suppresses out-of-plane magnetization. The perpendicular magnetic torque, on the other hand, is dominated by in-plane anisotropy. Complementary information (with net moment measurements) is gained from the perpendicular torques as they can directly probe diagonal magnetic susceptibility contributions which arise due to microstructural inhomogeneities in the film.

The ability to measure both components of magnetic torque is enabled by precise design and nanofabrication of sensors that grant first-order sensitivity to vibrational modes which respond independently to in- and out-of-plane torques. Two such modes are depicted in Fig. 6 for a nanomechanical resonator embedded with a split-beam photonic crystal cavity. Magnetic torques are generated at the thin-film element on the square paddle end of the torque sensor (opposite the split-beam). An out-of-plane RF field generates a y-torque, τ_y (Fig. 6a), while the RF field in the y-direction drives the τ_z mode (Fig. 6b). In terms of vector cross-products of the magnetic moment m_i and field H_i, and differential volume susceptibilities, χ_i, the torque expressions for each mechanical mode are:

$$\tau_y^{RF} = -m_x \mu_0 H_z^{RF} + H_z^{RF} \chi_z V \mu_0 H_x^{DC}, \tag{1}$$

$$\tau_z^{RF} = +m_x \mu_0 H_y^{RF} - H_y^{RF} \chi_y V \mu_0 H_x^{DC} \tag{2}$$

This coupled system of equations can be applied to paired measurements of hysteresis through the τ_y and τ_z to extract the differential susceptibility components [25]. In the limit of thin films, the analysis is simplified as χ_z approaches zero. When the torque is hysteretic at or near $H_x = 0$, the quantity $(\chi_y - \chi_z)H_x$ can be directly extracted as a function of field. The method can be further extended to the third orthogonal torque axis for full measurement of the diagonal magnetic susceptibility components (and necessary for comprehensive analysis of three-dimensional structures).

Einstein-de Haas Torques

The intrinsic connection between mechanical angular momentum and magnetic moment is described through the magnetomechanical ratio, g', which is generally determined through Einstein-de Haas (EdH) effect measurements. Historically, EdH measurements have been difficult on the account of very small EdH torque amplitudes in comparison to conventional cross-product torques generated by the magnetic moment and applied field. Wallis et al. reported a pioneering micromechanical measurement of the EdH effect in 2006 [26]. Nanoscale torque sensors, with very high mechanical resonance frequencies, offer a path of continued resurgence for EdH studies. The amplitude of the EdH effect scales with frequency, becoming comparable to or even exceeding cross-product torques when driven at radio frequencies. In the quasistatic regime, the EdH torque is also 90 degrees out of phase to conventional magnetic torques, allowing one to distinguish each component in quadrature measurements [27].

5   Micromagnetic Simulation

Micromagnetic simulations based on the Landau-Lifshitz-Gilbert equation, as used to model magnetization dynamics in geometrically confined microstructures, are very helpful for extending insight into the measurables from micromechanical torque experiments. Using open-source code such as Mumax3 [28, 29], the change in system energy can be calculated due to a change in applied fields, and torque calculations can be extracted from these results.

The most familiar occurrences of mechanical torque from magnetic systems are constant (DC) torques, while the torques measured in micromechanical studies all are AC torques; it can be helpful to connect the two scenarios. DC torques will rotate two bar magnets into anti-alignment (such as a compass needle in the Earth's magnetic field). Traditional torsion balances use this phenomenon to determine the anisotropies of bulk specimens by measuring the mechanical counter-torque to the equilibrium magnetic torque on a specimen. This is done as a function of the angle between a coordinate system fixed in the specimen's frame of reference and an external DC field. An AC torque measurement, in contrast, employs an AC drive ("dither") field perpendicular to the DC field direction to modulate the equilibrium torque. When the dither field amplitude is small in comparison to the DC field, the AC perturbation effectively measures the first derivative of the DC torque versus field direction.

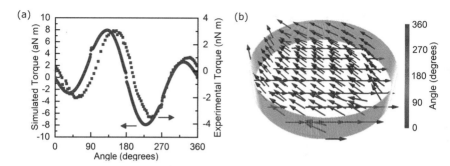

Fig. 7 (**a**) Magnetic torque studies of exchange-biased antiferromagnetic/ferromagnetic bilayers. The left axis (red circles) is the simulated torque for a 448 nm diameter permalloy disk with a height of 35 nm. The right axis (blue squares) is the measured torque from reference [30] of a NiFe (30 nm)/FeMn (15 nm) bilayer of 6 μm diameter at 20 mT. (**b**) The simulated representation of the spins in the exchange-biased permalloy system. The exchange bias direction is 0 degrees and at the base of the disk, with the magnetization direction of the ferromagnet in plane indicated by the colorbar scale

Micromagnetic simulations can similarly be scripted to calculate both DC and AC torque components for model specimens. As in other applications of finite element micromagnetic simulation, tests of scaling behavior versus cell size, and other controls such as testing for linear response to small-field amplitude or anisotropy parameter changes, are performed to ensure that physically meaningful conclusions are drawn. This is especially important when the type of simulation grid is not ideally suited to the problem, as, for example, in a prototypical case of a circular magnetic disk simulated on a rectangular grid.

These torque simulations can be used to gain insight into more complex magnetic material systems such as exchange-biased antiferromagnetic/ferromagnetic bilayers. Exchange bias introduces a unidirectional anisotropy, describable as a negative cosinusoidal variation of energy as a function of the angle between the exchange bias (EB) direction and the magnetization in the ferromagnet. The corresponding torsion spring constant, κ_{EB}, from the angular second derivative of energy, is positive/negative for magnetic moment aligned/anti-aligned with the EB direction. The out-of-plane torque measurements on a macroscopic circular disk of (nominally) isotropic ferromagnetic thin film can isolate the EB contribution and are shown in Fig. 7(a) (sample details in caption). In this case, the ferromagnetic film also contained an unintentional uniaxial anisotropy, revealed upon increasing the strength of the external magnetic field [30]. An illustrative simulation of the DC torque on a microdisk in a similar bilayer system qualitatively agrees with the experimental results, as shown in Fig. 7(a). The magnitude difference on the order of one billion (compare left- and right-axis scales) stems from the ratio of magnetic volumes. A representation of the spin configuration in the simulated microdisk is shown in Fig. 7(b).

6 Calibration and Sensitivity

Calibration methods based on the intrinsic Brownian motion of the sensor have been borrowed from the atomic force microscopy community and adapted from calibrating linear displacement and force to angular displacement and torque. The torque calibration follows from the fluctuation-dissipation theorem [31]. In the nanomechanical device, mechanical losses, characterized by the mechanical quality factor Q_m, couple energy to the environment as heat. Conversely, the external thermal bath couples energy into the nanomechanical mode as Brownian motion, resulting in the TM noise signal. This occurs at each mechanical mode of motion. Since the thermal motion is in the form of white noise, the angular displacement spectral density is given by:

$$
\begin{aligned}
S_{\theta,n}(f) &= \frac{k_B T f_n}{2\pi^3 Q_{m,n} I_{\text{eff},n}} \frac{1}{\left(f_n^2 - f^2\right)^2 + \left(\frac{f_n f}{Q_m}\right)^2} \\
&= \frac{2k_B T f_n^3}{\pi k_{\text{tor},n} Q_{m,n}} \frac{1}{\left(f_n^2 - f^2\right)^2 + \left(\frac{f_n f}{Q_{m,n}}\right)^2}
\end{aligned}
\tag{3}
$$

where k_B is the Boltzmann constant, T is the temperature, and f_n is the mechanical frequency of mode n. Both forms of the equation follow from $2\pi f_n = \sqrt{\frac{k_{\text{tor},n}}{I_{\text{eff},n}}}$. The effective moment of inertia, $I_{\text{eff},n}$, can be calculated analytically for simple structures or by using finite element modelling for more complex devices [32].

The angular spectral density is then related to the mean square angular displacement through:

$$
\langle\theta\rangle^2 = \int_0^\infty S_\theta(f) df.
\tag{4}
$$

This can be calibrated using the equipartition theorem equating the thermal energy of the mode with the energy of the torsional spring, $\frac{1}{2}k_B T = \frac{1}{2}k_{\text{tor},n}\langle\Theta\rangle^2$. If the mechanical device is driven in the linear regime, the driven torque displacements can be calculated by comparing these signals to the TM noise measurement. The conversion to torque occurs through $\tau_n = \kappa_{\text{tor},n}\theta_n$.

Next, the sensitivity of these devices can be extracted from the TM noise measurement. Once converted to angular displacement and torque, the minimal detectable levels are determined by the (white) noise floor in the experiment off of the mechanical resonance. However, one powerful property of using a mechanical resonator to measure magnetic torque is the fact that the signal is enhanced through the resonator's ability to store energy as characterized by its Q_m. As such, on-resonance displacement and torque sensitivities are improved by a factor of $\sqrt{Q_m}$

[33]. Record torque sensitivities at the level of 10^{-24}Nm $\sqrt{\text{Hz}}$ have been achieved in a dilution refrigerator used to suppress the TM noise. This sensitivity level also had the added restriction of a 20-ms optical sampling limitation required to prevent sensitivity degradation due to heating [34]. At atmospheric conditions (where the Q-enhancement is significantly reduced due to air damping), torque sensitivity has been measured in the 10^{-20}Nm $\sqrt{\text{Hz}}$ range [35]. It should be noted that both of these measurements used optical cavities to achieve their high sensitivity.

If the net magnetic moment of the sample is known, as well as the applied field magnitude, the magnetic moment sensitivity of the sensor can be determined through scaling to the TM noise amplitude [36]. The conservative assumption for all magnetic moments contributing to torque generally applies to mechanical modes with compliant torque axes perpendicular to a saturated sample. Often is the case of micromagnetic structures that the net moment is unknown, and micromagnetic simulations assist in providing a close approximation.

7 Nanomechanical Torque Magnetometry in Comparison to Other Methods: Summary

Nanoscale optomechanical torque magnetometry is ideally suited for harnessing the direct transfer of angular momentum between the magnetic and mechanical subsystems to investigate underlying spin mechanical effects (such as magnetomechanical ratios and spin-lattice coupling), as well as properties of the magnetic material from torques communicated through quasistatic and dynamic magnetization processes. The exquisite detection sensitivity allows for measurement of single nanoscale specimens, making it highly competitive with (as well as offering complementary information to) other existing methods. The construction of hybrid torque sensors is amenable to various fabrication platforms, allowing for the measurement of a wide range of materials and magnetic systems. Optical transduction methods can be designed to be generally noninvasive to the magnetic system.

The ability to clearly measure equilibrium magnetization responses (such as hysteresis) in single, nanoscale samples is a key feature of nanoscale torque magnetometry and not as easily realized using other techniques. Although possible in principle to acquire static magnetization using magneto-optical methods (such as MOKE), in practice, it is difficult to achieve due to the low laser fluences generally needed in order to avoid sample heating while at the same time attempting to detect sub-microradian polarization changes. Signal averaging times required approach $1/f$ noise, limiting the signal-to-noise ratio. DC magnetization detection is possible using local probing methods such as MRFM and nitrogen-vacancy center magnetometry, or X-ray circular magnetic dichrism (XMCD), though generally numerical reconstruction methods are needed on top of long averaging times (in other words, not a one-shot measurement possible through mechanical torque detection). SQUID measurements are commonly used for highly sensitive static magnetization measurements, though they offer limitations for operation in high temperature. Coupling to small specimens inductively for DC measurements is possible through superconducting electronics, though as size scales decrease, parasitic inductances hamper adequate signal-to-noise ratio.

Torque detection of slow dynamics (such as those induced through thermal activation or domain wall resonances) and AC susceptibility can be achieved at the mechanical resonance frequencies, while spin resonances at much higher frequencies can be measured through drive signal modulation or detection demodulation schemes. The ability to monitor simultaneously the dynamics alongside the underlying equilibrium magnetization in small specimens gives nanomechanical torque magnetometry an advantage not yet served by other methods.

As detection frequencies increase, Einstein-de Haas torques amplitudes will grow in magnitude and become a routine component of the measured data. Furthermore, a materials-dependent upper limit of mechanical detection frequencies will be imposed when the spin-lattice relaxation time becomes longer than the oscillation cycle of the sensor. These effects speak in particular to the most exciting aspect of nanoscale torque sensors, in which the method goes beyond simple miniaturization to providing access to new physical information.

References

1. J. Moreland, Micromechanical instruments for ferromagnetic measurements. J. Phys. D. Appl. Phys. **36**, R39 (2003)
2. J.E. Losby, V.T.K. Sauer, M.R. Freeman, Recent advances in mechanical torque studies of small-scale magnetism. J. Phys. D. Appl. Phys. **51**, 483001 (2018)
3. D.C. Mattis, *The Theory of Magnetism Made Simple* (World Scientific, 2006)
4. M. Rashid, M. Toroš, A. Setter, H. Ulbricht, Precession motion in levitated optomechanics. Phys. Rev. Lett. **121**, 253601 (2018)
5. P. Huillery, T. Delord, L. Nicolas, M. Van Den Bossche, M. Perdriat, G. Hetet, Spin-mechanics with levitating ferromagnetic particles. Phys. Rev. B **101**, 134415 (2020)
6. T. Wang, S. Lourette, S.R. O'Kelley, M. Kayci, Y.B. Band, D.F. Jackson-Kimball, A.O. Sushkov, D. Budker, Dynamics of a ferromagnetic particle levitated over a superconductor. Phys. Rev. Appl. **11**, 044041 (2019)
7. D. Karabacak, T. Kouh, K.L. Ekinci, Analysis of optical interferometry displacement detection in nanoelectromechanical systems. J. Appl. Phys. **98**, 124309 (2005)
8. M. Wu, A.C. Hryciw, C. Healey, J. Harishankar, M.R. Freeman, J.P. Davis, P.E. Barclay, Dissipative and dispersive optomechanics in a nanocavity torque sensor. Phys. Rev. X **4**, 021052 (2014)
9. W.K. Hiebert, D. Vick, V. Sauer, M.R. Freeman, Optical interferometric displacement calibration and thermomechanical noise detection in bulk focused ion beam-fabricated nanoelectromechanical systems. J. Micromech. Microeng. **20**(11), 115038 (2010)
10. C.P. Michael, M. Borselli, T.J. Johnson, C. Chrystal, O. Painter, An optical fiber-taper probe for wafer-scale microphotonic device characterization. Opt. Express **15**, 4745 (2007)
11. D. Taillaert, P. Bienstman, R. Baets, Compact efficient broadband grating coupler for silicon-on-insulator waveguides. Opt. Lett. **29**, 2749 (2004)

12. N. Liu, M. Belov, J.E. Losby, J. Moroz, A.E. Fraser, G. McKinnon, T.J. Clement, V.T.K. Sauer, W.K. Hiebert, M.R. Freeman, Time-domain control of ultrahigh-frequency nanomechanical systems. Nat. Nanotechnol. **3**, 715 (2008)
13. J.E. Losby, F. Fani Sani, D.T. Grandmont, Z. Diao, M. Belov, J.A.J. Burgess, S.R. Compton, W.K. Hiebert, D. Vick, K. Mohammad, E. Salimi, G.E. Bridges, D.J. Thomson, M.R. Freeman, Torque-mixing magnetic resonance spectroscopy. Science **350**, 798 (2015)
14. C. Reinhardt, T. Muller, A. Bourassa, J.C. Sankey, Ultralow-noise SiN trampoline resonators for sensing and optomechanics. Phys. Rev. X **6**, 021001 (2016)
15. R. Fischer, D.P. McNally, C. Reetz, G.G.T. Assumpção, T. Knief, Y. Lin, C.A. Regal, Spin detection with a micromechanical trampoline: Towards magnetic resonance microscopy harnessing cavity optomechanics. New J. Phys. **21**, 043049 (2019)
16. M.J. Burek, J.D. Cohen, S.M. Meenehan, N. El-Sawah, C. Chia, T. Ruelle, S. Meesala, J. Rochman, H.A. Atikian, M. Markham, D.J. Twitchen, M.D. Lukin, O. Painter, M. Lonvcar, Diamond optomechanical crystals. Optica **3**, 1404 (2016)
17. M. Mitchell, B. Khanaliloo, D.P. Lake, T. Masuda, J.P. Hadden, P.R. Barclay, Single-crystal diamond low-dissipation cavity optomechanics. Optica **3**, 963 (2016)
18. Z. Diao, J.E. Losby, J.A.J. Burgess, V.T.K. Sauer, W.K. Hiebert, M.R. Freeman, Stiction-free fabrication of lithographic nanostructures on resist-supported nanomechanical resonators. J. Vac. Sci. Technol. B **31**, 051805 (2013)
19. J.P. Davis, D. Vick, D.C. Fortin, J.A.J. Burgess, W.K. Hiebert, M.R. Freeman, Nanotorsional resonator torque magnetometry. Appl. Phys. Lett. **96**, 072513 (2010)
20. J.E. Losby, J.A.J. Burgess, C.M.B. Holt, J.N. Westwood, D. Mitlin, W.K. Hiebert, M.R. Freeman, Nanomechanical torque magnetometry of permalloy cantilevers. J. Appl. Phys. **108**, 123910 (2010)
21. J.A.J. Burgess, J.E. Losby, M.R. Freeman, An analytical model for vortex core pinning in a micromagnetic disk. J. Magn. Magn. Mater. **361**, 140 (2014)
22. J.A.J. Burgess, A.E. Fraser, F. Fani Sani, D. Vick, B.D. Hauer, J.P. Davis, M.R. Freeman, Quantitative magneto-mechanical detection and control of the Barkhausen effect. Science **339**, 1051 (2013)
23. G. Alzetta, E. Arimondo, C. Ascoli, A. Gozzini, Paramagnetic resonance experiments at low fields with angular-momentum detection. Il Nuovo Cimento **52**, 392 (1967)
24. C. Ascoli, P. Baschieri, C. Frediani, L. Lenci, M. Martinelli, G. Alzetta, R.M. Celli, L. Pardi, Micromechanical detection of magnetic resonance by angular momentum absorption. Appl. Phys. Lett. **69**, 3920 (1996)
25. G. Hajisalem, J.E. Losby, G. de Oliveira Luiz, V.T.K. Sauer, P.E. Barclay, M.R. Freeman, Two-axis cavity optomechanical torque characterization of magnetic microstructures. New J. Phys. **21**, 095005 (2019)
26. T.M. Wallis, J. Moreland, P. Kabos, Einstein–de Haas effect in a NiFe film deposited on a microcantilever. Appl. Phys. Lett. **89**, 122502 (2006)
27. K. Mori, M.G. Dunsmore, J.E. Losby, D.M. Jenson, M. Belov, M.R. Freeman, Einstein-de Haas effect at radio frequencies in and near magnetic equilibrium. Phys. Rev. B **102**, 054415 (2020)
28. A. Vansteenkiste, J. Leliaert, M. Dvornik, M. Helsen, F. Garcia-Sanchez, B. Van Waeyenberge, The design and verification of MuMax3. AIP Adv. **4**, 107133 (2014)
29. J. Leliaert, J. Mulkers, J. De Clercq, A. Coene, M. Dvornik, B. Van Waeyenberge, Adaptively time stepping the stochastic Landau-Lifshitz-Gilbert equation and nonzero temperature: Implementation and validation in MuMax3. AIP Adv. **7**, 125010-1–125010-13 (2017)
30. J. Rigue, A.M.H. de Andrade, M. Carara, A torque magnetometer for thin films applications. J. Magn. Magn. Mater. **324**, 1561 (2012)
31. H.B. Callen, T.A. Welton, Irreversibility and generalized noise. Phys. Rev. **83**, 34–40 (1951)
32. B.D. Hauer, C. Doolin, K.S.D. Beach, J.P. Davis, A general procedure for thermomechanical calibration of nano/micro-mechanical resonators. Ann. Phys. **339**, 181–207 (2013)
33. A.N. Cleland, *Foundations of Nanomechanics* (Springer-Verlag, 2003)

34. P.H. Kim, B.D. Hauer, C. Doolin, F. Souris, J.P. Davis, Approaching the standard quantum limit of mechanical torque sensing. Nat. Commun. **7**, 13165 (2016)
35. M. Wu, N.L.-Y. Wu, T. Firdous, F.F. Sani, J.E. Losby, M.R. Freeman, P.E. Barclay, Nanoscale optomechanical torque magnetometry and radiofrequency susceptometry. Nat. Nanotechnol. **12**, 127 (2017)
36. J.E. Losby, J.A.J. Burgess, Z. Diao, D.C. Fortin, W.K. Hiebert, M.R. Freeman, Thermomechanical sensitivity calibration of nanotorsional magnetometers. J. Appl. Phys. **111**, 07D305 (2012)

Part III
Imaging Techniques

Magneto-Optical Microscopy

Rudolf Schäfer and Jeffrey McCord

Abstract Magneto-optical microscopy, prominently represented by Kerr microscopy, is the most versatile in-lab imaging technique for magnetic microstructure analysis on any kind of magnetic material being microscopically ordered in magnetic domains. Domains can be studied on a wide range of lateral scales reaching from centimetres down to 200 nanometres, under static conditions as well as under dynamic magnetic field excitation ranging between quasistatic and beyond the microwave regime, at temperatures between 4 K and 800 K, and in applied fields exceeding the Tesla range in arbitrary directions. The surface magnetisation vector field of soft magnetic materials can be quantitatively measured, and the magnetic layers in superlattices may be imaged separately. By plotting the image intensity as a function of magnetic field, local magnetisation curves are obtained that can be interpreted straightforwardly as the underlying magnetisation process is imaged simultaneously. In this chapter the physical and technical basics of magneto-optical microscopy are reviewed together with examples that demonstrate the benefits of the technique.

Keywords Magnetic domains · Magnetic imaging · Magneto-optical microscopy · Magneto-optical Kerr microscopy · Magneto-optical magnetometry · Magneto-optical effects · Magneto-optical Kerr effect · Faraday effect · Magneto-optical Voigt effect · Magneto-optical gradient effect · Magnetic linear birefringence · Magnetic linear dichroism

R. Schäfer (✉)
Leibniz Institute for Solid State and Materials Research (IFW) Dresden, Dresden, Germany
e-mail: r.schaefer@ifw-dresden.de

J. McCord
Kiel University, Institute for Materials Science, Kiel, Germany
e-mail: jmc@tf.uni-kiel.de

© Springer Nature Switzerland AG 2021
V. Franco, B. Dodrill (eds.), *Magnetic Measurement Techniques for Materials Characterization*, https://doi.org/10.1007/978-3-030-70443-8_9

1 Introduction

Magneto-optical microscopy is based on physical effects that lead to an influence of the spontaneous magnetisation of matter on the emission and propagation of light. Such effects occur at visible light frequencies as well as at X-ray wavelengths. The former, which may be called 'conventional' effects [1], include the magneto-optical Faraday, Kerr, Voigt, and gradient effects that are applied in optical polarisation microscopes (often laxly called 'Kerr microscopes') for imaging, while for X-ray microscopy, tunable X-rays are required as they are available at synchrotron radiation sources. This chapter is devoted to the conventional effects and their application for domain imaging; for X-ray spectro-microscopy, we refer to Chapt. 2.c.vii.

While the history of magneto-optical domain observation dates back to the 1950s, it was in the nineteenth century already when the most important underlying effects have been discovered (see Fig. 1). In 1846, Michael Faraday reported that plane-polarised light is rotated by traversing glass if a magnetic field is applied along the propagation direction [2]. The Faraday effect can also be observed in transparent materials with spontaneous magnetisation like magnetic garnets, where it is caused by the magnetisation rather than by a magnetic field. As it is limited to optically transparent materials, Faraday microscopy can only rarely be employed. After its first application for domain imaging on garnet films by Fowler and Fryer [3] in 1956, the method has nevertheless played a decisive role for the development of the magnetic bubble memory in the 1970s. Also the Voigt effect, discovered as magnetic double refraction in vapour by Woldemar Voigt in 1898 [4] and later in paramagnetic liquids [5], played an established role for the observation of in-plane domains in garnets [6]. In recent years, the interest in transmission microscopy has raised again because garnet films have retrieved renewed popularity for magnonic and spintronic applications. Also ultrathin metallic films, being thinner than the penetration depth of light and deposited on transparent substrates, can favourably be studied in transmission geometry. Due to strong magnetic contrasts, Faraday and (transmission) Voigt microscopy could always be easily performed without the need for electronic contrast enhancement.

This is different for domain observation on non-transparent samples. The magneto-optical effect, corresponding to the Faraday effect but in reflection, was discovered on iron by John Kerr in the 1870s for the cases of the magnetisation being perpendicular [7] to the surface, called polar Kerr effect, and parallel [8] to the surface along the plane of incidence, called longitudinal effect. A third variant in which the magnetisation is in-plane but orthogonal to the plane of incidence was later found by Zeeman [9], now called transverse Kerr effect. The first Kerr images were reported independently by Williams et al. [10] and by Fowler and Fryer [11] on iron and cobalt crystals, respectively. As the Kerr effect on in-plane magnetised material is weak and technically difficult, for a long time, meaningful domain images could only be obtained in carefully optimised polarisation microscopes on favourable and well-prepared samples. Despite significant contrast enhancement by coating the samples with optical interference layers [12], Kerr microscopy was considered a niche technique for many years, only being used by a few specialists. This has changed with the introduction of digital image processing in the 1980s [13, 14]. In the standard difference technique, a background image (obtained, e.g. by saturating the specimen in a magnetic field) is digitally subtracted from an image containing domain information, leading to a difference image with pure magnetic contrast that can be enhanced by electronic means. After the introduction of electronic contrast enhancement, continuous methodical developments have boosted the capabilities of Kerr microscopy, which has evolved into a universal, affordable but yet very powerful in-lab domain imaging technique that allows for domain studies on virtually all magnetic materials. Image processing has also opened the path for the discovery of the Voigt effect in reflection and of the magneto-optical gradient effect [15] in 1990.

In this chapter the basics of magneto-optical domain imaging are reviewed. After introducing the conventional magneto-optical effects in Sect. 2, physical aspects of magneto-optical microscopy like domain contrast and resolution are addressed in Sect. 3. In Sect. 4 the state-of-the-art technology is presented, followed by Sect. 5 that collects various methodologies. Emphasise will be on Kerr microscopy, which is the most commonly applied technique. We will nevertheless also introduce the basics of the other effects—imaging on the basis of the Voigt and gradient effects certainly plays a niche role, but both effects can contribute to any domain contrast in Kerr microscopy. It is therefore worth being aware of them. Imaging by Faraday microscopy is of timeless importance anyway, and the application of the Faraday effect in magneto-optical indicator films experiences increasing attention. Previous reviews on magneto-optical microscopy can be found in refs. [1, 16, 17]. Section 2 in ref. [18] contains a general overview of domain observation techniques.

2 Magneto-Optical Effects

The phenomenological difference between the conventional magneto-optical effects is apparent from the right column of Fig. 1. For the Kerr, Voigt, and gradient effects, a four-phase domain pattern of an FeSi crystal, in which the surface domains are magnetised along two orthogonal easy axes, was imaged in reflection in an optical polarisation microscope. For each effect a typical domain contrast was adjusted by properly setting the optical components of the microscope and by choosing the appropriate incidence of light as indicated. In the Kerr effect, the four domain phases appear in up to four different grey levels as this effect depends linearly on the magnetisation vector. Since the Voigt effect has a quadratic dependence on magnetisation, the same domain pattern displays only two grey levels in Voigt microscopy, one for each magnetisation *axis* and independent of the magnetisation direction. Imaged in the gradient effect, which is sensitive to *changes* in magnetisation, the domain boundaries show up with a contrast that depends on the relative magnetisation directions of the neighbouring domains. Both gradient and Voigt effects require a compensator (i.e. a rotatable retardation plate

Effect (Discovery)	Optical description	M-dependence of contrast	First application for imaging (year)	Contrast appearance and (particular) experimental conditions
Faraday Effect (1846)	Circular birefringence & dichroism Transmission	$\sim M$	Fowler & Fryer (1956)	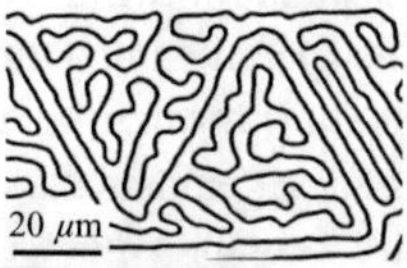 20 μm ‖ Polarisation Perpendicular incidence
Kerr Effect (1877)	Circular birefringence & dichroism Reflection	$\sim M$	Williams et al. (1951) Fowler & Fryer (1952)	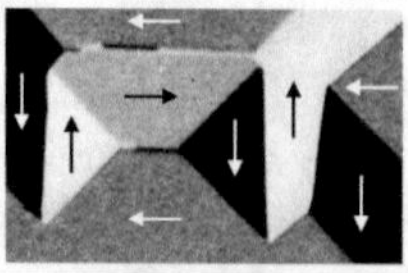 ‖ Polarisation Oblique incidence
Voigt Effect (1898)	Linear birefringence Transmission & reflection	$\sim M^2$	Transmission: Dillon (1958) Reflection: Schäfer & Hubert (1990)	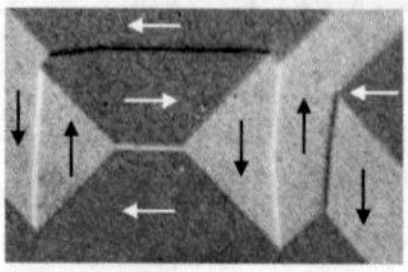 ∥ Polarisation Perpendicular incidence
Gradient Effect (1990)	Birefringence due to M-gradients Reflection & Transmission	$\sim \partial M/\partial r$	Schäfer & Hubert (1990)	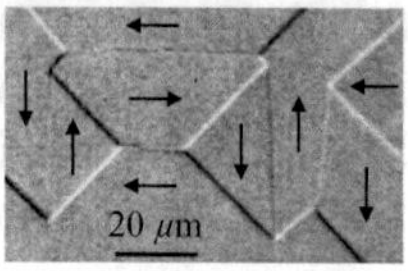 20 μm = Polarisation Perpendicular incidence

Fig. 1 Summary of the four conventional magneto-optical effects that can be applied for magnetic domain imaging. Listed are, from left to right, effect name with its year of discovery, optical description, sensitivity to the magnetisation vector M, authors and year of the first application to imaging, and the typical contrast appearance in an optical polarisation microscope. The domain images were taken from a single-crystal garnet film with perpendicular anisotropy (upper row) and on a (100) surface of an Fe 3 wt% Si sheet with a thickness of 0.5 mm (other rows). Indicated is the polarisation direction of the illuminating light and its plane of incidence. The arrows in the images indicate the local magnetisation directions of the domains. Identical domain patterns are shown for the Kerr and Voigt effect, while a very similar domain state was imaged for the gradient effect. The images on the FeSi material are adapted from ref. [1] with kind permission from Springer Verlag

that transforms elliptically polarised light into plane-polarised light) for optimal contrast adjustment. They are strongest at perpendicular incidence where a Kerr contrast of in-plane domains is not possible. If a transparent material with similar domain phases would be imaged in transmission, the same contrast features would be observed, but now in the Faraday, the transmission Voigt, and the transmission gradient effects. In Fig. 1, a perpendicularly magnetised garnet film at perpendicular incidence was chosen to represent the Faraday effect as this polar geometry is most frequently used in Faraday microscopy.

To apply the conventional effects for microscopy, the sample is illuminated by plane-polarised light, i.e. by a transverse electromagnetic wave, which moves along the propagation vector k with electric and magnetic field components that oscillate

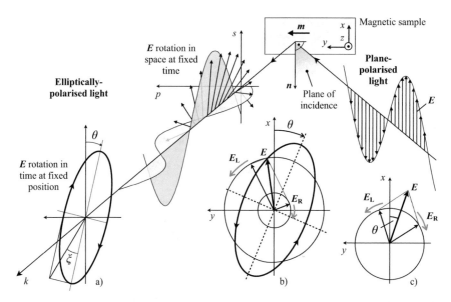

Fig. 2 (**a**) Visualisation of elliptically polarised light being generated by the phase-shifted superposition of two partial waves that are plane-polarised in the (s, k)- and (p, k)-planes. Shown is the spatial and time evolution of the electric field vector E. By assuming that the elliptical wave is caused by reflection of plane-polarised light from a magnetic specimen, the schematics illustrate the Kerr effect (in a strongly exaggerated way, though). (**b**) Representation of elliptical polarisation as superposition of right- and left-handed circularly polarised waves with different amplitudes and phases. (**c**) If the two circular waves are out of phase, but equal in amplitude, a rotated plane-polarised wave emerges by superposition. Adapted from ref. [1] with kind permission from Springer Verlag

within planes, both being perpendicular to each other (see textbooks on optics, e.g. refs. [19, 20]). The *electric* field of the wave can be described by the function

$$E = E^0 \exp[i(\boldsymbol{k} \cdot \boldsymbol{r} - \omega t)], \tag{1}$$

which is harmonic both in time, t, and position, r. The frequency and the amplitude of the oscillating field are represented by ω and E^0, respectively. Due to the interaction with the magnetisation of the specimen, the plane-polarised wave is transformed into rotated and/or elliptically polarised light as illustrated in Fig. 2a. Rotation and ellipticity are then converted into a domain contrast by properly setting an analyser and compensator as explained below in section "Domain Contrast". Like other optical properties of solids (dispersion, absorption, double refraction, optical activity, etc.), also the magneto-optical interactions can be treated within the frame of classical electromagnetic theory. The basic ideas of this phenomenological theory and their specific relevance for the magneto-optical effects are briefly outlined in the following subsections. For a comprehensive description, we refer to ref. [1].

Electromagnetic Basics

The starting point of electromagnetic theory is the Maxwell equations (div $D = \rho$; div $B = 0$; rot $E = -\dot{B}$; rot $H = j + \dot{D}$) that connect the various field vectors of electromagnetism. The magnetic and electric fields are represented by H and E, respectively, B is the magnetic induction, D is the dielectric displacement field, and j and ρ are the electric current and charge density, respectively. The field equations are linked by material-specific relations, most importantly by

$$D = \epsilon_0 \epsilon_r E \tag{2}$$

as well as by $B = \mu_0 \mu_r H$ and $j = \sigma E$. In those equations, which are given here for *isotropic* media, ϵ_0 and μ_0 are the dielectric constant and permeability of free space and ϵ_r, μ_r, and σ are the (relative) dielectric permittivity, magnetic permeability, and electric conductivity, respectively, of the material. While in free space the speed of the electromagnetic wave is given by $c_0 = 1/\sqrt{\epsilon_0\mu_0}$, it is defined as $v = 1/\sqrt{\epsilon_0\epsilon_r\mu_0\mu_r}$ in matter. The index of refraction of the material, n, relates the two velocities, i.e. $n = c_0/v = \sqrt{\epsilon_r\mu_r}$. The refractive index thus describes the fact that the light velocity in matter is different from that in free space and that it differs in different materials.

Electromagnetic theory applies to both transparent (dielectric) and absorbing (conductive) media. Electrical conductivity, allowing for light-induced electrical currents and thus being responsible for absorption in metallic materials, is considered by introducing a complex refractive index ($n = n' + in''$). The real part is the true refractive index, while the imaginary part, called *extinction coefficient*, accounts for absorption. This becomes obvious by entering n into the plane wave equation (1). With $|k| = n k_0 = \omega/v$ and $k_0 = \omega/c_0$ (with k_0 being the wave propagation number in free space), the plane wave (1) is expressed by

$$E = E^0 e^{i[\frac{\omega}{c_0}(n'+in'')z-\omega t]} = E^0 e^{-\frac{\omega}{c_0}n''z} e^{i[\omega\frac{n'}{c_0}z-\omega t]} . \tag{3}$$

According to the term $\exp(i[\omega\frac{n'}{c_0}z - \omega t])$, the wave advances in the z-direction with the speed c_0/n'. By propagating through the conductor, its amplitude, $E^0 \exp(-\omega n''z/c_0)$, is exponentially damped, i.e. the energy of the wave is more and more absorbed by the material. The power of the electromagnetic radiation (irradiance), which is proportional to the square of the amplitude, is then given by $I(r) = I_0 e^{-\alpha z}$. Here I_0 is the irradiance at the surface of the metal and $\alpha = 2\omega n''/c_0 = 4\pi n''/\lambda$ is the *absorption coefficient*. The energy density thus falls to $e^{-1} \approx 1/3$ of its value after the wave has propagated a distance of $1/\alpha$. This distance is known as the *skin depth*. For typical magnetic materials like iron, the skin depth is about 15 nm (with $n'' = 3$ and at a wavelength of 560 nm).

Four facts need to be taken into account: (1) If the refractive index is complex, also the wave vector and the field terms are complex due to the connecting relations. (2) In the case of optically *anisotropic* materials, the factor ϵ_r in Eq. (2) is a tensor!

This means that the direction of the D-vector, which represents the electric field vector within the material, may deviate from the direction of the incoming E-vector. (3) The components of the ϵ-tensor are frequency dependent; the wave propagation is thus dispersive. (4) The typical Larmor precession frequency of magnetic moments (<100 GHz) is much lower than the 500 THz frequency of visible light. Therefore at optical frequencies the magnetic moments are too 'slow' to follow the alternating magnetic field of the electromagnetic wave. All material-specific optical properties, including the magneto-optical effects, are therefore dominated by 'gyroelectric' terms and incorporated in the dielectric ϵ-tensor, while the 'gyromagnetic' terms and the permeability tensor μ do not play a role.

By properly combining the material and Maxwell equations, a wave equation for the E-field is obtained. With the plane wave ansatz (1) and by assuming visible light (i.e. $\mu = 1$), the wave equation can be written as a system of three linear equations for the three vector components E_i of the E-field amplitude [1]:

$$k_i(\mathbf{k} \cdot \mathbf{E}) - k^2 E_i + \sum_{j=1}^{3} k_0^2 \epsilon_{ij} E_j = 0 \,. \tag{4}$$

The non-trivial solutions for E to this equation system, the so-called *eigenmodes*, describe the propagable light waves in the medium. It turns out that not one but two waves of different velocity and possibly different absorption may in general propagate through the material. These can be plane- or circularly polarised waves. For a visualisation of circularly polarised light, we refer to Fig. 2a—the elliptical wave becomes circular if the two orthogonal field amplitudes are equal and shifted in phase by 90°. Also note that a plane-polarised wave may be seen as being composed of two circularly polarised waves of equal amplitude and opposite rotation sense as indicated in Fig. 2c. According to Eq. (4), the knowledge of the components ϵ_{ij} of the ϵ-tensor is required to obtain the eigenmodes. We will therefore start the discussion of the conventional magneto-optical effects by introducing the dielectric tensors in the following subsections. Although the Kerr effect applies to metallic or otherwise light-absorbing magnetic material, whereas the Faraday effect occurs in optically transparent media, both effects are treated together as both are predominantly *rotational* effects that follow the same phenomenology.

Kerr and Faraday Effect

For the Kerr and Faraday effect (and for materials with cubic crystal symmetry), the magneto-optic permittivity tensor is

$$\epsilon_{\mathrm{KF}} = \epsilon_{\mathrm{iso}} \begin{pmatrix} 1 & -iQ_{\mathrm{V}}m_z & iQ_{\mathrm{V}}m_y \\ iQ_{\mathrm{V}}m_z & 1 & -iQ_{\mathrm{V}}m_x \\ -iQ_{\mathrm{V}}m_y & iQ_{\mathrm{V}}m_x & 1 \end{pmatrix} \,, \tag{5}$$

with ϵ_{iso} representing an isotropic dielectric constant as defined below. If the off-diagonal elements of the tensor are zero, Eq. (2) would explicitly be written as $(D_1, D_2, D_3) = \epsilon_0 \epsilon_{iso}(E_1, E_2, E_3)$, i.e. the displacement vector would be along the same direction as the E-vector of the incident light. The light would then interact with the matter without rotation of its polarisation plane. Rotation (and ellipticity) of the out-coming light requires non-zero off-diagonal elements. They contain the direction cosines of the magnetisation vector $m = M/M_s$ (with $|m|^2 = 1$) along the three spacial directions defined in Fig. 2a and the Voigt constant Q_V, which is a frequency-dependent, complex material parameter. It is roughly proportional to the saturation magnetisation M_s and describes the strength of the effect. Usually the real part of Q_V is dominating with absolute values of the order of 10^{-2}. According to the tensor, the Kerr and Faraday effects are obviously linear in the components of the magnetisation vector as already demonstrated in Fig. 1.

As we will see below, any Kerr and Faraday rotation requires that the projection of the k-vector on the magnetisation vector m is non-zero. By solving the wave equation (4) under the condition of light propagation strictly *along* the magnetisation vector, it turns out [1] that the eigenmodes are two circularly polarised waves with opposite rotation sense, given by $E_y^0 = \pm i E_x^0$. Plane-polarised light entering the material is resolved into these two modes. Each mode 'feels' its own (complex) index of refraction n_+ and n_-, which depend on the Voigt parameter according to

$$n_\pm = \sqrt{\epsilon_{iso}}(1 \pm Q_V/2). \tag{6}$$

The factor $\sqrt{\epsilon_{iso}}$ represents the mentioned isotropic dielectric constant that is related to the magnetisation-independent, average refractive index $\bar{n} = \frac{1}{2}(n_+ + n_-) = \sqrt{\epsilon_{iso}}$. If Q_V is zero, n_+ and n_- would be identical, and any polarisation state could propagate, feeling generally the isotropic refractive index. As soon as Q_V appears, only the right and left circular normal modes can propagate. Since n_+ and n_- may differ in their real and imaginary parts, the two partial waves will advance with different velocities and absorption, respectively. Different velocities result in a phase shift and consequently in a rotation of the polarisation plane as illustrated in Fig. 2c. For this reason the Kerr and Faraday rotations may be classified as *circular birefringence* effects (i.e. a birefringence of circularly polarised light—note that, in general, a material that displays two different indices of refraction is said to be birefringent). The difference in absorption causes different amplitudes of the two circular modes, leading to ellipticity (compare Fig. 2b) and thus *circular dichroism*—again, a material that displays different absorption of partial waves is said to be dichroic. Both rotation and ellipticity are considered by introducing complex angles for the Faraday and Kerr effects as will be elaborated below. Note that the Faraday and Kerr effects are primarily *rotation* effects with some ellipticity superimposed.

The *microscopic origin* of the Faraday and Kerr effects can be traced back to light-induced electron transitions in the valence-band energy regime, e.g. from d-states to unoccupied p-states in case of a 3d ferromagnet (different to X-ray magnetic dichroism where X-ray light is absorbed by the excitation of core electrons, see chapter "X-Ray Magnetic Circular Dichroism and X-Ray Microscopy"). Prerequisite is the simultaneous presence of band exchange splitting, induced by the spontaneous magnetisation of magnetic materials, and spin-orbit coupling that also leads to band splitting. By considering now that incident, plane-polarised light may actually be seen as being composed of two circularly polarised waves of opposite helicity (compare Fig. 2c), it turns out that those two circular modes experience different absorption spectra according to the selection rules for electric dipole transitions. The complex dielectric constant can be related to the electronic transitions, thus connecting the microscopic picture with the previously presented macroscopic description. With $(n'_\pm + in''_\pm)^2 = \epsilon'_\pm + i\epsilon''_\pm$ and Eq. (6), the Voigt constant Q_V can then be expressed in terms of n_+ and n_- by $Q_V = 2(n_+ - n_-)/(n_+ + n_-)$. For details on this microscopic model, we refer to the literature [1, 21].

Faraday Effect

Both circular birefringence and dichroism and their superposition are illustrated in Fig. 3 for the case of perpendicular incidence on a perpendicularly magnetised medium in transmission, which is called *polar Faraday effect*. Although the Faraday rotation is reminiscent to the circular birefringence of optically active media [19], there is an important distinction: if the light passes the material again in reversed direction, the rotation does not cancel but is rather doubled in case of the Faraday effect. The reason for this irreversibility is that the Faraday rotation is tied to the *direction* of the magnetisation rather than to an optical axis. A reversal of the magnetisation relative to the propagation direction leads to an exchange of the preferred left and right circular modes along the *m*-axis. Oppositely magnetised

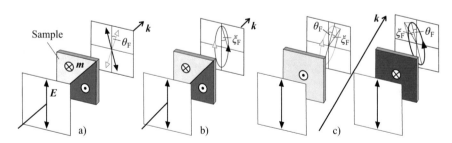

Fig. 3 Illustration of magnetisation-induced circular birefringence (**a**), circular dichroism (**b**), and superimposition of the two effects (**c**) in polar Faraday geometry with perpendicular incidence of plane-polarised light. The trace of light polarisation in the plane perpendicular to the propagation vector is shown. The two out-of-plane magnetised domains have different influences on the polarisation state as indicated by arrows with the same color as the domains. In (**c**) the Faraday rotation refers to the rotation of the major axis of the ellipse. Similar illustrations would apply to the polar Kerr geometry in reflection. Adapted from ref. [1] with kind permission from Springer Verlag

domains will thus cause a reversal of the Faraday rotation and of the handiness of ellipticity. Following the above arguments, the complex Faraday angle $\theta_F^c = \theta_F + i\xi_F$ can be calculated. For a polar geometry, the Faraday rotation (θ_F) and ellipticity (ξ_F) are given by [1]

$$
\begin{aligned}
\theta_F &= -l(\pi/\lambda_0)\,(n'_+ - n'_-) = -l(\pi/\lambda_0)\,\mathrm{Re}(\bar{n}\,Q_V) \\
\xi_F &= -l(\pi/\lambda_0)\,(n''_+ - n''_-) = -l(\pi/\lambda_0)\,\mathrm{Im}(\bar{n}\,Q_V),
\end{aligned}
\tag{7}
$$

with l being the covered distance in the material. A Faraday effect with the same phenomenology is also found if an in-plane magnetised material is illuminated at oblique light incidence with the propagation vector having a component along the magnetisation axis. The strength of this *longitudinal Faraday effect* is reduced compared to the polar effect as only a component of the k-vector is responsible for circular birefringence and dichroism. For an arbitrary orientation of magnetisation, it might be possible that also linearly polarised eigenmodes are excited (like for the Voigt effect, see section "Voigt Effect"), which get intermixed with the circular modes making an intuitive understanding of the resulting phenomenology complicated [22, 23].

A Faraday rotation of plane-polarised light does furthermore occur in isotropic dielectric media like glass in the presence of a magnetic field, given by $\theta_F = V B_z l$. The amount of rotation is proportional to the component of magnetic induction B_z in the propagation direction, to the length l of the traversed medium, and to the material parameter V, called the *Verdet constant*. In magneto-optical microscopy and magnetometry, the field-induced Faraday effect plays a parasitic role: Optical elements in the microscope, like the objective lens, can cause a Faraday rotation in the presence of magnetic fields that is superimposed to any light rotation being caused by the magnetism of the specimen. In quantitative Kerr microscopy, where saturated sample states of well-defined directions are created by external magnetic fields for calibration purposes, as well as in magnetometry, this disturbing Faraday effect has to be taken into account for a proper interpretation of results (see sections "MOKE Magnetometry" and "Quantitative Kerr Microscopy").

Kerr Effect

The magneto-optical *Kerr effects* are much weaker than the Faraday effects as the light only interacts with the magnetisation within its penetration depth. Nevertheless, also the Kerr effects can be interpreted in terms of circular birefringence and dichroism (possibly being superimposed by linear effects in case of oblique incidence), meaning that incident plane-polarised light is reflected as rotated and maybe elliptically polarised light. In analogy to the Faraday effect and again assuming *polar geometry*, the components of the complex Kerr angle $\theta_K^c = \theta_K + i\xi_K$ are calculated as [1]

$$\theta_K = \mathrm{Im}\Big(\frac{\bar{n}\, Q_V}{1 - \bar{n}^2}\Big) \quad \text{and} \quad \xi_K = -\mathrm{Re}\Big(\frac{\bar{n}\, Q_V}{1 - \bar{n}^2}\Big). \tag{8}$$

Interestingly, the role of the real and imaginary parts of the off-diagonal elements in the magneto-optical tensor is now opposite to the case of transmission [compare Eq. (7)]. The Kerr rotation thus corresponds to the Faraday ellipticity and the Kerr ellipticity to the Faraday rotation. Recalling that for non-absorbing materials the imaginary part of $\bar{n}\, Q_V$ is zero, it is obvious that for a transparent medium, no rotation can be expected on reflection—the reflected light will just be elliptically polarised with the major axis along the plane of polarisation of the incident light. Rotation in reflection rather requires a non-vanishing imaginary part and thus absorption. If transparent materials are supposed to be studied in reflection microscopes, it is therefore rational to place them on a metallic mirror, thus using effectively the Faraday effect.

Instead of using a circular basis for the description of elliptical polarisation (like in Fig. 2b, c), a cartesian decomposition as depicted in Fig. 2a is more suitable for the Kerr effect. To include the case of oblique incidence, it is useful to choose the (s, p, k) coordinate system for the reflection path. Here the direction of the light propagation is given by the wave vector axis k, the p-axis is in the plane of incidence (defined by the sample normal and the wave vector of obliquely incident light), and the s-direction is perpendicular to the plane of incidence. The two standard geometries of reflection and transmission of pure s- and p-polarised light are summarised in Fig. 4. Here n_0 is the refractive index of the environment (air or an immersion medium), $n_1 = n_1' + i n_1''$ is the complex index of the magnetic material, ϑ_0 is the angle of incidence (measured from the surface normal), and ϑ_1 is the (complex) angle of the light beam in the magnetic medium that is related to ϑ_0 by Snell's law:

$$n_0 \sin \vartheta_0 = n_1 \sin \vartheta_1. \tag{9}$$

Within that coordinate system, the change of the polarisation state on reflection is described by the following Jones-type [19] matrix equation:

$$\begin{pmatrix} E_p^{\mathrm{refl}} \\ E_s^{\mathrm{refl}} \end{pmatrix} = R \begin{pmatrix} E_p^{\mathrm{in}} \\ E_s^{\mathrm{in}} \end{pmatrix} \quad \text{with} \quad R = \begin{pmatrix} r_{pp} & r_{ps} \\ r_{sp} & r_{ss} \end{pmatrix}. \tag{10}$$

It connects the Jones vector of the incident light with that of the reflected light, both containing the p- and s-components of the electrical field amplitudes. The reflection matrix R includes the reflection coefficients r_{ij} that are defined by the corresponding ratios of the incident field amplitudes and the reflected amplitudes.

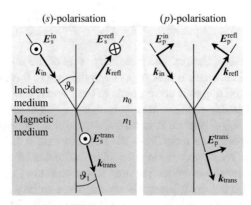

Fig. 4 Plane waves with (s)- and (p)-polarisation, represented by their incident, reflected and transmitted field amplitudes. On the left, the **E**-fields are normal to the plane of incidence; on the right they are parallel. The component of **E** normal to the plane of incidence undergoes a 180° phase shift upon reflection when the incident medium has a lower refractive index than the magnetic medium

The first index, i, labels the reflected component; the second index, j, is for the corresponding components of the incident light. Solving the Maxwell equations for the Kerr tensor (5) leads to the following reflection coefficients[1] [24, 25]:

$$r_{ss} = \frac{n_0 \cos \vartheta_0 - n_1 \cos \vartheta_1}{n_0 \cos \vartheta_0 + n_1 \cos \vartheta_1} \,,$$

$$r_{pp} = \frac{n_1 \cos \vartheta_0 - n_0 \cos \vartheta_1}{n_1 \cos \vartheta_0 + n_0 \cos \vartheta_1} - \frac{i\, 2 n_0 n_1 \cos \vartheta_0 \sin \vartheta_1 m_x Q_V}{(n_1 \cos \vartheta_0 + n_0 \cos \vartheta_1)^2} \,,$$

$$r_{sp} = \frac{i\, n_0 n_1 \cos \vartheta_0 (m_z \cos \vartheta_1 + m_y \sin \vartheta_1) Q_V}{(n_1 \cos \vartheta_0 + n_0 \cos \vartheta_1)(n_0 \cos \vartheta_0 + n_1 \cos \vartheta_1) \cos \vartheta_1} \,,$$

$$r_{ps} = \frac{i\, n_0 n_1 \cos \vartheta_0 (m_z \cos \vartheta_1 - m_y \sin \vartheta_1) Q_V}{(n_1 \cos \vartheta_0 + n_0 \cos \vartheta_1)(n_0 \cos \vartheta_0 + n_1 \cos \vartheta_1) \cos \vartheta_1} \,. \tag{11}$$

Note that the coefficient r_{pp} contains two components: the first fraction corresponds to the regular Fresnel coefficient for p-polarised light [19, 20], while the second fraction is magnetisation-dependent. Below we will call them r_{pp}^0 and r_{pp}^{mag}, respectively.

As an example let us examine the geometry of Fig. 2a again. Here the incident plane wave is s-polarised with the field amplitude $\mathbf{E} = (0, E_s^{in})$, which is then reflected as a rotated elliptical wave due to the magneto-optical interaction. The

---

[1]Please be aware of misprints in ref. [1], regarding the coefficients r_{pp} and r_{ps} in Eq. (2.92) and the coefficient r_{yy} in Eq. (2.85). Furthermore, in Eq. (2.97a) the minus sign in the third vector component has to go to the second component.

rotated ellipse may be seen as being composed of two phase-shifted, orthogonally polarised waves of different amplitudes. The high-amplitude partial wave is s-polarised along the same plane as the incident light, thus being described by the coefficient r_{ss} in Eq. (11). This coefficient corresponds to the regular Fresnel reflection coefficient [19, 20], not containing a magnetic, m_i-dependent term. The high-amplitude wave may therefore be interpreted as *regular component* R_N of the reflected light. The low-amplitude partial wave is p-polarised, being described by the coefficient r_{ps} in Eq. (11). It is purely caused by the magnetisation and may therefore be called *Kerr component* R_K. With Eq. (10) the reflected wave is now explicitly given by

$$\begin{pmatrix} E_p^{\text{refl}} \\ E_s^{\text{refl}} \end{pmatrix} = \begin{pmatrix} r_{pp} \ r_{ps} \\ r_{sp} \ r_{ss} \end{pmatrix} \begin{pmatrix} 0 \\ E_s^{\text{in}} \end{pmatrix} = \begin{pmatrix} r_{ps} E_s^{\text{in}} \\ r_{ss} E_s^{\text{in}} \end{pmatrix} = \begin{pmatrix} K_{ps} E_s^{\text{in}} \\ N_{ss} E_s^{\text{in}} \end{pmatrix} = \begin{pmatrix} R_K \\ R_N \end{pmatrix}. \quad (12)$$

The reflection coefficients, r_{ps} and r_{ss}, are labeled as K_{ps} and N_{ss} to emphasise their relevance as coefficients for the Kerr and regular amplitudes, respectively. The Kerr amplitude may actually be taken as *figure of merit* for the Kerr effects (see section "Domain Contrast").

The complex Kerr angle for s-polarised illumination is then defined by the ratio of the off-diagonal to the diagonal element of the reflection matrix:

$$\theta_K^s \equiv r_{ps}/r_{ss} \quad (13)$$

(for p-polarised incidence it would be $\theta_K^p \equiv r_{sp}/r_{pp}$). In our example we assume a magnetisation vector along the plane of incidence and along the sample's y-axis, i.e. $m_y = 1$ and $m_x = m_z = 0$. In this *longitudinal* Kerr configuration, the Kerr angle gets

$$(\theta_K^s)^{\text{long}} = (\frac{r_{ps}}{r_{ss}})^{\text{long}} = \frac{-i\, n_0 n_1 \cos \vartheta_0 \sin \vartheta_1 Q_V}{(n_1 \cos \vartheta_0 + n_0 \cos \vartheta_1) \cos \vartheta_1 (n_0 \cos \vartheta_0 - n_1 \cos \vartheta_1)}. \quad (14)$$

By making use of the Fresnel transmission coefficients (compare Fig. 4):

$$t_{pp}^{01} = \left(\frac{E_p^{\text{trans}}}{E_p^{\text{in}}} \right) = \frac{2n_0 \cos \vartheta_0}{n_0 \cos \vartheta_1 + n_1 \cos \vartheta_0}, \quad t_{ss}^{01} = \left(\frac{E_s^{\text{trans}}}{E_s^{\text{in}}} \right) = \frac{2n_0 \cos \vartheta_0}{n_0 \cos \vartheta_0 + n_1 \cos \vartheta_1} \quad (15)$$

and Snell's law (9), Eq. (14) can be written as

$$(\theta_K^s)^{\text{long}} = \frac{-i Q_V t_{pp}^{01} t_{ss}^{01} \sin \vartheta_0}{4 \cos \vartheta_0 \cos \vartheta_1} \frac{1}{r_{ss}^{01}}. \quad (16)$$

Recalling that the Kerr angle is the ratio between the Kerr coefficient and the coefficient of the normally reflected amplitude (here $N_{ss} = r_{ss}^{01}$), the Kerr coefficient K_{ps}^{long} is finally given by the first fraction in Eq. (16):

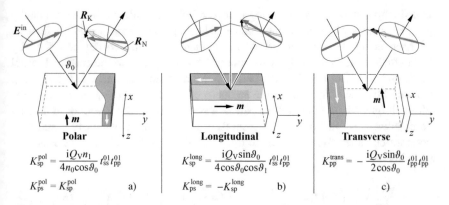

Fig. 5 Schematics of the three basic Kerr geometries. The illuminating light amplitude E^{in} results in the superposition of regular $\boldsymbol{R}_{\text{N}}$ and magneto-optical field amplitudes $\boldsymbol{R}_{\text{K}}$ on the reflection paths. Phase shifts, i.e. ellipticity, are ignored here for simplicity. Given are the Kerr coefficients in each case (see text for definition). After ref. [18], with kind permission of Springer Verlag

$$K_{\text{ps}}^{\text{long}} = \frac{-\mathrm{i}Q_{\text{V}} t_{\text{pp}}^{01} t_{\text{ss}}^{01} \sin\vartheta_0}{4\cos\vartheta_0 \cos\vartheta_1} \ . \tag{17}$$

The interpretation of this coefficient is straightforward: As indicated by the lower indices in $K_{\text{ps}}^{\text{long}}$, the incident light is s-polarised, resulting in a p-polarised Kerr amplitude. A portion of the incident light, given by t_{ss}^{01} [the upper index stands for transmission from medium (0) to (1)], penetrates into the magnetic material where it generates the Kerr amplitude by interaction with the magnetisation. Its size depends on the material parameter Q_{V} and the angle of incidence ϑ_0. The p-polarised Kerr amplitude has to leave the material to be detected. This would actually be described by the transmission coefficient t_{pp}^{10}. In Eq. (17), this coefficient is hidden in t_{pp}^{01}, which is related to t_{pp}^{10} by $t_{\text{pp}}^{10}/t_{\text{pp}}^{01} = n_1 \cos\vartheta_1 / n_0 \cos\vartheta_0$ according to the Fresnel transmission formulae (15).

For other polarisation and magnetisation directions, the Kerr coefficients can be derived in a similar way. In Fig. 5 they are explicitly listed for the three fundamental Kerr geometries, the polar, longitudinal, and transverse Kerr effects:

- In case of the *polar* effect (Fig. 5a), the magnetisation is aligned perpendicular to the sample surface. Oblique incidence will lead to the same Kerr amplitude for s- and p-polarised light. By entering the Fresnel transmission coefficients (15) into the formulae given in the figure, it becomes evident that $K_{\text{ps, sp}}^{\text{pol}} \sim \cos\vartheta_0$. Therefore the polar effect is strongest at perpendicular incidence. For $\vartheta_0 = 0$ a plane of incidence cannot be defined, and any polarisation direction of the incident beam will lead to the same signal. A reversal of the perpendicular magnetisation causes a reversal of the Kerr amplitude (and thus the Kerr rotation) as the signs of the Fresnel coefficients r_{ps} and r_{sp} are reversed for $m_{\pm z}$ [see Eq. (11)].

- The *longitudinal* effect (Fig. 5b) requires oblique incidence of light ($\sin \vartheta_0 \neq 0$) and a magnetisation that is oriented parallel to the sample surface and along the plane of incidence. The Kerr amplitude (and rotation) has opposite signs for *s*- and *p*-polarised light as m_y enters the coefficients r_{ps} and r_{sp} with opposite signs. And again, magnetisation reversal will cause opposite Kerr rotation due to a reversal of the signs of the coefficients r_{ps} and r_{sp} on reversal of m_y.

- For *transverse* orientation (Fig. 5c) the in-plane magnetisation is perpendicular to the plane of incidence, i.e. $m_x = \pm 1$ and $m_y = m_z = 0$. Then the off-diagonal reflection coefficients in Eq. (11) are zero so that a Kerr rotation cannot be expected. For *p*-polarised illumination, however, the coefficient r_{pp} becomes active which contains a magnetisation-dependent component r_{pp}^{mag}. This leads to the generation of a Kerr amplitude K_{pp} that is derived by the ratio of magnetic to non-magnetic component, i.e. by r_{pp}^{mag}/r_{pp}^0. As shown in Fig. 5c, this Kerr amplitude is again proportional to $\sin \vartheta_0$, but its polarisation direction is the same as that of the regularly reflected beam. The transverse Kerr effect therefore causes an *amplitude variation* of the light rather than a rotation. The intensity of the reflected wave changes sign on magnetisation reversal even if one works with unpolarised light and without an analyser on the detection side. The amplitude variation can be used for magneto-optical magnetometry, having the advantage that the signal is not sensitive to polar magnetisation components. To generate a measurable rotation also in transverse geometry, which is required for Kerr microscopy, the light has to be polarised at 45° to the plane of incidence. Then its perpendicular component is not affected by the magnetisation, while the parallel component receives the amplitude modulation [1]. Together this leads to a detectable rotation if a compensator is used—note that incident, plane-polarised light that is not strictly *s*- or *p*-polarised is reflected elliptically in any case [19]. In this configuration a superimposed polar sensitivity cannot be avoided, though.

The polar and in-plane Kerr effects differ in their strengths. To see this, let us write the longitudinal Kerr coefficient (17) as $K_{ps}^{long} = (i Q_V n_1 t_{ss}^{01} t_{pp}^{01} \tan \vartheta_1)/(4 n_0 \cos \vartheta_0)$ by using Snell's law (9). When comparing this coefficient with that of the polar effect (see Fig. 5), it is obvious that $|K_{sp}^{long}| = |K_{sp}^{pol}| \tan \vartheta_1$, i.e. the longitudinal and polar effects differ by the factor $\tan \vartheta_1$. As the refractive indices of metals are relatively large at optical frequencies, ϑ_1 and with it $\tan \vartheta_1$ are typically of the order of 0.1. The polar Kerr effect is consequently by an order of magnitude stronger than the longitudinal Kerr effect, making domain studies in perpendicular magnetised specimens much easier than on in-plane magnetised materials. For the latter, in fact, the size of the Kerr amplitude and with it the domain contrast scales with the angle of incidence as shown in Fig. 5. The incidence angle, on the other hand, depends on the numerical aperture of the objective lens. This aspect will be addressed in Sect. 4.

Along the previous discussion, we have shown that the Kerr and Faraday effects depend on the direction of magnetisation, light incidence and polarisation and can be rigorously deduced from the dielectric magneto-optical tensor, Maxwell' equations and the proper boundary conditions. The symmetry of the solutions, however, can also be derived on the base of a classical *Lorentz concept*. This becomes immediately evident by entering the tensor (5) into Eq. (2) and by rewriting the dielectric law as:

$$D = \epsilon_0 \epsilon_{KF} E = \varepsilon_0 n^2 [E + i Q_V (m \times E)]. \tag{18}$$

The E-vector of the incident light obviously interacts with the magnetisation vector m in a cross-product fashion in the same symmetry as a Lorentz force that acts on light-agitated electrons, thus revealing the gyroelectric nature of the Kerr and Faraday effects. The Lorentz concept is intuitive and helps in the understanding of the Kerr effect geometries. According to this concept, the E-vector of the light wave excites electrons in the specimen to an oscillatory motion parallel to its plane of polarisation. That motion acts as a source for the regularly emitted or transmitted light, which is polarised in the same plane as the incident light. At the same time, the Lorentz force induces a small component of vibrational motion perpendicular to the primary motion and to the direction of magnetisation. Because of Huygens' principle, the secondary motion creates secondary amplitudes, the Kerr and Faraday amplitudes, which point along the same direction as the vector $(m \times E)$. A magnetisation-dependent contribution to the D-vector is thus only generated if $m \times E$ is non-zero. With the Lorentz concept, one can also easily explain why a transverse Faraday effect does not exist: in transmission the beam direction does not change. Therefore the cross-product is either zero or points along the propagation direction. For a detailed treatment of the Lorentz concept, we refer to refs. [1] and [18].

The phenomenology of the magneto-optical Kerr effect (excluding the transverse effect) can finally be summarised by a simple rule:

The polar and longitudinal Kerr effects cause a (complex) rotation of light, which is proportional to the magnetisation component parallel to the reflected light beam and which depends linearly on the magnetisation.

According to the rule, the Kerr rotation is inverted when the magnetisation direction is inverted. Oppositely magnetised domains thus rotate plane-polarised light in opposite directions as illustrated in Fig. 6a for the longitudinal Kerr effect. The three Kerr effects can be combined in a quantitative way for general polariser and analyser settings as defined in Fig. 6b. Here the plane of the incoming light is specified by the polariser with setting α_{pol}, measured from the plane of incidence. On reflection, the light is rotated by the superposition of regular amplitudes R_N^p and R_N^s parallel and perpendicular to the plane of incidence, respectively, and the Kerr amplitudes R_K^{pol}, R_K^{lon} and R_K^{tra} for the polar, the longitudinal and the transverse cases [compare Eq. (12)], depending on the magnetisation components m_{pol}, m_{lon} and m_{tra}. Finally the light passes through the analyser (with the setting α_{an} measured from the axis perpendicular to the plane of incidence), leading to the total signal amplitude [12]:

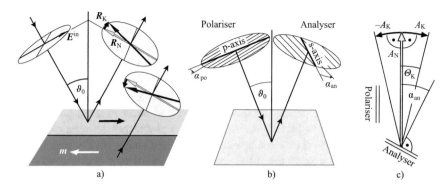

Fig. 6 (**a**) Longitudinal Kerr effect for p-polarised incidence, showing opposite Kerr amplitudes R_K for antiparallel domains. (**b**) Definition of polariser and analyser angles in a Kerr setup. (**c**) The interference of the (generalised) normally reflected component A_N and the Kerr amplitude A_K results in a magnetisation-dependent light rotation by the (small) angle Θ_K, which leads to the domain contrast by blocking with the analyser. The analyser should actually be set at a larger angle $\alpha_{an} > \Theta_K$ for optimum domain visibility. Possible phase shifts between A_N and A_K (Kerr ellipticity) are assumed to be eliminated by a compensator

$$A_{tot} = A_N \pm A_K \,, \text{ with}$$

$$A_N = R_N^s \sin \alpha_{po} \cos \alpha_{an} - R_N^p \cos \alpha_{po} \sin \alpha_{an} \,,$$

$$A_K = R_K^{pol} \cos (\alpha_{an} - \alpha_{po}) \, m_{pol} + R_K^{lon} \cos (\alpha_{an} + \alpha_{po}) \, m_{lon}$$

$$- R_K^{tra} \cos \alpha_{po} \sin \alpha_{an} \, m_{tra} \,. \tag{19}$$

This equation can also be used for a general magnetisation direction different to the conventional polar, longitudinal and transverse geometries. In section "Domain Contrast" we will show how the effective amplitude A_{tot} can be converted in a domain contrast by referring to Fig. 6c.

Voigt Effect

The Faraday and Kerr effects, presented in section "Kerr and Faraday Effect", are effects to first order in the components m_i of the magnetisation vector and in the magneto-optical parameter Q_V. This becomes obvious from the dielectric tensor (5) and the reflection coefficients (11) representing these effects. In contrast, the Voigt effect is a second-order effect that is even in the magnetisation. Often it is considered an independent effect with its own material coefficients B_1 and B_2, described by the following tensor:

$$\epsilon_V = \epsilon_{iso} \begin{pmatrix} B_1 m_x^2 & B_2 m_x m_y & B_2 m_x m_z \\ B_2 m_x m_y & B_1 m_y^2 & B_2 m_y m_z \\ B_2 m_x m_z & B_2 m_y m_z & B_1 m_z^2 \end{pmatrix} \,. \tag{20}$$

This *intrinsic Voigt effect*, however, is not the only conceivable second-order magneto-optical effect. As shown below, a quadratic effect with the same symmetry can also be caused by elementary gyroelectric interaction, derived from tensor (5) and determined by Q_V. In practice it is not easy to distinguish the two effects.

Independent of its origin, the prerequisite for the Voigt effect is a component of the magnetisation vector perpendicular to the propagation vector k. In Fig. 7 this is illustrated for perpendicular incidence on an in-plane magnetised medium, a geometry for which a Faraday or Kerr rotation would not be possible at all. While the latter are based on circular birefringence and dichroism, it turns out that for $k \perp m$ the solution of the wave equation (4) yields two *linearly polarised* eigenmodes with vibrational planes being aligned along and perpendicular to m. Incident light, plane-polarised along one of these two directions will thus not alter its polarisation in the magnetic medium. If the polarisation plane is at an angle to m, however, the polarisation state will be changed with the strongest effect occurring at an angle of 45°. This is the geometry depicted in Fig. 7. Both linear eigenmodes are experiencing different refractive indices $n_\parallel$ and $n_\perp$, given by $n_\parallel^2 = \epsilon_{iso} + B_1$ and $n_\perp^2 = \epsilon_{iso}(1 - Q_V^2)$ [1] (here both tensors, ϵ_{KF} and ϵ_V, have been considered in the solution of the wave equation because both can lead to quadratic effects as mentioned). Due to the different indices, the two modes will proceed in the medium with different velocities and with different attenuations (in case of a complex refractive index). The light thus experiences a *magnetic linear birefringence* proportional to $\mathrm{Re}(n_\parallel - n_\perp)$ and a *magnetic linear dichroism* proportional to $\mathrm{Im}(n_\parallel - n_\perp)$. Note that the word 'linear' refers to the polarisation mode of the light and not to the order of the effect. In case of linear birefringence, the two partial waves are retarded relative to each other so that the outgoing light is elliptically polarised with a handedness that depends on the relative orientation of the polarisation plane and magnetisation *axis* (Fig. 7a). Oppositely magnetised

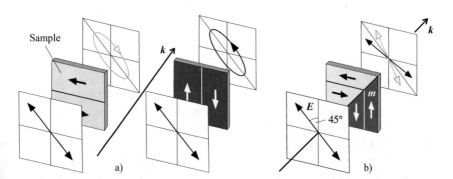

Fig. 7 Magnetically induced linear birefringence (**a**) and linear dichroism (**b**) effects, illustrated for perpendicular incidence on an in-plane magnetised material with 90° domains. The effects are strongest if the incident light is polarised at 45° relative to the domain axes. While linear birefringence leads to elliptical light of opposite rotation sense for the two domain axes, linear dichroism causes a magnetisation-dependent rotation of the polarisation plane

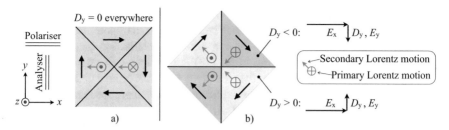

Fig. 8 Illustration of the secondary quadratic effect that has the same symmetry as the Voigt effect, but which is derived from gyroelectric interaction like the Kerr effect. (**a**) Horizontally polarised light creates an out-of-plane primary Lorentz motion in the vertically magnetised domains, which then generates a secondary motion that is polarised antiparallel to E_x. (**b**) A 45° rotation of the domain magnetisation relative to E_x leads to antiparallel D_y-components for domains magnetised along different axes

domains, following the same axis, can therefore not be distinguished in their birefringence effect. Linear dichroism results in a rotation of the emerging light (Fig. 7b) due to the different amplitudes of the partial wave. The phenomenology is thus opposite to that of magnetic *circular* birefringence and dichroism, where birefringence causes a rotation and dichroism leads to ellipticity (compare Fig. 3).

As mentioned, a second-order effect with the same symmetry as the intrinsic Voigt effect can as well be derived from gyroelectric interaction, i.e. by just considering the tensor ϵ_{KF} in Eq. (5). This effect is illustrated in Fig. 8 by using the Lorentz concept. Let us assume that x-polarised light is propagating in the z-direction, thus generating a zero-order displacement $D_x = \epsilon_{iso} E_x$ in the material. In the geometry of Fig. 8a, the magnetisation vectors are along and transverse to the polarisation axis. According to the gyroelectric tensor ϵ_{KF}, the D_x-field will not cause any Lorentz motion in the x-axis domains as here $\boldsymbol{m} \times \boldsymbol{E}$ is zero. In the two y-axis domains, however, Lorentz motions along the $\pm z$-directions are induced, leading to first-order displacements $D_z = -i\epsilon_{iso} Q_V m_y E_x$. They cause a first-order depolarising field $E_z = -D_z/\epsilon_{iso} = iQ_V m_y E_x$ to keep $D_z = 0$. The depolarising field then induces a second-order displacement $D'_x = i\epsilon_{iso} Q_V m_y E_z = -\epsilon_{iso}(Q_V m_y)^2 E_x$ that corresponds to a Lorentz motion along $-x$, opposite to the zero-order displacement. As for the transverse Kerr effect, such amplitude modulation cannot be converted to a domain contrast. If the sample is rotated by 45° (Fig. 8b), however, Lorentz motions are excited in all four domains, leading to D'_x-components along $-x$ in all domains and antiparallel D_y-components in neighbouring domains. The latter are finally responsible for antiparallel E_y-components in the reflected (or transmitted) light, which are phase shifted by 180°. When the phase of one of the two E_y vectors is matched with that of E_x by means of the compensator, a black/white microscopic contrast is created in neighbouring domains if the analyser is perpendicular to the polariser (see also section "Domain Contrast").

Compared to the first-order Kerr and Faraday effects, the second-order Voigt effect differs in four aspects that can be derived from the previous discussion:

- Except for the transverse Kerr effect, first-order effects primarily lead to a rotation of the light polarisation, which may be superimposed by some ellipticity. The magneto-optical amplitudes thus have a different polarisation plane compared to the incident light. According to Eq. (13) this indicates an effect in the off-diagonal elements (r_{sp} or r_{ps}) of the reflectivity tensor. A typical quadratic effect, on the other hand, contributes a modification of the regular reflectivity being manifested in the diagonal elements (r_{ss} and r_{pp}) of the tensor. It can thus be interpreted as (magnetically induced) linear birefringence with dominating elliptical polarisation, possibly being superimposed by some linear dichroism, i.e. rotation.

- While the Faraday and Kerr effects are direction-sensitive, the Voigt effect is *quadratic* in the magnetisation vector due to its axial sensitivity. Antiparallel domains can therefore not be distinguished in Voigt microscopy, while orthogonally magnetised domains with magnetisation axes at $\pm 45°$ relative to the polariser axis generate maximum domain contrast. For the Kerr and Faraday effects, the conditions are different: here maximum contrast is obtained for antiparallel domains, and the polariser has to be parallel or orthogonal to the domain magnetisation for the longitudinal effect or at an arbitrary angle for the polar effect.

- The rotational symmetry is also different: Due to its quadratic magnetisation dependence, the Voigt contrast changes sign every $90°$ when the sample or the polariser is rotated. This is different for the Faraday and Kerr contrast: due to the linear magnetisation dependency of these effects, the sample has to be rotated by $180°$ to achieve opposite, but equally intensive contrasts.

- The emerging light due to the Voigt effect is primarily elliptically polarised, requiring a compensator to obtain a useful signal. The domain contrast is inverted by rotating the compensator through its extinction position. For the longitudinal and polar Kerr and Faraday effects, being primarily rotation effects, the compensator may help to improve the domain contrast, but basically it is achieved by opening of the analyser. Here rotation of the analyser through the extinction position leads to contrast inversion (see section "Domain Contrast" for details).

As mentioned in Sect. 1, the Voigt effect plays some role for domain analysis in transparent media, while for reflection microscopy it may be seen as a niche effect compared to the Kerr effect. Nevertheless, there are three important facts that should be considered:

- Due to its axial sensitivity, domains in antiferromagnetic materials can be seen in the Voigt effect [26], opening new paths for Voigt microscopy in an emerging field of research. It should be noted, though, that domains in antiferromagnets may also cause magnetostrictive strain along the domain axes [27]. In an optical polarisation microscope, this can lead to a birefringence contrast with the same symmetry as the Voigt effect, making it difficult to separate Voigt- and strain-induced contrast.

- The Voigt effect offers the possibility to detect in-plane domains at perpendicular incidence (if they are magnetised along different axes), thus allowing to use objective lenses with low resolution that have small numerical apertures and insufficient angles of incidence to obtain reasonable Kerr signals (see Sect. 4).

- As the domain contrast due to the Voigt effect is usually weaker than that caused by the Kerr effect, pure Voigt microscopy can only be realised if contributions from the Kerr effect are avoided. This is possible by imaging in-plane domains at perpendicular incidence (compare the domain images in Fig. 1). In the general case of arbitrary angles between k-vector and magnetisation, there will be a superposition of linear and quadratic magneto-optical effects. Especially in MOKE magnetometry (section "MOKE Magnetometry") and quantitative Kerr microscopy (section "Quantitative Kerr Microscopy"), the distorting influence of the quadratic effect needs to be considered. We will see in section "Domain Contrast" that the Voigt effect is adjusted at crossed polariser and analyser with some opening of the compensator, while the Kerr and Faraday effects require uncrossing of polariser and analyser. If, in practice, a significant quadratic effect is superimposed in a Kerr experiment, the relative contribution can therefore be reduced or suppressed by choosing large analyser angles.

Gradient Effect

The gradient effect is a birefringence effect, which depends linearly on gradients (changes) in the magnetisation. It therefore stresses fine magnetisation modulations rather than large-scale domain features (see section "Voigt and Gradient Effect Microscopy" for examples). Magnetisation gradients are strongest across domain walls. Therefore the domain walls show up in a contrast if imaged in the gradient effect, which, however, is different from the wall contrast seen by Kerr microscopy. From the example in Fig. 1, it is evident that in the Kerr effect, which senses the magnetisation directly, the 180° walls appear in a colour that depends on the (in-plane) rotation sense of the wall magnetisation. Imaged in the gradient effect, the same walls appear homogeneously black or white, independent of the rotation sense. This 'boundary' contrast is primarily determined by the magnetisation jump from one to the other domain, i.e. by the magnetisation gradient *across* the wall and not by the wall itself. It consequently alternates for neighbouring walls when the surrounding domain magnetisation is inverted. The boundary contrast furthermore changes sign when the walls are rotated by 90° as schematically shown in Fig. 9, thus revealing the same rotational symmetry as the Voigt contrast which is detected at the same experimental conditions as the gradient effect.

The physical origin of the gradient effect can be ascribed to the same gyroelectric interaction ($D \propto m \times E$) that is also responsible for the Kerr or Faraday effect and which can be used to derive quadratic effects (see Fig. 8) [28–31]. This is illustrated in Fig. 9 for the case of a 180° domain wall at four orientations. For light polarisation along the x-axis at perpendicular incidence, the cross product

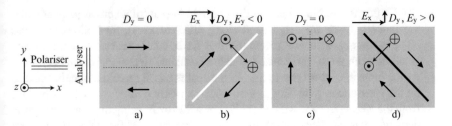

$D_y = 0$ E_x $D_y, E_y < 0$ $D_y = 0$ E_x $D_y, E_y > 0$

a)   b)   c)   d)

Fig. 9 Symmetry of the gradient contrast for a 180° wall that is rotated by 135° from (**a**) to (**d**). Like in Fig. 8, the oscillating Lorentz motion across the wall, excited by the gyroelectric interaction, is indicated. It causes a dielectric displacement vector with positive or negative D_y components.

$m \times E$, leading to an oscillating gyroelectric polarisation normal to the sample surface, varies between zero (if m is along the x-axis) and 1 (if m is along the y-axis). In neighbouring domains the oscillations are out-of-phase by 180° since the m_y-component, which causes the gyroelectric interaction, changes sign (i.e. has a gradient across the boundary). The pairs of out-of-phase oscillating electric dipoles produce a quadrupolar dielectric displacement with a component perpendicular to the domain boundary, which can be expressed by:

$$D_y = P_{gr} \left(-\partial m_y / \partial y \right) E_x, \tag{21}$$

with P_{gr} being a material constant that scales with that of the Kerr effect. The component D_y results in a component E_y in the emerging light. In the experiment, the component of this transverse field parallel to the analyser axis can be transformed to a contrast by phase matching like for the Voigt effect. Equation (21) contains just one component of a tensor that can be written as [15, 32]

$$\epsilon_G = \epsilon_{iso} P_{gr} \begin{pmatrix} -\dfrac{\partial m_x}{\partial y} - \dfrac{\partial m_y}{\partial x} & \dfrac{\partial m_x}{\partial x} - \dfrac{\partial m_y}{\partial y} \\ \dfrac{\partial m_x}{\partial x} - \dfrac{\partial m_y}{\partial y} & \dfrac{\partial m_x}{\partial y} + \dfrac{\partial m_y}{\partial x} \end{pmatrix} \tag{22}$$

under the condition of perpendicular incidence ($E_z = 0$). This tensor applies to both, strictly *in-plane* magnetised materials as well as samples with subsurface, perpendicular magnetisation gradients within the information depth, for which the condition div $m \approx 0$ needs to be fulfilled as reviewed in ref. [1].

Experimentally, contrasts due to the gradient effect can be seen under the same conditions as for the Voigt effect, i.e. by rotating the compensator while keeping the analyser and polariser crossed in a 'Kerr' microscope. At the end of section "Voigt Effect", it was noted that in a general magneto-optical microscopy experiment, there might be a superposition of linear and quadratic effects under given circumstances. The same is true for the gradient effect that may cause parasitic contrasts in Kerr

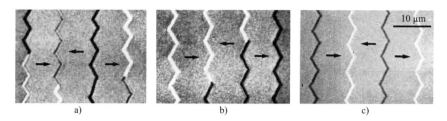

Fig. 10 Disturbing effect of the gradient effect in Kerr microscopy, demonstrated for V-line walls in an iron-silicon sheet with (100) surface orientation: (**a**) superposition of longitudinal Kerr and gradient contrast, (**b**) pure Kerr contrast after carefully aligning the compensator and (**c**) pure gradient contrast at perpendicular incidence of light. Images (**b**) and (**c**) show an identical wall pattern, while a similar structure is shown in (**a**). Adapted from ref. [15]

microscopy. An example, demonstrating this effect, is presented in Fig. 10a. Here a V-line pattern [18] was imaged in the longitudinal Kerr effect without worrying about the alignment of analyser and compensator. Rather than seeing an equal-amplitude black and white contrast of the wall segments, caused by opposite rotation senses of the wall magnetisation, the segments show up with different intensity due to the superposition of the gradient effect. In image (b) the gradient effect was eliminated by carefully aligning analyser and compensator, leading to the expected segment contrast of a similar pattern, while in image (c) the pure gradient contrast is shown. To isolate the gradient contrast from the Kerr contrast, perpendicular incidence of light is required in case of such in-plane magnetised materials. A Kerr contrast is not possible then, neither for the domains nor the walls. However, these are the same experimental conditions as for the Voigt effect. A separation of gradient and Voigt contrast is therefore not possible on in-plane materials. Furthermore, contrast for both effects is obtained by rotating the compensator while keeping analyser and polariser crossed, and both effects follow the same rotational symmetry (i.e. the contrast can be inverted by either rotating the sample by 90°, the polariser by 45°, or by rotating the compensator through its extinction position— see section "Voigt Effect"). This indicates a phase shift of the magneto-optical amplitude of the gradient contrast of roughly 90° relative to the phase of the Kerr amplitude, similar to that of the Voigt effect.

3 Physical Aspects of Magneto-Optical Microscopy

After having summarised the basics of the conventional magneto-optical effects in Sect. 2, let us now address some physical aspects that have an impact on wide-field magneto-optical microscopy before we move on with instrumentation in Sect. 4. These aspects include contrast formation, depth sensitivity, and lateral resolution.

Domain Contrast

Most important in magneto-optical microscopy is the *visibility* of magnetic domains, which depends on the achievable *contrast* and on the *signal-to-noise ratio* of the electronic detection system. In this section we will take a closer look at these parameters. The discussion concerns the Kerr contrast, but it equally well applies to the Voigt contrast by simply replacing the generalised Kerr amplitude A_K [defined in Eq. (19)] by a corresponding Voigt amplitude A_V.

By the example of the longitudinal Kerr effect, we have demonstrated in Fig. 6a that oppositely magnetised domains cause opposite Kerr amplitude vectors $\boldsymbol{R}_K$, which lead to opposite (complex) Kerr rotations by superposition with the regularly reflected amplitude $\boldsymbol{R}_N$. A domain contrast in the Kerr microscope is obtained if the reflected light from different domains is blocked differently with a rotatable polarising filter in the reflection path, called analyser. Different Kerr rotations are thus transferred to a detectable *difference in intensities*. The strongest contrast is expected between domains that create an opposite change of the polarisation state. In case of the Kerr effect, these are domains with antiparallel magnetisation (for the Voigt effect, orthogonally magnetised domains lead to the strongest contrast). If extinction is not possible due to elliptical contributions, a phase-shifting retardation plate (compensator) is required. For the moment, however, let us ignore this complication by assuming no phase shifts between R_N and R_K—ellipticity and the compensator will be addressed below.

The starting point of the discussion is the generalised reflected amplitude $A_{tot} = A_N + A_K$ defined in Eq. (19). The Kerr rotation is then given by the (small) angle $\theta_K = A_K/A_N$. According to Fig. 6c, it seems intuitive to block exactly the light that emerges from one of the two domains by choosing an analyser setting $\alpha_{an} = |\theta_K|$. This domain would then appear dark, whereas the opposite domain (together with all other domains magnetised at different angles) would appear more or less bright. In practice, however, it is common to open the analyser by a *few* degrees (see Fig. 6c) rather than by the intrinsic Kerr angle that is typically significantly less than a degree.

This counterintuitive fact can be explained by calculating the Kerr contrast. For an analyser opening angle α_{an}, the intensities of the two domains in Fig. 6 are given by the squares of their reflected amplitudes [18]:

$$I_1 = A_N^2 \sin^2(\alpha_{an} - \theta_K) + I_0 \quad \text{and} \quad I_2 = A_N^2 \sin^2(\alpha_{an} + \theta_K) + I_0 . \quad (23)$$

Here I_0 is a background intensity that may be caused by imperfect polarisation degrees of polariser and analyser, the use of a finite illumination aperture and by depolarising effects in the lenses and at the sample. The relative Kerr *signal*, which is the difference between the two domain intensities, is then approximately given by

$$S_K = I_2 - I_1 \approx 2 \sin(2\alpha_{an}) A_N A_K , \quad (24)$$

considering $\theta_K = A_K/A_N \ll 1$. Accordingly, a maximum Kerr signal can be expected at an analyser angle of $45°$, causing a very high brightness at the same time. A high Kerr signal, however, does not automatically mean high contrast for *visual* observation. The contrast is given by $C = (I_2 - I_1)/(I_1 + I_2)$. Optimisation with respect to the analyser angle α_{an} yields a maximum contrast of

$$C_{opt} = \frac{A_K A_N}{\sqrt{(A_K^2 + I_0)}\sqrt{(A_N^2 + I_0)}} \approx \frac{A_K}{\sqrt{(A_K^2 + I_0)}} , \qquad (25)$$

which is achieved already at an analyser angle $\alpha_{opt}^C = \arctan\sqrt{(A_K^2 + I_0)/(A_N^2 + I_0)}$ smaller than $45°$. As A_N^2 is much larger than I_0, the optimum contrast can be approximated by the second term in (25). It obviously depends on the background intensity and on the Kerr amplitude only—not on the regular amplitude or the Kerr rotation!

In contemporary Kerr microscopy, however, the 'visible' contrast according to Eq. (25) is not the crucial criterion for the visibility of domains. Today, Kerr microscopes are usually equipped with image processing (see Sect. 4) so that the contrast can be easily enhanced electronically. In fact, it turns out that in digitally enhanced microscopy, an analyser setting of $\alpha_{an} = \alpha_{opt}^C$ would lead to an insufficient image intensity. Rather than the contrast, the *signal-to-noise ratio* (SNR) now becomes decisive for good domain visibility. As the intensity increases with α_{an}^2, analyser opening angles that are larger than those necessary for optimum (visible) contrast are required to increase the signal according to Eq. (24). The best analyser setting actually depends on the relative proportion of various noise contributions:

- The detection electronics adds a temperature-dependent noise that is usually independent of the image intensity. It depends on the quantum efficiency of the detection sensor (CCD or CMOS chip, see Sect. 4), the number of 'binned' camera pixels and the integration time [17]. If this 'dark' noise would be the sole noise mechanism, the optimum angle α_{opt}^{SN} would in fact be $45°$, i.e. the angle for maximum signal according to Eq. (24).
- Fluctuations of the light create noise that is proportional to the image intensity.
- The quantised nature of light leads to shot noise, which varies with the square root of the photon number in the image. This unavoidable noise is of statistical nature and can therefore be reduced by averaging over a large number of photons.

Let us now discuss the case of pure (unavoidable) shot noise, i.e. we assume 'ideal' image processing and a stable light source, neglecting short integration and exposure times. The shot noise depends on the incident photon flux, F_{inc}, according to $N_{shot} = \sqrt{0.5 F_{inc}(I_1 + I_2)}$. With the absolute signal $S_{mo} = F_{inc}(I_1 - I_2)$, the SNR is written as $r_{SN} = S_{mo}/N_{shot} = \sqrt{F_{inc}}(I_1 - I_2)/\sqrt{0.5(I_1 + I_2)}$. Inserting the intensities (23) and minimising with respect to the analyser angle, we obtain for the maximum SNR [18]:

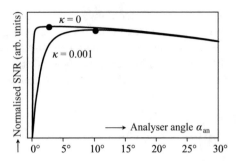

Fig. 11 Relative signal-to-noise ratio (SNR) as a function of the analyser angle for two extinction ratios of the polarisers ($\kappa = 0$ would be perfect polarisers with full extinction). The angles of best SNR are indicated by dots. Adapted from ref. [17] with kind permission from J. Phys. D

$$r_{SN}^{opt} = \frac{4A_K A_N \sqrt{F_{inc}}}{\sqrt{A_N^2 + I_0} + \sqrt{A_K^2 + I_0}} \approx 4A_K \sqrt{F_{inc}}, \tag{26}$$

which is achieved at an analyser opening of $\alpha_{opt}^{SN} = \arctan \sqrt[4]{(A_K^2 + I_0)/(A_N^2 + I_0)}$, again being beyond the setting of optimum visible contrast as expected. The maximum SNR is obviously determined by the Kerr amplitude and the photon flux, but again not by the Kerr rotation. Note that in practice the illumination intensity cannot be arbitrarily enhanced to avoid overheating of the specimen. If electronic noise or light fluctuations are added, the SNR will be reduced, and the optimum analyser angle will be shifted within the boundaries $\alpha_{opt}^C < \alpha_{opt}^{SN} < 45°$. The proportionality between r_{SN} and the magneto-optical amplitude A_K is preserved, however. Figure 11 shows the simulated signal-to-noise ratio as a function of the analyser opening angle for two different extinction ratios of polariser and analyser (i.e. different background intensities I_0), proving that analyser angles of several degrees (orders of magnitude higher than the the Kerr rotation) are indeed typical for optimisation.

So far we have assumed that A_K is in-phase with A_N, leading to rotated, *plane*-polarised light. In case of a phase shift, the Kerr angle gets complex, and the emerging light is elliptically polarised (see Fig. 2a for visualisation). Ellipticity may be intrinsically caused if the Voigt parameter Q_V has a non-vanishing imaginary part (see section "Kerr Effect"). Metallic reflection at oblique incidence always generates elliptical light if the polarisation plane of the incident light is not parallel or perpendicular to the plane of incidence. In magneto-optical microscopy, however, only linearly polarised light can be converted into a contrast by means of the analyser. So, phase shifts need to be tuned by the compensator, which may be placed before the analyser along the reflection path or after the polariser along the illumination path (see Fig. 16) in a magneto-optical experiment. In wide-field microscopy, Brace-Köhler-type compensators are preferred as they cause a homogeneous phase retardation across the whole field of view (compared to Babinet compensators with line-characteristics—see ref. [1] for details).

A Brace-Köhler compensator is a transparent, birefringent plate of suitable thickness and material (like mica, calcite, or some stretched plastic sheet). It has two distinguished, orthogonal in-plane axes, called fast and slow axis (Fig. 12a), along which plane-polarised light propagates with different velocities due to different refractive indices. The two rays are consequently retarded relative to each other with a phase difference of $\Delta\varphi = 2\pi N$. The variable N is the retardation number, expressed in fractions of the wavelength. For a quarter wave (or $\lambda_0/4$) plate, e.g. $N = \frac{1}{4}$ so that a phase difference of $\pi/2$ is introduced between the fast and slow beams. When the compensator is rotated around the k-axis, the value of retardation R is given by $R = \Delta\varphi \sin{(2\,|\alpha_{\text{pol}} - \alpha_{\text{comp}}|)}$. Here α_{comp} is the rotation angle of the compensator's fast axis, measured from the same axis as the polariser angle α_{pol} (see Fig. 12a). Apparently, the compensator will have no effect if the incident light is polarised parallel to either of the two principal axis (the fast axis, i.e. $\alpha_{\text{comp}} = \alpha_{\text{pol}}$, or the slow axis, i.e. $\alpha_{\text{comp}} = \alpha_{\text{pol}} + 90°$). A retardation always requires two components. As an example let us consider the regularly used quarter-wave plate: Here both field components along the two principal axes will have equal amplitudes if the entering light is plane-polarised at $45°$ to the axes. With $\alpha_{\text{pol}} = 0°$ and $\alpha_{\text{comp}} = 45°$, the phase shift will be $90°$, and the plane wave will be converted into circular light. Any other orientation of the quarter-wave plate will lead to elliptical light (Fig. 2) with an ellipticity that depends upon the degree of relative phase retardation R. Vice versa, any elliptically polarised wave, emerging from a magnetic sample, can be converted into plane-polarised light by a properly rotated compensator.

Now let us assume a Kerr experiment in which two oppositely magnetised domains emit elliptical light with opposite rotation and handedness as illustrated (strongly exaggerated, though) in Fig. 12b. The upper ellipse can be linearised by rotating the compensator by $90° + \theta$. The linear wave is then rotated by $\xi + \theta$. Rotating the analyser by $\alpha_{\text{ap}} = 90° + \xi + \theta$ will block the light from the corresponding domain. At the same setting, however, the light from the opposite domain will increase its ellipticity, thus leaving the analyser with some intensity, which finally leads to a domain contrast. For the practice of magneto-optical microscopy, three conclusions can be drawn:

- With a rotatable compensator, it is only possible to linearise the reflected light from one domain phase. Usually it is sufficient to compensate the ellipticity of the darkest domains. Then the above formulae for the contrast and signal remain valid if A_K is replaced by $|A_K|$ [12]. In practice, the compensator and the analyser are rotated simultaneously, leaving the polariser fixed, until an image of satisfactory contrast and brightness is obtained.
- Alternatively, the compensator can even be omitted. The intrinsic Kerr ellipticity in a certain domain phase can rather be compensated by illuminating the sample with elliptical light of opposite handedness and rotation. In microscopy, this can be achieved by carefully turning the polariser away from an exact p- or s-direction.

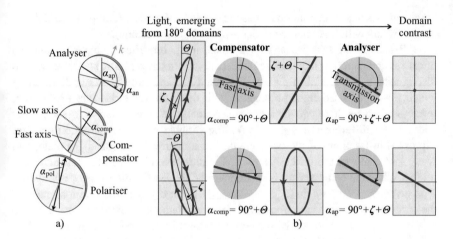

Fig. 12 (**a**) A Brace-Köhler compensator contains a phase-retardation plate with an optical axis parallel to the plate plane. For adjustment, the plate is rotated around the propagation (**k**)-axis of the light. The azimuth angles of polariser, compensator and analyser are defined relative to the vertical axis. Note that here the function of the 'polariser' is just to create polarised light for the compensator. In a Kerr experiment, there would be a sample placed between polariser and compensator. (**b**) Transformation examples of a quarter-wave compensator in combination with an analyser, demonstrating the geometry of a Kerr experiment with superimposed elliptical polarisation

- In case of magnetic linear birefringence (Voigt effect), two equally oriented ellipses with opposite handedness are created (Fig. 7a). If the analyser is at 90° to the polariser, both ellipses would pass with equal intensity. Adding a compensator with its fast axis turned away from the polariser axis will cause elliptical light of different ellipticity for the two domains, which passes the crossed analyser with different intensity thus creating a Voigt contrast [1]. The contrast can be inverted by rotating the compensator through its zero position, thus interchanging the degrees of ellipticity for the two domain axes. The experimental conditions are different for the Kerr contrast: as Kerr ellipticity is usually small, a Kerr contrast is mainly generated and inverted by rotating the analyser—the compensator just helps to optimise the contrast by linearising the light that emerges from one domain type.

Besides electronic enhancement, there is a 'natural' way to increase the Kerr contrast: The strongest Kerr signal can be expected if all light is absorbed in the sample, creating magneto-optical interaction, and if the generated Kerr amplitude can freely exit in turn [33]. Any light that is uselessly reflected from the surface cannot contribute to the Kerr amplitude. A *dielectric antireflection coating* [12, 34] creates that condition as demonstrated in Fig. 13c. This effect can be explained with the sketch in Fig. 13a: For a correct coating thickness, the phases of amplitudes $R_N^{(2)}$ and $R_N^{(1)}$ are shifted by 180°. Between $R_N^{(3)}$ and $R_N^{(2)}$, the phase difference is 360° (consider the 180° shift at internal reflection). The total (regular) reflection

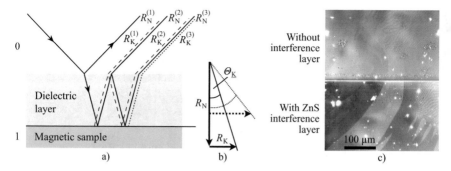

Fig. 13 Enhancement of domain contrast by interference coating: (**a**) schematic view of the effect of a dielectric interference layer on the regular and Kerr amplitudes (see text). (**b**) If the effective regular amplitude decreases while the Kerr amplitude is enhanced, the effective Kerr rotation increases. (**c**) Domain image of an amorphous ribbon for which the upper half was coated by ZnS while the lower half is without interference coating. The contrast enhancement is obvious. The sketch in (**a**) is taken from ref. [18] with kind permission from Springer Verlag

will be zero by destructive interference if $R_N^{(1)}$ is equal to the combined amplitude $(R_N^{(2)} + R_N^{(3)} + \ldots)$. At the same time, the created Kerr amplitudes $R_K^{(i)}$ add up by constructive interference, leading to an increased *effective* Kerr rotation (compare Fig. 13b). For a quantitative description, we refer to the review on antireflection coatings in Chapt. 2 of ref. [18].

In practice it is sufficient to minimise the reflectivity at normal incidence as large angles are hardly used. For metals a suitable dielectric is ZnS; for magnetic oxides MgF_2 or SiO_2 is favourable. The films are evaporated while optically controlling the reflectivity in the deposition chamber by using the same wavelength of light that will later be used for imaging. For a correct layer thickness, a metallic specimen displays a dark blue colour. Before the introduction of digital image processing, antireflection coatings were mandatory to obtain a reasonable Kerr contrast. Nowadays they can be abandoned. However, if an evaporation chamber is available in a laboratory, there is no reason not to apply coatings—one can only gain, especially if domains are to be imaged on materials with intrinsically small Kerr amplitudes.

Depth Sensitivity

Magneto-optical microscopy is *not* a surface-sensitive method, different, e.g., to scanning electron microscopy with polarisation analysis (SEMPA) or X-ray photo emission electron microscopy (XPEEM), which relies on the generation of secondary and photo-electrons, respectively, that mainly emerge from surfaces [18]. This is trivial for transparent materials like garnets, but also for metals an *information depth* of the order of 20 nm can be expected and in oxides it can even be higher. So, the Kerr effect is not a 'surface' effect—the term 'SMOKE' for surface-

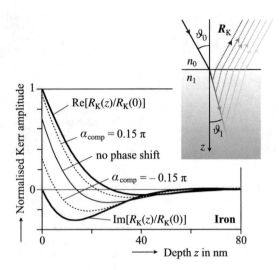

Fig. 14 Depth dependence of the longitudinal Kerr amplitude for bulk iron, normalised to the Kerr amplitude at the surface (with $\vartheta_0 = 30°$, $\lambda_0 = 633$ nm, homogeneous in-plane magnetisation along the plane of incidence). Between the real and imaginary part graphs, curves with three different compensator settings are added. The inset indicates the partial beams that are contributing to the Kerr amplitude. The plot is adapted from ref. [1] with kind permission from Springer Verlag

magneto-optical Kerr effect, which was coined for Kerr magnetometry on films [35], has to be taken carefully.

For a quantification of the depth sensitivity, the *phase* of the magneto-optical amplitude has to be taken into account. In case of the Kerr effect, it turns out that the total effective Kerr amplitude, which interferes with the effective regular amplitude (section "Kerr Effect"), can actually be seen as a superposition of Kerr amplitudes that are generated within the penetration depth of light (see inset in Fig. 14). These amplitude contributions are exponentially damped with increasing depth, and they differ in phase according to a complex amplitude penetration function [33]:

$$\varphi(z) = \exp(-2k_0 \, \mathrm{i} \cos \vartheta_1 \, n_1 z). \tag{27}$$

Like in section "Kerr and Faraday Effect", n_1 is the refractive index of the specimen, ϑ_1 is the angle of incidence within the sample, and $k_0 = 2\pi/\lambda_0$ is the wave propagation number with λ_0 being the vacuum wavelength of the (assumed) monochromatic, illuminating light. The light wave has to travel in and out again, hence the factor of 2. In Fig. 14 the depth dependency of the real and imaginary parts of the Kerr amplitude, which follow this function, is plotted for the example of the longitudinal Kerr effect on bulk iron, both being normalised to the Kerr amplitude at the surface.

As introduced in section "Domain Contrast", a Kerr contrast is obtained by converting the Kerr amplitude into a detectable signal, with the required light rotation being generated by interference of the regularly reflected light amplitude and the Kerr amplitude. A *detectable* rotation is only obtained, if the two components are in phase so that they can interfere to a plane-polarised wave. In case of a phase shift, a compensator needs to be added to shift the phases properly. With a rotatable compensator, however, the phase difference can be selected freely. For the plot in Fig. 14, for instance, the relative phase was chosen in such a way that the surface magnetisation is detected optimally, meaning that the Kerr amplitude generated right at the surface is allowed to interfere with the (effective) regular amplitude. The real part curve of the normalised amplitude in the figure applies to this case, whereby the total Kerr amplitude can be derived by integrating this curve over the depth. Interestingly, the magnetisation at a certain depth, where this curve passes zero, will be invisible, and magnetisation contributions from deeper in the material may even reduce the overall signal. The zero crossing of the real part curve may therefore be seen as 'information depth' parameter [33]. According to Eq. (27), it is found at $z_0 = \lambda_0/8n_1'$ where n_1' is the real part of the refractive index. For iron ($n_1' = 2.89$) the information depth z_0 is around 20 nm for visible light. The $1/e$-decay length of the absolute value of the signal can be calculated as $z_d = \lambda_0/4\pi n_1''$ with n_1'' being the imaginary part of the refractive index.

As mentioned, the relative phase angle between Kerr and regular amplitude can be arbitrarily adjusted by rotating the compensator. With regard to Fig. 14, this means that any linear combination between the real and the imaginary part of the Kerr amplitude can be realised. If, for instance, the phase sensitivity is adjusted at 90° to the phase of the surface amplitude, the imaginary curve in the figure will become the real part curve. Then there will be no surface sensitivity, and one will rather 'focus' on some subsurface magnetisation. So, in principle any Kerr amplitude can be selected to interfere with the regular amplitude, allowing to focus on certain depth magnetisations or to intentionally blind out the magnetisation in a certain depth by adjusting the Kerr amplitude out of phase with respect to the regular light—all, of course, within the absorption limit. In Fig. 14 examples for three different compensator settings are added. In section "Depth-Selective Imaging" we will see that this concept can be applied for layer-selective imaging of multilayer systems.

To summarise, for bulk material the Kerr contributions from different depths differ in phase, thus reducing the overall efficiency of the Kerr effect. In materials with large information depth like magnetic oxides, the decay length z_d is much larger than the position of the first zero z_0. Many oscillations in the depth-dependent Kerr signal can then occur, leading to a reduction of the (stronger) surface contributions so that the integral Kerr effect of oxides is usually quite weak. For details on this depth sensitivity concept, we refer to the review in [1] and to a series of articles in which it was developed [31, 33, 36–39].

Resolution

When the dimensions of the objects to be imaged in an optical microscope are reaching the order of the wavelength of visible light, the spatial resolution is limited by diffraction and can be expressed by the *resolving power R*, which corresponds to the smallest distance between two closely spaced objects so that they can still be distinguished [40]:

$$R = f \frac{\lambda}{NA} \quad \text{with} \quad NA = n_0 \sin \alpha. \tag{28}$$

Here λ is the wavelength of the light and NA the (dimensionless) numerical aperture of the objective lens. The angle α is half the angular aperture of the objective lens (i.e. half the angle of the cone of light from the object that is accepted by the lens), and n_0 is the refractive index of the medium between objective and sample ($n_0 = 1$ for air; $n_0 \approx 1.5$ for immersion oil). The higher α and n_0, the more orders of diffracted light are collected by the objective and the higher the resolution. The factor f is somewhat arbitrary as it depends on numerous parameters like the form of the magnetic object and the light intensity that emerges from it, the aperture, angle and shape of the incidence light bundle and the sensitivity of the detector (in digitally enhanced imaging, also the sampling interval of the camera sensor has an influence on the effective resolution). Typical values for f are 0.61 (Rayleigh criterion), 0.47 (Sparrow criterion) or 0.5 (Abbe criterion) [40]. Let us take the Abbe criterion as reference. It takes into account that on the observation side, the full aperture of the objective lens is effective in the collection of diffracted beams, even though the effective illumination is typically restricted to about 20% of the aperture for contrast reasons (see section "High-Resolution Microscopy"). Other influences on the resolution are not considered in the Abbe criterion.

In the table of Fig. 15a, the spatial resolution of a number of objective lenses, as they are commonly used in Kerr microscopy, is collected for blue and red light. Note that the magnification of the objective lenses is irrelevant as the resolution is only determined by NA and the wavelength. According to the table, a resolution of almost 150 nm can be expected by using blue light and an oil-immersion objective with highest numerical aperture. In fact, on perpendicular recording media, domains as narrow as 160 nm could be clearly resolved by using the blue spectral lines (404 nm and 435 nm) of a mercury arc high-pressure lamp and a 100× immersion lens with $NA = 1.25$ [41]. The use of deep ultraviolet (UV) light would further improve the resolution, but it requires the use of special lens material (like quartz glass) as the transmission of conventional optical glass drops rapidly below 400 nm and becomes zero below 320 nm. The gain in resolution, achievable in a UV Kerr microscope, does not justify the considerable expense.

Magnification	NA	$R_{460\,nm}$	$R_{640\,nm}$
5 ×	0.13	1769 nm	2462 nm
10 ×	0.25	920 nm	1280 nm
20 ×	0.50	460 nm	640 nm
50 ×	0.80	288 nm	400 nm
100 ×	0.90	256 nm	356 nm
100 × oil	1.45	159 nm	221 nm

a) b) c) d)

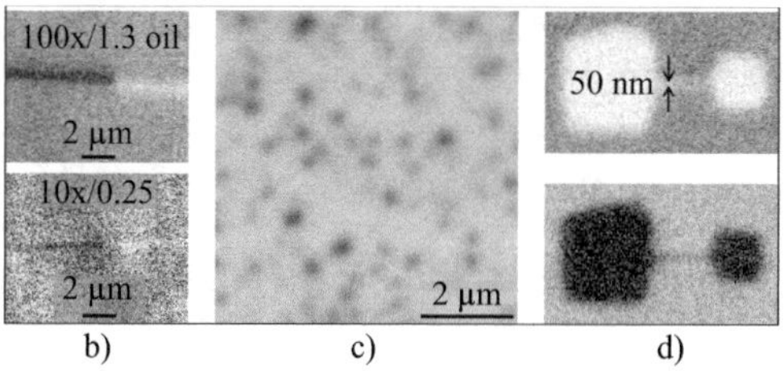

Fig. 15 (a) Spatial resolution according to the Abbe criterion for typical objective lenses used in wide-field Kerr microscopy. (b) Vortex domain wall on amorphous $Fe_{24}Co_{18}Ni_{40}Si_2B_{16}$ ribbon (20 μm thick), imaged in the longitudinal Kerr effect at highest resolution (upper image, 100×/1.3 oil objective) and at low resolution (lower image, 10×/0.25 lens). (c) Bubble domains on a [Pt/Co/Ir]Pt multilayer, imaged with a 100×/1.3 oil objective. (d) Remanent states of metallic film with perpendicular anisotropy, patterned to a 50 nm wide nanowire in the middle and again imaged with a 100×/1.3 oil objective. The images in (b) and (c) are adapted from ref. [42] with kind permission from APL. The sample for images (d) was provided by M. Moneck, CMU, Pittsburgh, USA

The resolving power or resolution limit should not be confused with the *visibility* of isolated magnetic objects. Virtually there is no limit for the visibility as long as the reflected intensity of the object is strong enough to make it stand out from any surrounding background. Here, perpendicularly magnetised structures, imaged in the (stronger) polar Kerr effect, have a clear advantage compared to in-plane structures that need the longitudinal Kerr effect. Although it is not 'resolved', the object can nevertheless be seen with a contrast profile that is reduced in intensity and broadened by diffraction compared to the profile of a larger object. In Fig. 15 this is demonstrated for three examples: In (a) a vortex domain wall [18] of an amorphous ribbon was imaged in the longitudinal Kerr effect with the sensitivity transverse to the wall. The in-plane surface rotation of the wall shows up with a black/white contrast that depends on the rotation sense of the wall magnetisation. With a width of 310 nm, the wall is well resolved in the upper image, which was obtained at a nominal resolution of about 160 nm, while in the lower image, the wall is still well visible although its width is just 1/3 of the resolution of 920 nm. In Fig. 15c isolated bubble domains with diameters ranging between 50 and 450 nm were imaged in the polar Kerr effect. Depending on their width, they show up with more or less intensity, but still visible. By measuring the integral contrast of the objects in (b) and (c) and normalising it to the maximum achievable contrast under the given conditions, the width of the objects can even be quantised although being well below resolution [42]. In fact it turns out that the image of a sub-resolution magnetic object becomes reduced in amplitude but becomes wider so that the integral contrast stays constant. This property of magneto-optical images is due to the fact that the magnetic contrast is to first order a linear function of the magnetisation, not a quadratic one as with other scattering objects. In Fig. 15d, finally, perpendicularly magnetised nanowires of 50 nm width are showing up in black and white contrast at positive and negative remanence. In ref. [43] it was shown that in a wide-field Kerr microscope, a magnetic signal can even be obtained from nanowires with a width of just 30 nm.

4 Technical Aspects of Magneto-Optical Microscopy

For magneto-optical domain imaging, either scanning optical microscopes or conventional wide-field microscopes can be used. In scanning microscopy [44], only a single spot is illuminated by a focused laser beam and recorded at a time, and an image is then formed by scanning the beam across the surface. In wide-field microscopy, the sample is illuminated by parallel light, and the whole field of view is seen immediately. The former offers some advantages for time-resolved imaging [45], but far more wide-field microscopes are in use for domain imaging. We therefore restrict this article to the latter; reviews on scanning Kerr microscopy can be found in refs. [1, 18]. Three types of wide-field polarisation microscopes are used for magneto-optical domain observation:

- Most frequently applied for Kerr, reflection Voigt and reflection gradient microscopy are *reflected light microscopes*, often referred to as incident light, epi-illumination or metallurgical microscopes. Here a single objective lens is applied for illumination and observation at the same time, and the microscope is operated in the bright-field mode. This arrangement is ready for domain observation up to highest spatial resolution and with high magnification.
- Bright-field microscopes with separate illumination and observation paths, allowing to obtain Kerr images on larger sample areas in the millimetre and centimetre regime at a strongly reduced resolution, though.
- Transmission microscopes are used for Faraday, transmission Voigt and transmission gradient microscopy. This is the microscope type that the layman usually has in mind when referring to an 'optical microscope'. Transmission microscopy is thoroughly covered in numerous textbooks on optics. We therefore refrain from elaborating on the technical aspects here. To make a transmission microscope a magneto-optical microscope, it needs to be equipped with electromagnets and image processing in the same way as reflection microscopes. Note that the transmission effects can also be used in reflection if the transparent specimen is placed on a metallic mirror in a reflection microscope. Also indicator films make use of that (see section "Indicator Films").

In the following the above-mentioned two fundamentally different configurations of wide-field magneto-optical imaging in reflection are discussed. Yet, the laid out technical microscope aspects can directly be related to domain imaging in transmission or Faraday effect imaging using indicator films. An additional section on shared aspects of observation, primarily the camera system of choice, is included. The principle aspects of time-resolved imaging modes are further discussed. Kerr microscopes also offer the possibility for magnetometry with high spatial resolution.

Microscopy

High-Resolution Microscopy

Achieving best results in reflected light magneto-optical microscopy requires a detailed understanding of the illumination path to adjust the desired modes of Kerr sensitivity. For instance, a comprehensive knowledge of the principles of the so-called Köhler illumination and the significance of the conjugate optical planes for the resulting Kerr images is needed to obtain the optimum results in terms of spatially even magneto-optical contrast across the entire field of view. To achieve optimal magneto-optical contrast, several aspects like the correct adjustment of illumination and the proper setting of the polarisation elements have to be taken into account. Historically, arc discharge lamps being directly attached to the microscope were used for their high radiance output levels. By transmitting the illumination with a fibre optic light guide into the microscope, a physical detachment of the light source and the microscope is achieved, by which a reduced input of heat into the microscope and thus an improvement in thermal stability of the microscope are obtained. The usage of alternative light sources, such as light emitting diodes (LEDs) and lasers, offers significant advantages over the traditionally used arc lamps. Both, LEDs and lasers, produce a greater and steadier intensity than arc lamps. For laser systems, as a result of laser speckles, image artefacts are introduced in the low-contrast magnetic domain images, which are amplified using difference imaging techniques. By applying different technical modulation methods, the effects are suppressed to a level, which makes laser illuminations suitable for magneto-optical imaging [14, 46, 47]. Yet, laser-based illuminations are currently mostly applied for wide-field *time-resolved* imaging setups, which will be described in section "Quantitative Imaging of Magnetisation Dynamics". For easy magneto-optical contrast adjustment, the laser-based systems make use of fibre illumination [46].

In recent years, fibre-coupled LEDs have become the standard for illumination in magneto-optical microscopy [17, 48, 49]. The spectral irradiance is similar or superior to high pressure arc lamps, as several watts of collimated output power are achievable with current LED illuminators. Most importantly, LEDs possess low noise. They also offer pulsed operation modes that permit an easy adaption of imaging schemes for advanced component-selective quasi-static (section "Separation of Effects") and time-resolved microscopy (section "Quantitative Imaging of Magnetisation Dynamics"). Unlike for lasers, speckle patterns are no issue for LED-based illuminations. With high efficiency coupling into optical fibres, high-power LEDs are now the illumination of choice for most magneto-optical microscope experiments.

Essential for obtaining correctly adjusted magneto-optical effects is the exact setting of the Köhler illumination, where the illuminating light source, e.g. the fibre-optic output, is fully defocused onto the magnetic sample, thus obtaining an even illumination of the specimen. The principle illuminating path in a Kerr microscope is displayed in Fig. 16a. The light source and the back focal plane of the objective lens lie in conjugate aperture planes (AP). Additionally, several conjugate image planes (IP) exist, most importantly the field diaphragm and the magnetic sample. For obtaining the best magnetic imaging results, the proper alignment of the fibre position in all three axes is most important. The possibly varying position of the back focal plane for different objective lenses is compensated by varying the optical fibre output along the imaging axis or by the application of an adjustable condenser lens in the illumination pathway. Due to the diameter of the illuminating fibre output, the specimen is illuminated in a narrow angular spread of incidence as indicated in Fig. 16b, thus leading to well-defined conditions for the magneto-optical sensitivity (see sect. 2). Practically, the setting of the desired mode of sensitivity is achieved by positioning the fibre output at different off-axis positions in the aperture plane.

It should be noted that for high numerical aperture and high magnification objective lenses, depolarisation effects occur that lead to an increase of the background intensity. This slightly decreases the signal-to-noise ratio and has consequences for the optimal analyser settings to achieve best magneto-optical contrast (see section "Domain Contrast"). Moreover, the contrast of the resulting magneto-optical images strongly depends on the optical transmission properties of the objectives, which determine the effective overall accessible intensity, thus being as important as the quantum efficiency of the camera system (see section "Camera Systems"). The scattering of light properties and the polarisation quality of the objective lenses affect the overall contrast (section "Domain Contrast"), in particular the SNR in magneto-optical imaging. Unwanted Faraday rotation in the objective lenses occurs with the application of high magnetic fields (see section "Faraday

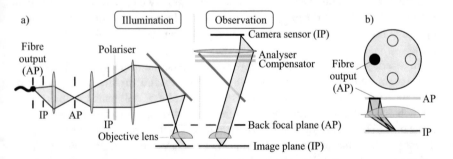

Fig. 16 (**a**) Ray diagram of a high-resolution Kerr microscope with the optical path of illumination and a simplified path of observation restricted to the primary image plane. Köhler illumination is assumed for the illumination path on the basis of a fibre-coupled light source. Indicated are the rays corresponding to the conjugate set of aperture planes (AP) for oblique incidence, the equivalent conjugate sets of field or image planes (IP) and exemplary locations of rotatable polariser, retarder plate and analyser. (**b**) Corresponding image of the back-focal plane of the objective lens and its conjugated focal planes for a single illumination source for the case of oblique incidence of light together with the resulting image rays illuminating the magnetic sample for imaging in the longitudinal magneto-optical mode. Different planes and directions of incidence can be realized by activating other LEDs in the indicated cross of the fibre outputs

Effect"), not only leading to an additional intensity change but also to a reduction of the SNR. Such contributions can be compensated and reduced, respectively, by readjusting the analyser [50] or by using advanced imaging schemes (see section "Separation of Effects"). Also, the use of special low permeability objective lenses is advantageous in magneto-optical imaging applications to avoid side effects from magnetic forces acting on the lens during (high) magnetic field applications, which can lead to unwanted artefacts and reduced quality of the images. This effect and also mechanical drift can be compensated by adjusting the sample position with nanometer resolution using piezoelectric positioning systems.

Overview Microscopy

Kerr microscopy of large samples at low magnification can be easily achieved for *polar* contrast configuration in the single-objective-lens, high-resolution microscope discussed above. Yet, due to the limited numerical aperture of low magnification objectives, which limits the achievable in-plane longitudinal contrast due to low angles of incidence [17], the imaging of in-plane magnetisation distributions in large-scale samples, where a viewing field of up to several millimetres or centimetres may be desirable, is not easily achievable. This limitation can be overcome by using an inclined microscope setup with completely separated, symmetrically arranged illumination and reflection paths [18]. With such an arrangement, significant longitudinal domain contrast with nearly optimal Kerr amplitude can be attained. A further advantage of such systems is that the optical polarising elements can be arranged between the lenses and the magnetic sample. This eliminates depolarizing effects, occurring at the lenses' surfaces, and the mentioned Faraday effect with the application of magnetic fields. Using zoom lenses, a variable field of view is achievable.

The advantages of strong domain contrast together with a large field of view in an overview microscope are achieved at a cost. Due to the inclination of the objective lens, only a narrow strip of the sample is in focus, the region of which being defined by the depth of field of the optical system. The under- and overfocus of the magnetic specimen may be surmounted by tilting the objective lens away from the illumination axis in such a way that a focused image at the camera sensor is obtained. Consequently, the resulting sample image is then distorted due to the essentially varying magnification factor across the field of view of the microscope. This issue can be eliminated by correcting for the distorted image perspective by real-time imaging processing. Constant magnification with constant focus across the full field of view is achieved with the use of telecentric lenses together with a Scheimpflug camera mount [17]. A principle sketch of an optimized telecentric Kerr microscopy system is shown in Fig. 17a. Even for a strongly inclined axis of observation, zero distortion magnetic images are acquired. The resulting domain image is still compressed perpendicular to the plane of incidences of light, and a linear operation to obtain an equalized image map is needed. Exemplary overview

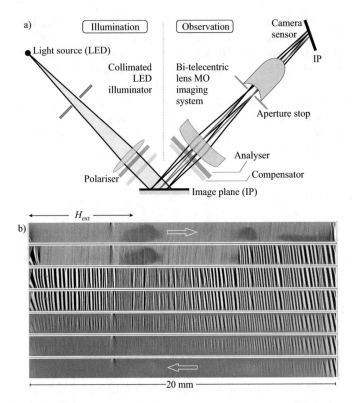

Fig. 17 (a) Optical path of a bi-telecentric overview Kerr microscope. Parallel illumination of the magnetic sample is achieved by a collimated high-power LED light source. The positions of rotatable polariser, compensator and analyser are indicated. The aperture stop is positioned at the focal plane of the front optical lens group. The conjugate image planes are tilted relative to the optical axis. Tilting the camera detector provides in-focus imaging of the sample by bringing the image plane in line with the camera sensor. (**b**) Magnetic domain formation in a magnetoelectric sensor element with varying magnetic field after saturation along the long axis of the sensor. The schematics of (**a**) are adapted from ref. [17]

Kerr microscopy images from the magnetisation reversal of a magneto-electric composite cantilever sensor are shown in Fig. 17b.

Camera Systems

Independent of using high-resolution or overview microscopes, the camera system is an essential part that determines the performance of every magneto-optical microscope as it significantly contributes to the overall SNR. Relevant are the maximum quantum efficiency and the noise characteristics of the camera. The quantum efficiency is wavelength-dependent, and the corresponding spectral response of the camera system needs to be considered in the selection of the illumination source. For long camera exposure times under low light conditions, cooling of the sensor chip becomes essential. Due to recent developments, highly sensitive, complementary metal oxide semiconductor (CMOS) sensors with a good SNR, a high dynamic range and high frame rates are favourable for most Kerr microscope applications. The high frame rates further enable advanced imaging modes by combining domain images of different sensitivities close in time (see section "Separation of Effects"). The achievable temporal resolution is determined by the minimum exposure time of the camera system (see also section "Time-Resolved Kerr Microscopy"). CMOS-based camera sensors offer both a so-called rolling shutter mode, where individual lines of the sensor array are exposed at different times, and a global shutter mode, where every pixel in the sensor is exposed simultaneously. The rolling mode has advantages in terms of lower read-out noise characteristics, while the global mode secures an exact time resolution. Image amplifiers, with and without gating, allow imaging at very low light levels and enable high temporal resolutions [51].

Time-Resolved Kerr Microscopy

The principles of high-resolution and overview microscopy discussed above can be adapted to the observation of dynamic magnetisation processes, which may include femtosecond processes for ultrafast spin-dynamics, resonant magnetisation dynamics on the sub- or nanosecond timescale and eddy-current dominated magnetisation dynamics in the micro- to millisecond range for thick samples [17]. The whole range of timescales is accessible using *time-resolved* Kerr microscopy, and the rather broad range can be addressed by implementing fast illumination and/or fast detection schemes in magneto-optical microscope setups. In the following, the emphasis will be on the features of the microscopes themselves for achieving high temporal resolution based on either using fast camera schemes, fast LEDs or laser-based illumination schemes. The aspects of fast magnetic field generation or alternating schemes of initiating magnetisation dynamics based on electrical current or optics will not be discussed.

Different imaging modes for time-resolved imaging do exist, which can be separated into real-time, single-shot and stroboscopic imaging methods. The applicability of the individual modes is limited by the frame rate of the camera system as well as by the time resolution of the illumination light source. Due to the technical limitations of both in relation to the needed or aimed-for temporal resolution and the actual scientific problem of interest, not all approaches are practically suitable for the imaging of dynamic magnetic domain processes. The three available main modes of imaging are specified as:

- Direct *real-time imaging* with continuous alternation of the magnetic domain states relies on the *steady* observation of the magnetisation processes as sketched in Fig. 18a. The domain evolution in a changing magnetic field is visualized directly in 'real time' with a temporal resolution determined by the frame rate of the camera system.

- *Single-shot imaging* of individual single magnetic events is possible through pulsed high-intensity illumination sources like pulsed arc flash lamps, LEDs and lasers. In pulsed light, imaging of *single* magnetic events with high temporal resolutions, determined by the illumination source, is possible. The principle of the method is sketched in Fig. 18b. The main difference to real-time observation is due to the existing restraints in the frame rate of the camera system, allowing for recording only a singular magnetic event within one fast magnetisation process. Further limitations of single-shot imaging result from the achievable SNR ratio, which is mostly limited by the illumination intensity and the sensitivity of the used camera system.

- *Stroboscopic imaging* is also based on the use of pulsed illumination sources. The method relies on phase matching of the magnetic excitation with the illumination pulse as sketched in Fig. 18c. Alternatively, camera systems with gated image

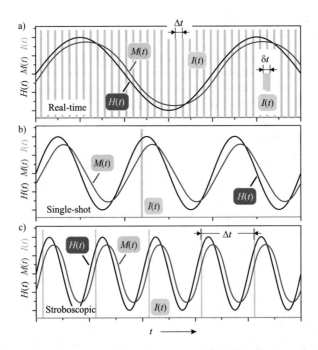

Fig. 18 (a) Real-time observation with a varying magnetic field $H(t)$, the magnetisation response $M(t)$ and a continuous illumination $I(t)$. Domain processes are probed with an exposure time ∂t. The time interval Δt is determined by the frame rate of the camera. (b) Single-shot Kerr microscopy with a single Kerr image taken in a time window ∂t defined by an illumination pulse with the duration ∂t at a delay time t. (c) Stroboscopic domain imaging with the composite magneto-optical micrographs obtained by summing up repeatable magnetic events with a temporal resolution ∂t defined by an illumination pulse train at a delay time t relative to the exciting magnetic field $H(t)$. A harmonic magnetic field excitation $H(t)$) is assumed for the examples

intensifiers can be used. Both enable the imaging of *repeatable* magnetic events up to very fast timescales. The overall image formation takes place by integration over the magnetic events by the camera system. The limiting temporal factor is the pulse width of the illumination source, respectively the gating time of the image intensifier.

A different view on the aspects of time-resolved magneto-optical microscopy arises looking at the different imaging options in relation to the needed temporal resolution to resolve magnetisation processes at different frequencies. The relevant points are:

- *Low-frequency dynamics* in the regime of an excitation frequency of a few to several Hertz can be visualized in real time with regular Kerr microscopy setups as the use of camera systems with exposure times of the order of 10 ms is standard. On the other hand, using pulsed LED illumination sources, a similar time resolution is achievable. The latter avoids problems with rolling shutters in the camera systems. The limiting factor for real-time imaging is the frame rate of the camera, rather than the exposure time. Giving sufficient light intensities, single-shot imaging down to 10 µs is possible with standard imaging systems as demonstrated for overview microscopy in ref. [17] (Fig. 19a). For lower light levels, stroboscopic imaging can easily be achieved relying on pulsed LED illuminations. Using high-speed cameras together with high magneto-optical contrast, continuous-detection imaging up to 500 Hz is possible [52] (see Fig. 21e).

- *Kilohertz magnetisation dynamics* can be imaged best by means of stroboscopic techniques using gated image intensifiers [53] or pulsed LED illuminations [54, 55] (Fig. 19b). Both offer variable repetition rates and continuously adjustable time resolutions. Yet, current LED sources are limited to temporal resolutions of about a few microseconds. As pointed out above, at least partial repeatability of the magnetisation processes is required for stroboscopic imaging. Yet, single-shot imaging is possible within this timescale using triggerable pulsed laser sources.

- *High-frequency magnetisation dynamics* in the megahertz regime can be imaged by means of stroboscopic techniques using gated image intensifiers [51] or pulsed laser systems. The image intensifier systems provide variable repetition rates and continuously flexible temporal resolutions into the sub-nanosecond regime. Selecting proper solid-state lasers, a similar flexibility is also available for laser-based systems, but providing advantages in terms of SNR compared to intensified cameras. Single-shot imaging is possible within the timescale down to 20 ns by using pulsed laser sources with the corresponding pulse widths (Fig. 19c).

- *Resonant frequency dynamics* in the gigahertz regime and beyond can be imaged stroboscopically only by using pulsed laser systems [56]. The temporal resolution is defined by the laser pulses, being in the picosecond or femtosecond regime. For achieving highest time resolution, special care must be taken to minimize timing jitter of the microscope system. The best flexibility is achieved with systems where the magnetic excitation and the laser pulses are running from the same clock signal and thus being locked in phase. Practically, jitter-free time-resolved Kerr microscopy with picosecond time resolution for direct

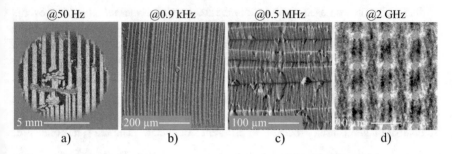

Fig. 19 (**a**) Single-shot Kerr image from an FeSi electrical steel sample with a camera exposure time of $10\,\mu s$ at the operational frequency of 50 Hz. (**b**) Stroboscopic image from a magnetic field modulated magneto-electric sensor device with a LED pulse width of $10\,\mu s$ at 876 Hz and (**c**) from an electrical field modulated magneto-electric sensor with a laser pulse width of 20 ns at 0.516 MHz. (**d**) Stroboscopic Kerr microscopy of standing magneto-static spin-wave modes in a CoFeB/Ru/CoFeB anti-dot array with a laser pulse width of 7 ps at 2 GHz magnetic field excitation

imaging under a continuous microwave field excitation is possible with such a microscope setup [17, 57] (Fig. 19d). A further advanced system, allowing for component-selective and quantitative time-resolved imaging, will be discussed in section "Quantitative Imaging of Magnetisation Dynamics".

Overall, Kerr microscopy methods provide the almost sole possibility for the analysis of fast magnetisation dynamics in the laboratory. Stroboscopic wide-field Kerr microscopy [56, 57] for the investigation of dynamic magnetic domain processes is compatible with all the Kerr microscopy imaging modes. Often, time-resolved Kerr microscopy is performed in the difference imaging mode, varying the phase between the rf-field excitation and the imaging laser pulse. Difference images are calculated from the difference of two magnetisation states at different time delays, by which only alterations of magnetisation become visible in the domain response images. Imaging sensitivities down to variations of $m = 0.01$ have been achieved in this way. A review on the realisation of stroboscopic dynamic imaging by Kerr microscopy is given in ref. [17].

Temperature-Dependent Microscopy

Adapting Kerr microscopy for temperature-dependent imaging requires the integration of cryostats for low temperatures and heating stages for elevated temperatures into the microscope systems. The magnetic sample is usually mounted on a cold or heating finger, respectively. The technology for low-temperature imaging was partly driven by the imaging of flux penetration into type-II superconductors, where the

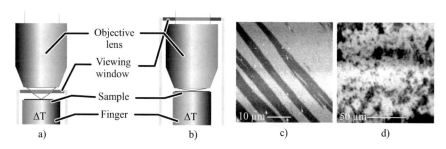

Fig. 20 (**a**) Detail of an imaging setup with the viewing window between the objective lens and the sample. (**b**) Setup with the window on top of the objective lens, where the objective lens is integrated into the cryostat or heating chamber. Exemplary Kerr microscopy images obtained (**c**) from the chirality reversal in an exchange spring soft/hard magnetic bi-layer film at 77 K [60] and (**d**) from a noncrystalline ferromagnet at 623 K [62].

observation is based on Faraday microscopy using detection films (see section "Indicator Films"), but it is as well-suited for regular Kerr microscopy. Applications of high-temperature Kerr observations are, for instance, the investigation of high-temperature domains in soft magnetic nanocrystalline materials [58]. Examples for low-temperature Kerr imaging include exchange bias phenomena in coupled ferromagnetic thin film systems [59, 60] and ferromagnetic semiconductors [61].

Two basically different approaches are used for temperature-dependent microscopy. In both approaches the sample is typically under vacuum, and thus special care needs to be taken for the integration of the objective lens. The elementary difference lies in the positioning of the objective lens outside or inside the vacuum chamber (Fig. 20). In one case, imaging is performed through an optical window between the magnetic object and the objective lens. Using this approach (Fig. 20a), long-distance objective lenses with an optical correction for the window thickness are required. This restricts the achievable spatial resolution. For the other method (Fig. 20b), the objective lens must be integrated into the cryostat or heater [63]. By this, regular high numerical aperture objective lenses with short working distances and higher resolving power can be used, and the highest resolution magneto-optical imaging becomes achievable. On the other hand, objective lenses with longer working distances are favoured due to the strong temperature gradient between sample and objective lens, and the microscope objective lenses must be vacuum-compatible. Both methods require a stress-free viewing window to minimize influences on the polarisation state of the light. To avoid imaging artefacts, especially relevant for the mostly applied difference imaging technique, the imaging system must be vibrational-decoupled from the supporting vacuum systems. Thermal drift with varying temperature can be compensated for by an active drift correction of the sample position.

Indicator Films

The indirect imaging of magnetic domains by probing the magnetic stray fields that are generated by magnetic microstructures at the sample surface can be achieved by placing a magneto-optically active detection layer on top of a specimen. For this method mainly rare earth doped ferrite garnet magneto-optical indicator films (MOIFs) are applied, which are micrometres in thickness. The films are nearly transparent and have a high Faraday rotation. Imaging is then performed in reflection in a high-resolution or overview microscope with a mirror layer on the bottom of the indicator film. As the magnetic stray field of the sample, which is eventually responsible for a modulation of the indicator film magnetisation and thus a Faraday domain contrast in the MOIF, emerges from magnetic poles in the specimen, MOIF microscopy may be seen as '*magnetic pole*' microscopy. The method is especially applicable for magnetic materials with no or very weak magneto-optical contrast and for specimens with optically non-transparent cover layers. For a review on the many aspects of magnetic imaging with MOIFs, we refer to ref. [64].

The principal scheme of MOIF imaging of the magnetic domains in a magnetic material with out-of-plane components of magnetisation is sketched in Fig. 21a. The possible schemes for an in-plane magnetised sample with charged domain walls are shown in Fig. 21b and c. Magnetic domains with an out-of-plane magnetic (field) contribution and charged domain walls become detectable. The contrast formation is due to an alternation of magnetisation in the MOIF in connection with the stray fields generated by the magnetic domains or domain walls of the sample under investigation. Examples for typical MOIF domain images are presented in Fig. 21d–f. The achievable spatial resolution is usually about a micrometre for in-plane magnetised MOIFs (Fig. 21d). It is mainly limited by the thickness of the MOIF in connection with the distance to the sample surface. While in-plane MOIFs rely on rotational magnetisation processes in the indicator film, wall motion occurs in the case of perpendicular MOIFs—the latter are thus more sensitive due to a higher permeability. Yet, for out-of-plane magnetised MOIFs, the spatial resolution is influenced by the actual magnetic domain size in the detection layer. Therefore, perpendicular MOIFs are especially advantageous for use at low spatial resolution in the 10 micrometre range, as for overview microscopy (section "Overview Microscopy"), so that the perpendicular eigen-domains of the indicator film are not resolved (Fig. 21e, f). The underlying domain pattern of the sample then shows up as a modulated dark-bright contrast [66]. As the domain contrast essentially arises from the (strong) polar Faraday effect, even single-shot dynamic imaging of electrical steel at 50 Hz power frequency in the presence of insulation coating becomes possible by MOIF microscopy [52] (see Fig. 21f and section "Time-Resolved Kerr Microscopy").

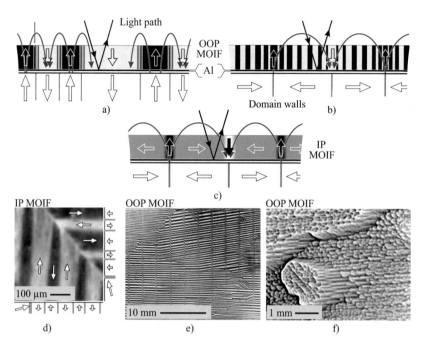

Fig. 21 Imaging magnetic domains by MOIF microscopy with an indicator film with (**a, b**) magnetic out-of-plane (OOP) anisotropy and (**c**) with in-plane (IP) anisotropy. The overall MOIF structure contains the magneto-optically active iron garnet film grown on a transparent substrate (not shown) and an aluminum mirror. Magnetic domain observation in reflection is achieved by placing the MOIF on top of the magnetic specimen. The magnetic response of the MOIF to the stray field, emerging from the sample, is imaged using the polar Faraday effect. (**d**) Exemplary MOIF image displaying magnetic domain contrast from a twinned magnetic shape memory crystal with slightly tilted magnetic anisotropy relative to the surface [65] (IP magnetised MOIF). (**e**) Overview images of the static magnetic domain distribution in a laser-scribed electrical steel sample (OOP magnetised MOIF) and (**f**) dynamic magnetic domain formation in an electrical steel sample (OOP magnetised MOIF) at an excitation frequency of 50 Hz obtained by time-resolved magneto-optical microscopy [52]

MOKE Magnetometry

The optical measurement of magnetisation curves by MOKE magnetometry is usually based on a laser beam that is focused on the sample with a spot size in the micrometre range and on the measurement of the reflected Kerr signal with an optical detector [1]. Space-resolved measurements require scanning over the sample in such magnetometres. MOKE magnetometry on the basis of wide-field Kerr microscopy is an elegant alternative. Plotting the intensity of a selected spot or area in the Kerr image as a function of magnetic field leads immediately to a magnetisation curve. The local character enables the analysis of different material properties in magnetically inhomogeneous magnetic materials. An example of a

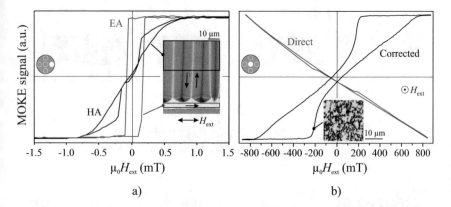

Fig. 22 (**a**) Longitudinal MOKE loops from different regions of an FeCoSiB(4 μm) film displaying easy axis (EA, blue) and hard axis (HA, black) magnetisation loop behavior [55, 67]. (**b**) Directly measured (blue) polar MOKE loop from an FePd (10 nm)/FePt(25 nm)/FePd(10 nm) film sample, including parasitic Faraday effects, and magnetisation loops with an active in situ correction (black) [50]. Exemplary domain images are shown in the insets

magnetostrictive film with local stress anisotropy, leading to a local variation of magnetic anisotropy alignment, is displayed in Fig. 22a. Compared to any other kind of magnetometry, the wide-field approach offers the advantage that the relevant domain images along the loop can be readily recorded, which is extremely helpful for the interpretation of hysteresis loops. Some critical points need to be considered, though:

- As the microscope's objective lens is exposed to the magnetic field, the loops may be strongly distorted by non-linear Faraday rotations that occur in the objective lens and that are superimposed on the Kerr signal. This is especially true if the magnetic field is aligned along the objective axis. Such (polar) Faraday contributions can be subtracted in the computer after measuring them separately by recording the Kerr intensity on a non-magnetic spot in the image. In ref. [50] an in situ compensation method was suggested based on a motorised analyser (Fig. 22b). For magnetometry in in-plane fields, there may as well be polar Faraday contributions in the objective that can arise by field inhomogeneities and by stray fields emerging from the edges of magnetised *bulk* samples. They can be compensated by recording the loops in the pure in-plane sensitivity mode [68] (see section "Separation of Effects").
- The magnetisation can only be measured *qualitatively*, not quantitatively as in vibrating sample or SQUID magnetometry. If the Kerr intensity, corresponding to the saturation magnetisation M_s, can be precisely measured by compensating any Faraday contribution, the magnetisation can be plotted in reduced units M/M_s.
- Any MOKE curve is caused by domain contributions that are recorded within the information depth of Kerr microscopy. For bulk specimens only the surface domains will therefore contribute. Surface domains, however, differ in general

drastically from interior domains so that MOKE magnetometry has to be applied with care in case of non-transparent materials. For thin films this is not a problem.

- In any case it needs to be considered that *local* hysteresis loops are recorded by MOKE magnetometry, which may be very different from curves at other locations and from an integral loop measured on the complete sample.

5 Advanced Methodology

Magneto-optical Kerr microscopy is a well-established technique for the observation of magnetic domains. Despite being around for so many years, recent developments have further improved the quality of domain imaging with major advances in the application and the separation of the intertwined magneto-optical effects. Advanced methodologies now allow for magnetisation component selectivity and quantitative domain imaging on different timescales.

The recent progress in Kerr microscopy is in particular due to advanced light sources. These allow complete control over the illumination conditions for obtaining different Kerr sensitivities and for having the concurrent possibility to reversibly switch between the various incidence-of-light conditions. These advances are reviewed in this section, starting from the separation of Kerr effects to real-time quantitative microscopy and time-resolved quantitative domain imaging of precessional magnetisation dynamics. Further advanced non-standard techniques like depth-selective Kerr microscopy and imaging modes involving the magneto-optical Voigt and gradient effect are discussed. These techniques offer additional possibilities, which are in particular of relevance for magnetic thin film investigations and the imaging of antiferromagnetic materials.

The presented advanced imaging modes are so far not necessarily used routinely for domain imaging. They open up opportunities for additional fields of investigations beyond standard Kerr microscopy.

Separation of Effects

At the end of section "Kerr Effect", we have summarised the phenomenology of the magneto-optical Kerr effect (except the transverse effect) by a simple rule. Formulated in terms of contrast, that rule reads: *The polar and longitudinal Kerr contrast is proportional to the magnetisation component parallel to the reflected light beam.* In Fig. 23 this is visualized for antiparallel in-plane and perpendicular domains. At oblique incidence (Fig. 23a, b), both domain types have vector components along the reflected beam, thus showing a contrast. The contrasts, however, have different dependencies on the direction of incidence: While the (polar) contrast of the perpendicular domains does not depend on the incidence direction, the longitudinal contrast of the in-plane domains changes sign when the direction

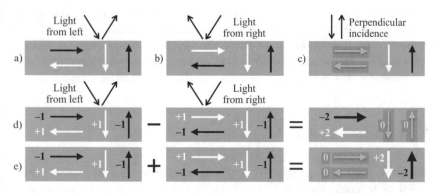

Fig. 23 Kerr contrast of in-plane and perpendicular domains with light being reflected to the right (**a**), to the left (**b**) and for perpendicular incidence (**c**). The longitudinal domain magnetisation is along the plane of incidence, which is in the plane of projection. (**d**) Subtraction of the two images with inverted oblique incidence leads to pure in-plane sensitivity, while addition (**e**) results in pure polar sensitivity. The numbers in (**d**) and (**e**) are supposed to indicate the domain intensities

of incidence is inverted as the vector components of the left and right domains along the reflected beam change sign. So, at oblique incidence there is always a superposition of longitudinal and polar contrast. At perpendicular incidence (c) the in-plane domains do not show up because there is no magnetisation component along the reflected beam. This is the condition for pure polar contrast.

The described dependencies are the basis for contrast separation [17, 47, 49, 69]. Technically this can be realised by running proper fibre outputs (see section "High-Resolution Microscopy") arranged in a crossed configuration (Figs. 24 and 16b) either in a static or pulsed mode. For the latter [49], the light pulses with a typical switching time in the microseconds range need to be synchronised with the camera exposure. If the frame rate of the camera is beyond ten frames per second, the human eye cannot follow the pulsating images, and the domain picture on the screen appears static. An alternative implementation for the component-selective imaging of magnetisation dynamics using pulsed laser sources [69] will be discussed in section "Quantitative Imaging of Magnetisation Dynamics". The following contrast modes can be achieved:

- *Biaxial in-plane sensitivity*. Oblique incidence along two orthogonal planes with *s*- and *p*-polarised light corresponds to longitudinal Kerr sensitivities along the two planes (with polar sensitivity superimposed). There are two technical ways to realise this biaxial mode of imaging:

  1. Choosing different colours for the LEDs that define the two planes of incidence and using some optical device that separates the two corresponding images [48]. It is convenient to place an image splitter between the microscope and camera, which splits the two partial images on the same camera chip, thus leading to the simultaneous display of the two orthogonal magnetisation components. In Fig. 24a an example of this *dichromatic mode* of biaxial imaging is presented. For further possibilities see ref. [17].

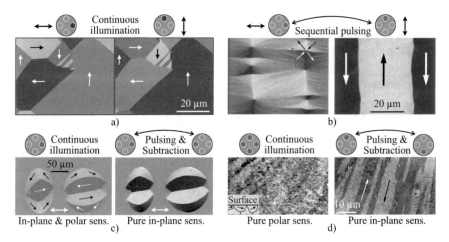

Fig. 24 (**a**) Biaxial imaging in the dichromatic mode with blue and red LED light. The horizontal and vertical magnetisation components of a 50 nm thick epitaxial iron film are imaged simultaneously in the same frame. (**b**) Biaxial imaging of cross-tie domain walls in a 40 nm thick permalloy film in the monochromatic mode using white light. Again the two magnetisation components are displayed, but now in separate images and quasi-simultaneously by alternately pulsing the indicated LEDs. (**c**) Pure-in-plane mode achievement by sequentially pulsing opposite LEDs. (**d**) Separation of perpendicular and in-plane magnetisation components in a nanocrystalline NdFeB material with a grain size of round 300 nm and an in-plane effective easy axis. In the polar image, the grains show up due to their up and down out-of-plane magnetisation components, while magnetostatic interaction domains [18, 49] are revealed in the image with pure in-plane sensitivity. The in-plane sensitivity axis is indicated by double arrows. Images (**b**) and (**c**) are adapted from ref. [49] with kind permission from AIP, and the in-plane image of (**d**) is adapted from ref. [70] with kind permission from Springer Verlag

2. Using monochromatic light and alternately activating the LEDs that define the two orthogonal planes of incidence in the pulsed mode [49]. After synchronisation with the camera exposure, the two partial images can be displayed quasi-simultaneously. Figure 24b shows an example of this *monochromatic mode* of biaxial imaging.

- *Pure in-plane sensitivity.* Subtracting two images, which are obtained with oblique incidence from opposite directions, leads to a cancellation of the polar contrast, thus leaving pure in-plane sensitivity (Fig. 23d). In Fig. 24a this is demonstrated for Permalloy film elements (note that 'pure in-plane sensitivity' is meaningless here as permalloy is magnetised strictly in-plane anyway). Interestingly, in this pure in-plane mode, it turns out that also the polar Faraday effect in the objective lens of the microscope is suppressed [68], which is very favourable for MOKE magnetometry (section "MOKE Magnetometry") and quantitative Kerr microscopy (section "Quantitative Kerr Microscopy"). To achieve the pure in-plane mode technically, opposite LEDs in the LED cross have to be run in the pulsed mode as indicated in Fig. 24c.

• *Pure polar sensitivity.* Adding two images, obtained with oblique incidence from opposite directions, results in a cancellation of in-plane contrast leaving pure polar sensitivity (Fig. 23e). Technically this can be achieved by running opposite LEDs in the static mode. If more light intensity is needed, also the transverse LEDs can be activated. As long as the LEDs are run symmetrically, any in-plane contrast will cancel, and the illumination is effectively perpendicular (alternatively also a centred LED could be used). In Fig. 24b an example is shown in which an identical domain pattern has been imaged in the pure polar and pure in-plane modes—the wide interaction domains, showing up in the pure in-plane mode, could not be guessed from the polar image.

The experimental conditions to separate the Voigt and gradient effects have already been discussed at the ends of sections "Voigt Effect" and "Gradient Effect", respectively.

Quantitative Kerr Microscopy

Quantitative magneto-optical Kerr microscopy is an advanced technique that is so far rarely applied for magnetic materials investigations [18, 71]. The technique is based on the merging of two images of the domain structure of interest, obtained with preferably nearly orthogonal Kerr sensitivities. It involves a calibration procedure with multiple calibration images, by which the angular magneto-optical sensitivity function in the Kerr microscope must be determined. By fitting two sinusoidal calibration functions (Fig. 25a) with complimentary in-plane sensitivities that are ideally phase-shifted close to $180°$, two linear equations for the two magnetisation components m_x and m_y are computed. By matching the magneto-optical contrast of two complementary domain images with the computed sensitivity curves obtained under the same contrast conditions, quantitative vectorial magnetisation images are obtained. Conventionally, the two sensitivity functions are recorded at the beginning and at the end of the experiment, respectively, before and after the domain pattern of interest is recorded. Only one pair of complementary domain images and thus one single magnetic vector field can be obtained under the same contrast conditions as the calibration curves, limiting the method to the analysis of static magnetic domain patterns.

Advanced approaches that allow for real-time quantitative magnetic domain observations, respectively, the quantitative imaging of continuous magnetisation processes, have evolved, overcoming the drawback of a restriction to static domain analysis and allowing for the quantitative analysis of a sequence of domain states during the magnetisation reversal. These quantitative methods rely on the methods of contrast separation described in section "Separation of Effects". One approach is based on the dichromatic imaging mode, using two LEDs of different wavelengths,

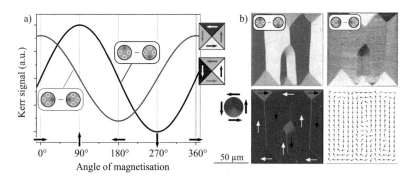

Fig. 25 (**a**) Phase-shifted complimentary magneto-optical sensitivity functions of nearly orthogonal sensitivity, together with resulting intensities of the magnetisation directions. Illumination schemes in the aperture plane for obtaining the corresponding sensitivities, showing typical LED arrangements, are indicated by the insets (compare to Fig. 24). (**b**) Vertical and horizontal pure in-plane Kerr sensitivity images together with the corresponding quantitative image and derived vector distribution of a magnetic domain state in a Permalloy thin film element [72]

which are positioned orthogonally in the back focal plane of the microscope [17, 48]. By that, simultaneous Kerr sensitivities along the vertical and horizontal axes are attained with a single exposure, ensuring simultaneous imaging of both in-plane magnetisation components. The images of complementary sensitivity can then be recorded independently. In this way, the two (still needed) phase-shifted calibration curves can be applied simultaneously to a sequence of image pairs and not only to a single pair of static images as in the aforementioned approaches for quantitative imaging. The precise spatial alignment of the domain images of different sensitivity is however problematic. Furthermore, the method does not permit for easy elimination of possible polar magneto-optical contrast contributions. This can be achieved by using a refined method [72], which is based on utilizing the pulsed LED scheme introduced in section "Separation of Effects". A result obtained with this method is displayed in Fig. 25b. By properly switching the LEDs in synchronisation with different camera exposures, domain images with orthogonal sensitivity can be acquired. Moreover, images with pure in-plane sensitivity can be obtained, by which also parasitic, magnetic field-dependent Faraday rotations in the microscope optics are eliminated [68] during calibration as well as in the evaluated domain images. This leads to a substantially improved accuracy in the quantitative imaging results, which is a significant step forward in quantitative magnetic domain imaging.

A related methodology based on a single illumination source is applied for time-resolved magneto-optical imaging. It should be noted that the temporal resolution of the dichroitic method is limited by the image exposure time, whereas for the pulsed LED scheme, it is further determined by the frame rate of the camera system.

Quantitative Imaging of Magnetisation Dynamics

The methods of quantitative imaging (section "Quantitative Kerr Microscopy") and, especially, magneto-optical contrast selectivity (section "Separation of Effects") are of utmost importance for dynamic Kerr microscopy. If precessional magnetisation processes are dynamically excited, timely varying in- and out-of-plane magnetisation components will occur even in a soft magnetic film, leading to entangled magneto-optical contrast contributions that involve the superposition of time-varying in-plane and out-plane Kerr signals. As the superimposing magneto-optical signals might also occur with a different time dependency, the separation of in-plane and out-of-plane magnetisation components, including vectorial dynamic wide-field imaging, is essential for the proper analysis of precessional magnetisation processes.

A dynamic wide-field vector imaging approach for the separation of dynamic in-plane and out-of-plane magnetisation response information and for achieving dynamic quantitative response imaging capabilities was demonstrated in ref. [69]. A principle imaging scheme, which is also relevant for regular fast time-resolved wide-field Kerr imaging setups, is shown in Fig. 26. The microscope observation path is not included in the schematics. The three main components of the setup are a high-stability Kerr microscope, a variable pulsed laser illumination and a high-frequency magnetic field excitation source. The rf signal generator, the pulsed laser light source and the sampling oscilloscope are synchronized to the same clock as a prerequisite for making component-selective imaging up to the GHz regime possible. Minimisation of laser speckles is achieved by the integration of a rotating ground glass diffusor and a vibrational diffusor. In contrast to the system described in section "Separation of Effects", here a lens with computer-controlled position adjustment is applied to reproducibly separate the Kerr effects (Fig. 26b). To untangle the polar and longitudinal Kerr contributions, the diametric counterparts of the two longitudinal sensitivities are measured in the same dynamic measurement cycle. Using a calibration procedure similar to the one described in section "Quantitative Kerr Microscopy" together with a procedure similar to the one presented in section "Separation of Effects", vectorial- and component-selective data can be obtained. Especially, the extraction of the dynamic out-of-plane information enables the analysis of the dynamic in-plane magnetisation response, and subtraction of the diametric vertically and horizontally measured modes yields the pure in-plane magnetisation response. An example of the quantitative component selection for the precessional response of a magnetic domain structure in a soft magnetic strip, including simultaneous in-plane and out-of-plane response, is shown in Fig. 27. Dominant in-plane magnetisation precession in the central domains and a spin-wave like response [73] become evident in the magnetic vector response images.

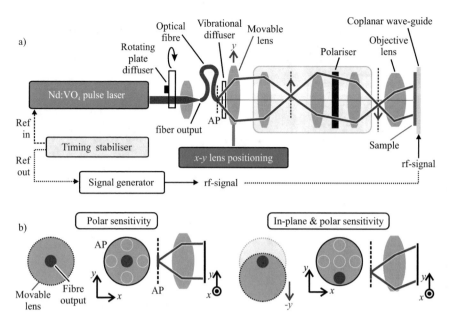

Fig. 26 (**a**) Principle sketch of the illumination of a quantitative time-resolved Kerr microscopy setup with a pulsed laser emitting a train of pulses at a fixed repetition rate. A rotating plate diffusor and a vibrational diffusor are used for despeckling. The laser is fed into an optical fibre, which output position is used for setting the needed magneto-optical contrast conditions. For this, the position is set by a movable focussing lens mounted on a motorised x-y stage for computer-controlled variable angle of incidence of light. The specimen's magnetisation is excited by a sinusoidal microwave magnetic field generated by a coplanar wave guide. Signal generator and laser probe are locked to the same clock. The phase relation between field excitation and laser pulse is adjusted for time-slicing. (**b**) Relative position of the optical fibre output and the resulting incident of light for perpendicular illumination and oblique illuminations. (Details of the system are described in ref. [69])

Voigt and Gradient Effect Microscopy

Along the presented methods of magneto-optical effect separation, Voigt effect microscopy offers additional complimentary input for the investigation of magnetic domains. Yet, the magneto-optical Voigt effect (section "Voigt Effect") is rarely used for the imaging of magnetic domain structures, as the obtained signal is weak compared to that of the magneto-optical Kerr effects. Furthermore, in Voigt microscopy the contrast is selected in polar incidence geometry for in-plane magnetised materials, where the longitudinal Kerr contrast ideally vanishes (note that in practice the elimination of residual Kerr contrasts is not easily attained). As the Voigt effect is even in magnetisation, it provides also access to antiferromagnetic domains not accessible by the linear magneto-optical effects. First experiments demonstrating the imaging of antiferromagnetic states in NiO(001) films at room temperature have been published [26].

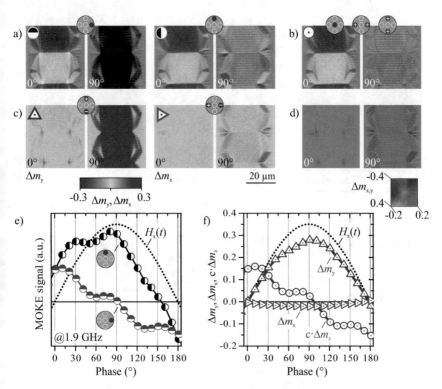

Fig. 27 (a) Oblique incidence and (b) polar dynamic Kerr images from a FeCoB thin film strip (thickness = 160 nm) excited in a harmonic 1.9 GHz magnetic field. (c) Derived magnetisation component (Δm_x, Δm_y) and (d) vector in-plane $\Delta m_y m_x$ magnetisation response images for a distinct phase angle of magnetic field excitation. The phase angles for all shown images correspond to 0° (left) and 90° (right). Comparison of (e) measured magneto-optical and (f) component-selective time response of a central domain. The measured magneto-optical response with oblique incidence is a superposition of longitudinal (in-plane; Δm_x, Δm_y) and polar (out-of-plane; $c \cdot \Delta m_z$) magnetisation response. (Data from ref. [69])

The Voigt contrast is possibly supplemented by the magneto-optical gradient contrast (section "Gradient Effect"), which is most visible at domain walls as an alternating domain boundary contrast. In Fig. 28 the connection between Kerr, Voigt and gradient effect is visualised for a soft magnetic thin film element [54]. A uniform and alternating gradient contrast is seen for all the magnetic domain boundaries. Neither of the domain wall transitions of any type is visible in the Voigt and gradient microscopy images, as the gradient contrast is independent from the internal structure of the domain walls. In practice, for the adjustment of Voigt and gradient contrast, starting with maximum extinction at crossed polariser and analyser settings and perpendicular plane of incidence, only the compensator is rotated until sufficient contrast conditions are obtained.

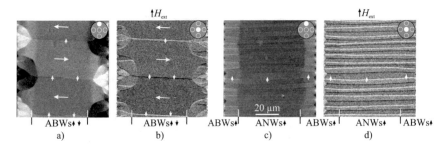

Fig. 28 Complementary (**a**) Kerr and (**b**) Voigt microscopy image of a magnetic domain state with asymmetric Bloch walls (ABWs) after demagnetising along the easy axis of magnetisation in a single layer $Co_{40}Fe_{40}B_{20}$ film with a thickness of 160 nm. (**c**) Kerr and (**d**) Voigt effect image of a magnetic domain state with a mixture of asymmetric Bloch and Néel (ANWs) walls at a bias field $H_{ext} = 100$ A/m. The regions with Bloch and Néel walls, the magnetic field direction and exemplary magnetic field directions are indicated

Depth-Selective Imaging

In section "Depth Sensitivity" it was shown that the total Kerr signal is due to a superposition of Kerr amplitudes that are generated within the penetration depth of light and that differ in amplitude and phase according to the complex amplitude penetration function of Eq. (27). The phase difference can be applied to obtain *depth-selective* information on the magnetisation, most favourably for magnetic sandwich systems in which magnetic films are interspaced by non-magnetic layers as shown schematically in Fig. 29a. By turning the compensator, the phase of the Kerr amplitude can be tuned relative to the regularly reflected amplitude so that the zero crossing of the information depth curve appears somewhere within one of the two layers. Properly adjusted, the integral Kerr contribution from this layer vanishes, and only the magnetisation from the other layer remains visible. By proper phase selections, the magnetisation of both layers can thus be imaged separately. This depth selection method was first applied in refs. [74, 75], where the two iron films of an Fe/Cr/Fe sandwich sample have been imaged separately. Figure 29b shows another example (which at the same time demonstrates the sensitivity of Kerr microscopy to ultra-thin magnetic films).

In practice, the compensator and analyser are rotated 'simultaneously' until the desired selectivity is obtained. To find the proper settings, it is helpful if the individual layers of the stack have characteristic properties that help to distinguish them. Different coercivities, for instance, will result in selective domain activities in an applied AC magnetic field of different amplitude. Making the domain activity in one of the layers invisible will then define the settings to block out the magnetisation of that layer. This procedure, however, will be difficult if more than two magnetic films are involved as only one film at a time can be made invisible. Besides the phase, also tuning the wavelength or the angle of incidence [77] can in principle be used for depth-selective imaging as both parameters enter the penetration function.

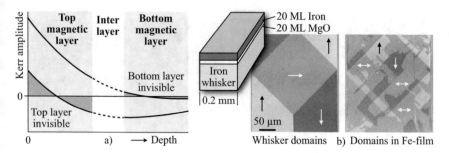

Fig. 29 (**a**) Thickness dependence of the Kerr amplitude (schematically, compare Fig. 14) for a trilayer system in which two magnetic films are interspaced by a non-magnetic film. By proper phase selection, the total Kerr amplitude of one of the layers can be nullified so that only a signal from the other layer will be left. (**b**) Kerr micrographs on a bulk iron whisker that is covered by 20 monolayer (ML) thick MgO and iron films. An identical domain pattern in the whisker and iron film is imaged selectively (adapted from ref. [76] with kind permission from APS)

While the latter can hardly be employed due to a weak dependence of the Kerr signal on the incidence angle in the case of metallic films, applying a tunable light source might be feasible. For Kerr microscopy, however, this has not been demonstrated yet.

X-ray spectro-microscopy [1] can also be used for depth-selective microscopy by tuning the photon energy to elemental absorption lines. In the case of multilayers, the chemical composition of the individual layers has to be different to image them separately. The Kerr approach, relying on interference effects rather than on the electronic properties of the material, can also be applied for layers of the same composition as demonstrated in Fig. 29b. For a thorough review on magnetic microscopy of layered structures, we refer to ref. [1].

6 Summary and Outlook

Recent developments in advanced magneto-optical microscopy techniques and the availability of commercial microscope systems have led to a revival of the method. *Kerr* microscopy especially is still one of the most versatile *in-lab* techniques for the investigation of magnetic domains and magnetic domain wall behaviour.

Together with its magnetometric capabilities and the possibilities of easy sample manipulation during observation, Kerr microscopy has become the 'workhorse' in many magnetism research laboratories. It enables magnetic microscopy of magnetic domain features on length scales from centimetres to even below the boundaries of spatial resolution. It facilitates magnetic imaging down to picosecond timescales and allows for imaging from cryogenic to elevated temperatures. Using and separating the various magneto-optical effects by advanced illumination schemes have led to significant progress in Kerr microscopy. Direct imaging modes and indicator film techniques are available for the investigation of the magnetisation behaviour of all kinds of magnetic materials and functional magnetic devices.

With all the recent and continuing developments, magneto-optical microscopy will continue to be one of the most important techniques for magnetic domain analysis. Thus, it will continue to make considerable contributions to the research on magnetic materials and the investigation of magnetic devices. Continuing progress in the field is taking place to meet the imminent scientific challenges. The limits of Kerr microscopy have not been reached yet by a long way. The development and use of applied magneto-optics are nowhere near its end.

Acknowledgments We thank Ivan Soldatov (Dresden), Babak Mozooni (Kiel) and Rasmus B. Holländer (Kiel) for helping with the development and implementation of advanced Kerr microscopy modes. J.M. thanks the German Science Foundation for support through the grants DFG Mc9/9-1 and DFG Mc9/9-2. J.M. acknowledges additional funding from the Collaborative Research Centre SFB 1261 (DFG Mc9/19-1).

References

1. W. Kuch, R. Schäfer, P. Fischer, F.U. Hillebrecht, *Magnetic Microscopy of Layered Structures* (Springer, Berlin, 2015)
2. M. Faraday, Phil. Trans. R. Soc. (London) **136**, 1 (1846)
3. C.A. Fowler, E.M. Fryer, Phys. Rev. **104**(2), 552 (1956)
4. W. Voigt, Nachr. Kgl. Ges. Wiss. Göttingen, Math.-Phys. Kl. **4**, 355 (1898)
5. A. Cotton, H. Mouton, Ann. Chim. Phys. **11**(8), 145 (1907)
6. J.J.F. Dillon, J. Appl. Phys. **29**(9), 1286 (1958)
7. J. Kerr, Philos. Mag. **3**(5), 321 (1877)
8. J. Kerr, Philos. Mag. **5**, 161 (1878)
9. P. Zeeman, Leiden Commun. **29**, 3 (1896)
10. H.J. Williams, F.G. Foster, E.A. Wood, Phys. Rev. **82**(1), 119 (1951)
11. C.A. Fowler, E.M. Fryer, Phys. Rev. **86**(3), 426 (1952)
12. J. Kranz, A. Hubert, Z. Angew. Phys. **15**, 220 (1963)
13. F. Schmidt, W. Rave, A. Hubert, IEEE Trans. Magn. **21**, 307 (1986)
14. B. Argyle, B. Petek, D. Herman Jr., J. Appl. Phys. **61**(8), 4303 (1987)
15. R. Schäfer, A. Hubert, physica status solidi (a) **118**(1), 271 (1990)
16. R. Schäfer, in *Handbook of Magnetism and Advanced Magnetic Materials*, vol. 3., ed. by H. Kronmüller, S.S.P. Parkin (Wiley, London, 2007), p. 1513
17. J. McCord, J. Phys. D: Appl. Phys. **48**(33), 333001 (2015)
18. A. Hubert, R. Schäfer, *Magnetic Domains: The Analysis of Magnetic Microstructures* (Springer, Berlin, 1998)
19. E. Hecht, *Optics* (Addison Wesley, San Francisco, 2002)
20. M. Born, E. Wolf, *Principles of Optics* (Pergamon Press, Oxford, 1980)
21. P. Bruno, Y. Suzuki, C. Chappert, Phys. Rev. B **53**(14), 9214 (1996)
22. A.V. Sokolv, *Optical Properties of Metals* (Blackie and Son, London, Glasgow, 1967)
23. S. Višňovský, *Optics in Magnetic Multilayers and Nanostructures* (Taylor & Francis, Boca Raton, 2006)
24. J. Zak, E. Moog, C. Liu, S. Bader, J. Magn. Magn. Mater. **89**(1–2), 107 (1990)
25. C.Y. You, S.C. Shin, J. Appl. Phys. **84**(1), 541 (1998)
26. J. Xu, C. Zhou, M. Jia, D. Shi, C.L.H. Chen, G. Chen, G. Zhang, Y. Liang, J. Li, W. Zhang, Y. Wu, Phys. Rev. B **100**, 134413 (2019)

27. W. Roth, J. Appl. Phys. **31**(1), 2000 (1960)
28. V. Kamberský, Phys. Stat. Sol. (a) **123**(1), K71 (1991)
29. V. Kamberský, Phys. Stat. Sol. (a) **125**(2), K117 (1991)
30. V. Kamberský, J. Magn. Magn. Mater. **104–107**, 311 (1992)
31. V. Kamberský, L. Wenzel, A. Hubert, J. Magn. Magn. Mater. **189**(2), 149 (1998)
32. A. Thiaville, A. Hubert, R. Schäfer, J. Appl. Phys. **69**(8), 4551 (1991)
33. G. Traeger, L. Wenzel, A. Hubert, Phys. Stat. sol. (a) **131**(1), 201 (1992)
34. P.H. Lissberger, J. Opt. Soc. Am. **51**, 957 (1961)
35. S.D. Bader, J. Magn. Magn. Mater. **100**, 440 (1991)
36. A. Hubert, G. Traeger, J. Magn. Magn. Mater. **124**(1–2), 185 (1993)
37. G. Träger, L. Wenzel, A. Hubert, V. Kamberský, IEEE Trans. Magn. **29**, 3408 (1993)
38. L. Wenzel, V. Kamberský, A. Hubert, Phys. Stat. Sol. (a) **151**(2), 449 (1995)
39. L. Wenzel, A. Hubert, V. Kamberský, J. Magn. Magn. Mater. **175**(1), 205 (1997)
40. R. Wayne, *Light and Video Microscopy* (Academic, New York, 2010)
41. F. Schmidt, A. Hubert, J. Magn. Magn. Mater. **61**(3), 307 (1986)
42. I. Soldatov, W. Jiang, S. te Velthuis, A. Hoffmann, R. Schäfer, Appl. Phys. Lett. **112**, 262404 (2018)
43. E. Nikulina, O. Idigoras, P. Vavassori, A. Chuvilin, A. Berger, Appl. Phys. Lett. **100**(14), 142401 (2012)
44. C. Wright, N. Heyes, W. Clegg, E. Hill, Microsc. Anal. **46**, 21 (1995)
45. M. Freeman, W. Hiebert, in *Spin Dynamics in Confined Magnetic Structures I*, ed. by B. Hillebrands, K. Ounadjela (Springer, Berlin, 2002), pp. 93–126
46. B.E. Argyle, J.G. McCord, J. Appl. Phys. **87**(9), 6487 (2000)
47. B.E. Argyle, J.G. McCord, in *Magnetic Storage Systems Beyond 2000* (Springer, Berlin, 2001), pp. 287–305
48. T. Hofe, N. Urs, B. Mozooni, T. Jansen, C. Kirchhof, D. Bürgler, E. Quandt, J. McCord, Appl. Phys. Lett. **103**, 142410 (2013)
49. I. Soldatov, R. Schäfer, Rev. Sci. Instr. **88**(33), 073701 (2017)
50. I. Soldatov, R. Schäfer, J. Appl. Phys. **122**, 153906 (2017)
51. D. Chumakov, J. McCord, R. Schäfer, L. Schultz, H. Vinzelberg, R. Kaltofen, I. Mönch, Phys. Rev. B **71**(1), 014410 (2005)
52. R. Schäfer, I. Soldatov, S. Arai, J. Magn. Magn. Mat. **474**, 221 (2019)
53. S. Flohrer, R. Schäfer, J. McCord, S. Roth, L. Schultz, G. Herzer, Acta Mat. **54**(12), 3253 (2006)
54. N.O. Urs, B. Mozooni, P. Mazalski, M. Kustov, P. Hayes, S. Deldar, E. Quandt, J. McCord, AIP Adv. **6**(5), 055605 (2016)
55. N.O. Urs, E. Golubeva, V. Röbisch, S. Toxvaerd, S. Deldar, R. Knöchel, M. Höft, E. Quandt, D. Meyners, J. McCord, Phys. Rev. Appl. **13**(2), 024018 (2020)
56. A. Neudert, J. McCord, D. Chumakov, R. Schäfer, L. Schultz, Phys. Rev. B **71**(13), 134405 (2005)
57. B. Mozooni, T. von Hofe, J. McCord, Phys. Rev. B **90**(5), 054410 (2014)
58. R. Schäfer, A. Hubert, G. Herzer, J. Appl. Phys. **69**, 5325 (1991)
59. T. Hauet, S. Mangin, J. McCord, F. Montaigne, E.E. Fullerton, Phys. Rev. B **76**(14), 144423 (2007)
60. J. McCord, Y. Henry, T. Hauet, F. Montaigne, E.E. Fullerton, S. Mangin, Phys. Rev. B **78**(9), 094417 (2008)
61. I. Soldatov, N. Panarina, C. Hess, L. Schultz, R. Schäfer, Phys. Rev. B **90**(10), 104423 (2014)
62. S. Flohrer, R. Schäfer, C. Polak, G. Herzer, Acta Mat. **53**, 2937 (2005)
63. M. Lange, S. Guénon, F. Lever, R. Kleiner, D. Koelle, Rev. Sci. Instr. **88**(12), 123705 (2017)
64. R. Grechishkin, S. Chigirinsky, M. Gusev, O. Cugat, N.M. Dempsey, in *Magnetic Nanostructures in Modern Technology* (Springer, Berlin, 2008), pp. 195–224
65. Y. Lai, N. Scheerbaum, D. Hinz, O. Gutfleisch, R. Schäfer, L. Schultz, J. McCord, Appl. Phys. Lett. **90**(19), 192504 (2007)

66. H. Richert, H. Schmidt, S. Lindner, M. Lindner, B. Wenzel, R. Holzhey, R. Schäfer, Steel Res. Int. **87**, 232 (2016)
67. N.O. Urs, I. Teliban, A. Piorra, R. Knöchel, E. Quandt, J. McCord, Appl. Phys. Lett. **105**(20), 202406 (2014)
68. D. Markó, I. Soldatov, M. Tekielak, R. Schäfer, J. Magn. Magn. Mat. **396**, 9 (2015)
69. R.B. Holländer, C. Müller, M. Lohmann, B. Mozooni, J. McCord, J. Magn. Magn. Mat. **432**, 283 (2017)
70. R. Schäfer, in *Nanoscale Magnetic Materials and Applications*, ed. by J. Liu, E. Fullerton, O. Gutfleisch, D. Sellmyer (Springer, Boston, 2009), pp. 275–307
71. W. Rave, R. Schäfer, A. Hubert, J. Magn. Magn. Mat. **65**(1), 7 (1987)
72. I. Soldatov, R. Schäfer, Phys. Rev. B **95**(1), 014426 (2017)
73. R.B. Holländer, C. Müller, J. Schmalz, M. Gerken, J. McCord, Sci. Rep. **8**(1), 1 (2018)
74. M. Rührig, Mikromagnetische untersuchungen an gekoppelten weichmagnetischen schichten (micromagnetic investigations on coupled soft magnetic films). Ph.D. Thesis, Universität Erlangen-Nürnberg (1993)
75. R. Schäfer, J. Magn. Magn. Mater. **148**, 226 (1995)
76. R. Schäfer, R. Urban, D. Ullmann, H. Meyerheim, B. Heinrich, L. Schultz, J. Kirschner, Phys. Rev. B **65**, 144405 (2002)
77. J. Hamrle, J. Ferré, M. Nývlt, Š. Višňovský, Phys. Rev. B **66**, 224423 (2002)

X-Ray Magnetic Circular Dichroism and X-Ray Microscopy

Joachim Gräfe

Abstract X-ray magnetic circular dichroism (XMCD) and X-ray microscopy are ubiquitous techniques at modern synchrotron facilities, showing their importance for modern magnetism research. Indeed, XMCD and X-ray microscopy yield unique insights into the microscopic magnetic structure of materials that cannot be achieved with other techniques. Here, a beginner's introduction to the fundamental X-ray absorption effects is given, and important spectral properties of magnetic materials are discussed. Subsequently, various X-ray microscopies and their advantages and disadvantages are discussed, and the possibilities of such measurements are reviewed.

Keywords X-rays · Spectroscopy · Microscopy

1 Introduction

Magnetic measurements using soft X-rays generated by synchrotrons became a vital part of magnetism research with the advent of X-ray magnetic circular dichroism and X-ray magnetic linear dichroism [1, 2]. The discovery of these effects allowed X-rays to be used in a way that was previously only known for neutrons [3]. These X-ray techniques allow an unprecedented insight into the microscopic magnetic structure that no other measurement can achieve. In the following, a brief introduction to the capabilities is given that is only meant to give a taste of the almost endless possibilities of magnetic X-ray measurements. First, X-ray absorption spectroscopies will be discussed to introduce the fundamental effects and the information that can be gained from the spectral properties of materials. Second, an overview of different X-ray microscopies is given to showcase the ultimate capabilities of these techniques to unravel microscopic magnetic structures.

J. Gräfe (✉)
Max Planck Institute for Intelligent Systems, Stuttgart, Germany
e-mail: graefe@is.mpg.de

© Springer Nature Switzerland AG 2021
V. Franco, B. Dodrill (eds.), *Magnetic Measurement Techniques for Materials Characterization*, https://doi.org/10.1007/978-3-030-70443-8_10

2   X-Ray Absorption Spectroscopy

The main idea of using X-ray absorption experiments for materials characterization is to achieve some sort of absorption difference or contrast for X-rays depending on material parameters of interest. Commonly, the absorption is quantified by the energy-dependent X-ray absorption coefficient μ that can be determined from a transmission measurement according to Beer-Lambert law:

$$I = I_0 \cdot e^{-\mu d}.$$

with the initial intensity I_0 and the attenuated intensity I after transmission through a material layer of thickness d. Compared to visible light, the advantages of soft X-rays are strongly element-specific as well as chemically and magnetically sensitive absorption contrasts that are not accessible otherwise. However, this advantage comes at the cost of a more complicated experiment that relies on monochromatic soft X-rays that are only available from synchrotron radiation sources. In the following, the discussion is limited to a phenomenological description, and the reader is referred elsewhere [4–6] for a fundamental theory of X-ray absorption.

Elemental and Chemical Contrast

When an X-ray photon is absorbed, a core electron is excited into a state in an outer shell as illustrated in Fig. 1a. Thereby, the energy splitting between the different

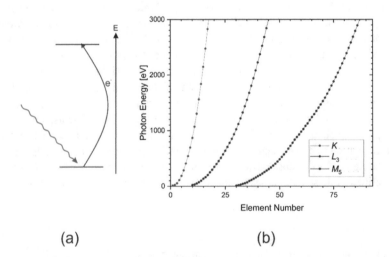

(a)   (b)

Fig. 1 (a) Schematic of the X-ray photon absorption process. (b) X-ray absorption energies of the K, L_3, and M_5 edges depending on the element number. Data taken from [7]

Fig. 2 (**a**) Schematic of the x-ray absorption process when the electron excitation involves final states that are split by the chemical environment of the atom. (**b**) illustration of the resulting absorption edge with multiple peaks corresponding to the split final state as indicated by the gray arrows

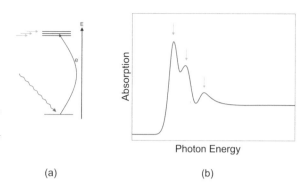

(a)   (b)

electron shells is effectively probed. The transition probability for the electron depends on the number of electrons in the initial state and the number of available final states, i.e., a core electron is used to probe unoccupied density of states. Hence, the X-ray absorption coefficient μ will be large if the photon energy matches the energy difference between the initial and final state and latter has a large unoccupied density of states. Therefore, X-ray absorption is very element-specific, and, thus, the energy dependence of μ can be used to determine the elemental composition of a sample by *X-ray absorption spectroscopy* (XAS). Because the X-ray absorption drastically increases when the photon energy becomes large enough to excite a core electron, these transitions are also called absorption edges. The energy of the K, L_3, and M_5 edges depending on the element number is shown in Fig. 1b [7], showing that the elemental composition can easily be identified by the characteristic position of these absorption edges. The nomenclature of the absorption edges goes back traditionally to the shell or principal quantum number of the excited core electron, e.g., K corresponds to the excitation of a 1 s electron, while the subscripted number indicates the orbital quantum number, e.g., L_2 and L_3 correspond to the excitation of a 2p electron. Most magnetic elements exhibit an absorption edge in the soft X-ray range up to a few keV [8].

The energy of the transition from core state to final state not only depends on the element but is also sensitive to the oxidation state and the chemical surrounding of the atom. This leads to small, but noticeable, changes to the photon absorption energies and the resulting X-ray absorption spectrum as illustrated in Fig. 2. Thus, the fine structure of the absorption edge, the so-called near-edge X-ray absorption fine structure (NEXAFS), yields extensive information about the sample's microscopic structure, e.g., the oxidation state, valence distribution, and coordination sphere of the atoms [4, 9].

In the context of magnetic materials, the NEXAFS signature is widely used to investigate magnetic oxides [10–13], but is not limited to these. Probably, the most prominent example is Fe_xO_y, as shown as an example in Fig. 3 [10]. The characteristic splitting of approximately 2 eV at the Fe L_3 edge between Fe^{2+} and Fe^{3+} species, indicated by gray arrows in Fig. 3, and the relative intensities of the peaks can be used to distinguish various iron oxides, ranging from FeO over Fe_3O_4 and γ-Fe_2O_3 to α-Fe_2O_3.

Fig. 3 Fe L_3 and L_2 edges of an Fe_xO_y compound measured at different temperatures. Gray arrows indicated the X-ray absorption energies for the Fe^{2+} and Fe^{3+} species in the compound at the L_3 and L_2 edges. Adapted from [10]

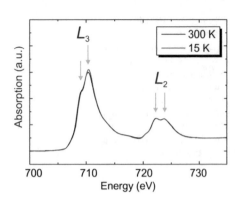

Magnetic Contrast

While NEXAFS spectroscopy already yields a wealth of information, X-ray absorption spectroscopy is the most powerful for magnetic materials characterization when employing magnetic contrast effects. Both *X-ray magnetic circular dichroism (XMCD)* and *X-ray linear circular dichroism* (XMLD) rely on polarized X-ray photons to achieve a difference in the X-ray absorption coefficient μ depending on the microscopic magnetic moment. The XMCD effect uses circular polarized X-ray photons and is linearly proportional to the magnetization, i.e., sensitive to uncompensated magnetic moments [3, 5, 14]. On the other hand, the XMLD effect uses linear polarized X-ray photons and is quadratically proportional to the magnetization, i.e., sensitive to compensated magnetic moments [15]. While XMLD measurements are an excellent tool to investigate anti-ferromagnets, XMCD measurements are more widespread as the effect is larger and the spectra are easier to interpret [3, 5, 14, 15].

XMCD was first theoretically predicted by Erskine and Stern [16] and experimentally demonstrated by Schütz et al. [2] in 1987. First experiments were based on hard X-rays, because of experimental limitations at the time, but since have moved to the soft X-ray regime. As discussed before, X-ray absorption transitions probe the unoccupied density of states. Thus, 2p → 3d and 3d → 4f transitions, i.e., $L_{2,3}$ and $M_{4,5}$ edges, are most interesting as the electronic occupation of the 3d and 4f states determines the magnetism of the transition and rare-earth elements, respectively. More precisely, the spin-dependent occupation of the 3d and 4f states determines their magnetism. Therefore, XMCD needs to provide X-ray photon absorption that results in spin-dependent 2p → 3d and 3d → 4f transitions, which is schematically illustrated in Fig. 4a.

To this end, circular polarized X-ray photons are used as they can transfer their angular momentum to preferentially excite spin-up or spin-down core electrons, depending on their helicity (indicated in green in Fig. 4a). The photon's angular momentum does not directly couple to the electron's spin but is mediated by spin orbit coupling [4, 14]. The electron's spin is conserved during the transition from

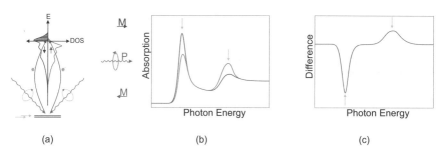

Fig. 4 Simplified schematic of the (**a**) spin- and polarization-dependent X-ray absorption transition and the (**b**) resulting X-ray absorption spectrum at the $L_{2,3}$ edge. (**c**) XMCD spectrum that results as a difference between the two XA spectra for positive and negative helicity. Gray arrows indicate the position of the L_2 and L_3 edge

initial to final state as a spin flip is forbidden [4]. Thus, the transition probability, i.e., the absorption coefficient μ, depends on the number of unoccupied states with appropriate spin (indicated in red/blue in Fig. 4a). Therefore, the peak intensity in the XAS will be different for the two photon helicities due to the different number of unoccupied spin-up and spin-down states, in turn, that allows deduction of the spin-dependent number of electrons [17], as illustrated in Fig. 4b. The difference in absorption $\Delta\mu_{XMCD}$ is proportional to the scalar product of the photon polarization P and the magnetization M:

$$\Delta\mu_{XMCD} \propto \vec{P} \cdot \vec{M}$$

where the effect can amount to up to 50% [1]. The resulting difference spectrum is called XMCD spectrum and is illustrated in Fig. 4c.

An XMCD spectrum yields a lot of additional insight into the microscopic structure of a magnetic material as illustrated in Fig. 5 [10]. While the XAS (cf. Fig. 3) only exhibited two features at the L_3 edge, the XMCD spectrum shows three. As the XMCD is not only sensitive to the element and chemical surrounding of the atom but also to the orientation of its magnetic moment, it distinguishes between magnetic ions on different sublattices with different spin orientation. In the example shown in Figs. 3 and 5, these are the typical sublattices of magnetic materials featuring a spinel structure [10, 18]. Thus, the XMCD spectrum allows to distinguish not only Fe^{2+} and Fe^{3+} species but also whether these are located on a lattice site with parallel or antiparallel magnetic coupling to its neighboring magnetic atoms.

It is noteworthy that the peaks in the XMCD spectrum sketched in Fig. 4c exhibit opposing sign at the L_3 and L_2 edge. This is due to the spin splitting of the $2p_{3/2}$ and $2p_{1/2}$ states that gives rise to the energy splitting between the L_3 and L_2 edge and that, in turn, results in a different sign of the spin orbit coupling for the two states. Thus, the photon helicity that leads to a preferential excitation of spin-up core electrons at the L_3 edge will lead to a preferential excitation of spin-down

Fig. 5 XMCD spectrum at
the Fe L_3 and L_2 edges of the
Fe_xO_y compound shown in
Fig. 3 measured at different
temperatures. Gray arrows
indicated the position of the
different Fe species,
depending on their oxidation
state and the parallel (down)
or antiparallel (up) alignment
with the magnetic majority
moment. (Adapted from [1,
10])

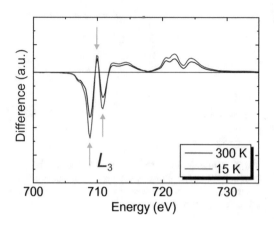

core electrons at the L_2 edge while still probing the same density of 3d states [14].
Therefore, the XMCD effect is not limited to determining the spin moment, but
it is also uniquely capable of determining the orbital moment. To achieve this,
it is necessary to also probe the orbital moment of the unoccupied final states,
i.e., to allow deduction of the orbital moment of the occupied states. Due to the
dipole selection rules, only certain transitions are allowed. Consequently, the orbital
moment can be calculated by comparing the photon absorption at the L_3 and L_2
edge.

Sum Rule Analysis

As hinted before, XMCD measurements not only result in a qualitative insight
into the presence and orientation of magnetic components in a sample but can
also be used for quantitative evaluation of the magnetic moment. First theoretically
derived by Thole et al. [19], extended by Carra et al. [17, 20], and experimentally
demonstrated by Chen et al. [21], the so-called sum rules allow the calculation of
magnetic spin and orbital moment. While the theoretical derivation is complicated
and was long disputed in literature, the principle analysis is quite simple and
depends on integrating the area under the XAS and XMCD spectra [3].

The necessary integrals for the sum rule analysis of transition metals are
illustrated in Fig. 6. First, the non-magnetic XAS needs to be analyzed. This can be
measured independently either using linear polarized X-ray photons or by averaging
the spectra for both helicities. The integral area underneath the absorption spectrum
(cf. Fig. 6a) is used for the so-called *charge rule C*:

$$C = \frac{I}{N}$$

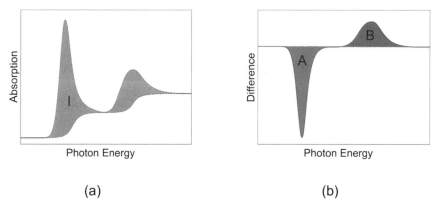

Fig. 6 Schematic illustration of the XMCD sum rule analysis. The areas in (**a**) show the integral I for the determination of the charge and (**b**) shows the areas A and B for the determination of the orbital and spin magnetic moments

with the absorption intensity I and the number of holes N, i.e., unoccupied states, that is usually taken from theory. Simplified, C corresponds to the total number of core electrons that can be excited. The so-called NEXAFS step function below the resonant absorption signal is subtracted before the area I is calculated. This step results from core electrons that are directly emitted as photo electrons and do not exhibit a resonant transition into a bound final state. Hence, they do not probe the unoccupied density of states. The step function is usually approximated with an error function:

$$I_{\text{step}}(E) = H \cdot \left[\frac{1}{2} + \frac{1}{2} \cdot \text{erf}\left(\frac{E - P}{W} \right) \right]$$

where I_{step} is the step intensity; H is the step height; E is the photon energy; P is the photo emission energy, i.e., the position of the step; and W is a parameter for the line width [14]. When only the integral for the sum rules is of interest, the choice of the parameter W is not relevant.

Subsequently, the orbital and spin magnetic moments can be calculated from the integral area underneath the XMCD spectrum (cf. Fig. 6b). The spin rule for the spin moment m_s reads:

$$\frac{m_{\text{s}}}{\mu_B} = -\frac{A - 2 \cdot B}{C}$$

with the Bohr magneton μ_B, and the orbital rule for the orbital moment m_o reads:

$$\frac{m_{\text{o}}}{\mu_B} = -\frac{2}{3} \frac{A + B}{C}.$$

Overall, the sum rule analysis routinely achieves measurement uncertainties below 10%. However, several issues, like self-absorption, unequal polarization, and improper background determination, can lead to inaccuracies. The reader is referred to more thorough discussions of sum rule analysis elsewhere [3, 5, 13, 14, 22, 23].

3 X-Ray Microscopy

In the previous part, integral measurements of X-ray contrast and microscopic information derived from spectral properties were discussed. However, all X-ray absorption effects can also be used for X-ray microscopy, i.e., imaging with element, chemical, and magnetic contrast. In addition to these unique contrasts, X-ray microscope also achieves superior resolution in comparison to conventional microscopy, due to the much shorter wavelength in the range of a few nm. In the following, a few major techniques for imaging with X-rays are discussed.

Photoemission Electron Microscopy

Photoemission electron microscopy (PEEM) was one of the first X-ray microscopies, pioneered in the 1930s and refined in the 1960s and 1970s by Engel and others [24]. PEEM benefited from the developments of electron optics for *transmission electron microscopy* (TEM) and is the only X-ray microscopy that does not rely on X-ray optics but purely on electron optics. This is a major advantage as the development of powerful X-ray optics has long been a challenge, due to the small change in refractive index that can be achieved with conventional lenses [6, 24–26].

A sketch of the working principle of PEEM is shown in Fig. 7. The unfocused X-ray beam is directed onto the sample where X-ray absorption processes result in the emission of photoelectrons. The emitted photoelectrons are focused onto a screen or 2D image detector using electron optics where they result in a microscopic image of the sample. The magnification is solely achieved by the electron optics. The main advantages of PEEM are the relatively simple setup, the lack of X-ray optics, and the direct imaging. A disadvantage is that the X-ray beam always illuminates a large area of the sample; thus, only a small fraction of the photons contributes to the image at high magnifications. An additional disadvantage for investigating magnetic materials with PEEM is that electrons, influenced by both the applied and the sample's magnetic stray field, are used for imaging, leading to image distortions. Furthermore, the use of electrons for imaging limits the magnetic fields that can be applied to the sample without interfering with the electron optics. Due to the PEEM relying on the emission of photoelectrons, it is in intrinsically surface-sensitive technique.

Fig. 7 Schematic of a photo electron emission microscope (PEEM). The unfocused X-ray beam is directed onto the sample where X-ray absorption processes result in the emission of photo electrons. The emitted photo electrons are focused onto a 2D image detector using electron optics where they result in a microscopic image of the sample. The magnification is solely achieved by the electron optics

Transmission X-Ray Microscopy

With the advent of micro- and nanostructuring techniques, X-ray optics became possible by moving to *Fresnel zone plates* (FZP), i.e., lenses that rely on diffraction instead of refraction. Unfortunately, the resolution r of such zone plates depends on the width of the outermost zone w with:

$$r = \frac{1.22}{w}.$$

Thus, fabricating high-quality and high-resolution X-ray lenses is still a major endeavor [25–28]. Regardless, Schmahl et al. [27] demonstrated a full-field *transmission X-ray microscopy* (TXM) using FZP-type X-ray lenses in 1974. The working principle of a TXM is similar to a conventional microscope and is sketched in Fig. 8. A condenser lens is used to illuminate the sample with the incoming X-ray beam, and an objective FZP lens projects the photons onto a screen or 2D image detector. While the resolution of the objective FZP determines the image resolution, the requirements for the condenser FZP are less stringent, and a capillary condenser can be used instead [6, 9]. Apart from the focusing FZP, the experimental setup of TXM is simple, and it directly generates an image. Unlike a PEEM, TXM is a photon-in, photon-out technique and is insensitive to magnetic fields generated by the sample or the sample environment. Similar to PEEM, the photon efficiency of TXM is reduced at higher magnifications. Hence, TXM is best suited for investigations requiring a larger field of view at moderate magnifications, like imaging of the evolution of domain patterns at the mesoscale. In contrast to PEEM,

Fig. 8 Schematic of a transmission X-ray microscope (TXM). Analogous to a conventional microscope, the sample is illuminated with a condenser lens, and subsequently the beam is projected onto a screen using an objective lens. Due to the small refractive index of X-rays, diffractive Fresnel zone plates (FZP) need to be used as objective lenses

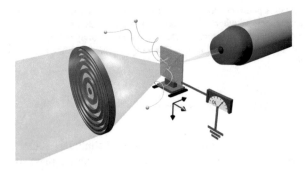

Fig. 9 Schematic of a scanning X-ray microscope (SXM). A Fresnel zone plate (FZP) is used to focus the X-ray beam to the resolution limiting illumination spot size. An order separating aperture (OSA, not shown) is used to eliminate higher diffraction orders of the FZP. The sample is scanned through the X-ray spot and arbitrary detectors can be used to measure the X-ray absorption contrast. The image is subsequently generated in a computer

the TXM is not surface-sensitive; however, it requires samples that are X-ray-transparent, i.e., thinner than 1 μm due to the small attenuation length of soft X-rays.

Scanning X-Ray Microscopy

Another FZP lens-based technique is *scanning X-ray microscopy* (SXM), whose working principle is sketched in Fig. 9. It was pioneered by Horrowitz and Howell in 1972 [29]. Similar to the TXM, an FZP is used to focus the X-ray beam to the smallest possible spot. An order separating aperture (OSA) is used to eliminate higher diffraction orders of the FZP. For an SXM, the resulting illumination spot determines the imaging resolution. Like in other scanning microscopy techniques, the sample is scanned through the X-ray spot using piezo-controlled nanopositioners

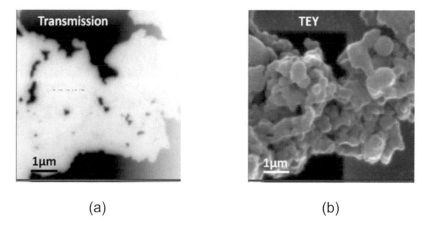

(a)   (b)

Fig. 10 SXM images with simultaneous use of (**a**) a transmission and (**b**) a total electron yield (TEY) detector of agglomerated Fe_xO_y nanoparticles. (Reproduced from [30])

[6, 9]. As the image is generated by scanning the sample, an SXM does not rely on spatially resolving detectors, and, indeed, multiple detectors can be used simultaneously, e.g., an *avalanche photodiode* (APD) to detect transmitted photons and a sensitive ammeter to determine the drain current of electrons ejected from the surface, i.e., the total electron yield (TEY) [30]. As the pixels of the image are measured individually, the two-dimensional image of the sample needs to be generated by a computer.

At large magnifications, SXM has an advantage over the previously discussed methods, because all photons are used in the smallest focal spot. However, for the same reasons, SXMs have a disadvantage when scanning a large field of view at low magnification as the need to make up the image from individual pixels of the focal spot size makes the measurements slow. Like TXM, SXM benefits from being a photon-in, photon-out technique that is insensitive to magnetic fields. In combination with the freedom of using arbitrary detectors, this makes SXM the most versatile platform for complicated sample environments, e.g., to apply RF fields or currents [31–34] or even gasses and liquids to the sample during measurement [35, 36]. However, this makes SXM an experimentally more complex technique. Depending on the detector that is used, SXM can either be surface- or bulk-sensitive, or both at the same time.

An example of such a measurement, making use of two detectors simultaneously, is shown in Fig. 10 [30]. Agglomerated Fe_xO_y particles are imaged using both the transmission and TEY mode of a SXM. While the former yields only little contrast as the agglomerated particles are optically dense, the surface-sensitive TEY image results in an almost three-dimensional representation that resembles scanning electron micrographs.

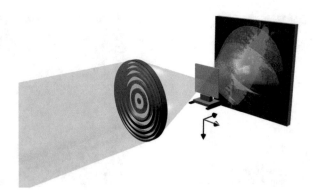

Fig. 11 Schematic of a scanning X-ray microscope (SXM) with 2D image detector for ptychography. The X-ray beam is focused onto the sample with a Fresnel zone plate (FZP) and the sample is scanned across the beam. For each position, the full diffraction pattern behind the sample is recorded, and the real space image is recovered by a computer algorithm after the measurement

Ptychography

Due to the limitations in FZP fabrication, the resolution achieved by an SXM is not wavelength-limited and is in the range of 20 nm [37]. One way to overcome this limitation is through coherent diffractive imaging (CDI) techniques that not only use the photon absorption but also evaluate the photon scattering in the sample [38, 39]. Using CDI resolution enhancement techniques, resolutions below 10 nm have been achieved [9, 37, 38, 40]. Ptychography is such a CDI technique that extends an SXM by a 2D detector behind the sample to record the photon scattering patterns as illustrated in Fig. 11. For ptychography, the diffraction pattern behind the sample is captured. Each subsequent illumination has an overlap with the previous, so that some of the diffraction patterns stay the same and some change. In combination with the known displacement of the sample during the measurement, the amplitude and phase of the sample's transmission and the illumination probe function can be recovered using an advanced computer algorithm [37–43]. In this image scheme, the resolution is not limited by the resolution of the FZP but by the maximum diffraction angle that is captured with the 2D detector behind the sample. Therefore, the requirements regarding the resolution of the FZP are not as stringent as for an SXM. Indeed, a larger illumination spot allows for faster imaging, because a smaller number of individual diffraction patterns need to be captured for a larger field of view. However, the lesser requirements for the FZP lens come at the cost of complex algorithmic data processing that needs a lot of computing power. Furthermore, all CDI techniques are photon hungry, i.e., a lot of coherent photons is necessary to get a sufficient signal-to-noise ratio of the diffraction patterns.

4 Summary

Soft X-ray absorption techniques are a very powerful tool for investigating magnetic materials. Although these techniques rely on synchrotron radiation, they provide a unique insight into the microscopic properties of the materials. X-ray absorption not only is element-specific but also depends on the oxidation, coordination, and magnetization state of the probed atoms. This can be leveraged for spectroscopic measurements that yield information about the collective properties and their distribution, but it can also be used as contrast for X-ray microscopic imaging.

X-ray imaging can be achieved with various approaches categorized as photon-in/electron-out or photon-in/photon-out techniques. While the former are experimentally easier, the latter are insensitive to magnetic fields and more suited for imaging of magnetic samples. Especially, scanning X-ray microscopy provides a very versatile platform for advanced experiments with challenging sample environments. Finally, the resolution of X-ray microscopies can be improved with coherent diffractive imaging techniques, like ptychography, which enhance the image by using diffraction as well as absorption information.

In conclusion, X-rays provide a unique access to nanoscopic sample properties and provide crucial information for pushing magnetic material science and engineering forward.

Acknowledgments The author would like to acknowledge fruitful discussions on how to explain XMCD with Eberhard J. Goering. Furthermore, the author is grateful to Nick Träger for generating magnificent 3D renderings.

References

1. G. van der Laan, B.T. Thole, Strong magnetic x-ray dichroism in 2p absorption spectra of 3d transition-metal ions. Phys. Rev. B **43**(16), 13401–13411 (1991)
2. G. Schütz et al., Absorption of circularly polarized x rays in iron. Phys. Rev. Lett. **58**(7), 737–740 (1987)
3. G. van der Laan, *25 Years of Magnetic X-Ray Dichroism* (Springer International Publishing, Cham, 2013)
4. J. Stöhr, NEXAFS spectroscopy, in *Springer Series in Surface Sciences*, vol. xv, 1st edn., (Springer, Berlin ; New York, 1996), p. 403
5. J. Stöhr, H.C. Siegmann, *Magnetism : From Fundamentals to Nanoscale Dynamics. . Springer Series in Solid-State Sciences*, vol xvii (Springer, Berlin ; New York, 2006), p. 820
6. P. Fischer, *Magnetic Imaging with Polarized Soft X-Rays* (Springer International Publishing, Cham, 2013)
7. C. Albert, D.T.A. Thompson, E.M. Gullikson, M.R. Howells, J.B. Kortright, A.L. Robinson, J.H. Underwood, K.-J. Kim, J. Kirz, I. Lindau, P. Pianetta, H. Winick, G.P. Williams, J.H. Scofield, *X-Ray Data Booklet*, 2nd edn. (Center for X-Ray Optics and Advanced Light Source, Berkeley, CA, 2001)
8. D. Attwood, *Soft X-Rays and Extreme Ultraviolet Radiation: Principles and Applications* (Cambridge University Press, Cambridge, 1999)

9. A.P. Hitchcock, Soft X-ray spectromicroscopy and ptychography. J. Electron Spectrosc. Relat. Phenom. **200**, 49–63 (2015)

10. D. Nolle et al., Structural and magnetic deconvolution of FePt/FeOx-nanoparticles using x-ray magnetic circular dichroism. New J. Phys. **11**(3), 033034 (2009)

11. T. Tietze et al., XMCD studies on co and li doped ZnO magnetic semiconductors. New J. Phys. **10**(5), 055009 (2008)

12. E.J. Goering et al., Absorption spectroscopy and XMCD at the Verwey transition of Fe3O4. J. Magn. Magn. Mater. **310**(2, Part 1), e249–e251 (2007)

13. E. Goering et al., Vanishing Fe 3d orbital moments in single-crystalline magnetite. Europhysics Letters (EPL) **73**(1), 97–103 (2006)

14. J. Stöhr, Exploring the microscopic origin of magnetic anisotropies with X-ray magnetic circular dichroism (XMCD) spectroscopy. J. Magn. Magn. Mater. **200**(1–3), 470–497 (1999)

15. G. van der Laan, *Anisotropic X-ray magnetic linear dichroism*, in *Magnetism and Synchrotron Radiation: Towards the Fourth Generation Light Sources*, (Springer International Publishing, Cham, 2013)

16. J.L. Erskine, E.A. Stern, Calculation of the ${M}_{23}$ magneto-optical absorption spectrum of ferromagnetic nickel. Phys. Rev. B **12**(11), 5016–5024 (1975)

17. P. Carra et al., X-ray circular dichroism and local magnetic fields. Phys. Rev. Lett. **70**(5), 694–697 (1993)

18. E. Goering, Large hidden orbital moments in magnetite. Phys. Status Solidi B **248**(10), 2345–2351 (2011)

19. B.T. Thole et al., X-ray circular dichroism as a probe of orbital magnetization. Phys. Rev. Lett. **68**(12), 1943–1946 (1992)

20. P. Carra et al., Magnetic X-ray dichroism: General features of dipolar and quadrupolar spectra. Phys. B Condens. Matter **192**(1), 182–190 (1993)

21. C.T. Chen et al., Experimental confirmation of the X-ray magnetic circular dichroism sum rules for iron and cobalt. Phys. Rev. Lett. **75**(1), 152–155 (1995)

22. E. Goering, X-ray magnetic circular dichroism sum rule correction for the light transition metals. Philos. Mag. **85**(25), 2895–2911 (2005)

23. P. Audehm et al., Pinned orbital moments—A new contribution to magnetic anisotropy. Sci. Rep. **6**, 25517 (2016)

24. J. Feng, A. Scholl, Photoemission electron microscopy (PEEM), in *Science of Microscopy*, ed. by P. W. Hawkes, J. C. H. Spence, (Springer New York, New York, NY, 2007), pp. 657–695

25. K. Keskinbora et al., Ion beam lithography for Fresnel zone plates in X-ray microscopy. Opt. Express **21**(10), 11747–11756 (2013)

26. U.T. Sanli et al., Overview of the multilayer-Fresnel zone plate and the kinoform lens development at MPI for intelligent systems. SPIE Optics + Optoelectronics **9510** (2015) SPIE

27. B. Niemann, D. Rudolph, G. Schmahl, Soft X-ray imaging zone plates with large zone numbers for microscopic and spectroscopic applications. Opt. Commun. **12**(2), 160–163 (1974)

28. K. Keskinbora et al., Rapid prototyping of Fresnel zone plates via direct Ga+ ion beam lithography for high-resolution X-ray imaging. ACS Nano **7**(11), 9788–9797 (2013)

29. P. Horowitz, J.A. Howell, A scanning X-ray microscope using synchrotron radiation. Science **178**(4061), 608–611 (1972)

30. D. Nolle et al., High contrast magnetic and nonmagnetic sample current microscopy for bulk and transparent samples using soft X-rays. Microsc. Microanal. **17**(05), 834–842 (2011)

31. H. Stoll et al., Imaging spin dynamics on the nanoscale using X-ray microscopy. Front. Phys. **3**, 26 (2015)

32. F. Groß et al., Nanoscale detection of spin wave deflection angles in permalloy. Appl. Phys. Lett. **114**(1), 012406 (2019)

33. S. Chung et al., Direct observation of Zhang-Li torque expansion of magnetic droplet solitons. Phys. Rev. Lett. **120**(21), 217204 (2018)

34. N. Ohmer et al., Phase evolution in single-crystalline LiFePO4 followed by in situ scanning X-ray microscopy of a micrometre-sized battery. Nat. Commun. **6**, 6045 (2015)

35. J. Lim et al., Origin and hysteresis of lithium compositional spatiodynamics within battery primary particles. Science **353**(6299), 566–571 (2016)
36. W.E. Gent et al., Coupling between oxygen redox and cation migration explains unusual electrochemistry in lithium-rich layered oxides. Nat. Commun. **8**(1), 2091 (2017)
37. D.A. Shapiro et al., Chemical composition mapping with nanometre resolution by soft X-ray microscopy. Nat. Photonics **8**, 765 (2014)
38. X. Zhu et al., Measuring spectroscopy and magnetism of extracted and intracellular magneto-somes using soft X-ray ptychography. Proc. Natl. Acad. Sci. **113**(51), E8219–E8227 (2016)
39. P. Thibault, M. Guizar-Sicairos, A. Menzel, Coherent imaging at the diffraction limit. J. Synchrotron Radiat. **21**(5), 1011–1018 (2014)
40. P. Thibault et al., High-resolution scanning X-ray diffraction microscopy. Science **321**(5887), 379–382 (2008)
41. J. Donatelli et al., CAMERA: the center for advanced mathematics for energy research applications. Synchrotron Radiat. News **28**(2), 4–9 (2015)
42. S. Marchesini et al., SHARP: A distributed GPU-based ptychographic solver. J. Appl. Crystallogr. **49**(4), 1245–1252 (2016)
43. P. Thibault et al., Probe retrieval in ptychographic coherent diffractive imaging. Ultramicroscopy **109**(4), 338–343 (2009)

Transmission, Scanning Transmission, and Scanning Electron Microscopy

Akira Sugawara

Abstract The properties of magnetic materials are closely related to their microstructure and micromagnetic structures like magnetic domains and spin textures. Electron microscopy is a powerful technique to characterize such features at nanometer resolution. This chapter describes structural characterization and magnetic imaging techniques using (scanning) transmission electron microscopy, scanning electron microscopy, and related techniques.

Keywords Microstructures · Structural analysis · Electron diffraction · Transmission electron microscopy · Scanning electron microscopy · Scanning transmission electron microscopy · Lorentz microscopy · Electron holography · Differential phase contrast imaging · Magnetic domain imaging · Magnetic domain wall · Micromagnetics · Magnetization reversal · Electron phase microscopy

1 Introduction

Overview

Electron microscopes are tools to observe objects at high resolution by using electrons as the primary beam. Electron microscopes magnify images of specimens by converging and diverging electron beams with electron lenses that consist of beam-bending electromagnets. Electrons have wave-particle duality. The wavelength of an electron wave depends on the accelerating voltage. For example, the wavelength is 2.51 pm when the voltage is 200 kV. Such short wavelengths enable us to observe the structure of matter at the atomic scale. This chapter describes three types of electron microscopes and associated observation methods using them: transmission

A. Sugawara (✉)
Center for Exploratory Research, Hitachi Ltd., Saitama, Japan
e-mail: akira.sugawara.ne@hitachi.com

© Springer Nature Switzerland AG 2021
V. Franco, B. Dodrill (eds.), *Magnetic Measurement Techniques for Materials Characterization*, https://doi.org/10.1007/978-3-030-70443-8_11

electron microscopy (TEM), scanning transmission electron microscopy (STEM), and scanning electron microscopy (SEM).

The main use of such electron microscopy is structural analysis. To improve the performance of magnetic materials, it is necessary to control microstructure characteristics such as the phase, interface structure, shape, size, and crystallographic orientation. The structural information obtained by electron microscopy is feedback for further development. Another important use is magnetic microscopy. Magnetic domain structures and magnetic textures can be determined by analyzing the interaction between the electron beam and magnetic induction, or spin states within a material.

Electron-Solid Interaction

Elastic Scattering

Electron propagation in an electromagnetic field is described by the following Schrodinger equation, where A is the vector potential and $\varnothing$ is the electric potential:

$$H = \frac{1}{2m}(p - eA)^2 + e\varnothing.$$

For transmission electron diffraction and microscopy, elastic forward scattering of fast electrons in the electric potential within a solid plays a central role. Elastic scattering of an electron by a single atom results from the attractive Coulomb interaction between the atomic nucleus and the electron. The scattering amplitude $f(\mathbf{k})$ is a function of the scattering vector $\mathbf{k}$, which is called the *atomic scattering factor*. The electron propagation within a crystalline specimen is a superposition of such scattering waves due to individual atoms arranged periodically. The resultant wave is represented by the *structure factor* $F_{\mathbf{g}}$, written in the following form:

$$F_{\mathbf{g}} = \sum f_i \, \exp\left(-2\pi \, \mathrm{i} \, \mathbf{g} \cdot \mathbf{r}_i\right),$$

where $\mathbf{g}$ is a reciprocal lattice vector, and f_i and $\mathbf{r}_i$ are the atomic scattering factor and coordinate vector, respectively, of the ith atom in a unit cell. Bragg diffraction occurs when the scattering vector coincides with g, causing $|F_g|$ to become nonzero. The structure factor is used in the same form in X-ray and neutron crystallography to determine crystal structure precisely on the basis of kinematical diffraction theory. Electron diffraction is not suitable, however, for that purpose, because the intensity of electron diffraction deviates from $|F_g|^2$, which varies in a complex manner depending on the specimen thickness and the angular deviation of the incident beam from the exact zone (high-symmetry) axis due to multiple scattering [1].

For magnetic analysis, electron phase shift is the central problem. Electrons subject to magnetic induction B are deflected by the Lorentz force, which results in a tilted wavefront, even though the amplitude does not change. If the gradient of the electron phase ϕ is determined by electron holography or transport-of-intensity equation (TIE) analysis, the magnetization state can be determined by [2, 3]:

$$\frac{\partial \phi}{\partial \boldsymbol{r}} = \frac{e}{\hbar} \int B(\boldsymbol{r}, z) \, dz.$$

Phase shift also occurs because of the electrostatic potential in a solid. As long as the measurement is not performed at atomic resolution, the mean inner potential V_0 (the zero-order Fourier component of the crystal potential) determines the phase shift:

$$\phi = \frac{2\pi}{\lambda E} \frac{E + E_0}{E + 2 E_0} \int V_0(\boldsymbol{r}, z) \, dz,$$

where λ is the electron wavelength, and E and E_0 are the kinetic and rest mass energies, respectively [3]. The magnetic and electrostatic contributions can be separated experimentally in an appropriate manner, as described later. Note that electron phase microscopy is capable of determining the magnetic induction (**B**-field) but not the magnetization (**M**-field). The influence of a demagnetization field should thus be considered.

Inelastic Scattering

Inelastic scattering is an event in which an electron loses part of its kinetic energy by, for example, inner-shell excitation, secondary electron emission, intraband transitions, or plasmon and phonon excitation. Inelastic scattering disrupts coherent image formation, but it provides useful information to analyze the material structure.

Thermal diffuse scattering by phonons causes high-angle scattering. Such electrons are used for high-angle annular dark-field (HAADF) imaging by STEM. Inelastic electron scattering associated with excitation of inner-shell electrons in a specimen is used to determine the chemical species and composition. Characteristic X-ray radiation occurs when an outer-shell electron fills a vacancy generated by inner-shell excitation of atoms by high-energy electrons. The energy (wavelength) of the photon is the difference between the electron energy levels that is specific for elements. The composition of the local area irradiated by electrons is determined by measuring the intensity profile as a function of the X-ray energy (wavelength). Electron energy-dispersive X-ray spectroscopy (EDS or EDX) using a compact solid-state detector is combined with TEM, STEM, and SEM. Wavelength-dispersive X-ray spectroscopy (WDS or WDX) using analyzer crystals is generally combined with SEM because of their large size of the spectrometer. Finally, electron energy-loss spectroscopy (EELS) uses a spectrometer capable of measuring the energy distribution of transmitted electrons. Edge energies in the range of 100–2500 eV are used to identify atoms. Furthermore, the energy-loss near-edge structure provides valuable information about the electronic states and local configurations of atoms.

Secondary and Auger Electron Emission

Electron emission from a solid surface occurs when it is irradiated by primary electrons. Secondary electrons are then emitted from the valence states of the specimen's constituent elements. Secondary electron microscopy is used as a surface-sensitive imaging method because of the short escape depth of secondary electrons. Auger electron emission is related to hole generation by excitation and relaxation of a core electron by an intra-atom transition of two outer-shell electrons. Auger electron spectroscopy is used for surface composition analysis because of the element-specific kinetic energy.

Diffraction and Imaging through Multiple Electron Lenses

An electron lens is a beam-bending electromagnet composed of a coil and a pair of pole pieces with a hole. Microscope images can thus be enlarged (or shrunk) by diverging (or converging) the electron beam. The focal length varies with lens current. For STEM and SEM, the lens is also used to form a small spot on the specimen surface by focusing the beam.

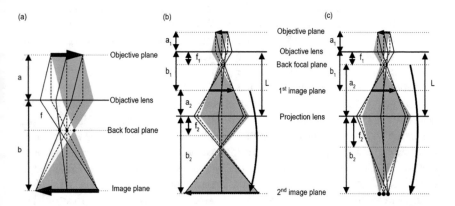

Fig. 1 Ray diagrams for an electron beam passing through an electron lens: (**a**) diffraction and image formation through a single lens, (**b**) transfer of image through a two-lens system, and (**c**) transfer of electron diffraction pattern through a two-lens system

The objective lens forms a diffraction pattern at the back focal plane and an image at the image plane, as shown in Fig. 1(a). Because of the electron diffraction in a crystalline specimen, the electron wave exiting from the specimen splits into several beams traveling in different directions. When the electron wave passes through a perfect objective lens (with no aberration), beams that travel in the same direction but exit from different positions on the specimen converge at a point on the diffraction plane, which is the focal length away from the lens. On the other hand, beams exiting from the same point on the specimen but propagating in different directions converge at a point on the image plane. The following lens formula determines the relationship among the object distance a, the image distance b, and the lens's focal length f:

$$\frac{1}{a} + \frac{1}{b} = \frac{1}{f}.$$

The lens's magnification M is then defined as:

$$M = b/a.$$

By stacking such electromagnetic lenses and tuning the individual currents, we can obtain a broad range of magnification (e.g., 200–1,500,000 ×). We can also conveniently switch between the TEM imaging mode and electron diffraction modes through a two-lens system, as shown in Fig. 1(b, c), respectively. The distance between the objective plane, on which the specimen is located, and the objective lens is a_1. The distance between the objective lens and the projection lens is L. The diffraction pattern and the image are formed at the back focal plane and first image plane, respectively. The focal length of the objective lens, f_1, is varied by tuning objective lens current. Accordingly, the distance between the objective lens and the first image plane, b_1, is varied so as to satisfy $\frac{1}{a_1} + \frac{1}{b_1} = \frac{1}{f_1}$. The transfer through the projection lens is described in similar manner. It is convenient to fix the distance between projection lens and the second image plane, b_2, when the downstream lenses are set to transfer the image on the first image plane to the detector. The objective plane of the projection lens can be controlled by changing the focal length of the projection lens f_2 according to the relation $\frac{1}{a_2} + \frac{1}{b_2} = \frac{1}{f_2}$, where a_2 is the distance between the objective lens and its objective plane. In order to form a specimen image on the second image plane, the image on the first image plane needs to be transferred. That is the distance between the first image plane and the projection lens $a_2 = L - b_1$ (Fig. 1(b)). On the other hand, a diffraction pattern is transferred from the back focal plane to the second image plane when the projection lens's focal length f_2 to satisfy $a_2 = L - f_1$ (Fig. 1(c)). Such combinations of lens currents are preset by manufacturers so that users can switch modes conveniently. Such modes cannot be used, however, for magnetic microscopy, because the specimen magnetization may saturate in such a large magnetic field as that of an objective lens excited with a large current. It is thus important to understand the multistage lens system to implement magnetic imaging modes on individual apparatuses.

Real electron lenses have aberrations that cause image and spot blurring. The electron wave traveling through the lens's outer zone converges at a point closer to the lens because of spherical aberration. Another important aberration is chromatic aberration, which causes waves with different wavelengths to converge at different points. Aberration also causes electron phase shifts as a function of the electron wave number. As a result, the electron image obtained as a superposition of scattered waves experiencing different phase shifts is different from that of the electron wave exiting a specimen [1].

Reduction of such aberrations improves the resolution and image distortion of electron microscopes. In light optics, spherical aberration can be suppressed to nearly zero by combining convergent and divergent lenses. An electromagnetic lens with rotational symmetry, however, acts only as a converging lens, and improvement of the aberration coefficient was a long-standing issue. The technical breakthrough was the development of a spherical aberration (C_s) corrector composed of multipole lenses [4, 5]. A C_s corrector produces a negative aberration coefficient and consequently acts as a divergent lens. As a result, the overall C_s coefficient of the combination of an objective lens and an aberration corrector can be reduced nearly to zero.

2   Transmission Electron Microscopy and Scanning Transmission Electron Microscopy

Instrumentation

Figure 2 shows a schematic diagram of a transmission electron microscope (TEM). Scanning transmission electron microscopy (STEM) can also be functionalized on a TEM by combining appropriate attachments, although dedicated STEM machines are also available. The electron path is evacuated to ensure a long travel distance for the electrons. A high-energy beam is generated in an electron gun consisting of an electron source and an accelerating tube. The illumination condition of the electron beam is tuned through an illumination system on the upstream side of the specimen. Electron scattering occurs when the beam passes through the specimen. The scattered wave is then transferred through objective and projection lenses to the detector.

An electron source with high brightness, a small probe size, and a small energy spread is suitable to perform high-resolution imaging, electron phase imaging, and elemental analysis in the nanometer region. A field emission (FE) electron source satisfies this requirement. In this source electrons are extracted by the tunneling effect when a high electric field is applied to the source's tip. A cold field emission (CFE) source produces electrons from a sharp W needle at room temperature.

Fig. 2 Schematic diagram of a transmission electron microscope (TEM): (1) electron source, (2) accelerating tube, (3) condenser lenses, (4) condenser aperture, (5) objective lens, (6) specimen holder, (7) goniometer, (8) objective aperture, (9) selected-area aperture, (10) intermediate and projection lenses, (11) image detector

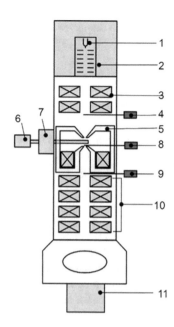

A Schottky-type source (also called a thermal FE source) produces electrons by heating emitter materials such as ZnO in a high electric field. Thermionic emitters like a hot W filament and LaB_6 are also used as inexpensive, stable sources. The electrons emitted from the source gain kinetic energy in the accelerating tube. An illumination system, including condenser lenses, the upper half of an objective lens, beam deflectors, and an aperture, controls the area, size, and incident and spread angles of the electron beam with respect to the specimen. A specimen is illuminated by a parallel broad beam in TEM, whereas it is illuminated by a focused fine beam spot in STEM, combined with a beam scan deflector unit and an optional probe-forming C_s corrector.

The specimen is usually located in the middle of the objective lens's pole-piece gap to obtain high spatial resolution and large magnification. For magnetic observation, however, the specimen has to be at a field-free position as described later. An imaging and projection system consists of the objective lens (lower half), intermediate and projection lenses, objective and selected-area apertures, and an optional image-forming C_s corrector. In the case of TEM, a real image of the specimen is formed at the first image plane and transferred to an image detector through multiple lenses, as described in the previous section. A typical image detector is a digital slow-scan charge-coupled device (CCD) or complementary metal-oxide-semiconductor (CMOS) camera specialized for TEM image acquisition, which replaces photographic plates. In the case of STEM, the beam focused on the specimen diverges on the downstream side and forms a disc-shaped intensity distribution at the STEM detector. The STEM signal source is mainly the electron current arriving at the annular detector electrodes as a function of the beam position.

Images are displayed by mapping the electron intensity accordingly. The electron energy loss and X-ray intensity are also used as sources for elemental mapping.

Specimen

The specimen has to be thin enough to transmit electrons for TEM and STEM. Such specimens are prepared by mechanical polishing followed by electrochemical and ion-beam etching. Focused ion beam (FIB) etching is used when a specific region of a sample needs to be thinned, such as a nanopatterned device on a Si wafer or the boundaries between grains of interest in a permanent magnet. After setting the specimen on a specimen holder, the holder is inserted into the microscope column. The specimen can be moved and tilted with respect to the beam using a goniometer.

The domain structure and magnetization reversal may be modified in thin specimens. One reason for this is the change in shape. For example, a cross-section specimen prepared from a thin film takes a wire shape. This changes the magnetization configuration significantly as a result of changes in the demagnetization factor. Another reason is defects induced during the thinning process and in an oxidation layer. For example, thinning may degrade the magnetocrystalline anisotropy of hard magnetic specimens.

Structural Analysis

Electron Diffraction

The transmission electron diffraction (TED) pattern formed at the back focal plane corresponds to the Fourier transform of the specimen's exit wave field. When the incident beam is tuned parallel to a zone axis of the specimen, a net electron diffraction pattern (in which the spots are arranged two-dimensionally) is obtained. The material phase is identified by comparing the diffraction spot arrangement and reciprocal lattice of the candidate materials. The most remarkable feature of TED is crystal structure determination in a local area. This is performed by placing a selective-area aperture at the image plane or illuminating the local area with a focused beam. A typical example is phase identification of a specific grain in a polycrystalline specimen. On the other hand, a ring diffraction pattern is obtained from a randomly oriented polycrystalline specimen, such as one prepared by vapor growth, while a halo pattern is obtained from an amorphous specimen.

Note that Bragg diffraction also occurs for inelastically scattered electrons, which are regarded as a spherical wave propagating from a point emitter (atom). A net intensity distribution, called a Kikuchi pattern [6], results from the diffraction. This phenomenon is observed in both TED and electron backscattering in SEM, and it is used to analyze the crystallographic orientations of specimens.

TEM Imaging

TEM imaging is performed in several modes. A bright-field image is obtained when an objective aperture is placed at the back focal plane to pass the direct beam only. This mode is conveniently used to observe microstructures in low and medium magnification, in which diffraction contrast results from the grain structure, thickness variation, local crystal bending, and absorption contrast associated with inelastic scattering. A dark-field image is obtained when the aperture is placed to pass a specific diffraction spot(s). Areas generating the diffraction spot(s) show bright contrast in the image. This mode is used to determine the crystallographic features of structural defects such as dislocations and planar defects. The resolution of bright- and dark-field images is suppressed to be low, because the small aperture in each case filters off high-spatial frequency information.

High-resolution TEM imaging is used to observe the local atomic structure at, for example, interfaces and defects. Such atomic images result from the interference of multiple diffraction beams excited by aligning the zone axis parallel to the incident beam. Interpretation of the image contrast is not easy because of the dynamical diffraction (multiple scattering) effect, as described earlier. Because of the dynamical effect, the exit wave may not represent the specimen's projected potential. Another problem is transfer through an objective lens with spherical aberration. The image blurring due to such spherical aberration determines the image resolution. On the other hand, the image formed by a perfect lens under an in-focus condition exhibits no contrast. The spherical aberration and defocus, which give extra electron phase shift as a function of the scattering vector, are therefore essential for contrast formation. The optimum defocus derived by Scherzer is $\Delta f_s = -1.15 \sqrt{C_s \lambda}$ [7]. Image simulation based on the multi-slice method is commonly used to examine the influence of the specimen thickness, defocus, and illumination condition [8].

In the case of a C_s-corrected TEM, an alternative measure to "resolution" is the *information limit*, which is determined by the wave coherency, chromatic aberration, illumination angle and stability of the lens current, and acceleration voltage. Therefore, only TEMs equipped with FE sources can benefit from C_s correctors [4].

Bright-Field/HAADF STEM Imaging

STEM imaging is classified into several methods according to the electron scattering angle. When the beam is focused on the specimen, it diverges on the downstream side and forms a disc-shaped intensity distribution at the detector plane. Bright-field imaging is performed by collecting small-angle scattered electrons near the direct beam. Dark-field imaging is performed in a similar manner when the incident beam is tilted at twice the Bragg angle. These imaging modes relate to the similar TEM imaging modes by reciprocal theory.

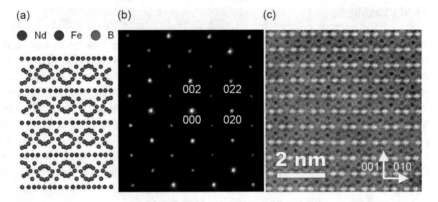

Fig. 3 Crystal structure of $Nd_2Fe_{14}B$ viewed along the [100] direction: (**a**) a structure model, (**b**) a TED pattern, and (**c**) a high-resolution HAADF-STEM image (courtesy of T. Kanemura and M. Shirai at Hitachi High-Tech)

Annular dark-field imaging is performed by using annular detectors to collect electrons scattered in a specific angle range. High-angle annular dark-field (HAADF) imaging uses electrons whose scattering angle is typically in the range of 50–200 mrad. Because the intensity is dominated by thermal diffuse scattering, which is enhanced as the atomic number (Z) increases, this is also called Z-contrast imaging. Interpretation of the image contrast is much simpler than in the case of TEM, because it is insensitive to the diffraction conditions, including the effect of multiple scattering. On the other hand, annular bright-field (ABF) images formed by the outer ring of the direct beam disc are used for high-resolution imaging of low-Z elements. HAADF-STEM has become a major method for sub-angstrom structural imaging since Cs-corrected STEM became widespread in the 2010s. Figure 3 shows a structure model, a TED pattern, and an atomic-resolution HAADF-STEM structural image obtained by a C_s-corrected TEM-STEM for $Nd_2Fe_{14}B$, which is the main phase in Nd-Fe-B magnets.

Compositional Mapping Using EDX/EELS

The magnetic properties of alloys and compounds can be modified by substituting elements without changing the crystal symmetry. In such cases, it is helpful to examine the local constituent elements and composition by energy-dispersive X-ray spectroscopy (EDX) and electron energy loss spectroscopy (EELS). Elemental mapping is effectively performed with STEM. Atomic-column X-ray imaging is achieved with a small- and large-current scanning probe formed using C_s corrector. Energy-filtered imaging is also possible in TEM with in-column or post-column energy filters.

Magnetic (Scanning) Transmission Electron Microscopy

Basics

The magnetic induction (**B**-field) within and around a magnetized specimen causes an electron phase gradient, as described previously, although it does not change the wave amplitude. No magnetic contrast is observed at the in-focus image plane. To observe magnetic structure, we have to measure the electron phase directly or convert such phase information into a visible intensity distribution. This section focuses on three representative methods: (1) Fresnel-mode and Foucault-mode Lorentz imaging, (2) differential phase contrast (DPC) Lorentz imaging, and (3) electron holography.

High-magnification structural imaging and diffraction are usually performed when a specimen is placed in a high magnetic field (~ 2 T) in the objective lens's pole-piece gap and excited with a large current. Because the magnetic domain structure is not usually preserved in such a high magnetic field, the specimen has to be located in a (nearly) field-free position. There are several approaches to satisfy such a condition:

1. Placing the specimen in the pole-piece gap, while setting the lens current to zero. Another lens on the downstream side (like an intermediate lens or Lorentz mini-lens) is used as an image-forming lens. The magnification and resolution are generally poor, because the additional lens's focal length is much larger than that of the high-resolution objective lens.
2. Placing the specimen at a field-free position between the objective lens and condenser lenses by installing an extra specimen stage. The conventional objective lens is then used as an image-forming lens with a long focal length. A spatial resolution of 0.23 nm has been achieved by combining a 1.2-MV CFE source and a C_s corrector [9].
3. Using a custom pole-piece design to shield or cancel out the magnetic field at the specimen position [10]. By combining a custom pole-piece canceling the field to zero at the specimen position and a C_s-corrected STEM, a 92-pm spatial resolution has been achieved [11].

The resolution of phase microscopy is measured by determining a small phase gradient between two positions. It is not determined simply by the spatial resolution of image transfer through the lens but by the precision of phase determination [12]. Because the phase difference is proportional to the specimen thickness, thick specimens are frequently used to improve the phase sensitivity. The phase microscopy, however, determines only the projected phase gradient over the beam path in the specimen. The uniformity of the magnetization distribution along the beam path should also be considered.

It is also possible to examine the changes in magnetization states, such as domain wall (DW) motion or magnetization rotation, in response to a varying external field and temperature. In situ magnetization observation is performed with a magnetizing

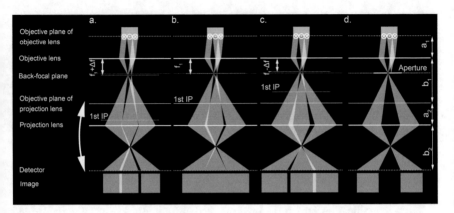

Fig. 4 Electron-beam ray diagrams for Fresnel-mode Lorentz imaging in the (**a**) under-focus, (**b**) in-focus, and (**c**) over-focus conditions and for (**d**) Foucault-mode Lorentz imaging

device. Some of the experimental examples described in this chapter were obtained by such in situ experiments.

Lorentz Microscopy

Lorentz microscopy is a method to observe a magnetic configuration by converting phase information to a visible intensity distribution. The Fresnel-mode Lorentz method is performed under a defocus condition, whereas the Foucault mode is based on dark-field imaging. Figure 4 shows ray diagrams of these methods on the basis of a two-lens system, as in the previous section. For simplicity, stripe domains separated by 180° DWs are assumed. Domains with opposite magnetization deflect the beam in opposite directions. As in the case of Bragg diffraction, beams deflected by the Lorentz force to the same direction converge to a spot at the back focal plane. On further propagation, the deflected beams reach the first image plane of the objective lens with a tilted wave front.

Figures 4(a–c) show the ray diagrams for the under-focus, in-focus, and over-focus conditions, respectively. The focal length of the objective lens (f_1) is varied, while that of the projection lens (f_2) is fixed to transfer the image formed at $z = a_1 + b_1$ to the detector. When the first image plane is shifted downstream (upstream) of $z = a_1 + b_1$ by the increase (decrease) in f_1, the under-focus (over-focus) image at $z = a_1 + b_1$ is transferred to the detector. In Fig. 4(a), the left DW shows bright contrast due to beam convergence, while the right DW shows dark contrast due to beam divergence. The in-focus image in Fig. 4(b) shows flat contrast. In Fig. 4(c), the left DW shows dark contrast, while the right DW shows bright contrast due to beam divergence. The converging and diverging contrast reverses when the sign of the defocus reverses. This makes it easy to distinguish whether contrast formation has a magnetic or diffractive origin by reversing the defocus sign. The width of a DW image is nearly proportional to the defocus, so it should not be regarded as the DW width. Fresnel-mode imaging is thus suitable for DW imaging. The imaging-mode can be performed on most commercial TEM's under-defocus condition in "low-magnification mode," of which objective lens current is nearly zero. Focusing is done by tuning the first projective (intermediate) lens. Its magnification and resolution thus may not be sufficient enough because the objective lens is deactivated. On the other hand, a Foucault-mode Lorentz image is obtained from the waves propagating in a specific direction as a result of the Lorentz deflection in domains. The direction is selected by using an aperture at the back focal plane, as shown in Fig. 4(d). Because only beams passing through the aperture contribute to the image intensity, Foucault-mode imaging is suitable for domain imaging. This imaging mode usually requires customized lens configuration: a special imaging lens (frequently called as "Lorentz lens" or "mini-lens") is placed in between a specimen and a selective-area aperture, or another aperture is placed on the downstream side of the first projective (intermediate) lens.

Figures 5(a–c) show under-focus, in-focus, and over-focus Fresnel-mode Lorentz images, respectively, of a nanopatterned permalloy rectangle. The DW contrast seen in the bright and dark lines reverses when the defocus sign reverses, whereas no contrast is observed in the in-focus image. Raw Fresnel-mode images are easily interpreted when the DW contrast is clearly involved. Fresnel-mode imaging is based on beam convergence/divergence due to propagation of a tilted wave front. The phase can be retrieved by numerically solving a differential equation describing the wave propagation, called the *transfer of intensity equation*(TIE):

$$\frac{2\pi}{\lambda} \frac{\partial I(r, z)}{\partial z} = -\frac{\partial}{\partial r} \left(I(r, z) \frac{\partial}{\partial r} \phi(r, z) \right).$$

The problem is reduced to image calculation using only three images: the over-focus, in-focus, and under-focus images [13]. Figure 5(d) shows a phase map calculated from Figs. 5(a–c) by applying TIE analysis. The phase's gradient represents the magnetic induction. The local induction direction and magnitude are expressed using a hue-saturation-value (brightness) color wheel, as shown in Fig. 5(e). Figure 5(f) shows the flux distribution around a vortex center where 90° walls cross. Phase retrieval by TIE analysis does not require a microscope equipped with an FE source. It is also suitable to obtain a large field of view. TIE analysis is especially useful to observe the intuitive magnetization variation, as when the magnetization direction changes gradually, in cases like a helical spin order [14] and a skyrmion in FeGe [15].

Uniaxial (or nearly so) magnetic materials are suitable for Foucault-mode Lorentz imaging because of the small angular distribution of the Lorentz deflection. An anisotropic Nd-Fe-B magnet is one of the most popular permanent magnets because of its large magnetization and coercivity. The dominant DW is of a 180° Bloch-wall-type, and the deviation of the easy axis is typically 15°. Figure 6 shows a series of Foucault-mode Lorentz images with a varying in-plane field for a sintered

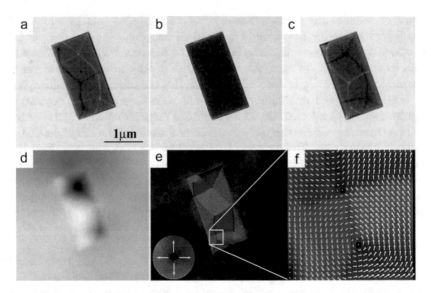

Fig. 5 Fresnel-mode Lorentz imaging and phase retrieval by TIE analysis for a nanopatterned permalloy rectangle: (**a**) under-focus, (**b**) in-focus, and (**c**) over-focus Fresnel images; (**d**) an electron phase image calculated by TIE analysis; (**e**) a B-field distribution color map; and (**f**) the magnified configuration around a vortex (lower) and an antivortex (upper) [13]. [Reprinted figure with permission from V.V. Volkov, Y. Zhu, and M. De Graef, Micron 33, 411 (2002). Copyright 2002 by Elsevier.]

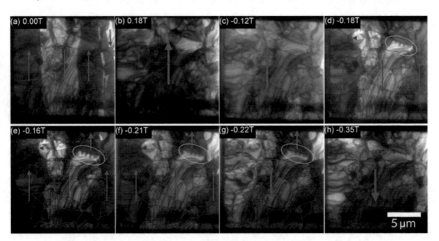

Fig. 6 In situ observation of the magnetization reversal within a polycrystalline anisotropic Nd-Fe-B magnet while varying the applied field [16]. The external field is (**a**) 0 T, (**b**) 0.18 T, (**c**) −0.12 T, (**d**) −0.18 T (**e**) −0.16 T, (**f**) −0.21 T, (**g**) −0.22 T and (**h**) −0.35 T. [Reprinted figure with permission from A. Sugawara, T. Shimakura, H. Nishihara, T. Akashi, Y. Takahashi, N. Moriya, M. Sugaya, Ultramicroscopy, 197, 105–111 (2019). Copyright 2019 by Elsevier.]

anisotropic Nd-Fe-B magnet. A specially designed magnetizing holder was used to apply an in-plane field up to $\mu_0 H = -0.5$ T [16]. The as-prepared specimen took a thermally demagnetized state. Stripe domains with 180° Bloch DWs are observed parallel to the easy direction within individual grains. The DWs moved continuously within the grains when the field was increased. Eventually, a saturation state was established as the DWs were swept out of the specimens. The DW nucleation followed by motion from the saturation phase occurred more discontinuously. The first reversed domain formation occurred at $\mu_0 H = -0.12$ T, but it was localized in a single grain. Significant propagation of reversal across several grains occurred at $\mu_0 H = -0.18$ and -0.23 T. Most of the domain boundaries coincide with grain boundaries. Few DWs are observed within the grains, except for head-on domains aligned on the c-plane grain boundaries marked by the dotted ovals. The DWs were therefore pinned at grain boundaries. Although coercivity of the bulk material is typically greater than 1 T, that of the thin, electron-transparent specimen was degraded by surface effects, including ion damage during specimen preparation.

Differential Phase Contrast Lorentz Imaging

Differential phase contrast (DPC) Lorentz imaging is performed using a STEM apparatus. It uses a circular detector (or annular detectors) divided typically into four circumferential segments. A magnetized specimen produces an asymmetric current distribution flowing into the individual segments as a function of the magnitude and direction of magnetic induction within the individual domains, as shown in Fig. 7 [17].

Figure 8 show a series of DPC Lorentz images of a permalloy ring with a 2.5-/3-μm inner/outer diameter as a function of the in-plane magnetic field. The magnetic field (170 kA/m) was applied parallel to the electron beam by using the

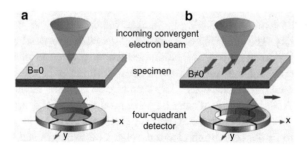

Fig. 7 Principle of DPC Lorentz imaging using a four-quadrant annular detector [17]. Electron current distribution is (**a**) symmetric when the electron beam travels through vacuum and non-magnetic specimens (where B=0), while it is (**b**) asymmetric when the beam travels through ferromagnetic specimens (where B≠0). [Reprinted figure with permission from T. Uhlig and J. Zweck, Phys. Rev. Lett. 93, 047203 (2004). Copyright 2004 by the American Physical Society.]

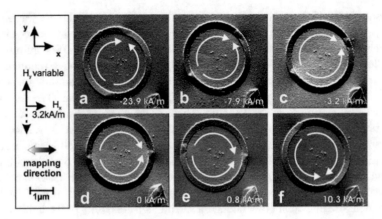

Fig. 8 Variation of the magnetic induction distribution within a permalloy ring as a function of the in-plane external field [17]. The external field is (**a**) −23.9 kA/m, (**b**) −7.9 kA/m, (**c**) −3.2kA/m, (**d**) 0 kA/m (**e**) 0.8 kA/m, (**f**) 10.3 kA/m. [Reprinted figure with permission from T. Uhlig and J. Zweck, Phys. Rev. Lett. 93, 047203 (2004). Copyright 2004 by the American Physical Society.]

objective lens, and the in-plane field component was controlled through specimen tilt. An "onion" state composed of a pair of half-ring domains with opposite circulation directions occurred when a sufficiently large field was applied to the specimen. Head-to-head and tail-to-tail DWs were formed accordingly. The pair of DWs moved on the ring as a function of the applied field, so that the net magnetization parallel to the applied field varied. It was also found that the domain configuration switched abruptly through an intermediate state with a vortex (flux closure) structure.

Electron Holography

Holography is a two-stage imaging method to obtain amplitude and phase information separately. The principle of holography was first proposed by Gabor [18], whose original motivation was aberration correction of electron microscopes. The first step is recording the interference fringes, forming a hologram, between a reference wave and an object wave. The second step is phase reconstruction from the hologram. A highly coherent electron wave produced by FE sources is required. The apparatus uses a Moellenstedt-type [19] electron biprism composed of a thin conductive wire in the center and electrode plates on the right and left sides. The biprism splits the electron wavefront and tilts the two split beams in opposite directions in the electrostatic field. By placing the biprism upstream of the first image plane, superposition of the object and reference waves occurs at that plane. The biprism is usually placed at the position of the selective-aperture. As in the case of the Foucault-mode Lorentz imaging, a customized lens configuration is frequently required for high-magnification and resolution observation to optimize

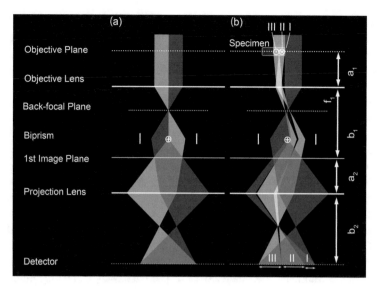

Fig. 9 Electron beam ray diagram for electron holography: (a) a reference hologram and (b) an object hologram

the geometry among a specimen, a biprism, and imaging and projection lenses. The interference pattern is then transferred to the detector through the projection lens. Figure 9(a) shows the ray diagram for the case in which both beams travel in vacuum. Although widely spaced fringes associated with the defocused image of the prism wire edge are superimposed, the intensity distribution is basically sinusoidal. Such a hologram is used as a reference hologram to subtract the background phase map.

When the beam on the left passes through a magnetic specimen with a uniform thickness, beam deflection due to the Lorentz force occurs as depicted in Fig. 9(b). The fringe spacing in zone I is the same as that in the reference hologram, because the interference fringe between unperturbed waves occurs there. The fringe spacing in zone II (III) is narrower (wider) than that in zone I, because the relative angle between the object and reference waves is larger (smaller) there. On the other hand, the electrostatic phase shifts the peak positions of the fringes without changing the spacing. Information about the phase gradient due to the magnetic nature and the phase shift due to the electrostatic potential is thus embedded in the hologram.

The phase reconstruction is performed through numerical computation. A digital image recording system with high linearity and a wide dynamic range, like a slow-scan CCD or CMOS camera, serves as the basis for numerical phase reconstruction. The Fourier transform of the reference hologram simply returns a zero-order peak and two first-order peaks centered at the spatial frequency of the hologram fringes. In contrast, the Fourier transform of the object hologram returns a scattering profile around the spatial frequency center, because the fringe spacing and direction are modified by the magnetic phase gradient and electrostatic phase shift. A zone centered at the wave number of the carrier fringe is called a "sideband." After recentering one of the following sidebands by applying a mask, typically with a radius of 1/3 of the spatial frequency, the inverse Fourier transform is applied to the sideband in complex form. The amplitude and phase are then separated from the resultant complex image. The formulation and experimental procedure of reconstruction using a fast Fourier transform (FFT) are described in a handbook article [20].

To determine the **B**-field distribution, the electrostatic contribution to the phase has to be excluded. Procedures for this include subtraction of the phase map obtained by flipping the specimen, subtraction of the phase map obtained for ferromagnetic and paramagnetic states by varying the temperature [21], and subtraction of saturation in the opposite direction by applying a strong in-plane pulse field [12]. Once the magnetic phase distribution is determined, the B-field is mapped by calculating the local phase gradient. A color wheel is frequently used as in Figs. 5(e–f). In addition, the equiphase lines obtained by applying a sine or cosine function to the magnetic phase map represent the magnetic flux. The direction and spacing of the equiphase lines indicate the respective direction and magnitude of the **B**-field. Before numerical phase reconstruction using computers became popular in the 1990s, reconstruction was performed by laser interferometry from holograms recorded on photographic plates. Such early results included experimental verification of the Aharonov-Bohm effect [22]. Interferograms corresponding to the equiphase line distribution were frequently used for data representation at the time [2].

(Ga,Mn)As is a ferromagnetic semiconductor obtained by substituting Mn into Ga sites in GaAs. Its magnetic anisotropy is controlled from in-plane to out-of-plane anisotropy. The temperature dependence of the in-plane anisotropy is also interesting. The anisotropy axes rotate in the film plane with temperature: bidirectional anisotropy ([100] and [010]) is dominant at low temperature, whereas unidirectional anisotropy ([110]) is dominant near the Curie temperature. The magnetization direction can be measured by electron phase measurement as a function of temperature even in the absence of an external field. Such measurement was performed for $Ga_{0.96}Mn_{0.04}As$, which had a saturation magnetization of only 0.02 T and Curie temperature of approximately 60 K, using a liquid-He specimen-cooling holder [21]. Figure 10 shows Fresnel-mode Lorentz images and flux distribution images obtained by electron holography. Only a near-180° DW parallel to $[1\bar{1}0]$ was visible at (a) 30.5 K. As the temperature was lowered to (b) 25.8 K, a small-angle DW became visible along [110] according to the B-field rotation toward the [100] and [010] directions. Near-90° DWs were found to cross at (c) 9.8 K, and the B-fields were nearly parallel to [100] and [010]. The results also confirm that the magnetization increases as the temperature decreases, as evidenced by an increase in the equiphase line density.

Computed tomography (CT) is widely used to obtain cross-sectional images of a three-dimensional object by acquiring many images from different directions. Tomographic reconstruction of scalar quantities such as density has been well

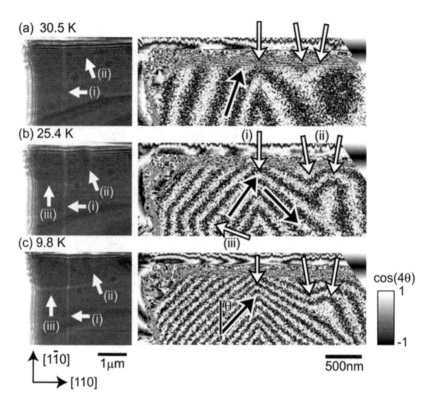

Fig. 10 Variation of the magnetization state of (Ga,Mn) as a function of temperature, shown via (left) Fresnel-mode Lorentz images and (right) magnetic flux distribution images obtained by electron holography, at (**a**) 30.5, (**b**) 25.4, and (**c**) 9.8 K. The white arrows in the flux images show the DWs, and the black arrows indicate the local magnetic flux direction [21]. [Reprinted figure with permission from A. Sugawara, H. Kasai, A. Tonomura, P.D. Brown, R.P. Campion, K.W. Edmonds, B.L. Gallagher, J. Zemen, T. Jungwirth, Phys. Rev. Lett. 100, 047202 (2008). Copyright 2008 by the American Physical Society.]

established. On the other hand, vector field tomography to determine, for example, a three-dimensional magnetization distribution is still under development [23].

The 3D magnetization configuration within a pillar containing a stack of Fe (30 nm)/Cr (10 nm)/Fe (20 nm) discs of 180-nm diameter was examined by electron holography and tomography [24]. The specimen was prepared from a multilayer specimen by using a focused ion beam. Tomographic data was acquired by using a special holder that could rotate the specimen pillar by 360° around two independent rotation axes in a 1-MV CFE-TEM. Circulating flux structures (vortices) were formed in both Fe discs. Although the circulation directions were parallel, the vertical magnetizations in the vortex cores were antiparallel, as shown in Fig. 11. Interestingly, the vortex centers were slightly misaligned with each other, probably to reduce the magnetostatic energy penalty due to the tail-to-tail

Fig. 11 3D view of the tail-to-tail vortex cores in a stack of Fe (30 nm)/Cr (10 nm)/Fe (20 nm) discs with a diameter of 180 nm [24]. [Reprinted figure with permission from T. Tanigaki, Y. Takahashi, T. Shimakura, T. Akashi, R. Tsuneta, A. Sugawara, D. Shindo, "Three-dimensional observation of magnetic vortex cores in stacked ferromagnetic discs", Nano Lett. 15 (2015) 1309–1314. Copyright 2015 American Chemical Society.)]

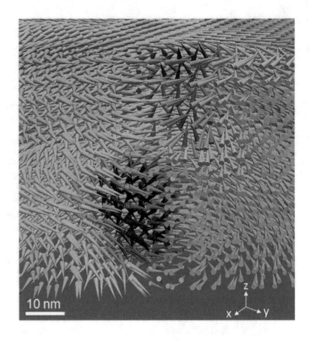

magnetization configuration. This tendency was also confirmed by Landau-Lifshitz-Gilbert micromagnetic simulation.

3 Scanning Electron Microscopy

Instrumentation

A scanning electron microscope (SEM) maps various types of position-dependent signals, such as secondary electrons, backscattered electrons, Kikuchi diffraction patterns, and emitted X-rays, which are excited by scanning a focused electron beam on the specimen surface [25]. To increase or activate such scattering events in the subsurface region, a relatively low electron acceleration voltage is favored (typically from a few hundred volts to tens of kilovolts). SEM is used more frequently than TEM, because much larger specimens can be observed without time-consuming specimen preparation in many cases. It is possible to observe, for example, a large bulk specimen or large-area thin-film specimen as deposited or as nanofabricated on a substrate. Although the resolution is better for the in-lens configuration (i.e., when the specimen is located in the pole-piece gap), as in the case of TEM, the out-lens configuration is still suitable for observing large magnetic specimens. The latter configuration also allows us to give various measurement attachments access to the specimen, because of the large free space around it.

Structural Analysis

Secondary Electron (SE) and Backscattered Electron (BSE) Imaging

The main use of SEM is surface observation by secondary electron (SE) imaging. An SE image is obtained by mapping the SE intensity emitted from the surface. The kinetic energy of most SEs is smaller than 20 eV, so their escape depth is on the order of nanometers. SE images mainly provide information about surface morphology and composition.

A backscattered electron (BSE) image is obtained by mapping the intensity of electrons scattered elastically at angles of nearly 180°. Because the scattering amplitude increases with the atomic number Z, BSE images are sensitive to the composition distribution.

The resolution of SE imaging is nearly the same as the size of the primary electron beam. For example, the resolution of a commercial SEM equipped with a CFE source at 30 kV reaches down to 0.4 nm for the in-lens configuration. On the other hand, the resolution of BSE imaging is not as high as that of SE imaging, because the scattering events occur deeper below the surface.

Electron Backscattering Diffraction (EBSD)

The electron backscattering diffraction (EBSD) method determines a two-dimensional mapping of crystal structure and orientation. For example, it is used to analyze the phase and orientation distributions of anisotropic permanent magnets [26] and rolled electromagnetic steels [27]. As described in the previous section, Kikuchi patterns due to the diffraction of inelastically scattered electrons are observed in the backscattering configuration in SEM. The crystal phase and orientation at the beam position are determined through pattern matching between the observed and simulated Kikuchi patterns of candidate materials. Another important feature of this method is its high sensitivity (down to $\sim 10^{-4}$) to strain. Two-dimensional phase, orientation, and strain mapping of polycrystalline, multiphase materials can be obtained by repeating acquisition with the scanned beam.

WDX Imaging (EPMA)

SEMs are frequently equipped with apparatuses for EDX and wavelength-dispersive X-ray spectroscopy (WDX). WDX is capable of separating elements with excitation energies close to each other by applying Bragg diffraction using analyzer crystals. It is usually combined with SEM because of large size of the spectrometer. An apparatus specialized for the WDX function is also called an *electron probe microanalyzer* (EPMA). This method is useful for analyzing compounds including multiple rare-earth elements. Examples include sintered Nd-Fe-B hard magnets with low concentrations of heavy rare-earth elements such as Dy and Tb substituted to enhance the coercivity.

Magnetic Microscopy

Type-I and Type-II Contrast

Magnetic domain contrast occurs frequently in SEM observation of ferromagnetic materials. Such contrast formation is due to deflection of secondary or backscattered electrons in the stray field near the specimen surface. Type-I contrast is observed for secondary electrons [28], while type-II contrast is observed for backscattered electrons [29].

Scanning Electron Microscopy with Polarization Analysis (SEMPA)

Low-energy secondary electrons escaping from the subsurface region carry information about the electron state of the conduction band slightly below the Fermi level. These secondary electrons are also spin-polarized in the case of magnetic materials containing 3D transition metals. Spin-up and spin-down secondary electrons are measured separately by using a spin analyzer. The magnetic domain is imaged by mapping the spin polarization, which is the difference between the spin-up and spin-down electrons [30]. A Mott detector is a spin analyzer applying asymmetric backscattering of the secondary electrons impinging on a heavy-metal target such as Au. Another type of high-efficiency analyzer using a single-crystalline oxygen-passivated Fe film recently became available. A technical difficulty of SEMPA is that measurement has to be performed in ultrahigh vacuum so as not to degrade the spin polarization of secondary electrons by surface contamination.

An important feature of SEMPA is its high sensitivity to surface magnetization (i.e., the **M**-field). For example, it is difficult to measure perpendicularly magnetized thin films by electron phase microscopy using TEM, because the **B**-field is almost canceled when the demagnetization factor is nearly one. Figure 12 shows the in-plane magnetization distribution of a perpendicularly magnetized Co/Ir bilayer [31]. The spin configuration within the DW was found to be Neel-type whirling, suggesting that the round domain is a Neel-type skyrmion, which is also difficult to measure by TEM-based electron phase microscopy because of the demagnetization field.

Fig. 12 SEMPA images obtained for Co/Ir (111): (**a**) a vertical magnetization component, (**b**) a horizontal magnetization component, and (**c**) a composite image [31]. [Reprinted figure with permission F. Kloodt-Twesten, S. Kuhrau, H.P. Oepen, R. Frömter, Phys. Rev. B 100, 100402(R) (2019). Copyright 2019 by the American Physical Society.]

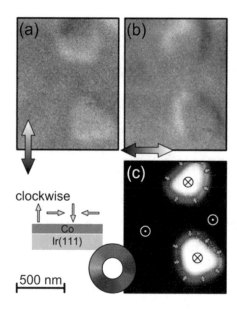

4 Summary

This chapter has described structural and magnetic characterization of magnetic materials by using TEM, STEM, and SEM. The performance of structural analyses at the atomic scale has been improved significantly in recent years by applying the benefits of C_s correctors. Magnetic characterization with sub-nanometer resolution is also becoming possible. A barrier to use the magnetic imaging described in this chapter is the customized configuration of the microscopes. Some research institutes accept proposals to use such microscopes as shared facilities to enhance research opportunities [e.g., [32]].

The technical difficulties arising in recent years are due to a lack of phase sensitivity. Small **B**-field components, like perpendicularly magnetized layers of small-magnetization materials, have come to be used in novel, compact, efficient spintronics devices (the ultimate form is anti-ferromagnetic spintronics). One novel technique capable of accessing the spin state (**M**-field) at the atomic scale is electron magnetic circular dichroism (EMCD) [33], although it has not yet reached a practical stage. Another challenging issue is improving temporal resolution to observe fast magnetic switching. Electron phase microscopy using an ultrafast TEM is expected to lead to new fields [34].

Acknowledgments AS thanks Kanemura T and Shirai M for providing a high-resolution image obtained using Hitachi High-Tech, HF5000. AS also thanks Akashi T, Kasasi H, Takahashi Y, Tanigaki T, Yoshida T, Fukunaga K, Kohashi T, Harada K, Tonomura A, Osakabe N, Shinada H, Otsuka N, Venables JA, Nakamura Y, and Nittono O for helpful discussion and supervision. A special thanks is devoted to Scheinfein MR for providing me a research opportunity of magnetic microscopy in connection with micromagnetic simulations.

References

1. J.M. Zuo, J.C. Spence, *Advanced Transmission Electron Microscopy: Imaging and Diffraction in Nanoscience* (Springer, New York, 2017)
2. A. Tonomura, Electron-holographic interference microscopy. Adv. Phys. **41**, 59–103 (1992)
3. H. Hopster, H. P. Oepen (eds.), *Magnetic Microscopy of Nanostructures* (Springer-Verlag, Berlin Heidelberg, 2005)
4. M. Haider, H. Rose, S. Uhlemann, E. Schwan, B. Kabius, K. Urban, A spherical-aberration-corrected 200 kV transmission electron microscope. Ultramicroscopy **75**, 53–60 (1998)
5. O.L. Krivanek, N. Dellby, A.R. Lupini, Towards sub-Å electron beams. Ultramicroscopy **78**, 1–11 (1999)
6. S. Nishikwa, S. Kikuchi, Diffraction of cathode rays by mica. Nature **121**, 1019–1020 (1928)
7. O. Scherzer, The theoretical resolution limit of the electron microscope. J. Appl. Phys. **20**, 20 (1949)
8. E.J. Kirkland, *Advanced Computing in Electron Microscopy* (Springer, New York, 2010)
9. Y. Takahashi, T. Akashi, T. Shimakura, T. Tanigaki, T. Kawasaki, H. Shinada, N. Osakabe, Resolution assessment of an aberration corrected 1.2-MV field emission transmission electron microscope. Microsc. Microanal. **21**(S3), 1865–1866 (2015)
10. M.A. Schofield, M. Beleggia, Y. Zhu, G. Pozzi, Characterization of JEOL 2100F Lorentz-TEM for low-magnification electron holography and magnetic imaging. Ultramicroscopy **108**, 625–634 (2008)
11. N. Shibata, Y. Kohno, A. Nakamura, et al., Atomic resolution electron microscopy in a magnetic field free environment. Nat. Commun. **10**, 2308 (2019)
12. T. Tanigaki, T. Akashi, A. Sugawara, et al., Magnetic field observations in CoFeB/Ta layers with 0.67-nm resolution by electron holography. Sci. Rep. **7**, 16598 (2017)
13. V.V. Volkov, Zhu, Y.M. De Graef, Micron **33**, 411 (2002)
14. M. Uchida, Y. Onose, Y. Matsui, Y. Tokura, Real-space observation of helical spin order. Science **311**, 359–361 (2006)
15. X.Z. Yu et al., Real-space observation of a two-dimensional skyrmion crystal. Nature **465**, 901–904 (2010)
16. A. Sugawara, T. Shimakura, H. Nishihara, T. Akashi, Y. Takahashi, N. Moriya, M. Sugaya, A 0.5-T pure-in-plane-field magnetizing holder for in-situ Lorentz microscopy. Ultramicroscopy **197**, 105–111 (2019)
17. T. Uhlig, J. Zweck, Direct observation of switching processes in Permalloy rings with Lorentz microscopy. Phys. Rev. Lett. **93**, 047203 (2004)
18. D. Gabor, A new microscopic principle. Nature **161**, 777–778 (1948)
19. G. Möllenstedt, H. Düker, Beobachtungen und Messungen an Biprisma-Interferenzen mit Elektronenwellen. Z. Physik **145**, 377–397 (1956)
20. R.E. Dunin-Borkowski, A. Kovács, T. Kasama, M.R. McCartney, D.J. Smith, Electron holography, in *Springer Handbook of Microscopy. Springer Handbooks*, ed. by P. W. Hawkes, J. C. H. Spence, (Springer, Cham, 2019)
21. A. Sugawara, H. Kasai, A. Tonomura, P.D. Brown, R.P. Campion, K.W. Edmonds, B.L. Gallagher, J. Zemen, T. Jungwirth, Domain walls in the (Ga, Mn)As diluted magnetic semiconductor. Phys. Rev. Lett. **100**, 047202 (2008)
22. A. Tonomura, N. Osakabe, T. Matsuda, T. Kawasaki, J. Endo, S. Yano, H. Yamada, Phys. Rev. Lett. **56**, 792 (1986)
23. C. Phatak, M. Beleggia, M. De Graef, Ultramicroscopy **108**, 503–513 (2008)

24. T. Tanigaki, Y. Takahashi, T. Shimakura, T. Akashi, R. Tsuneta, A. Sugawara, D. Shindo, Three-dimensional observation of magnetic vortex cores in stacked ferromagnetic discs. Nano Lett. **15**, 1309–1314 (2015)
25. N. Erdman, D.C. Bell, R. Reichelt, Scanning electron microscopy, in *Springer Handbook of Microscopy. Springer Handbooks*, ed. by P. W. Hawkes, J. C. H. Spence, (Springer, Cham, 2019)
26. T.T. Sasaki, T. Ohkubo, K. Hono, Structure and chemical compositions of the grain boundary phase in Nd-Fe-B sintered magnets. Acta Mater. **115**, 269–277 (2016)
27. D. Schuller, D. Hohs, R. Loeffler, T. Bernthaler, D. Goll, G. Schneider, AIP Adv. **8**, 047612 (2018)
28. D.C. Joy, J.P. Jakubovics, Direct observation of magnetic domains by scanning electron microscopy. Philos. Mag. **17**, 61–69 (1968)
29. D.J. Fathers, J.P. Jakubovics, D.C. Joy, D.E. Newbury, H. Yakowitz, A new method of observing magnetic domains by scanning electron microscopy. Phys. Status Solidi A **20**, 535–544 (1973)
30. K. Koike, K. Hayakawa, Scanning electron microscope observation of magnetic domains using spin-polarized secondary electrons. Jpn. J. Appl. Phys. **23**, L187–L188 (1984)
31. F. Kloodt-Twesten, S. Kuhrau, H.P. Oepen, R. Frömter, Measuring the Dzyaloshinskii-Moriya interaction of the epitaxial co/Ir(111) interface. Phys. Rev. B **100**, 100402(R) (2019)
32. The Atomic Scale Electromagnetic Field Analysis Platform, https://www9.hitachi.co.jp/atomicscale_pf/en/. Accessed 30 March 2020
33. P. Schattschneider, S. Rubino, C. Hébert, J. Rusz, J. Kunes, P. Novák, E. Carlino, M. Fabrizioli, G. Panaccione, G. Rossi, Detection of magnetic circular dichroism using a transmission electron microscope. Nature **441**, 486–488 (2006)
34. G.M. Caruso, F. Houdellier, S. Weber, M. Kociak, A. Arbouet, High brightness ultrafast transmission electron microscope based on a laser-driven cold-field emission source: Principle and applications. Adv. Phys.: X **4**, 1 (2019)

Part IV
Field Sensing and Neutron Scattering in Magnetism

Magnetic Field Sensing Techniques

Philip Keller (ORCID)

Abstract We provide an overview of some of the most widely used magnetic field sensing techniques: Hall sensors, fluxmeters, fluxgates, anisotropic magnetoresistive (AMR) and giant magnetoresistive (GMR) sensors, and nuclear magnetic resonance (NMR) magnetometers. For each technology, we summarize the history, principle of operation, benefits and limitations, typical applications, key specifications, and recent developments. We conclude with an overview of magnetometer calibration, from the definition of magnetic field units to practical considerations.

Keywords Magnetometer overview · Hall sensor · Fluxmeter · Voltage integrator · Fluxgate · AMR sensor · GMR sensor · NMR magnetometer · Magnetometer calibration · Magnetic units

1 Magnetic Field Sensors: Overview

This chapter is complementary to the others, focusing on the measurement of magnetic fields rather than the characterization of magnetic materials. Some of the instruments described here underlie material characterization methods and are also used to monitor the magnetic environment to which material samples are subjected.

There are well over a dozen different techniques to measure magnetic fields, as summarized in Fig. 1. We can learn much from this diagram; notably, we see that the biotope of magnetic field measurement techniques is indeed rich and diverse, ranging from classical to very high-tech. Also note the impressive measurement range: the horizontal axis spans 14 orders of magnitude! Under the right circumstances, the accuracy is also astounding, easily in the parts per million; however, such high accuracy is only achievable at relatively high field strengths.

New techniques are still being developed; an example are sensors based on the planar Hall effect [2]. However, this chapter will limit itself to the most developed

P. Keller (✉)
Metrolab Technology S.A., Plan-les-Ouates, Geneva, Switzerland
e-mail: keller@metrolab.com

© Springer Nature Switzerland AG 2021
V. Franco, B. Dodrill (eds.), *Magnetic Measurement Techniques for Materials Characterization*, https://doi.org/10.1007/978-3-030-70443-8_12

Fig. 1 Classification of important field sensing techniques, by measurement range and accuracy [1]

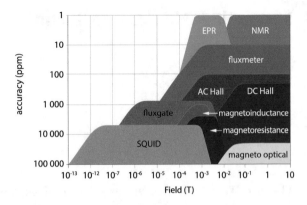

and widely used technologies: Hall sensors, fluxmeters, fluxgates, magnetoresistive sensors (AMR and GMR), and NMR magnetometers.

For each of these techniques, we will briefly describe its history and interest, an intuitive description of its principle of operation, benefits and limitations, typical applications, key specifications, and recent developments.

The chapter will conclude with a section discussing the calibration of magnetic field sensors—a critical topic for any of these techniques.

2 Hall Sensors

The Hall effect is named after the man who discovered it in 1879, Edwin Hall. Mass production of Hall sensors became viable with the advent of solid-state electronics; since then, their use has proliferated in industry as well as science. As can be seen in Fig. 1, they cover a wide measurement range with good accuracy and are the magnetic field sensor of choice in many situations.

Hall Sensors: Principles

The principle of operation of a traditional Hall sensor is illustrated in Fig. 2a. To a first approximation, the Hall voltage V_H is proportional to the bias current I as well as the normal component of the magnetic flux density, $B_\perp$:

$$V_H = \frac{R_H}{t} I B_\perp \tag{1}$$

where R_H is the so-called Hall coefficient and t is the thickness of the plate [[3], p. 69].

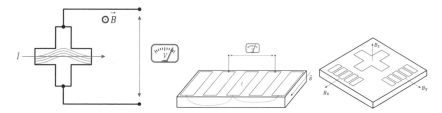

Fig. 2 Principle of operation of a Hall sensor: (a) In a traditional planar Hall sensor, the bias current I is deflected by a normal magnetic field, **B**, generating a voltage V_H across the voltage terminals. (b) In a vertical Hall sensor, the bias current I is injected in the middle terminal and is split into two channels flowing in opposite directions. Since these are deflected in opposite directions by a magnetic field **B** parallel to the plate, a differential voltage V_H between the two voltage terminals is created. (c) Combining a traditional planar Hall sensor with two vertical Hall sensors creates an integrated three-axis Hall sensor

The cross is a common, but not the only, shape for Hall sensors [[3] pp. 193–197]. In recent years, the so-called vertical Hall sensor (see Fig. 2b) has become important, because it measures an in-plane component of *B*, rather than the normal component $B_\perp$. As illustrated in Fig. 2c, the combination of two vertical Hall sensors and a traditional planar Hall sensor makes an integrated three-axis sensor that measures all three components of the magnetic field.

Such integrated Hall sensors can be readily implemented using complementary metal-oxide-semiconductor (CMOS) technology in silicon (Si). But silicon is not the only material used for Hall sensors; in fact, it is far from being the best. Important criteria in the choice of a material are the bandgap and the mobility of the charge carriers. The former affects the carrier density and temperature dependence and the latter the sensitivity [[3], pp. 250–252]. n-doped materials are used, since the mobility of electrons is higher than that of holes. Indium antimonide (InSb) and indium arsenide (InAs) feature very high mobility, but also small bandgaps. Gallium arsenide (GaAs) is another good material, with a much lower mobility than InSb and InAs, but still roughly 5.5x that of Si, and with a much higher bandgap than InSb and InAs.

There are also different approaches for the circuit supplying the bias current. Physically, it is the current that is important, but as a practical matter, it is often easier to supply a constant voltage than a constant current. Since the input resistance of a Hall sensor does not vary much, this may be an acceptable simplification. Another option is to use an AC bias current, allowing a significant noise reduction through synchronous detection; this explains the distinction between "AC Hall" and "DC Hall" in Fig. 1.

Finally, the low-field sensitivity can be improved by up to 10x by using a magnetic flux concentrator—a pair of soft-iron rods placed above and below the Hall sensor. An alternative geometry, compatible with integrated circuit technology, involves a flux concentrator with a gap placed over a differential pair of Hall sensors; these in fact detect the flux leakage in the gap [[3], pp. 292–305]. Disadvantages of a flux concentrator are that it disturbs the field being measured and introduces additional offset and hysteresis errors.

Hall Sensors: Benefits and Limitations

One of the primary benefits of Hall sensors is that they are relatively simple solid-state electronic devices, easily integrated into larger electronic systems and benefitting from the constant evolution of semiconductor manufacturing techniques. The range of sensitivities is wide and can be easily adjusted by changing the bias current. They can be used over a wide temperature range, including cryogenic temperatures. They are also fast, with bandwidths that can reach into the hundreds of kHz. Last but not least, they can measure all three components of B.

Hall sensors also have significant limitations. One of the most important is the offset voltage at zero field, caused, for example, by misalignment of the voltage terminals. The gain and offset are both temperature dependent, primarily because of the variation in number of charge carriers. Hall sensors also exhibit nonlinearity, caused by a multitude of effects specific to the geometry and material used. All these factors drift with time, due to aging of the semiconductor material. Like any resistor, Hall sensors have a noise spectrum with $1/f$ and white-noise components, limiting the measurement resolution, especially for DC measurements. Depending on the material, the planar Hall effect, at its maximum when B is in-plane and at 45° to the current flow, can introduce significant systematic errors, especially in three-axis sensors. The voltage leads are highly susceptible to induced noise from the environment. And finally, the quantum Hall effect limits the resolution at low temperatures and high fields.

Hall Sensors: Applications and Specifications

In describing the typical applications of Hall sensors, a sharp distinction must be drawn between commercial and scientific applications. Hall sensors, produced by the millions by mainstream IC manufacturers, have now crept into every conceivable consumer item, from flip-phones to automobiles. Typical functions are the detection of proximity, shaft rotation speed, user controls, and current. Similar sensors are widely used in industry, for example, in robotic systems.

In contrast, Hall sensors for scientific and metrological applications represent a much smaller market, with relatively few suppliers and typical annual production quantities in the thousands, rather than millions. Hall magnetometers are, however, general-purpose instruments, flexible, and easy to use. They are suitable for almost any type of task—measurement, mapping, or monitoring—as long as the precision is sufficient. Traditionally, Hall systems measured only a single component of B, but three-axis instruments are becoming the norm, using either an integrated sensor, as described above, or an assembly of three single-axis sensors. They can readily be organized into arrays, for rapid mapping or monitoring of a volume. Specialized instruments cover very wide temperature ranges, including cryogenic applications. The probes can also be made to be very compact, allowing them to fit into small gaps. Instruments may be packaged as traditional benchtop units, handheld units, or industrial-control modules.

Specifications of a Hall sensor include its sensitivity, offset voltage, input and output resistance, typical and maximum values of bias current/voltage, temperature coefficients of both the gain and offset, and linearity. A Hall magnetometer is generally judged by its measurement range, accuracy, resolution, bandwidth, probe size, and operational temperature range.

Hall Sensors: Recent Developments

One of the most important developments in recent years is the widespread application of the spinning-current technique, illustrated in Fig. 3 and described in greater detail in [[3] pp. 284–286]. Another important recent development is the proliferation of integrated devices, with built-in signal processing and analog-to-digital conversion, or multiple sensors to form gradiometers and magnetic-moment sensors, or even large arrays of Hall sensors on a single chip to form a "magnetic camera."

A third and final recent development is the use of "two-dimensional electron gas" (2DEG) Hall sensors, referring to stacks of semiconductors that confine the conduction electrons to a ~10-nm-thick quantum well. As we can easily see from Eq. 1, this reduction of the thickness of the Hall plate (by several orders of magnitude!) will have a direct effect on the sensitivity. Such structures, already described in [[3] pp. 257–259], are now in commercial use. A competing development, with the same goal, is the use of graphene to confine the conduction electrons to a single atomic layer [4].

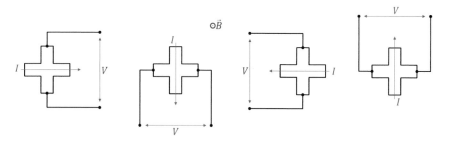

Fig. 3 The spinning-current technique involves switching the voltage and current leads of a Hall element. The benefit of this technique is threefold: suppression of the zero-field offset, for example, caused by misalignment of the voltage leads; suppression of the planar Hall effect; and chopping the signal, which suppresses low-frequency ($1/f$) noise

These exciting developments demonstrate that derivatives of this nineteenth-century discovery are still bubbling with innovation and that Hall sensors promise to remain a mainstay magnetic field sensing technique for decades to come.

3 Fluxmeters

Fluxmeters are another nineteenth-century invention, even older than Hall sensors: see the fascinating online "Historical Scientific Instrument Gallery" [5]. They are still in common use today, valued for their precision and flexibility—but the functionality and the styling have fortunately evolved.

Fluxmeters: Principles

A fluxmeter consists of two parts: a coil adapted to the application and a general-purpose voltage integrator. Flux changes in the coil induce voltage changes; these are integrated by the voltage integrator to yield the total flux change.

This is a straightforward application of Faraday's law of induction:

$$V = -\frac{d\Phi}{dt}, \tag{2}$$

where V is the induced voltage and $-\frac{d\Phi}{dt}$ is the time rate of change of magnetic flux. Integrating from t_1 to t_2:

$$\Delta\Phi = -\int_{t_1}^{t_2} V\, dt. \tag{3}$$

Expressed in terms of the flux density, B, the flux enclosed by the coil is given by the surface integral over the surface area of the coil:

$$\Phi = \int_S B \cdot da. \tag{4}$$

If the effective area of the coil is a constant, A, and the angle between the coil axis and B is θ, we can combine these last two equations and compute the change in flux density:

$$\Delta B = \frac{\Delta\Phi}{A\cos\theta} = -\frac{1}{A\cos\theta}\int_{t_1}^{t_2} V\, dt. \tag{5}$$

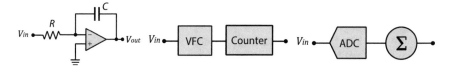

Fig. 4 Different types of voltage integrators. From left to right: analog integrator; voltage-to-frequency converter (VFC) and counter; fully digital integrator

From this last equation, we see the need of a voltage integrator. Figure 4 illustrates three different approaches for its implementation, each with its limitations:

- Analog integrator: the low end of the bandwidth is limited by the size of the capacitor, and there are numerous sources of error, such as leakage currents and temperature dependence. In addition, the output generally needs to be digitized anyway. A clever variant on this design is the "digitally compensated analog integrator": the analog integrator is used as a null detector, thus continually discharging the capacitor and avoiding many of its limitations [6].
- Voltage-to-frequency converter (VFC) and counter: the chief limitation is the VFC frequency range, which limits the resolution. The technique is, of course, also completely dependent on the VFC linearity. The beauty of this approach lies in the elegant combination of analog and digital, avoiding all quantification noise.
- Analog-to-digital converter (ADC) with digital adder: this method has to limit the high end of the bandwidth to satisfy the Nyquist criterion. It also depends critically upon the linearity of the ADC and can suffer from quantification noise. However, ADCs are continually improving in speed and resolution, so the long term favors this solution.

As for the coil, there is an astounding variety of designs, catering to different types of measurements. Here is a sampling:

- A surrounding coil is a coil that encloses all the flux in a magnetic system, for example, when wrapped around a magnet pole.
- A field coil is a coil of a known area that only samples the flux in a given volume. In a time-varying field, a field coil will measure voltage changes, even without moving.
- A moving coil is a field coil that is moved, generally from the region of high magnetic flux density to a field-free region. The measured flux change is $\Delta\Phi = AB_\perp$, where A is the area of the coil and $B_\perp$ is the component of $\boldsymbol{B}$ perpendicular to the coil area.
- By taking partial integrals along the way, a moving coil actually maps the field along its path. The path can be adapted to the geometry of the system: linear, circular, etc. Sometimes—for example, with a rotor—it is easier to move the magnetic system than the coil; the result is of course the same.
- In very small gaps, a moving wire can be used instead of a moving coil. In this case, the "coil" is actually the entire loop of wire, running from the integrator

through the magnet back to the integrator. As the wire sweeps through the magnet, partial integrals can be recorded to produce a map.

- A flip coil is a field coil with a mechanical arrangement to flip it $180°$. The measured flux change is $\Delta\Phi = 2AB_\perp$, where A is the area of the coil and $B_\perp$ is the component of B perpendicular to the coil area.
- A rotating coil, or harmonic coil, is like a flip coil, except that it rotates a full $360°$ and the mechanical arrangement allows measuring partial integrals at intermediate angles. The resulting measurements constitute a circular field map, which can be decomposed into circular harmonics, providing an "n-pole" model of the field, i.e., a superposition of dipole, quadrupole, octupole, ... n-pole components [7].
- Usually, harmonic coils are designed with additional coils—so-called bucking coils—to improve the sensitivity for higher-order harmonics. For example, when measuring a dipole magnet, a figure-eight shaped coil can be used to "buck out" the dipole component: flux changes in the top loop of the "8" cancel out those in the bottom loop, thus suppressing the dipole component.
- A Helmholtz coil is commonly used to measure one component of the magnetic moment of a permanent magnet. To measure all three components, a three-axis Helmholtz coil can be used [6].
- A potential coil is a long rigid rod with a coil wound along its entire length, designed to measure differences in magnetic potential. By analogy to the electric potential, the magnetic scalar potential ψ is defined as $H = -\nabla\psi$ (valid only in regions with no free current!), where H is the magnetizing field. In free space, this reduces to $B = \mu_0 H = -\mu_0\nabla\psi$. The tip of the probe is placed on the point where ψ *is* to be measured; the coil must be long enough that the other end is in a field-free region. The long coil integrates $B_\perp$ from the measurement point to the field-free region and therefore measures ψ. The tip can then be moved to another point, and the difference is $\Delta\psi$. Dividing $\Delta\psi$ by the distance between the two points (and μ_0) provides an estimate of ΔB [6].

Fluxmeters: Benefits and Limitations

As can be seen from the list in the previous section, the geometry of the coil can be readily adapted to an amazing variety of measurement problems. The sensitivity can also be easily adjusted, by changing the number of windings in the coil. A static field coil is insensitive to a DC field, allowing it to measure minute fluctuations in a strong field—for example, the effect of eddy currents. Since a coil is a passive device, fluxmeter measurements can be used in cryogenic environments as well as at very high temperatures. And finally, sometimes fluxmeter measurements are much more efficient and accurate than other techniques; an example is measuring the integrated field along a particle trajectory.

Some of the limitations of fluxmeters are just the flip side of their benefits: the need to adapt the coil to each measurement task, for example, or the insensitivity to

DC fields. Another significant disadvantage is that even the simplest measurements require proficiency in mathematics and physics, as well as very careful experimental technique to avoid problems related, for example, to drift or integrator input impedance.

Fluxmeters: Applications and Specifications

Many systems include fluxmeters without calling them that, for example, to measure AC currents. In this section, we will only consider metrological applications. We already hinted at these while enumerating the different coil types; however, the use of fluxmeters is very different in industrial and scientific applications.

In industry, fluxmeters with standardized coils are used for a wide variety of measurements, for example, measuring the strength and magnetic moment of permanent magnets, measuring the flux in a gap, measuring the potential loss in a yoke, or determining the working point of a magnet [6].

In science, the use of fluxmeters is almost exclusively restricted to the characterization of accelerator magnets. The equipment is very different than that used for industrial applications: coils are often custom-built for a specific magnet, and the integrator needs to be able to rapidly produce partial integrals, for example, synchronized by a rotational encoder. The resolution requirements are generally much higher than in industrial applications, on the order of at least 0.01%.

The key specifications of the coil generally involve its area and geometry. For bucking coils, the key parameter is harmonic rejection. Parameters for the integrator include flux resolution, gain, drift, linearity, bandwidth, and input impedance. For scientific applications, time resolution is also important. Key parameters for the fluxmeter system as a whole are the resolution and accuracy; the latter depends heavily on the accurate calibration of the coil area.

Fluxmeters: Recent Developments

Scientific applications continue to drive the evolution of fluxmeter systems. Two recent developments are the increased use of printed circuit board (PCB)-based coils and the more extensive use of wire-based techniques for very small gaps. For example, harmonic measurements are now also performed using a single stretched wire [8].

Like Hall sensors, the fluxmeter is a nineteenth-century discovery that is still very much alive. Especially the techniques in the scientific realm are abstruse and difficult to master, but there continue to be many problems for which a fluxmeter is simply the best tool. Users in search of a field measurement solution owe it to themselves to seriously evaluate this option, despite the hurdles.

4 Fluxgate Magnetometers

Fluxgate magnetometers are low-field devices, primarily used to measure perturbations in the earth's field. Invented in 1936 and first applied during the Second World War for submarine detection, they are slightly more recent than Hall sensors and fluxmeters. After the war, surplus units were recycled for geomagnetism research. They continue to be the standard against which all other low-field techniques are measured.

Fluxgates: Principles

The principle of operation of a fluxgate magnetometer is summarized in Fig. 5. The magnetic permeability of the core—the slope of the B vs. H curve — is modulated as it goes into and out of saturation: unsaturated, the core has the high permeability of soft iron; saturated, it suddenly drops to the low permeability of free space. This means that the flux density B in the core due to an external field H is also modulated; one can consider that the flux due to the external field is switched off as the core saturates and back on as the core desaturates—hence the name "fluxgate." Clever demodulation schemes allow the magnitude of this gated field to be measured by the sense coil.

The actual performance depends on many factors:

– Core material: To maximize the signal, the core material should be as magnetically "soft" as possible, with a very narrow hysteresis loop and a sharp saturation point. Ideally, to avoid Barkhausen noise during the switch, the core should consist of a single magnetic domain. Other characteristics of the core material, such as temperature coefficient, isotropy, and long-term stability, determine other key sensor characteristics.

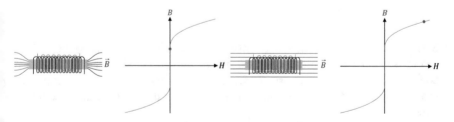

Fig. 5 Principle of operation of a fluxgate magnetometer, consisting of a soft-iron core, a drive coil, and a sense coil. The drive coil drives the core through the entire hysteresis cycle: unsaturated—positive saturation—unsaturated—negative saturation. In the unsaturated state, the core concentrates external flux; however, when saturated, that flux is expelled. The sense coil measures these flux changes

- Core geometry: Clever core geometries dramatically improve sensitivity by cancelling the direct transmission of drive to sense signal. The Vacquier design uses two parallel cores, with the drive coil wound in opposite directions and the sense coil wound around the pair, so that the flux generated by the two drive coils cancels out. The Förster design also uses two parallel cores, but here the drive coils are wound in the same direction, and the sense coil is wound in one direction on one core and the opposite direction on the other core [9].
- Another core geometry is a ring, with the drive coil wound toroidally and the sense coil around the entire ring—a straightforward extension of the Vacquier design, with the additional advantage that the core is a single piece, thus minimizing problems with offset errors.
- Multi-axis: the direction of the sense coil determines the sensitive direction of the sensor; two orthogonal sense coils on the same ring make an XY sensor; and with two rings, one can make an XYZ sensor. Note that in this design, the axes are centered on a single point.
- Synchronous detection: A single cycle of the drive voltage causes two saturation/desaturation signals from the sense coil: once in the forward and once in the reverse direction. Synchronous second harmonic detection can therefore be used to improve the signal-to-noise ratio.
- Feedback: Dynamic range and linearity can be greatly improved by using the fluxgate signal as a null detector. It controls a current that is fed back into either the sense coil or a separate feedback coil, cancelling the external field.
- Frequency response: A higher drive frequency allows a higher sensor bandwidth, but needs to be traded off against the sensitivity.
- Demodulation electronics: The demodulator must be carefully designed to optimize noise, offset, bandwidth, and the temperature coefficient.
- Calibration: To achieve good accuracy, good calibration technique is paramount. Given the great sensitivity of fluxgate sensors, it is especially difficult to create a good zero reference field.

Fluxgates: Benefits and Limitations

The key benefit of the fluxgate is its low noise—in the $pT/\sqrt{Hz}$—allowing sub-nanotesla precision with excellent linearity. Unlike fluxmeters, fluxgates can measure DC fields, and three-axis devices are commonplace. Their relatively low cost and low power consumption make them suited for field work.

For many applications, the key limitation of a traditional fluxgate is its size: smaller than a shoebox, but definitely bigger than a matchbox! The measurement volume is correspondingly large—typically $10,000\times$ that of an integrated three-axis Hall sensor—which makes it unsuitable for high gradients. Relative gain and offset errors are higher than with other techniques. For applications involving magnets, the measurement range—usually not much above the earth's field—is also a significant limitation. The frequency response is generally limited to 1 kHz and might be limited to a few Hz to achieve high precision. The drive signal causes a high-frequency perturbation of the field being measured. Finally, the orthogonality error of multi-axis units is non-negligible.

Fluxgates: Applications and Specifications

Fluxgates are used in diverse fields such as geology, archeology, space exploration, defense, oil exploration, security, science, and medicine—always to detect low-level magnetic perturbations. Concretely, this might mean searching for iron and mineral deposits, searching for archeological artifacts, mapping the magnetic field of a planet, detecting weapons or moving vehicles, surveying the magnetic environment before installing an MRI scanner, or checking the magnetic shielding for a physics experiment.

As with any sensor, the key specifications of fluxgates include gain, offset, linearity, and bandwidth, including all related errors. Manufacturers usually specify the noise density and how much of the drive signal feeds through to the output. Power consumption and size are also significant parameters.

Fluxgates: Recent Developments

There is a clear trend toward miniaturization of fluxgates; there is now even a commercially available fluxgate-on-a-chip, destined for accurate electric current measurement [10].

However, the magnetic noise power is inversely proportional to the magnetic core volume, so, to preserve the unique benefits of fluxgates, dramatic improvements are required. [11] provides a good summary of possible approaches to noise reduction: new magnetic materials with more uniform Barkhausen jumps; optimized core geometry and excitation mode; and operating the core close to the Curie point.

Looking back at Fig. 1, we are reminded of how important fluxgates are. Magnetoresistive techniques continue to gain in popularity, but they have neither the accuracy nor the sensitivity. SQUIDs have better sensitivity, but, as a super-conducting device, cannot compete in terms of cost and usability. We are forced to conclude that, despite their considerable limitations, fluxgates are the method of choice for many applications.

5 Magnetoresistive Sensors

As the name implies, magnetoresistive sensors rely on a change in resistance to detect a magnetic field. This section will discuss the most common of these, the anisotropic magnetoresistive (AMR) and giant magnetoresistive (GMR) sensors.

AMR Sensors: Principles

The magnetoresistive effect was discovered by Lord Kelvin in 1857, when he noticed a slight change in the resistance of iron when placed in a magnetic field. However, it took until 1971 for Robert P. Hunt of Ampex to propose the first magnetoresistive sensor, for reading magnetic tapes [12]. This proposal was picked up by IBM, first for magnetic-stripe readers, then in tape drives, and finally, starting 1990, in read heads for hard disks.

The magnetoresistive effect in ferromagnetic materials is caused by the scattering of the conduction electrons, from the 4s subshell, by the electrons in the 3d subshell. The shape of the 3d orbit causes it to orient itself in a magnetic field, thus modulating the scattering cross-section [13].

The principle of a modern AMR sensor is illustrated in Fig. 6. The ferromagnetic material is a thin film of permalloy ($Ni_{80}Fe_{20}$) deposited on a silicon substrate. Its resistance varies as:

$$R = R_0 \left(1 + \delta_H \cos^2\alpha \right), \tag{6}$$

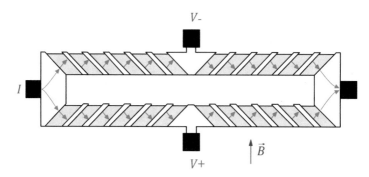

Fig. 6 Typical configuration of an AMR sensor. The permalloy bars are longitudinally magnetized, and a perpendicular external field rotates the magnetization vector toward the transverse direction. The resistance to current flowing through the bar is at a maximum when the magnetization is parallel to the current flow and at a minimum when perpendicular. The Wheatstone bridge arrangement renders the output insensitive to the absolute resistance, thus minimizing the temperature dependence while providing good sensitivity to resistance changes. The "barber pole" shunts redirect the current locally to ±45°, in order to exploit the linear region of the response curve

where α is the angle between the magnetization and the current, R_0 is the resistance with no field, and δ_H is the fractional variation in resistance in an applied field H [13]. Note that the sensitivity $\frac{dR}{d\alpha}$ is zero at $\alpha = 0°$ and $\pm 90°$; the "barber pole" shunts shown in Fig. 6 redirect the current by $\pm 45°$ in order to utilize the most sensitive and linear portions of the response curve. Also note that δ_H for the anisotropic magnetoresistance of permalloy is no more than 1.5–3%; the Wheatstone bridge arrangement shown in Fig. 6 maximizes the sensitivity to these small changes while minimizing sensitivity to overall changes, for example, due to temperature variations.

Further refinements [14] include mechanisms to:

- Re-magnetize the permalloy after exposure to a strong external field. This is accomplished by a strong "set" pulse in a coil, or an on-chip strap, surrounding the entire sensor. A "reset" pulse, of opposite polarity, can be used to magnetize the permalloy in the opposite direction, thus dynamically reversing the sensor's output polarity.
- Cancel the zero offset of the Wheatstone bridge. The offset can be determined by comparing a given field measurement after a "set" pulse with one after a "reset" pulse. The offset can then be compensated in a number of ways: a trim resistor; averaging the two measurements; subtracting the offset from the output signal; or a current in an offset coil/strap.
- Compensate the effects of nearby iron, with an additional offset current in the offset coil/strap.
- Auto-calibrate the sensor, by driving a known calibration current in the offset coil/strap.
- Improve the linearity, by measuring the current in the offset coil/strap that is required to cancel the external field and using the sensor as a null detector.

GMR Sensors: Principles

The ever-increasing areal density of hard disks created the need for even smaller sensors. It was natural that attention should turn to the 1988 discovery of giant magnetoresistance (GMR) by Albert Fert and, independently, Peter Grünberg, work which earned them the 2007 Nobel Prize in Physics.

As shown in Fig. 7, GMR is observed in thin sandwich structures of ferromagnetic and non-ferromagnetic conductors. Several different layer structures are used in sensors [15]:

- Unpinned sandwich: a simple sandwich with a relatively thick copper inner layer (3–5 nm) that weakens the magnetic coupling between the ferromagnetic outer layers. The sandwich material is patterned into narrow stripes, with a small in-plane current running the length of the stripe. With no external field, the field caused by the current makes the ferromagnetic layers counter-align. An external field will override this alignment, causing it to transition to an aligned or low-

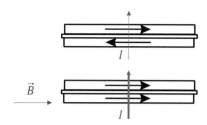

Fig. 7 Basic principle of GMR. A very thin (<10 nm) non-magnetic conductor, such as copper, is sandwiched between two very thin ferromagnetic layers. Depending on the thickness of the middle layer, the magnetization of the ferromagnetic layers will spontaneously counter-align. The counter-alignment can, however, be overcome by an external field, causing the layers to become aligned. It was found that the resistance of the aligned configuration is significantly lower than the counter-aligned one—curiously enough, regardless of whether the current is in the plane (CIP) or perpendicular to the plane (CPP—as shown here). The observed change in resistance is at least an order of magnitude greater than the "natural" anisotropic magnetoresistance—hence the name, giant magnetoresistance

resistance configuration. Typical δ_H values are 4–9%, and the saturation field is 3–6 mT.

– Antiferromagnetic multilayers: stacks of alternating magnetic and non-magnetic layers. The non-magnetic layers are very thin (1.5–2 nm), causing the surrounding magnetic layers to spontaneously counter-align, due to antiferromagnetic coupling. Again, an external field will override this alignment, causing the entire stack to switch to the low-resistance configuration. The use of multiple layers amplifies the GMR effect and provides better linearity and lower hysteresis than the unpinned sandwich. Typical δ_H values are 12–16%, and the saturation field is around 25 mT.

– Spin valve: similar to the unpinned sandwich, with the addition of an antiferromagnetic layer. This layer pins the magnetization direction of the adjacent ferromagnetic layer through exchange bias; the direction of the magnetization is fixed by cooling the material in a strong magnetic field. The magnetization in the unpinned layer changes by rotation rather than domain wall motion, thus minimizing hysteresis. Typical δ_H values are 4–20%, and the saturation field is 1–8 mT.

GMR sense elements usually consist of long serpentine patterns, thus increasing the resistance and minimizing power consumption. As with AMR, four GMR elements are arranged in a Wheatstone bridge structure. However, unlike "barber pole"-biased AMR elements, GMR elements do not distinguish between positive and negative fields, and it is therefore impossible to make a bridge with four active elements. Instead, two elements are covered by a magnetic shield, serving as reference resistors, with the same temperature coefficient as the active elements. The magnetic shield also serves as a flux concentrator, whose geometry can be adapted to provide different sensitivity and saturation characteristics with the same basic sensor [15].

Magnetoresistive Sensors: Benefits and Limitations

The key benefits of magnetoresistive sensors are their sensitivity, small size, and low cost. In addition, they provide DC response and large bandwidth; this was especially important for the original application of tape read heads, since the tape speed became much less critical.

These sensors are, however, limited to relatively low fields [14]. AMR sensors cover a range similar to fluxgates, with an upper limit of under 1 mT. GMR sensors can have higher saturation fields, up to roughly 10 mT. The noise density, at best 100 pT/$\sqrt{\text{Hz}}$, is clearly worse than a traditional fluxgate [16].

The magnetization of AMR sensors is destroyed by a strong external field and needs to be restored by a powerful "set" pulse that temporarily perturbs the field being measured. It is also important to repeat that GMR, unlike a "barber pole"-biased AMR element, does not distinguish between positive and negative fields.

Magnetoresistive Sensors: Applications and Specifications

The field of applications for magnetoresistive sensors is vast and diverse. As described previously, their initial development was driven by tape and disk read heads. Other applications include compassing, navigation, position and attitude sensing, detection of vehicles or other iron objects, current sensing, fault detection in pipes or plates, subsurface geophysical exploration, reading magnetic ink, detecting nerve impulses, and detecting magnetic nanoparticles used as biological markers.

The typical specifications of a magnetoresistive sensor are the field range, sensitivity, resolution, linearity, hysteresis, number of axes, axis orthogonality, and axis cross-talk. In addition, one has all the specifications associated with any electronic component, such as size, packaging, supply voltage, power, and operating temperature range.

Magnetoresistive Sensors: Recent Developments

The commercial success of AMR and GMR has inspired much research into other magnetoresistive technologies. Sensors based on tunnel magnetoresistance (TMR) are already commercially available, especially for low-power magnetic switches and angle sensors, hard-disk read heads, and magnetoresistive random-access memory (MRAM) [15]. In this variation of GMR, the conductor is replaced by an insulator, enabling only a tunnel current to flow. Because of high intrinsic resistance, power-efficient and very small elements—several μm square—can be produced. In addition, these devices regularly achieve values of δ_H as high as 30%.

Colossal magnetoresistance (CMR) is a form of magnetoresistance of certain materials, associated with a ferromagnetic-to-paramagnetic phase transition [17]. Some of the materials were known as far back as 1950, and interest was revived in the 1990s. At this time, these discoveries have not yet found commercial application.

Extraordinary magnetoresistance (EMR) describes the magnetoresistance of certain semiconductor-metal structures; the δ_H is truly extraordinary — as much as 750,000% [18]. This is achieved not by magnetoresistive material properties, but by the Hall effect and a cleverly designed device geometry, which deflects the charge carriers away from a metallic shunt. As such, these structures do not rival AMR and GMR for low fields : the δ_H cited above was achieved at 4 T!

Given their commercial importance, and consequently the R&D being lavished on the technology, one can assume that magnetoresistive sensors will continue to gain in popularity, also for metrological applications. They already offer tremendous advantages in terms of size, power, cost, and bandwidth. As the noise performance improves, they are becoming serious rivals for fluxgates in many applications.

6   NMR Magnetometers

Nuclear magnetic resonance (NMR) was experimentally demonstrated in 1945. It was only in the late 1970s, however, that the first NMR magnetometers arrived on the market, first used by accelerator physicists and soon by the then-nascent manufacturers of magnetic resonance imaging (MRI).

The importance of NMR has always been its precision and accuracy: NMR and the related technique EPR (electron paramagnetic resonance) have the undisputed top spot in Fig. 1.

NMR Magnetometers: Principles

If a nucleus has spin, it tends to align itself to an external magnetic field. However, by giving it exactly the right additional amount of energy, the nucleus can be induced to flip into the opposite spin state. The resonance occurs when a radio-frequency (RF) field applied to a sample is just the right frequency—called the Larmor frequency—to induce this spin-flip. EPR is a similar effect, with an electron rather than a nucleus.

The energy difference between the aligned and counter-aligned nuclear states depends linearly on the field strength, B, with a proportionality constant called the gyromagnetic ratio. It is approximately 42.5 MHz/T for protons (hydrogen nuclei), 6.5 MHz/T for deuterium, and 40 MHz/T for fluorine. EPR has a much higher gyromagnetic ratio, approximately 28 GHz/T.

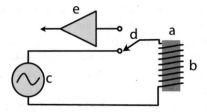

Fig. 8 Schematic diagram of an NMR magnetometer with pulsed-wave detection: (a) NMR sample; (b) excitation/sense coil; (c) RF generator; (d) transmit/receive switch; (e) detector. No modulator is required for pulsed-wave detection

As illustrated in Fig. 8, an NMR magnetometer has five main elements [19]:

– NMR sample: The sample material must have a nuclear spin; many common isotopes, such as ^{12}C or ^{16}O, have zero spin and are transparent to NMR. The material must also exhibit a sharp resonance; in a molecule, the NMR resonance is broadened by interactions with the other nuclei and electrons. Finally, the nuclei must "relax" to their initial spin-aligned state in a "reasonable" amount of time: too short prevents the detection of the resonance; too long slows the measurement rate. The most readily available sample material is water: NMR resonance of ^{1}H, or a proton.

– Excitation/sense coil: The coil that supplies the RF field to induce the spin-flip must be more or less perpendicular to the field being measured. Unlike Hall sensors, imperfect alignment does not change the measurement result; it simply causes some loss of sensitivity.

– RF generator: The key parameters are bandwidth ($\approx$1 MHz to 1 GHz for ^{1}H), stability ($\approx$ parts per billion/day), and suppression of spurious frequencies ($<\approx$ 80 dB). Producing such a generator economically is one of the major challenges in designing an NMR magnetometer.

– Detector: There are several basic approaches for detecting the NMR resonance:

  – Continuous wave (CW): at resonance, the NMR sample absorbs energy, increasing the apparent resistance of the excitation coil. One can either directly detect the slight voltage dip across the excitation coil, or one can incorporate the excitation coil in a marginally stable oscillator whose amplitude dips at resonance.

  – An alternate CW technique is the inductive bridge, where a pick-up coil perpendicular to the excitation coil detects when, at resonance, spins are rotated by 90°.

  – Pulsed-wave (PW): the sample is excited with a short, wide-band pulse whose frequency corresponds approximately to the Larmor frequency. In a second step, the excitation coil is used to detect the precession frequency of the spins during the relaxation time.

– Modulator: The CW techniques require some sort of modulation, to detect a change when crossing the resonance. The most obvious solution is to modulate the frequency, but it is often more practical to add a small coil to modulate the field being measured, taking care to synchronize the measurement with the modulation zero-crossing. The fact that the PW technique does not require any modulation whatsoever is a major advantage.

NMR Magnetometers: Benefits and Limitations

The key benefit of NMR is its very high resolution: one can typically resolve a 1 Hz change in the resonant frequency; in a 1 T = 42.5 MHz field, that results in roughly 0.02 ppm resolution. The accuracy, however, is generally limited to roughly 5 ppm, by various effects including sample shape and susceptibility of surrounding material. In addition, NMR is not subject to temperature drift, aging, or angular error.

The most constraining limitation of NMR is that the technique does not tolerate strong field gradients: destructive interference between the slightly different resonant frequencies across the sample kills the signal. The NMR signal also disappears at low fields (~10 mT), because the energy gap between the spin-flipped and non-flipped states becomes so small that it is swamped by thermal transitions. Finally, the measurement rate is limited by the finite relaxation time, to typically well under 100 Hz.

NMR Magnetometers: Applications and Specifications

The most widespread use of NMR magnetometers is the mapping of highly uniform fields, such as those produced by MRI magnets. This application is commercially so important because the required field uniformity cannot be achieved by construction; instead, every such magnet must be mapped and corrected using magnetic shims.

The accuracy and stability of NMR magnetometers also make them a perfect secondary standard for calibrating other magnetometers. Hall magnetometers as well as fluxmeters are generally calibrated against NMR (see Sect. 7).

Finally, NMR is used to monitor and/or regulate magnetic fields that must be extremely reproducible. Examples include accelerator and spectrometry magnets.

The key specifications of an NMR magnetometer are its precision, accuracy, measurement range, and measurement rate.

NMR Magnetometers: Recent Developments

The most recent generation of NMR magnetometers has extended the range of proton-sample probes from 2.1 to 24 T, with RF generators that go to over 1 GHz.

The resolution has also improved, by better than a factor of 10, due to PW detection and digital signal processing (DSP). DSP, by improving the signal-to-noise ratio, has also significantly improved the tolerance to field gradients.

Recent years have also seen incremental improvements in low-field performance. As seen in Fig. 1, EPR should provide a dramatic improvement. In the past, some manufacturers have indeed offered EPR probes; unfortunately, there is no known stable EPR sample material with sufficient spin density and narrow resonance width.

The efforts to push back the limitations of NMR magnetometers continue, and the range of applications that can benefit from this technique is becoming ever broader. However, the use of NMR magnetometers is ultimately limited by the number of applications that specifically require ppm-level resolution; the vast majority is better served by simpler techniques, such as Hall sensors.

7 Calibration

Regardless of a magnetometer's sensor technology, it must be calibrated. This section summarizes the most important notions concerning magnetometer calibration.

Calibration: Magnetic Field Units

Calibration relates a magnetometer's output to the unit of magnetic flux density, as defined by international and national standards bodies. Therefore, we start our story with the definition of this measurement unit.

The unit for magnetic flux density in the International System of Units ("SI"), the tesla (T), is a derived unit. Its definition in terms of the SI base units can be deduced from the Lorentz force law; in the absence of an electric field:

$$\boldsymbol{F} = q\boldsymbol{v} \times \boldsymbol{B} \tag{7}$$

where $\boldsymbol{F}$ is the force exerted on the charge q travelling at velocity $\boldsymbol{v}$ through a magnetic field $\boldsymbol{B}$. The dimensional analysis of this equation yields:

$$\frac{[kg]\,[m]}{[s]^2} = [C]\,\frac{[m]}{[s]}\,[T] \tag{8}$$

or:

$$[T] = \frac{[kg]}{\frac{[C]}{[s]}[s]^2} = \frac{[kg]}{[A]\,[s]^2} \tag{9}$$

This is the expression in SI base units. It is equivalent, by Faraday's law of induction (c.f. Eq. 5), to:

$$[T] = \frac{[V][s]}{[m]^2} \equiv \frac{[Wb]}{[m]^2} \qquad (10)$$

where the weber (Wb) is the SI unit of flux.

One also still sees magnetic flux density expressed in gauss (G), using the old centimeter-gram-second (CGS) system of electromagnetic units. This was accepted in polite SI society, because the conversion to the official SI units was exactly:

$$[T] = 10^4 [G]. \qquad (11)$$

However, since the 2019 redefinition of the SI base units [20], this is no longer the case, and the gauss should be finally allowed to enjoy its well-earned retirement [21].

Magnetometers measure the magnetic flux density B. However, when discussing bulk material properties such as hysteresis, this observed quantity is conceptually split into the external magnetizing field H and the internal material magnetization M:

$$B = \mu_0 (H + M), \qquad (12)$$

where

$\mu_0 \cong 4\pi \times 10^{-7} \dfrac{V \cdot s}{A \cdot m}$ is the permeability of free space. (Note that since the 2019 redefinition of the SI units, the value of μ_0 is no longer exact, but a measured value.) The units of H are A/m, as can be deduced from Ampere's law :

$$\nabla \times H = J_f, \qquad (13)$$

where J_f is the free current density. Dimensional analysis yields:

$$\frac{1}{[m]} H = \frac{[A]}{[m]^2}, \text{ or } H = \frac{[A]}{[m]}. \qquad (14)$$

To interpret historical literature, we also need to understand the equivalent formulation in CGS units. In this system of units, the permeability of free space was unity:

$$B = H + 4\pi M \qquad (15)$$

Nonetheless, H had its own unit, the oersted (Oe). The conversion from Oe to A/m is easy to deduce from the flux density generated by 1 Oe in free space:

$$H [Oe] = 1 \ Oe \sim 1 \ G = 10^{-4} \ T \sim \mu_0 H \ [A/m] \qquad (16)$$

Consequently,

$$[Oe] = \frac{10^{-4}}{4\pi \times 10^{-7}} \ [A/m] \cong 79.577 \ [A/m] \qquad (17)$$

Note that, since the 2019 redefinition of the SI units, this conversion is doubly inexact: neither the conversion from G to T nor the value of μ_0 is the exact value anymore. Like his colleague the gauss, the oersted is now definitively retired.

Calibration: NMR as a Secondary Standard

Calibrating a magnetometer involves placing the probe in a region with a precisely known magnetic flux density. In principle, this requires a dimensionally accurate magnet—perhaps a Helmholtz geometry—and an equally accurate current source. To eliminate perturbations—especially from the earth's field—the system needs to be placed in a magnetically shielded room, with absolutely no ferromagnetic material, perhaps in combination with a compensation coil or a compensated coil design [22].

In practice, only national standards laboratories can manage such a demanding setup; most practitioners use a conventional, high-quality laboratory electromagnet and power supply, with an NMR magnetometer (see Sect. 6) as a secondary standard. Besides resolution that far exceeds that of the sensor being calibrated, the NMR measurements have the advantage of being easily traceable to international standards, only requiring an accurate clock and precise knowledge of the gyromagnetic ratio. Measurements of the gyromagnetic ratio of water are regularly consolidated by the Committee on Data for Science and Technology [23] and published by, for example, the USA's National Institute of Science and Technology [24]. If the actual sample material is not water, the (generally small) deviation of the gyromagnetic ratio can readily be determined via NMR spectroscopy [25].

NMR can be used as a direct calibration reference down to ~10 mT (see Sect. 6). The upper field strength that can be calibrated in this way is in practice not limited by the NMR magnetometer, but by the availability of magnets with sufficient uniformity, stability, and gap size; above roughly 2 T, these need to be superconducting magnets. Since ramping the field in these magnets is very slow, one generally has data for only a limited number of field strengths, and especially the sensor nonlinearity becomes difficult to quantify.

Calibration: Low-Field Sensors

To calibrate sensors at low fields, when a direct NMR reference is no longer feasible, one often uses a sinusoidal excitation current, with a fluxmeter as reference measurement (see Sect. 3). To avoid hysteresis effects, this requires an air-core magnet, such as a Helmholtz coil or solenoid. The fluxmeter is, in turn, calibrated with an NMR reference, thus introducing an additional link in the traceability chain and increasing the error budget correspondingly.

Because of the sensitivity of low-field sensors, the determination of the zero-field point of the sensor is particularly critical. The problem is compounded by the fact that the fluxmeter reference provides no zero reference. The answer usually comes in the form of an elaborate "Zero Gauss Chamber" (pardon the antiquated term!), built out of multiple layers of mu-metal.

Calibration: Three-Axis Sensors

Since almost all sensors—except NMR!—have a field-sensitive direction, good calibration technique requires a special jig to reproducibly hold the sensor with the field properly aligned. The jig must obviously be non-ferromagnetic and, for low-field sensors with AC excitation, non-conducting to avoid distortions due to eddy currents.

The mechanical design becomes especially difficult with three-axis sensors, where each of the three axes must be in turn aligned with the field. The orthogonality of the jig is critical, to be able to detect and correct the non-orthogonality of the sensor itself. Another complication, this one very practical, is that the length of a probe may not fit crosswise into the gap of a dipole or solenoid; only Helmholtz geometries are basically immune to this issue, but they are limited in field strength.

Some have argued that calibrating only the sensor's primary axes cannot produce accuracy of better than roughly 0.1%, since the errors will cumulate in the regions between the axes [26]. Instead, they propose that the sensor be rotated in all directions and that its output be modeled with a spherical-harmonic fit. This method, although theoretically well-founded, has not caught on in practice, partly because of the mechanical and analytical complexity and partly because it requires a magnet with a very large gap.

In this section, we have attempted to outline only some of the most common issues encountered when calibrating magnetic field sensors. Each type of sensor comes with its own set of additional complications. We hope that it has become clear that a major part of the value added by commercial instrument manufacturers is the professional calibration of their products.

8 Magnetic Field Sensors: Conclusion

Over 150 years' worth of brilliant physicists and engineers have lavished their talents on the problem of magnetic field sensing. We have attempted an intuitive explanation of what we consider the most popular techniques. Neither the list nor the explanations are exhaustive: even for these "top five," this chapter only scratches the surface; the literature for each is vast and sophisticated. We encourage enlightened users with the appetite for more to delve into this all-you-can-eat buffet.

References

1. L. Bottura (2002) Field Measurement Methods. Presentation at CERN Accelerator School (CAS) on Superconductivity, Erice, May 8–17, 2002. http://m.home.cern.ch/m/mtauser/www/archives/events/erice/MM.pdf. Accessed 12.03.2007
2. A. Grosz, V. Mor, E. Paperno, et al., Planar hall effect sensors with Subnanotesla resolution. IEEE Magn. Lett. **4** (2013). https://doi.org/10.1109/LMAG.2013.2276551
3. R.S. Popovic, *Hall Effect Devices*, 2nd edn. (Institute of Physics Publishing, Bristol and Philadelphia, 2004)
4. J. Dauber, A.A. Sagade, M. Oellers, et al., Ultra-sensitive hall sensors based on graphene encapsulated in hexagonal boron nitride. Appl. Phys. Lett. **106** (2015) https://arxiv.org/pdf/1504.01625.pdf. Accessed 10.08.2019
5. University of Nebraska-Lincoln, Department of Physics and Astronomy (2019) Historical Scientific Instrument Gallery—Magnetism. https://www.unl.edu/physics/historical-scientific-instrument-gallery/magnetism. Accessed 25.07.2019
6. E. Steingroever, G. Ross, *Magnetic Measurement Techniques* (MAGNET-PHYSIK Dr. Steingroever GmbH, Köln, 2008)
7. L. Walckiers (1992) The Harmonic-Coil Method. Proceedings of the Magnetic measurement and alignment workshop, Montreux 1992, pp. 138–166. https://cds.cern.ch/record/245405/files/p138.pdf. Accessed 25.07.2019
8. G. le Bec, J. Chavanne, C. Penel (2001) Stretched-wire measurements of multipole magnets at the ESRF. Presented at IMMW 17, ALBA, Barcelona, Spain, Sep 2011. https://public.cells.es/workshops/immw17.cells.es/presentations/Thu01-IMMW17-LeBec.pdf. Accessed 25.07.2019
9. K. Evans, *Fluxgate Magnetometer Explained—Mar. 2006* (Invasens, Cheltenham, 2006) http://invasens.co.uk/FluxgateExplained.PDF. Accessed 26.07.2019
10. Texas Instruments (2016) DRV425 Fluxgate Magnetic-Field Sensor. http://www.ti.com/lit/ds/symlink/drv425.pdf. Accessed 26.07.2019
11. Korepanov V, Marusenkov A (2008) Modern Flux-Gate Magnetometers Design. In proceedings of International Scientific Conference on Magnetism—Geomagnetism—Biomagnetism MGB—2008, Sežana, Slovenija, Nov 7–8 2008, pp. 31–36. http://www.viviss.si/download/viviss/ZBORNIK%20MGB/Korepanov_paper_31_36.pdf. Accessed 26.07.2019
12. R.P. Hunt, A magnetoresistive readout transducer. IEEE Trans. Magn. **6**(1), 150–154 (1971) https://www.researchgate.net/publication/3109306_A_Magnetoresistive_Readout_Transducer. Accessed 07.08.2019
13. Measurement Specialties, Inc. (2017) Basics of Magnetoresistive (MR) Sensors. https://www.te.com/content/dam/te-com/documents/sensors/global/basics-magnetoresistive-mr-sensors-white-paper.pdf. Accessed 07.08.2019
14. M.J. Caruso, T. Bratland, C.H. Smith, et al. (1998) A New Perspective on Magnetic Field Sensing. https://aerospace.honeywell.com/en/~/media/aerospace/files/technical-articles/anewperspectiveonmagneticfieldsensing_ta.pdf. Accessed 07.08.2019

15. C.H. Smith, R.W. Schneider (1999) Low-Field Magnetic Sensing with GMR Sensors. Presented at Sensors EXPO, Baltimore, USA, May, 1999. https://www.nve.com/Downloads/lowfield.pdf. Accessed 08.08.2019

16. N.A. Stutzke, S.E. Russek, D.P. Pappas, et al., Low-frequency noise measurements on commercial magnetoresistive magnetic field sensors. J. Appl. Phys. **97**(10Q107) (2005) https://ws680.nist.gov/publication/get_pdf.cfm?pub_id=31842. Accessed 10.08.2019

17. A.P. Ramirez, Colossal magnetoresistance. J. Phys. Condens. Matter **9**(39), 8171–8199 (1997) https://iopscience.iop.org/article/10.1088/0953-8984/9/39/005. Accessed 10.08.2019

18. S.A. Solin, T. Thio, D.R. Hines, et al., Enhanced room-temperature geometric magnetoresistance in inhomogeneous narrow-gap semiconductors. Science **289**, 1530–1532 (2000) http://www.ruf.rice.edu/~rau/phys600/ERM21.pdf. Accessed 09.08.2019

19. P. Keller, NMR magnetometers. Magn. Technol. Int. **2011**, 68–71 (2011) https://www.ukimediaevents.com/publication/9c6782fc. Accessed 10.08.2019

20. Bureau International des Poids et Mesures (2019) Information for users about the redefinition of the SI. https://www.bipm.org/utils/common/pdf/SI-statement.pdf. Accessed 06.08.2019

21. R.B. Goldfarb, The permeability of vacuum and the revised international system of units. IEEE Magn. Lett. **8**, 1–3 (2017) https://ieeexplore.ieee.org/stamp/stamp.jsp?tp=&arnumber=8241772. Accessed 06.08.2019

22. H. Harcken (2011) MNPQ-Projekt: Ein neuartiges kompensiertes Spulensystem für die Messung magnetischer Momente mit geringem Streufeldeinfluss und die streufeldarme Erzeugung magnetischer Felder. https://www.ptb.de/cms/fileadmin/internet/fachabteilungen/abteilung_2/2.5_halbleiterphysik_und_magnetismus/2.51/KompSpule.pdf. Accessed 06.08.2019

23. Committee on Data for Science and Technology (CODATA). http://www.codata.org. Accessed 06.08.2019

24. NIST, Physical Measurement Laboratory, Fundamental Physical Constants. https://www.nist.gov/pml/fundamental-physical-constants. Accessed 06.08.2019

25. P. Keller (2017) Values of Gyromagnetic Ratios. https://www.metrolab.com/wp-content/uploads/2018/07/Values-of-Gyromagnetic-Ratios-04.07.2018.pdf. Accessed 06.08.2019

26. F. Bergsma (2003) Calibration of Hall sensors in three dimensions. Presented at IMMW13, SLAC, May 19–22, 2003. https://cds.cern.ch/record/1072471/files/cer-002727968.pdf. Accessed 06.08.2019

Introduction to Neutron Scattering as a Tool for Characterizing Magnetic Materials

Cindi L. Dennis

Abstract This chapter will introduce the concepts of neutron scattering, including the benefits and limitations, especially in comparison to other experimental scattering techniques. It will also describe recent expansion of neutron scattering capabilities, including polarization analysis and time resolution, and their potential. Example applications demonstrating the power of neutron scattering as a characterization tool will be discussed.

Keywords Neutron scattering · Small angle neutron scattering · Neutron reflectivity · Neutron diffraction · Polarization analysis · Time resolution · Characterization · Magnetic materials

1 Introduction

Fundamentally, there are different ways to probe materials, including wavelength-specific electromagnetic radiation (e.g., visible light, infrared, x-rays, etc.) or particles (e.g., electrons and neutrons) or forces (e.g., electrostatic, electromagnetic, etc.), to name three common ones. These probes can then be used in diverse ways to determine different properties of a material, such as optical or force microscopy to image the surface of a material, electron or x-ray diffraction to learn about composition and atomic-level crystal structure, magnetometry to understand magnetic changes in a material, and light/x-ray/neutron scattering to learn about micro- and meso-scale structure as well as atomic structure. These measurements can be made as a function of applied magnetic field, temperature, stress or strain, etc., to better understand the behavior of a material, either fundamentally or for its specific application. Choosing the right technique to answer a specific question about a material is critical; understanding the potential and the limitations of a

C. L. Dennis (✉)
National Institute of Standards and Technology, Gaithersburg, MD, USA
e-mail: cindi.dennis@nist.gov

V. Franco, B. Dodrill (eds.), *Magnetic Measurement Techniques for Materials Characterization*, https://doi.org/10.1007/978-3-030-70443-8_13

given technique is crucial to making an informed choice. Many different techniques have been discussed in other chapters of this book. Here, the focus is on neutron scattering, specifically the elastic coherent techniques of neutron diffraction (ND), neutron reflectivity (NR), and small angle neutron scattering (SANS).

This chapter is not an exhaustive treatise on neutron scattering. The interested reader is referred to the following chapter and to a number of excellent references, either generally [1] and [2] and [3] or for specific techniques like polarization analysis [4, 5] or time resolution [6, 7]. Instead, this chapter will introduce the basic physics concepts, using a few limited equations for understanding experimental data. Then, specific instrumentation and measurement methods, including what to look for in the data, will be discussed with examples.

2 Neutron Sources

Unlike with electromagnetic radiation (e.g., visible light or x-rays) or electrons or forces, there are no easy lab-scale sources of neutrons. Instead, neutrons are generated though one of two sources: a steady neutron flux that is a by-product of controlled fission in a nuclear reactor, such as the split core reactor at the National Institute of Standards and Technology's (NIST) Center for Neutron Research [8], or pulses of neutrons from a spallation source, such as those generated from the impact of protons accelerated at a steel target filled with mercury as at the Spallation Neutron Source at Oak Ridge National Laboratories [9].

Neutron fluxes are typically higher at spallation sources than from reactors because all the neutrons in a pulse can be used. (The time of flight yields the velocity of the individual neutrons, which is necessary for the scattering wave vector Q to be determined, as explained below.) In contrast, neutrons from a reactor are a steady stream with a range of energies/velocities and therefore typically must be monochromated so that the velocity of the neutrons is known with reasonable uncertainty. This throws away a significant amount of intensity. In either case, the total flux is still significantly lower than an equivalent x-ray source.

The neutrons generated come in two "temperatures": thermal neutrons which have a very short wavelength (e.g., 2 Å, but it depends on the temperature of the moderator and/or monochromator that the neutrons pass through) and "cold" neutrons with longer wavelength (e.g., 7 Å, which result from passing thermal neutrons through a cold source composed of, for example, liquid hydrogen or solid deuterium oxide ("heavy water") to slow the neutrons down further). The difference in wavelength allows the neutrons to probe different length scales in materials. Reactors produce a larger flux of cold neutrons; spallation sources produce more thermal neutrons.

Although the Fukushima nuclear disaster in 2011 has reduced the ranks of power generation nuclear reactors globally, there are still a number of research reactors and spallation sources available around the world. A non-exhaustive list includes the NIST Center for Neutron Research (NCNR) and the Spallation Neutron Source

(SNS) and High Flux Isotope Reactor (HFIR) in the USA, the Open-Pool Australian Lightwater reactor (OPAL) in Australia, the Japan Proton Accelerator Research Complex (J-PARC) in Japan, ISIS Neutron and Muon Source in the UK, the Institut Laue-Langevin (ILL) in France, Forschungs-Neutronenquelle Heinz Maier-Leibnitz (FRM II) in Germany, and the (under construction as of 2020) European Spallation Source (ESS) in Sweden. All of these facilities have teams of instrument scientists who operate specific instruments (beamlines) at each location and host user scientists from around the world for specific experiments, awarded through beamtime proposals. More information about procedures for applying for beamtime can be found on each facility's website.

3 Physics of Neutron Scattering

The idea of scattering is straightforward. You direct a beam of something, electromagnetic radiation or particles, at the material of interest. The beam then interacts with the material that it passes through, causing the beam to deviate from a straight line. The physics can be simple or complicated, depending on the specific process causing the scattering and whether or not it is elastic or inelastic. To aid in understanding, we compare three specific types of beams and their interactions: x-rays, electrons, and neutrons (see Fig. 1).

X-rays scatter off the electrons surrounding the nucleus of an atom through an electromagnetic interaction. The more electrons there are, the more efficiently the x-ray is scattered (see Fig. 2). Therefore, if you have a material composed of heavy atoms (like yttrium and copper) and light atoms (oxygen and barium), the heavy atoms dominate the signal. As a result, if you are looking for the position of an oxygen atom in a high-temperature superconductor filled with heavy atoms of yttrium and copper, it will be challenging to see it.

Charged particles like electrons also interact with the electrons surrounding the nucleus, but through an electrostatic interaction. They are either attracted or repelled depending on the sign of the charge on the particle. As a result, electron transparent materials must be very thin (<100 nm) to avoid significant attenuation of the signal (see Fig. 2).

Neutrons, in contrast, interact with a material differently, for three reasons. First, the neutron has no charge. The neutrons interact with the material primarily through the nuclear force, which is very short range (as compared to the electrostatic/electromagnetic forces). As a result, most neutrons pass unattenuated through materials. This enables complex sample environments, since the neutrons will go through the container holding the sample readily. Unfortunately, the end result is that significantly more material is required and much longer counting times (as the flux of neutrons is not very high to begin with) in order to get good statistics for neutron scattering, as compared to the same x-ray/electron scattering technique. Second, neutrons have a magnetic moment. Therefore, neutrons can also interact with the material through dipole-dipole interactions if there are unpaired electrons in

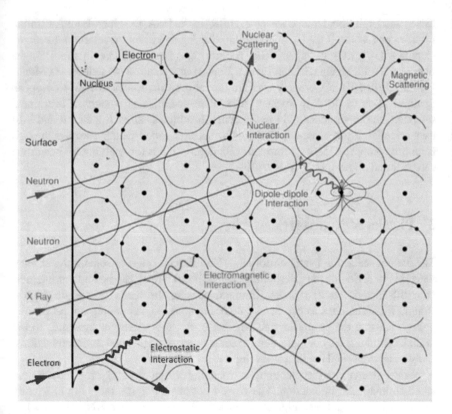

Fig. 1 Beams of neutrons, x-rays, and electrons interact with a material through different mechanisms. (Used with permission from [3]. Modified colors for better visibility)

the material. The end result is that neutrons can probe not only the nuclear structure of a material but also the magnetic structure.

Third, neutrons have a unique scattering cross section [10] which depends heavily upon isotope, as compared to x-ray and electrons, which have a more straightforward dependence of scattering on the atomic number. For example, hydrogen (H) has a small neutron scattering cross section of 1.7583 barns (1.7583×10^{-24} cm^2), but deuterium (D) is much larger at 5.592 barns. The end result is that neutrons are uniquely placed to study hydrogen-containing compounds, like polymers, something that x-rays and electrons do not see as well. In addition, the non-monotonic behavior of the penetration depth (see Fig. 2) of neutrons allows the use of "contrast matching" to pick and choose what is studied. This means that an isotopic substitution of elements, such as deuterium for hydrogen, can significantly change what is seen (see Fig. 3). A good example of this is in iron oxide magnetic nanoparticles dispersed in water. The nuclear scattering length density of magnetite is closely matched at 6.975×10^{-6}/Å^2 to that of "heavy water" D_2O at 6.335×10^{-6}/Å^2, but is significantly different from that of "light water" or H_2O

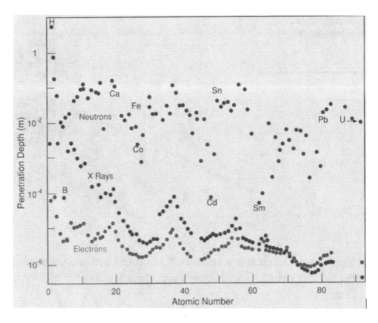

Fig. 2 Penetration depth of neutrons, electrons, and x-rays by element (Used with permission from [3])

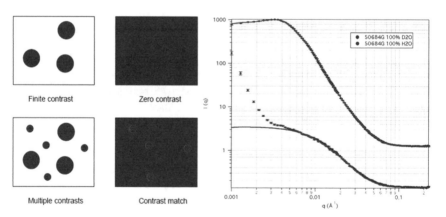

Fig. 3 (Left) Schematic of contrast matching for an iron oxide nanoparticle dispersed in water. (Right) SANS data showing nuclear and magnetic scattering of the iron oxide nanoparticles in light (H_2O) and heavy (D_2O) water

at $-0.561 \times 10^{-6}/\text{Å}^2$. As a result, changing the isotope of hydrogen in the water enables the user to make visible or invisible the nuclear scattering from the iron oxide. Therefore, the magnetic scattering can be determined, even though it is only 4% of the signal intensity (see Fig. 3).

Fig. 4 Scattering schematic for SANS. (Courtesy of NCNR)

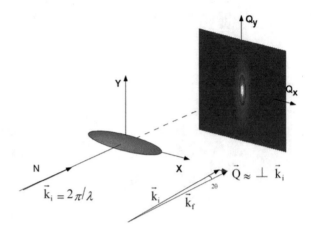

From quantum mechanics, we know that neutrons have wave-particle duality. This means we can associate a neutron wavelength λ with the velocity **v** of the neutron, through:

$$\frac{h\vec{k}}{2\pi} = m\vec{v} \tag{1}$$

where h is Planck's constant, **k** is the neutron wave vector ($k = 2\pi/\lambda$), and m is the mass of the neutron. Since the wavelength of the neutron is much larger than the range of the nuclear force, the nucleus acts as a point scatterer and scatters the neutrons isotropically. (This is not the case for x-rays, nor is it the case for dipole-dipole "magnetic" neutron scattering.) In all neutron scattering experiments, the result is the intensity I of neutrons scattered per incident neutron as a function of the momentum transfer (h**Q**/2π) and energy transfer (ε):

$$I(\mathbf{Q}, \varepsilon) = NF(\mathbf{Q}, \varepsilon)\, S(\mathbf{Q}, \varepsilon) \tag{2}$$

where N is the number of scatters per unit volume, $\mathbf{Q} = \mathbf{k_i} - \mathbf{k_f}$ and $\mathbf{k_i}$ is the incident wave vector and $\mathbf{k_f}$ is the scattered wave vector (see Fig. 4), and F(**Q**, ε) is the form factor and S(**Q**, ε) is the structure factor. (These "factors" will be briefly discussed later.) Interestingly, the anisotropy inherent in magnetic scattering manifests itself as a simple rule: magnetic scattering can only be seen when the scattering vector **Q** is perpendicular to the magnetic field of an unpaired electron or, more generally, the sample magnetization. With this equation, there are three types of scattering:

1. Elastic coherent scattering ($\varepsilon = 0$).
2. Inelastic coherent scattering ($\varepsilon \neq 0$).[1]

---

[1] A triple-axis spectrometer (at a reactor) and a chopper spectrometer (at a spallation source) are examples of techniques which provide information about the collective motions of atoms (e.g., phonons or magnons) as not only **k** and **k'** must be determined to get **Q** but also ε.

3. Incoherent scattering.

Only elastic coherent scattering will be discussed here, of which ND, NR, and SANS are example techniques. They provide information about the equilibrium structure.

One final comment: there exist multiple neutron scattering instruments which make similar measurements, as they are tailored specifically to the science they are designed to study because there is always a trade-off between intensity and resolution. For example, SANS (see Fig. 4) has a small diffraction angle for studying large objects, but also a reduction in monochromatization—only enough to match the angular resolution to maximize neutron flux. In contrast, back scattering has very good energy resolution, but uses large analyzer crystals so it has poor Q resolution.

4 Neutron Diffraction

Neutron diffraction (see Fig. 5) is analogous to x-ray diffraction and follows Bragg's law:

$$n\lambda = 2dsin(\theta) \tag{3}$$

where 2θ is the scattering angle, d is the lattice constant, and n is an integer. λ, the wavelength of the probe, is typically chosen to be on the order of the lattice spacing. Therefore, I(Q) is typically dominated by S(Q) where S(Q) is the structure factor and is primarily dependent upon the placement of the specific atoms. Diffraction provides information about the dimensions of the crystal lattice and the positions of atoms within it, the symmetry of the crystal, and the extent of thermal vibrations in various directions. It is a very useful tool for studying stress and strain as well as texture in materials. However, unlike x-ray diffraction, the magnetic moment of the neutron means that it is possible to see magnetic lattices as well as atomic lattices. The atomic and magnetic scattering can be separated, for example, by changing the orientation of a saturating magnetic field from parallel to perpendicular to Q. When H and Q are parallel, only nuclear scattering is seen. When H and Q are perpendicular, the magnetic scattering is added to the nuclear scattering, changing the peak intensities. For example, when measuring a ferromagnet such as MnPtGa (see Fig. 6), the atomic lattice spacing between the Mn atoms and the magnetic lattice spacing between the Mn atoms is the same, since the Mn moment is parallel to its neighbor. When decreasing the temperature so that MnPtGa changes to a canted anti-ferromagnet, the atomic lattice spacing between the Mn atoms is double that of the magnetic lattice spacing, since the Mn atoms are coupled anti-ferromagnetically (see Fig. 6). (Recall: doubling the lattice spacing gives rise to a peak at the half order position.) The latter is an alternative approach to changing the magnetic field orientation. Therefore, neutron diffraction also enables a direct probe of the magnetic structure and has been particularly useful for anti-ferromagnetic and spin-canted materials.

BT-8 Residual Stress Diffractometer

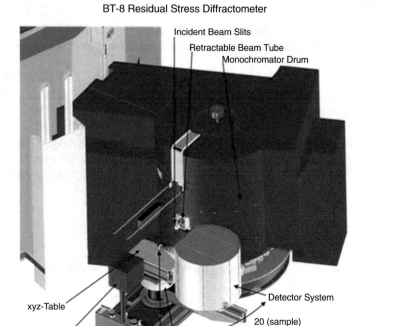

Fig. 5 Schematic of a neutron diffractometer. (Courtesy of NCNR)

5   Neutron Reflectivity

Neutron reflectivity is analogous to x-ray reflectivity (see Fig. 7), where x-rays/neutrons are reflected off a smooth surface at very small angles to study the surface structure. For light, this works only for very smooth surfaces at the top and follows Fresnel's law. For neutrons, the "surface" will actually be all interfaces within a few thousand nanometers of the topmost layer. This makes neutron reflectivity a very powerful method for probing interfaces, especially buried interfaces. As most modern devices have shrunk in size and have at least one interface, if only from a protective coating, there are many avenues for research here. For example, a recent review summarized the studies on the self-assembly process of magnetic nanoparticles on a surface and the impact of the surfactant, surface preparation, and applied magnetic field (see Fig. 8).

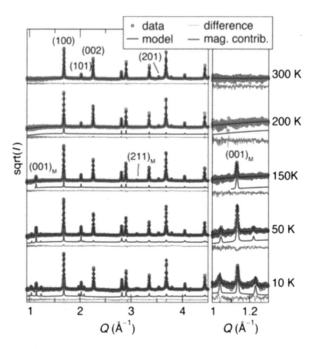

Fig. 6 Experimental ND data demonstrating the transition of MnPtGa from a ferromagnet, to a canted anti-ferromagnet, to the eventual formation of a spin-density-wave state at low temperatures. (Used with permission from [11])

Fig. 7 Schematic of a neutron reflectometer (Courtesy of NCNR)

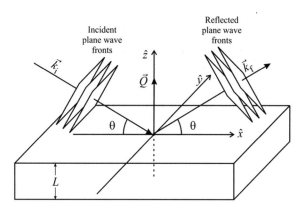

When examining NR data, the spacing of the oscillations and/or peaks and the height and shape of these features are indicative of the layer thickness and the interface quality. Generally, the less dampening of the oscillations there is, the cleaner the interface is; the closer the oscillations are in Q, the thinner the layers are in practice.

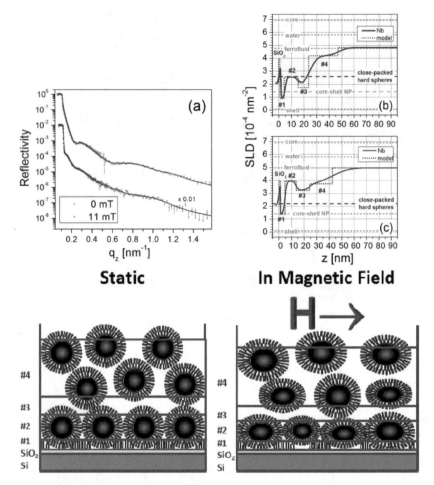

Fig. 8 (Top) (**a**) NR data on iron oxide nanoparticles coated with oleic acid and their self-assembly onto a Si/SiO₂ surface as a function of applied magnetic field. The nuclear scattering length density profiles as a function of distance z from the Si surface at applied magnetic fields of (**b**) 0 mT and (**c**) 11 mT, obtained from the model fits of the NR data in (**a**). (Bottom) Model showing hexagonal close packed structure in the initial layer followed by increasing disorder in the layers above. The magnetic field increases the disorder and elongation in the field direction. (Used with permission from [12])

6 Small Angle Neutron Scattering

Small angle neutron scattering also obeys Bragg's law (Eq. 3), but now the size d is that of larger objects, such as pores in rock or nanoparticles in solution or blocks in a co-polymer. Therefore, longer wavelengths are needed and/or smaller angles. Then,

I(Q) is typically dominated by F(Q) where F(Q) is the form factor and is dependent mainly upon the shape of the object. Whereas ND provides information about the atomic structure, SANS provides information about the shape and size of nanoscale to microscale structure.

Within SANS, there are specific instruments that cover specific dimensions (see Fig. 9 for examples from NCNR). 30 m SANS covers dimensions approximately 1–700 nm, ultra-SANS (USANS) covers approximately 100–20,000 nm, and very-SANS (VSANS) covers approximately 1–2000 nm.

Data reduction [13] and data modeling programs [14] are readily available and account for differences in data formatting between instruments at different facilities. Data reduction takes into account the total flux of neutrons incident on the sample (by removing the sample, so that the beam is "open"), what the background is for the instrument configuration (by a blocked beam), and what the transmission of the sample is (by measuring the central beam intensity at the detector). The scattering measurement blocks the main transmitted beam so that the scattered intensity can be measured. While data are collected on a 2D detector, they are rarely analyzed that way. Instead, the data are averaged (see Fig. 10). This could be a circular average where only the radius from the center matters. It could be a sector average, if there is anisotropy in the data such as shown in Fig. 10, for example, due to an applied magnetic field or structural anisotropy such as chaining. Finally, it could be an annular average, where the data at a specific radius are plotted as a function of angle.

Interpretation of SANS data requires specific models, and the details of the model matter greatly. Therefore, secondary information is always necessary in order to make good guesses about what model and parameters to use. For example, the same dataset (see Fig. 11) can be modeled with a polydisperse sphere or with a parallelepiped. The shapes are quite different, and to most experimentalists, the fits are equally good. However, the exact detail of the transition to background is quite different between the two models and is significant to the interpretation.

However, in looking at SANS data, you are looking for the location in Q of peaks or turn-overs or asymmetry in the data. For example, the turn-overs in Fig. 12 of modeled polydisperse spheres are located at different Q values (where $Q \approx 2\pi/d$), which are indicative of the differing size of the objects; in this case, the size is a diameter. The difference in intensity is less important, as the total intensity is spread out over the whole Q range, including low Q values not accessible for the specific instrument. In addition, when looking at SANS data, recall that the selection rules dictate that the magnetic scattering only appears when the sample magnetization is perpendicular to Q. Therefore, magnetic scattering has a characteristic sinusoidal-like pattern, as seen in Fig. 10.

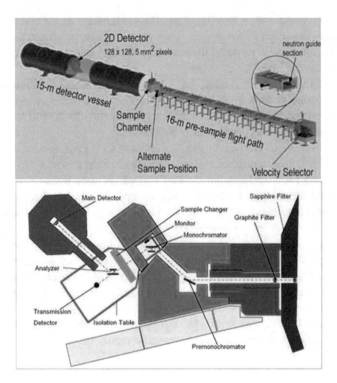

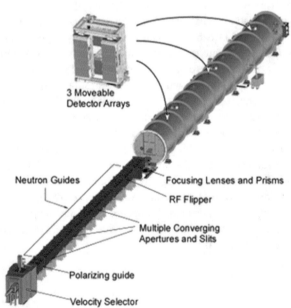

Fig. 9 (Top) SANS instrument schematic. (Middle) USANS instrument schematic. (Bottom) VSANS instrument schematic. (Courtesy of NCNR)

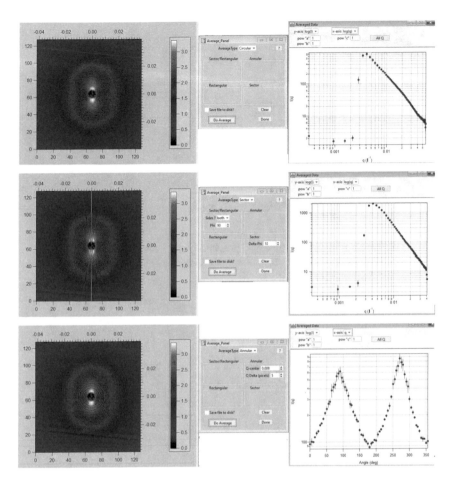

Fig. 10 Image of 2D detector data collected on iron oxide nanoparticles in water with (top) circular averaged·I(Q), (middle) sector averaged I(Q), and (bottom) annular averaged I(Q)

7 Polarization Analysis

Normally, the neutron source is unpolarized, so the neutron moments point in all directions. However, very interesting physics can be done if the neutron moments have a large degree of alignment or polarization. (Polarization analysis can occur using any of the scattering techniques, but it is only discussed here for SANS and NR, see Fig. 13, due to space constraints. The basic principles apply for all methods.) While this polarization can originate from a large magnetic field applied

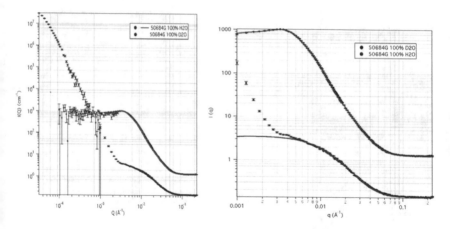

Fig. 11 Iron oxide nanoparticles in water, fit with a (left) spherical model and (right) parallelepiped model. Note the difference in the shape of the transition to background of the two models

Fig. 12 Model SANS data of
spheres of different radii

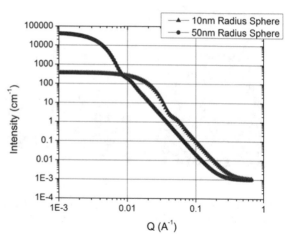

to the neutron beam, in practice, a polarizing device is used.[2] For example, a supermirror (e.g., alternating layers of Fe and Si which reflect away the undesired spin state), or a helium-3 spin filter (which adsorbs the undesired spin state), or a Heusler crystal (which has a scattering length of zero for the undesired spin state) is placed in the beam (see Fig. 9). This polarization is then maintained via a small guide magnetic field (on the order of 1 mT).

Experimentally, once you have a polarized beam, the neutrons are labeled as spin up (parallel to the guide field) or spin down (antiparallel to the guide field). When the polarized neutrons interact with a material, their spin direction can change.

---

[2]At NCNR, the incident polarization on the two SANS instruments is done with a V-shaped supermirror cavity.

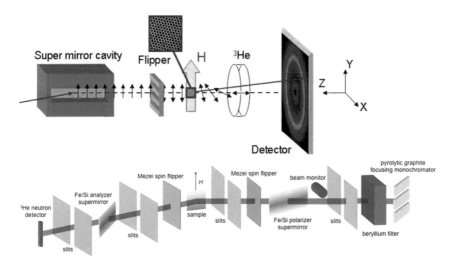

Fig. 13 (Top) Schematic of SANS including polarization of the neutron beam. (Bottom) Schematic of NR including polarization of the neutron beam (Courtesy of NCNR)

For magnetic materials, the non-spin flip scattering (spin-up to spin-up scattering and spin-down to spin-down scattering) represents nuclear scattering; the spin flip scattering (spin-up to spin-down scattering and spin-down to spin-up scattering) represents magnetic scattering only. The components of the magnetization parallel and perpendicular to the applied magnetic field can be separated when the polarization is perpendicular to the scattering wave vector, and the vector magnetization within the material can be unambiguously determined.

The next step is to be able to flip the neutron polarization at will, to create spin-up and spin-down configurations. A flipper is used to change the direction of the moment of the neutron relative to the guide field. This is typically done by having a coil of wire to cancel the guide field and a second coil to create a field perpendicular to reverse the moment of the neutron. (While this is actually a quantum mechanical effect, like many things in magnetism, a simple classical analogue works: the perpendicular field causes a torque to act on the neutron and rotate it.) After the polarized beam passes through the material, the polarization of the neutron beam must be determined. This is often done in SANS with polarized helium-3 cells, in which neutrons of one spin state are absorbed, while neutrons with the opposite orientation travel through relatively unimpeded [15].

Data reduction of polarized neutrons is more complicated than the unpolarized data [15]. Additional corrections include the efficiency of the polarizers (supermirror and/or He-3 cell) and, if appropriate, the decay of the He-3 cell polarization as a function of time. Polarization analysis permits the unique determination of the magnetization vector within a sample. This has already demonstrated enormous utility for determining the vector magnetization when studying magnetic nanoparticles and thin films. At low fields, it has shown the internal magnetic domain structure (see Fig. 14), enabling correlation of the magnetic domains with

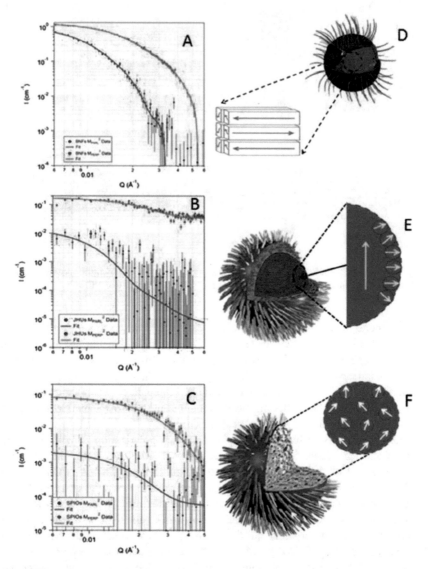

Fig. 14 Internal magnetic domain in iron oxide nanoparticles as determined with polarization analysis in SANS. (**a**), (**b**), and (**c**) show the parallel and perpendicular magnetic scattering and their fits for three different iron oxide nanoparticles while (**d**), (**e**), and (**f**) are schematics of the domain structure from (**a**), (**b**), and (**c**) respectively. (Used with permission from [16])

the AC field response and therefore the heat generation for the cancer treatment hyperthermia [16]. It has also been used to demonstrate surface spin canting in compositionally and structurally uniform magnetic nanoparticles under a nominally saturating magnetic field (see Fig. 15) and how the surface spins change as a function of applied magnetic field and temperature [17].

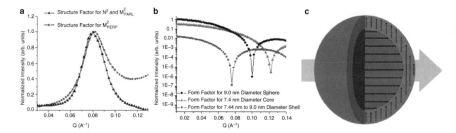

Fig. 15 Surface spin canting on iron oxide nanoparticles as determined with polarization analysis in SANS. (**a**) is the structure factor and (**b**) is the form factor for the nuclear and magnetic (parallel and perpendicular) scattering. (**c**) is a schematic of the magnetic structure within the iron oxide nanoparticle based on (**a**) and (**b**). (Used with permission from [17])

Early studies (see Fig. 16) of bilayers of Fe/Gd thin films demonstrated the power of polarized NR, as they were able to identify the antiferromagnetic coupling between the layers and demonstrate the diverse range of magnetic orientations that resulted as a function of layer thickness and temperature.

8 Time Resolution

This is naturally done at a spallation source, where all the neutrons are time stamped in order to determine their energies. It is also possible to time stamp neutrons from a reactor. This leads to two forms of time resolution: time-of-flight (TOF) and time-resolved (TR). TOF is limited by the spread of the neutron beam as it travels, resulting in overlap of different time signals, thereby determining the size of the bins. Time-resolved puts a chopper in the beam (see Fig. 17), breaking up the neutron flux into small packets. These can then be time synchronized to something else, like an oscillating electric or magnetic field or sinusoidal stress. The neutrons can then be binned according to when they interacted with the sample with respect to the AC stimulation, permitting tracking of the time-dependent response. TR is limited by the spread of the beam packet (due to having a polychromatic beam even after monochromating) during flight from the sample to the detector as well as detector electronics. Another limitation is that the chopper throws away neutrons, further reducing your signal. Early work has confirmed expected behavior in magnetic nanoparticles (see Fig. 18).

9 Summary

In summary, neutrons are a unique and powerful tool with which to characterize materials. Advantages included *in situ* measurements and the ability to highlight

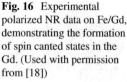

Fig. 16 Experimental polarized NR data on Fe/Gd, demonstrating the formation of spin canted states in the Gd. (Used with permission from [18])

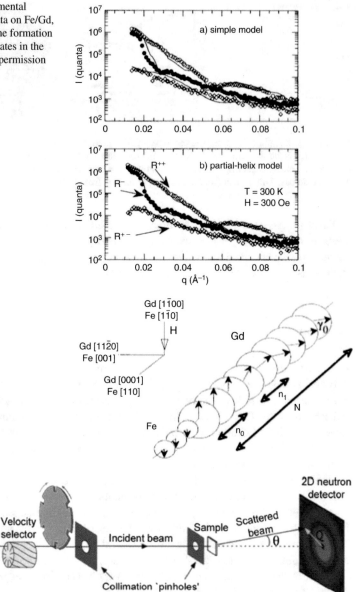

Fig. 17 Time-resolved SANS instrument schematic (Used with permission from [7])

different aspects of the sample with contrast matching. Furthermore, neutrons also have exquisite sensitivity to the vector magnetization on the nanoscale. However, although access is readily available via user proposal system, neutrons do require specialized facilities. Therefore, if it is possible to measure the same property with

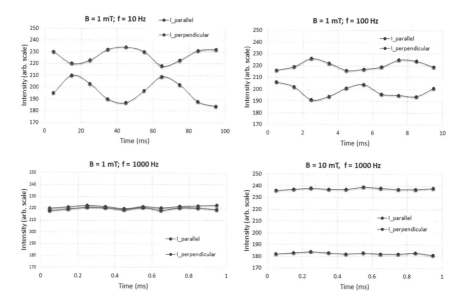

Fig. 18 TOF data on magnetic nanoparticles showing reversal behavior at low frequencies, and then blocked behavior at low fields and high frequencies, before reversal behavior reappears at high fields and high frequencies. (Used with permission from [19])

a lab-scale technique, that is preferable (especially as neutron experiments often require more material and longer counting times than other scattering methods). However, there are a number of cases where neutrons are the best method available, such as vector magnetization of an ensemble of nanoparticles or magnetization reversal mechanisms in magnetic thin films, to name just two. This then opens up new avenues of research and new insights into materials.

Acknowledgments C.L.D. thanks Julie Borchers, Kathryn Krycka, Andrew Jackson, William Ratcliffe, Markus Bleuel, Charlie Glinka, John Barker, Steve Kline, and Paul Butler for many useful discussions over the years.

References

1. T. Chatterji, *Neutron Scattering from Magnetic Materials* (Elsevier, New York, 2006)
2. B. Hammouda, *Probing Nanoscale Structures – The SANS Toolbox* (NIST, Gaithersburg, MD, 2016)
3. R. Pynn, *Neutron Scattering: A Primer* (Los Alamos Neutron Science Center)
4. K.L. Krycka, J.A. Borchers, Y. Ijiri, R. Booth, S.A. Majetich, Polarization-analyzed small-angle neutron scattering. II. Mathematical angular analysis. J. Appl. Crystallogr. **45**, 554–565 (2012)
5. A. Michels, J. Weissmueller, Magnetic-field-dependent small-angle neutron scattering on random anisotropy ferromagnets. Rep. Prog. Phys. **71**, 066501 (2008)

6. A. Wiedenmann, U. Keiderling, K. Habicht, M. Russina, R. Gaehler, Dynamics of field-induced ordering in magnetic colloids studied by new time-resolved small-angle neutron-scattering techniques. Phys. Rev. Lett. **97**, 057202 (2006)

7. C. Glinka, M. Bleuel, P. Tsai, D. Zakutna, D. Honecker, D. Dresen, F. Mees, S. Disch, Sub-millisecond time-resolved small-angle neutron scattering measurements at NIST. J. Appl. Crystallogr. **53**, 598–604 (2020)

8. J. Rush and R. Cappelletti, The NIST Center for Neutron Research: Over 40 years serving NIST/NBS and the nation, NIST Special Publication 1120 (2011)

9. "How SNS Works," 2020. [Online]. Available: https://neutrons.ornl.gov/content/how-sns-works

10. V.F. Sears, Neutron scattering lengths and cross sections. Neutron News **3**(3), 29–37 (1992)

11. J. Cooley, J. Bocarsly, E. Schueller, E. Levin, E. Rodriguez, A. Huq, S. Lapidus, S. Wilson, R. Seshadri, Evolution of noncollinear magnetism in magnetocaloric MnPtGa. Phys. Rev. Mater. **4**, 044405 (2020)

12. K. Theis-Bröhl, A. Saini, M. Wolff, J. Dura, B. Maranville, J. Borchers, Self-assembly of magnetic nanoparticles in Ferrofluids on different templates investigated by neutron reflectometry. Nano materials **10**, 1231 (2020)

13. S. Kline, Reduction and analysis of SANS and USANS data using IGOR pro. J. Appl. Crystallogr. **39**, 895–900 (2006)

14. "SASVIEW," 2020. [Online]. http://www.sasview.org/

15. K. Krycka, W. Chen, J. Borchers, B. Maranville, S. Watson, Polarization-analyzed small-angle neutron scattering. I. Polarized data reduction using Pol-Corr. J. Appl. Crystallogr. **45**, 546–553 (2012)

16. C. Dennis, K. Krycka, J. Borchers, R. Desautels, J. van Lierop, N. Huls, A. Jackson, C. Gruettner, R. Ivkov, Internal magnetic structure of nanoparticles dominates time-dependent relaxation processes in a magnetic field. Adv. Funct. Mater. **25**, 4300–4311 (2015)

17. K. Krycka, R. Booth, C. Hogg, Y. Ijiri, J. Borchers, W. Chen, S. Watson, M. Laver, T. Gentile, L. Dedon, S. Harris, J. Rhyne, S. Majetich, Core-Shell magnetic morphology of structurally uniform magnetite nanoparticles. Phys. Rev. Lett. **104**, 207203 (2010)

18. O. McGrath, N. Ryzhanova, C. Lacroix, D. Givord, C. Fermon, C. Miramond, G. Saux, S. Young, A. Vedyayev, Observation and interpretation of a partial Gd twisted spin state in an epitaxial Gd/Fe bilayer. Phys. Rev. B **54**, 6088 (1996)

19. C. Dennis, C. Glinka, J. Borchers, C. Grüttner, R. Ivkov, Effect of applied AC magnetic field on response of magnetic nanoparticles. J. Appl. Phys.. submitted

Neutron Scattering in Magnetism: Fundamentals and Examples

Javier Campo and Víctor Laliena

Abstract In this chapter, an up-to-date survey of theoretical concepts and experimental results in the field of the neutron scattering techniques applied to magnetism, and their relevance to experiments in real materials, is presented. Main emphasis of the chapter is to enlarge the use of these techniques among the researchers (physicist, chemists, engineers, etc.) in the area of magnetism and magnetic materials. For that reason we do not enter deeply in topics as neutron production, neutron optics, detectors, instrumentation, etc., but we focused the chapter on the neutron scattering applications in such a field.

Along this chapter, together with some basic concept on quantum mechanics, solid-state physics, magnetism, and symmetry, we introduce the main theoretical concepts, peculiarities, and language, of the neutron scattering techniques, and we particularize it to the crystalline matter. In a second part, we briefly introduce the most commonly employed techniques (powder diffraction, single crystal diffraction, polarization analysis, inelastic, SANS, reflectivity, etc.) and how these techniques are applied to understand different magnetic phenomena and magnetic materials, trying to cover a large variety of topics with interest in magnetism. Every time, the examples showed here show how neutrons can enlighten problems that are quasi-impossible to be solved with other techniques.

Keywords Neutron Scattering · Magnetic excitations · Magnetic structures · Magnetic energy levels · Polarized neutrons · Magnetic crystallography

J. Campo (✉) · V. Laliena
Aragón Nanoscience and Materials Institute (CSIC – Zaragoza University), Spanish National Research Council (CSIC), Zaragoza, Spain
e-mail: javier.campo@csic.es; laliena@unizar.es

© Springer Nature Switzerland AG 2021
V. Franco, B. Dodrill (eds.), *Magnetic Measurement Techniques for Materials Characterization*, https://doi.org/10.1007/978-3-030-70443-8_14

1 Introduction

Since the 1950s, when the first experiments using neutron beams were reported, the neutron scattering techniques are of paramount importance in the study of any type of magnetism and magnetic materials. It is due to the fact that the neutron carries a spin, and therefore a magnetic moment, that can interact, via the dipolar magnetic interaction, with the different sources of magnetism in the matter (unpaired electrons, magnetic nuclei, etc.). In this chapter, it is not intended to deduce from the scattering theory all the equations governing the scattering, magnetic or nuclear, of neutrons by matter, but only the mathematical expressions, and their approximations, necessary to understand the experiments are described.

Different sets of *canonical* experiments, related to different neutron scattering techniques, will be explained bearing in mind that the goal of the chapter is to motivate the readers by showing how the neutron techniques can help in their scientific problems in the area of magnetism.

The chapter is organized as follows. In Sect. 2 the fundamentals of thermal neutron scattering theory will be explained, giving only the minimal information to understand the meaning of the formulae and the approximations employed to obtain them. In Sect. 3 the scattering theory will be particularized for crystalline matter, and in Sect. 4 a brief deduction of the scattering of polarized neutrons is given.

In Sect. 5 the concepts of magnetic space groups and the formalism of the propagation vector to describe magnetic structures are introduced. Then the powder diffraction techniques, explaining a little bit the Rietveld method, are presented in Sect. 6. Applications of single-crystal neutron diffraction with unpolarized neutrons are explained in Sect. 7. In Sects. 8 and 9, the use of polarized neutrons in single crystals, by exploring the potentiality of Blume-Maleev equations, is highlighted. The small-angle neutron scattering (SANS) techniques are also often employed to solve important problems in magnetism: physics of magnetic nanoparticles, magnetic clusters, skyrmions, etc. In Sect. 10 examples of the use of these techniques in cubic helimagnets to characterize skyrmionic phases will be shown. Some ideas of magnetic inelastic scattering are described in Sect. 11, and finally other neutron scattering techniques (neutron spin echo, neutron reflectivity, and quasielastic neutron scattering) initially developed to study soft matter and that now are more and more employed in magnetism will be explained in Sect. 12.

An introduction to the application of neutron techniques to magnetic materials, with a focus on the physics behind the instrumentation and general data analysis for SANS, reflectivity, and diffraction, has been described in the previous chapter.

The chosen examples in this chapter comprise several materials showing different magnetic behaviors: spin glasses, cubic helimagnets, skyrmions, random fields, solitons, single-molecule magnets, organic magnets, multiferroics, battery electrodes, molecular rotors, etc.

Before starting, let us describe briefly the conventions used in this chapter. Although in general it is uncommon in physical literature, we found it convenient to adhere to the crystallographic and spectroscopy conventions and define the wave number as the inverse of the wavelength and the frequency as the inverse of time period, without the 2π factors that are present in the more common convention. As a drawback, the 2π factors appear in the exponential factors of the Fourier transforms, but they disappear from many expressions, as, for instance, from the normalization factors of the Fourier transform. The symbol h denotes the Planck constant,[1] $h \approx 6.626 \times 10^{-34}$ Js, and, as usual, $\hbar = h/2\pi$. Energy and momentum are related to frequency and wave number by $E = h\nu$ and $p = hk$, respectively.

2 Fundamentals of Neutron Scattering

Properties of Neutrons

The neutron is a baryon, an elementary particle formed by three quarks bound by the strong interaction. Its mass, $m = 1.67 \times 10^{-27}$ kg $= 939.6$ MeV/c^2, is almost equal to the proton mass, and it has an intrinsic angular momentum, or spin, of magnitude $\hbar/2$. The neutrons are thus fermions and obey the Fermi-Dirac statistics. Neutrons are stable within the nuclei, but free neutrons are unstable, decaying via the weak interaction into a proton, an electron, and an antineutrino, with a mean life of ~ 881 s. Although it is a neutral particle, its constituent quarks are charged, and thus the neutron interacts electromagnetically. Its dipolar electric moment does vanish, or, if it does not, it is extremely small: at the time of writing, the latest experimental result is quoted as $(0.0 \pm 1.1_{\text{stat}} \pm 0.2_{\text{sys}}) \times 10^{-26}$ $|e| \cdot$ cm, where the subscripts *stat* and *sys* stand for the statistical and systematic contributions to the uncertainty and $e < 0$ is the electron charge [1]. Therefore, the electromagnetic interactions of neutrons are dominated by its dipolar magnetic moment, described by the quantum operator

$$\vec{\mu}_n = -\gamma \mu_N \frac{1}{2} \vec{\sigma}, \tag{1}$$

where $\mu_N \approx 5.05 \times 10^{-27}$ J/T is the nuclear magneton, $\gamma = 3.826$ is the neutron *g-factor*, and $\vec{\sigma}$ are the Pauli matrices acting on the neutron spin space.

Neutrons are produced in nuclear reactions and as a product of the fission of heavy nucleus, such ^{252}Cf, ^{235}U, etc. It is difficult to have intense neutron sources. At present, such intense sources are of two types: *nuclear reactors*, in which the neutrons are a by-product of the fission of the nuclear fuel, and *spallation sources*, in which neutrons are produced when a beam of high energy protons, accelerated by linear accelerators (LINAC), impacts on a heavy metal target.

According to their energy, in meV, neutrons are customarily classified into cold (0.1–10), thermal (10–100), hot (100–500), and epithermal (>500). The rest energy

---

[1] Occasionally, the symbol h will be used to denote one of the Miller indices, but the context will resolve any ambiguity.

of the neutron is many orders of magnitude higher than the kinetic energy of these kinds of neutrons, which are thus non-relativistic particles, with energy given by $E = h^2 k^2 / 2m$, where $k = 1/\lambda$ is the wave number and λ the wavelength. This has a very interesting consequence for the study of condensed matter systems, since the wavelength of neutrons with energy of the order of meV, which corresponds to the excitation energies of many systems, is in the range of the Å, which corresponds to the inter-particle distances in the systems. In comparison, the energy of the X-rays with such wavelengths is in the range of the keV, orders of magnitude higher than the excitation energies of condensed matter systems.

Being a spin $1/2$ particle, the quantum state of a free neutron is completely characterized by the three components of its wave vector, $\vec{k}$, which form a continuous set, and a two-component spinor which characterizes the spin state, or *polarization state* of the neutron, denoted by $|\sigma\rangle$. Each spinor is the eigenstate of the projection of the spin operator $\vec{\sigma}$ onto the direction determined by the expectation value of the spin operator in this polarization state, $\langle \sigma | \vec{\sigma} | \sigma \rangle$, which is a unit vector. The polarization state of the neutron is completely characterized by this unit vector which is called the polarization vector of the neutron. In a beam, different neutrons may be in different polarization states. The polarization vector of the beam, $\vec{P}$, is defined as the average of the polarization vectors of all neutrons. Oviously, its modulus satisfies $P \leq 1$. The polarization states of the beam are statistically described by a 2×2 *density matrix* operator,

$$\varrho = \frac{1}{N} \sum_n |\sigma_n\rangle \langle \sigma_n|, \tag{2}$$

where n labels the neutrons in the beam, N is the number of neutrons, and $|\sigma_n\rangle$ stands for the normalized spin-wave function (spinor) of the n-th neutron. The density matrix is Hermitian and satisfies $\mathrm{Tr}\,\varrho = 1$ and $\mathrm{Tr}(\varrho\vec{\sigma}) = \vec{P}$. The only 2×2 Hermitian matrix that satisfies the two conditions is

$$\varrho = \frac{1}{2} \left(I_2 + \vec{P} \cdot \vec{\sigma} \right), \tag{3}$$

where I_2 is the two-dimensional unit matrix. The probability of finding a neutron of the beam in some arbitrary polarization state, $|\sigma\rangle$, is given by the expectation value of the density matrix in this state, $\langle \sigma | \varrho | \sigma \rangle$, as can be readily checked. Notice that the expectation values on *all* possible states, $|\sigma\rangle$, fully determine the density matrix. The eigenvectors of ϱ are those of $\vec{P} \cdot \vec{\sigma}$, and its two eigenvalues are given by $(1 \pm P)/2$. A beam with $\vec{P} = 0$ is called an *unpolarized* beam. In it, all polarization states have the same probability.

Notice finally that an orthonormal basis on the neutron spin space is always associated to a given direction, defined by some unit vector $\hat{u}$. Therefore, if the first vector of the basis is the eigenvector of $\hat{u} \cdot \vec{\sigma}$ with positive eigenvalue, the second vector is the eigenvector with negative eigenvalue. Hence, choosing an orthonormal basis in spin space is tantamount to choosing a *quantization* axis, and the vectors of the basis are the eigenstates of the projection of the spin operator onto the quantization axis. If we choose the quantization axis along the beam polarization direction, the vectors of the basis are the eigenstates of the density matrix.

The Scattering Problem

In a scattering experiment, a probe, in our case a neutron beam, impinges a target, and the resulting scattered beam is detected at a large distance from the target. The spatial and energy distributions of the scattered neutrons provide a considerable amount of physical information about the target. The incoming neutron is described by a plane wave with wave vector $\vec{k}_i$ and a spin state $|\sigma_i\rangle$, which is one of the eigenstates of the incident beam density matrix, ϱ_i. If the beam is unpolarized, any orthonormal basis in spin space can be taken. If the state of target before the interaction is $|\varphi_i\rangle$, the initial state of the composite neutron-target system is written as $|\vec{k}_i\sigma_i; \varphi_i\rangle$. A long time after the scattering took place, there will be a certain probability that the composite system is found in the state $|\vec{k}_f\sigma_f; \varphi_f\rangle$. The spin state of the scattered neutron can be referred to a quantization axis different from that of the incoming beam, that is, the orthonormal basis $\{|\sigma_f\rangle\}$, need not be the same as the orthonormal basis $\{|\sigma_i\rangle\}$.

In the scattering experiment, the initial neutron beam is prepared according to some convenient specification, and the scattered neutrons are detected by detectors conveniently distributed around the target. The *differential scattering cross section* is defined as the number of neutrons scattered in the direction $\hat{k}_f$ per unit time; solid angle, $d\Omega_{\hat{k}_f}$; and energy, dE_f, normalized by the neutron flux, which is the number of incident neutrons per unit of time and beam cross section area [2]. If $\mathcal{H}_{\text{int}}$ is the interaction Hamiltonian operator for the neutron and the target system, the differential scattering cross section in the lowest order pertubation theory is given by

$$\left(\frac{d^2\sigma}{d\Omega_{\hat{k}_f} dE_f}\right)_{\sigma_f} = \frac{k_f}{k_i}\left(\frac{2\pi m}{h^2}\right)^2 |\langle\vec{k}_f\sigma_f; \varphi_f|\mathcal{H}_{\text{int}}|\vec{k}_i\sigma_i; \varphi_i\rangle|^2\delta(E_{\varphi_i} - E_{\varphi_f} + h\nu),$$

(4)

where E_{φ_i} and E_{φ_f} are the energies of the target before and after the interaction and $h\nu = E_i - E_f$ is the difference between the energies of the incident and scattered neutron. Although the above formula cannot be rigorously derived in a few lines, since it requires the application of scattering theory [2], the origins of each ingredient entering (4) can be easily understood. The Dirac delta function, δ, enforces the conservation of energy in the whole neutron-target system a long time after the interaction took place, when the interaction energy is negligible. The matrix element of $\mathcal{H}_{\text{int}}$ is the lowest-order perturbation theory approximation to the probability amplitude that the composite neutron-target system is found in the state

$|\vec{k}_f \sigma_f; \varphi_f\rangle$ after the interaction took place. Its modulus squared divided by h gives the probability per unit time of finding this final state (compare, for instance, with the *Fermi Golden rule* [3]). The density of final neutron states (per unit energy and per unit solid angle) is mk_f/h^2, what accounts for the k_f factor and one m/h^2 factor. And the incident neutron flux is proportional to the neutron velocity, hk_i/m, what introduces the k_i in the denominator and another m/h factor.

The initial target pure state, $|\varphi_i\rangle$, is unknown. It is assumed that the target is in equilibrium, so that its quantum state is actually mixed, given by the density matrix corresponding to the Boltzmann distribution. This means that we have to average the right-hand side of expression (4) over the initial target states, $|\varphi_i\rangle$, with the Boltzmann weight $P_{\varphi_i} = \exp(-E_{\varphi_i}/k_\mathrm{B}T)/\mathcal{Z}$, where T is the temperature, k_B the Boltzmann constant, and $\mathcal{Z}$ the canonical partition function that ensures the normalization $\sum_{\varphi_i} P_{\varphi_i} = 1$. The final state of the target, $|\varphi_f\rangle$, is not observed, so that we have to sum over all of them. Analogously, we have to average over the polarization state of the incident neutron, $|\sigma_i\rangle$, which has a probability $p_{\sigma_i} = \langle \sigma_i | \varrho_i | \sigma_i \rangle$, and, if the polarization state of the scattered neutron is not observed, we have to sum over $|\sigma_f\rangle$. To be general, we assume here that it is observed. Finally, there are properties of the target that influence the scattering result that are usually not under control. An example is the isotopic distribution of the nuclei. These properties have to be treated as a disorder, and we have to average over them with the proper disorder distribution. We denote this average by a horizontal bar over the quantities that are averaged. Taking all this into account, (4) becomes

$$
\left(\frac{d^2\sigma}{d\Omega_{\hat{k}_f}\, dE_f} \right)_{\sigma_f} = \frac{k_f}{k_i} \left(\frac{2\pi m}{h^2} \right)^2
$$

$$
\times \sum_{\sigma_i \varphi_i \varphi_f} p_{\sigma_i} P_{\varphi_i} \overline{|\langle \vec{k}_f \sigma_f; \varphi_f | \mathcal{H}_{\mathrm{int}} |\vec{k}_i \sigma_i; \varphi_i \rangle|^2} \delta(E_{\varphi_i} - E_{\varphi_f} + h\nu).
$$

(5)

Neutron interactions at low energy do not depend on neutron momentum. Hence, the interaction Hamiltonian contains no derivative with respect to the neutron coordinates. Since the neutron wave functions are plane waves, the matrix elements entering the cross section can be partially evaluated as

$$
\langle \vec{k}_f \sigma_f; \varphi_f | \mathcal{H}_{\mathrm{int}} |\vec{k}_i \sigma_i; \varphi_i \rangle = \int d^3 r\, e^{i2\pi \vec{q}\cdot\vec{r}} \langle \sigma_f | \langle \varphi_f | \mathcal{H}_{\mathrm{int}}(\vec{r}; \ldots) |\varphi_i\rangle |\sigma_i\rangle,
$$

(6)

where $\vec{r}$ is the neutron position and the ellipsis stands for the target degrees of freedom on which the interaction hamiltonian depends: the position and momentum operators of the i-th target particle, $\vec{r}_i$ and $\vec{p}_i$, respectively, and any other operators, like the spin of the i-the particle, I_i. We have introduced the *scattering vector*, $\vec{q} = \vec{k}_i - \vec{k}_f$, which is proportional to the momentum transfer vector.

It is convenient to define the *amplitude* operator

$$\mathcal{A}_{\vec{q}} = \frac{2\pi m}{h^2} \int d^3r \, e^{i2\pi \vec{q}\cdot\vec{r}} \, \mathcal{H}_{\text{int}}(\vec{r}; \vec{r}_1 \vec{p}_1 I_1; \ldots, \vec{r}_{N_P} \vec{p}_{N_P} I_{N_P}), \tag{7}$$

which is a matrix in the neutron spin degrees of freedom and an operator in the Hilbert space of the target. It has the dimensions of a length. In terms of $\mathcal{A}_{\vec{q}}$, the cross section reads:

$$\left(\frac{d^2\sigma}{d\Omega_{\hat{k}_f} dE_f}\right)_{\sigma_f} = \frac{k_f}{k_i} \sum_{\sigma_i \varphi_i \varphi_f} p_{\sigma_i} P_{\varphi_i}$$

$$\times \langle \varphi_i | \langle \sigma_i | \mathcal{A}_{\vec{q}}^{\dagger} | \sigma_f \rangle | \varphi_f \rangle \langle \varphi_f | \langle \sigma_f | \mathcal{A}_{\vec{q}} | \sigma_i \rangle | \varphi_i \rangle \, \delta(E_{\varphi_i} - E_{\varphi_f} + h\nu). \tag{8}$$

We can write an interesting expression for the cross section by inserting in (8) the Fourier representation of the Dirac delta function

$$\delta(E_{\varphi_i} - E_{\varphi_f} + h\nu) = \frac{1}{h} \int_{-\infty}^{\infty} dt \, e^{-i2\pi(E_{\varphi_i} - E_{\varphi_f} + h\nu)t/h}, \tag{9}$$

and using the relations $e^{-iE_{\varphi_i}t/\hbar} |\varphi_i\rangle = \mathcal{U}(t) |\varphi_i\rangle$, $\langle \varphi_f | e^{iE_{\varphi_f}t/\hbar} = \langle \varphi_f | \mathcal{U}^{\dagger}(t)$, and

$$\mathcal{U}^{\dagger}(t) \, \mathcal{A}_{\vec{q}} \, \mathcal{U}(t) = \mathcal{A}_{\vec{q}}(t), \tag{10}$$

where $\mathcal{U}(t)$ is the time evolution operator of the target in the Schrödinger picture and $\mathcal{A}_{\vec{q}}(t)$ is the Heisenberg operator that coincides with $\mathcal{A}_{\vec{q}}$ at $t = 0$. Using the notation $\langle \mathcal{O} \rangle$ for the thermal quantum average of any operator $\mathcal{O}$,

$$\langle \mathcal{O} \rangle = \sum_{\varphi} P_{\varphi} \langle \varphi | \mathcal{O} | \varphi \rangle, \tag{11}$$

we get

$$\left(\frac{d^2\sigma}{d\Omega_{\hat{k}_f} dE_f}\right)_{\sigma_f} = \frac{k_f}{k_i} \int_{-\infty}^{\infty} \frac{dt}{h} e^{-i2\pi\nu t} \sum_{\sigma_i} p_{\sigma_i} \overline{\left\langle \langle \sigma_i | \mathcal{A}_{\vec{q}}^{\dagger}(0) | \sigma_f \rangle \langle \sigma_f | \mathcal{A}_{\vec{q}}(t) | \sigma_i \rangle \right\rangle}. \tag{12}$$

The sum over σ_i gives the density matrix of the incident beam, ϱ_i, so that the above equation can be written as

$$\left(\frac{d^2\sigma}{d\Omega_{\hat{k}_f} dE_f}\right)_{\sigma_f} = \frac{k_f}{k_i} \int_{-\infty}^{\infty} \frac{dt}{h} e^{-i2\pi\nu t} \, \text{Tr} \, \overline{\langle \mathcal{A}_{\vec{q}}^{\dagger}(0) \, \mathcal{P}_{\sigma_f} \, \mathcal{A}_{\vec{q}}(t) \, \varrho_i \rangle}, \tag{13}$$

where $\mathcal{P}_{\sigma_f} = |\sigma_f\rangle \langle \sigma_f|$ is the projector onto the neutron final spin state and Tr stands for the trace in neutron spin space.

We see that the cross section contains a big amount of information about the target. The interaction Hamiltonian and $\mathcal{A}_{\vec{q}}$ depends on the degrees of freedom (coordinates, spin, etc.) of the target particles. Equation (13) shows that the scattering cross section is related to the Fourier transform of the time correlation function of the target degrees of freedom involved in the interaction with the neutron.

We find it convenient to use the following notation for the Fourier transform of the time correlation function of two operators A and B acting on the target Hilbert space:

$$\langle AB \rangle_{\nu} = \int_{-\infty}^{\infty} \frac{dt}{h} e^{-i2\pi \nu t} \langle A(0)B(t) \rangle, \tag{14}$$

so that we have the compact expression:

$$\left(\frac{d^2\sigma}{d\Omega_{\hat{k}_f} dE_f} \right)_{\sigma_f} = \frac{k_f}{k_i} \operatorname{Tr} \overline{\langle \mathcal{A}_{\vec{q}}^{\dagger} \mathcal{P}_{\sigma_f} \mathcal{A}_{\vec{q}} \varrho_i \rangle}_{\nu}. \tag{15}$$

The sum over a complete set of orthonormal final polarization states gives the intensity of the scattered beam, relative to that of the incoming beam, propagating along $\hat{k}_f$ with energy between E_f and $E_f + dE_f$. Taking into account that $\sum_{\sigma_f} \mathcal{P}_{\sigma_f} = I_2$, this intensity is given by

$$I_{\vec{q}\,\nu} = \frac{k_f}{k_i} \operatorname{Tr} \overline{\langle \mathcal{A}_{\vec{q}}^{\dagger} \mathcal{A}_{\vec{q}} \varrho_i \rangle}_{\nu}. \tag{16}$$

The density matrix of the scattered beam, ϱ_f, can be obtained from Eq. (12) and from it the polarization vector of the scattered beam, which is given by

$$\vec{P}_f = \frac{1}{I_{\vec{q}\,\nu}} \operatorname{Tr} \overline{\langle \mathcal{A}_{\vec{q}}^{\dagger} \vec{\sigma} \mathcal{A}_{\vec{q}} \varrho_i \rangle}_{\nu}. \tag{17}$$

For $t \to \infty$ the correlation between two operators disappears, and we have

$$\lim_{t \to \infty} \langle \mathbb{A}(0)\mathbb{B}(t) \rangle = \lim_{t \to \infty} \langle \mathbb{A}(0) \rangle \langle \mathbb{B}(t) \rangle = \langle \mathbb{A} \rangle \langle \mathbb{B} \rangle, \tag{18}$$

where we used the invariance of the system under time translation and the fact the Heisenberg operators coincide with the Schrödinger operators at $t = 0$. If the product $\langle \mathbb{A} \rangle \langle \mathbb{B} \rangle$ does not vanish, then $\langle \mathbb{A}\mathbb{B} \rangle_{\nu}$ is singular, since the integrand of the right-hand side of (14) does not vanish for $t \to \infty$. We may isolate the singularity by substracting $\langle \mathbb{A} \rangle \langle \mathbb{B} \rangle$ from the integrand, what makes the integral convergent (non-singular), and adding the remaining part, which is

$$\int_{-\infty}^{\infty} \frac{dt}{h} e^{-i2\pi \nu t} \langle \mathbb{A} \rangle \langle \mathbb{B} \rangle = \delta(h\nu) \langle \mathbb{A} \rangle \langle \mathbb{B} \rangle, \tag{19}$$

so that we have

$$\langle \mathbb{A}\mathbb{B} \rangle_v = \int_{-\infty}^{\infty} \frac{dt}{h} e^{-i2\pi vt} \left[\langle \mathbb{A}(0)\mathbb{B}(t) \rangle - \langle \mathbb{A} \rangle \langle \mathbb{B} \rangle \right] + \delta(hv) \langle \mathbb{A} \rangle \langle \mathbb{B} \rangle . \tag{20}$$

Applying Eq. (20) to (15), we separate the scattering into *elastic* and *inelastic*:

$$\left(\frac{d^2\sigma}{d\Omega_{\hat{k}_f} dE_f} \right)_{\sigma_f} = \left(\frac{d^2\sigma^{(e)}}{d\Omega_{\hat{k}_f} dE_f} \right)_{\sigma_f} + \left(\frac{d^2\sigma^{(i)}}{d\Omega_{\hat{k}_f} dE_f} \right)_{\sigma_f} \tag{21}$$

where

$$\left(\frac{d^2\sigma^{(e)}}{d\Omega_{\hat{k}_f} dE_f} \right)_{\sigma_f} = \delta(E_i - E_f) \langle \sigma_f | \overline{\langle \mathcal{A}_{\vec{q}} \rangle \varrho_i \langle \mathcal{A}_{\vec{q}}^{\dagger} \rangle} | \sigma_f \rangle . \tag{22}$$

is the *elastic scattering cross section* and

$$\left(\frac{d^2\sigma^{(i)}}{d\Omega_{\hat{k}_f} dE_f} \right)_{\sigma_f} = \frac{k_f}{k_i} \mathrm{Tr} \overline{\left\langle (\mathcal{A}_{\vec{q}}^{\dagger} - \langle \mathcal{A}_{\vec{q}}^{\dagger} \rangle) \mathcal{P}_{\sigma_f} (\mathcal{A}_{\vec{q}} - \langle \mathcal{A}_{\vec{q}} \rangle) \varrho_i \right\rangle}_v . \tag{23}$$

is the *inelastic scattering cross section*. We used the properties of the trace operation to write the elastic cross section in the form of Eq. (22).

It is customary to present the integrated elastic cross section by integrating (22) over dE_f:

$$\left(\frac{d\sigma^{(e)}}{d\Omega_{\hat{k}_f}} \right)_{\sigma_f} = \langle \sigma_f | \overline{\langle \mathcal{A}_{\vec{q}} \rangle \varrho_i \langle \mathcal{A}_{\vec{q}}^{\dagger} \rangle} | \sigma_f \rangle . \tag{24}$$

One has to bear in mind that $k_f = k_i$ in the above equation.

The probability of finding a neutron in the scattered beam with polarization state σ_f is given by $\langle \sigma_f | \varrho_f | \sigma_f \rangle$, where ϱ_f is the spin density matrix of the scattered beam. From Eq. (24) we see that for the elastic scattered beam ϱ_f is given by

$$\varrho_f = \frac{1}{I_{\vec{q}}} \overline{\langle \mathcal{A}_{\vec{q}} \rangle \varrho_i \langle \mathcal{A}_{\vec{q}}^{\dagger} \rangle}, \tag{25}$$

where

$$I_{\vec{q}} = \sum_{\sigma_f} \langle \sigma_f | \overline{\langle \mathcal{A}_{\vec{q}} \rangle \varrho_i \langle \mathcal{A}_{\vec{q}}^{\dagger} \rangle} | \sigma_f \rangle \tag{26}$$

is proportional to the intensity of the elastically scattered beam in the $\hat{k}_f$ direction. Therefore, the polarization vector of the elastically scattered beam is given by

$$\vec{P}_f = \frac{1}{I_{\vec{q}}} \operatorname{Tr} \left(\overline{\langle \mathcal{A}_{\vec{q}} \rangle \varrho_i \langle \mathcal{A}_{\vec{q}}^{\dagger} \rangle} \vec{\sigma} \right). \tag{27}$$

Let us stress that ϱ_f and $\vec{P}_f$ depend on the direction at which the scattered beam is observed, given by $\hat{k}_f$. The above equations will be used later in the introduction to polarization analysis.

Neutron Interactions at Low Energy

Let us study the interaction Hamiltonian of low-energy neutrons with matter. A free neutron interacts with an atom through two kinds of processes: the nuclear interaction with the nucleus and the electromagnetic interaction of the neutron magnetic moment with the charge of the nucleus and the electrons.

The nuclear force is very complex and strong, so that it cannot be treated perturbatively in general. However, it is also extremely short ranged, its range being orders of magnitude shorter than the wavelength of slow neutrons (with energies smaller than a few eV). Therefore, slow neutrons are effectively scattered off a point like nucleus, and in this case the scattering is essentially isotropic: only the s-wave gives a substantial contribution to the amplitude of the scattered wave, which therefore is given by

$$\psi_S(\vec{r}) = -\frac{b}{r} e^{i2\pi kr}, \tag{28}$$

where k is the neutron wave number, $\vec{r}$ is the vector position of the neutron relative to the nucleus, and b is the so-called *scattering length*, which in general is a complex number. The minus sign is conventionally chosen so that the real part of b is positive for a repulsive force. The imaginary part of b is only significant in the vicinity of capture resonances, that is, for nuclei that strongly absorb slow neutrons. All the complexity of the nuclear force is embodied in the scattering length, which can be obtained from measurements. This allows us to construct an effective potential, the Fermi *pseudopotential* [4], which gives the exact result for scattering in the lowest order of perturbation theory (the first Born approximation):

$$V_F(\vec{r}) = \frac{h^2}{2\pi m} b \, \delta(\vec{r}), \tag{29}$$

where m is the neutron mass and $\delta(\vec{r})$ the Dirac delta function. Indeed, an elementary application of perturbation theory gives (28) and $\sigma \sim |b|^2$ from the potential (29).

It is worthwhile to stress that (29) is not the complicated true nuclear interaction Hamiltonian, which is strong and non-central. The Fermi pseudo-potential is a simple way to describe effectively the scattering of slow neutrons with a single parameter, b. The scattering lengths may be in principle computed from the nuclear interaction Hamiltonian, but this is a formidable problem. Fortunately, they are available from measurements.

The facts that the size of the atom is ~ 1 Å $= 10^{-10}$ m, the size of the nuclei is $\sim 10^{-14}$ m, and the short range of the neutron-nuclei interaction, have as immediate consequence the high penetration power of neutrons in matter, except if there are absorbing isotopes. We can say that *from the point of view of neutrons, the matter is empty*, because they only *see* the nuclei which are $\sim 10^4$ times smaller than the atoms. This consequence facilitates the experiments with neutron beams using very complex sample environments, which are easily traversed by the neutrons. We will explain later that neutrons are also able to *see* the unpaired electrons at the atoms.

The nuclear interactions depend strongly on the spin of the nucleus and the neutron, and the above discussion holds for given total spin states of the neutron-nucleus system. That is, b in the above equations depends on the total spin of the neutron-nucleus system, which can take two values: $J^{\pm} = I \pm 1/2$, where I is the total spin of the nucleus.[2] Therefore, we have actually two scattering lengths, b^+ and b^-, corresponding to the two composite states with total spin J^+ and J^-, respectively. This dependence on the total spin can be taken into account by introducing a *scattering amplitude operator*, $\mathbb{B}$, which, due to the rotational symmetry, can be written as [5]

$$\mathbb{B} = A + B \frac{1}{2} \vec{\sigma} \cdot \vec{\mathcal{I}}, \tag{30}$$

where the components of matrix vector $\vec{\sigma}$ are the Pauli matrices that act on the neutron spin degrees of freedom, $\vec{\mathcal{I}}$ is the nucleus spin operator, and

$$A = \frac{1}{2I+1}[(I+1)b^+ + Ib^-], \qquad B = \frac{2}{2I+1}(b^+ - b^-). \tag{31}$$

Thus, we have to substitute the complex number b in (29) by the operator in spin space $\mathbb{B}$. If the target contains N nucleus, labelled by the index n and located at $\vec{R}_n$, the nuclear interaction Hamiltonian is

$$\mathcal{H}_{\text{int}}^{(n)} = \frac{h^2}{2\pi m} \sum_{n=1}^{N} \mathbb{B}_n \delta(\vec{r} - \vec{R}_n), \tag{32}$$

with $\mathbb{B}_n$ given by (30).

The electromagnetic interactions are long ranged but very weak, and thus can be treated perturbatively in the Born approximation. In the scales involved in the problem, the neutron can be considered electromagnetically as a neutral point particle with a magnetic moment. Any other residual electromagnetic interaction

---

[2]Note that since we are dealing with *s*-wave scattering, the total spin of the system is conserved.

originating from the internal structure of the neutron (higher-order multipoles) is much smaller than the magnetic dipolar interaction. Since the nuclear magneton is two orders of magnitude smaller than the Bohr magneton, the interaction of the neutron magnetic moment with the nucleus is about two orders of magnitude weaker than the interactions with the electrons. The neutron-atom electromagnetic interaction is thus dominated by the interaction of the neutron magnetic moment with the electrons [6–8]. Generically, the electromagnetic interaction is introduced by substituting the momentum operator of the particle, $\vec{p}$, by $\vec{p} - q\vec{A}(\vec{r})$, where q is the electric charge of the particle and $\vec{A}(\vec{r})$ is the electromagnetic vector potential at the position of the particle. To this we have to add the dipole-dipole interaction between the intrinsic magnetic moments of the electron and the neutron, which is of relativistic origin and is derived from the Dirac equation [8]. The dipole-dipole interaction potential can be written alternatively as the energy of the electron intrinsic magnetic moment in the magnetic field created by the intrinsic magnetic moment of the neutron or as the energy of the neutron magnetic moment in the magnetic field created by the intrinsic magnetic moment of the electron. We will use the former form.

The vector potential at a point $\vec{r}_j$ by the magnetic moment of the neutron $\vec{\mu}_n$, located at point $\vec{r}$, is

$$\vec{A}_n(\vec{r}_j - \vec{r}) = \frac{\mu_0}{4\pi} \nabla_j \times \frac{\vec{\mu}_n}{|\vec{r}_j - \vec{r}|}. \tag{33}$$

It originates a magnetic field

$$\vec{B}_n(\vec{r}_j - \vec{r}) = \frac{\mu_0}{4\pi} \nabla_j \times \left(\nabla_j \times \frac{\vec{\mu}_n}{|\vec{r}_j - \vec{r}|} \right) = \frac{\mu_0}{4\pi} \nabla \times \left(\nabla \times \frac{\vec{\mu}_n}{|\vec{r}_j - \vec{r}|} \right), \tag{34}$$

where ∇ and ∇_j stand for the gradient with respect to $\vec{r}$ and $\vec{r}_j$, respectively, and $\mu_0 = 4\pi \times 10^{-7}$H/m is the vacuum permeability.

Let us write the Hamiltonian, $\mathcal{H}$, of the system composed of a free neutron of momentum $\vec{k}$ and position $\vec{r}$ and a system of N_e electrons and N_n nuclei. Let us denote by $\vec{p}_j$ and $\vec{r}_j$ the j-th electron momentum and position, respectively, and by $\vec{R}_n$, $\vec{P}_n$, and M_n the n-th nucleus momentum, position, and mass, respectively. We have

$$\mathcal{H} = \sum_{n=1}^{N_n} \frac{1}{2M_n} P_n^2 + \sum_{j=1}^{N_e} \frac{1}{2m_e} \left[\vec{p}_j - e\vec{A}(\vec{r}_j) \right]^2 + U_s(\vec{r}_j, \vec{R}_n)$$

$$+ \frac{1}{2m} k^2 + \mathcal{H}_{\text{int}}^{(n)} - \sum_{j=1}^{N_e} \vec{\mu}_j \cdot \vec{B}_n(\vec{r}_j - \vec{r}), \tag{35}$$

where U_S contains the interaction of the electron-nuclei system, $\vec{\mu}_j$ is the magnetic moment operator of the j-th electron, and $\vec{A}(\vec{r}_j)$ is the total vector potential at the position of the j-th electron and is a superposition of the internal vector potential of the electron-nuclei system, $\vec{A}_S$, any applied external field, $\vec{A}_{ext}$, and the vector potential due to the neutron, $\vec{A}_n$:

$$\vec{A}(\vec{r}_j) = \vec{A}_S(\vec{r}_j) + \vec{A}_{ext}(\vec{r}_j) + \vec{A}_n(\vec{r}_j - \vec{r}). \tag{36}$$

Notice that the vector potential is absent from the neutron kinetic energy term since it is a neutral particle, and, as discussed before, it is neglected in the nuclei terms. The last term in (35) is the dipole-dipole interaction between the electrons and the neutron. One has to bear in mind, when dealing with (35), that momentum and position of the same particle do not commute: they are operators that satisfy canonical commutation relations.

Let us write $\mathcal{H}$ in the form

$$\mathcal{H} = \mathcal{H}_S + \frac{1}{2m}k^2 + \mathcal{H}_{int}^{(n)} + \mathcal{H}_{int}^{(m)} \tag{37}$$

where $\mathcal{H}_S$ is the Hamiltonian of the electron-nuclei system in absence of any neutron, and

$$\mathcal{H}_{int}^{(m)} = \frac{\mu_0 \mu_B}{2\pi} \vec{\mu}_n \cdot \sum_{j=1}^{N_e} \left[-\frac{1}{\hbar} \frac{\vec{r}_j - \vec{r}}{|\vec{r}_j - \vec{r}|^3} \times \vec{p}_j + \nabla \times \left(\nabla \times \frac{\vec{s}_j}{|\vec{r}_j - \vec{r}|} \right) \right] \tag{38}$$

is the *magnetic interaction* Hamiltonian, neglecting the terms quadratic in $\vec{A}$, which are quadratic in e and thus two orders of magnitude smaller than the linear terms. In deriving $\mathcal{H}_{int}^{(m)}$ we have used $\vec{p}_j = -i\hbar\nabla_j$ and $\nabla_j \cdot \vec{A}_n = 0$. As usual, $\mu_B = -e\hbar/2m_e \approx 9.27 \times 10^{-24}$ J/T is the Bohr magneton, and we set $\vec{\mu}_j = -2\mu_B \vec{s}_j$, where $\vec{s}_j$ is the spin operator of the j-th electron in units of $\hbar$, whose eigenvalues are $\pm 1/2$. We will see that the first and second terms within the square brackets of (38) give the magnetic scattering due to orbital and spin magnetic moments, respectively.

Nuclear Scattering

For nuclear scattering (32), the $\mathcal{A}_{\vec{q}}$ operator (7) takes the simple form

$$\mathcal{A}_{\vec{q}}^{(n)} = \sum_{n=1}^{N_n} \left(A_n + \frac{1}{2} B_n \vec{I}_n \cdot \vec{\sigma} \right) e^{i2\pi \vec{q} \cdot \vec{R}_n}. \tag{39}$$

We will consider here the case of real A_n and B_n (no absorption). We have to compute the time correlation function entering the right-hand side of (13)

$$\mathrm{Tr}\, \overline{\langle \mathcal{A}_{\vec{q}}^{(n)\,\dagger}(0)\, \mathcal{P}_{\sigma_f}\, \mathcal{A}_{\vec{q}}^{(n)}(t)\, \varrho_i\rangle}, \tag{40}$$

The result is obtained straightforwardly after introducing the physically reasonable assumptions enumerated in the following list.

1. The nuclear spin dynamics are very weakly correlated to the nuclear motion, so that the correlation

$$\langle \mathcal{I}_n^\alpha(0)\mathcal{I}_{n'}^\beta(t)e^{-i2\pi\vec{q}\cdot\vec{R}_{n'}(0)}e^{i2\pi\vec{q}\cdot\vec{R}_n(t)}\rangle \tag{41}$$

factorizes as

$$\langle \mathcal{I}_n^\alpha(0)\mathcal{I}_{n'}^\beta(t)\rangle\langle e^{-i2\pi\vec{q}\cdot\vec{R}_{n'}(0)}e^{i2\pi\vec{q}\cdot\vec{R}_n(t)}\rangle. \tag{42}$$

2. The nuclei are not polarized, so that $\langle \vec{\mathcal{I}}\rangle = 0$.
3. The spins of different nuclei are also weakly coupled and therefore essentially uncorrelated, so that $\langle \mathcal{I}_n^\alpha(t)\mathcal{I}_{n'}^\beta(0)\rangle = \langle \mathcal{I}_n^\alpha(t)\mathcal{I}_n^\beta(0)\rangle\,\delta_{nn'}$.
4. The dynamics of the nuclear spins are much slower than the orbital dynamics, since the interactions of the nuclear spins are in general very weak. We then neglect the time dependence of $\langle \mathcal{I}_n^\alpha(t)\mathcal{I}_n^\beta(0)\rangle$.
5. The nuclear spin motion is nearly random, and therefore

$$\langle \mathcal{I}_n^\alpha \mathcal{I}_n^\beta\rangle = \frac{1}{3}\langle \vec{\mathcal{I}}^2\rangle\delta_{\alpha\beta} = \frac{1}{3}I_n(I_n+1)\delta_{\alpha\beta}. \tag{43}$$

6. The isotopic probability distributions of nuclei n and n' are independent, which means

$$\overline{A_n A_{n'}} = \overline{A_n}\,\overline{A_{n'}}\,(1-\delta_{nn'}) + \overline{A_n^2}\,\delta_{nn'} = \overline{A_n}\,\overline{A_{n'}} + (\overline{A_n^2} - \overline{A}_n^2)\delta_{nn'}. \tag{44}$$

From these assumptions we have

$$\sum_{\alpha,\beta}\frac{1}{4}\overline{B_n B_{n'}\langle \mathcal{I}_n^\alpha(t)\mathcal{I}_{n'}^\beta(0)\rangle}\,\mathrm{Tr}(\sigma_\alpha \mathcal{P}_{\sigma_f}\sigma_\beta \varrho_i) = \frac{1}{4}\overline{B_n^2 I_n(I_n+1)}\,\delta_{nn'}\,\langle \sigma_f|\varrho_i|\sigma_f\rangle, \tag{45}$$

where we used $\sum_{\alpha\beta}\delta_{\alpha\beta}\sigma_\alpha\varrho_i\sigma_\beta = 3\varrho_i$.

It is convenient to define the *coherent* and *incoherent* scattering length as, respectively, $b_{\mathrm{coh},n} = b_n = \overline{A_n}$ and

$$b_{\mathrm{inc},n} = \left[(\overline{A_n^2} - \overline{A}_n^2) + (1/4)\overline{B_n^2 I_n(I_n+1)}\right]^{1/2}. \tag{46}$$

The average is over isotope abundance in the target.

Taking into account the above considerations, it is straightforward to obtain

$$
\mathrm{Tr}\ \overline{\langle \mathcal{A}_{\vec{q}}^{(n)\,\dagger}(0) P_{\sigma_f} \mathcal{A}_{\vec{q}}^{(n)}(t)\varrho_i\rangle} =
$$
$$
\sum_{nn'}(\overline{b_n b_{n'}} + b_{\mathrm{inc},n}^2 \delta_{nn'})\langle e^{-i2\pi\vec{q}\cdot\vec{R}_{n'}(0)} e^{i2\pi\vec{q}\cdot\vec{R}_n(t)}\rangle\ \langle \sigma_f | \varrho_i | \sigma_f\rangle . \tag{47}
$$

By definition, the density matrix of a beam is the matrix whose expectation value in a spin state gives the probability of such state in the beam. Equation (47) shows that, in the case of nuclear scattering by unpolarized nuclei, the density matrix of the scattered beam is the same as the density matrix of the incident beam, so that there is no change in polarization. Since polarization is uninteresting at this point, in the remaining of this section, we will sum over the polarization states of the scattered neutrons.

Equation (47) allows us to separate the nuclear scattering cross section into coherent and incoherent components

$$
\frac{d^2\sigma^{(n)}}{d\Omega_{\hat{k}_f}\,dE_f} = \frac{d^2\sigma_{\mathrm{coh}}^{(n)}}{d\Omega_{\hat{k}_f}\,dE_f} + \frac{d^2\sigma_{\mathrm{inc}}^{(n)}}{d\Omega_{\hat{k}_f}\,dE_f}, \tag{48}
$$

where

$$
\frac{d^2\sigma_{\mathrm{coh}}^{(n)}}{d\Omega_{\hat{k}_f}\,dE_f} = \frac{k_f}{k_i}\sum_{nn'} \overline{b_n}\,\overline{b_{n'}}\langle e^{-i2\pi\vec{q}\cdot\vec{R}_n(0)} e^{i2\pi\vec{q}\cdot\vec{R}_{n'}(t)}\rangle_v \tag{49}
$$

is the *coherent nuclear scattering cross section* and

$$
\frac{d^2\sigma_{\mathrm{inc}}^{(n)}}{d\Omega_{\hat{k}_f}\,dE_f} = \frac{k_f}{k_i}\sum_{n} b_{\mathrm{inc},n}^2 \langle e^{-i2\pi\vec{q}\cdot\vec{R}_n(0)} e^{i2\pi\vec{q}\cdot\vec{R}_n(t)}\rangle_v \tag{50}
$$

is the *incoherent nuclear scattering cross section*. Remember that $\langle \mathbb{AB}\rangle_v$ is defined in Eq. (14). Notice that, in spite of the appearance due to the condensed notation, the two operators within brackets in Eq. (50) are not inverse, since they are taken at different times.

Notice the big difference between coherent and incoherent scattering. The former gives the Fourier transform of the time correlation function between pairs of nucleus [9]. Indeed, in the case of a *monoatomic* target, it is usual to write the scattering cross section as

$$
\frac{d^2\sigma}{d\Omega_{\hat{k}_f}\,dE_f} = \frac{\sigma_{\mathrm{coh}}}{4\pi}\frac{k_f}{k_i} N_{\mathrm{n}} S_{\mathrm{coh}}(\vec{q}, v) + \frac{\sigma_{\mathrm{inc}}}{4\pi}\frac{k_f}{k_i} N_{\mathrm{n}} S_{\mathrm{inc}}(\vec{q}, v) \tag{51}
$$

where $\sigma_{coh} = 4\pi b_{coh}^2$, $\sigma_{inc} = 4\pi b_{inc}^2$, and S_{coh} and S_{inc} are called the coherent and incoherent *scattering functions*,[3] given, respectively, by

$$S_{coh}(\vec{q}, v) = \frac{1}{N_n} \sum_{nn'} \langle e^{-i2\pi\vec{q}\cdot\vec{R}_n(0)} e^{i2\pi\vec{q}\cdot\vec{R}_{n'}(t)} \rangle_v \qquad (52)$$

and

$$S_{inc}(\vec{q}, v) = \frac{1}{N_n} \sum_{n} \langle e^{-i2\pi\vec{q}\cdot\vec{R}_n(0)} e^{i2\pi\vec{q}\cdot\vec{R}_n(t)} \rangle_v. \qquad (53)$$

The coherent scattering function is the Fourier transform of the *time-dependent pair correlation function* $G(\vec{r}, t)$ [9],

$$S_{coh}(\vec{q}, v) = \int \frac{dt}{h} e^{-i2\pi vt} \int d^3r\, e^{i2\pi\vec{q}\cdot\vec{r}}\, G(\vec{r}, t), \qquad (54)$$

which is defined by[4]

$$G(\vec{r}, t) = \frac{1}{N_n} \sum_{nn'} \int d^3r' \langle \delta[\vec{r} - \vec{r}' - \vec{R}_n(0)]\delta[\vec{r}' - \vec{R}_{n'}(t)] \rangle. \qquad (55)$$

Analogously, the incoherent scattering function is the Fourier transform of the time-dependent self-correlation function $G_s(\vec{r}, t)$ [9]

$$S_{inc}(\vec{q}, v) = \int \frac{dt}{h} e^{-i2\pi vt} \int d^3r\, e^{i2\pi\vec{q}\cdot\vec{r}}\, G_s(\vec{r}, t), \qquad (56)$$

where

$$G_s(\vec{r}, t) = \frac{1}{N_n} \sum_{n} \int d^3r' \langle \delta[\vec{r} - \vec{r}' - \vec{R}_n(0)]\delta[\vec{r}' - \vec{R}_n(t)] \rangle. \qquad (57)$$

It is also convenient to define the *intermediate functions* as

$$I_{coh}(\vec{q}, t) = \int d^3r\, e^{i2\pi\vec{q}\cdot\vec{r}}\, G(\vec{r}, t) \qquad (58)$$

---

[3]The scattering function is also called response function, dynamic scattering function, dynamic structure factor, or scattering law.

[4]These expressions have to be manipulated carefully, since the Heisenberg operators $\vec{R}_n(0)$ and $\vec{R}_{n'}(t)$ do not commute for $t \neq 0$. For instance,

$$\exp\{-i2\pi\vec{q}\cdot\vec{R}_n(0)\}\exp\{i2\pi\vec{q}\cdot\vec{R}_{n'}(t)\} \neq \exp\{i2\pi\vec{q}\cdot[\vec{R}_{n'}(t) - \vec{R}_n(0)]\}.$$

$$I_{\text{inc}}(\vec{q}, t) = \int d^3r \, e^{i2\pi \vec{q} \cdot \vec{r}} \, G_s(\vec{r}, t) \tag{59}$$

These ideas are not specific to monoatomic targets, not even to nuclear scattering. It is a general feature that coherent scattering provides information about the time-dependent correlations between pairs of degrees of freedom in the target, while incoherent scattering provides information about the time self-correlation function of single degrees of freedom in the target.

Elastic Nuclear Scattering

After summing over the neutron final spin states, an expression for the elastic nuclear scattering is obtained from [c.f. Eq. (22)].[5]

$$\text{Tr}\left[\overline{\langle \mathcal{A}_{\vec{q}}^{(n)\,\dagger}\rangle \langle \mathcal{A}_{\vec{q}}^{(n)}\rangle} \varrho_i\right] = \sum_{nn'} (b_n b_{n'} + b_{\text{inc},n}^2 \delta_{nn'}) \langle e^{-i2\pi \vec{q} \cdot \vec{R}_n}\rangle \langle e^{i2\pi \vec{q} \cdot \vec{R}_{n'}}\rangle. \tag{60}$$

In fluids, the particle motion is translational, or diffusive, and takes place in a region of macroscopic dimensions, much larger than $1/q$, and thus the expectation value $\langle \exp\{i2\pi\vec{q} \cdot \vec{R}_n\}\rangle$ vanishes unless $\vec{q} = 0$. But $\nu = 0$ and $\vec{q} = 0$ mean that there is no scattering at all, so that in liquids all scattering is inelastic. In solids, however, the particle motion takes place around an equilibrium position and is limited to a region comparable to $1/q$, and therefore $\langle \exp\{i2\pi\vec{q} \cdot \vec{R}_n\}\rangle \neq 0$. Hence, elastic scattering in solids is very important and often dominant.

Expression (60) allows us to obtain the *coherent and incoherent nuclear elastic* cross sections, which are given by

$$\frac{d\sigma_{\text{coh}}^{(n,\,e)}}{d\Omega_{\hat{k}_f}} = \left| \sum_n b_n \langle e^{i2\pi \vec{q} \cdot \vec{R}_n}\rangle \right|^2, \tag{61}$$

and

$$\frac{d\sigma_{\text{inc}}^{(n,\,e)}}{d\Omega_{\hat{k}_f}} = \sum_n b_{\text{inc},n}^2 \left| \langle e^{i2\pi \vec{q} \cdot \vec{R}_n}\rangle \right|^2, \tag{62}$$

respectively. Notice again the big difference between coherent and incoherent elastic scatterings. The coherent scattering is characterized by the strong interference between the scattered waves, which produces peaks in the scattered intensity along specific directions, determined by the sample atomic structure, that frequently are very sharp. This interference is completely absent in the incoherent scattering that is

---

[5]Equivalently the elastic nuclear scattering can be obtained from the $t \to \infty$ limit of Eq. (47).

entirely due to the presence of disorder, which cancels the interference terms. Thus, the contribution of the incoherent scattering is often merely a smooth background.

Inelastic Nuclear Scattering

The inelastic scattering cross section is obtained by subtracting the elastic scattering from the whole cross section. The explicit expressions are

$$
\frac{d^2\sigma_{\text{coh}}^{(n,i)}}{d\Omega_{\hat{k}_f}\,dE_f} = \frac{k_f}{k_i} \int \frac{dt}{h}\, e^{-i2\pi\nu t} \sum_{nn'} b_n b_{n'} \\
\times \left[\left\langle e^{-i2\pi\vec{q}\cdot\vec{R}_n(0)} e^{i2\pi\vec{q}\cdot\vec{R}_{n'}(t)} \right\rangle - \left\langle e^{-i2\pi\vec{q}\cdot\vec{R}_n} \right\rangle\!\left\langle e^{i2\pi\vec{q}\cdot\vec{R}_{n'}} \right\rangle \right]
\tag{63}
$$

for the *coherent nuclear inelastic* cross section and

$$
\frac{d^2\sigma_{\text{inc}}^{(n,i)}}{d\Omega_{\hat{k}_f}\,dE_f} = \frac{k_f}{k_i} \int \frac{dt}{h}\, e^{-i2\pi\nu t} \sum_{n} b_{\text{inc},n}^2 \\
\times \left[\left\langle e^{-i2\pi\vec{q}\cdot\vec{R}_n(0)} e^{i2\pi\vec{q}\cdot\vec{R}_n(t)} \right\rangle - \left\langle e^{-i2\pi\vec{q}\cdot\vec{R}_n} \right\rangle\!\left\langle e^{i2\pi\vec{q}\cdot\vec{R}_n} \right\rangle \right]
\tag{64}
$$

for the *incoherent nuclear inelastic* component.

In fluids, the elastic scattering is absent, and the time-dependent pair correlations and self-correlations are related to mean inter-particle distances, velocity correlations, and other properties of the fluid motion [10].

In solids, the nuclei perform oscillations about some equilibrium position. Let $\vec{R}_n = \vec{R}_n^e + \vec{u}_n$, where $\vec{R}_n^e = \langle\vec{R}_n\rangle$ denotes the equilibrium position and $\vec{u}_n$ is the quantum operator that describes the displacements from the equilibrium position. Then (63) can be written as

$$
\frac{d^2\sigma_{\text{coh}}^{(n,i)}}{d\Omega_{\hat{k}_f}\,dE_f} = \frac{k_f}{k_i} \int \frac{dt}{h}\, e^{-i2\pi\nu t} \sum_{nn'} b_n b_{n'} e^{i2\pi\vec{q}\cdot(\vec{R}_{n'}^e - \vec{R}_n^e)} T_n(-\vec{q})\,T_{n'}(\vec{q}) \\
\times \left[\frac{\left\langle e^{-i2\pi\vec{q}\cdot\vec{u}_n(0)} e^{i2\pi\vec{q}\cdot\vec{u}_{n'}(t)} \right\rangle}{T_n(-\vec{q})\,T_{n'}(\vec{q})} - 1 \right],
\tag{65}
$$

where

$$
T_n(\vec{q}) = \left\langle e^{i2\pi\vec{q}\cdot\vec{u}_n} \right\rangle
\tag{66}
$$

is the so called Debye–Waller factor.

If we assume that the displacements from the equilibrium position are small, the dynamics of $\vec{u}_n$ is approximately described by a set of coupled harmonic oscillators.

Then, $\vec{u}_n(t)$ can be expanded in terms of normal modes, the phonons, which describe the collective excitations of the system, as

$$\vec{u}_n(t) = \sum_r \sqrt{\frac{h}{2M_n \nu_r}} \vec{\xi}_{r,n} \left(e^{-i2\pi \nu_r t} a_r + e^{i2\pi \nu_r t} a_r^\dagger \right), \tag{67}$$

where M_n is the mass of the n-th nucleus, $r = 1, \ldots, N_p$ labels the phonon branches, N_p is the number of different phonon branches, and ν_r, $\vec{\xi}_{r,n}$, a_r and $a_r^\dagger$ are the frequency, the polarization vector, and the annihilation and creation operators of the corresponding phonon, respectively. The polarization vectors are real and dimensionless and satisfy the normalization and closure conditions

$$\sum_n \vec{\xi}_{r,n} \cdot \vec{\xi}_{r',n} = \delta_{rr'}, \qquad \sum_r \xi_{r,n}^\alpha \xi_{r,n'}^\beta = \delta_{\alpha\beta} \delta_{nn'}. \tag{68}$$

The numerical factor in the right-hand side of (67) arises from the canonical commutation relations. Notice that expression (67) is valid for crystalline as well as for amorphous solids.

The term within brackets in (65) is an even function of $\vec{q}$. We expand it in powers of $\vec{q}$, and we get only even powers of the components of $\vec{q}$. The term of order $2p$ is called the p-phonon term and is interpreted as a sum of scattering events in which p phonons are involved, some absorbed and some emitted. It is not difficult to get the 1-phonon term, which is the most important since it provides the phonon spectrum. To second order in $\vec{q}$, we have

$$\frac{\left\langle e^{-i2\pi \vec{q} \cdot \vec{u}_n(0)} e^{i2\pi \vec{q} \cdot \vec{u}_{n'}(t)} \right\rangle}{T_n(-\vec{q}) T_{n'}(\vec{q})} - 1 = 4\pi^2 \left\langle [\vec{q} \cdot \vec{u}_n(0)][\vec{q} \cdot \vec{u}_n(t)] \right\rangle + O(q^4). \tag{69}$$

From Eqs. (69) and (67), we see that we have to compute $\langle a_r a_{r'} \rangle$, $\langle a_r a_{r'}^\dagger \rangle$, $\langle a_r^\dagger a_{r'} \rangle$, and $\langle a_r^\dagger a_{r'}^\dagger \rangle$. The phonons form an ideal Bose gas of non-conserved particles; thus it has zero chemical potential, and, at temperature T, the thermal quantum average of an operator $\mathcal{O}$ is given by

$$\langle \mathcal{O} \rangle = \frac{1}{Z} \sum_{n_1=1}^\infty \cdots \sum_{n_{N_p}=1}^\infty \langle n_1, \ldots, n_{N_p} | \mathcal{O} | n_1, \ldots, n_{N_p} \rangle \exp\left(-\sum_{r=1}^{N_p} n_r h\nu_r / k_B T \right), \tag{70}$$

where Z is the partition function and $|n_1, \ldots, n_{N_p}\rangle$ is the state that contains n_r phonons of type r. Clearly we have $\langle a_r a_{r'} \rangle = \langle a_r^\dagger a_{r'}^\dagger \rangle = 0$ and

$$\langle a_r a_{r'}^\dagger \rangle = \langle n_r \rangle \delta_{rr'}, \qquad \langle a_r^\dagger a_{r'} \rangle = [1 + \langle n_r \rangle] \delta_{rr'}, \tag{71}$$

where in the last equation we used the commutation relation $[a_r, a_{r'}] = \delta_{rr'}$ and $\langle n_r \rangle$ is the avearge value of the number of type r phonons, given by the Bose-Einstein statistics:

$$\langle n_r \rangle = \frac{1}{\exp(h\nu_r/k_B T) + 1}. \tag{72}$$

Then we have

$$\langle [\vec{q} \cdot \vec{u}_n(0)][\vec{q} \cdot \vec{u}_{n'}(t)] \rangle = \sum_r \left[(\vec{q} \cdot \vec{\xi}_{r,n})(\vec{q} \cdot \vec{\xi}_{r,n'})^*(1 + \langle n_r \rangle) e^{i2\pi\nu_r t} \right.$$
$$\left. + (\vec{q} \cdot \vec{\xi}_{r,n})^*(\vec{q} \cdot \vec{\xi}_{r,n'}) \langle n_r \rangle e^{-i2\pi\nu_r t} \right]. \tag{73}$$

Inserting Eqs. (69) and (73) into (65), we obtain, to 1-phonon level

$$\frac{d^2\sigma_{coh,1ph}^{(n,i)}}{d\Omega_{\hat{k}_f} dE_f} = \frac{k_f}{k_i}(2\pi)^2 q^2 \sum_r |\hat{q} \cdot \vec{\upsilon}_r(\vec{q})|^2 \frac{1}{2\nu_r} [(1 + \langle n_r \rangle)\delta(\nu - \nu_r) + \langle n_r \rangle\delta(\nu + \nu_r)], \tag{74}$$

where

$$\vec{\upsilon}_r(\vec{q}) = \sum_n e^{i2\pi\vec{q}\cdot\vec{R}_n^e} \frac{b_n}{\sqrt{M_n}} T_n(\vec{q}) \vec{\xi}_{r,n}. \tag{75}$$

The first term in Eq. (74) corresponds to a neutron loosing an amount of energy $h\nu_r$ by creating one r-type phonon, and the second term to a neutron gaining an amount of energy $h\nu_r$ by absorbing one r-type phonon. In the case of an amorphous solid, one has to average over the disorder in equilibrium positions and harmonic forces between nuclei, and the phonon spectrum is a continuum whose structure is difficult to disentangle. In this case, the more interesting quantity is the density of states. For a crystal, however, the phonon spectrum is separated into phonon branches, and the coherent 1-phonon neutron scattering allows to obtain the dispersion relation of each branch, as we will see.

An analogous phonon expansion can be developed for the incoherent inelastic cross section. For a monoatomic system, it can be expressed solely in terms of the density of states. The 1-phonon contribution to the incoherent scattering has the relatively simple expression

$$\frac{d^2\sigma_{inc,1ph}^{(n,i)}}{d\Omega_{\hat{k}_f} dE_f} = N_n \frac{k_f}{k_i} \frac{b_{inc}^2}{2M} (2\pi)^2 q^2 |T(\vec{q})|^2 \frac{Z(\nu)}{\nu} \frac{1}{2} \left[\coth\left(\frac{h\nu}{2k_B T}\right) + 1 \right], \tag{76}$$

where $Z(\nu) = \sum_r \delta(\nu - \nu_r)$ for $\nu > 0$ is the density of states, and we have defined $Z(-\nu) = Z(\nu)$.

Expressions (74) and (76) are valid in the harmonic approximation. Anharmonic corrections are usually important only at high temperature, but some phenomena like thermal expansion and thermal conductivity in solids cannot be explained without anharmonicity [11]. In the harmonic approximation, the Debye–Waller factor takes the simple form

$$T_n(\vec{q}\,) = \mathrm{e}^{-\vec{q}\cdot\tilde{\beta}_n\vec{q}}, \tag{77}$$

where $\tilde{\beta}_n$ is a 3×3 matrix that depends on the temperature and on the frequencies and polarization vectors of phonons.

Magnetic Scattering

Expressions for the magnetic scattering cross section can be obtained by applying the same ideas as in nuclear scattering. The derivations and the expressions are much more difficult since the magnetic interaction Hamiltonian is much more complex than its nuclear counterpart. Furthermore, the atomic wave functions of the electrons enter the final results, due to the long-range nature of the magnetic interactions.

Let us denote by $\mathcal{A}_{\vec{q}}^{(m)}$ the amplitude operator corresponding to magnetic scattering, defined by Eq. (7). Since the interaction Hamiltonian is given by Eq. (38), we need the two integrals:

$$\int d^3 r\, \mathrm{e}^{\mathrm{i}2\pi\vec{q}\cdot\vec{r}}\, \frac{\vec{r}_j - \vec{r}}{|\vec{r}_j - \vec{r}|^3} = -\mathrm{i}4\pi \mathrm{e}^{\mathrm{i}2\pi\vec{q}\cdot\vec{r}_j}\, \frac{\hat{q}}{q}, \tag{78}$$

$$\int d^3 r\, \mathrm{e}^{\mathrm{i}2\pi\vec{q}\cdot\vec{r}}\, \nabla \times \left(\nabla \times \frac{\vec{s}_j}{|\vec{r}_j - \vec{r}|}\right) = 4\pi \mathrm{e}^{\mathrm{i}2\pi\vec{q}\cdot\vec{r}_j}\, \hat{q} \times (\vec{s}_j \times \hat{q}). \tag{79}$$

Then we can write $\mathcal{A}_{\vec{q}}^{(m)} = p\, \vec{\mathcal{M}}_{\perp\vec{q}} \cdot \vec{\sigma}$, where the operator

$$\vec{\mathcal{M}}_{\perp\vec{q}} = -2\mu_{\mathrm{B}} \sum_j \mathrm{e}^{\mathrm{i}2\pi\vec{q}\cdot\vec{r}_j} \left[\hat{q} \times (\vec{s}_j \times \hat{q}) + \frac{\mathrm{i}}{\hbar q}(\vec{p}_j \times \hat{q})\right], \tag{80}$$

is called the *magnetic interaction vector operator* and acts only on the electron degrees of freedom. The factor $-2\mu_{\mathrm{B}}$ in $\vec{\mathcal{M}}_{\perp\vec{q}}$ is introduced by convenience, so that, as we will show later, $\vec{\mathcal{M}}_{\perp\vec{q}}$ is the component perpendicular to the scattering vector of the Fourier transform of the magnetization density operator. With these definitions, we have $p = (m/m_{\mathrm{p}})\gamma r_{\mathrm{e}}/2\mu_{\mathrm{B}}$, where $r_{\mathrm{e}} = e^2/(4\pi\epsilon_0 m_{\mathrm{e}} c^2)$, with ϵ_0 and c being the vacuum permittivity and the velocity of light, respectively. It is customary to substitute by one the ratio between the neutron and proton mass, which is approximately 1.0014. The quantity r_{e} is the so-called classical radius of the electron and has the value $r_{\mathrm{e}} \approx 2.818 \times 10^{-15}$m. This is of the same order of magnitude as the nuclear scattering lengths, which means that *the magnetic scattering of thermal neutrons is of the same order of magnitude as the nuclear scattering*. The constant $p = 0.2695$ provides the conversion of magnetic moments, given in Bohr magnetons μ_{B}, to scattering lengths in units of 10^{-14} m.

Clearly, $\vec{\mathcal{M}}_{\perp\vec{q}}$ is orthogonal to $\vec{q}$ and can be written as

$$\vec{\mathcal{M}}_{\perp\vec{q}} = \hat{q} \times (\vec{\mathcal{M}}_{\vec{q}} \times \hat{q}) = \vec{\mathcal{M}}_{\vec{q}} - \left(\hat{q} \cdot \vec{\mathcal{M}}_{\vec{q}}\right) \hat{q}, \qquad (81)$$

or, in components

$$\mathcal{M}_{\perp\vec{q}}^{\alpha} = \sum_{\beta} \left(\delta_{\alpha\beta} - \frac{q_{\alpha}q_{\beta}}{q^2}\right) \mathcal{M}_{\vec{q}}^{\beta}, \qquad (82)$$

where

$$\vec{\mathcal{M}}_{\vec{q}} = -2\mu_{\mathrm{B}} \sum_{j} e^{i2\pi\vec{q}\cdot\vec{r}_j} \left[\vec{s}_j + \frac{i}{\hbar q}(\vec{p}_j \times \hat{q})\right] \qquad (83)$$

is called the *magnetic structure factor operator*. Notice that any operator proportional to $\vec{q}$ can be added to $\vec{\mathcal{M}}_{\vec{q}}$ without changing $\vec{\mathcal{M}}_{\perp\vec{q}}$. This fact can be used to simplify some expressions of the matrix elements of $\vec{\mathcal{M}}_{\perp\vec{q}}$. We want to stress the following important fact: *only the component of $\vec{\mathcal{M}}_{\vec{q}}$ perpendicular to $\vec{q}$ contributes to the magnetic neutron scattering.* We anticipate here that $\vec{\mathcal{M}}_{\vec{q}}$ is the Fourier transform of the magnetic moment density, or magnetization density, operator. Thus, not surprisingly, the magnetic scattering of neutrons is closely linked to the magnetization density.

Using the general formula (15), we obtain the following expression for the magnetic scattering cross section in the case that the scattered neutron polarization state, σ_f, is observed:

$$\left(\frac{d^2\sigma^{(\mathrm{m})}}{d\Omega_{\hat{k}_f}\, dE_f}\right)_{\sigma_f} = p^2 \frac{k_f}{k_i} \sum_{\alpha\beta} \rho_{\alpha\beta}^i(\sigma_f) \left\langle \mathcal{M}_{\perp\vec{q}}^{\beta\,\dagger}(0)\, \mathcal{M}_{\perp\vec{q}}^{\alpha}(t) \right\rangle_{\nu}, \qquad (84)$$

where $\mathcal{M}_{\perp\vec{q}}^{\alpha}$ is the α-th component of the $\vec{\mathcal{M}}_{\perp\vec{q}}$ operator, and

$$\rho_{\alpha\beta}^i(\sigma_f) = \langle \sigma_f | \sigma_{\alpha} \varrho_i \sigma_{\beta} | \sigma_f \rangle. \qquad (85)$$

After integrating over dE_f, the elastic contribution to the magnetic scattering cross section is

$$\left(\frac{d\sigma^{(\mathrm{m,\,e})}}{d\Omega_{\hat{k}_f}}\right)_{\sigma_f} = p^2 \sum_{\alpha\beta} \rho_{\alpha\beta}^i(\sigma_f)\langle \mathcal{M}_{\perp\vec{q}}^{\beta}\rangle^* \langle \mathcal{M}_{\perp\vec{q}}^{\alpha}\rangle, \qquad (86)$$

and the inelastic contribution is

$$\left(\frac{d^2\sigma^{(m,\,i)}}{d\Omega_{\hat{k}_f}\,dE_f} \right)_{\sigma_f} = p^2 \frac{k_f}{k_i} \sum_{\alpha\beta} \rho_{\alpha\beta}^i(\sigma_f)$$

$$\times \int \frac{dt}{h}\, e^{-i2\pi\nu t} \left[\langle \mathcal{M}_{\perp\vec{q}}^{\beta\,\dagger}(0)\mathcal{M}_{\perp\vec{q}}^{\alpha}(t)\rangle - \langle \mathcal{M}_{\perp\vec{q}}^{\beta}\rangle^*\langle \mathcal{M}_{\perp\vec{q}}^{\alpha}\rangle \right] \tag{87}$$

If the polarization states of the scattered neutrons are not resolved, we have to sum over σ_f in the above expressions. Since

$$\sum_{\sigma_f} \rho_{\alpha\beta}^i(\sigma_f) = \delta_{\alpha\beta} - \sum_{\gamma} \epsilon_{\alpha\beta\gamma}\, P_i^{\gamma}, \tag{88}$$

where $\epsilon_{\alpha\beta\gamma}$ is the tridimensional totally antisymmetric tensor and $\vec{P}_i$ is the polarization vector of the incident beam, we obtain

$$\frac{d\sigma^{(m,\,e)}}{d\Omega_{\hat{k}_f}} = p^2 \left[\langle \vec{\mathcal{M}}_{\perp\vec{q}}\rangle^* \cdot \langle \vec{\mathcal{M}}_{\perp\vec{q}}\rangle - \left(\langle \vec{\mathcal{M}}_{\perp\vec{q}}\rangle^* \times \langle \vec{\mathcal{M}}_{\perp\vec{q}}\rangle \right) \cdot \vec{P}_i \right] \tag{89}$$

for the *magnetic elastic* contribution and

$$\frac{d^2\sigma^{(m,\,i)}}{d\Omega_{\hat{k}_f}\,dE_f} = p^2 \frac{k_f}{k_i} \int \frac{dt}{h}\, e^{-i2\pi\nu t} \left\{ \langle \vec{\mathcal{M}}_{\perp\vec{q}}^{\dagger}(0) \cdot \vec{\mathcal{M}}_{\perp\vec{q}}(t)\rangle - \langle \vec{\mathcal{M}}_{\perp\vec{q}}\rangle^* \cdot \langle \vec{\mathcal{M}}_{\perp\vec{q}}\rangle \right.$$

$$\left. + \left[\langle \vec{\mathcal{M}}_{\perp\vec{q}}^{\dagger}(0) \times \vec{\mathcal{M}}_{\perp\vec{q}}(t)\rangle - \langle \vec{\mathcal{M}}_{\perp\vec{q}}\rangle^* \times \langle \vec{\mathcal{M}}_{\perp\vec{q}}\rangle \right] \cdot \vec{P}_i \right\} \tag{90}$$

for the *magnetic inelastic* contribution.

Physical Meaning of the Operator $\vec{\mathcal{M}}_{\perp\vec{q}}$

We show in the following that $\vec{\mathcal{M}}_{\perp\vec{q}}$ is the component of the Fourier transform of the magnetization density operator perpendicular to the scattering vector, $\vec{q}$ [12]. Let us write $\vec{\mathcal{M}}_{\perp\vec{q}} = \vec{\mathcal{M}}_{S\perp\vec{q}} + \vec{\mathcal{M}}_{L\perp\vec{q}}$, where $\vec{\mathcal{M}}_{S\perp\vec{q}}$ and $\vec{\mathcal{M}}_{L\perp\vec{q}}$ are the spin and orbital contributions to $\vec{\mathcal{M}}_{\perp\vec{q}}$:

$$\vec{\mathcal{M}}_{S\perp\vec{q}} = -2\mu_B \sum_j e^{i2\pi\vec{q}\cdot\vec{r}_j}\, \hat{q} \times (\vec{s}_j \times \hat{q}), \tag{91}$$

$$\vec{\mathcal{M}}_{L\perp\vec{q}} = -2\mu_B \sum_j e^{i2\pi\vec{q}\cdot\vec{r}_j}\, \frac{i}{\hbar q}(\vec{p}_j \times \hat{q}). \tag{92}$$

It is clear that

$$\vec{\mathcal{M}}_{S\perp\vec{q}} = \hat{q} \times (\vec{\mathcal{M}}_{S\vec{q}} \times \hat{q}),\tag{93}$$

where $\vec{\mathcal{M}}_{S\vec{q}}$ is the Fourier transform of

$$\vec{\mathbb{M}}_S(\vec{r}) = \sum_j -2\mu_B\,\vec{s}_j\,\delta(\vec{r}-\vec{r}_j),\tag{94}$$

which is the spin magnetic moment density operator, i.e., the magnetization density operator due to spin.

Let us turn to the orbital part. Since $\vec{q}\cdot\vec{r}_j$ commutes with $\vec{p}_j \times \hat{q}$, because the former operator is proportional to the component of $\vec{r}_j$ parallel to $\vec{q}$ and the latter is the component of $\vec{p}_j$ perpendicular to $\vec{q}$, we have

$$\vec{\mathcal{M}}_{L\perp\vec{q}} = -i\frac{\mu_B}{\hbar q}\sum_j \left[e^{i2\pi\vec{q}\cdot\vec{r}_j}(\vec{p}_j \times \hat{q}) + (\vec{p}_j \times \hat{q})e^{i2\pi\vec{q}\cdot\vec{r}_j} \right].\tag{95}$$

The above equation is equivalent to

$$\vec{\mathcal{M}}_{L\perp\vec{q}} = \frac{i}{q}\int d^3r\, e^{i2\pi\vec{q}\cdot\vec{r}}\vec{\mathbb{D}}(\vec{r}) \times \hat{q},\tag{96}$$

where, bearing in mind that $\mu_B = e\hbar/2m_e$,

$$\vec{\mathbb{D}}(\vec{r}) = -\frac{e}{2m_e}\sum_j \left[\delta(\vec{r}-\vec{r}_j)\vec{p}_j + \vec{p}_j\delta(\vec{r}-\vec{r}_j) \right].\tag{97}$$

The operator $\vec{\mathbb{D}}(\vec{r})$ is $-e$ times the electron probability density current, and therefore it is the electric current density operator. The solenoidal part of $\vec{\mathbb{D}}$ is, by definition, the rotational of the orbital magnetic moment density, $\vec{\mathbb{M}}_L(\vec{r})$, and the irrotational part, written as $\nabla\Phi$, is the conduction current:

$$\vec{\mathbb{D}}(\vec{r}) = \nabla \times \vec{\mathbb{M}}_L(\vec{r}) + \nabla\Phi.\tag{98}$$

The Fourier transform of $\vec{\mathbb{D}}(\vec{r})$ is related to the Fourier transforms of $\vec{\mathbb{M}}_L(\vec{r})$ and Φ by

$$\int d^3r\, e^{i2\pi\vec{q}\cdot\vec{r}}\vec{\mathbb{D}}(\vec{r}) = -i\vec{q} \times \int d^3r\, e^{i2\pi\vec{q}\cdot\vec{r}}\vec{\mathbb{M}}_L(\vec{r}) - i\vec{q}\int d^3r\, e^{i2\pi\vec{q}\cdot\vec{r}}\Phi(\vec{r}).\tag{99}$$

Inserting the above equation into (96), and noticing that the term with $\Phi(\vec{r})$ does not contribute to it, since it is proportional to $\vec{q}$, we obtain $\vec{\mathcal{M}}_{L\perp\vec{q}} = \hat{q} \times (\vec{\mathcal{M}}_{L\vec{q}} \times \hat{q})$, where

$$\vec{\mathcal{M}}_{L\vec{q}} = \int d^3r \mathrm{e}^{\mathrm{i}2\pi\vec{q}\cdot\vec{r}} \vec{\mathbb{M}}_L(\vec{r})$$ (100)

is the Fourier transform of the orbital magnetic moment density operator.

Collecting the spin and orbital contributions to $\vec{\mathcal{M}}_{\perp\vec{q}}$, we obtain

$$\vec{\mathcal{M}}_{\perp\vec{q}} = \hat{q} \times (\vec{\mathcal{M}}_{\vec{q}} \times \hat{q}),$$ (101)

where the magnetic structure factor operator, $\vec{\mathcal{M}}_{\vec{q}} = \vec{\mathcal{M}}_{L\vec{q}} + \vec{\mathcal{M}}_{S\vec{q}}$, is the Fourier transform of the total magnetization density operator, $\vec{\mathbb{M}}(\vec{r})$, which is the sum of the orbital, $\vec{\mathbb{M}}_L(\vec{r})$, and spin, $\vec{\mathbb{M}}_S(\vec{r})$, magnetization density operators. Therefore, $\vec{\mathcal{M}}_{\perp\vec{q}}$ is the perpendicular component of the Fourier transform of the total magnetization density operator to the direction of the scattering vector.

Let us stress again the important fact that *only the component of the magnetic structure factor operator perpendicular to the scattering vector contributes to the magnetic neutron scattering*.

Matrix Elements of $\vec{\mathcal{M}}_{\perp\vec{q}}$

The computations of these matrix elements are complex, and we will give here only some hints to understand the origin of the main ingredients that enter the matrix elements.

The matrix elements of $\vec{\mathcal{M}}_{\perp\vec{q}}$ between the states $|\varphi_i\rangle$ and $|\varphi_f\rangle$ cannot be computed generically without making assumptions about the system. For definiteness, let us consider a system in which the magnetic moments are due to the unpaired electrons localized at ions. This is the case of magnetic moments originated by unpaired p electrons belonging to molecular orbitals in organic radicals, and of transition metals and rare earths, in which the unpaired electrons that contribute to the magnetic moment belong to a d and to an f sub-shell, respectively.

Let N be the number of ions, labelled by the index n, and let N_n be the number of electrons of ion n-th. The position of the ion nucleus is denoted by $\vec{R}_n$, and the position of its j-th electron with respect to the ion center by $\vec{r}_{nj}$, so that the position of the nj-th electron is $\vec{R}_n + \vec{r}_{nj}$. The spin operator of the nj-th electron is $\vec{s}_{nj}$, and the spin quantum number corresponding to the spin z component is denoted by $m_{s,nj}$. Let $e_{nj} = \{\vec{r}_{nj}, m_{s,nj}\}$ denote the complete set of electron degrees of freedom. The $\vec{\mathcal{M}}_{\perp\vec{q}}$ operator is given by

$$\vec{\mathcal{M}}_{\perp\vec{q}} = \sum_{n=1}^{N} \mathrm{e}^{\mathrm{i}2\pi\vec{q}\cdot\vec{R}_n} \vec{\mathcal{M}}_{\perp\vec{q}}^{(n)},$$ (102)

where $\vec{\mathcal{M}}_{\perp\vec{q}}^{(n)} = \sum_{j=1}^{N_n} \vec{\mathcal{M}}_{\perp\vec{q}}^{(nj)}$, with

$$\vec{\mathcal{M}}_{\perp\vec{q}}^{(nj)} = -2\mu_B e^{i2\pi\vec{q}\cdot\vec{r}_{nj}}\left[\hat{q}\times(\vec{s}_{nj}\times\hat{q}) + \frac{i}{\hbar q}(\vec{p}_{nj}\times\hat{q})\right]. \tag{103}$$

The state of the system can be written as

$$|\varphi\rangle = \sum_{\mu_1}\cdots\sum_{\mu_N}\phi_{\mu_1,\dots,\mu_N}|\psi_{\mu_1,1}\rangle\cdots|\psi_{\mu_N,N}\rangle, \tag{104}$$

where $\phi_{\mu_1,\dots,\mu_N}$ depends only on the nuclei coordinates $\vec{R}_1\dots\vec{R}_N$ and the electronic wave function $\psi_{\mu_n,n}$ depends on $e_1,\dots,e_{N_n}$ and is localized around the n-th ion position, $\vec{R}_n$. It is antisymmetric under electron exchange, and there is negligible overlapping between $\psi_{\mu_n,n}$ and $\psi_{\mu'_n,n'}$ if $n\neq n'$. Since the electronic states do not overlap, there is no need to antisymmetrize the wave function with respect to the exchange of electrons at different ions, in the sense that antisymmetrization does not change the result obtained with the above wave function. The index μ_n labels all the possible electronic states of the n-th ion, so that two different states differ by the functions ϕ.

The matrix element of $\vec{\mathcal{M}}_{\perp\vec{q}}$ between states $|\varphi_i\rangle$ and $|\varphi_f\rangle$ is

$$\langle\varphi_f|\vec{\mathcal{M}}_{\perp\vec{q}}|\varphi_i\rangle = \sum_n\sum_{\{\mu_1,\dots,\mu_N\}_n}\sum_{\mu_n,\mu'_n}\int d^3R_1\cdots d^3R_N$$
$$\times\phi_{\mu_1,\dots,\mu'_n,\dots,\mu_N}^{(f)*}e^{i2\pi\vec{q}\cdot\vec{R}_n}\langle\psi_{\mu'_n,n}|\vec{\mathcal{M}}_{\perp\vec{q}}^{(n)}|\psi_{\mu_n,n}\rangle\phi_{\mu_1,\dots,\mu_N}^{(i)}, \tag{105}$$

where $\{\mu_1,\dots,\mu_N\}_n$ stands for the set of $N-1$ indices $\mu_1,\dots,\mu_N$ excluding μ_n. In the above equation, we used the orthonormality of the ionic wave functions, $\langle\psi_{\mu,n}|\psi_{\mu',n}\rangle = \delta_{\mu\mu'}$. It is apparent that we only need the matrix element of $\vec{\mathcal{M}}_{\perp\vec{q}}^{(n)}$ between the electronic states localized at ion n-th.

We consider the electronic structure of the ion described by a Hartree-Fock scheme in the central field approximation [13]. The multi-electronic state, $|\psi_{\mu,n}\rangle$, is a suitable antisymmetric superposition of products of single electron states, which are computed from a self-consistent central potential common to all the electrons. Therefore, the single-electron states are labelled by four numbers: the principal quantum number, n_p; the orbital angular momentum ℓ; the z component of the orbital angular momentum, m; and the spin z component, m_s. Recall that $-\ell\leq m\leq\ell$ and m_s take only the two values $\pm1/2$. The single electron state is written as $|n_p\ell mm_s\rangle$, and its wave function is given by

$$\langle\vec{r}|n_p\ell mm_s\rangle = f_{n_p\ell}(r)Y_\ell^m(\hat{r})\chi_{m_s}, \tag{106}$$

where $f_{n_p\ell}(r)$ is the radial wave function, $Y_\ell^m(\hat{r})$ is the ℓm spherical harmonic evaluated on the polar angles θ and φ determined by the unit vector $\hat{r}$, and χ_{m_s} is the m_s component of the two-component spinor that describes the electron spin state.

Due to the rotational symmetry of the single-electron Hamiltonian, the energy is independent of m and m_s. An $n_p\ell$ state is called a *sub-shell*. At this zero-th order of approximation, the electrons fill the $n_p\ell$ sub-shells with lowest energy. The sub-shell $n_p\ell$ can be occupied at most by $2(2\ell + 1)$ states. A state of n_e electrons occupying the $n_p\ell$ sub-shell is specified by the $2n_e$ numbers $m_1 m_{s1}, \ldots, m_{n_e} m_{s n_e}$ and is given by the antisymmetric combination

$$\frac{1}{\sqrt{n_e!}} \sum_P (-1)^P |n_p\ell m_{P1} m_{sP1}\rangle_1 \cdots |n_p\ell m_{Pn_e} m_{sPn_e}\rangle_{n_e}, \tag{107}$$

where P is a permutation of $1, \ldots, n_e$, $(-1)^P$ is the sign of P, and $|\cdot\rangle_i$ describes the state of the i-th electron. Notice that the single-electron states in sub-shell $n_p\ell$ all have the same radial wave function.

The difference between the true interaction Hamiltonian and the self-consistent field, which contains the electrostatic interaction between the electrons, the spin-orbit coupling, and the crystal field energy, is treated as a perturbation. It lifts the degeneracy of the $n_p\ell$ sub-shell. The way in which the $n_p\ell$ sub-shell is split into energy eigenstates depends on which term of the perturbations dominates. Usually, the electrostatic energy is much higher than the spin-orbit coupling and than the crystal field energy. In this case the perturbation Hamiltonian at first order does not depend on the spin and, given that it is rotationally symmetric, commutes with the total orbital angular momentum (and with the total spin, obviously). Hence, the splitting of sub-shell $n_p\ell$ can be characterized by the four quantum numbers L, M_L, S, and M_S, corresponding to total orbital angular momentum, z component of total orbital angular momentum, total spin, and total spin z component,[6] in addition to $n_p\ell$. This is called the Russell-Saunders, or LS, spin-orbit coupling scheme, which we assume in the following.

The completely filled sub-shells have no net orbital angular momentum and no net spin and thus do not contribute to the magnetic moment. Analogously, paired electrons, which combine their orbital angular momentum and their spin in such a way that both cancel, do not contribute to the magnetic moment. Consider then an ion with n_u unpaired electrons occupying a partially filled sub-shell $n_p\ell$. These electrons are described by an LS state, $|n_p\ell L S M_L M_S\rangle$, which is an anitsymmetric linear combination of states of the form

$$|n_p\ell m_1 m_{s1}\rangle_1 \cdots |n_p\ell m_{n_u} m_{s n_u}\rangle_{n_u} \tag{108}$$

---

[6]Notice that the electrostatic energy operator commutes with the spin of each electron separately (but not with the single electron orbital angular momentum). At first sight, it seems that the multi-electron states can be characterized by the single-electron spins, not just by the total spin. However, the individual spins are not observables since the electrons are indistinguishable, and the single-electron spin states are not good quantum numbers. Indeed, the single-electron spin states are mixed in the antisymmetrization process, as can be seen in Eq. (107).

From now on we drop the quantum number n_p from the expressions, and we will write the radial wave function as $f(r)$. Let us stress that in general there are many ways of combining n_u orbital angular momentum and spin into ℓLS multiplets. This means there are many LS multiplets in $n\ell$ sub-shell that can be occupied by n_u electrons. Which one is occupied is decided by the minimization of the electrostatic repulsion between the electrons.

At this level of approximation, an LS state is $(2L + 1)(2S + 1)$ fold degenerate. This degeneracy is lifted by the remaining interactions. If the spin-orbit coupling dominates over the crystal field energy, rotational symmetry is preserved, and then the states are split into multiplets characterized by the magnitude and z component of the total angular momentum operator, J and M, respectively. Thus the states are written as $|\ell LSJM\rangle$, and it remains a $2J + 1$ degeneracy, due to the rotational symmetry of the isolated ion, which is removed by the crystal field interaction. These states are of course given by appropriate linear combinations of states of the form (108), If, on the contrary, the crystal field dominates over the spin-orbit coupling, the $(2L + 1)(2S + 1)$ degeneracy may be completely lifted, and in this case the expectation value of the three components of the orbital angular momentum vanishes, so that they do not contribute to the expectation value of the magnetic moment, which is entirely due to spin. In this case, it is said that the orbital angular momentum is *quenched*.

Usually, the energy differences between different LS multiplets are of the order of 1 eV or higher, and thus neutrons with energies in the thermal range cannot induce transitions between them. This is an important point, since in these cases the neutron interaction can only change the values of M_L and M_S, that is, the ion spatial orientation. Hence, we have to consider only the matrix elements of $\vec{\mathcal{M}}^{(n)}_{\perp\vec{q}}$ between the states of the same LS multiplet, with constant L and S (and of course constant n_p and ℓ). Evidently, the matrix element of $\vec{\mathcal{M}}^{(nj)}_{\perp\vec{q}}$ is a linear combination of single particle matrix elements,

$$\langle \ell m' m'_s | \vec{\mathcal{M}}^{(nj)}_{\perp\vec{q}} | \ell m m_s \rangle. \tag{109}$$

The contribution of filled shells or paired electrons cancels, and we have to consider only the contribution of unpaired electrons. Since the electrons are indistinguishable, the single-electron matrix element (109) does not depend on j. To avoid heavy notation, from now on we drop the indices nj from the expressions, and we denote by $\vec{r}$ the electron coordinate with respect to the center of the ion.

Let us consider first the contribution to (109) of the orbital part of $\vec{\mathcal{M}}^{(nj)}_{\perp\vec{q}}$. Since it does not depend on spin, we get an overall factor $\delta_{m_s m'_s}$. It remains

$$\langle \ell m' | e^{i2\pi\vec{q}\cdot\vec{r}} (\hat{q} \times \nabla) | \ell m \rangle = \int d^3 r \, f(r) Y_\ell^{m'*}(\hat{r}) e^{i2\pi\vec{q}\cdot\vec{r}} (\hat{q} \times \nabla) f(r) Y_\ell^m(\hat{r}). \tag{110}$$

Let us give only a few hints about how to compute this, in order to recognize the origin of some factors entering the magnetic cross section, mainly the *magnetic form factor.* A thorough analysis with many detailed calculations can be found in the book of Lovesey [14]. First, notice that by integrating by parts and using the fact that $(\hat{q} \times \nabla) \exp(i2\pi \vec{q} \cdot \vec{r}) = 0$, we can remove the derivative from the radial function $f(r)$. We get

$$
\langle \ell m' | e^{i2\pi \vec{q} \cdot \vec{r}} (\hat{q} \times \nabla) | \ell m \rangle = \frac{1}{2} \int d^3 r f^2(r) e^{i2\pi \vec{q} \cdot \vec{r}} \left[Y_\ell^{m'*}(\hat{r})(\hat{q} \times \nabla) Y_\ell^m(\hat{r}) \right.
$$
$$
\left. - Y_\ell^m(\hat{r})(\hat{q} \times \nabla) Y_\ell^{m'*}(\hat{r}) \right]. \tag{111}
$$

The derivative acts now only on the angular part of the wave function. We can extract a factor $1/r$ by defining $\hat{\nabla}$ so that $\nabla = (1/r)\hat{\nabla}$.

The spherical symmetry of the single electron wave functions makes it convenient to exploit the algebra of spherical tensors, working in terms of the spherical components of $\hat{q} \times \hat{\nabla}$, which is a vector operator.[7] We also expand the plane wave state $\exp(i2\pi \vec{q} \cdot \vec{r})$ in the basis of spherical waves as

$$
e^{i2\pi \vec{q} \cdot \vec{r}} = 4\pi \sum_{L=0}^{\infty} \sum_{M=-L}^{L} i^L j_L(2\pi q r) Y_L^M(\hat{r}) Y_L^{M*}(\hat{q}), \tag{112}
$$

where $j_L(x)$ is the spherical Bessel function of order L. Inserting this expansion into (111), in which we take the M'-th spherical component, we get

$$
\langle \ell m' | e^{i2\pi \vec{q} \cdot \vec{r}} (\hat{q} \times \nabla)_{M'} | \ell m \rangle =
$$
$$
2\pi \sum_{L=0}^{\infty} \sum_{M=-L}^{L} i^L Y_L^{M*}(\hat{q}) \int dr r^2 \frac{1}{r} f^2(r) j_L(2\pi q r) I_{M'LM}^{\ell m m'}(\hat{q}), \tag{113}
$$

where

$$
I_{M'LM}^{\ell m m'}(\hat{q}) = \int d^2 \hat{r} \, Y_L^M \left[Y_\ell^{m'*}(\hat{q} \times \hat{\nabla})_{M'} Y_\ell^m - Y_\ell^m (\hat{q} \times \hat{\nabla})_{M'} Y_\ell^{m'*} \right] \tag{114}
$$

is an angular integral, with $d^2 \hat{r} = \sin\theta d\theta d\varphi$. The $1/r$ factor in the radial integral comes from the definition of $\hat{\nabla}$ given above.

The angular integral can be evaluated straightforwardly using properties of spherical tensor operators. It gives rather complicated, and not very illuminating, linear combination of the spherical harmonics of $Y_L^M(\hat{q})$, with coefficients that are linear combinations of products of Clebsch-Gordan coefficients. Its product with the spherical harmonics entering (113) can be again reduced to a linear combination of simple spherical harmonics.

---

[7]Let us recall that the spherical components of a vector operator $\vec{V}$ are denoted by V_M, with $M = -1, 0, 1$, and defined as $V_{\pm 1} = \mp (V_x \pm i V_y)/\sqrt{2}$ and $V_0 = V_z$.

For the radial part, we use the relation

$$j_L(x) = \frac{x}{2L+1}\left[j_{L-1}(x) + j_{L+1}(x)\right],\tag{115}$$

and the definition

$$\bar{J}_L(q) = \int dr\, r^2 f^2(r) j_L(2\pi q r),\tag{116}$$

and we can write the expression

$$\langle \ell m'|e^{i2\pi \vec{q}\cdot\vec{r}}(\hat{q}\times\nabla)_M|\ell m\rangle = 4\pi^2 q \sum_{L=1}^{\infty}\frac{i^L}{2L+1}(\bar{J}_{L-1} + \bar{J}_{L+1})C_L(\hat{q}),\tag{117}$$

where $C_L(\hat{q})$ is a linear combination of spherical harmonics. For the spin part, the computation is similar. One difference is that there is no gradient term; hence the $1/r$ factor is absent in the radial integral, and only a single $\bar{J}_L$ comes from the radial part.

Therefore, the matrix element of the M-th spherical component of $\vec{\mathcal{M}}_{L\perp\vec{q}}^{(n)}$ between LSJ states has the form

$$\langle \ell LSJM'_J|\mathcal{M}_{L\perp\vec{q}}^{(n)\,M}|\ell LSJM_J\rangle = \sum_{L'=1}^{\infty}(\bar{J}_{L'-1} + \bar{J}_{L'+1})A_{L'}(\hat{q})\tag{118}$$

$$\langle \ell LSJM'_J|\mathcal{M}_{S\perp\vec{q}}^{(n)\,M}|\ell LSJM_J\rangle = \sum_{L'=0}^{\infty}\bar{J}_{L'}B_{L'}(\hat{q})\tag{119}$$

where the coefficients $A_{L'}$ and $B_{L'}$ obviously also depend on ℓ, L, S, J, M_J, M'_J, and M, and, we stress it again, are linear combinations of spherical harmonics. The symmetries of the Clebsch-Gordan coefficients imply that they vanish for sufficiently large L', so that the sums are in practice finite and suitable for numerical computations. It is however interesting to simplify $A_{L'}$ and $B_{L'}$ by making some approximation that provides more physical insight than expressions (118) and (119). This is the aim of the *dipole approximation*.

Dipole Approximation

The dipole approximation is obtained by truncating of the sums in (118) and (119) to $L' \le 1$ and keeping in $A_{L'}$ and $B_{L'}$ only the spherical harmonics $Y_{L''}^{M''}$ with $L'' \le 2$ [15]. It can be shown that, in the dipole approximation, the matrix elements of $\vec{\mathcal{M}}_{\vec{q}}^{(n)}$ between $LSJM_J$ states are equal to the matrix elements of the operator

$$\vec{\mathcal{M}}_{\vec{q}}^{(n,D)} = -\mu_B \left\{ \left[\bar{J}_{0n}(q) + \bar{J}_{2n}(q) \right] \vec{\mathbb{L}}_n + 2\bar{J}_{0n}(q) \vec{\mathbb{S}}_n \right\}, \tag{120}$$

where $\vec{\mathbb{L}}_n$ and $\vec{\mathbb{S}}_n$ are the total orbital angular momentum operator and the total spin operator, respectively, for the n-th ion. According to the Wigner-Eckart theorem, all vector operators are proportional to each other and, therefore, to the total angular momentum operator $\vec{\mathbb{J}}$, on a given $|LSJM_J\rangle$ subspace. Consequently, if the total angular momentum is approximately conserved, so that J and M are good quantum numbers, as it happens in the case of rare earth ions, we have

$$\vec{\mathcal{M}}_{\vec{q}}^{(n,D)} = -\mu_B g_n F_n(q) \vec{\mathbb{J}}_n, \tag{121}$$

where g_n is the Landé factor of the n-th ion,

$$g_n = 1 + \frac{J_n(J_n+1) - L_n(L_n+1) + S_n(S_n+1)}{2J_n(J_n+1)}, \tag{122}$$

and the magnetic form factor for the n-th ion is given by

$$F_n(q) = \bar{J}_{0n}(q) + \frac{J_n(J_n+1) + L_n(L_n+1) - S_n(S_n+1)}{3J_n(J_n+1) + L_n(L_n+1) - S_n(S_n+1)} \bar{J}_{2n}(q). \tag{123}$$

In the case of transition metal ions, the crystal field removes the degeneracy in the total orbital angular momentum, mixing states with different M_L quantum numbers, although it preserves the degeneracy in the spin quantum number M_S, since the crystal field Hamiltonian is independent of the spin. Hence, S and M_S remain good quantum numbers. If the degeneracy in the M_L quantum numbers is completely lifted, it can be easily shown that the expectation value of the total orbital angular momentum operator, $\vec{\mathbb{L}}$, does vanish in these states. As we said before, this phenomenon is called the *quenching* of the orbital angular momentum. And, since the neutron does not have enough energy to overcome the crystal field splitting energy, the only matrix element of $\vec{\mathbb{L}}$ that contributes to the magnetic neutron scattering vanishes. Hence, the scattering is only due to spin, and the magnetization operator is given by

$$\vec{\mathcal{M}}_{\vec{q}}^{(n,D)} = -\mu_B 2\bar{J}_{0n}(q) \vec{\mathbb{S}}_n. \tag{124}$$

Actually, the spin-orbit interaction couples the states split by the crystal field, and the quenching of the orbital angular momentum is not complete. In that case, $\vec{\mathbb{L}}$ is proportional to $\vec{\mathbb{S}}$ in the ground-state subspace, so that $\vec{\mathbb{L}} = (g-2)\vec{\mathbb{S}}$, with g close to 2, and we have

$$\vec{\mathcal{M}}_{\vec{q}}^{(n,D)} = -\mu_B g_n F_n(q) \vec{\mathbb{S}}_n, \tag{125}$$

with

$$F_n(q) = \frac{1}{g_n} \left[g_n \bar{J}_{0n}(q) + (g_n - 2) \bar{J}_{2n}(q) \right].\tag{126}$$

The dipole approximation is valid at low q. The spherical Bessel function $j_L(x)$ vanishes as $x \to 0$ as x^L, and therefore for $q \to 0$ we have $\bar{J}_L(q) \sim (qr_{max})^L$, where r_{max} is the maximum of the radial function, $f(r)$, which is proportional to the Bohr radius. Then, the dipole approximation is justified if q is small in comparison to $1/r_{max}$.

Magnetic Scattering Cross Section in the Dipole Approximation

The explicit form of the matrix elements of $\vec{\mathcal{M}}_{\perp \vec{q}}$ can be obtained from (102) with the dipole approximation for $\vec{\mathcal{M}}_{\vec{q}}^{(n)}$, using Eq. (121) or (125), depending on the case. To be generic, we write

$$\vec{\mathcal{M}}_{\vec{q}}^{(n,D)} = F_n(q) \vec{\mathbb{M}}_n,\tag{127}$$

where $\vec{\mathbb{M}}_n$ denotes the n-th ion magnetic moment density operator, which is given by $-\mu_B g_n$ times either the total angular momentum operator or the total spin operator, depending on the case. $F_n(q)$ and g_n are the corresponding *magnetic form factor* and *Landé factor*, respectively, of the n-th ion.

From the general equation (84) and the definition (102), we immediately get

$$\left(\frac{d^2 \sigma^{(m)}}{d\Omega_{\vec{k}_f} dE_f} \right)_{\sigma_f} = p^2 \frac{k_f}{k_i} \sum_{\alpha\beta} \rho^i_{\alpha\beta}(\sigma_f) \times$$
$$\sum_{nn'} F_n^*(q) F_{n'}(q) \langle e^{-i2\pi\vec{q}\cdot\vec{R}_n(0)} e^{i2\pi\vec{q}\cdot\vec{R}_{n'}(t)} \mathbb{M}_{n\perp\vec{q}}^{\beta\,\dagger}(0) \mathbb{M}_{n'\perp\vec{q}}^{\alpha}(t) \rangle_v.\tag{128}$$

The magnetic moment dynamics and the nuclear dynamics are weakly coupled, so that to a good approximation the time-dependent correlation function in (128) factorizes into nuclear and magnetic moment factors, and we can write

$$\left(\frac{d^2 \sigma^{(m)}}{d\Omega_{\vec{k}_f} dE_f} \right)_{\sigma_f} = p^2 \frac{k_f}{k_i} \sum_{\alpha\beta} \rho^i_{\alpha\beta}(\sigma_f) \sum_{nn'} F_n^*(q) F_{n'}(q)$$
$$\times \int \frac{dt}{h} e^{-i2\pi vt} \langle e^{-i2\pi\vec{q}\cdot\vec{R}_n(0)} e^{i2\pi\vec{q}\cdot\vec{R}_{n'}(t)} \rangle \langle \mathbb{M}_{n\perp\vec{q}}^{\beta\,\dagger}(0) \mathbb{M}_{n'\perp\vec{q}}^{\alpha}(t) \rangle.\tag{129}$$

3 Neutron Scattering by a Crystal

Let us consider now the specific case of diffraction by a crystalline material. Let us define a crystal by its unit cell $(\vec{a}, \vec{b}, \vec{c})$ and the atoms contained in it, labelled by $d = 1, \ldots, N_{uc}$. The reciprocal lattice is generated by the basis $(\vec{a}^*, \vec{b}^*, \vec{c}^*)$, defined as usually in crystallography as

$$\vec{a}^* = \frac{(\vec{b} \times \vec{c})}{V}, \qquad \vec{b}^* = \frac{(\vec{c} \times \vec{a})}{V}, \qquad \vec{c}^* = \frac{(\vec{a} \times \vec{b})}{V}, \qquad (130)$$

where $V = (\vec{a} \times \vec{b}) \cdot \vec{c}$ is the volume of the direct lattice unit cell. We denote by $\vec{H}$ a node of the reciprocal lattice with $\vec{H} = h\vec{a}^* + k\vec{b}^* + l\vec{c}^*$, where $h, k, and l$ are integer numbers (the Miller indices).

Nuclear Scattering

The coherent scattering cross section can be obtained by applying the general results for solids obtained in section "Nuclear Scattering", taking into account that the equilibrium position of an atom can be written as $\vec{R}_{ld}^{\,e} = \vec{R}_l + \vec{r}_d$, where $\vec{R}_l = l_a\vec{a} + l_b\vec{b} + l_c\vec{c}$, with l_a, l_b, and l_c integers, represents the position of the origin of the l-th cell and $\vec{r}_d$ is the position of d-th atom relative to the origin of coordinates of the unit cell. Then, we have to substitute the summation in n entering the equations of section "Nuclear Scattering" by summations in l and d.

For the coherent elastic scattering, given by Eq. (61), we have to compute

$$\left| \sum_n b_n \langle e^{i2\pi\vec{q}\cdot\vec{R}_n} \rangle \right|^2 = \sum_{l,\,l'} e^{i2\pi\vec{q}\cdot(\vec{R}_{l'} - \vec{R}_l)} |N_{\vec{q}}|^2 \qquad (131)$$

where

$$N_{\vec{q}} = \sum_d^{N_{uc}} b_d \, T_d(\vec{q}) \, e^{i2\pi\vec{q}\cdot\vec{r}_d} \qquad (132)$$

is the so-called *nuclear structure factor* of the unit cell and N_{uc} is the number of atoms inside it. Using the Poisson summation formula

$$\sum_{l,\,l'} e^{i2\pi\vec{q}\cdot(\vec{R}_{l'} - \vec{R}_l)} = \frac{N_c}{V} \sum_{\vec{H}} \delta(\vec{q} - \vec{H}), \qquad (133)$$

where N_c is the number of crystal unit cells in the system, we obtain

$$\frac{d\sigma_{coh}^{(n,\,e)}}{d\Omega_{\hat{k}_f}} = \frac{N_c}{V} \sum_{\vec{H}} \delta(\vec{q} - \vec{H}) |N_{\vec{q}}|^2. \qquad (134)$$

The above expression tells us that the scattered neutrons will follow directions such a way that $\vec{q} = \vec{H}$, *i.e.,* the Bragg reflection will occur only when the scattering vectors correspond to the nodes of the reciprocal lattice. These nodes are related to the characteristic distances between crystal planes (d_{hkl}) at the direct space. The amount of neutrons scattered for each of these directions is proportional to the modulus square of the complex number $N_{\vec{H}}$, the *nuclear structure factor* evaluated at $\vec{q} = \vec{H}$.

Sometimes it is convenient to define the atoms of the crystal unit cell through the number of atoms in its asymmetric unit cell (N_{au}), each one indexed with d, with coherent scattering length b_d, and located at $\vec{r}_d$. Let $\tilde{g}_p = \{\mathcal{R}_p | \vec{t}_p\}$ be the elements of the space group of the crystal ($\tilde{g}_p \in \mathcal{G}$), with rotational and translational parts, respectively, $\mathcal{R}_p$, and $\vec{t}_p$. The d-orbit in the unit cell is defined by the atoms at the p-site related to the d-site by $\vec{r}_p = \tilde{g}_p \vec{r}_d$. The number of atoms on a given orbit is less or equal than the order of the group $\mathcal{G}$, and all of them have the same scattering length b_d. The position of the same atom d at another unit cell, l, is $\vec{R}_{ld} = \vec{R}_l + \vec{r}_d$. The nuclear structure factor in (132) is rewritten as

$$N_{\vec{q}} = \sum_{d}^{N_{au}} b_d \sum_{p}^{d-orbit} T_{pd}(\vec{q})\,e^{i2\pi\vec{q}\cdot\tilde{g}_p\vec{r}_d}, \tag{135}$$

where $\vec{q}$ labels the nuclear Bragg reflexion ($\vec{q} = \vec{H}$) and p runs over the symmetry operators of the space group of the crystal that generates the atoms of the d-orbit. The factor $T_{pd}(\vec{q})$ is the Debye–Waller factor of the d-th atom at its equivalent position p. In the harmonic approximation, it is given by

$$T_{pd}(\vec{q}) = e^{-\vec{q}\cdot\mathcal{R}_p\tilde{\beta}_d\mathcal{R}_p^{\dagger}\vec{q}}. \tag{136}$$

The elastic incoherent cross section by a crystal is immediately obtained from Eq. (62) by summing on all the atoms at the unit cell or by summing first on the asymmetric unit and then on each d-orbit (p)

$$\frac{d\sigma_{inc}^{(n,\,e)}}{d\Omega_{\hat{k}_f}} = N_c \sum_{d}^{N_{uc}} b_{inc,d}^2 |T_d(\vec{q})|^2 = N_c \sum_{d}^{N_{au}} b_{inc,d}^2 \sum_{p}^{d-orbit} |T_{pd}(\vec{q})|^2. \tag{137}$$

Scattering by Phonons

The coherent inelastic 1-phonon scattering by a crystal can be readily obtained from the general equation (74), setting $\vec{R}_{ld}^e = \vec{R}_l + \vec{r}_d$ and taking into account that, due to the lattice translational symmetry, the phonons r are characterized by a wave vector, $\vec{p}$, which belongs to the first Brillouin zone in reciprocal space and a branch index, s. The number of branches is equal to the number of atoms in the unit cell. The polarization vector is given by Fultz [16]

$$\vec{\xi}_{s,ld}(\vec{p}) = \frac{1}{\sqrt{N_c}}\,e^{i2\pi\vec{p}\cdot\vec{R}_l}\vec{e}_{s,d}(\vec{p}), \tag{138}$$

and thus $\vec{v}_r(\vec{q}\,)$, defined by Eq. (75), reads

$$\vec{v}_{s\vec{p}}(\vec{q}\,) = \frac{1}{\sqrt{N_c}} \sum_{ld} e^{i2\pi\vec{q}\cdot\vec{R}_l}\, e^{i2\pi\vec{q}\cdot\vec{r}_d}\, \frac{b_d}{\sqrt{M_d}}\, T_d(\vec{q}\,)\,\vec{e}_{s,d}(\vec{p}\,). \tag{139}$$

Inserting this into Eq. (74) and using (133), we obtain

$$\frac{d^2\sigma_{\text{coh,1ph}}^{(n,i)}}{d\Omega_{\hat{k}_f}\, dE_f} = \frac{k_f}{k_i}\, q^2 \frac{(2\pi)^2}{2V} \sum_{s,\vec{p}} \left| \sum_d \frac{b_d}{\sqrt{M_d}} T_d(\vec{q}\,) e^{i2\pi\vec{q}\cdot\vec{r}_d}\, \hat{q}\cdot\vec{e}_{s,d}(\vec{q}\,) \right|^2$$

$$\times \frac{1}{v_{s\vec{p}}} \sum_{\vec{H}} \Big[\delta(\vec{q}-\vec{p}-\vec{H})\delta(h\nu - h\nu_{s\vec{p}})\big(1 + \langle n_{s\vec{p}}\rangle\big)$$

$$+ \delta(\vec{q}+\vec{p}-\vec{H})\delta(h\nu + h\nu_{s\vec{p}})\langle n_{s\vec{p}}\rangle \Big]. \tag{140}$$

This equation provides a method to measure the *phonon dispersion relation*, that is, the phonon frequency as a function of its wave vector, for each branch. The technique is as follows. A monochromatic beam, with fixed wave vector $\vec{k}_i$, is scattered by a single crystal, and the intensity of the scattered beam is measured at a fixed direction, $\hat{k}_f$, as a function of the neutron energy E_f (or as a function of wave number, k_f). The Dirac delta functions of (140) impose severe constraints to the scattering: for given k_f, there will be a discrete set of phonon wave vectors that will satisfy the constraint imposed by the wave vector Dirac delta function, namely, those that differ from the scattering vector by a reciprocal lattice vector: $\vec{p} = \vec{k}_i - \vec{k}_f + \vec{H}$. But then, in general, none of the frequencies $\nu_{s\vec{p}}$ will satisfy the constraint imposed by the frequency Dirac delta functions: $\nu_{s\vec{p}} \neq |E_i - E_f|/h$, since the index s is discrete. It is only for specific values of k_f that both constraints will be satisfied: sharp intensity peaks will appear only if $\vec{p} = \vec{k}_i - \vec{k}_f + \vec{H}$ and $\nu_{s\vec{p}} = |E_i - E_f|/h$. These peaks provide the phonon dispersion relations. The polarization vectors can be obtained from the intensities of the peaks. Neutron instruments suitable to measure the phonon dispersion relations are the *triple axis* and *time-of-flight* spectrometers.

The expression for the inelastic scattering is not sensitive to the atom arrangements, and thus it is given by (74).

Magnetic Scattering

If some of the ions in the unit cell have localized unpaired electrons, the neutrons suffer magnetic scattering. Let d_m label the magnetic ions within a cell. The position of the ld_m ion is given by $\vec{R}_l + \vec{r}_{d_m} + \vec{u}_{ld_m}$, where $\vec{u}_{ld_m}$ is the quantum operator that describes the displacements from the equilibrium position. The magnetic form factors depend only on d_m, and not on the cell to which the ion belongs. We have to substitute the sums in n and n' in Eq. (129) by sums in $l d_m$ and $l' d'_m$. We obtain

$$
\left(\frac{d^2\sigma^{(m)}}{d\Omega_{\hat{k}_f} dE_f}\right)_{\sigma_f} = p^2 \frac{k_f}{k_i} \sum_{\alpha\beta} \rho^i_{\alpha\beta}(\sigma_f) \sum_{l d_m l' d'_m} e^{i2\pi\vec{q}\cdot(\vec{R}_{l'}-\vec{R}_l)} e^{i2\pi\vec{q}\cdot(\vec{r}_{d'_m}-\vec{r}_{d_m})} F^*_{d_m}(q) F_{d'_m}(q)
$$
$$
\int \frac{dt}{h} e^{-i2\pi\nu t} \left\langle e^{-i2\pi\vec{q}\cdot\vec{u}_{l d_m}(0)} e^{i2\pi\vec{q}\cdot\vec{u}_{l' d'_m}(t)} \right\rangle \left\langle \mathbb{M}^{\beta\,\dagger}_{l d_m \perp \vec{q}}(0) \mathbb{M}^{\alpha}_{l' d'_m \perp \vec{q}}(t) \right\rangle. \tag{141}
$$

The elastic scattering component is extracted from the factorization of the time correlations in the $t \to \infty$ limit. Defining the vector $\vec{M}_{l d_m} = \langle \mathbb{M}_{l d_m}\rangle$, which is real, and denoting by $\vec{M}_{l d_m \perp \vec{q}}$ its component perpendicular to $\vec{q}$, we have

$$
\left(\frac{d\sigma^{(m,\,e)}}{d\Omega_{\hat{k}_f}}\right)_{\sigma_f} = p^2 \sum_{\alpha\beta} \rho^i_{\alpha\beta}(\sigma_f) \sum_{l d_m l' d'_m} e^{i2\pi\vec{q}\cdot(\vec{R}_{l'}-\vec{R}_l)}
$$
$$
\times\, e^{-i2\pi\vec{q}\cdot\vec{r}_{d_m}} T^*_{d_m}(\vec{q}) F^*_{d_m}(\vec{q}) e^{i2\pi\vec{q}\cdot\vec{r}_{d'_m}} T_{d'_m}(\vec{q}) F_{d'_m}(\vec{q}) M^{\beta\,*}_{l d_m \perp \vec{q}} M^{\alpha}_{l d_m \perp \vec{q}}. \tag{142}
$$

Notice that, in the most general case, $\vec{M}_{l d_m}$ needs not be commensurate with the nuclear lattice, but it can always be expanded in Fourier series as

$$
\vec{M}_{l d_m} = \sum_{\{\vec{K}\}} \vec{S}_{\vec{K} d_m} e^{-i2\pi\vec{K}\cdot\vec{R}_l}, \tag{143}
$$

where the complex vectors $\vec{S}_{\vec{K} d_m}$ are the *Fourier coefficients* and $\{\vec{K}\}$ are the set of *propagation vectors*. The propagation vectors relate the magnetic moment at the l-th unit cell with respect to the zero-th unit cell. In the case of $\vec{K} = 0$, which means that the nuclear and magnetic periodicities are the same, then the Fourier coefficients $\vec{S}_{\vec{K} d_m}$ are equal to the magnetic moment, $\vec{M}_{d_m}$, at site d_m, and do not depend on the lattice index l.

Moreover, because the magnetic moment $\vec{M}_{l d_m}$ must be a real vector, it imposes the condition that summation extends to pairs of propagation vectors $\vec{K}$ and $-\vec{K}$ and the relation $\vec{S}_{-\vec{K} d_m} = \vec{S}^*_{\vec{K} d_m}$ between the Fourier coefficients.

Inserting expansion (143) into (142) and using Eq. (133) for the sum in $l l'$, we get

$$
\frac{d\sigma^{(m,\,e)}_{\mathrm{coh}}}{d\Omega_{\hat{k}_f}} = \frac{N_c}{V} \sum_{\vec{H}\vec{K}} \delta(\vec{q} - \vec{H} - \vec{K}) |\vec{M}_{\perp\vec{q}}|^2, \tag{144}
$$

where we have defined the *magnetic interaction vector*, $\vec{M}_{\perp\vec{q}}$, and the *magnetic structure factor*, $\vec{M}_{\vec{q}\vec{K}}$, for the crystal unit cell[8] as:

---

[8]Other authors define the magnetic structure factor for the crystal by keeping the sum on the unit cells, l and l', in Eq. (142) in its definition.

$$\vec{M}_{\perp\vec{q}} = \hat{q} \times (\vec{M}_{\vec{q}\vec{K}} \times \hat{q}), \tag{145}$$

$$\vec{M}_{\vec{q}\vec{K}} = p \sum_{d_m}^{N_{uc}} F_{d_m}(\vec{q}) T_{d_m}(\vec{q}) \, e^{i2\pi\vec{q}\cdot\vec{r}_{d_m}} \vec{S}_{\vec{K}d_m}. \tag{146}$$

To have elastic magnetic scattered neutrons (magnetic Bragg peaks) the Dirac delta in (144) imposes that the scattering vector must fulfill the condition $\vec{q} = \vec{H} \pm \vec{K}$ ($\vec{H}$ is a node of the nuclear reciprocal lattice and $\vec{K}$ is the magnetic propagation vector).

The intensity of magnetic peaks is very sensitive to temperature, since the expectation value of the magnetic moment $\vec{M}_{ld_m}$ is. By increasing temperature $\vec{M}_{\vec{q}\vec{K}}$ decreases until it vanishes above the magnetic ordering temperature, in which case only the nuclear peaks survive. The behavior with temperature is one way of distinguishing magnetic and nuclear scattering.

Magnetic Inelastic Scattering

To discuss magnetic inelastic scattering, it is convenient to introduce more notation. Let us define

$$N_E = \langle e^{-i2\pi\vec{q}\cdot\vec{u}_{ld_m}} \rangle \langle e^{i2\pi\vec{q}\cdot\vec{u}_{l'd'_m}} \rangle, \quad M_E^{\alpha\beta} = \langle \mathbb{M}_{ld_m\perp\vec{q}}^{\alpha\,\dagger} \rangle \langle \mathbb{M}_{l'd'_m\perp\vec{q}}^{\beta} \rangle, \tag{147}$$

and

$$N_I = \langle e^{-i2\pi\vec{q}\cdot\vec{u}_{ld_m}(0)} e^{i2\pi\vec{q}\cdot\vec{u}_{l'd'_m}(t)} \rangle - \langle e^{-i2\pi\vec{q}\cdot\vec{u}_{ld_m}} \rangle \langle e^{i2\pi\vec{q}\cdot\vec{u}_{ld_m}} \rangle, \tag{148}$$

$$M_I^{\alpha\beta} = \langle \mathbb{M}_{ld_m\perp\vec{q}}^{\alpha\,\dagger}(0) \mathbb{M}_{l'd'_m\perp\vec{q}}^{\beta}(t) \rangle - \langle \mathbb{M}_{ld_m\perp\vec{q}}^{\alpha\,\dagger} \rangle \langle \mathbb{M}_{l'd'_m\perp\vec{q}}^{\beta} \rangle. \tag{149}$$

The integrand of the time integral in Eq. (141) is split into four terms

$$s\langle e^{-i2\pi\vec{q}\cdot\vec{u}_{ld_m}(0)} e^{i2\pi\vec{q}\cdot\vec{u}_{l'd'_m}(t)} \rangle \langle \mathbb{M}_{ld_m\perp\vec{q}}^{\alpha\,\dagger}(0) \mathbb{M}_{l'd'_m\perp\vec{q}}^{\beta}(t) \rangle$$
$$= N_I M_I^{\alpha\beta} + N_E M_I^{\alpha\beta} + N_I M_E^{\alpha\beta} + N_E M_E^{\alpha\beta}. \tag{150}$$

Correspondingly, the magnetic scattering cross section is split into four terms, one for each of the above terms. The $N_E M_E^{\alpha\beta}$ term gives the elastic scattering discussed before. The $N_I M_E^{\alpha\beta}$ term gives the so-called *magnetovibrational scattering*, which is elastic in the magnetic sector and inelastic in the nuclear part, with emission or absorption of phonons. This is essentially identical to the pure inelastic nuclear scattering, with the form factor modified by the magnetic factors. The $N_E M_I^{\alpha\beta}$ term gives scattering elastic in the nuclear sector and inelastic in the magnetic sector, with energy transfer between the neutron and the magnetic degrees of freedom. Finally, the $N_I M_I^{\alpha\beta}$ term gives inelastic scattering both in the nuclear and magnetic sectors, with energy exchange between the neutron and the magnetic and nuclear degrees of freedom.

Scattering by Spin-Waves

Let us consider the inelastic magnetic scattering by an ordered magnetic material with no transfer of energy with phonons (i.e., the term $N_E M_I^{\alpha\beta}$). In this case, the neutrons exchange energy only with the elementary magnetic excitations around the ordered magnetic state, which are called *magnons* [17]. They are described as follows. Let $\vec{\mathbb{J}}_{ld_m}$ denote the spin or total angular momentum operator, whatever matters, of the ld_m-th magnetic ion. Let us choose $\vec{M}_{ld_m}$ as the local quantization axis for the ld_m-th ion, and let $\mathbb{J}^z_{ld_m}$ the projection of $\vec{\mathbb{J}}_{ld_m}$ along $\vec{M}_{ld_m}$ and $\mathbb{J}^x_{ld_m}$ and $\mathbb{J}^y_{ld_m}$ the projections onto two mutually perpendicular directions in the plane orthogonal to $\vec{M}_{ld_m}$. Let $\mathbb{J}^\pm_{ld_m}$ be the corresponding ladder operators. The magnetic ground state of the ion is given by $|JM_J\rangle$ with $M_J = J$. At low temperature, M_J cannot depart much from its maximum value J. Evidently, this cannot be the case if $J = 1/2$; hence we assume $J > 1/2$. Let us introduce the departure number, $n = J - M_J$, to characterize the magnetic states $|n\rangle = |J\,J - n\rangle$. We perform a Holstein-Primakoff transformation [18], introducing the operators a_{ld_m} and $a^\dagger_{ld_m}$ which satisfy the bosonic commutation relations

$$[a_{ld_m}, a_{l'd'_m}] = 0, \quad [a^\dagger_{ld_m}, a^\dagger_{l'd'_m}] = 0, \quad [a_{ld_m}, a^\dagger_{l'd'_m}] = \delta_{ll'}\,\delta_{d_m d'_m}. \tag{151}$$

At each ion, they act on the magnetic states as the creation and annihilation operators of an harmonic oscillator:

$$a_{ld_m}|n\rangle = n^{1/2}\,|n - 1\rangle, \qquad a^\dagger_{ld_m}|n\rangle = (n + 1)^{1/2}\,|n + 1\rangle. \tag{152}$$

By introducing this bosonic operators, the space of states has been artificially enlarged to the space of states with any $n > 0$, but only the states with $n \le 2J$ have physical meaning, and states with $n > 2J$ do not influence the physical results. Indeed, the angular momentum operators are expressed in terms of the bosonic operators as

$$\begin{aligned}
\mathbb{J}^+_{ld_m} &= \left(2J - a^\dagger_{ld_m}a_{ld_m} - 1\right)^{1/2} a_{ld_m}, \\
\mathbb{J}^-_{ld_m} &= \left(2J - a^\dagger_{ld_m}a_{ld_m}\right)^{1/2} a^\dagger_{ld_m}, \\
\mathbb{J}^z_{ld_m} &= J - a^\dagger_{ld_m}a_{ld_m}.
\end{aligned} \tag{153}$$

The system Hamiltonian, expressed in terms of the bosonic operators, is expanded in powers of a_{ld_m} and $a^\dagger_{ld_m}$. It is easy to see from the above equations that this expansion is actually an expansion in powers of $1/2J$. The lowest order is provided by the quadratic terms. In the linear, or harmonic, approximation, only the quadratic terms are taken into account, and the magnetic dynamics is described by a system of coupled harmonic oscillators. In this case we have

$$\mathbb{J}_{ld_m}^+ = \sqrt{2J}\, a_{ld_m}, \quad \mathbb{J}_{ld_m}^- = \sqrt{2J}\, a_{ld_m}^\dagger, \quad \mathbb{J}_{ld_m}^z = J - a_{ld_m}^\dagger a_{ld_m}. \tag{154}$$

This approximation is good for large J, and in this case the neglected terms can be systematically taken into account by perturbation theory.

In the linear approximation, the eigenstates of the system, which are called *magnons*, form a Bose-Einstein gas of non-conserved particles, with zero chemical potential. Moreover the correlation $N_E M_1$ of Eq. (150) can be readily computed in a way completely analogous to that of phonons. Let us consider for simplicity a crystal with only one magnetic ion per unit cell, so that the magnetic ions form a Bravais lattice (and we can remove the d_m index), and assume that the system is ferromagnetically ordered. Thus, the only non-vanishing Fourier mode of the magnetic moment is that with $\vec{K} = 0$, which is denoted in the following by S. To linear order, the solution to the equations of motion is given by

$$a_l(t) = \frac{1}{\sqrt{N_c}} \sum_{\vec{p}} e^{i2\pi(\vec{p}\cdot\vec{R}_l - v_{\vec{p}} t)}\, b_{\vec{p}}, \quad a_l^\dagger(t) = \frac{1}{\sqrt{N_c}} \sum_{\vec{p}} e^{-i2\pi(\vec{p}\cdot\vec{R}_l - v_{\vec{p}} t)}\, b_{\vec{p}}^\dagger, \tag{155}$$

where $\vec{p}$ belongs to the first Brillouin zone of the magnetic ion lattice, the operators $b_{\vec{p}}$ and $b_{\vec{p}}^\dagger$ satisfy the commutation relation $[b_{\vec{p}}, b_{\vec{p}'}^\dagger] = \delta_{\vec{p}\,\vec{p}'}$, and $v_{\vec{p}}$ are the magnon frequencies.

Using the fact that $\vec{\mathbb{M}}_l = g\mu_B \vec{\mathbb{J}}_l$, Eqs. (154) and (155), and defining $M = g\mu_B J$, we easily obtain $\langle \mathbb{M}_l^x \rangle = \langle \mathbb{M}_l^y \rangle = 0$, and

$$\langle \mathbb{M}_l^z \rangle = M\Big[1 - (1/J N_c)\sum_{\vec{p}}\langle n_{\vec{p}}\rangle\Big], \tag{156}$$

where $\langle n_{\vec{p}}\rangle$ is the average value of the number of magnons with wave vector $\vec{p}$, given by the Bose-Einstein distribution (72). For the time correlation functions, we have $\langle \mathbb{M}_l^z(0)\mathbb{M}_{l'}^\alpha(t)\rangle = \langle \mathbb{M}_l^\alpha(0)\mathbb{M}_{l'}^z(t)\rangle = 0$, with $\alpha = x, y$, and

$$\langle \mathbb{M}_l^z(0)\mathbb{M}_{l'}^z(t)\rangle = M^2\Big(1 - \frac{2}{J N_c}\sum_{\vec{p}}\langle n_{\vec{p}}\rangle\Big) = \langle \mathbb{M}_l^z\rangle\langle \mathbb{M}_{l'}^z\rangle + O(1/J^2), \tag{157}$$

so that this correlation, called the *longitudinal* correlation, does not contribute to inelastic scattering.[9] For the remaining components, called the *transverse* correlations, we have

$$\langle \mathbb{M}_l^x(0)\mathbb{M}_{l'}^x(t)\rangle = \langle \mathbb{M}_l^y(0)\mathbb{M}_{l'}^y(t)\rangle, \quad \langle \mathbb{M}_l^x(0)\mathbb{M}_{l'}^y(t)\rangle = -\langle \mathbb{M}_l^y(0)\mathbb{M}_{l'}^x(t)\rangle, \tag{158}$$

---

[9]Notice that this correlation is independent of time. The $1/J^2$ terms have to be neglected in the linear approximation for consistency. Higher-order corrections guarantee that the second equality in (157) holds to all orders in the $1/J$ expansion.

with

$$\langle M_l^x(0)M_{l'}^x(t)\rangle = \frac{M^2}{2JN_c}\sum_{\vec{p}}\left[e^{-i2\pi\phi_{ll'}(\vec{p})}\left(\langle n_{\vec{p}}\rangle + 1\right) + e^{i2\pi\phi_{ll'}(\vec{p})}\langle n_{\vec{p}}\rangle\right], \qquad (159)$$

$$\langle M_l^x(0)M_{l'}^y(t)\rangle = \frac{iM^2}{2JN_c}\sum_{\vec{p}}\left[e^{-i2\pi\phi_{ll'}(\vec{p})}\left(\langle n_{\vec{p}}\rangle + 1\right) - e^{i2\pi\phi_{ll'}(\vec{p})}\langle n_{\vec{p}}\rangle\right], \qquad (160)$$

where $\phi_{ll'}(\vec{p}) = \vec{p}\cdot(\vec{R}_{l'} - \vec{R}_l) - \nu_{\vec{p}}\,t$.

Consider the scattering of an unpolarized beam. We do not observe the polarization state of the scattered neutrons, so that $\sum_{\sigma_f}\rho_{\alpha\beta}^i(\sigma_f) = \delta_{\alpha\beta}$. Then, the cross section depends only on the projection of the time correlation function onto the component perpendicular to $\vec{q}$, which is

$$\sum_{\alpha\beta}\left(\delta_{\alpha\beta} - \frac{q_\alpha q_\beta}{q^2}\right)\left[\langle M_l^\beta(0)M_{l'}^\alpha(t)\rangle - \langle M_l^\beta\rangle\langle M_{l'}^\alpha\rangle\right] = \left(1 + \frac{q_z^2}{q^2}\right)\langle M_l^x(0)M_{l'}^x(t)\rangle, \qquad (161)$$

where we used $1 - q_x^2/q^2 + 1 - q_y^2/q^2 = 1 + q_z^2/q^2$.

Now, to obtain the inelastic magnetic scattering cross section by a ferromagnet to $1/J$ order (1-magnon scattering), we can insert (161) into (149) and this into (129). The sums in l and l' are obtained using (133), and we get

$$\frac{d^2\sigma^{(m)}}{d\Omega_{\hat{k}_f}\,dE_f} = p^2\,\frac{k_f}{k_i}\,\frac{1}{V}\,\frac{M^2}{2J}\left(1 + \frac{q_z^2}{q^2}\right)F^2(q)\,T^2(\vec{q})$$

$$\times \sum_{\vec{p}\,\vec{H}}\left[\delta(\vec{q} - \vec{p} - \vec{H})\,\delta(h\nu - h\nu_{\vec{p}})\left(\langle n_{\vec{p}}\rangle + 1\right) \qquad (162)$$

$$+ \delta(\vec{q} + \vec{p} - \vec{H})\,\delta(h\nu + h\nu_{\vec{p}})\,\langle n_{\vec{p}}\rangle\right].$$

This expression is similar to expression (140) for the scattering by phonons, and the *magnon dispersion relations*, which provide the dependence of the frequency $\nu_{\vec{p}}$ with the wave vector $\vec{p}$, can be obtained by applying the same ideas. One difference between magnon and phonon scattering is the factor $1 + q_z^2/q^2$, which takes values between 1 and 2. If the scattering vector is parallel to $\hat{z}$, the factor is 2, and if it is perpendicular to $\hat{z}$, it is 1. This corresponds to the fact that the deviation of the magnetic moment with respect to its equilibrium value is perpendicular to $\hat{z}$. Therefore, it has two components perpendicular to $\hat{q}$ if this vector is parallel to $\hat{z}$ and only one component perpendicular to $\hat{q}$ if it is perpendicular to $\hat{z}$.

The magnon dispersion relation allows us to investigate the nature of the magnetic interactions. For instance, in a ferromagnetic system described by a Heisenberg model, we have $h\nu_{\vec{p}} = 2J[K(0) - K(\vec{p})]$, where $K(\vec{p})$ is the Fourier transform of the exchange interaction coupling.

4 Polarization Analysis

The scattering of a polarized beam of neutrons together with the analysis of the polarization of the scattered beam is a powerful tool that allows us to extract much more information about the target than the simpler scattering of unpolarized neutrons. Since, as we are going to see, polarization analysis introduces correlations between nuclear and magnetic scattering amplitudes which are absent in the case of unpolarized beams, we have to consider both nuclear and magnetic scattering at once. To this end, let us write the scattering amplitude operator as a matrix in spin space:

$$\mathcal{A}_{\vec{q}} = \vec{\mathcal{V}}_{\vec{q}} \cdot \vec{\sigma} + \mathcal{W}_{\vec{q}} \, I_2, \tag{163}$$

where $\vec{\sigma}$ and I_2 are, respectively, the Pauli matrices and the 2×2 identity matrix acting on the neutron spin degrees of freedom, and

$$\vec{\mathcal{V}}_{\vec{q}} = p\vec{\mathcal{M}}_{\perp\vec{q}} + \frac{1}{2} \sum_n e^{i2\pi\vec{q}\cdot\vec{R}_n} B_n \vec{\mathcal{I}}_n, \qquad \mathcal{W}_{\vec{q}} = \sum_n e^{i2\pi\vec{q}\cdot\vec{R}_n} A_n. \tag{164}$$

In the above expressions, $\vec{R}_n$, $\vec{\mathcal{I}}_n$, A_n, and B_n denote, respectively, the position, the spin, and the two scattering lengths of the n-th nucleus [c.f. Eq. (30)].

Let us consider elastic scattering. The density matrix and the polarization vector of the scattered beam are obtained by substituting (163) into Eqs. (25) and (27), respectively. The sum over σ_f defining $I_{\vec{q}}$ and the trace entering Eq. (27) can be easily computed applying several times the well-known identity for the Pauli matrices

$$\sigma_\alpha \sigma_\beta = I_2 \, \delta_{\alpha\beta} + i \sum_\gamma \epsilon_{\alpha\beta\gamma} \sigma_\gamma, \tag{165}$$

and using $\mathrm{Tr}\, \sigma_\alpha = 0$ and $\mathrm{Tr}\, \sigma_\alpha \sigma_\beta = 2\delta_{\alpha\beta}$. In the above equation, $\epsilon_{\alpha\beta\gamma}$ is the totally antisymmetric tensor in three dimensions. Then, the average over the disorder can be computed as in section "Nuclear Scattering". The assumptions are as follows: (1) there is no correlation between the probability distributions of isotopes in different nuclei; (2) the nuclear spins are uniformly randomly oriented in space; (3) the spin orientations of different nuclei are uncorrelated; and (4) the disorder in magnetic properties is independent from the disorder in nuclear properties.

Let us introduce the following quantities, appropriate for a generic scattering target

$$W_{\vec{q}} = \overline{\langle \mathcal{W}_{\vec{q}} \rangle} = \sum_n b_n \langle e^{i2\pi\vec{q}\cdot\vec{R}_n} \rangle, \tag{166}$$

$$\vec{V}_{\vec{q}} = \overline{\langle \vec{\mathcal{V}}_{\vec{q}} \rangle} = p\langle \vec{\mathcal{M}}_{\perp\vec{q}} \rangle, \tag{167}$$

where $b_n = \bar{A}_n$ is the coherent scattering length of the n-th nucleus. Notice that by definition $\vec{V}_{\vec{q}}$ is perpendicular to $\vec{q}$. Sometimes $W_{\vec{q}}$ and $\vec{V}_{\vec{q}}$ are referred to, respectively, as generic nuclear structure factor and magnetic interaction vector for the whole system. Retaining only the coherent component from Eq. (27), we get

$$I_{\vec{q}} = W_{\vec{q}}\, W_{\vec{q}}^* + \vec{V}_{\vec{q}} \cdot \vec{V}_{\vec{q}}^* + \left(W_{\vec{q}}\, \vec{V}_{\vec{q}}^* + \vec{V}_{\vec{q}}\, W_{\vec{q}}^*\right)\cdot \vec{P}_i - i\left(\vec{V}_{\vec{q}} \times \vec{V}_{\vec{q}}^*\right)\cdot \vec{P}_i \quad (168)$$

and

$$I_{\vec{q}}\vec{P}_f = \left(W_{\vec{q}}\, W_{\vec{q}}^* - \vec{V}_{\vec{q}} \cdot \vec{V}_{\vec{q}}^*\right)\vec{P}_i + \vec{V}_{\vec{q}}\left(\vec{P}_i \cdot \vec{V}_{\vec{q}}^*\right) + \left(\vec{P}_i \cdot \vec{V}_{\vec{q}}\right)\vec{V}_{\vec{q}}^* \\ + i\left(\vec{V}_{\vec{q}}\, W_{\vec{q}}^* - W_{\vec{q}}\, \vec{V}_{\vec{q}}^*\right)\times \vec{P}_i + W_{\vec{q}}\, \vec{V}_{\vec{q}}^* + W_{\vec{q}}^*\, \vec{V}_{\vec{q}} + i\left(\vec{V}_{\vec{q}} \times \vec{V}_{\vec{q}}^*\right), \quad (169)$$

where $\vec{P}_f$ and $\vec{P}_i$ are, respectively, the final and initial beam polarizations.

The theoretical foundations of the scattering of polarized neutrons were developed in [12, 19–26] and can be summarized in the Blume-Maleev Eqs. (168) and (169) [24, 27] which open new possibilities to design experiments with polarized neutron beams.

Two consequences of Eq. (169) are immediately apparent: (1) if there is no magnetic scattering, the scattered beam has the same polarization vector as the incident beam; and (2) the magnetic scattering can polarize an unpolarized beam, since the last three terms are independent of $\vec{P}_i$.

Let us consider the elastic scattering by a crystal. As before, the ion position is written as $\vec{R}_l + \vec{r}_d + \vec{u}_{ld}$, and Eqs. (166) and (167) become

$$W_{\vec{q}} = N_{\vec{q}} \sum_l e^{i2\pi \vec{q} \cdot \vec{R}_l}, \qquad \vec{V}_{\vec{q}} = \sum_{\vec{K}} \vec{M}_{\perp \vec{q}} \sum_l e^{i2\pi (\vec{q}+\vec{K}) \cdot \vec{R}_l}. \quad (170)$$

where $N_{\vec{q}}$ and $\vec{M}_{\perp \vec{q}}$ are given by Eqs. (132) and (145). Then each term of (168) and (169) contains sums in l and l' of terms of the form (133). Thus, each term of these equations is a sum of terms that contain a Dirac delta function $\delta(\vec{q} - \vec{H} \pm \vec{K})$. This is Bragg scattering, and the delta functions define the Bragg peaks at $\vec{q} = \vec{H} \pm \vec{K}$. We get the first Blume-Maleev equation for the scattered intensity in a crystal as

$$I_{\vec{q}} = |N_{\vec{q}}|^2 + |\vec{M}_{\perp \vec{q}}|^2 + \left(N_{\vec{q}}\vec{M}_{\perp \vec{q}}^* + N_{\vec{q}}^*\vec{M}_{\perp \vec{q}}\right)\cdot \vec{P}_i + i\left(\vec{M}_{\perp \vec{q}}^* \times \vec{M}_{\perp \vec{q}}\right)\cdot \vec{P}_i \quad (171)$$

where the vectors $N_{\vec{q}}\vec{M}_{\perp \vec{q}}^* + N_{\vec{q}}^*\vec{M}_{\perp \vec{q}}$ and $i\vec{M}_{\perp \vec{q}}^* \times \vec{M}_{\perp \vec{q}}$ are called, respectively, *nuclear-magnetic interference vector* and *chiral vector*. The second Blume-Maleev equation for the scattered polarization in a crystal is

$$I_{\vec{q}}\vec{P}_f = \left(|N_{\vec{q}}|^2 - |\vec{M}_{\perp \vec{q}}|^2\right)\vec{P}_i + \left(\vec{P}_i \cdot \vec{M}_{\perp \vec{q}}^*\right)\vec{M}_{\perp \vec{q}} + \left(\vec{P}_i \cdot \vec{M}_{\perp \vec{q}}\right)\vec{M}_{\perp \vec{q}}^*$$

$$-\mathrm{i}\left(N_{\vec{q}}\vec{M}^*_{\perp\vec{q}} - N^*_{\vec{q}}\vec{M}_{\perp\vec{q}}\right) \times \vec{P}_i + N_{\vec{q}}\vec{M}^*_{\perp\vec{q}} + N^*_{\vec{q}}\vec{M}_{\perp\vec{q}}$$

$$-\mathrm{i}\left(\vec{M}^*_{\perp\vec{q}} \times \vec{M}_{\perp\vec{q}}\right). \tag{172}$$

Notice that, by definition, $N_{\vec{q}}$ vanishes if $\vec{K}$ is not a vector of the nuclear reciprocal lattice, and the nuclear scattering does not contribute to the magnetic satellites.

5 Magnetic Crystallography

Let us consider that the nuclear unit cell, as defined in section "Nuclear Scattering", contains some magnetic atom located at site d_m of the asymmetric unit cell with magnetic moment $\vec{M}_{d_\mathrm{m}}$. The magnetic moment is an *axial vector*, and therefore its transformation rule includes the determinant of the operator (det). Moreover, the symmetry group that we need to consider is not the space group of the crystal but its magnetic space group where the *time reversal* operation ($\mathcal{I}'$) can play a role (this operation reverses the components of the magnetic moment and does not change the atomic coordinates).

The magnetic space groups ($\mathcal{M}$) are obtained as the outer direct product of the crystallographic group $\mathcal{G}$ and the group $\mathcal{K}$ formed by the *identity*, $\mathcal{I}$, and the *time reversal*, $\mathcal{I}'$, operations $\mathcal{M} \subset \mathcal{G}\bigotimes\mathcal{K}$ where $\mathcal{K} = \{\mathcal{I},\ \mathcal{I}'\}$. So a magnetic group has *primed* operations, in which the time reversal acts after an operation of the crystallographic group, and *unprimed* operations. Starting from the 230 space groups $\mathcal{G}$, also known as Fedorov groups, it is possible to build the 1651 magnetic space groups, or Shubnikov groups, $\mathcal{M}$, as follows: Type (I) 230 *colorless* groups without *primed* operations ($\mathcal{M} = \mathcal{G}$). Type (II) 230 *grey* or paramagnetic groups composed with the same operations *primed* and *unprimed* ($\mathcal{M} = \mathcal{G} + \mathcal{G}\mathcal{I}'$). The other 1191 called *Black-White* groups are built as $\mathcal{M} = \mathcal{H} + (\mathcal{G} - \mathcal{H})\mathcal{I}'$ where $\mathcal{H}$ are the index 2 subgroups of $\mathcal{G}$ as constituting the *unprimed* elements and the rest of operators $(\mathcal{G} - \mathcal{H})$ multiplied by the time reversal $\mathcal{I}'$. These *Black-White* groups can be split into type (III) 674 groups in which $\mathcal{H}$ has the same translation group $\mathcal{T}$ as $\mathcal{G}$ and type (IV) 517 groups in which the translation group $\mathcal{T}$ contains translations associated with the time reversal (*primed* translations). Shubnikov groups can be generalized to magnetic *superspace* groups in order to describe the symmetry of the incommensurate magnetic structures.

A paramagnetic crystal is described with a *grey* group, but when it orders magnetically, the time reversal symmetry is broken. Therefore, in the ordered phase, its Shubnikov group will not include the $\mathcal{I}'$ operation alone, but it can appear combined with any other transformation.

Let $\tilde{g}_p = \{\delta_p | \mathcal{R}_p | \vec{t}_p\}$ be a symmetry operation of the magnetic space group $\mathcal{M}$. A magnetic atom of the d_m-orbit located at site p, related to the site d_m by $\vec{r}_p = \tilde{g}_p \vec{r}_{d_\mathrm{m}}$, will have a magnetic moment $\vec{M}_p = \delta_p \det(\mathcal{R}_p)\mathcal{R}_p\vec{M}_{d_\mathrm{m}}$ where δ_p is -1 or $+1$ for *primed* or *unprimed* $\tilde{g}_p$ operations, respectively.

Therefore the magnetic structure factor defined in (146) considering N_{au} as the number of magnetic atoms at the asymmetric unit cell, indexed with d_m and located at $\vec{r}_{d_m}$, and its p-th equivalents by symmetry (d_m-orbit), reads:

$$\vec{M}_{\vec{q}\vec{K}} = p \sum_{d_m}^{N_{au}} F_{d_m}(q) \sum_{p}^{d_m-orbit} T_{pd_m}(\vec{q})\delta_p \det(\mathcal{R}_p)\mathcal{R}_p \vec{S}_{\vec{K}d_m} e^{i2\pi\vec{q}\cdot\vec{g}_p\vec{r}_{d_m}} \qquad (173)$$

where the d_m-index runs over the magnetic atoms associated to the propagation vector $\vec{K}$ (notice that the Fourier coefficient with index $\vec{K}$ contributes only to the K-satellite). The sum over p concerns the symmetry operators of the crystal that belong to the propagation vector group, and $F_{d_m}(q)$ is the magnetic form factor for the d_m-th atom. The constant p in front of the right-hand side of (173) was defined in section "Magnetic Scattering". In the case of a propagation vector different from zero, Eq. (144) tells us that two magnetic satellites will be observed around each Bragg nuclear reflection.

Describing Magnetic Structures

It is possible to describe magnetic structures using the magnetic content of the nuclear unit cell (magnetic moments or Fourier coefficients) and a propagation vector which describe the relation between the orientations of the magnetic moments located in equivalent magnetic atoms but in different nuclear unit cells. The propagation vector appears naturally in the Landau theory of the second-order phase transitions as the modulation that minimizes the magnetic energy in the ordered state. The *Irreducible Representations* (*irrep*) of the symmetry group of the propagation vector $\mathcal{G}^{\vec{K}} \in \mathcal{G}$, knowing that it transforms as a polar vector, are also employed to classify a magnetic structure. Usually, only one propagation vector pair ($\{\vec{K}, -\vec{K}\}$) survives as ground state describing the majority of the magnetic structures. However also multi-$\vec{K}$ structures and structures with one $\vec{K}$ and its harmonics are possible.

Based on the translation properties of a magnetic structure, it is possible to define *commensurate* and *incommensurate* structures when, respectively, the components of the $\vec{K} = (p\,q\,r)$ in the nuclear reciprocal lattice basis are rational numbers ($\{p, q, r\} \in \mathbb{Q}$) or not. In the first case, it is possible to find a magnetic unit cell that is a *multiple* of the nuclear one.

Let us now consider initially the simplest case in Eq. (143), that is, $\vec{K} = (0\,0\,0)$. In this case the Fourier coefficient $\vec{S}_{\vec{K}d_m}$ is a real vector equal to the magnetic moment at site d_m, $\vec{M}_{ld_m}$. This $\vec{K}$ means that the magnetic and the crystallographic unit cells are the same. All the ferromagnetic (FM) structures, and a large part of the antiferromagnetic (AF) ones, occur with this propagation vector. Notice that AF with $\vec{K} = 0$ is possible if there are more than one similar atom at the unit cell.

For a propagation vector, $\vec{K}$, on the Brillouin zone border, i.e., $\vec{K} = \frac{1}{2}\vec{H}$, the Fourier coefficient $\vec{S}_{\vec{K}d_m}$ is real and the magnetic moment, along different unit cells, change its sign, but not its modulus, according to:

$$\vec{M}_{ld_m} = (-1)^{\vec{R}_l \cdot \vec{K}} \vec{S}_{\vec{K}d_m}, \tag{174}$$

where $\vec{R}_l$ was defined in section "Nuclear Scattering" and $\vec{K}$ has the form $(\frac{1}{2}\,\frac{1}{2}\,\frac{1}{2})$, $(\frac{1}{2}\,0\,\frac{1}{2})$, etc.

Let us consider now that $(\{\vec{K}, -\vec{K}\})$ are at the interior of the Brillouin zone. If the Fourier coefficients $\vec{S}_{\vec{K}d_m}$ are real or its imaginary and real parts are parallel, then they can describe an *amplitude modulated* or *sinusoidal* structure:

$$\vec{M}_{ld_m} = \vec{S}_{\vec{K}d_m} \cos\{2\pi(\vec{K}\cdot\vec{R}_l + \phi_{\vec{K}d_m})\}. \tag{175}$$

Also with $(\{\vec{K}, -\vec{K}\})$ at the interior of the Brillouin zone, in the general case where the Fourier coefficients $\vec{S}_{\vec{K}d_m}$ are complex vectors, an *helicoidal* or a *cycloidal* structure can be described if both the real and imaginary parts of $\vec{S}_{\vec{K}d_m}$ are orthogonal. In the first case, the propagation vector is also orthogonal to both components, but in the second case, the propagation vector is contained in the plane formed by both components. Let us see how evolves the magnetization along different unit lattices l:

$$\vec{M}_{ld_m} = 2\mathrm{Re}\left[\vec{S}_{\vec{K}d_m}\right] \cos\{2\pi(\vec{K}\cdot\vec{R}_l + \phi_{\vec{K}d_m})\}$$
$$- 2\mathrm{Im}\left[\vec{S}_{\vec{K}d_m}\right] \sin\{2\pi(\vec{K}\cdot\vec{R}_l + \phi_{\vec{K}d_m})\} \tag{176}$$

If the real and imaginary part vectors have the same modulus, then the helicoid, or the cycloid has a circular envelope; if not the envelope is elliptical.

The experimental finding of the propagation vector is the first step to determine a magnetic structure. Usually this is done with powder neutron diffraction techniques, which allow a quick inspection of the 3D reciprocal space projected in 1D. The indexation of the magnetic peaks will produce the magnetic lattice and therefore the propagation vector. Once this is done, it is very convenient to use symmetry arguments to reduce the number of possible magnetic structures. Finally, for each possible model, it is necessary to try and fit the magnetic intensities measured with neutrons.[10] A complete monograph on these procedures can be found in [28].

At this point it is very important to remember that the experiments produce the scattered intensities, $I_{\vec{q}}$, which are real positive numbers, because they are proportional to the modulus square of the nuclear structure factor, or the magnetic interaction vector, which in absence of spatial inversion symmetry are complex

---

[10]The FullProf Suite has the necessary software to accomplish these tasks.

quantities. It means that the experiment does not give a complete information on the nuclear or magnetic structure, because different structures differing in a global phase factor will produce the same intensities pattern. One example of this is explained latter.

6 Neutron Powder Diffraction

The first commercial X-ray powder diffractometer was introduced by Philips company in 1947. In the 1950s and 1960s, the use of these apparatuses increased, mainly due to mineralogists and metallurgists, as a tool to perform primary structural studies (phase identification, defects, vacancies, etc.). In 1969 Rietveld [29] developed a method for the whole pattern analysis of the powder neutron diffraction data. Later, Cox, Young, Thomas, and others [30–32] firstly applied the method to analyze synchrotron and conventional X-ray data. It is important to mention here that the Rietveld method is a refinement process that requires an approximation to the correct structure in advance. Nowadays powder diffraction is a very popular technique that can be found in many laboratories to characterize any kind of crystalline matter and plays a central role in structural physics, chemistry, or materials science. It has produced important advances in the knowledge and understanding of high-T_c superconductors, zeolites, fullerenes, permanent magnets, etc. Extensive monographs on powder diffraction can be found in [33, 34]. Here, we will explain some particularities or advantages of the neutron powder diffraction that justify their use.

For a nuclear structure, which includes the determination of the positions of any light element (H, Li, C, N, O, S, etc.), or light elements in the presence of any heavy element (U, Pb, Bi, Ta, etc.), the use of neutron diffraction is the most suitable. For example, in [35] by using powder neutron diffraction in the multiferroic $(NH_4)_2FeCl_5·H_2O$, the anomaly observed in the specific heat at 79 K was assigned to an order-disorder phase transformation due to the hydrogen atoms in the NH_4*** cation, which produces a change of the symmetry from the space group $Pnma$ at room temperature to $P2_1/c$ for $T < 79$ K. As an example of the second case, high-resolution *in situ* powder neutron diffraction was employed to determine the changes in the UO_2 structure during oxidation allowing the determination and refinement of the structure of β-U_3O_7 [36] at 483 K. Also, if some next-neighbor elements are present (Mn, Fe, Co, Ni, etc.) simultaneously in the structure (e.g., Fe_nNi_m alloys), they can be easily distinguishable by using neutron diffraction.

In situ or in operando experiments, employing complex sample environments (very low and high temperatures mK to 3000 K, high pressure up to 30 GPa, high magnetic field up to 20 T), are easy and commonly performed. Due to the large penetration depth of neutron beams and their non-destructiveness, the determination of residual stress or textures measurements in bulk materials is another typical engineering application of neutron powder diffraction [37].

As was explained before, neutrons interact with matter via the strong nuclear force with the nuclei but also via the magnetic dipolar interaction with the unpaired electrons, both interactions with a comparable intensity. That is the reason why neutron powder diffraction is the primary technique to determine magnetic structures, as will be explained in detail in this section. In certain cases, it is necessary to go further using single-crystal techniques, with polarized or unpolarized neutrons, as it will be explained later in Sects. 8 and 9.

A *powder* is a polycrystalline sample formed by a huge number of microscopic single crystals randomly oriented. Under these conditions, there are thousands of crystallites[11] in diffraction conditions when the powder is illuminated with an appropriate wave. The three-dimensional reciprocal space of the powder formed by these crystallites is a set of concentric spherical shells, with radiuses equal to the modulus of each node at the reciprocal lattice (indexed with the vector $\vec{H}$). The intersection of these shells with the *Ewald Sphere* gives rise to the powder diffraction as a set of *Debye-Scherrer Cones*, and the intersections of these with a 2D detector give rise to the typical *Debye-Scherrer Rings*. The integration around these rings reduces the 2D data to a 1D diffraction pattern where the abscissae axis is the 2θ angle ($n\lambda = 2d\sin\theta$) and the ordinates axis corresponds to the diffracted intensity. Therefore, the 3D reciprocal lattice is condensed into a 1D in powder diffraction pattern. This leads to both accidental and exact peak overlap and sometimes complicates the determination of individual peak intensities. Also, if the powder is not randomly oriented and therefore it has any preferred orientation, or textures, then the powder leads to biased peak intensities.

In reactor-based neutron sources, where the diffractometers operate usually at constant wavelength, the scattering variable is the 2θ angle. However in spallation neutron sources, due to their pulsed nature, the diffractometers are based on *time of flight* (ToF) techniques being the scattering variable the incident neutron wavelength λ_i of a polychromatic beam.

The principle of ToF is based on the fact that neutrons have a mass (m) and therefore the de Broglie equation relates the wavelength of the neutron, λ_i, to its velocity, v_i according to $\lambda_i = h/mv_i$, where h is Planck's constant. Therefore, it is possible to determine the wavelength of the neutron by measuring the time t_i that a neutron employs to travel a fixed distance L. Combining these ideas with the Bragg equation yields:

$$\lambda_i = \frac{ht_i}{mL} = 2d_i \sin\theta_0, \qquad t_i = 505.55685(40)\, L\, d_i \sin\theta_0. \qquad (177)$$

Therefore the d-spacing is proportional to the time of flight (t_i), and the diffraction patters are collected at fixed θ_0. To measure the t_i, all we need is to measure the arrival time of the neutron at the detector, knowing that the starting time of the

---

[11]Notice that 1 cm^3 contains $\sim 10^9$ crystallites with size $\sim 10\,\mu$m or $\sim 10^{12}$ crystallites with size $\sim 1\,\mu$m.

neutron is fixed by the use of chopper, in reactor sources, or by using the pulses structure, in spallation sources. In Eq. (177) the units are μs, m, and Å for t_i, L, and d_i, respectively.

Some Ideas About the Rietveld Method

A powder diffraction pattern is a histogram of *counts* measured at the detector (y_i, at the ordinates axis) for each value of the scattering variable (x_i at the abscissae axis), which can be scattering angles ($2\theta_i$), time of flight (t_i), or energies E_i. The Rietveld method consists in minimizing the weighted squared difference between the observations y_i and the calculated y_{ical} against a set of different parameters ($\vec{\xi}$) related mainly with the atomic structure and microstructure of the sample and the optics of the instrument (monochromators, collimators, slits, pulse structure, etc.):

$$\chi^2 = \sum_{i=1}^{n} w_i \left\{ y_i - y_{ical}(\vec{\xi}) \right\}^2, \tag{178}$$

where w_i is the inverse of the variance (σ_i^2) of the observation y_i and y_{ical} represents the calculated *counts* as follows:

$$y_{ical} = \sum_{\phi} S_\phi \sum_{\vec{q}} I_{\phi,\vec{q}} \ \Omega(x_i - x_{\phi,\vec{q}}) + y_{bi}. \tag{179}$$

The index ϕ runs over all the different phases, nuclear or magnetic, present at the powder with scale S_ϕ; the vector $\vec{q}$ labels a nuclear or magnetic Bragg reflection ($\vec{q} = \vec{H}$ or $\vec{q} = \vec{H} \pm \vec{K}$ where $\vec{H}$ is a node of the reciprocal lattice and $\vec{K}$ is a magnetic propagation vector); $\Omega(x_i - x_{\phi,\vec{q}})$ is the peak-shape normalized function centered at the Bragg position $\vec{q}$ of phase ϕ, which includes the instrumental and sample microstructural effects; and $I_{\phi,\vec{q}}$ is the intensity of that Bragg reflection, which includes the modulus squared either of the nuclear structure factor $|N_{\vec{q}}|^2$ (Eq. 135) or the magnetic interaction vector $|\vec{M}_{\perp\vec{q}}|^2$ (Eq. 145), among other different corrections (asymmetry, transmission, preferred orientation, efficiencies, etc.). The background is represented by y_{bi}.

The Rietveld method needs an initial, nuclear and/or magnetic, structural model as input. Essentially, it enters at the calculations of the nuclear structure factor (atomic coordinates, occupancies, *Debye–Waller* factors, etc.) or at the magnetic interaction vector (vectorial magnetic atomic moments). Also the Bragg positions $x_{\phi,\vec{q}}$ are a function of the unit cell ($a, b, c, \alpha, \beta, \gamma$).

Selected Examples

Magnetic Structure of MnO

Up to year 1949, the only method to detect antiferromagnetism in a compound was via indirect experiments, i.e., by observing some anomalies in the specific heat and susceptibility. However, based on the pioneering paper of Halpern and Johnson [12], Shull and Smart, at the Oak Ridge National Laboratory, initiated a research program to investigate at room temperature the magnetic properties of powders of MnF_2, $MnSO_4$, MnO, and α-Fe_2O_3 with neutron scattering. They found no magnetic Bragg signals in MnF_2 and $MnSO_4$ but a short-range order in MnO and coherent magnetic scattering in α-Fe_2O_3, being the ordering temperatures 122 K and 950 K, respectively [38]. In the case of MnO, they also collected neutron scattering data at 80 K and observed the same Bragg nuclear peaks observed at 300 K and, additionally, also new strong peaks not indexed with the chemical unit cell but with a doubled cell in each crystallographic direction, in agreement with an AF longe-range order (LRO) with a propagation vector $\vec{K} = (\frac{1}{2} \frac{1}{2} \frac{1}{2})$. This was one of the first magnetic structures determined with neutron scattering.

Antiferromagnetic Structures with $\vec{K} = 0$ in $A_2FeX_5 \cdot H_2O$ (A=K, Rb, X=Cl, Br)

The family of low anisotropy antiferromagnets $A_2FeX_5 \cdot H_2O$, where (A=K, Rb, X=Cl, Br) was well known in the 1980s [39] due to the dimensionality crossover observed in its magnetic behavior [40] and also because its experimental accessibility to the *spin flop* phase transition at relatively low magnetic fields [41] (see Sect. 7). Moreover, the diamagnetically site diluted compounds $A_2Fe_{1-x}In_xX_5 \cdot H_2O$ were a good physical realization of the theoretical *ramdom field Ising model* [42] by its equivalence with the *diluted antiferromagnet in presence of a field* model, as proved by Fishman and Aharony [43]. Moreover, this AF family showed a significant *remanent magnetization*, with singular scaling properties, when cooled in presence of very low fields. All these facts increased the interest in these compounds [44–48].

However, the magnetic structure of these compounds was unknown till Palacio et al. collected neutron powder diffraction at the D1B and D2B instruments at the ILL with several main objectives: (1) to determine the magnetic super-exchange pathways by refining the O and H atomic positions with high precision at low temperature, (2) to determine and classify the magnetic structure, and (3) to measure the temperature evolution of the magnetic moment for each sub-lattice in order to compare it with the observed remanent magnetization to see if both were proportional [49].

The analysis of the data accomplished all these goals. The system crystallizes at the *Pnma* space group. The unit cell of $A_2FeX_5 \cdot D_2O$ contains four discrete $[FeX_5 \cdot H_2O]^{2-}$ octahedra connected by hydrogen bonds. In each octahedron three halogen atoms, the oxygen atom, and the iron atom are in special positions $4c$, and the other atoms are in general positions. The octahedra are arranged in planes perpendicular to the $\vec{b}$-axis. The hydrogen bonds connect octahedra related by the inversion operator to form chains along the $\vec{b}$-axis. The structure is schematized in Fig. 6.

The propagation vector found was $\vec{K} = (0\,0\,0)$ giving ac-ferromagnetic planes, antiferromagnetically coupled along the $\vec{b}$-axis with all the magnetic moments parallel to the $\vec{a}$-axis (easy axis), being the super-exchange pathway along the $\vec{b}$-axis (J_1) the most intense. The symmetry analysis helped to classify the magnetic structure as the *irrep* $\Gamma_{1u}(A_x\,0\,C_z)$ of *Pnma* which is compatible with an A_x, AF, mode along $\vec{a}$-axis and an C_z mode, also AF, along the $\vec{c}$-axis, even if accidentally the refinement did not produce any appreciable magnetic moment along the $\vec{c}$-axis. The magnetic structure had the symmetry of the $Pn'm'a'$ Shubnikov group. As we will see later, the mode C_z will play an important role in the magnetic structure of the ammonium derivative (see section "Modulated Magnetic Structures in Multiferroics"). Unexpectedly the magnetic moment found at low temperature was 3.9(1) μ_B and 4.06(5) μ_B for, respectively, K and Rb derivatives, which is around 20% lower than that expected for Fe^{3+}. This question will be analyzed later in Sect. 8. The temperature evolution of the sub-lattice magnetic moment was found to be not proportional to the observed remanent magnetization, excluding therefore the possibility of a *bulk effect* as the origin of that remanence.

The Problem of the *Global Phase* in $Ca_3Co_{2-x}Fe_xO_6$

The spin-chain compounds $Ca_3Co_{2-x}Fe_xO_6$ ($x = 0.2$ and 0.4) crystallize in the rhombohedral space group $R\bar{3}c$ where Co^{3+} ions are located at Wyckoff position $6a$ ($0\,0\,\frac{1}{4}$), with $S = 2$, and $6b$ ($0\,0\,0$), with $S = 0$, separated by Ca^{2+}. The Co^{3+} ions in the adjacent spin chains are shifted by 1/6 or 1/3 along the $\vec{c}$-axis. The compound with $x = 0$ was reported to exhibit a *partially disordered antiferromagnetic* (PDA) state, where 2/3 ferromagnetic Ising spin-chain orders with AF interchain interaction while the remaining 1/3 are disordered with zero net magnetization. After the Rietveld refinement of neutron powder diffraction, it was possible to locate the Fe^{3+} at the trigonal prims sites $6a$. At low temperatures additional magnetic Bragg peaks were indexed in the nuclear lattice with a propagation vector $\vec{K} = (0\,0\,1)$ which indicates that the centering translation $\vec{t} = (0\,0\,\frac{1}{2})$ was lost in the magnetic structure. The Γ_2 *irrep* of $\mathcal{G}^{\vec{K}}$, common in both sites, labels the magnetic structure and corresponds with ferromagnetic ac-planes, with the magnetic moment parallel to $\vec{c}$-axis, coupled AF along the $\vec{b}$-axis.

In [50] it is demonstrated that two different magnetic structures, (1) amplitude modulated structure with a propagation vector $\vec{K} = (0\,0\,1)$ in $R\bar{3}c$ and (2) a PDA structure, with a propagation vector $\vec{K} = (0\,0\,0)$ in $P\bar{1}$, are able to fit the same neutron diffraction pattern because the Fourier coefficients for each model only differ in a global phase factor that cannot be determined by the experiment (see section "Magnetic Scattering"). Both models assure that the total magnetic moment in the unit cell is zero and their Fourier coefficient is:

$$\vec{S}_{\vec{K}, pd_m} = m_z \, (0 \, 0 \, c) e^{-i2\pi \psi_{\vec{K}, pd_m}}$$

$$\psi_{(0\,0\,1), pd_m} - \psi_{(0\,0\,0), pd_m} = \frac{\pi}{6} \qquad (180)$$

where m_z is the refined magnetic moment and $c = 1$ or $c = 2\sqrt{3}$ for, respectively, $\vec{K} = (0\,0\,1)$ or $\vec{K} = (0\,0\,0)$ and the phase factor $\psi_{\vec{K}, pd_m}$ for each model differs in $\frac{\pi}{6}$ for all the atoms in the orbit of both sites $6a$ and $6b$. Both models correspond to a magnetic moment parallel to the $\vec{c}$-axis, which changes its modulus along the z coordinate of the unit cell following the sequences $\ldots \frac{\sqrt{3}}{2}, 0, -\frac{\sqrt{3}}{2}, \frac{\sqrt{3}}{2}, 0, -\frac{\sqrt{3}}{2} \ldots$ for $\vec{K} = (0\,0\,0)$ and $\ldots 1, -\frac{1}{2}, -\frac{1}{2}, 1, -\frac{1}{2}, -\frac{1}{2} \ldots$ for $\vec{K} = (0\,0\,1)$.

Incommensurate Spiral Magnetic Structure in LiFeAs$_2$O$_7$

Iron diarsenate was intensively studied because of its redox properties as cathode material in Li-ion batteries. However it was also interesting to understand how the Fe ions interacted, through magnetic exchange pathways involving two oxygens, originating a LRO at temperatures $\sim$35 K. For that reason, and to refine the crystallographic structure regarding the light elements, Rousse et al. [51] performed high-resolution powder neutron diffraction experiments at the instrument D1A at several temperatures and thermo-diffraction from 2 K to 50 K at the high flux D1B instrument, which is optimized for magnetic measurements with its good resolution at low angles.

The refined nuclear structure confirmed the space group as $C2$ and allowed to locate with high accuracy the Li atoms, which are in the free channels along the $(0\,0\,1)$ direction formed by each FeO$_6$ octahedron surrounded by six diarsenate As$_2$O$_7$. Unlike the phosphate analogue, the patterns collected in LiFeAs$_2$O$_7$ below 35 K showed new extra Bragg peaks indexed with a propagation vector $\vec{K} = (0.709\ 0\ 0.155)$, which is not a high symmetry point of the Brillouin zone, representing a incommensurate structure.

With these conditions, the symmetry analysis does not provide any constraint on the Fourier coefficients. The authors excluded the possibility of any spiral structure with axis parallel to neither any crystallographic axis nor the propagation vector. However, by using *simulating annealing* techniques, the authors found two possible orientations for the spiral axis, related by the twofold rotation axis, indistinguishable by the diffraction pattern, and a magnetic moment for Fe ion 4.28(3) μ_B. This magnetic structure is intermediate between a pure helical and pure cycloidal structure.

The magnetic spiral structure can be explained by considering a triangular topology of AF interactions between the Fe ions (J_1, J_2, and J_3), which produces some frustration responsible from the non collinearity of the spins, which fulfil the conditions $J_2 \approx 0.522J_1$ and $J_3 \approx 6.932J_1$.

Long Period Helical Structures

TM_3S_6 (T=Transition metal, M = Nb and Ta), an intercalated systems of $2H-MS_2$, are described by the chiral space group $P6_322$, and they *apparently* showed a variety of magnetic behaviours: paramagnetism for $TiNb_3S_6$, VNb_3S_6, $TiTa_3S_6$, and VTa_3S_6; AF for $FeNb_3S_6$, $CoNb_3S_6$, $NiNb_3S_6$, $CoTa_3S_6$, and $NiTa_3S_6$; and ferromagnetism for $CrNb_3S_6$, $MnNb_3S_6$, $CrTa_3S_6$, $MnTa_3S_6$, and $FeTa_3S_6$ [52–54].

The monoaxial helimagnet $CrNb_3S_6$ has received attention recently due to the observation of magnetic chiral soliton lattice directly probed by Lorentz transmission electron microscopy (LTEM) [55] and by the theoretical predictions about its magnetic phase diagrams [56–59]. However thermal neutron diffraction studies firstly reported that $CrNb_3S_6$ was a ferromagnet because the magnetic peaks were observed at the same diffraction angle as the nuclear ones [60].

The pitch angle of a chiral helimagnetic structure, mainly determined by the ratio of the exchange interaction (J) and the Dzyaloshinskii-Moriya interaction (D), is usually very small, and therefore the period can be hundreds of angstroms[12]. In some cases, the angular resolution of conventional neutron diffractometers is not high enough to separate nuclear Bragg from magnetic satellite peaks. As a consequence, some compounds with helical ordering may be easily misinterpreted as ferromagnets.

In this example, it is highlighted that ultra high-resolution neutron powder diffraction experiments made it possible to detect incommensurate satellite peaks related to very small propagation vectors in the compounds $CrNb_3S_6$ and $CrTa_3S_6$, previously reported as ferromagnets [61].

For that purpose, the experiments were performed at the SuperHRPD diffractometer, in the Materials and Life Science Facility (MLF), at 10 K and paramagnetic phase at 160 K and 180 K for, respectively, $CrNb_3S_6$ and $CrTa_3S_6$. Figure 1 shows the results around the nuclear $(0\,0\,4)$ and $(1\,0\,3)$ reflections for $CrNb_3S_6$ and $CrTa_3S_6$ obtained by the backward detector bank. The magnetic satellites for both compounds (Fig.1) were successfully separated and indexed by $\vec{K} = (0\,0\,\delta)$, with $\delta = 0.0206(2)$ for $CrNb_3S_6$, giving a helimagnetic period of 510 Å, slightly longer than that reported with LTEM. This discrepancy was assigned to the sample quality because a fewer defects in Cr sites may change the helical periodicity governed by D/J. The new observed peaks for $CrTa_3S_6$ were indexed with $\delta = 0.0538(1)$, which produced a period of 225 Å.

7   Single Crystal Neutron Diffraction

As it was explained before, the NPD techniques provide unique information about the magnetic periodicities, or magnetic correlations, in magnetically long-range ordered (LRO) materials. However the information given by the powder diffraction

---

[12]In helical structures, driven by *magnetic frustration*, the period is usually shorter.

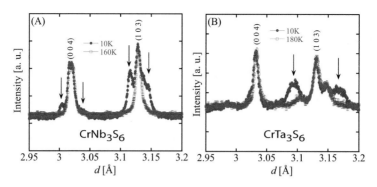

Fig. 1 Neutron powder diffractograms around the nuclear $(0\,0\,4)$ and $(1\,0\,3)$ reflections in (a) $CrNb_3S_6$ and (b) $CrTa_3S_6$. The vertical arrows indicate observed magnetic satellite peaks

techniques is sometimes incomplete because, basically, we have only access to a projection of the 3D reciprocal space onto one dimension (1D), i.e., only the modulus of the reciprocal lattice $\vec{H}$ node vector is measurable, and therefore different reflections $I_{\vec{H}}$ associated to $\vec{H}$ vectors with the same modulus will overlap at the same point in the powder diffractogram. However the diffraction in single crystals will permit the full exploration of the 3D reciprocal space and the individual measurement of each reflexion $I_{\vec{H}}$. Moreover, diffraction in single crystals will specially be suitable when an external axis needs to be defined (e.g., a magnetic field). In the next examples, these ideas will be illustrated.

Modulated Magnetic Structures in Multiferroics

The magneto-electric effect, i.e., the cross control of magnetization with electric fields and the electric polarization by magnetic fields, was predicted a long time ago by Pierre Curie [62]. Later on, in the 1960s and 1970s, materials with this effect attracted great interest. However, it was after the discovery of magneto-electric multiferroicity in the perovskite $TbMnO_3$ by Kimura et al. [63] when these materials became interesting from the point of view of its spintronic applications, because of its enhanced coupling. In their seminal paper, several magnetic models to explain the observed magneto electricity properties in $TbMnO_3$ were suggested, and, as a consequence, it was imperative to revisit the previous knowledge about its published magnetic structures.

Magnetic Structures of TbMnO₃ At room temperature the crystal structure of $TbMnO_3$ can be described with the space group *Pbnm*. The heat capacity shows three anomalies at $T_{Mn} \approx 41$ K, $T_L \approx 27$ K, and $T_{Tb} \approx 7$ K. Previous neutron scattering experiments seemed to indicate that the Mn sublattice ordered with a sinusoid modulated AF structure between T_{Mn} and T_L, with the magnetic moments parallel to the $\vec{b}$ axis, and an incommensurate propagation vector $\vec{K} = (0\ 0.295\ 0)$, which slightly decreased its value to $\vec{K} = (0\ 0.27\ 0)$. In this region, X-ray diffraction revealed that the modulated magnetic order was accompanied with a lattice modulation. In the region between T_L and T_{Tb}, the magnetic propagation vector locked at constant value $\vec{K} = (0\ 0.27\ 0)$, and below T_{Tb} the Tb ions ordered magnetically with a different propagation vector. It is below T_L when TbMnO$_3$ develops ferroelectric behavior with an electrical polarization parallel to the $\vec{c}$ axis. It is clear that the knowledge of the magnetic structure is not complete, and for that reason new neutron diffraction experiments in single crystals were needed. Initially Kajimoto and co-workers [64, 65], from experiments in single crystals, suggested the existence of a *complex non-collinear* magnetic order below T_L together with the incommensurate lattice modulation.

Moreover, the experiments done by Kenzelmann and co-workers [66] contributed to clarify the magnetic models at every region of the $T - B$ phase diagram. Their works agreed with the order observed previously for the region between T_{Mn} and T_L. They classified the magnetic structure under the Γ_3 *irrep* of $\mathcal{G}^{\vec{K}}$ without observing high-order Fourier components. However, in the region from T_L to T_{Tb}, they described the magnetic structure by using Γ_2 and Γ_3 *irreps* forming a elliptical spiral perpendicular to the $\vec{b}$ axis and with a significant magnetization along the $\vec{a}$ axis in the Tb sub-lattice. The simultaneous presence of both *irreps* indicates the breaking of inversion symmetry and therefore the presence of electrical polarization below T_L. Below T_{Tb} a new propagation vector, $\vec{K}_{Tb} = (0\ 0.425\ 0)$, associated with Tb ions appeared together with several strong odd high-order harmonics, indicating strong distortions at the sinusoidal modulated Tb magnetic structure.

Multiferroicity in the Molecular Compound $(NH_4)_2FeCl_5 \cdot H_2O$ The magnetic structures of the low anisotropy AF series A$_2$FeX$_5$·H$_2$O at zero magnetic field were discussed in section "Antiferromagnetic Structures with $\vec{K} = 0$ in A$_2$FeX$_5$·H$_2$O (A=K, Rb, X=Cl, Br)". However, the ammonium derivative (NH$_4$)$_2$FeCl$_5$·H$_2$O, which has been described as the first molecular multiferroic [67], develops a slightly different one [68, 69]. In fact, this compound shows two anomalies in the heat capacity curves at $T_{FE} = 6.9$ K and $T_N = 7.2$ K associated with different magnetic phases. Single-crystal neutron diffraction at zero magnetic field allowed to determine an incommensurate propagation vector $\vec{K} = (0\ 0\ 0.2288)$ at 2K. Moreover, none of the models described by an unique *irrep* of the $\mathcal{G}^{\vec{K}}$, with $\mathcal{G} = P2_1/c$, did reproduce the collected magnetic reflections. However the simultaneous use of 2 *irrep*, with a phase difference between the Fourier coefficients of the respective *irrep* of $\pi/2$, fitted it. It was a hint of a symmetry reduction from $2/m \rightarrow m$, also needed to explain the observed ferroelectricity in this compound. Within this model, the Fe magnetic moments develop two pairs of AF coupled cycloids contained in the a, c-plane and propagating along the $\vec{c}$-axis with a phase shift of 41.5°. The same super-exchange pathway along the $\vec{b}$-axis (J_1) that produces the AF coupling in the cycloids and the reduced magnetic moment of 3.8 μ_B measured for the Fe^{+3} ions were also observed in the K and Rb derivatives.

Spin-Flop Transition and Magnetic Structures in the Hybrid Multiferroic $(NH_4)_2FeCl_5 \cdot H_2O$

In the preceding paragraphs, the magnetic structures of the series of easy axis Heisenberg AF $A_2FeX_5 \cdot H_2O$, at zero magnetic field, determined from NPD experiments, have been shown. However, due to the low anisotropy of the ion Fe^{3+} in these compounds, its *Spin–Flop* transitions occur at relatively low fields ($\vec{B} \leq 5T$)) [39]. It facilitated the determination of the magnetic structures inside the *Spin–Flop* region, by using single-crystal neutron diffraction techniques with an applied magnetic field. In fact, the magnetic moment flopped from the $\vec{a}$ axis at zero field to the $\vec{c}$ axis for fields $\vec{B} \leq 3T$ keeping the same *irreps* Γ_{1u} of the space group *Pnma* [70].

In crystals of the ammonium derivative $(NH_4)_2FeCl_5 \cdot H_2O$, which shows a multiferroic behaviour, the evolution of its cycloidal magnetic structure was also studied, by applying a magnetic field parallel to its $\vec{a}$ easy axis at low temperatures ($T = 2K$), with the help of neutron diffraction [71]. These experiments showed the change of the propagation vector from $\vec{K}_0 = (0\ 0\ 0.23)$ to $\vec{K}_{3.5} = (0\ 0\ \frac{1}{4})$ and $\vec{K}_{5.5} = (0\ 0\ 0)$ for fields $B = 0$, $2.5 \leq B \leq 5$, and $5 < B$ T. The electrical polarization vectors related to the magnetic structures determined in these experiments are in agreement with the macroscopic measurements done in Ref. [67]. The knowledge of the magnetic structures, incommensurate and distorted commensurate cycloidal in the *ac*-plane and collinear in the $\vec{c}$ axis (corresponding to the spin–flop transition), leads to the conclusion that the mechanism responsible for multiferroicity in $(NH_4)_2FeCl_5 \cdot H_2O$ is changing from the so-called *spin-current* mechanism to the *spin-dependent p-d hybridization* mechanism. The magnetic structures in the three different magnetic field regimes are displayed in Fig. 2.

Magnetic Ordering by Dipolar Interactions in Single-Molecule Magnet Mn_{12}-Acetate

In the 1990s the molecular magnetic clusters like Mn_{12} or Fe_8 or Mn_4 and others were intensively studied because they show a single-molecule magnet (SMM) behaviour including quantum tunnelling of the magnetization vector [72–77]. All of them behave as super-paramagnetic entities with a large magnetic moment, blocking temperatures below 10 K and a large magnetic anisotropy. Here we will focus on the Mn_{12}-acetate cluster.

Mn_{12}-acetate is a cluster containing 12 Mn atoms arranged in such a way that four ions with Mn^{4+} with S=3/2 are in a core surrounded by 8 Mn^{3+} with S=2 coupled AF [78, 79] with those of the core giving a total spin equal to 10 and therefore a magnetic moment of $20\,\mu_B$. It crystallizes in the tetragonal space group $I\bar{4}$ with four molecules in the unit cell with a quasi-null intermolecular exchange due to weak van der Waals forces.

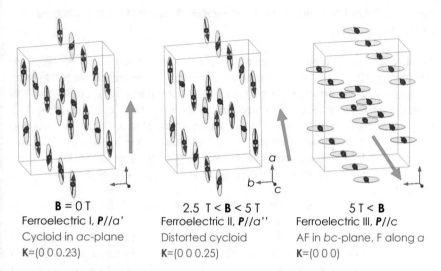

B = 0 T	**2.5 T < B < 5 T**	**5 T < B**
Ferroelectric I, **P**//a'	Ferroelectric II, **P**//a''	Ferroelectric III, **P**//c
Cycloid in ac-plane	Distorted cycloid	AF in bc-plane, F along a
K=(0 0 0.23)	**K**=(0 0 0.25)	**K**=(0 0 0)

Fig. 2 Magnetic structures of $(NH_4)_2FeCl_5 \cdot H_2O$ for the three regimes of the magnetic field applied parallel to the $\vec{a}$ axis at $T = 2$ K. The propagation vector $\vec{K}$ is changing from $\vec{K}_0 = (0\,0\,0.23)$ to $\vec{K}_{3.5} = (0\,0\,\frac{1}{4})$ and $\vec{K}_{5.5} = (0\,0\,0)$ for fields $0, 2.5 \leq B \leq 5$, and $5 < B$ T. The green arrow represents the direction of the electrical polarization vector

Mn$_{12}$-acetate was theoretically predicted at low temperatures ($\sim$0.5 K) to be a ferromagnetic LRO due to strong dipolar magnetic interactions between the high-spin molecules [80]. Therefore the verification of this prediction was a very nice experiment to be done with the use of neutron diffraction in a deuterated single crystal ($0.5 \times 0.5 \times 1$ mm^3) of Mn$_{12}$-acetate.

However, the spin reversal via quantum tunnelling in Mn$_{12}$-acetate becomes extremely slow at low temperatures ($\sim$2 months at T = 2 K), and, for the time scales 10^2–10^4 seconds of a typical neutron diffraction experiment, the spins are unable to attain thermal equilibrium below a blocking temperature $T_B \sim 3$ K, higher than the ordering temperature $T_C \sim 0.5$ K. This problem could be circumvented by the application of an orthogonal magnetic field to the anisotropy axis $\vec{c}$ of Mn$_{12}$-acetate ($\vec{B}_\perp$) that promotes quantum tunnelling of the molecular spins [81].

The experiment was performed at the instrument D10 of ILL, as proposed in the last paragraph, by using a cryomagnet to apply vertical magnetic fields up to 6 T and with a dilution fridge to go down to $\sim$30 mK. The examination of several appropriate magnetic reflections of a deuterated crystal of Mn$_{12}$-acetate, aligned with it $\vec{c}$ axis in the horizontal plane and its $(\bar{1}\,1\,0)$ axis parallel to $\vec{B}_\perp$, as a function of T and $\vec{B}_\perp$ was enough to determine its magnetic phase diagram, which shows a ferromagnetic phase transition line at $\sim$0.9 K at zero field in agreement with the predictions (Fig. 3). These results verified that for large enough $\vec{B}_\perp$, the molecule of Mn$_{12}$-acetate becomes diamagnetic because its ground state is something like $|\phi\rangle = |10, +10\rangle - |10, -10\rangle$ giving $S_T = 0$

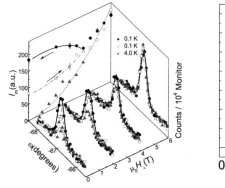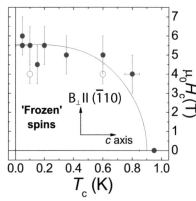

Fig. 3 Evolution of the reflection (2 2 0) for different magnetic fields $\vec{B}_\perp$ and temperatures (left) and the experimental magnetic phase diagram of Mn_{12}-acetate (right). From ref. [81]

Trapping the Different Magnetic Phases in the Chiral Molecular Magnet $[Cr(CN)_6][Mn(S)\text{-pnH}(H_2O)]\cdot(H_2O)$

The magnetic-field-induced effects can be anticipated to be very large if the structurally chiral compound orders magnetically (e.g., ferrimagnetism, ferromagnetism, weak ferromagnetism, conical, helical, etc.) below a certain temperature. Despite this interest, very few molecular candidates likely to exhibit magnetic and nuclear chirality have been synthesized until now.

Inoue and co-workers have recently developed a synthetic approach based on cyano-bridged bimetallic complexes by networking magnetic "bricks" that have more than three connections. In these compounds, cyanometallate magnetic anions $[M(CN)_n]^{m-}$ and a second magnetic metal ion M' coordinated to a chiral ligand alternate as building blocks to develop a chiral lattice [82, 83]. Here we analyze the magnetic structures of the chiral magnetic molecular compound of formula $[Cr(CN)_6][Mn(S)\text{-pnH}(H_2O)]\cdot(H_2O)$ where (S)-pn=(S)-1,2-diaminopropane was prepared following Inoue's strategy.

This compound presents three nuclear phases; each one is magnetic at low temperature with critical temperatures of $T_C = 38$ K, 39 K, and 73 K for phases I, II, and III, respectively, all remarkably high temperatures for a metal-organic molecular magnet. The three phases can be transformed into one another, differing mainly in the position of the chiral carbon and in the water content, by heating and cooling and applying vacuum conditions. Remarkably enough, all three nuclear phases (and therefore also the magnetic ones) are reversible and can be reached in the same experiment by using neutron LAUE diffraction [84].

For the three phases, the space group is $P2_12_12_1$. In phase I the Mn^{2+} ions are linked to four Cr^{3+} ions by four of the six cyanide groups in the $[Cr(CN)_6]^{3-}$ ion, thus forming a bimetallic patchwork that is arranged almost perpendicularly to the $\vec{c}$ axis to form a two-dimensional chiral network. The Mn^{2+} ions are also linked to one water molecule and to the amino acid ligand, which induces the chirality and completes the octahedron [85].

As particularities of the neutron LAUE diffraction method, we can mention that (1) it gives access, in a very fast way, to a large portion of the reciprocal space and therefore it allows some "dynamic" crystallographic studies and (2) LAUE methods use a quasi "white" neutron beam, which means that the number of neutrons arriving to the sample increases, and therefore it can be employed in very small crystals.

In a small crystal of $[Cr(CN)_6][Mn(S)\text{-pnH}(H_2O)]\cdot(H_2O)$ this technique determined that the propagation vector is $\vec{K} = (0\ 0\ 0)$ for the three magnetic phases, so the nuclear and the magnetic unit cells are the same. All three phases show two interpenetrating magnetic sublattices, one of Cr and another one of Mn AF coupled. Magnetic phases I and II can be described by the *irrep* Γ_4 of $P2_12_12_1$, which leaves the $\vec{a}$ axis ferromagnetic, while phase III is described by a Γ_3, leaving the $\vec{b}$ axis ferromagnetic. The magnitudes of the moments are slightly less than the spin-only moments for Cr^{3+} and Mn^{2+} but are in good agreement with similarly coordinated ions in the other Cr and Mn compounds.

8   Polarized Neutron Diffraction

The experiments described till now were performed with non-polarized neutron beams, so that the intensity diffracted was $I_{\vec{q}} = N_{\vec{q}}N_{\vec{q}}^* + \vec{M}_{\perp\vec{q}} \cdot \vec{M}_{\perp\vec{q}}^*$ ($\vec{q} = \vec{H} \pm \vec{K}$ where $\vec{H}$ is a node of the reciprocal lattice and $\vec{K}$ is the magnetic propagation vector that plays only in the magnetic part). In such expression, the nuclear structure factor, $N_{\vec{q}}$, and the magnetic interaction vector, $\vec{M}_{\perp\vec{q}}$, appear not coupled.

However, if we polarize the neutron beam, by using, for example, ^{3}He spin filters or Heusler-Alloy monochromators, the first Blume-Maleev equation (Eq. 171) states that, on top of the modulus square of $N_{\vec{q}}$ and $\vec{M}_{\perp\vec{q}}$, the scattered intensity, $I_{\vec{q}}$, incorporates the initial polarization vector of the neutron beam, $\vec{P}_i$, multiplying the *nuclear-magnetic interference vector* $(N_{\vec{q}}\vec{M}_{\perp\vec{q}}^* + N_{\vec{q}}^*\vec{M}_{\perp\vec{q}})$ and the *chiral vector* $i(\vec{M}_{\perp\vec{q}}^* \times \vec{M}_{\perp\vec{q}})$, which is non-null for non-collinear magnets.

In the following examples, the initial polarization $\vec{P}_i$ of the neutron beam can be changed between two states, up ($\uparrow$) and down ($\downarrow$), or parallel and antiparallel, and the detector counts for all the scattered neutrons for each initial polarization state up, $I_{\vec{q}}^{\uparrow}$, or down, $I_{\vec{q}}^{\downarrow}$. These scattered intensities include both *non spin-flip* and *spin-flip* processes (e.g., $I_{\vec{q}}^{\uparrow} = I_{\vec{q}}^{\uparrow\uparrow} + I_{\vec{q}}^{\uparrow\downarrow}$ and $I_{\vec{q}}^{\downarrow} = I_{\vec{q}}^{\downarrow\downarrow} + I_{\vec{q}}^{\downarrow\uparrow}$). These are the ideas of the linear neutron polarimetry initially developed experimentally by Moon, Riste, and Koheler [86].

The Fourier transform of the magnetic structure factors, $\vec{M}_{\vec{q}}$, which can be experimentally determined by using polarized neutron diffraction (PND) is directly related to the magnetization density in the unit cell. In these experiments a key point is the maximization of the induced magnetization in the sample along the magnetic field axis, which is parallel to the initial beam polarization. For that reason, in most cases, high magnetic fields and low temperatures will be used for paramagnetic samples. Also, in these experiments, the previous knowledge of the crystalline structure (i.e., the nuclear structure factors $N_{\vec{q}}$) obtained from a non-polarized neutron experiment is needed.

While the nuclear contribution to the diffraction intensities is independent of the polarization of the neutron beam, the magnetic contribution varies depending on whether the incident beam is polarized parallel ($\uparrow$) or anti-parallel ($\downarrow$) to the applied field. In a PND experiment, the measured quantity is the *flipping ratio*, $R_{\vec{q}} = I_{\vec{q}}^{\uparrow}/I_{\vec{q}}^{\downarrow}$, where $I_{\vec{q}}^{\uparrow}$ and $I_{\vec{q}}^{\downarrow}$ were defined before and they adopt the following expression when the polarization of the neutron beam is parallel to a magnetic field applied along the $\vec{z}$ axis:

$$I_{\vec{q}}^{\uparrow(\downarrow)} = |N_{\vec{q}}|^2 \pm (N_{\vec{q}} \vec{M}_{\perp \vec{q}}^{z*} + N_{\vec{q}}^* \vec{M}_{\perp \vec{q}}^{z}) + |\vec{M}_{\perp \vec{q}}|^2 \pm i(\vec{M}_{\perp \vec{q}}^{x*} \vec{M}_{\perp \vec{q}}^{y} \mp \vec{M}_{\perp \vec{q}}^{x} \vec{M}_{\perp \vec{q}}^{y*}). \quad (181)$$

where the upper or down sign in $\pm$ or $\mp$ is taken for up ($\uparrow$) or down ($\downarrow$) initial polarization, respectively. Assuming that the magnetization is parallel (or quasi-parallel) to the polarization of the beam $\vec{P}_i$ or the interaction vector is real ($\vec{M}_{\perp \vec{q}} = \vec{M}_{\perp \vec{q}}^*$), the *flipping ratio* $R_{\vec{q}}$ is given by

$$R_{\vec{q}} = \frac{I_{\vec{q}}^{\uparrow}}{I_{\vec{q}}^{\downarrow}} = \frac{|N_{\vec{q}}|^2 + (N_{\vec{q}} \vec{M}_{\perp \vec{q}}^{z*} + N_{\vec{q}}^* \vec{M}_{\perp \vec{q}}^{z}) + |\vec{M}_{\perp \vec{q}}|^2}{|N_{\vec{q}}|^2 - (N_{\vec{q}} \vec{M}_{\perp \vec{q}}^{z*} + N_{\vec{q}}^* \vec{M}_{\perp \vec{q}}^{z}) + |\vec{M}_{\perp \vec{q}}|^2}, \quad (182)$$

where $\vec{q}$ is the scattering vector of the Bragg reflection[13] and $N_{\vec{q}}$ and $\vec{M}_{\perp \vec{q}}$ are the nuclear structure factor and magnetic interaction vector, respectively. When the crystal structure is non-centrosymmetric, both $N_{\vec{q}}$ and $\vec{M}_{\perp \vec{q}}$ are complex, and the magnetic structure factors $\vec{M}_{\vec{q}}$ cannot be directly extracted from Eq. 182 by using *direct methods* (Fourier transform, maximum entropy). Instead, the spin density can be modeled to provide a best fit to the flipping ratios by using *indirect methods*. Two approaches have been utilized to model the spin-density distribution: (a) magnetic wave function approach [87] and (b) multipolar-expansion approach [88].

Magnetic Wave Function Approach In this approach the magnetization density $m(\vec{r}) = |\langle \vec{\mathbb{M}}(\vec{r}) \rangle|$ coming from each ion can be described in terms of an orbital $|\psi_i\rangle$ centered in that ion

$$m(\vec{r}) = \sum_{i \in atoms} S_i \, |\langle \vec{r} \mid \psi_i \rangle|^2 \text{and} |\psi_i\rangle = \sum_j \alpha_{ij} |\varphi_j\rangle. \quad (183)$$

---

[13]If these experiments are done in a magnetically ordered crystal, then the nuclear and magnetic lattices need to be the same.

The S_i is the atomic spin populations, and the $|\psi_i\rangle$ are the atomic magnetic wave functions, which are linear combinations of standard atomic orbitals, and the coefficients α_{ij} fulfill the condition $\sum_j |\alpha_{ij}|^2 = 1$. The atomic spin populations, S_i, the coefficients α_{ij}, and the radial exponents of the Slater atomic orbitals are the parameters of the model.

Multipolar Expansion Approach In this approach the spin density is partitioned into separate atomic contributions which are expanded in the basis of real spherical harmonics $\mathcal{Y}_l^m$, which are also referred to as multipoles [88]

$$m(\vec{r}) = \sum_{i \in atoms} \sum_l \mathcal{R}_i^l(|\vec{r} - \vec{r}_i|) \sum_{m=-l}^{m=l} p_i^{lm} \mathcal{Y}_l^m(\theta_i, \varphi_i), \tag{184}$$

where p_i^{lm} is the population coefficients of the real spherical harmonics $\mathcal{Y}_l^m(\theta_i, \varphi_i)$ and $\mathcal{R}_i^l$ is the Slater radial functions defined as:

$$\mathcal{R}_i^l(r) = \frac{\zeta_l^{n_l+3}}{(n_l + 2)!} r^{n_l} e^{-\zeta_l r}. \tag{185}$$

The ideas described here were applied to understand the magnetic interaction mechanisms in the purely organic magnet family p-X-C_6F_4CNSSN and in the low anisotropy AF family $A_2FeX_5 \cdot H_2O$.

Spin Density Experiments in p-O_2N-C_6F_4CNSSN and $A_2FeX_5 \cdot H_2O$ (A=K, Rb, X=Cl, Br)

Magnetic Interaction Mechanism in $p-X-C_6F_4CNSSN$ ($X = O_2N$, CN, Br, I) To enhance the magnetic interaction, and therefore the temperature for magnetic LRO, sulfur-based organic magnets were proposed [89] due to the larger radially extension of the sulfur atomic orbitals compared with those of the oxygen in p-NPNN ($T_C = 0.65$ K) [90]. This strategy proved to be very successful, and it is today the phase β of p-NC-C_6F_4CNSSN, which undergoes a phase transition to a canted AF state below 36 K, being the metal-free compound with the highest reported magnetic LRO temperature [91].

The spin density studies in a crystal of p-O_2N-C_6F_4CNSSN, obtained with polarized neutron diffraction and supported by *ab initio* density functional theory (DFT) calculations, were crucial to provide a fundamental understanding of the magnetic behavior of the dithiadiazolyl radical family and the magnetism of p-orbitals, which can be employed to design new metal-free sulfur-based magnets [92].

The polarized neutron diffraction studies revealed that the spin density is almost entirely located on the sulfur and nitrogen atoms of the dithiadiazolyl ring in a p_z orbital-type distribution with the $\hat{z}$ axis perpendicular to the ring and a small

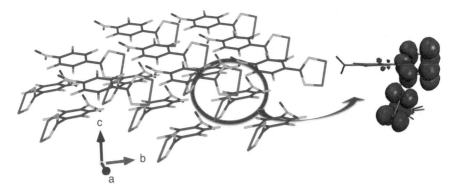

Fig. 4 Relative space localization of two molecules of p-O_2N-C_6F_4CNSSN. The magnetic interaction between these two molecules is supposed to be ferromagnetic due to the small magnetic orbital overlapping

negative spin density on the C atom arising out of spin polarization. These studies also show that the spin density is displaced away from both atoms and bonds within the dithiadiazolyl ring, in agreement with the anti-bonding nature of the SOMO determined by *ab initio* DFT calculations. Both the experimental and theoretical studies also indicate small negative spin densities in the plane of the heterocyclic ring.

This exhaustive and complete study of the spin density in the prototype system p-O_2N-C_6F_4CNSSN helped to better understand the magnetic interaction mechanism, and therefore the magnetic behavior, in all the members of the series p-X-C_6F_4CNSSN (X=O_2N, CN, Br, I), for which the packing is controlled by the substituent in the *para* position. This is due to the fact that the relative positions in the space of the interacting magnetic molecular orbitals, which have been determined to be located in the ring of the ·CNSSN, control the nature and strength of the magnetic interactions.

Figure 4 shows the relative positions of the magnetic orbitals of the rings of ·CNSSN from two different molecules of p-O_2N-C_6F_4CNSSN. Because of the quasi-null overlapping of the magnetic orbitals, the magnetic interactions between those molecules must be ferromagnetic.

Enhancement of the Magnetic Interaction Constant by Spin Delocalization in the Family $A_2FeX_5 \cdot H_2O$ ($A = K$, Rb, $X = Cl$, Br) As mentioned in the preceding sections, these series of low anisotropy AF showed magnetic LRO at relatively high temperatures ($T_C \geq 10$ K) [39] compared with other hydrated salts of transition metal ions $Cs_2FeCl_5 \cdot 4H_2O$ ($T_C = 0.185$ K [93]) or $Cs_2MnCl_4 \cdot 2H_2O$ ($T_C = 1.8$ K [94]) with similar coordination and with similar, or even shorter, magnetic super-exchange pathways. In $A_2FeX_5 \cdot H_2O$ there are five different possible magnetic interactions which are propagated through two super-exchange pathways, which consist of double bridges of the sort $Fe - X \cdots X - Fe$ or $Fe - O - H \cdots X - Fe$. These super-exchange pathways, which will be explained in detail later in this chapter, are surprisingly effective in transmitting the magnetic interaction that results in relatively high transition temperatures.

The knowledge of the spin density distribution in the $A_2FeX_5 \cdot H_2O$ series allows the determination of whether the relatively high magnetic transition temperatures are due to an important spin delocalization from the iron ion toward the ligand atoms. The existence of such a spin delocalization, which would strengthen the interaction between magnetic orbitals localized in different octahedra, is also supported by a reduced magnetic moment observed with neutron powder diffraction experiments at the Fe^{3+} site (3.9(1) μ_B and 4.06(5)μ_B for $K_2FeCl_5 \cdot H_2O$ and $Rb_2FeCl_5 \cdot H_2O$, respectively), far below the $5\mu_B$ expected for the saturated magnetic moment of the Fe^{3+} [95].

Polarized neutron diffraction experiments, done at the D23 instrument from the ILL, on $K_2FeCl_5 \cdot H_2O$ and $Rb_2FeBr_5 \cdot H_2O$ were performed to corroborate the hypothesis [96]. In this case, the *flipping ratios* were analyzed with the maximum entropy method [97] and with a multipolar expansion of the spin magnetization, and it was demonstrated that the entity $[FeX_5 \cdot H_2O]$ is much more effective at delocalizing the spin, $\sim 20\%$, than other $3d$ transition metal complexes aqua or halide coordinated [98–107]. As for iron compounds, in a Fe^{2+} compound, FeF_2, there is a 10% spin delocalization from the iron ion [108], and in a $[FeCl_4]^-$ complex, the spin population of the chlorine atoms is around 0.15 μ_B [109]. Also Campo and co-workers [96] showed that the spin delocalization is larger for Br than for Cl atoms, which explains why the transition temperatures in the bromide compounds are higher than in their chloride analogs. It was found also that, inside the distorted octahedra $[FeX_5 \cdot H_2O]$, the shorter the distance between the halide (X) and the Fe ion, the higher the spin delocalization at the halide, except in the case of the halide forming part of the pathway $Fe - O - H \cdots X - Fe$, which define the most intense magnetic interaction constant, J_1, in these compounds.

This spin delocalization reflects the fact that the magnetic molecular orbitals are spread over the ligand atoms, thus enhancing the magnetic interactions through a super-exchange magnetic interaction mechanism. This enhancement of the magnetic interactions is the cause of the surprisingly high transition temperatures in this series of compounds.

Determination of Atomic Site Susceptibility Tensors

In the last section, it was assumed that the magnetic field-induced magnetization of the sample ($\vec{M}$) and the applied magnetic field ($\vec{B}$) were parallel, i.e., the magnetic anisotropy was negligible and therefore $\vec{M} = \chi\vec{B}$. However this condition is not every time satisfied. In these cases it is necessary to take into account the anisotropic susceptibility for each magnetic crystallographic site, the bulk magnetization being the sum vector of all of these. The formalism for calculating the magnetic scattering in these cases was developed by Gukasov and Brown in Ref. [110]. For example, for one magnetic ion located on site d_m, the local induced magnetization will be

$\vec{M}_{d_m} = \tilde{\chi}^{d_m} \vec{B}$, where $\tilde{\chi}^{d_m}$ is the symmetric local susceptibility tensor for site d_m with the number of independent components given by the local symmetry of the site d_m. An equivalent magnetic atom located at site p related to the site d_m by $\vec{r}_p = \tilde{g}_p \vec{r}_{d_m}$, where the magnetic group symmetry operator is $\tilde{g}_p = \{\delta_p | \mathcal{R}_p | \vec{t}_p\}$, would have magnetization $\vec{M}_p = \tilde{\chi}^p \vec{B} = \mathcal{R}_p \tilde{\chi}^{d_m} \mathcal{R}_p^{-1} \vec{B}$. Therefore, the magnetization distribution at the unit cell, considering only one magnetic atom at $\vec{r}_{d_m}$ and its equivalents by symmetry (d_m-orbit), reads

$$\vec{M}(\vec{r}) = \sum_p^{d_m-orbit} m_{d_m}(\vec{r} - \tilde{g}_p \vec{r}_{d_m}) \mathcal{R}_p \tilde{\chi}^{d_m} \mathcal{R}_p^{-1} \vec{B}, \tag{186}$$

where $m_{d_m}(\vec{r})$ is the magnetization distribution (assumed spherical) around the atom d_m and p runs over the symmetry operators of the space group of the crystal that generates the atoms of the d_m-orbit. The expression (186) can be easily generalized for several magnetic sites in the asymmetric unit cell. Therefore the magnetic structure factor becomes

$$\vec{M}_{\vec{q}} = \sum_{d_m}^{N_{au}} F_{d_m}(q) \sum_p^{d_m-orbit} \mathcal{R}_p \tilde{\chi}^{d_m} \mathcal{R}_p^{-1} \vec{B} \, e^{i2\pi \vec{q} \cdot \tilde{g}_p \vec{r}_{d_m}}, \tag{187}$$

where the d_m index runs over the magnetic atoms and the $F_{d_m}(q)$ is the magnetic form factor for the d_m-th atom. At most, six free parameters define the local susceptibility tensors $\tilde{\chi}^{d_m}$. For some cases, when the magnetic atom occupies an special Wyckoff position, it is possible to add symmetry constraints, reducing the number of parameters, by imposing $\mathcal{R}_p \tilde{\chi}^{d_m} \mathcal{R}_p^{-1} \equiv \tilde{\chi}^{d_m}$ for those operations of the space group $\tilde{g}_p$ for which $\tilde{g}_p \vec{r}_{d_m} = \vec{r}_{d_m}$. The relationship between the bulk and site susceptibilities tensor is $\tilde{\chi}^{bulk} = \sum_{d_m} \sum_p \mathcal{R}_p \tilde{\chi}^{p d_m}$. The bulk susceptibility tensor transforms with the full symmetry of the crystallographic class.

The magnetic structure factor expressed in Eq. 187, parametrized with the components of the site susceptibility tensors, can be used to model the *flipping ratios* (Eq.182) which can be measured with polarized neutron diffraction experiments. These ideas were used to reinvestigate the magnetic structures of $Nd_{3-x}S_4$ and $U_3Al_2Si_3$ [110, 111].

The compound $Nd_{3-x}S_4$ crystallizes in the cubic space group $I\bar{4}3d$ in which the twelve Nd atoms occupy the Wyckoff position $12a$ with a local tetragonal symmetry $\bar{4}$ (3 sets of 4 Nd atoms with tetragonal axis parallel to each of the 3 main axis). Unpolarized neutron diffraction experiments with magnetic field parallel to one of the main axis demonstrated that there is a large difference between the magnetic moments found for the Nd located in the set with its tetragonal axis parallel to the field and the other two sets of $12a$. However, for an arbitrary orientation of the crystal with the magnetic field, no symmetry constraints on each Nd magnetic moment could be applied. The approach of the site susceptibility tensor, for this particular space group and Wyckoff position, implies that only two parameters are needed regardless of the field direction $\tilde{\chi}_{11}^{12a} = \tilde{\chi}_{22}^{12a} \neq \tilde{\chi}_{33}^{12a}$ and $\tilde{\chi}_{12}^{12a} = \tilde{\chi}_{23}^{12a} = \tilde{\chi}_{13}^{12a} = 0$. By measuring *flipping ratios*, it was possible to determine these two parameters: $\tilde{\chi}_{11}^{12a} = 1.45(5)\mu_B$ and $\tilde{\chi}_{33}^{12a} = 0.55(5)\mu_B$. These parameters represent an oblate susceptibility tensor for each one of the three sets of Nd in $12a$, each having its short axis parallel to the main axis on which it lies. The non-collinear distribution of Nd magnetic moments is not incompatible, however, with the bulk cubic symmetry of the crystal, since the vector sum of all magnetic moments remains parallel to the field and invariant with respect to the field orientation.

$U_3Al_2Si_3$ crystallizes in the tetragonal body-centered space group $I\bar{4}$, and the U atoms are sited in three different Wyckoff positions: U1 and U2 in $2a$, with tetragonal local symmetry, and U3 in $8c$, with triclinic symmetry. The system orders magnetically below $T_C = 36$ K, with propagation vector $\vec{K} = 0$, and previous neutron powder diffraction experiments seem to indicate, with good agreement factors, a collinear AF order for U1 and U2 sub-lattices and ferromagnetic order for the U3 one. However, this model, in which U1 and U2 are in similar Wyckoff positions, produces very different magnetic moment, for each of these uranium atoms. Nevertheless, the local susceptibility approach shows clearly that, in the paramagnetic region, there is a strong elongation of the magnetic susceptibility tensors of U3 (prolate) along one of the (110)-type axes, which increases as the temperature decreases. In the ordered phase, a non-collinear structure emerges with magnetic moments pointing along the same (110)-type axes, giving a net ferromagnetic moment in the ab-plane. On the contrary, the U1 and U2 atoms show a rather low local susceptibility, suggesting that these atoms have small moments, characterized by nearly spherical magnetic ellipsoids, indicating low local anisotropy. The absence of ordering in the U1 and U2 sub-lattices at low temperatures must be attributed partly to the low moment and partly to the fact that the first magnetic neighbor of either U1 or U2 atoms is far away.

9 Spherical Neutron Polarimetry

In the previous section, the polarization of the incoming and scattered neutron beam, respectively, $\vec{P}_i$ and $\vec{P}_f$, was fixed to be parallel or antiparallel to a given direction ($\hat{z}$) following the experimental method developed by Moon et al. [86]. However, this method cannot distinguish a rotation of the neutron polarization vector from a change in its modulus. This is the reason why the method of three-dimensional analysis of polarization (or spherical neutron polarimetry, SNP) was developed. Basically, the idea is to implement experimentally the second Blume-Maleev equation (172) [112–114].

However, it was in 1989 when Tasset and his colleagues developed at the ILL the instrument called CRYOPAD, which allowed the three-dimensional analysis of the polarization at large scattering angles [115, 116]. In CRYOPAD the neutron beam can be initially polarized, $\vec{P}_i$, in the three orthogonal directions $\hat{x}$, $\hat{y}$, and $\hat{z}$, (righ-hand coordinate system). The scatterer (sample) is placed in zero magnetic field, and the three components of the polarization vector of the scattered neutrons, $\vec{P}_f$, are measured for each initial $\vec{P}_i$ and for a given scattering vector $\vec{q}$. This conforms the polarization tensor $\tilde{P}_{fi}$, where the indices f and i refer to the *final* and *initial* polarizations. For example, it is possible to measure the diagonal components $\tilde{P}_{fi}^{xx}$, $\tilde{P}_{fi}^{yy}$, and $\tilde{P}_{fi}^{zz}$, the same determined in the linear polarization analysis, and the non-diagonal components ($\tilde{P}_{fi}^{yx}$, etc.), related to a polarization rotation in the scattering event. The SNP, sometimes also called zero-field polarimetry, allows the determination of complex magnetic structures. Domain structures and the effect of external influences can also be studied [117–121]. Another advantage of SNP (also valid for linear neutron polarimetry) is that the measured data, to a good approximation, are not affected by extinction effects. In the next subsections, several applications of this technique will be highlighted.

Proving the Magneto-Electric Coupling in the Molecular Multiferroic (NH₄)₂FeCl₅·H₂O

As explained in the previous sections, the compound $(NH_4)_2FeCl_5 \cdot H_2O$ has been described as a molecular *improper* multiferroic [35, 67, 69, 71, 122]. However, from the macroscopic characterization and the unpolarized neutron diffraction experiments, it was not possible to elucidate the nature of the magnetic phase observed between $T_{FE} = 6.9\,K$ and $T_N = 7.2\,K$. In fact, it was supposed, by analogy with the archetypical compound TbMnO₃ where two order parameters condense successively at T_{FE} and T_N [66], that it should correspond with a sinusoidal modulated magnetic phase, either parallel to the $\vec{a}$ or to the $\vec{c}$ axes, respectively, labelled with the irreps Γ_1 or Γ_2 of the little group of the propagation vector $\vec{K}_0 = (0\,0\,0.23)$ in $P2_1/c$. This would correspond to a transition on warming from a cycloidal ($T < T_{FE}$) to a collinear sinusoidal AF structure ($T_{FE} < T < T_N$), the latter being compatible with the absence of electric polarization. Furthermore, the symmetry at the multiferroic phase below T_{FE} was unclear.

Rodríguez-Velamazán [123] did SNP experiments in $(NH_4)_2FeCl_5 \cdot H_2O$ under an applied electric field ($\vec{E}$), to explore the possibility of controlling the *hand* of the cycloidal magnetic domains and, therefore, the electric polarities. These experiments allowed the determination of the absolute magnetic configuration and domain population of this system, under different applied external fields $\vec{E}$. In fact, for this compound, the term $\tilde{P}_{fi}^{yx}$ gave the information about the domain population. The experiment showed that the system is completely switchable, which represents a direct evidence of its magneto-electric coupling. Moreover, SNP data were used to refine the previously proposed magnetic structure of the ground state ($T < T_{FE}$), obtaining a more accurate magnetic structure model. In fact the group $P11'(\alpha\beta\gamma)0s$ is the only compatible symmetry with the measured polarization tensor $\tilde{P}_{fi}$. Finally, the existence of a collinear sinusoidal magnetic phase with moments parallel to the $\vec{a}$ axis in the temperature region $T_{FE} < T < T_N$ was confirmed by the sign of the term $\tilde{P}_{fi}^{yy}$, which is negative.

Elucidating the Magnetic Order in GdB$_4$

The family of compounds RB$_4$ attracted a huge interest in the past for their interesting physical properties (LRO, mixed valence, heavy fermion behavior, etc.). For example, the presence of magneto-electric coupling in GdB$_4$ was proposed, assuming that the inversion center was combined with the time reversal operation and a zero propagation vector $\vec{K} = 0$ with AF behavior [124, 125]. Also, in resonant X-ray scattering experiments at the Gd-L_3 edge, some interference between magnetic and anisotropic charge contributions was observed, which was compatible with several magnetic models for GdB$_4$ [126]. Blanco et al., in an elegant and easy way, shed light on these questions by using SNP [127]. In the next paragraphs, this instructive example will be summarized.

This compound crystallizes in the tetragonal space group $P4/mbm$, in which 4 Gd ions occupy the Wyckoff position $4g$ with point symmetry $m2m$, whereas the 16 B atoms are in $4e$, $8j$, and $4h$. From macroscopic measurements (heat capacity, magnetic susceptibility, etc.), it was well known that the system ordered AF below $T_N < 42$ K and that all the magnetic moments resided on the ab-plane. The resonant X-ray experiments did not discern between collinear and non-collinear arrangements of the magnetic moments [128], but electrical resistance experiments seemed to indicate some non-collinear model [129]. Despite all these experimental findings, the magnetic structure was not determined with neutron scattering because of the strong absorption of both B and Gd natural especies. However, Blanco et al. grew a ^{11}B-enriched single crystal to undertake SNP experiments, at very short wavelength in order to minimize the absorption effects and to find out the magnetic structure unambiguously [127]. At low temperature, they found a propagation vector $\vec{K} = 0$ and measured the polarization tensors $\tilde{P}$ for different set of reflections ($h0l$ and $hk0$), finding that the off-diagonal terms were null within the experimental error. Another significant feature was the non-observation of scattering for the lines $hh0$ and $h\bar{h}0$ (i.e., $\tilde{P}_{fi}^{xx} \sim \tilde{P}_{fi}^{yy} \sim \tilde{P}_{fi}^{zz} \sim 1.0$), which was only compatible with a non-collinear model of AF pairs parallel to the [110] and [1$\bar{1}$0] directions. The complete analysis of all the polarization tensors, using the magnetic symmetry models derived from the space group $P4/mbm$ with $\vec{K} = 0$, gave unequivocally the Shubnikov group $P4/m'b'm'$, with a Gd magnetic moment equal to 7.14(17) μ_B, as model for the magnetic order in GdB$_4$.

10 Small-Angle Neutron Scattering in Magnetism

Small-angle neutron scattering (SANS) is a widely employed technique for microstructure determinations in several disciplines, such as materials science, physics, chemistry, biology, etc. The reason is that SANS is one of the few experimental methods available to probe mesoscopic length-scales in the real space (from 1 to 300 nm) and in bulk materials. In addition, the spin of the neutron confers to the SANS the possibility to study magnetic phenomena and materials in the nano- and mesoscopic scales with high sensitivity. SANS is often utilized complementarily to other magnetic image and surface techniques like Lorentz transmission electron microscopy, magnetic or atomic force microscopy, Kerr effect magnetization, scanning tunneling microscopy with spin polarization, etc.

Magnetic SANS techniques are employed to study different magnetic materials: magnetic nanoparticles, magnetic steels, hard and soft magnets, magnetic oxides, ferrofluids, metallic alloys, magnetic skyrmions, chiral solitons and helical order in non centro-symmetric magnets, vortex in type II superconductors, etc. In a recent review [130], detailed information on the applications of magnetic SANS in magnetic materials can be found. In this section only several examples, related to magnetic textures, will be explained in more detail. Additionally, in the previous chapter, other examples related to the use of SANS to study magnetic materials have been described.

The expressions governing the SANS cross sections for unpolarized or polarized neutrons are basically the Blume-Malayev equations (168) and (169) introduced in the previous sections and adapted to the common SANS configurations ($\vec{B} \parallel \vec{k}_i$ or $\vec{B} \perp \vec{k}_i$). In Fig. 5 these configurations are schematized for typical experiments on skyrmion lattice studies.

In addition, the discrete nature of the matter has no relevance for the length-scales accessible with SANS, and therefore the magnetization on each atom $\vec{M}_{d_m}$ is replaced by a continuous vector field $\vec{M}(\vec{r})$. The magnetic SANS signal will reflect the variation of the magnetization, in modulus and orientation, at the nano-scale length. Then the *magnetic structure factor* vector ($\vec{M}(\vec{q})$) employed in SANS is the Fourier transform of the field $\vec{M}(\vec{r})$. In a similar way, the scalar *nuclear structure factor* ($N(\vec{q})$) is defined by using a *scattering length density* $b(\vec{r})$ instead of the atomic b_n. Therefore we get:

$$\vec{M}(\vec{q}) = \int d^3r \; \vec{M}(\vec{r}) \, e^{-i2\pi \vec{q} \cdot \vec{r}}, \tag{188}$$

$$N(\vec{q}) = \int d^3r \; b(\vec{r}) \, e^{-i2\pi \vec{q} \cdot \vec{r}}. \tag{189}$$

The sample scattering volume appears at SANS cross sections via a constant $K = V^{-1} p^2$, where p was defined in section "Magnetic Scattering".

In the next subsection, an example related to vortex matter will be presented in order to illustrate the power of the magnetic SANS.

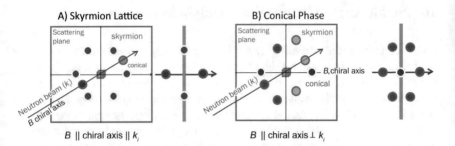

Fig. 5 Scheme showing the typical SANS configurations to observe the skyrmion lattice (**a**) and the conical phase (**b**) in cubic chiral helimagnets. Blue (**a**) and red (**b**) spots are observed at the neutron detector related, respectively, with the hexagonal packing of the SKL and the conical phase with propagation vector parallel to the field ($\vec{B} \parallel \vec{K}$)

Stroboscopic SANS Experiments on Skyrmionic Lattices (SKL)

Skyrmion textures, which are topologically protected nanometric vortex-like magnetic objects that have been discovered in cubic magnets without inversion symmetry (MnSi [131], $Fe_{1-x}Co_xSi$ [132], FeGe [133], Cu_2OSeO_3 [134, 135]), were theoretically predicted a long time ago by the Bogdanov seminal studies [136, 137]. However, the stability of the skyrmion lattices (SKL) has been studied only in a limited way, mostly concerning with radial or elliptic stability [136, 138]. Hence, solitonic configurations that are considered stable or metastable may actually be unstable due to some mode that is non-homogeneous along the magnetic field direction. Closely related to the stability analysis is the idea that some metastable state may become the equilibrium state by virtue of the thermal fluctuations. These two related problems, the stability of stationary points and the effect of thermal fluctuations at relatively low temperatures, are addressed in Ref. [139], and, as a result, it is shown that the low T phase diagram $B - T$ contains an unexpected new SKL phase.

Nakajima et al. [140, 141] by using stroboscopic SANS measurements in single crystals of MnSi at the *time-of-flight* instrument TAIKAN located at the Spallation Neutron Source (MLF) in Japan explored the low-temperature region of the phase diagram. By appropriately synchronizing the neutron and electrical current pulses (which are employed to create temperature ramps in the sample) with the neutron 2D-detector, it is possible to study the time evolution of SANS patterns with a resolution of 30 ms. The employed setup allowed to cool down the sample temperature with a rate of 100 K s^{-1}.

The skyrmion lattices are usually visible in SANS experiments under the configuration for which the magnetic field is parallel to the incoming neutron beam ($\vec{B} \parallel \vec{k}_i$). This is schematized in Fig. 5a and shows the six-fold typical pattern of a hexagonal lattice. However, the two spots defining the conical state are observed when the propagation vector of the helix, which is often parallel to the magnetic field, is perpendicular to the beam ($\vec{B} \parallel \vec{K} \perp \vec{k}_i$). See Fig. 5b. When the SKL is observed in $\vec{B} \parallel \vec{k}_i$, no signal appears in $\vec{B} \perp \vec{k}_i$ and reciprocally for the conical state. The crystal studied by Nakajima et al. was aligned with its (0 0 1) axis parallel to the horizontal magnetic field and the (1 1 0) in the vertical.

Nakajima et al. explored the phase diagram by doing measurements in equilibrium and also by doing fast quenching from the paramagnetic state, passing through the *A-phase*, to the low T region. At some values in the $T - B$ plane, SANS data were also collected. Metastable SKL states were captured at low T, in agreement with [139], and also a magnetic phase transition from triangular to square SKL was identified with decreasing magnetic field at low temperatures.

11 Magnetic Inelastic Scattering

The previous sections have been dedicated to describing different applications of neutron scattering based on the measurement of the nuclear or the magnetic elastic coherent part of the double differential cross sections, i.e., the incoming neutron exchanges momentum with the sample, but it does not exchange its energy. However, there are also nuclear and magnetic inelastic coherent parts, which are very small but which wear very important information about the collective excitations of atomic or magnetic origin (e.g., phonons, spin-waves, etc.).

In magnetically ordered systems, each magnetic moment deviates from its equilibrium position along a particular direction. However, these deviations cannot be arbitrary because the magnetic moments are coupled. Basically it produces collective coherent deviations, also called spin-waves (see section "Scattering by Spin-Waves"). As illustrated in section "Scattering by Spin-Waves", by using the second quantization picture, it is possible to obtain the single quantum (the *magnon*), and its energy $E(\vec{q})$ (dispersion relations), where $\vec{q}$ is a vector of the Brillouin zone that labels each magnon. By using linear approximations, valid when the deviations with respect to the magnetic moment are small (low temperatures and/or large enough magnetic moment), it is easy to demonstrate that $E(\vec{q})$ is related to the Fourier transform of the magnetic exchange interaction constants J_{l,d_m} between the different magnetic atoms d_m at different lattices l. The number of branches, acoustic or optic, depends on the number of independent magnetic atoms in the unit cell, and their degeneracy is related to the multiplicity of each magnetic orbit.

It has been found that this formalism is also appropriate to reasonably describe the spin-waves in $3d$ transition-metal magnets. In AF materials it is found that the dispersion relations are linear in q for small q, and, depending on the ion anisotropies, several gaps are opened at $\vec{q} = 0$.

On the other hand, the spin-wave dispersion relations are considered as the fingerprint to understand the superconductivity. For example, in related materials like the *pnictides* and *chalcogenides*, it seems that there exists some recognizable pattern in the next nearest neighbor magnetic interactions even if the first neighbor interactions are completely different.

Moreover, when the crystal field produced by electrostatic interactions with surrounding atoms is predominant, the energy levels of a magnetic atom are minimally dispersive. In these cases, it is possible to use inelastic neutron scattering to determine those energy levels assuming that they are accessible by magnetic dipolar transitions.[14]

An intermediate case, between isolated magnetic centers and extended magnets, happens for magnetic clusters formed by several atoms magnetically coupled inside of a molecule (e.g., SMM, metallo-proteines, phthalocianines, etc. . .). These objects can be prepared in a highly controlled manner, via chemical routes, and often they allow the study of new magnetic quantum phenomena.

The next two examples will illustrate how to determine the hierarchy between different magnetic interaction constants in the family $A_2FeX_5 \cdot H_2O$ (A=K, Rb, X=Cl, Br) and how to determine the parameters in the spin Hamiltonian in SMM (e.g., Mn_{12}-acetate and Fe_8). Both systems have been described in the preceding sections.

Understanding Magnetic Interactions in $K_2FeCl_5 \cdot D_2O$

The aim of this example is to go a step further in the understanding of the magnetic interaction mechanisms in this series of compounds by determining the magnetic coupling constants [142]. The analysis of the five magnetic coupling constants, particularly the comparison between magnetic interactions including and not including oxygen atoms in the super-exchange pathways, contributes to the understanding of the relative importance of two influential factors on the strength of the magnetic interaction: (1) the spin population on the ligand atoms (see the example in section "Spin Density Experiments in p-O_2N-C_6F_4CNSSN and $A_2FeX_5 \cdot H_2O$ (A=K, Rb, X=Cl, Br)") and (2) the distances between the two ligand atoms in the magnetic super-exchange pathways transmitting the magnetic interaction.

Therefore, in this example, after determining all the possible magnetic interactions from an analysis of the nuclear structure, previously determined by using single-crystal neutron diffraction experiments, which allowed us to locate accurately the position of the oxygen and the hydrogen atoms, the study of the magnetic coupling constants determined by inelastic triple axis neutron spectroscopy (TAS) is reported.

The unit cell of $K_2FeX_5 \cdot D_2O$ has been described in section "Antiferromagnetic Structures with $\vec{K} = 0$ in $A_2FeX_5 \cdot H_2O$ (A=K, Rb, X=Cl, Br)". In this structure there are five possible magnetic interactions between an iron ion and its first shell of neighbour Fe ions. Figure 6 A represents the unit cell of $K_2FeCl_5 \cdot D_2O$, whereas

---

[14] $\Delta M_S = \pm 1$.

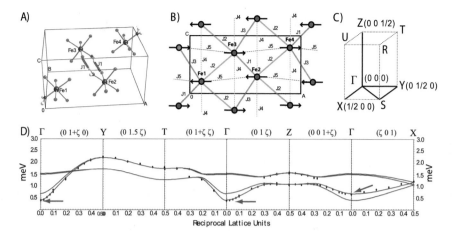

Fig. 6 (**a**) Structure of $K_2FeCl_5 \cdot D_2O$. The potassium atoms have been omitted for clarity. The halogen labels are represented in all the octahedra together with the label for the four iron ions: $Fe1(x, y, z)$, $Fe2(\frac{1}{2} + x, \frac{1}{4}, \frac{1}{2} - z)$, $Fe3(\frac{1}{2} - x, \frac{3}{4}, \frac{1}{2} + z)$, and $Fe4(-x, \frac{3}{4}, -z)$. (**b**) Scheme of the five magnetic interactions projected on the ac-plane. Thick lines represent interactions connecting a Fe ion with two other Fe ions, one with $y + \frac{1}{2}$ and the other with $y - \frac{1}{2}$. The magnetic structure is also shown. Orange circles correspond to Fe ions with $y = \frac{1}{4}$, whereas pink circles $y = \frac{3}{4}$. (**c**) High symmetry points of the first BZ together with the different directions that have been measured (thick lines). (**d**) Experimental points measured for each direction of high symmetry at the BZ. The red line is the fit of the points to a linear spin-wave model including uniaxial and rhombic anisotropies

Fig. 6b shows a scheme of the five magnetic interactions projected on the ac-plane together with the magnetic structure.

The inelastic neutron scattering experiment was performed on the three-axis spectrometer IN12 at the ILL. In order to determine the magnon dispersion curves, constant-q scans along the $(0\ 1\ \zeta)$, $(0\ 1 + \zeta\ 0)$, $(0\ \frac{3}{2}\ \zeta)$, $(0\ 1 + \zeta\ \zeta)$, $(0\ 0\ 1 + \zeta)$, and $(\zeta\ 0\ 1)$ directions were performed at 1.6 K, temperature much lower than the Néel temperature of 14.06 K. The measured directions in the reciprocal space were selected in order to have a correct description of the first Brillouin zone (BZ). See Fig. 6c.

A gap in the energy is observed in the low-energy magnon branch at the center of the zone (Γ). Moreover, an additional feature in the dispersion curves is that this energy gap appears to be split in the different Brillouin zones, around 0.38 meV and 0.58 meV in the $(0\ 1\ 0)$ and the $(0\ 0\ 1)$ zones, respectively. This different gap can be observed in Fig. 6d indicated with green arrows. Therefore, a model with a rhombic magnetic anisotropy term is needed in order to correctly fit the different gaps. The Hamiltonian which includes this rhombic magnetic anisotropy is:

$$\mathcal{H} = - \sum_{l,l',d_m,d'_m} J_{ld_m,l'd'_m} \vec{\mathbb{M}}_{ld_m} \cdot \vec{\mathbb{M}}_{l'd'_m} - \sum_{ld_m} \sigma_{d_m} A_{d_m} \mathbb{M}^z_{ld_m}$$

$$- \sum_{ld_m} E_{d_m} \left[\left(\mathbb{M}^x_{ld_m} \right)^2 - \left(\mathbb{M}^y_{ld_m} \right)^2 \right]. \tag{190}$$

The magnetic exchange coupling constants obtained, after fitting all the experimental dispersion curves to the relation $E(\vec{q})$ derived from Hamiltonian expressed in Eq. (190), are $J_1 = -0.113(3)$, $J_2 = -0.037(1)$, $J_3 = -0.032(3)$, $J_4 = -0.023(2)$, and $J_5 = -0.025(2)$ meV. The values of the variable A_{d_m} and E_{d_m} were 0.073 meV and 0.0072 meV, respectively, which gives an axial anisotropy $D = 0.0146$ meV. The experimental and fitted dispersion curves are depicted in Fig. 6d.

The results of this study can be related to other studies of several magnetic phenomena of this AF series: (1) The energy gap at the Γ point in the magnon dispersion curves is in good agreement with the energy gap calculated from an analysis of the boundaries of the magnetic phase diagram [41]. (2) The physical interpretation of the two different energy gaps at the Γ point implies that the hardest magnetic axis is the $\vec{b}$-axis. The same conclusion is deduced from the magnetic structure in the spin-flop phase in which the magnetic moments are collinear to the $\vec{c}$-axis [47]. (3) The magnetic structures determined from the values of the magnetic coupling constants correspond to the experimental magnetic structure. (4) Estimation of the Neel temperatures using a mean field approach results in values close to the experimental ones. (5) The values of the magnetic interactions also explain the magnetic dimensionality crossover observed in the heat capacity and magnetic measurements [40].

Neutron Spectroscopy in Magnetic Molecular Clusters

The interest of SMM, like Mn_{12}-acetate, has already been introduced in section "Magnetic Ordering by Dipolar Interactions in Single-Molecule Magnet Mn_{12}-Acetate". In these materials, the knowledge of the magnetic anisotropy of the clusters, which is assumed to be much smaller than the exchange energy, is central to understanding how the quantum tunneling of the magnetization (QTM) can occur. With uniaxial anisotropy, the energy levels of a cluster depend on the orientation of the spin S relative to its anisotropy axis. However, for QTM to be allowed, the rotational symmetry about the z axis must be broken by small high-order spin terms. The general Hamiltonian in expression (191), where quadratic and quartic terms in $\tilde{S}$ are included, describes appropriately the energy levels in Mn_{12}-acetate and Fe_8 SMM:

$$\mathcal{H} = D\left[\mathbb{S}^2_z - \frac{1}{3}S(S+1) \right] + E\left[\mathbb{S}^2_x - \mathbb{S}^2_y \right] + B^0_4 \mathbb{O}^0_4 + B^2_4 \mathbb{O}^2_4 + B^4_4 \mathbb{O}^4_4 \tag{191}$$

where the operators $\mathbb{O}_4^0$, $\mathbb{O}_4^2$, and $\mathbb{O}_4^4$ are defined as:

$$\mathbb{O}_4^0 = 35\,\mathbb{S}_z^4 - [30S(S+1) - 25]\mathbb{S}_z^2 - 6S(S+1) + 3S^2(S+1)^2$$

$$\mathbb{O}_4^2 = \frac{1}{4}\left\{\left[7\mathbb{S}_z^2 - S(S+1) - 5\right]\left(\mathbb{S}_+^2 + \mathbb{S}_-^2\right) + \left(\mathbb{S}_+^2 + \mathbb{S}_-^2\right)\left[7\mathbb{S}_z^2 - S(S+1) - 5\right]\right\}$$

$$\mathbb{O}_4^4 = \frac{1}{2}\left(\mathbb{S}_+^4 + \mathbb{S}_-^4\right)$$

Thus, to validate the different theories of QTM, the knowledge of the transversal anisotropy parameters (B_4^0, B_4^2, B_4^4) in the spin Hamiltonian is very important, whose precise determination requires a suitable experimental technique. The advantage of inelastic neutron scattering over other techniques, like high field electron paramagnetic resonance (EPR), is that it is not necessary to perturb the system by applying a magnetic field and it can give a detailed picture of the scheme levels from a straightforward analysis.

The cluster $[Fe_8O_2(OH)_{12}(tacn)_6]^{8+}$ (in short Fe8) is a molecule formed by eight Fe^{3+} ions in a planar butterfly geometry giving a $S = 10$ due to the AF coupling between the Fe ions. Cacciuffo et al. employed INS techniques in non-deuterated powder of Fe8 at low temperatures ($1 \leq T \leq 10$ K) at the time of flight spectrometer IN5 of ILL working with a resolution of $19\,\mu$eV with the aim to determine the different parameters defining the spin-Hamiltonian of this cluster [143].

The ground level, $|S = 10; M = \pm 10\rangle$, is the only one populated at the lowest temperature. As the temperature increases, other levels start to be populated, being the next one the $|S = 10; M = \pm 9\rangle$. The uniaxial and rhombic anisotropy parameters (D, E) mainly control the energy distance between the lowest levels, whereas the other parameters (B_4^0, B_4^2, B_4^4) are involved in some mixing of the upper energy levels ($|S; M\rangle$ with $|M| \leq 6$) less distant among them. Usually these levels are populated at the highest temperatures. In fact, the position of the peaks at the energy transfer axis gives directly the eigenvalues of the Hamiltonian (191) (which depends on the parameters), whereas the intensity of the peaks is related to the wave function of the eigenvectors. Therefore, in principle, it is possible to obtain the values of the parameters by fitting the peak positions.

Cacciuffo et al. found $D = -25.2(2)\,\mu$eV and $E = -4.02(3)\,\mu$eV, and to fit correctly all the measured peaks, it was needed to include the quartic terms of the Hamiltonian giving the values $B_4^0 = 0.87(6)10^{-4}\,\mu$eV, $B_4^2 = 0.1(1)10^{-4}\,\mu$eV, and $B_4^4 = 7.4(6)10^{-4}\,\mu$eV. These values produced an energy barrier $A/k_B = 32.8(1)$ K slightly higher than the one reported from ac susceptibility experiments and demonstrated the importance of the $\mathbb{O}_4^4$ operator, which is related to transverse anisotropy, in the tunneling processes.

In Mn_{12}-acetate, Mirebeau et al. also measured the energy levels by using the INS techniques [144]. Their main result was the precise determination of the coefficient B_4^4 in the non-diagonal transverse term (in Mn_{12} the tetragonal symmetry does not allow the presence of the term $\mathbb{O}_4^2$ in the Hamiltonian). The presence of the

$\mathbb{O}_4^4$ term was searched for by many experimentalists, since only components that do not commute with S_z could induce tunneling with $\Delta M = \pm 4$, as observed experimentally in the magnetization experiments.

In the 2000s the extrinsic interactions, arising from disorder, were proposed as the dominant sources of QTM in Mn_{12} nanomagnets [145]. However, Carbonera et al. presented a detailed study of the effects that crystalline disorder has on the QTM of Mn_{12}-benzoate SMM. The authors were able to isolate the influence of long-range crystalline disorder thanks to the absence of interstitial molecules in the crystal structure of this molecular compound. For this, they compared inelastic neutron scattering results obtained at IN5 under two extreme situations: a crystalline sample and a nearly amorphous material. Results showed that crystalline disorder weakly affects the anisotropy and other parameters in the spin Hamiltonian and therefore has a little effect on the quantum tunnelling of these materials. It follows that disorder is not a necessary ingredient for the existence of QTM [146].

12 Other Neutron Scattering Techniques

This section is intended to briefly explain other neutron scattering techniques that are increasingly being employed to solve problems in magnetism. They give access to the very small energy transfers and therefore allows the study of slow dynamics. An example on the use of neutron reflectometry technique to characterize magnetic surfaces will be also illustrated.

Quasielastic Neutron Scattering

In the previous sections, it has been explained that the scattering law can be split in coherent and incoherent parts. The first one is related to the correlations in time and in space of pair of scatterers, whereas the second is about self-correlations of scatterers. It means that the coherent part is related to the structure (elastic) and the collective movements or excitations (inelastic). However, the incoherent part gives information about the individual dynamics of nuclei (vibrations, rotations, or translations) and magnetic moments.

Due to the large incoherent cross section of hydrogen, compared with other isotopes, the scattering cross sections measured in hydrogenated samples are dominated by the incoherent part, and therefore neutron scattering techniques are ideal to characterize the individual movements of hydrogen in these samples. When the movement can be decoupled in vibrations, rotations, and translations, then it is possible gain access to the quasi-elastic scattering function whose width is related to the characteristic times of the movements in the range 10^{-10}–10^{-12} s. The intensity of the elastic part of this incoherent signal gives information about the geometry of the movements. More detailed information on quasi-elastic neutron scattering can be found in [147]. Here a simple example on the use of these techniques, involving magnetic molecules, will be commented.

Rodríguez-Velamazán et. al [148] applied quasi-elastic neutron scattering techniques, together with solid state ^{2}H-NMR, to characterize, as a function of temperature, the rotational movement of the pyrazine rings in the spin crossover molecular compound {Fe(pyrazine)[Pt(CN)$_4$]}. In this compound, the Fe ion can switch between high and low spin states at temperatures $\sim$285 K and $\sim$309 K for decreasing or increasing the temperature, respectively. The structural studies at low temperatures, in the low spin state, showed some anomalous elongation at the Debye–Wallers factors at the hydrogen atoms of the pyrazine rings. This fact could be an indication of some structural or dynamical disorder present in the materials. In this sense, quasi-elastic neutron scattering techniques can help to elucidate the origen of the disorder.

They demonstrate experimentally that the bistability between rotation and blocking of the pyrazine rings connecting the Fe ions arranged in different planes is accompanied with the change between high and low spin states. The rotation in the high spin state is described as a 4-fold jump motion about the nitrogen axis coordinated with the Fe ion. In the low spin state, the rotation is suppressed. They also observed that when benzene molecules are incorporated in the system, the pyrazine movements are restricted.

Neutron Spin Echo Techniques

In 1972 F. Mezei published his pioneering work establishing the grounds for the neutron spin echo techniques [149]. The very basic idea behind these techniques is the Larmor precession of the polarization of the neutron beam when neutrons travel along a distance L with an uniform magnetic field perpendicular to the polarization. If the neutron travels again through a second region with the same length L but with opposite magnetic field, then the dephasing precession angle turned in the first region is deturned in the second one, giving a zero total dephasing precession angle. However, if a sample is put in between both opposite magnetic field regions, some neutrons will exchange energy with the sample and therefore will change their velocities before entering the second region producing a different dephasing precession angle. As a result, the sample could produce an easily measurable non-zero dephasing angle for all neutrons.

It can be easily demonstrated that these techniques allow to measure directly the intermediate scattering function $I(\vec{q}, t)$, defined in Eq. (58), which measures the time correlations of the sample. Typically, the accessible time ranges go from picoseconds to microseconds, which correspond to hundredths of μeV to several neV energy exchanges within the sample. Moreover, the energy transfer can be determined with very high precision (10^{-5}) without using severe conditions at the monochromator. Detailed descriptions of these techniques can be found in [150, 151].

In the previous paragraphs, we assumed that the neutron spin does not change in the scattering process, which is true for non-magnetic nuclear interaction. When applying these ideas to magnetic materials, where the magnetic interactions can change the polarization state, we need to consider the magnetic field that the neutrons will feel when traveling through the sample, which is basically the *magnetic interaction vector* $\vec{M}_{\perp\vec{q}}$. It is also necessary to bear in mind how to control the effects of beam depolarization by the different magnetic domain distributions.

The next example will illustrate one of the most employed applications of NSE in magnetism, which is the measurement of relaxation times in spin glasses. In fact, the time correlations of the magnetization close to the glassy transition (T_g) have been the object of discussions and controversy over the last few decades. This has often been characterized by a stretched exponential function $\sim \exp\{-(t/\tau)^\beta\}$, but when self-similarity (i.e., fractal behaviour) occurs, this function does not describe adequately the structural or magnetic relaxation. Then a modified function, which phenomenologically describes the Monte Carlo calculations by incorporating a power law, $q(t) = \langle S_i(0)S_i(t)\rangle \propto t^{-x} \exp\{-(t/\tau)^\beta\}$, is employed. The implications of such a non-exponential function open the necessity of to think about global models that may lead to hierarchical relaxation. In this context Weron [152] proposed a generalized function, $\varphi(t) = \{1 + k(t/\tau)^\beta\}^{-1/k}$, where k is a parameter accounting for effective interactions and β is related to the self-similarity concept that can be adapted easily to spin glass dynamics and, in the limit $k \to 0$, reduces to the exponential function.

The validity of the model based in the generalized relaxation function was investigated in the archetypal metallic spin glasses $Cu_{1-x}Mn_x$ ($x = 0.1, 0.16, 0.35$) and $Au_{1-x}Fe_x$ ($x = 0.14$) with neutron spin echo experiments in [153] accessing to time scales from 0.002 to 1 ns. The authors measured the intermediate function ($I(\vec{q}, t)$) at several temperatures and q values. All the spectra were independent of q and well described with the Weron function with consistent physically meaning parameters, for both samples. In turn, the k parameters obtained from the fits to the Weron function are related to the sub-extensivity parameter $\tilde{q}$ of models with non-extensive entropy for complex and multifractal systems giving a $\tilde{q} \sim 5/3$ parameter at the spin glass transition temperature T_g. It seems that the spin glass system evolves from an extensive Boltzmann-Gibbs thermodynamics at high temperatures to a sub-extensive dynamics at T_g.

Neutron Reflectivity

Since the pioneering work of Felcher et al. in the early 1980s [154] about the neutron reflectivity grounds, these techniques have reached a mature state, and nowadays neutron reflectometers are fundamental instruments in all types of neutron sources, pulsed and steady-state sources. Due to their great power, these techniques are applied to the study and characterization of a wide range of phenomena and materials, mainly in soft matter and nanomagnetism areas. In fact, there are scientific topics in nanomagnetism that can only be addressed with neutron reflectivity, e.g., magnetization depth profiling, magnetic structures in multilayers and heterostructures, interdiffusion, magnetic domains, magnetic nanoparticles, study of patterned nanostructures, etc. In this section we will focus only on some applications of these techniques in thin films. For a more extended review, we refer to the nice chapter of Toperverg and Zabel in [155].

X-ray reflection by matter can be understood by means of the refractive index of the matter for X-ray photons. In electromagnetism theory the refractive index is related directly to the electron density of the material, which is at the origin of the interaction potential of the X-ray scattering, and the wavelength of the X-ray photons. The fact that this refractive index is less than unity implies that there exists a critical incidence angle below which there is total reflectivity. For larger angles, the X-rays partially penetrate the surface.

For neutron reflectivity the idea is similar. A refractive index for neutrons, related to the scattering length density, which is involved in the neutron-nuclei interaction, and with the magnetic moment distribution in the sample, also involved in the magnetic interaction with the neutron spin, can be built. However, the magnetic part of the refraction index can change its sign depending on the relative orientation of the magnetic moments of the material with the neutron beam polarization.

After the discovery of giant magnetoresistance effect, the neutron reflectivity techniques have been widely employed to study interlayer AF exchange coupling in magnetic superlattices, which are the base of different spintronic devices. Usually these multilayers are characterized from magnetic hysteresis measurements done with MOKE, SQUID, or VSM magnetometry experiments. However, these macroscopic characterization techniques do not unveil the angle between adjacent ferromagnetic layers nor the magnetic correlation in the perpendicular direction or the domain size distributions. These parameters are at the origin of poor transport properties in these systems and can be understood after polarized neutron reflectivity (PNR) experiments [156].

As a pioneer work, V. Lauter et al. [157] studied with PNR techniques the archetypical Fe/Cr multilayers, which have a strong AF interlayer exchange coupling and in-plane uniaxial anisotropy, to shed light on the role of the different ingredients, such as spin configurations in the flopped phase, role of the anisotropy, border effects, etc.

These authors determined unambiguously the magnetic structure of each ferromagnetic layer n (with $1 \leq n \leq 12$) in $[^{57}Fe(67 \text{ Å})/Cr(9 \text{ Å})]_{12}$, at the spin-flop phase by applying a magnetic field along the easy axis. The magnetic structure broke down into lateral domains, with a mean size of 2800 Å, deviated each one an angle $\pm\phi_n$ with respect to the magnetic field direction. The sequence of angle deviations along each layer n fulfils $\pm\phi_n = \mp\phi_{13-n}$, and the angle at the end layers $\pm\phi_1 = \mp\phi_{12} = \pm40°$ is smaller than in the interior layers $\pm\phi_6 = \mp\phi_7 = \pm63°$, which is related to the fact that in the end layers, there is one missing neighbor. The canting angle in the interior reaches the value of the canting angle in the bulk sample for the same magnetic field. The resulting configuration has no net perpendicular component to the magnetic field, and, therefore, it is stable. The splitting into lateral domains at each layer is interpreted as a minimization of the dipolar energy via demagnetization at each individual layer.

13 Conclusions

In this chapter, we showed the fundamentals of neutron scattering and discussed some contributions of neutron scattering to solve a huge variety of problems in the field of magnetism. After more than 70 years from the first neutron scattering experiment, these techniques are widely employed to characterize magnetic materials together with other complementary techniques. Moreover, after this long history, with new neutron sources, and in particular pulsed sources, one can expect that these techniques will experience a substantial improvement with the availability of higher neutron fluxes and lower backgrounds. The use of new optical systems, faster detectors, and new techniques exploiting new physical phenomena will allow all together the use of smaller samples, to design new experiments or to do faster *in operando* experiments, etc.

In the majority of the neutron sources, the experiments related to magnetism are the most demanded by the scientist, and often the diffraction techniques (powder, single crystal, and small angle) are the most requested with the goal to determine the evolution of some magnetic structure as a function of one or several parameters (temperature, magnetic field, pressure, concentration, etc.).

Acknowledgments Much of the work reported here was carried out during funding by the Spanish Ministry of Science and Innovation grant PGC-2018-099024-B-I00 or its equivalents before. JC is grateful for the hospitality received at the different neutron sources and in particular to the Institut Laue-Langevin and the Laboratoire Leon Brillouin. We are indebted to many discussions with our colleagues and co-workers, but in particular with Fernando Palacio, Javier Luzón, Garry McIntyre, José Alberto Rodríguez-Velamazán, Ángel Millán, Oscar Fabelo, Laura Cañadillas, Juan Rodríguez-Carvajal, Katsuya Inoue, Yusuke Kousaka, Yuko Hosokoshi, Kazuki Ohishi, Jun Kishine, Yoshiiko Togawa, Cristina Sáenz de Pipaón, Clara González, Clara Rodríguez, Miguel Pardo-Sáenz, Milagros Tomás, and Larry Falvello.

References

1. C. Abel et al., Phys. Rev. Lett. **124**, 081803 (2020)
2. R.G. Newton, *Scattering Theory of Waves and Particles* (Dover Publications, Mineloa, New York, 2002)
3. A. Galindo, P. Pascual, *Quantum Mechanics II* (Springer, Berlin, 1990)
4. E. Fermi, Ric. Sci. **7** 13 (1936)
5. J. Schwinger, E. Teller, Phys. Rev. **52**, 286 (1937)
6. F. Bloch, Phys. Rev. **50**, 259 (1936)
7. F. Bloch, Phys. Rev. **51**, 994 (1937)
8. J. Schwinger, Phys. Rev. **51**, 544 (1937)
9. L. Van Hove, Phys. Rev. **95**, 249 (1954)

10. N.H. March, M.P. Tosi, *Introduction to Liquid State Physics* (World Scientific, Singapore, 2002)
11. N.W. Ashcroft, N. David Mermin, *Solid State Physics* (Harcourt, San Diego, 1976)
12. O. Halpern, M.H. Johnson, Phys. Rev. **55**, 898 (1939)
13. E.U. Condon, H. Odabasi, *Atomic Structure* (Cambridge University Press, Cambridge, 1980)
14. S.W. Lovesey, *Theory of Neutron Scattering from Condensed Matter*, vol. 2 (Oxford University Press, Clarendon, Oxford, 1984)
15. D.F. Johnston, Proc. Phys. Soc. **88**, 37 (1966)
16. B. Fultz, Progr. Mater. Sci. **55**, 247–352 (2010)
17. F. Bloch, Z. Phys. A **61**, 206–219 (1939)
18. T. Holstein, H. Primakoff, Phys. Rev. **58**, 1098 (1940)
19. S.V. Maleev, Zh. Eksperim. i Teor. Fiz. **33**, 129 (1958)
20. S.V. Maleev, Zh. Eksperim. i Teor. Fiz. **40**, 1224 (1961)
21. A.W. Saenz, Phys. Rev. **119**, 1542 (1960)
22. Yu.A. Izyumov, S.V. Maleev, Zh. Eksperim. i Teor. Fiz. **41**, 1644 (1961)
23. Yu.A. Izyumov, Usp. Fiz. Nauk. **80**, 41 (1963)
24. M. Blume, Phys. Rev. 130:1670, 1963.
25. M. Blume, Phys. Rev. **133**, A1366 (1964)
26. M.R.I. Schermer, M. Blume, Phys. Rev. **166**, 554 (1968)
27. V.G. Barýaktar S.V. Maleev, P.A. Suris, Sov. Phys. Solid State **4**, 2533 (1963)
28. V.E. Naish Yu.A. Izyumov, R.P. Ozerov, (CRC Press, Taylor & Francis, New York, 1991)
29. H.M. Rietveld, Appl. Crystallogr. **2**, 65–71 (1969)
30. R.A.Young, M.P. Eackie, R.B. Von Dreele, Appl. Crystallogr. **10**, 262–269 (1977)
31. G. Malmros, J.O. Thomas, Appl. Crystallogr. **10**, 7–11 (1977)
32. J.B. Hastings D.E. Cox, W. Thomlinson, C.T. Prewill, Nucl. Instrum. Methods **208**, 573–578 (1983)
33. R.A. Young, (ed.), *The Rietveld Method* (Oxford University Press, Oxford, 1993)
34. W.I.F. David, K. Shankland, M.L.B, B. CH, (eds)., *Structure Determination From Powder Diffraction Data* (Oxford Science Publication, Oxford, 2002)
35. J.A. Rodríguez-Velamazán, O. Fabelo, A. Millán, J. Campo, R.D. Johnson, L. Chapon. Sci. Rep. **5**, 14475 (2015)
36. L. Desgranges, G. Baldinozzi, G. Rousseau, J.C. Niepce, G. Calvarin, Inorg. Chem. **48**, 7585–7592 (2009)
37. M.T. Hutchings, P.J. Withers, T.M. Holden, T. Lorentzen, (CRC Press, Taylor & Francis, London, 2005)
38. C.G. Shull, J. Samuel Smart, Phys. Rev **76**, 1257 (1949)
39. R.L. Carlin, F. Palacio, Coord. Chem. Rev. **65**, 141 (1985)
40. J.A. Puértolas, R. Navarro, F. Palacio, J. Bartolomé, D. González, R.L. Carlin, Phys. Rev. B **31**(1), 516–526 (1985)
41. F. Palacio, A. Paduan-Filho, R.L. Carlin, Phys. Rev. B **21**(1), 296–298 (1980)
42. Y. Imry and S.-K. Ma, Phys. Rev. Lett. **35**, 1399 (1975)
43. S. Fishman, A. Aharony, J. Phys. C: Solid State Phys. **12**(18), L729–L733 (1979)
44. F. Palacio, J. Campo, M.C. Morón, A. Paduan, C.C. Becerra, Int. J. Mod. Phys. B **12**(18), 1781–1793 (1998)
45. F. Palacio, M. Gabas, J. Campo, C.C. Becerra, A. Paduan-Filho, V.B. Barbeta, Phys. Rev. B **56**(6), 3196 (1997)
46. C.C. Becerra, V.B. Barbeta, A. Paduan-Filho, F. Palacio, J. Campo, M. Gabás, Phys. Rev. B **56**(6), 3204 (1997)
47. J. Campo, F. Palacio, M.C. Moron, C.C. Becerra, A. Paduan, J. Phys. Condens. Matter **11**(22), 4409–4425 (1999)
48. C.C. Becerra, V.B. Barbeta, A. Paduan, F. Palacio, J. Campo, M. Gabás, Phys. Rev. B **56**(6), 3204–3211 (1997)
49. M. Gabas, F. Palacio, J. Rodríguez Carvajal, D. Visser, J. Phys. Condens. Matter **7**, 4725 (1995)

50. A. Jain, S.M. Yusuf, J. Campo, L. Keller, Phys. Rev. B **79**(18), 184428 (2009)
51. G. Rousse, J. Rodríguez-Carvajal, C. Wurm, C. Masquelier, Phys. Rev. B **88**(21), 214433 (2013)
52. F. Hulligur, E. Pobitschka, J. Solid State Chem. **1**, 117 (1970)
53. S.S.P. Parkin, R.H. Friend, Philos. Mag. B **41**, 65 (1980)
54. S.S.P. Parkin, R.H. Friend, Philos. Mag. B **41**, 95 (1980)
55. Y. Togawa, T. Koyama, K. Takayanagi, S. Mori, Y. Kousaka, J. Akimitsu, S. Nishihara, K. Inoue, A. Ovchinnikov, J. Kishine, Phys. Rev. Lett. **108**(10), 107202 (2012)
56. V. Laliena, J. Campo, J.-I. Kishine, A.S. Ovchinnikov, Y. Togawa, Y. Kousaka, K. Inoue, Phys. Rev. B **93**(13), 1420 (2016)
57. V. Laliena, J. Campo, Y. Kousaka, Phys. Rev. B **95**(22), 1420 (2017)
58. V. Laliena, Y. Kato, G. Albalate, J. Campo, Phys. Rev. B **98**(14), 144445 (2018)
59. V. Laliena, J. Campo, Y. Kousaka, Phys. Rev. B **94**(9), 960 (2016)
60. B. van Laar, H.M. Rietveld, D.J.W. Ijdo, J. Solid State Chem. **3**, 154 (1971)
61. Y. Kousaka, T. Ogura, J. Zhang, P. Miao, S. Lee, S. Torii, T. Kamiyama, J. Campo, K. Inoue, J. Akimitsu, J. Phys. Conf. Ser. **746**, 012061 (2016)
62. P. Curie, J. Phys. **3**(III), 393–415 (1894)
63. T. Kimura, T. Goto, H. Shintani, K. Ishizaka, T. Arima, Y. Tokura, Nature **426**(6962), 55–58 (2003)
64. R. Kajimoto, H. Yoshizawa, H. Shintani, T. Kimura, Y. Tokura, Phys. Rev. B **70**(1), 916 (2004)
65. T. Kimura, G. Lawes, T. Goto, Y. Tokura, A.P. Ramirez, Phys. Rev. B **71**(22), 415 (2005)
66. M. Kenzelmann, A.B. Harris, S. Jonas, C. Broholm, J. Schefer, S.B. Kim, C.L. Zhang, S.W. Cheong, O.P. Vajk, J.W. Lynn, Phys. Rev. Lett. **95**(8), 87206 (2005)
67. M. Ackermann, D. Brüning, T. Lorenz, P. Becker, L. Bohat, N. J. Phys. **15**, 123001 (2013)
68. J.A. Rodríguez-Velamazán, O. Fabelo, A. Millan, J. Campo, R.D. Johnson, L. Chapon, Sci. Rep. **5**, 14475 (2015)
69. W. Tian, H. Cao, J. Wang, F. Ye, M. Matsuda, J.Q. Yan, Y. Liu, V.O. Garlea, H.K. Agrawal, B.C. Chakoumakos, B.C. Sales, R.S. Fishman, J.A. Fernández-Baca, Phys. Rev. B **94**, 214405 (2016)
70. J. Campo, F. Palacio, A. Paduan-Filho, C.C. Becerra, M.T. Fernández Díaz, J. Rodríguez Carvajal, Phys. B **234**, 622 (1997)
71. J.A. Rodríguez-Velamazán, O. Fabelo, J. Campo, A. Millán, J. Rodríguez-Carvajal, L.C. Chapon, Phys. Rev. B **95**, 174439 (2017)
72. R. Sessoli, D. Gatteschi, A. Caneschi, M.A. Novak, Nature (London) **365**, 141 (1993)
73. D. Gatteschi, A. Caneschi, L. Pardi, R. Sessoli, Science **265**, 1054 (1994)
74. F. Luis, J. Bartolomé, J. Fernández, J. Tejada, J. Hernández, X. Zhang, R. Ziolo, Phys. Rev. B **55**, 11449 (1997)
75. J. Hernádez, X. Zhang, F. Luis, J. Bartolomé, J. Tejada, R. Ziolo, Europhys. Lett. **35**:301, 1996.
76. L. Thomas, F. Lionti, R. Ballou, D. Gatteschi, R. Sessoli, B. Barbara, Nature (London) **383**, 145 (1996)
77. J.R. Friedman, M.P. Sarachik, J. Tejada, R. Ziolo, Phys. Rev. Lett. **76**, 3830 (1996)
78. T. Lis, Acta Crystallogr. Sect. B **36**, 2042, (1980)
79. R.A. Robinson, P.J. Brown, D.N. Argyriou, D.N. Hendrickson, S.M.J. Aubin, J. Phys. Condens. Matter **12**(12), 2805–2810 (2000)
80. J. Fernández, J. Alonso, Phys. Rev. B **62**, 53 (2000)
81. F. Luis, J. Campo, J. Gómez, G. J. McIntyre, J. Luzón, D. Ruiz-Molina, Phys. Rev. Lett. **95**, 227202 (2005)
82. K. Inoue, K. Kikuchi, M.Ohba, Angew. Chem. **42**, 4810–4813 (2003)
83. H. Higashikawa, K. Okuda, J. Kishine, N. Masuhara, K. Inoue, Chem. Lett. **36**, 1022–1023 (2007)
84. C. González, J. Campo, F. Palacio, G.J. McIntyre, Y. Yoshida, K. Inoue, K. Kikuchi (2020) submitted for publication
85. Y. Yoshida, K. Inoue, K. Kikuchi, M. Kurmoo, Chem. Mater. **28**, 7029–7038 (2016)

86. R.M. Moon, T. Riste, W.C. Koehler, Phys. Rev. **181**, 920 (1969)
87. E. Ressouche, J.X. Boucherle, B. Gillon, P. Rey, J. Schweizer, J. Am. Chem. Soc. **115**, 3610 (1993)
88. P.J. Brown, A. Capiomont, B. Gillon, J. Schweizer, J. Magn. Magn. Mater. **14**, 289 (1979)
89. J.M. Rawson, A.J. Banister, I. Lavender, Adv. Heterocycl. Chem. **62**, 137 (1995)
90. M. Kinoshita, P. Turek, M. Tamura, K. Nozawa, D. Shiomi, Y. Nakazawa, M. Ishikawa, M. Takahashi, K. Awaga, T. Inabe, Y. Maruyama, Chem. Lett. **7**, 1225 (1991)
91. F. Palacio, G. Antorrena, M. Castro, R. Burriel, J.M. Rawson, J.N.B. Smith, N. Bricklebank, J. Novoa, C. Ritter. Phys. Rev. Lett. **79**, 2336 (1997)
92. J. Luzón, J. Campo, F. Palacio, G.J. McIntyre, J.M. Rawson, R.J. Less, C.M. Pask, A. Alberola, R.D. Farley, D.M. Murphy, A.E. Goeta, Phys. Rev. B **81**, 144429 (2010)
93. R.L. Carlin, R. Burriel, Phys. Rev. B **27**, 3012 (1983)
94. T. Smith, S.A. Friedberg, Phys. Rev. **177**, 1012 (1969)
95. M. Gabás, F. Palacio, J. Rodríguez-Carvajal, D. Visser, J. Phys. Condens. Matter **7**, 4725 (1995)
96. J. Luzón, J. Campo, F. Palacio, G.J. McIntyre, A. Millan, Phys. Rev. B **78**, L279 (2008)
97. R. Papoular, B. Gillon, Europhys. Lett. **13**, 429 (1990)
98. J.B. Forsyth, in *Electron and Magnetization Densities in Molecules and Crystals*, ed. by P. Becker (Plenum Press, New York, 1980), pp. 791–819
99. B.E.F. Fender, B.N. Figgis, J.B. Forsyth, P.A. Reynolds, E. Stevens, Proc. R. Soc. Lond. A **404**, 127 (1986)
100. B.E.F. Fender, B.N. Figgis, J.B. Forsyth, Proc. R. Soc. Lond. A **404**, 139 (1986)
101. R.J. Deeth, B.N. Figgis, J.B. Forsyth, E.S. Kucharski, P.A. Reynolds, Proc. R. Soc. Lond. A **421**, 153 (1989)
102. B.N. Figgis, J.B. Forsyth, E.S. Kucharski, P.A. Reynolds, F. Tasset, Proc. R. Soc. Lond. A **428**, 113 (1990)
103. R.J. Deeth, B.N. Figgis, J.B. Forsyth, E.S. Kucharski, P.A. Reynolds, Proc. R. Soc. Lond. A **436**, 417 (1992)
104. B.N. Figgis, P.A. Reynolds, Mol. Phys. **80**, 1377 (1993)
105. G.S. Chandler, B.N. Figgis, R.A. Phillips, P.A. Reynolds, R. Mason, G.A. Wiliians, Proc. Royal Soc. Lond. A **384**, 31 (1982)
106. B.N. Figgis, P.A. Reynolds, R. Mason, Proc. R. Soc. Lond. A **384**, 49 (1982)
107. F.A. Wedgwood, Proc. R. Soc. Lond. A **349**, 447 (1976)
108. P.J. Brown, B.N. Figgis, P.A. Reynolds, J. Phys. Condens. Matter **2**, 5297 (1990)
109. B.N. Figgis, P.A. Reynolds, R. Mason, Inorg. Chem. **23**, 1149 (1984)
110. A. Gukasov, P.J. Brown, J. Phys. Condens. Matter **14**, 8831 (2002)
111. A. Gukasov, P. Rogl, P. Brown, M. Mihalik, A. Menovsky, J. Phys. Condens. Matter **14**, 8841 (2002)
112. M. Th Rekveldt, J. Phys. Colloques. C1 **32**, 579 (1971).
113. G.M. Drabkin, A.I. Okorokov, V. Runov, V. Pisńa, Zh. Eksp. Teor.Fiz. **15**, 458 (1972)
114. A.I. Okorokov et al., Zh. Eksp. Teor. Fiz. **69**, 590 (1975)
115. F. Tasset, Phys. B **156–157**, 627 (1989)
116. F. Tasset, P.J. Brown, E. Lelievre-Berna, T. Roberts, S. Pujol, J. Allibon, E. Bourgeat-Lami, Phys. B **267–268**, 69 (1999)
117. P.J. Brown, J.B. Forsyth, F. Tasset, Proc. R. Soc. Lond. A **442**, 147 (1993)
118. F. Tasset, Phys. B **297**, 1 (2001)
119. P.J. Brown, Phys. B **297**, 198 (2001)
120. P.J. Brown, J. Crangle, K.-U. Neumann, J.G. Smith, K.R.A. Ziebeck, J. Phys. Condens. Matter **9**, 4729 (1997)
121. P.J. Brown, Phys. B **14** (1993)
122. A.J. Clune, J. Nam, M. Lee, K.D. Hughey, W. Tian, J.A. Fernández-Baca, R.S. Fishman, J. Singleton, J.H. Lee, J.L. Musfeldt, Npj Quantum Mater. **1** (2019)
123. J.A. Rodríguez-Velamazán, O. Fabelo, J. Campo, J. Rodríguez Carvajal, N. Qureshi, L.C. Chapon, Sci. Rep. **8**, 10665 (2018)

124. S.W. Lovesey, J. Fernández-Rodríguez, J.A. Blanco, P.J. Brown, Phys. Rev. B **70**, 172414 (2004)
125. S.W. Lovesey, E. Balcar, K.S. Knight, J. Fernández, Phys. Rep. **411**, 233 (2005)
126. S. Ji, C. Song, J. Koo, K.-B. Lee, Y.J. Park, J.Y. Kim, J.H. Park, H.J. Shin, J.S. Rhyee, B.H. Oh, Phys. Rev. Lett. **91**, 257205 (2003)
127. J. Blanco, P. Brown, A. Stunault, K. Katsumata, F. Iga, S. Michimura, Phys. Rev. B **73**, 212411 (2006)
128. J. Fernández-Rodríguez, J.A. Blanco, P.J. Brown, K. Katsumata, A. Kikkawa, F. Iga, S. Michimura, Phys. Rev. B **72**, 52407 (2005)
129. B.K. Cho, J.S. Rhyee, J.Y. Kim, M. Emilia, O.C. Canfield, J. Appl. Phys. **97**, 10923 (2005)
130. S. Mühlbauer, D. Honecker, E.A. Périgo, F. Bergner, S. Disch, A. Heinemann, S. Erokhin, D. Berkov, C. Leighton, M.R. Eskildsen, A. Michels, Rev. Mod. Phys. **91**, 1174 (2019)
131. S. Mühlbauer, B. Binz, F. Jonietz, Ch. Pfleiderer, A. Rosch, A. Neubauer, R Georgii, P. Böni, Science **323**, 915 (2009)
132. W. Münzer, A. Neubauer, T. Adams, S. Mühlbauer, C. Franz, F. Jonietz, R. Georgii, P. Böni, B. Pedersen, M. Schmidt, A. Rosch, Ch. Pfleiderer, Phys. Rev. B **81**, 41203R (2010)
133. X.Z. Yu, Y. Onose, N. Kanazawa, J.H. Park, J.H. Han, Y. Matsui, N. Nagaosa, Y. Tokura, Nature **465**, 901 (2010)
134. C. Pappas, E. Lelievre-Berna, P. Falus, P. M. Bentley, E. Moskvin, S. Grigoriev, P. Fouquet, B. Farago, Phys. Rev. Lett. **102**, 197202 (2009)
135. S. Seki, X.Z. Yu, S. Ishiwata, Y. Tokura, Science **336**, 198–201 (2012)
136. A.N. Bogdanov, A. Hubert, J. Magn. Magn. Mater. **138**, 255 (1994)
137. A.N. Bogdanov, D.A. Yablonskii, Sov. Phys. JETP **68**, 101 (1989)
138. A.N. Bogdanov, A. Hubert, J. Magn. Magn. Mater. **195**, 182–192 (1999)
139. V. Laliena, J. Campo, Phys. Rev. B **96**, 1420 (2017)
140. T. Nakajima, H. Oike, A. Kikkawa, E.P. Gilbert, N. Booth, K. Kakurai, Y. Taguchi, Y. Tokura, F. Kagawa, T.H. Arima, Sci. Adv. **3**, e1602562 (2017)
141. T. Nakajima, Y. Inamura, T. Ito, K. Ohishi, H. Oike, F. Kagawa, A. Kikkawa, Y. Taguchi, K. Kakurai, Y. Tokura, T.-H. Arima, Phys. Rev. B **98**, 014424 (2018)
142. J. Campo, J. Luzon, F. Palacio, G. McIntyre, A. Millan, A. Wildes, Phys. Rev. B **78**(5), 054415 (2008)
143. R. Caciuffo, G. Amoretti, A. Murani, R. Sessoli, A. Caneschi, D. Gatteschi, Phys. Rev. Lett. **81**(21), 4744–4747 (1998)
144. I. Mirebeau, M. Hennion, H. Casalta, H. Andres, H. U. Güdel, A.V. Irodova, A. Caneschi, Phys. Rev. Lett. **83**, 628 (1999)
145. E.M. Chudnovsky, D.A. Garanin, Phys. Rev. Lett. **87**(18), 187203 (2001)
146. C. Carbonera, F. Luis, J. Campo, J. Sánchez-Marcos, A. Camón, J. Chaboy, D. Ruiz-Molina, I. Imaz, J. van Slageren, S. Dengler, M. González, Phys. Rev. B **81**(1), 014427 (2010)
147. M. Bée, *Principles and Applications in Solid State Chemistry, Biology and Materials Science* (IOP Publishing, Bristol, 2020)
148. J. Alberto Rodríguez-Velamazán, Miguel A. González, José A. Real, Miguel Castro, M. Carmen Muñoz, Ana B. Gaspar, R. Ohtani, M. Ohba, K. Yoneda, Y. Hijikata, N. Yanai, M. Mizuno, H. Ando, S. Kitagawa, J. Am. Chem. Soc. **134**(11), 5083–5089 (2012)
149. F. Mezei, Z. Phys. A: Hadrons Nucl. **255**(2), 146–160 (1972)
150. F. Mezei, The principles of neutron spin echo, in *Neutron Spin Echo*. Lecture Notes in Physics, ed. by F. Mezei, vol. 128 (Springer, Berlin, 1980)
151. F. Mezei, C. Pappas, T. Gutberlet, *Neutron Spin Echo Spectroscopy: Basics, Trends and Applications*. Lecture Notes in Physics (Springer, Berlin, 2002)
152. K. Weron, J. Phys. Condens. Matter **3**(46), 9151–9162 (1991)
153. R.M. Pickup, R. Cywinski, C. Pappas, B. Farago, P. Fouquet, Phys. Rev. Lett. **102**(9), 97202 (2009)
154. G.P. Felcher, R.O. Hilleke, R.K. Crawford, J. Haumann, R. Kleb, G. Ostrowski, Rev. Sci. Instrum. **58**(4), 609–619 (1987)

155. D.L. Price, F. Fernandez-Alonso, *Neutron Scattering: Magnetic and Quantum Phenomena*. Experimental Methods in the Physical Sciences (Academic Press, London, 2015)
156. H. Zabel, Mater. Today **9**(1–2), 42–49 (2006)
157. V. Lauter-Pasyuk, H.J. Lauter, B.P. Toperverg, L. Romashev, V. Ustinov, Phys. Rev. Lett. **89**(16), 309 (2002)

Part V
High Frequency Magnetization Dynamics

Radio-Frequency (RF) Permeameter

Shingo Tamaru

Abstract Although radio-frequency (RF) permeability measurements have been commonly used for a long time, it is still one of the fast-evolving fields of magnetic characterization, whose development is driven by various electronic systems that require high-performance RF magnetic components. There are several different types of RF permeameters designed to measure the permeability (equivalently the magnetic susceptibility) of magnetic materials having different sample shapes and sizes over a wide range of bandwidths and measurement conditions. Thus, it is important for researchers and engineers to understand the operating principle and measurement capabilities of each RF permeameter to fully utilize available measurement systems. This chapter initially explains the differences between ferromagnetic resonance and RF permeability measurements, as well as requirements imposed on RF permeameters. Next, certain RF permeameters that are commonly used for the characterization of bulk or sheet magnetic samples for RF applications are introduced. Then, a detailed description about one particular type of RF permeameter, which was recently developed by the author of this chapter, is presented. This system, called transformer coupled permeameter (TC-Perm), has achieved significant enhancements in terms of bandwidth and sensitivity when compared to the conventional RF permeameters; consequently, it can measure the permeability of a single particle in a magnetic powder. Descriptions of various RF permeameters, as well as strategies for improving the bandwidth and sensitivity taken in the development of TC-Perm, are expected to serve as a guide for readers to solve problems encountered in the RF permeability measurement of magnetic materials.

Keywords High-frequency permeability · Magnetic susceptibility · Ferromagnetic resonance · Magnetization dynamics · Electromagnetic waves ·

S. Tamaru (✉)
National Institute of Advanced Industrial Science and Technology (AIST),
Tsukuba, Ibaraki, Japan
e-mail: shingo.tamaru@aist.go.jp

Microwave · Inductor · Transformer · Noise suppressor · Soft magnetic material · Powder metallurgy

1 Introduction

Modern electronic systems use various radio-frequency (RF) magnetic components such as inductors, transformers, filters, isolators, attenuators, and noise suppressors. These components often play an important role in the RF circuitry and sometimes limit the overall system performance, because the characteristics of RF magnetic components are generally further from the ideal than any other passive components. Let us consider an inductor as an example. The relation between the voltage v and current i of an inductor is given as $v = L di/dt$, where L is the inductance. It is ideal for L to be a real constant value over the entire operating frequency range. However, this holds only at sufficiently low frequencies; the value of L shows complicated (in most cases deteriorating) behavior as the frequency increases, primarily due to various physical phenomena occurring in the magnetic core. The deviation of L from ideality usually starts from a much lower frequency than in other passive components such as resistors or capacitors. Therefore, it is crucial to study high-frequency dynamics of magnetic materials to improve the characteristics of the RF magnetic components and eventually the overall performance of the electronic systems that use them.

Ferromagnetic resonance (FMR) and RF permeability are the measurements that have been commonly used for the study of magnetization dynamics over the past several decades [1–5]. The systems for measuring FMR and RF permeability are called FMR spectrometer and RF permeameter, respectively. Both these systems measure the magnetic susceptibility $\chi = \mu_r - 1$, where μ_r is the relative permeability, of the material under test (MUT) based on a common principle; however, two major differences exist between them. One is that RF permeameters need to cover a much wider bandwidth than FMR spectrometers, especially on the low-frequency side. The other is that RF permeameters require accurate calibration to obtain the absolute quantity of χ, whereas FMR spectrometers do not.

To illustrate these differences, let us discuss the behavior of a soft magnetic thin film when an in-plane DC bias field H_B and in-plane RF magnetic field H_{RF} orthogonal to H_B are applied as shown in Fig. 1 as the simplest example. H_B saturates the magnetic film, while H_{RF} excites the motion of magnetization. In this film, the total magnetization behaves as a single magnetic moment, whose dynamical motion is described by the Landau-Lifshitz-Gilbert (LLG) equation[6, 7]:

$$\frac{d\vec{M}}{dt} = -\gamma \vec{M} \times \vec{H} + \frac{\alpha}{M_S} \vec{M} \times \frac{d\vec{M}}{dt} \tag{1}$$

where $\vec{M}$ is the magnetization, M_S is the saturation magnetization of the film, $\vec{H}$ is the magnetic field, α is the Gilbert damping parameter, γ is the gyromagnetic ratio given as $\gamma = g q_e \mu_0 / 2 m_e$, g is the Landé g-factor, q_e is the elementary charge

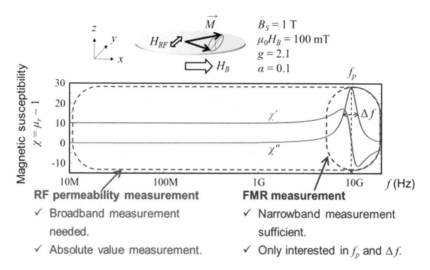

Fig. 1 Magnetic susceptibility χ of a soft magnetic thin film under the applications of an in-plane DC bias field H_B and in-plane RF magnetic field H_{RF} orthogonal to H_B as a function of frequency calculated using Eq. (2) with the magnetic parameters given in the figure. Also shown are the measurement bandwidths and quantities of interest for FMR and RF permeability measurements

(magnitude of the electron charge), μ_0 is the permeability of vacuum, and m_e is the electron mass. In Eq. (1), the first and second terms represent the precessional motion and energy damping, respectively. This equation can be analytically solved by decomposing both $\vec{M}$ and $\vec{H}$ into the static and dynamic terms and assuming that the dynamic terms are much smaller than their static counterparts. For the thin film as shown in Fig. 1, χ is expressed as follows [8, 9]:

$$\chi = \frac{\omega_M(\omega_0 - i\alpha\omega + \omega_M)}{(\omega_0 - i\alpha\omega)(\omega_0 - i\alpha\omega + \omega_M) - \omega^2}, \tag{2}$$

where $\omega = 2\pi f$, $\omega_0 = \gamma H_B$, and $\omega_M = \gamma M_S$. Figure 1 plots χ calculated using Eq. (2) with the magnetic parameters given in the figure, which exhibits certain important features. First, χ shows an FMR response at around 10 GHz, i.e., χ' (real part of χ) crosses zero, while χ'' (imaginary part of χ) becomes maximum. The FMR frequency f_p is given by the Kittel equation for thin films [10]:

$$2\pi f_p = \gamma \sqrt{H_B (H_B + M_S)}. \tag{3}$$

In real magnetic films, the total magnetic field consists of not only H_B but also the anisotropy field H_k and exchange field H_{ex}; thus, H_B in Eq. (3) should in general be replaced with $H_{tot} = H_k + H_{ex} + H_B$. In the frequency range above f_p, χ converges to zero as the frequency increases because the magnetization cannot respond to H_{RF}. In the frequency range below f_p, the magnetization can follow H_{RF} with a smaller phase delay as the frequency decreases, and in the low-frequency limit, χ converges to a real constant given as [11]:

$$\lim_{f \to 0} \chi = \frac{M_S}{H_B}. \tag{4}$$

The major differences between the FMR spectrometer and RF permeameter can be clearly illustrated by this plot. Let us first consider the FMR spectrometer. The primary objective of the FMR measurement is to gain insight into the magnetic properties and internal magnetic state from the FMR peak. The line width Δf of the χ'' peak is proportional to α, which sensitively reflects the energy dissipation processes [7, 12]. The frequency and bias field dependence of f_p provides information about the value of g and the internal magnetic state, because it permits the estimation of H_{tot}, which contains H_k and H_{ex}. A clear FMR peak can be detected at relatively high frequencies (usually above 1 GHz), because most magnetic materials become significantly inhomogeneous when H_{tot} is very weak, which smears out the FMR response at low frequencies. For these reasons, FMR spectrometers are designed to apply a finite H_B to saturate the MUT and measure uncalibrated χ just around the FMR peak in the gigahertz frequency range, from which only f_p and Δf are extracted as quantities of interest. In other words, FMR is a relatively narrow band (in certain systems fixed frequency) measurement, and the χ profile outside the FMR peak and the absolute magnitude of χ are ignored in most FMR measurements.

Conversely, the RF permeability measurement is meant to obtain the absolute quantity of χ over the entire frequency range relevant to the usage of the MUT. This is because the value of χ itself is an important physical quantity in many technological or engineering problems. More specifically, the real and imaginary parts of χ have different physical meanings, i.e., χ' and χ'' represent the energy stored in the form of magnetic fields and the energy dissipation rate, respectively. Inductor and transformer applications require a large χ' and small χ'' for a higher-quality factor, whereas attenuator and noise suppressor applications require the opposite for efficient energy absorption. These contrary requirements for χ lead to a different operating frequency range for each RF magnetic component. Inductors and transformers are mostly used in the kilohertz or megahertz frequency range, which is much lower than f_p, to take advantage of a real constant value of χ given by Eq. (4) [13]. One good example is the switching mode power supply (SMPS). Typical switching frequencies of commercial SMPS lie in low-frequency (LF) and medium-frequency (MF) bands (30 kHz–3 MHz), which is partly limited by the inductors and transformers [14]. Demand for a more compact and energy-efficient SMPS is rapidly growing with the advent of renewable energy, electric vehicles, ubiquitous computing and network infrastructures, and so on, which has promoted the development of various components used in the SMPS, including inductors and transformers, to increase the switching frequency. In real electronic systems, usually no H_B is applied to the magnetic components, and the shape of the magnetic material is not a thin film. In such situations, the magnetic material may take inhomogeneous and complicated magnetization distributions such as multi-domain structure and magnetic vortices. The speeds of domain wall motion and vortex gyration are inherently much slower than the precessional motion of the magnetization; thus, χ in the relatively low-frequency range (below gigahertz) is dominated by these slow physical processes. This necessitates careful material engineering in the development of inductors and transformers used in SMPS to suppress such undesirable magnetic distributions and attain a desired χ in the target frequency range. In addition, RF magnetic components are also used as a decoupling inductor or noise filter to filter out noise in the power line generated by the SMPS. Because such noise is harmonics of the switching frequency of the SMPS, the noise power is distributed from the switching frequency up to high-frequency (HF), very-high-frequency (VHF), and even ultra-high-frequency (UHF) bands (3 MHz–3 GHz). These considerations imply that a very broadband RF permeability measurement covering from LF to UHF bands (30 kHz–3 GHz) is required for the research and development of magnetic materials for inductors and transformers [13–15]. Another example is the noise suppressor that guards the analog circuits in the RF frontend and intermediate frequency domains in wireless communication devices by absorbing electromagnetic noise generated by pulse currents in the digital domain and SMPS, which are distributed in VHF, UHF, and even super-high-frequency (SHF) bands. A strong demand for higher data bandwidth in wireless communications inevitably increases the carrier frequencies. Currently, the 5G wireless network service is ramping up, which initially uses the lower SHF band (3 GHz–6 GHz), but will eventually migrate to the higher SHF band (24 GHz–30 GHz) or even higher frequencies. Noise suppressors need to work even at such high frequencies in accordance with the advancements in wireless technology. RF permeameters need to cover these bands to evaluate magnetic materials for the noise suppressor application too [14, 16, 17].

2 Overview of RF Permeameters

As explained in the previous section, broadband coverage spanning from very low frequencies up to microwave frequencies and accurate calibration to obtain the absolute quantity of χ are indispensable prerequisites for the RF permeameter. In addition, there are other measurement requirements imposed on the RF permeameter. One is high sensitivity, which is a universal requirement for any measurement. Another is the capability to apply H_B during measurement, because valuable information about magnetic properties of the MUT can be obtained by measuring χ as a function of H_B. No single RF permeameter can satisfy all these measurement requirements; thus, the use of multiple RF permeameters is often in order. It is important to understand the measurement principle, strengths, and weaknesses of each RF permeameter to decide which systems to use and to appropriately interpret the measurement results.

RF Permeameters Based on Transmission and Reflection of Propagating Electromagnetic Waves

Broadly speaking, RF permeameters are classified into two groups. One group of RF permeameters are based on the transmission and reflection of propagating electromagnetic waves, which are introduced in this subsection. Figure 2 shows three variations of RF permeameter jigs belonging to this group [18]. Figures 2a–c illustrate the coaxial transmission line [19, 20], rectangular waveguide [21, 22], and free space type [23, 24] jigs, which are referred to as types 1–1, 1–2, and 1–3 in this chapter, respectively. These jigs are connected to a vector network analyzer (VNA) to measure the scattering parameters (S-parameters) of the jig with the MUT inserted. Although these RF permeameters have different electromagnetic wave propagation paths, all of them follow the same measurement theory that was originally developed by Nicolson and Ross and later refined by Weir, which is henceforth referred to as NRW method [25–27]. The basic idea of the NRW method is explained using a schematic of the type 1–1 jig shown in Fig. 2d. The MUT having a relative permittivity ε_r and relative permeability μ_r is inserted into the coaxial transmission line. The characteristic impedance Z_m and phase velocity v of the MUT section are changed to $Z_m = Z_0\sqrt{\mu_r/\varepsilon_r}$ and $v = c/\sqrt{\varepsilon_r\mu_r}$, respectively, where Z_0 is the native characteristic impedance of the coaxial transmission line without the insertion of the MUT (nominally 50 Ω) and c is the speed of light. The reflection coefficient Γ at the incident surface of the MUT and the phase

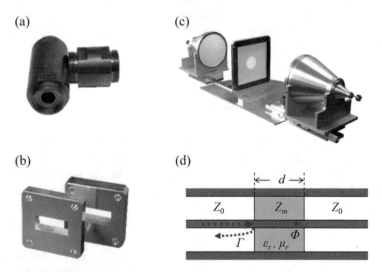

Fig. 2 Measurement jigs used for the NRW method. (**a**) Coaxial transmission line type jig (type 1–1), (**b**) rectangular waveguide type jig (type 1–2), and (**c**) free space type jig (type 1–3). These photos are courtesy of EM Labs, Inc. [18]. (**d**) Model for the derivation of the NRW method based on the type 1–1 jig

rotation ϕ of the electromagnetic wave propagating through the MUT are formulated as $\Gamma = (Z_m - Z_0)/(Z_m + Z_0)$ and $\phi = 2\pi fd/v$, respectively, where d is the thickness of the MUT. Γ and ϕ can be obtained from the reflection and transmission coefficients, i.e., S_{11} and S_{21}, respectively, from which both ε_r and μ_r can eventually be calculated. Readers interested in the rigorous mathematical expressions as well as the derivation of the NRW method taking into account the multiple reflections between both right and left surfaces of the MUT can refer to the original papers [25–27].

The strengths of the NRW method are as follows [28]. First, this method provides both ε_r and μ_r. Second, RF permeameters based on this method can be used up to very high frequencies. In fact, the NRW method with either the type 1–2 or 1–3 jig is the only commercially available solution for measuring μ_r above 40 GHz at this time. The three different types of the jigs shown in Figs. 2a–c are designed for different bandwidths. The type 1–1 jig is compatible with the Amphenol Precision Connector-7 mm (APC-7) standard, which is used on equipment operating from DC up to 18 GHz. The type 1–2 jigs conform to the Electronic Industries Alliance (EIA) standard waveguide dimensions and can be used to frequencies ranging from 0.96 to 110 GHz, although the entire frequency range is divided into 12 subbands and a different size of jig is required for each subband. The type 1–3 jigs that cover 2.6–26.5, 26.5–110, and 18–110 GHz are commercially available; however, the sizes of these jigs, especially the one for the lowest band, are very large (over 2 m in height for the lowest frequency band), because the diameter of the dielectric lenses must be at least several times larger than the wavelength to have a sufficiently collimated microwave beam.

Although the NRW method-based systems are very powerful techniques for measuring the electromagnetic properties of the MUT, they also have certain limitations as mentioned below. Obviously, the type 1–1 jig requires the MUT to be accurately shaped into a toroid to fit into the jig, because a clearance acts as a series capacitance to the MUT, which significantly alters the electric field distribution, especially when ε_r is large, which results in measurement errors. These measurements rely not on the difference taken under two different conditions but on the absolute values of S-parameters measured by a VNA; consequently, they are prone to be affected by drifts and calibration errors of the VNA, uncertainties in the MUT shape and position, and variations in connector mating and cable routing. All these types require the sample size to be sufficiently large to fill the entire wave propagation path, which implies that a high sensitivity measurement on a tiny sample is impossible in principle. Although the type 1–1 jig can technically work from very low frequencies, the measurement accuracy deteriorates as the frequency decreases, because ϕ is proportional to the frequency and thus becomes very small at low frequencies. None of these types of measurement systems can apply H_B to the MUT during measurement because of the following reasons. For the type 1–1 jig, the application of H_B breaks the coaxial symmetry. For the type 1–2 and 1–3 jigs, the jig and sample size are too large to fit into a typical electromagnet.

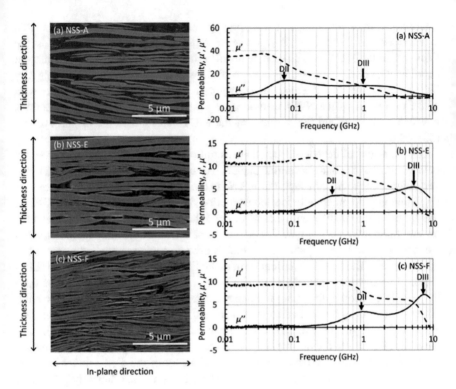

Fig. 3 SEM micrograph of the cross section of three different NSS samples and its permeability measured by the NRW method using the type 1–1 jig over 10 MHz–9 GHz. The compositions of (**a**) NSS-A, (**b**) NSS-E, and (**c**) NSS-F are Fe-9.8% Si-6.4%Al, Fe-14%Al, and Fe-50%Co, respectively [29]

To compensate for the low sensitivity of the NRW method using the type 1–1 jig at low frequencies, a measurement system dedicated for low frequencies (<1 GHz) is also commercially available [28, 30, 31]. In this system, a toroidal-shaped sample is inserted into a short-terminated end of a coaxial jig that forms a half-turn coil. The impedances of the jig with and without the sample are measured using an impedance analyzer, and the change of the impedance is converted into μ_r. This system can be used for frequencies as low as 1 kHz depending on the impedance analyzer used and the MUT thickness. However, this system cannot apply H_B either for the same reason as for the type 1–1 jig, i.e., the loss of coaxial symmetry.

Figure 3 shows an example of the permeability measurement by the NRW method [29]. In this work, noise suppression sheets (NSS) were prepared by dispersing magnetic powder of different particle sizes (ranging from a few tens to a few hundreds of micrometers) and materials (either FeSiAl or CoFe) into the polymer binder and solidifying it into a sheet with a thickness of 0.1 mm. They were cut into a toroidal shape with an outer diameter of 7.0 mm and inner diameter of 3.04 mm and inserted into an APC-7 coaxial transmission line jig (type 1–1),

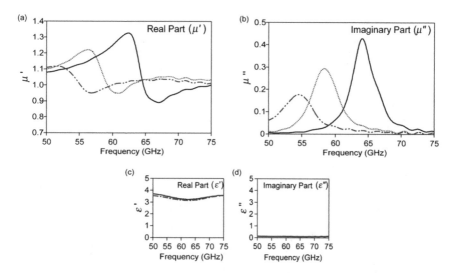

Fig. 4 (a) Real part of μ_r, (b) imaginary part of μ_r, (c) real part of ε_r, and (d) imaginary part of ε_r of ε-$Ga_xFe_{2-x}O_3$ with $x = 0.51$ (solid line), 0.56 (dotted line), and 0.61 (broken line) [17]

and its S-parameters were measured using a VNA (Keysight ENA5080A) over 10 MHz–9 GHz. The results clearly show two dispersion peaks in $\mu^{\prime\prime}$, which are labeled as DII and DIII. They are speculated to originate from FMR responses due to an internal effective field and magnetoelastic effects at the particle surface, respectively; however, their mechanisms are not fully understood yet.

Figure 4 shows another example of the permeability measurement by the NRW method [17]. In this work, a series of ε-$Ga_xFe_{2-x}O_3$ ($x = 0.51, 0.56, 0.61$) nanoparticle samples were prepared by a combination method of reverse-micelle and sol-gel techniques. Their ε_r and μ_r were measured by the NRW method using the free space type jig (type 1–3) over 50–75 GHz. The μ_r plots clearly show the FMR response at different frequencies for different compositions due to a high crystalline anisotropy field H_k, while ε_r plots exhibit no particular feature.

RF Permeameters Based on Inductive Response

The other group of RF permeameters is based on the inductive response of the MUT. The fundamental difference between this group and the other group introduced in the previous subsection is that this group eliminates the effect of electric fields and extracts only the magnetic response of the MUT. Figures 5a–c show three types of commercially available RF permeameters belonging to this group, which are referred to as types 2–1, 2–2, and 2–3 in this chapter, respectively. In the type 2–1 permeameter, the MUT is placed on top of a through transmission line, which is

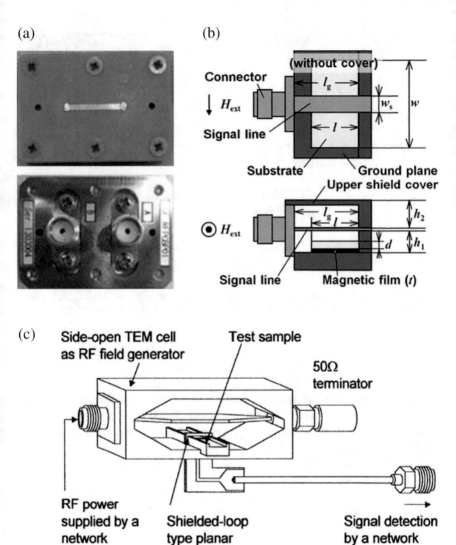

Fig. 5 Three types of RF permeameter jigs based on the inductive response of the MUT. (**a**) Through MSL jig (type 2–1) [32], (**b**) short-terminated MSL jig (type 2–2) [33], and (**c**) TEM cell and shielded loop coil jig (type 2–3) [34]

either a coplanar waveguide (CPW) or microstrip line (MSL), and a VNA measures S_{21} [32, 35–37]. This is basically the standard "through mode FMR" jig structure. In the type 2–2 permeameter, the MUT is inserted into the short-terminated end of an MSL where the magnetic field is maximum, while the electric field is minimum, and a VNA measures S_{11}. This is basically the standard "reflection mode FMR" jig structure [33, 38–44]. In the type 2–3 permeameter, the MUT is inserted into a transverse electromagnetic (TEM) cell, and a shielded loop coil wound around the MUT picks up magnetic fluxes while rejecting electric fluxes. A VNA sends a stimulus signal into the TEM cell and detects the signal induced in the pickup coil in the form of S_{21} [34, 45–47]. All these types of RF permeameters are designed to apply H_B and measure S-parameters under two magnetic fields, of which one is the field of interest (the measurement field denoted as H_B^m) and the other is a field that is sufficient to saturate the magnetic sample (the saturation field denoted as H_B^s). The difference in S_{11} or S_{21} between these two fields is converted into χ by applying a calibration algorithm appropriate for each measurement configuration while eliminating the effect of the electric fields.

Although these types of RF permeameters can theoretically work from very low frequencies up to microwave frequencies, because none of these types of jigs have a cutoff frequency, each of them has a practical bandwidth limit determined by various technical reasons. Regarding the high-frequency limit, the type 2–1 jig can theoretically work at very high frequencies, up to 100 GHz and beyond, if a properly designed transmission line is formed on top of a low loss substrate. Currently, broadband RF permeameters that can be used up to 40 GHz are commercially available [32, 36, 37]. The high-frequency limit of the type 2–2 jig is somewhat lower (10–30 GHz) because of the limitation in the jig structure that it must accommodate the MUT between the signal line and the ground plane of the MSL [33, 44]. The high-frequency limit of the type 2–3 jig is set by the self-resonant frequency of the pickup coil, which is 9 GHz at present [34]. The low-frequency limit is set by the fluctuation of the receiver sensitivity of the VNA, defined as the trace noise, in the type 2–1 and 2–2 jigs. This is because the signal of interest is a counter-electromotive force (CEMF) induced by the motion of the magnetization in the MUT; therefore, its amplitude is proportional to the frequency, while the stimulus signal that excites the motion of magnetization is constant over the entire measurement bandwidth. In the type 2–1 and 2–2 jigs, both the signal of interest and stimulus signal acting as a background are sent to the receiver of the VNA. This necessitates the VNA to extract a very small signal of interest from a much larger background signal that generates trace noises in proportion to it at low frequencies. Because of this problem, the signal-to-noise ratio (SNR) in the type 2–1 and 2–2 jigs rapidly deteriorates as the frequency decreases, resulting in a practical low-frequency limit of 10–100 MHz depending on the sample volume. The low-frequency limit of the type 2–3 jig is somewhat lower than that of the type 2–1 or 2–2 jigs because the stimulus signal sent to the TEM cell is inductively coupled to the pickup coil, resulting in the leakage of the stimulus signal from the TEM cell to the pickup coil proportional to the frequency. Therefore, both the signal of interest and the background signal reaching the receiver of the VNA become smaller at the same rate as the frequency decreases in the type 2–3 jig, allowing the VNA to extract the signal of interest from a small background signal, leading to significant improvement in the SNR at low frequencies. The low-frequency limit of the type 2–3 jig is ultimately set by the noise figure of the receiver of the VNA.

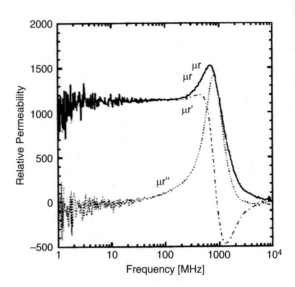

Fig. 6 Permeability profile of a 0.2-μm-thick $Co_{85}Nb_{12}Zr_3$ amorphous film, measured by the TEM cell and shielded loop coil jig (type 2–3) [34]

These types of RF permeameters are basically designed to measure film or sheet-shaped samples. The sensitivity of these inductive response-based techniques is inversely proportional to the square of the signal line width. In the type 2–1 jig, it is possible to make the signal line width of the CPW narrow, resulting in significant enhancement in the sensitivity, although the sensitivity at low frequencies is not very high owing to the aforementioned problem. In the type 2–2 and 2–3 jigs, the typical signal line widths range from a few millimeters up to a few centimeters, and it is practically impossible to make it narrower because the jigs of these types must accommodate the MUT inside, which limits the sensitivity of these types of RF permeameters.

Figure 6 shows an example of the permeability measurement result obtained by using the TEM cell and shielded loop coil jig (type 2–3) over 1 MHz–9 GHz [34]. The sample was a 0.2-μm-thick $Co_{85}Nb_{12}Zr_3$ amorphous film with a lateral size of 4 mm × 4 mm. This figure shows an FMR response at approximately 850 MHz and a low-frequency permeability of 1200. The SNR of the measurement result is reasonably good down to 1 MHz, owing to the type 2–3 jig structure in which both the signal of interest and the leakage of the stimulus signal from the TEM cell to the pickup coil become smaller at the same rate as the frequency decreases.

3 Transformer Coupled Permeameter

According to the discussion in the previous section, it may appear that an appropriate combination of multiple RF permeameters can satisfy most of the measurement requirements. In reality, each RF permeameter generally requires a different sample shape and size. Furthermore, the details of the signal detection mechanism and the calibration algorithm vary for each type of RF permeameter. Thus, it is often the case that measurement results obtained by using different RF permeameters do not accurately overlap each other around the transition frequency, causing serious difficulty in interpreting the results. The toughest requirement is a high sensitivity over the entire measurement bandwidth, which cannot be achieved by any combination of the conventional RF permeameters, despite a strong need in the development of RF magnetic components as outlined below.

Figure 7 shows a typical manufacturing flow of RF magnetic components. Most of the RF magnetic components are made of magnetic powder. Their manufacturing flow begins from an ingot of the raw magnetic material. It is either grounded or atomized into powder, which undergoes various processes such as annealing, mixing, and attrition. It is then solidified into the final component form by either dispersing into a polymer binder or sintering. The characterization, i.e., RF permeability measurement, is performed at the final stage of the manufacturing flow. Thus, the effect of each process performed when the magnetic material is in the powder form on the high-frequency characteristics can only be inferred from the measurement result obtained in the final component form. It is clearly advantageous to perform the characterization of the magnetic material in the earlier stages, preferably on a single particle in the magnetic powder, to clarify the effect of each process. However, as mentioned above, no RF permeameter commercially available at this time has a sufficiently high sensitivity to measure the permeability of a single magnetic particle. Because of these problems, a broadband and high sensitivity RF permeameter has been strongly sought for.

The author has developed an RF permeameter to overcome the problems associated with the commercially available conventional RF permeameters; consequently, significant enhancement in the measurement bandwidth and sensitivity when compared to the abovementioned conventional RF permeameters has been realized [48]. This RF permeameter is referred to as transformer coupled permeameter (TC-Perm). This chapter presents the details of TC-Perm, including its operation principle, system configuration, calibration algorithm, and sensitivity demonstration. Although this system is still under development and therefore has not been commercialized yet, it is hoped that the description about TC-Perm will help readers solve various problems encountered during the high-frequency characterization of magnetic materials.

System Overview of TC-Perm

Figure 8(a) is a schematic of the jig structure of TC-Perm. Two short-terminated CPWs (CPW1, CPW2) sandwich the MUT. These two CPWs are insulated by a 35-μm-thick Kapton tape, thereby forming a loosely coupled transformer. H_B is applied parallel to the CPWs. Figure 8(b) shows a block diagram of the system configuration and cross section of the jig. CPW1 is connected to port 1 (P1), while CPW2 is connected to receiver 2 (R2) of a VNA through a 10 dB attenuator. The

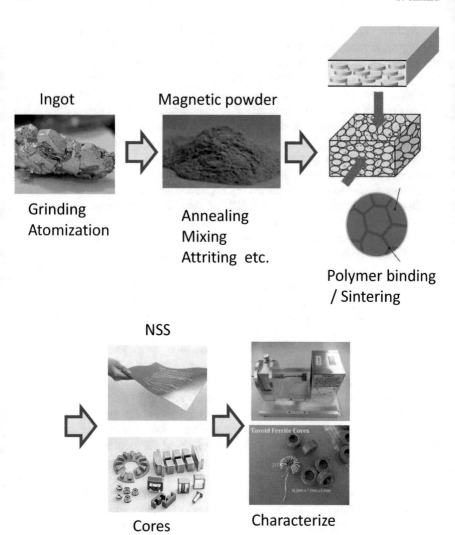

Fig. 7 Typical manufacturing flow of RF magnetic components. Photos and illustrations are courtesy of Tokin Corp. (NSS), Sumitomo Electric (compacted magnetic cores), and KEYCOM Corp. (permeameter)

reason for this connection is as follows. The VNA model used in this system (Keysight N5222A) has a directional coupler at ports 1 and 2 to separate the incoming reflected signal from the outgoing stimulus signal, and the reflected signal is sent to the receiver. The coupling strength of the directional coupler is proportional to the frequency, which weakens the signal reaching the receiver as the frequency decreases. Therefore, the signal from CPW2 should bypass the directional coupler and be directly fed to the receiver. The input impedance of the receiver was

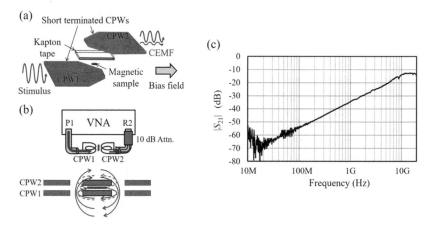

Fig. 8 (**a**) Schematic of the jig structure of TC-Perm. Two short-terminated CPWs (CPW1, CPW2) sandwich the MUT. These CPWs are insulated by Kapton tape and thus form a loosely coupled transformer. H_B is applied parallel to the CPWs. (**b**) Block diagram of the system configuration and cross section of the jig. A VNA sends a stimulus signal to CPW1, which generates H_{RF} and excites the motion of magnetization in the MUT. Magnetic fluxes, one from the MUT (shown as blue broken lines) and the other directly from CPW1 (shown as red solid lines), both thread CPW2 and induce electrical signals, which are detected by the VNA. (**c**) Magnitude of the transmission coefficient $|S_{21}|$ of the entire signal propagation path including the two cables and jig as a function of frequency [48]

measured to be approximately 65 Ω in the particular VNA used in this work; thus a 10 dB attenuator is inserted to suppress reflections from R2. The VNA sends a stimulus signal from P1 to CPW1, which generates H_{RF} and excites the motion of magnetization in the MUT. Magnetic fluxes, one from the MUT and the other directly from CPW1, both thread CPW2 and induce electrical signals, which are sent to R2 through the 10 dB attenuator and measured by the VNA in the form of S_{21}.

This jig structure provides the following benefits for achieving high sensitivity. First, it is easy to make the signal line of the two CPWs narrow. Because the signal amplitude is inversely proportional to the square of the signal line width as explained previously, this leads to a dramatic enhancement of the sensitivity. In the jig of TC-Perm developed in this work, the signal line width is approximately 200 μm, which is one to two orders of magnitude narrower than the typical signal line widths of the conventional permeameters using an MSL (type 2–2) or TEM cell (type 2–3). Second, the two CPWs are positioned very close to the MUT, thus maintaining tight couplings. Third, the stimulus signal leaking from CPW1 to CPW2 becomes smaller as the frequency decreases below approximately 10 GHz, as shown in Fig. 8(c), because these two CPWs are inductively coupled. This implies that TC-Perm does not have to extract the signal of interest from the much larger stimulus signal, similar to the RF permeameters using a TEM cell (type 2–3), which significantly enhances the sensitivity at low frequencies when compared to the conventional RF permeameters using a through (type 2–1) or short-terminated (type 2–2) transmission line, in which a very large stimulus signal reaches the receiver of the VNA along with the signal of interest. In the present TC-Perm system, the low-frequency limit is set by the VNA; thus, it may be extended to an even lower frequency by changing the VNA. On the other hand, the high-frequency limit is currently set by the jig, which shows a notch at approximately 22 GHz in S_{21}. Therefore, the measurement bandwidth is fixed between 10 MHz and 20 GHz in this section unless specified otherwise. The cause of the notch is speculated to be unwanted reflections somewhere in the jig structure; however, the actual cause has not been identified yet. Further extension of the measurement bandwidth by improving the jig structure and system configuration remains one of the important future tasks in the development of TC-Perm.

As in the case of the conventional RF permeameters based on the inductive response of the MUT, the VNA measures S_{21} of the entire signal propagation path from P1 all the way through R2 under two bias magnetic fields, H_B^m and H_B^s, in TC-Perm as well, which are denoted as S_{21}^m and S_{21}^s, respectively. Throughout this section, the saturation field is fixed at $\mu_0 H_B^s = 1.5$ T. The following two conditions are assumed. The first is that the reflection from R2 is negligibly small, which is guaranteed by the insertion of the 10 dB attenuator. The second is that the change in S_{21} between the two bias fields is much smaller than the average, i.e., $\left| S_{21}^m - S_{21}^s \right| \ll \left| \left(S_{21}^m + S_{21}^s \right) / 2 \right|$. If both these conditions are satisfied, it can be shown that the normalized S_{21} change defined as:

$$\Delta S_{21} = \frac{\overline{S_{21}^m} - \overline{S_{21}^s}}{\overline{S_{21}}}, \tag{5}$$

where $\overline{S_{21}^m}$, $\overline{S_{21}^s}$, and $\overline{S_{21}}$ are the averages of the corresponding S-parameter, is independent of the transmission characteristics of the cables connecting the jig and VNA and the calibration errors of the VNA. In general, the transmission characteristics of the cables can fluctuate for each measurement due to various factors such as different cable bending or routing, torque when tightening connectors, and a change in the ambient temperature. The introduction of ΔS_{21} ensures that the measurement results obtained by TC-Perm are robust against these fluctuation factors and eliminates the need for VNA calibration. ΔS_{21} is basically proportional to the difference in χ between H_B^m and H_B^s; however, it is still influenced by other parameters such as the frequency dependence of the jig's sensitivity, residual signal of the jig, and difference in the signal strength for each sample. The calibration algorithm for removing these influences is presented in the following subsection.

Calibration Algorithm of TC-Perm

Our approach for the calibration uses a reference sample as shown in Fig. 9(a): a 100-nm-thick permalloy (Py) coupon directly sputter deposited on Kapton tape. It

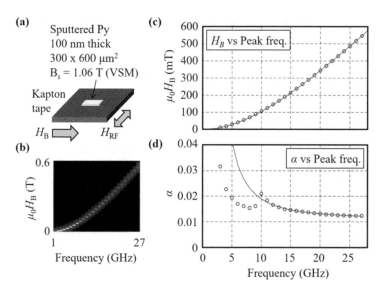

Fig. 9 (**a**) Schematic of the reference sample. (**b**) 2D mapped amplitude of the FMR spectra measured on the reference sample as a function of frequency and H_B. (**c**) H_B of the FMR peak as a function of frequency. The open circle and solid line are the experimentally observed peak and the fit of Eq. (3) to the experimental result, respectively. (**d**) α as a function of frequency. The open circle and solid line are the experimentally observed damping determined by the algorithm given in Ref. [49] and the fit of Eq. (6) to the experimental result for frequencies above 12 GHz, respectively [48]

has a lateral dimension of approximately 300×600 μm by masking the outside of the deposition area during sputtering. The saturation induction $B_S = \mu_0 M_S$ was estimated as 1.06 T using a vibrating sample magnetometer (VSM) measurement on a blanket film. This film showed a very soft magnetic property with the coercivity of less than 1 mT in the VSM measurement; thus, the anisotropy is ignored in the following discussion. This reference sample is attached on top of the signal line of CPW1 with its long axis along the CPW direction.

First, the FMR is measured for this reference sample using the technique described in Ref. [49]. Fig. 9(b) shows the 2D mapped amplitude of the FMR spectra as a function of frequency and H_B, which exhibits a clean Kittel mode, i.e., coherent precession as a single magnetic moment. This indicates that its magnetization dynamics are described by the LLG equation and its derivatives, i.e., Eqs. (1)–(3). By fitting Eq. (3) to the measured peak shown in Fig. 9(c), g is determined to be 2.055 for the reference sample. α at each frequency is determined by the algorithm described in Ref. [49], which is plotted in Fig. 9(d). The result shows an unusual dependence of α on frequency, i.e., a peak appearing at approximately 10 GHz, which corresponds to $\mu_0 H_B \approx 100$ mT. The physical origin of this peak is not understood at this time. One possible explanation is that the Kapton tape may have some surface roughness with a specific length scale, which causes two magnon scattering at 10 GHz. Because the purpose of the FMR measurement in this work is to determine the value of α for H_B under which the permeability of the reference sample is measured for the theoretical calculation of χ, the behavior of α around and below this peak is ignored in the following discussion. An empirical equation for the dependence of α on H_B is given as:

$$\alpha = \frac{A}{\mu_0 H_B} + B, \tag{6}$$

where A and B are constants that represent the extrinsic and intrinsic damping terms, respectively. The frequencies are converted to H_B by Eq. (3), and then Eq. (6) is fitted to the measured α shown in Fig. 9(d) above 12 GHz (corresponding to $\mu_0 H_B = 150$ mT), from which $A = 8.5 \times 10^{-4}$ and $B = 1.07 \times 10^{-2}$ are obtained. Next, TC-Perm is calibrated using the magnetic parameters obtained from the VSM and FMR measurements at $\mu_0 H_B^m = 400$ mT. The theoretical expression of χ is given in Eq. (2). Because TC-Perm measures S_{21} under H_B^m and H_B^s and calculates ΔS_{21} using Eq. (5), ΔS_{21} should reflect the difference of χ between these two bias fields, i.e.,

$$\chi_d(f) = \chi\left(H_B^m, f\right) - \chi\left(H_B^s, f\right). \tag{7}$$

The calibration algorithm is based on the following equation:

$$\Delta S_{21} = \frac{\chi_d(f)}{C_m S(f)} + R(f), \tag{8}$$

where C_m is the ratio of the signal magnitude relative to that of the reference sample, which depends on the product of the total magnetization of the MUT and the coupling strengths between the MUT and the CPWs, $S(f)$ is the inverse of the dependence of the sensitivity of the jig on frequency, and $R(f)$ is the residual signal that the jig exhibits in response to the change in H_B even without a magnetic sample.

For the calibration, ΔS_{21} is measured with and without inserting the reference sample into the jig under $\mu_0 H_B^m = 400$ mT, the results of which are shown in Figs. 10(a, b), respectively. Fig. 10(c) shows the theoretically calculated χ_d using Eqs. (2) and (7) with $\mu_0 H_B^m = 400$ mT and $\mu_0 H_B^s = 1.5$ T. $C_m = 1$ when the reference sample is measured by definition. By plugging all these values into Eq. (8), $S(f)$ can be determined as shown in Fig. 10(d). For the actual calibration, $S(f)$ is smoothed out by taking a moving average to reduce the random noise contained in the calibration measurements, shown as green lines overlapping on top of each of $\text{Re}[S(f)]$ and $\text{Im}[S(f)]$ in Fig. 10(d).

To validate this calibration algorithm, the reference sample was measured under different values of $\mu_0 H_B^m$. Fig. 11 shows the comparison of χ_d obtained experimentally and calibrated by the above algorithm with the theoretical calculation using Eqs. (2) and (7). For $\mu_0 H_B^m = 600$ mT, the experimentally obtained χ_d shows a small deviation from the theoretical calculation at higher frequencies, as shown in Fig. 11(a). One possible reason for this deviation is the contributions of the perpendicular components of H_{RF} and magnetization precession. The width

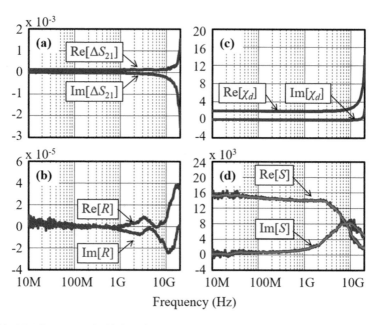

Fig. 10 (a) ΔS_{21} measured with the reference sample inserted into the jig under $\mu_0 H_B^m = 400$ mT. (b) Residual signal of the jig $R(f)$ measured with no sample inserted into the jig under $\mu_0 H_B^m = 400$ mT. (c) $\chi_d(f)$ calculated using Eqs. (2) and (7). (d) Inverse of the frequency dependence of the sensitivity of the jig $S(f)$ obtained by plugging the results shown in (a), (b), and (c) and $C_m = 1$ into Eq. (8). The green lines overlapping on top of both Re[S] and Im[S] are the moving average of $S(f)$ [48]

of the reference sample of 300 μm is somewhat larger than the signal line width of 200 μm, and thus the edges of the reference sample should be exposed to perpendicular components of H_{RF} with an anti-symmetric profile. Furthermore, the precession trajectory of magnetization becomes closer to circular as H_B increases, which should give rise to a finite coupling between these perpendicular and spatially nonuniform components of H_{RF} and the magnetization. On the other hand, the theoretical calculation based on Eq. (2) assumes that H_{RF} is completely in-plane and uniform over the entire sample width. This oversimplification may result in the deviation seen in the figure. Further studies to clarify and eliminate the source of the deviation are planned to extend the range of H_B over which TC-Perm can be used.

For $\mu_0 H_B^m = 400$ and 200 mT, the experimentally obtained and theoretically calculated χ_d show good quantitative agreement over the entire frequency range. Under $\mu_0 H_B^m = 200$ mT, not only the overall χ_d profile but also the shape of the FMR response appearing at approximately 14.4 GHz agrees quite well. These results confirm that the calibration algorithm developed in this work provides a quantitatively accurate χ_d over the entire frequency range under $\mu_0 H_B^m$ that is at least equal to or weaker than 400 mT.

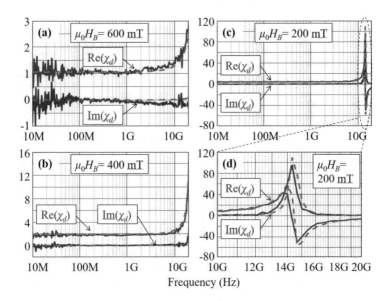

Fig. 11 (**a, b, c**) χ_d measured for the reference sample under $\mu_0 H_B^m = 600$, 400, and 200 mT, respectively. (**d**) Magnified plot of (**c**) between 10 GHz and 20 GHz to highlight the FMR peak. In these plots, the blue and red lines are the real and imaginary parts of χ_d measured by TC-Perm after calibration. The orange and green broken lines are the real and imaginary parts of the theoretical χ_d calculated using Eqs. (2) and (7) under the condition corresponding to each plot [48]

Sensitivity Demonstration of TC-Perm

A single Py flake particle comprising a noise suppression sheet was measured using TC-Perm to demonstrate its sensitivity. A bulk Py was ground and attrited into flake powder, from which one particle was picked. Figure 12(a) shows an optical micrograph of the particle. It has lateral dimensions of approximately $100 \times 180 \,\mu m$ and a thickness of approximately 0.7 μm. B_S is assumed to be the same as that of the bulk (1 T). This particle was attached on top of CPW1 with its short axis along the CPW.

First, χ_d between 10 and 100 MHz was measured under $\mu_0 H_B^m = 200$, 400, 600, and 800 mT. The real parts of χ_d calibrated by plugging the measured $\Delta S_{21}(f)$, $C_m = 1$, $S(f)$ obtained in the previous subsection, and $R(f) = 0$ (because $R(f)$ is confirmed to be negligibly small at low frequencies) into Eq. (8) are plotted in Fig. 12(b). Under a H_B^m sufficiently strong to saturate the magnetization, χ_d in the low-frequency limit can be calculated by using Eq. (4) as:

$$\lim_{f \to 0} \chi_d(f) = M_s \left(\frac{1}{H_B^m} - \frac{1}{H_B^s} \right). \tag{9}$$

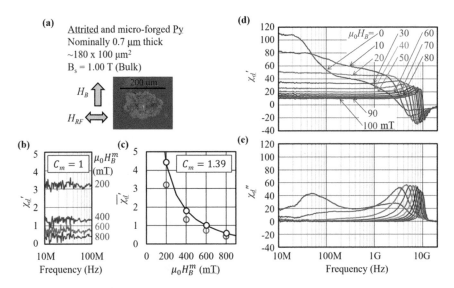

Fig. 12 (a) Optical micrograph of the Py flake particle measured using TC-Perm. (b) Real part of χ_d over 10 MHz–100 MHz under $\mu_0 H_B^m = 200, 400, 600,$ and 800 mT, calibrated with $C_m = 1$. (c) Comparison of χ_d in the low-frequency limit between the experimental and theoretical values. The black solid line is the theoretical values calculated using Eq. (9). The colored and black open circles show the average of the experimental values of χ_d calibrated with $C_m = 1$ and 1.39, respectively. (d, e) Real and imaginary parts of χ_d under $\mu_0 H_B^m$ varied from 0 mT to 100 mT measured using TC-Perm. These results are calibrated with $C_m = 1.39$ [48]

The average of χ_d experimentally measured under each $\mu_0 H_B^m$ and the theoretical value of χ_d calculated using Eq. (9) are shown as colored circles and a black solid line, respectively, in Fig. 12(c). The experimentally obtained χ_d is smaller than the theoretical values, because the MUT in this measurement gives a somewhat smaller signal than the reference sample. By adjusting the value of C_m to 1.39, the experimentally obtained χ_d moves to the black circles and agrees very well with the theoretical values.

This Py particle that was measured under $\mu_0 H_B^m$ varied from 0 to 100 mT with a step of 10 mT and was calibrated by the algorithm described in the previous subsection with $C_m = 1.39$. Figs. 12(d, e) show the real and imaginary parts of χ_d, respectively. The SNR in this measurement is very high over the entire frequency range and all values of $\mu_0 H_B^m$. These figures clearly indicate that the particle is not saturated; therefore, its magnetization dynamics are dominated by either domain wall motion or vortex gyration, giving rise to large values of χ_d at low frequencies when $\mu_0 H_B^m$ is weaker than 20 mT. When $\mu_0 H_B^m$ is equal to or stronger than approximately 20 mT, the particle is almost saturated, giving rise to a fairly clean FMR spectrum, whose resonant frequency shifts higher as a stronger $\mu_0 H_B^m$ is applied to the particle.

4 Summary

This chapter initially presented the measurement requirements that must be satisfied by the RF permeameter, which include the broadband coverage spanning from very low frequencies up to microwave frequencies, calibrated measurement to provide the absolute quantity of χ, high sensitivity, and capability to apply H_B during measurement. Next, various RF permeameters were introduced, which belong to either of the two groups, of which one is based on the transmission and reflection of propagating electromagnetic waves (NRW method) and the other is based on the inductive response of the MUT. Each of these RF permeameters has its own strengths and weaknesses; thus, the use of multiple RF permeameters is often in order. Measurement capabilities and sample requirements for each system were explained to help readers choose the appropriate RF permeameters for their measurement needs. Then, a detailed description about TC-Perm, which was recently developed by the author of this chapter, was presented. It was shown that TC-Perm achieved a significant enhancement in the sensitivity and bandwidth, thus making it possible to measure a single particle in magnetic powder, which should be a powerful tool for tracking down the effects of various processes performed on the magnetic powder on the high-frequency characteristics.

References

1. J.H.E. Griffiths, Anomalous high-frequency resistance of ferromagnetic metals. Nature **158**(4019), 670–671 (1946)
2. G. Eichholz, G.F. Hodsman, High-frequency permeability of ferromagnetic materials. Nature **160**(4061), 302–303 (1947)
3. M.H. Johnson, G.T. Rado, Ferromagnetism at very high frequencies II. Method of measurement and processes of magnetization. Phys. Rev. **75**(5), 841 (1949)
4. G.T. Rado, Ferromagnetic phenomena at microwave frequencies. Adv. Electron **2**, 251–295 (1950)
5. P.H. Haas, A radio-frequency Permeameter. J. Res. NBS **51**, 5 (1953)
6. L. Landau, and E. Lifshits, On the theory of the dispersion of magnetic permeability in ferromagnetic bodies. Perspectives in Theoretical Physics. Pergamon, 51–65 (1992)
7. T.L. Gilbert, A phenomenological theory of damping in ferromagnetic materials. IEEE Trans. Magn. **40**(6), 3443–3449 (2004)
8. G. Bate, D.J. Craik, *Magnetic oxides part 2* (Wiley Interscience, New York, 1975), p. 698
9. S. Tamaru, J.A. Bain, R.J.M. Veerdonk, T.M. Crawford, M. Covington, M.H. Kryder, Measurement of magnetostatic mode excitation and relaxation in permalloy films using scanning Kerr imaging. Phys. Rev. B **70**(10), 104416 (2004)
10. C. Kittel, On the theory of ferromagnetic resonance absorption. Phys. Rev. **73**(2), 155 (1948)
11. Y. Shimada, J. Numazawa, Y. Yoneda, A. Hosono, Absolute value measurements of thin film permeability. IEEE Transl. J. Magn. Jpn. **6**, 987–993 (1991)
12. M.C. Hickey, J.S. Moodera, Origin of intrinsic Gilbert damping. Phys. Rev. Lett. **13**(137601), 102 (2009)
13. M.K. Kazimierczuk, *High-Frequency Magnetic Components* (John Wiley & Sons, 2009)

14. D.J. Perreault, et al. Opportunities and challenges in very high frequency power conversion. 2009 Twenty-Fourth Annual IEEE Applied Power Electronics Conference and Exposition. IEEE, (2009)
15. M. Yamaguchi, K.H. Kim, S. Ikeda, Soft magnetic materials application in the RF range. J. Magn. Magn. Mater. **304**(2), 208–213 (2006)
16. S. Yoshida, H. Ono, S. Ando, F. Tsuda, T. Ito, Y. Shimada, M. Yamaguchi, K. Arai, S. Ohnuma, T. Masumoto, High-frequency noise suppression in downsized circuits using magnetic granular films. IEEE Trans. Magn. **37**(4), 2401–2403 (2001)
17. A. Namai, S. Kurahashi, H. Hachiya, K. Tomita, S. Sakurai, K. Matsumoto, T. Goto, S. Ohkoshi, High magnetic permeability of ε-$Ga_xFe_{2-x}O_3$ magnets in the millimeter wave region. J. Appl. Phys. **107**.9, 09A955 (2010)
18. EM labs, Inc. https://www.emlabs.jp/en/index.php
19. O. Acher, J.L. Vermeulen, P.M. Jacquart, J.M. Fontaine, P. Baclet, Permeability measurement on ferromagnetic thin films from 50 MHz up to 18 GHz. J. Magn. Magn. Mater. **136**(3), 269–278 (1994)
20. D. Ba, P. Sabouroux, Epsimu, a toolkit for permittivity and permeability measurement in microwave domain at real time of all materials: Applications to solid and semisolid materials. Microw. Opt. Technol. Lett. **52**(12), 2643–2648 (2010)
21. H. Bayrakdar, Complex permittivity, complex permeability and microwave absorption properties of ferrite–paraffin polymer composites. J. Magn. Magn. Mater. **323**(14), 1882–1885 (2011)
22. N.N. Al-Moayed, M.N. Afsar, U.A. Khan, S. McCooey, M. Obol, Nano ferrites microwave complex permeability and permittivity measurements by T/R technique in waveguide. IEEE Trans. Magn. **44**(7), 1768–1772 (2008)
23. D.K. Ghodgaonkar, V.V. Varadan, V.K. Varadan, Free-space measurement of complex permittivity and complex permeability of magnetic materials at microwave frequencies. IEEE Trans. Instrum. Meas. **39**(2), 387–394 (1990)
24. J. Munoz, M. Rojo, A. Parrefio, J. Margineda, Automatic measurement of permittivity and permeability at microwave frequencies using normal and oblique free-wave incidence with focused beam. IEEE Trans. Instrum. Meas. **47**(4), 886–892 (1998)
25. A.M. Nicolson, G.F. Ross, Measurement of the intrinsic properties of materials by time-domain techniques. IEEE Trans. Instrum. Meas. **19**(4), 377–382 (1970)
26. W.B. Weir, Automatic measurement of complex dielectric constant and permeability at microwave frequencies. Proc. IEEE **62**(1), 33–36 (1974)
27. A.-H. Boughriet, C. Legrand, A. Chapoton, Noniterative stable transmission/reflection method for low-loss material complex permittivity determination. IEEE Trans. Microwave Theory Tech. **45**(1), 52–57 (1997)
28. Keysight Technologies, Basics of measuring the dielectric properties of materials. Application Note (2014)
29. T. Igarashi, K. Kondo, T. Oka, S. Yoshida, Design and evaluation of noise suppression sheet for GHz band utilizing magneto-elastic effect. J. Magn. Magn. Mater. **444**, 390–393 (2017)
30. Keysight Technologies, Materials measurement: Magnetic materials. Application Brief (2014)
31. V. Radonić, N. Blaž, L. Živanov, Measurement of complex permeability using short coaxial line reflection method. Acta. Physica. Plolnica A **117**(5), 820–824 (2010)
32. T. Kimura, T. Kimura, S. Yabukami, T. Ozawa, Y. Miyazawa, H. Kenju, Y. Shimada, Permeability measurements of magnetic thin film with microstrip probe. J. Magn. Soc. Jpn. **38**.3-1, 87–91 (2014)
33. S. Takeda, M. Naoe, Size optimization for complex permeability measurement of magnetic thin films using a short-circuited microstrip line up to 30 GHz. J. Magn. Magn. Mater. **449**, 530–537 (2018)
34. M. Yamaguchi, Y. Miyazawa, K. Kaminishi, K. Arai, A new 1 MHz-9 GHz thin-film Permeameter using a side-open TEM cell and a planar shielded-loop coil. Trans. Magn. Soc. Jpn. **3**(4), 137–140 (2003)
35. Y. Ding, T.J. Klemmer, T.M. Crawford, A coplanar waveguide permeameter for studying high-frequency properties of soft magnetic materials. J. Appl. Phys. **96**(5), 2969–2972 (2004)

36. Y. Chen, X. Wang, H. Chen, Y. Gao, N.X. Sun, Novel ultra-wide band (10 MHz–26 GHz) permeability measurements for magnetic films. IEEE Trans. Magn. **54.11**, 6100504 (2018)
37. S. Yabukami, K. Kusunoki, H. Uetake, H. Yamada, T. Ozawa, R. Utsumi, T. Moriizumi, Y. Shimada, Permeability measurements of thin film using a flexible microstrip line-type probe up to 40 GHz. J. Magn. Soc. Jpn. **41**, 25–28 (2017)
38. Y. Liu, L. Chen, C.Y. Tan, H.J. Liu, C.K. Ong, Broadband complex permeability characterization of magnetic thin films using shorted microstrip transmission-line perturbation. Rev. Sci. Instrum. **76**(6), 063911 (2005)
39. S.N. Starostenko, K.N. Rozanov, A.V. Osipov, A broadband method to measure magnetic spectra of thin films. J. Appl. Phys. **103.7**, 07E914 (2008)
40. J. Wei, H. Feng, Z. Zhu, Q. Liu, J. Wang, A short-circuited coplanar waveguide to measure the permeability of magnetic thin films: Comparison with short-circuited microstrip line. Rev. Sci. Instrum. **86**(11), 114705 (2015)
41. T. Sebastian, S.A. Clavijo, R.E. Diaz, Improved accuracy thin film permeability extraction for a microstrip permeameter. J. Appl. Phys. **113**(3), 033906 (2013)
42. J. Neige, M. Ledieu, T. Le Bihan, E. Estrade, A.-L. Adenot-Engelvin, P. Belleville, N. Vukadinovic, Probing high-frequency dynamics of a single ferromagnetic flake using a broadband micro-loop permeameter. Appl. Phys. Lett. **104**, 162402 (2014)
43. V. Bekker, K. Seemann, H. Leiste, A new strip line broad-band measurement evaluation for determining the complex permeability of thin ferromagnetic films. J. Magn. Magn. Mater. **270**(3), 327–332 (2004)
44. M. Ledieu, O. Acher, New achievements in high-frequency permeability measurements of magnetic materials. J. Magn. Magn. Mater. **258**, 144–150 (2003)
45. M. Yamaguchi, S. Yabukami, K.I. Arai, A new permeance meter based on both lumped elements transmission line theories. IEEE Trans. Magn. **32**(5), 4941–4943 (1996)
46. S.Y. Shin, K.H. Shin, J.S. Kim, Y.H. Kim, S.H. Lim, G. Sa-gong, High frequency permeameter with semi-rigid pick-up coil. J. Magn. Magn. Mater. **304**(1), e480–e482 (2006)
47. F. Hettstedt, U. Schürmann, R. Knöchel, and E. Quandt, Double Coil Permeameter for the Characterization of Magnetic Materials." 2009 German Microwave Conference. IEEE, (2009)
48. S. Tamaru, N. Kikuchi, T. Igarashi, S. Okamoto, H. Kubota, S. Yoshida, Broadband and high-sensitivity permeability measurements on a single magnetic particle by transformer coupled permeameter. J. Magn. Magn. Mater. **501**, 166434 (2020)
49. S. Tamaru, S. Tsunegi, H. Kubota, S. Yuasa, Vector network analyzer ferromagnetic resonance spectrometer with field differential detection. Rev. Sci. Instrum. **89**(5), 053901 (2018)

Ferromagnetic Resonance

Tim Mewes and Claudia K. A. Mewes

Abstract Ferromagnetic resonance spectroscopy (FMR) is a versatile characterization technique that probes the magnetization dynamics in the frequency domain. Despite its long history, there continues to be significant improvement regarding the sensitivity and frequency range achievable in ferromagnetic resonance experiments. In this chapter we will briefly introduce the basic principles of ferromagnetic resonance spectroscopy and the long history of this technique. We will discuss the equation of motion for the magnetization that leads to the ferromagnetic resonance condition in the form of the Kittel equation and the Smit-Beljers equation, followed by a discussion of the resonance lineshape. We will cover four different experimental methods that are used to detect ferromagnetic resonance: cavity-based, shorted waveguide, transmission line, and electrically detected ferromagnetic resonance. We will give three examples that highlight some of the unique capabilities of broadband ferromagnetic resonance spectroscopy.

Keywords Ferromagnetic resonance · FMR · Resonance condition · Magnetization dynamics · Landau–Lifshitz–Gilbert equation of motion · Anisotropy · Damping · Gyromagnetic ratio

1 Introduction

Ferromagnetic resonance spectroscopy or, in short, ferromagnetic resonance (FMR) is an experimental technique to study the dynamic response of ferromagnetic materials. When the magnetization of a material is perturbed from its equilibrium, the magnetization precesses around the direction of the internal magnetic field and within a short period of time will relax and realign itself again along the

T. Mewes (✉) · C. K. A. Mewes
The University of Alabama, Tuscaloosa, AL, USA
e-mail: tmewes@ua.edu; ckmewes@ua.edu

© Springer Nature Switzerland AG 2021
V. Franco, B. Dodrill (eds.), *Magnetic Measurement Techniques for Materials Characterization*, https://doi.org/10.1007/978-3-030-70443-8_16

direction of the internal field. In ferromagnetic resonance experiments, a small, rapidly oscillating external electromagnetic field is applied to the sample to sustain the precession. If the frequency of this electromagnetic field coincides with the precession frequency of the ferromagnet, the material undergoes resonance. As will be discussed in detail in Sect. 2, the resonance frequency can be tuned using a static external magnetic field. Typical precession frequencies for magnetic materials are in the range from 1 GHz to 300 GHz, i.e., in the microwave part of the electromagnetic spectrum. Ferromagnetic resonance thus probes the magnetization dynamics at ns and sub-ns timescales, while the corresponding photon energies range from 1 meV down to 1 μeV.

As S.V. Vonsovskii pointed out [1], V.K. Arkad'yev first observed ferromagnetic resonance in 1911 [2]. In 1923 J. Dorfmann provided important new insights regarding the theoretical understanding of these observations and suggested experimentally testing the influence of strong external magnetic fields on ferromagnetic resonances [3]. In 1946 significant experimental progress was made by James Griffiths [4], who is often credited for the experimental discovery of ferromagnetic resonance [5]. Shortly thereafter, Charles Kittel explained Griffiths' observations [6, 7], laying the groundwork for the interpretation of countless ferromagnetic resonance experiments since then.

2 Ferromagnetic Resonance Condition

The magnetization dynamics of the magnetization M of ferromagnetic materials can be described by the Landau–Lifshitz–Gilbert (LLG) equation of motion

$$\frac{dM}{dt} = -\gamma_0 \, M \times H_{\text{eff}} + \frac{1}{M_s} M \times \alpha \frac{dM}{dt} \tag{1}$$

here $\gamma_0 = |\gamma|\mu_0$ is the gyromagnetic ratio γ rescaled by the vacuum permeability μ_0. For a free electron, the gyromagnetic ratio is $\gamma = -g\frac{\mu_B}{\hbar}$. With an electron g-factor of $g \approx 2$, one has $\frac{|\gamma|}{2\pi} = 28\,\text{GHz/T}$, which sets the frequency scale for ferromagnetic resonance phenomena.

The effective field H_{eff} is the sum of all fields acting on the magnetic moments. This effective field can be expressed as the variational derivative of the free energy density with respect to the magnetization M [8]. The second term on the right-hand side of Eq. (1) was first introduced in this form in 1955 by Thomas L. Gilbert [9, 10] as a phenomenological term to account for damping. However, since then this form of the damping term has been linked to a number of different physical mechanisms and has been derived from first principles [11, 12]. This dissipative term causes any precession of the magnetization around the effective field direction to diminish quickly, thereby causing the magnetization to align with the effective field direction.

Kittel Equation

In ferromagnetic resonance experiments, a small time varying field $\boldsymbol{h}(t)$ is applied perpendicular to the quasi-static field that saturates the sample. Thus, in ferromagnetic resonance experiments, the total externally applied field $\boldsymbol{H}$ is

$$\boldsymbol{H} = H\hat{\mathbf{z}} + h \cdot e^{i\omega t}\hat{\mathbf{x}}, \tag{2}$$

if one assumes the quasi-static field is applied along the z-axis and that the time-varying field $\boldsymbol{h}(t) = h \cdot e^{i\omega t}\hat{\mathbf{x}}$ is sinusoidal and applied along the x-axis. The dynamic magnetic susceptibility tensor χ characterizes the linear response of the magnetization and depends on the angular frequency $\omega = 2\pi f$ and can be written as follows [13]:

$$\chi(\omega) = \chi'(\omega) - i\,\chi''(\omega) \tag{3}$$

Here χ' is the real part and χ'' the imaginary part of the complex dynamic susceptibility. By assuming that the amplitude of the time-varying field is small compared to the amplitude of the static field, i.e., $h \ll H$ one can show the time-averaged power $\langle P \rangle$ absorbed by the sample is [14–16]

$$\langle P \rangle = \frac{1}{2}\mu_0\,\omega\,h^2 \cdot \chi''(\omega); \tag{4}$$

it is proportional to the imaginary part of the complex susceptibility $\chi''(\omega, H)$. The last notation indicates the complex susceptibility depends not only on the microwave frequency but also on the magnetic field. As will be discussed in more detail in Sect. 3, all experimental implementations of FMR spectroscopy take advantage of this increased absorption of ferromagnetic samples at their resonance frequency.

In order to determine the natural frequency of the dynamical system described by the Landau–Lifshitz–Gilbert equation of motion without damping, it is instructive to include the effect of the demagnetizing field $\boldsymbol{H}_{\text{demag.}}$ using an ellipsoidal demagnetizing tensor:

$$N = \begin{pmatrix} N_{xx} & 0 & 0 \\ 0 & N_{yy} & 0 \\ 0 & 0 & N_{zz} \end{pmatrix} \tag{5}$$

with $N_{xx} + N_{yy} + N_{zz} = 1$. By solving Eq. (1) in the small angle limit, one arrives at the following natural resonance frequency f_{res} of the system [6, 7]:

$$\begin{aligned} f_{\text{res}} &= \frac{\gamma_0}{2\pi}\sqrt{H_1 \cdot H_2} \\ &= \gamma_0'\sqrt{\left(H + (N_{yy} - N_{zz})M_z\right) \cdot \left(H + (N_{xx} - N_{zz})M_z\right)}, \end{aligned} \tag{6}$$

where $\gamma_0' = \frac{\gamma_0}{2\pi}$ is the reduced gyromagnetic ratio. H_1 and H_2 can be interpreted as effective stiffness fields characterizing the torque exerted on the magnetization [17, 18]. Equation (6) describes the resonance condition and is colloquially also known as the Kittel equation.

Smit-Beljers Equation

A more general equation including all contributions to the free energy density e and therefore to the effective field was developed by Smit and Beljers [19]:

$$\left(\frac{f_{\text{res}}}{\gamma_0'}\right)^2 = \frac{1}{M_s^2 \sin^2 \theta}\left[\frac{\partial^2 e}{\partial \theta^2}\frac{\partial^2 e}{\partial \phi^2} - \left(\frac{\partial^2 e}{\partial \theta \partial \phi}\right)^2\right]. \tag{7}$$

For this equation we have also dropped our earlier assumptions about the direction of the external magnetic field and allowed for an arbitrary orientation as shown in Fig. 1. Smit-Beljers' expression for the resonance condition has to be evaluated at the equilibrium points of the magnetization, that is, at the polar angle θ_0 and azimuthal angle ϕ_0 of the magnetization for which $\frac{\partial e}{\partial \theta}\big|_{\theta_0} = \frac{\partial e}{\partial \phi}\big|_{\phi_0} = 0$. It is noteworthy that an alternate form of Eq. (7) has been developed that avoids the singularity for $\theta = 0$. See reference [20] for more details regarding this approach.

The resonance condition in the Smit-Beljers' form highlights the tremendous potential of magnetic resonance techniques for characterizing different free energy contributions in ferromagnetic materials. By determining the resonance condition as a function of the microwave frequency f and the sample orientation with respect

Fig. 1 Geometry used in Eq. (7) to describe the ferromagnetic resonance condition. The quasi-static magnetic field $\boldsymbol{H}$ (blue) and the microwave field $\boldsymbol{h}_{mw}$ (green) are perpendicular to each other. The equilibrium angles of the magnetization $\boldsymbol{M}$ (red) are denoted with θ_0 and ϕ_0

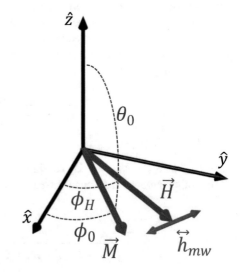

to the quasi-static field, one can map the energy landscape of the sample. On the other hand, with adequate prior knowledge regarding the symmetry of the contributions to the free energy of the system, a few measurements along high-symmetry orientations are often sufficient to determine anisotropy constants with excellent accuracy [21–23]. Furthermore, the resonance condition also indicates the possibility to determine the gyromagnetic ratio, and hence the g-factor, using FMR measurements. However, results published in the literature should be evaluated carefully in particular in view of more recent developments regarding the use of broadband frequency-dependent FMR measurements for the precise determination of the g-factor [24, 25]. When fitting experimental data, the different parameters entering the resonance condition turn out to be correlated. This can make their accurate determination difficult, in particular when data is only collected over a narrow frequency range.

Ferromagnetic Resonance Lineshape

Besides the wealth of information that can be obtained from the resonance condition, additional information can be gleaned from the fact that the power absorbed by the sample is proportional to the imaginary part of the complex susceptibility $\chi''(\omega, H)$. To get further insight, one can solve Eq. (1) for high-symmetry cases in the limit of small precession amplitudes to obtain an expression for the imaginary part of the magnetic susceptibility [13, 26, 27]:

$$\chi''(f) = \frac{\gamma_0' M_s \alpha f}{(f_{\text{res}} - f)^2 + f^2 \alpha^2}. \tag{8}$$

The observation that the imaginary part of the dynamic susceptibility has a Lorentzian lineshape $L(f) \propto \frac{\Delta f/2}{(f_{\text{res}} - f)^2 + (\Delta f/2)^2}$ also holds for more complicated cases. Besides the resonance frequency f_{res}, the lineshape is characterized by one additional parameter, its frequency full width at half maximum:

$$\Delta f_{\text{FWHM}} = 2\alpha f. \tag{9}$$

The linewidth increases linearly with the driving frequency f, and the proportionality factor is twice the Gilbert damping parameter α. An example for the expected imaginary part of the susceptibility for NiFe driven at a frequency of $f = 10\,\text{GHz}$ is shown in Fig. 2. Therefore, a careful study of the resonance linewidth can provide valuable insights regarding the damping term of the Landau–Lifshitz–Gilbert equation of motion.

However, as will be discussed in more detail in Sect. 3, it is customary for ferromagnetic resonance experiments to use the field-swept peak-to-peak linewidth ΔH_{pp}. This means not only to sweep the field instead of the frequency but also to use the peak-to-peak width instead of the full width at half maximum. The

Fig. 2 This graph shows the frequency dependence of the imaginary part χ'' of the dynamic magnetic susceptibility given by Eq. (8) for NiFe with a saturation magnetization $M_s = 800\,\text{kA/m}$, a driving frequency $f = 10\,\text{GHz}$, and a Gilbert damping parameter $\alpha = 0.007$. The frequency linewidth (full width at half maximum) Δf_{FWHM} is indicated in red

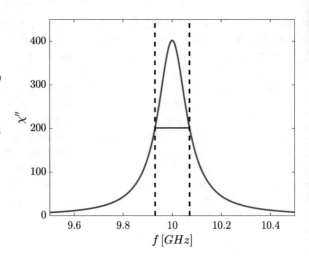

peak-to-peak linewidth is defined as the field separation between the two extrema of the derivative of the power absorbed by the sample with respect to field, i.e., the separation between the inflection points of χ''. For a Lorentzian lineshape, the two linewidths are connected as follows:

$$\Delta H_{\text{PP}} = \frac{\Delta H_{\text{FWHM}}}{\sqrt{3}}. \tag{10}$$

For the field-swept peak-to-peak linewidth due to Gilbert damping one has, using Suhl's approach [28]:

$$\Delta H_{\text{PP}}^{\text{GILBERT}} = \frac{1}{\sqrt{3}} \frac{\alpha}{|\partial f/\partial H|} \frac{\gamma_0'}{M_s} \left(\frac{\partial^2 e}{\partial \theta^2} + \frac{1}{\sin^2 \theta} \frac{\partial^2 e}{\partial \phi^2} \right). \tag{11}$$

Using this approach one can obtain an approximate expression for the field-swept linewidth [12, 29, 30]

$$\Delta H_{\text{PP}}^{\text{GILBERT}} \approx \frac{2}{\sqrt{3}} \frac{\alpha}{\gamma_0'} \frac{f}{\cos \beta}, \tag{12}$$

where the angle β between the magnetization and the external magnetic field takes into account the field dragging effect [31]. When the external magnetic field is sufficiently large and aligned with an easy or hard axis of the ferromagnet, the magnetization will align with the field, and the expression further simplifies to

$$\Delta H_{\text{PP}}^{\text{GILBERT}} = \frac{2}{\sqrt{3}} \frac{\alpha}{\gamma_0'} f \tag{13}$$

The same expression can be obtained using the frequency linewidth given by Eq. (9) and $\Delta f = \Delta H \frac{\partial f}{\partial H}$ [32, 33].

3 Experimental Methods

The power absorbed by a sample driven by a small microwave field according to Eq. (4) is proportional to the imaginary part of the dynamic magnetic susceptibility. The dynamic magnetic susceptibility has a peak at the resonance frequency of the sample, as shown by Eq. (8). Therefore any experimental method capable of detecting the microwave power absorbed by the sample can, in principle, be used to perform ferromagnetic resonance spectroscopy.

Resonant Microwave Cavity-Based FMR

Originally, samples were placed in resonant microwave cavities to perform ferromagnetic resonance spectroscopy. The resonant cavity determines the microwave frequency at which the precession of the magnetization can be driven. The microwave power in these systems was typically provided using a Klystron [34] or Gunn diode [35, 36]. By using a circulator [37] or directional coupler [38], the power reflected from the sample-loaded cavity is monitored, while the external magnetic field is swept through the resonance condition.

Because the change in reflected power at resonance is small, a small modulation field with a modulation frequency of a few hundred Hz is typically added to the quasi-static field. By using a lock-in amplifier, the signal-to-noise ratio can be drastically improved. The use of a lock-in detection results in the measurement of the first derivative $\partial \chi'' / \partial H$ of the imaginary part of the susceptibility with respect to the field H. This also explains the custom to use the peak-to-peak linewidth ΔH_{PP} to characterize the linewidth (see Eq. (10)). This linewidth can be determined using two easily identifiable points on the measured FMR response; see Fig. 3. For small modulation field amplitudes $\Delta H_{\mathrm{mod}} \ll \Delta H_{\mathrm{PP}}$, the measured signal is proportional to the modulation amplitude. Therefore, increasing the modulation amplitude generally increases the signal and consequently the signal-to-noise ratio. However, as for any modulation-based technique, care must be taken to avoid signal distortion due to overmodulation [39, 40]. For more details regarding cavity-based FMR instrumentation, see references [16, 41] and references therein.

Shorted Waveguide FMR

One of the challenges of cavity-based FMR is the need to use different cavities to perform measurements at different microwave frequencies. Therefore numerous approaches have been used to enable measurements over a broad frequency range. While not widely reported, shorted waveguides have been used for this at least since the early 1980s [42–44].

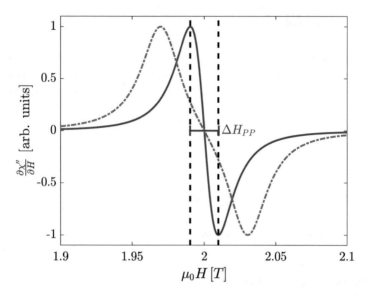

Fig. 3 Expected field-swept FMR signal $\partial \chi'' / \partial H$ (blue line) for NiFe with a saturation magnetization $M_s = 800\,\text{kA/m}$, a driving frequency $f = 10\,\text{GHz}$, and a Gilbert damping parameter $\alpha = 0.007$. The field-swept peak-to-peak linewidth ΔH_{PP} for an infinitesimally small modulation field is indicated in red. As a dash-dotted magenta line, the expected FMR signal for the same sample is shown for an overmodulation of $\mu_0 \Delta H_{\text{mod}} = 0.03\,\text{T}$

Shorted waveguide FMR takes advantage of the standing wave pattern that forms in a waveguide with a shorted end [45], as shown schematically in Fig. 4a. By placing the shorted end of the waveguide at the center of an electromagnet, the sample can be exposed to a quasi-static field and microwave field simultaneously. By placing the sample either on the short itself or on the (narrow) sidewalls of waveguide, different field geometries can be achieved for thin film samples. By using field modulation and lock-in detection, good signal-to-noise ratios can be achieved over the entire frequency band of the waveguide.

Other Transmission Line-Based FMR

Over the years various other transmission lines have been utilized to excite the ferromagnetic resonance of the sample. These include slotted lines [46–48], microstrip lines [49–52], and coplanar waveguides [26, 53–57]. Compared to microstrip lines, coplanar waveguides (CPWs) offer the advantage of several design parameters that can be used to tune the characteristic impedance to match the output impedance of the microwave source [58, 59]; see Fig. 5. By reducing the width s of the central conductor, higher current densities and hence larger microwave field amplitudes can be achieved. However, there is a trade-off between the increased excitation of

Fig. 4 (**a**) Schematic diagram of the magnetic component h of the standing wave field in a waveguide that is shorted with a conducting plate on the right-hand side (gray area). The sample can be placed on the short itself or on the sidewall depending on the desired field geometry. (**b**) X-band (8–12 GHz) waveguide and a copper shorting plate with the sample in the center of it

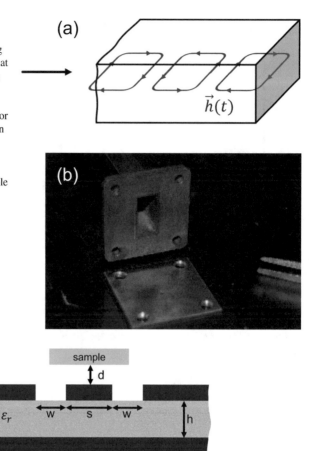

Fig. 5 Schematic diagram of a coplanar waveguide with an additional ground plane on the back of the substrate. The substrate is characterized by its relative permittivity ε_r and thickness h. The coplanar waveguide signal trace has a width s and is separated from the ground planes by a gap w. The sample is placed at a height d above the coplanar waveguide

the sample, the inhomogeneity of the microwave field [60, 61], and radiative losses [62]. With the proper design, a very broad frequency range can be covered using a single coplanar waveguide, with operation frequencies extending up to 65 GHz being recently reported [63–65]. A schematic diagram of a typical broadband FMR setup is shown in Fig. 6. All instruments are usually computer controlled to enable full automation. The choice of microwave source depends on the desired and/or achievable frequency range for a particular setup. Maintaining the characteristic impedance of the source throughout the microwave assembly is vital for maximizing the power delivered to the sample and transmitted to the detector. Similarly, the choice of detector depends on the overall bandwidth of the setup. Zero-biased Schottky detectors can generally achieve good results.

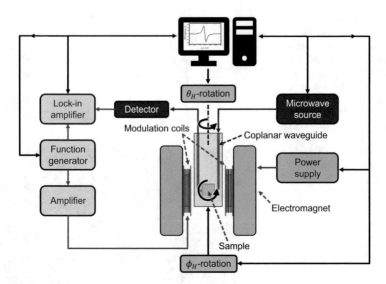

Fig. 6 Schematic diagram of a field-swept broadband coplanar waveguide based FMR setup. The absorption at resonance is detected in the transmitted power using lock-in detection based on field modulation or microwave frequency modulation (not shown). Motors enable in-plane angle ϕ_H and polar angle θ_H dependent measurements at arbitrary frequencies

Similar to cavity-based FMR, a lock-in amplifier is typically used to improve the signal-to-noise ratio compared to a DC detection of the transmitted power. Due to the broadband nature of the coplanar waveguide, frequency modulation of the microwave signal is also possible [40, 66] and results in the same lineshape as field modulation. The potential benefit of frequency modulation is significantly higher modulation frequencies achievable with this approach compared to modulation coils, where the coil inductance restricts the usable modulation frequencies. However, due to the limited frequency deviation that can be achieved with typical microwave sources, this is currently not a widely used approach. While most microwave sources also offer amplitude modulation, which enables lock-in detection with a lineshape that reflects χ'' instead of its first derivative, this approach typically suffers from a large signal offset away from the resonance.

The planar structure of the coplanar waveguide enables easy sample exchange as well as automated rotation of the sample to vary the in-plane angle of the applied field ϕ_H. This allows precise measurements of magnetic anisotropies [21, 23, 65] similar to cavity-based studies [67–69]. Furthermore, as indicated in Fig. 6, the coplanar waveguide structure can be rotated around its axis, changing the polar angle θ_H of the field relative to the surface normal. This allows detailed investigation of higher-order perpendicular anisotropies [70].

Temperature control is another common extension of ferromagnetic resonance systems [71–73] and can be implemented for broadband FMR systems in almost the same way as in cavity-based system, for example, using a closed-cycle helium cryostat and a heater [64, 74, 75].

Fig. 7 Schematic diagram for vector network analyzer FMR setup. A full two-port measurement of the scattering parameters as a function of frequency and field enables determination of both the imaginary and real parts of the complex susceptibility of the sample

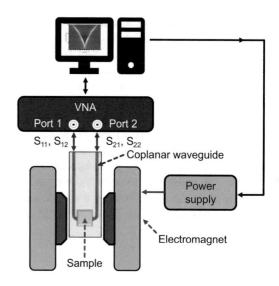

Vector Network Analyzer FMR

While vector network analyzer (VNA) FMR uses the same transmission lines as the aforementioned techniques, it differs by measuring the phase and amplitude of the transmitted and reflected signals in the form of scattering parameters (S-parameters) instead of measuring only the transmitted power [26, 33, 49, 76–81]. Only few components are needed for a typical setup, as shown in Fig. 7. Contrary to field-swept FMR, VNA FMR is usually done at a fixed field while sweeping the frequency. This is the typical mode of operation for a VNA. However, field-swept VNA FMR measurements have also been reported [82, 83]. One advantage of VNA FMR compared to other methods is that it enables the determination of both the real and imaginary parts of the complex susceptibility [26, 79, 84]. However, to achieve this a careful full two-port calibration of the VNA FMR system is required. The goal of this calibration is to move the reference planes of the VNA as close as possible to the sample to eliminate all contributions to the measured scattering parameters that do not originate from the sample itself. The most reliable measurements appear to be possible when measuring all four scattering parameters S_{11}, S_{21}, S_{12}, and S_{22} [26, 55].

Electrically Detected FMR

Over the last two decades, electrical detection of ferromagnetic resonance has been observed in a number of different magnetic structures [85–91]. The unifying aspect of these various electrical detection techniques is the generation of a DC voltage at the ferromagnetic resonance [90]. There are numerous physical mechanisms that have been utilized for this, including spin-transfer torque [89], spin-pumping [92, 93], spin rectification [94, 95], spin Hall [96], and inverse spin Hall effects [97–99]. The generated DC voltages often can be readily detected with a voltmeter. However, just like for the other FMR techniques, lock-in detection can often significantly improve the signal-to-noise ratio [89, 91, 100, 101]. The high sensitivity of electrically detected FMR not only offers many potential technological applications but also provides researchers with the ability to investigate ferromagnetic resonance phenomena of very small sample volumes down to individual nanomagnets [87, 89]. However, the multitude of potential mechanisms that can cause the observed DC voltages can make the interpretation of the results challenging. Additional valuable insights can often be gained by complementary characterization using the FMR techniques discussed in the preceding sections.

4 Examples

In this section we want to give a few examples that highlight some of the capabilities of broadband ferromagnetic resonance spectroscopy.

Kittel's resonance condition (6) in its simplest form suggests one advantage of FMR compared to many other magnetic characterization methods regarding the determination of the saturation magnetization M_s. Because the resonance condition only involves the microwave frequency f, the reduced gyromagnetic ratio γ_0', and the externally applied field, one should be able to determine the saturation magnetization with high accuracy. In field-swept FMR experiments, the microwave frequency is fixed and known to very high accuracy; thus the only limiting factor is the accuracy with which one can measure the applied field.

Saturation Magnetization and Perpendicular Anisotropy

The reason this is possible is the sensitivity of the ferromagnetic resonance condition to the internal field acting on the magnetization, which includes the demagnetizing field. For thin films the situation simplifies even further. For example, if one saturates the sample along the film normal (assumed to be along $\hat{\mathbf{z}}$), one has $N_{xx} = N_{yy} = 0$ and $N_{zz} = 1$, and hence the resonance condition reads:

$$f_{\text{res}} = \gamma_0' (H - M_s) . \tag{14}$$

However, there is one caveat to this: any internal field with the same functional form as the demagnetizing field will be indistinguishable from it. For the thin film

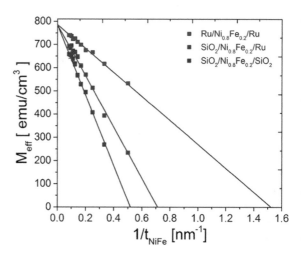

Fig. 8 Effective magnetization as a function of the inverse film thickness for $SiO_2/Ni_{0.8}Fe_{0.2}/SiO_2$ (blue symbols), $SiO_2/Ni_{0.8}Fe_{0.2}/Ru$ (red symbols), and $Ru/Ni_{0.8}Fe_{0.2}/Ru$ (green symbols). The lines are a linear fit to each data set. For details, see ref. [22]. Reprinted from Mohammadi et al. [22], with the permission of AIP Publishing

geometry, this is the case for a second-order uniaxial perpendicular anisotropy K_2 with a contribution to the free energy of the system of the form $e_{u,2} = -K_2 \cos^2 \theta$ that may be present in the sample. This is the reason for the introduction of the effective magnetization M_{eff} in ferromagnetic resonance spectroscopy

$$M_{eff} = M_s - \frac{2K_2}{\mu_0 M_s}, \tag{15}$$

that enters the resonance condition (14) instead of M_s. It is worth noting that for samples with a fourth-order perpendicular anisotropy K_4 of the form $e_{u,4} = -\frac{K_4}{2} \cos^4 \theta$, the effective magnetization entering the resonance condition will be different when measuring the in-plane and out-of-plane configuration (for details see [102, 103]).

Figure 8 shows an example from reference [22] demonstrating the use of broadband FMR to obtain a very precise value for the bulk saturation magnetization of $Ni_{0.8}Fe_{0.2}$ and the interfacial contribution to the second-order perpendicular anisotropy $K_{2,i}$. Due to the interfacial nature of the perpendicular anisotropy of the films in this study, one expects the effective magnetization measured by FMR to scale with the inverse of the film thickness t_{NiFe}. From the slopes in Fig. 8, one can therefore determine the interfacial contribution to the anisotropy from the different interfaces, which is strongest for the SiO_2 interface. On the other hand, for this material system, one does not expect any perpendicular anisotropy in the bulk. Therefore, by extrapolating the effective magnetization to infinite thick films or $1/t_{NiFe} = 0$, the saturation magnetization of $Ni_{0.8}Fe_{0.2}$ can be determined to be $M_s = 786 \pm 2 \, \text{kA/m}$ (or in cgs units, $M_s = 786 \pm 2 \, \text{emu/cm}^3$).

In-plane Anisotropies

Depending on the symmetry of the sample under investigation, additional anisotropy contributions can readily be included in the resonance condition (7). While obtaining analytical expressions for the resonance field is limited to high-symmetry orientations, a numerical evaluation is straightforward. As an example, in Fig. 9 the dependence of the resonance field on the external applied field is shown for an exchange bias system IrMn/CoFe studied in reference [23]. Broadband measurements were performed along the in-plane direction parallel and antiparallel to the exchange bias direction. The exchange bias field H_{eb} can be easily estimated as half the field separation of these two curves. However, in exchange bias systems, a uniaxial anisotropy is often observed in addition to the unidirectional anisotropy responsible for the exchange bias effect [21, 104, 105]. In order to precisely determine this anisotropy contribution, in-plane angle-dependent measurements can be used, as shown in Fig. 10. As described in detail in reference [23], the figure shows two approaches to analyze the data. The first one, shown as the green curve in part (a) and (b) of this figure, uses an analytical model that can be derived by assuming the magnetization is aligned with the external magnetic field [106, 107]. The second approach, shown in red, numerically minimizes the free energy to obtain the equilibrium angles required for the evaluation of the Smit-Beljers relation (7). While the two fits of the experimental data are indistinguishable in part (a) of

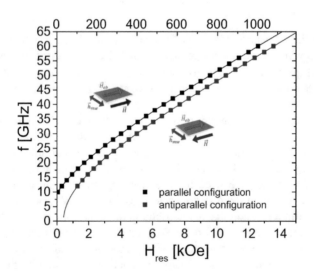

Fig. 9 Microwave frequency f versus resonance field H_{res} (Kittel plot) for a 6 nm thick CoFe exchange-biased layer. Black (red) symbols show broadband FMR data with the external magnetic field applied parallel (antiparallel) to the exchange bias direction. The corresponding solid lines are the result of a simultaneous fit to the Kittel equation of the system for both orientations. For more details see ref. [23]. Reprinted figure with permission from Mohammadi et al. [23]. Copyright (2017) by the American Physical Society

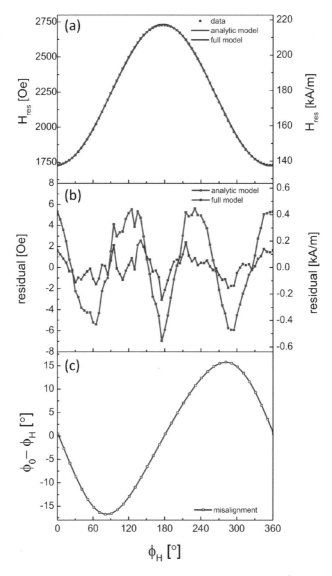

Fig. 10 Dependence on the in-plane angle of the applied field ϕ_H of (**a**) the resonance field H_{res} of a 6 nm CoFe exchange-biased layer, where the figure includes the experimental data (blue symbols), a fit using the analytic model (green line), and a fit using the full model (red line), (**b**) the residuals of the fit using the analytical model (green line and symbols) and the full model (red line and symbols), and (**c**) the misalignment of the in-plane angle of the magnetization ϕ_0 from the direction of the applied field ϕ_H calculated using the full model. For more details see ref. [23]. Reprinted figure with permission from Mohammadi et al. [23]. Copyright (2017) by the American Physical Society

the figure, close inspection of the residuals reveals systematic deviations of the analytical model that may erroneously be interpreted as a threefold anisotropy. However, as part (c) of the figure shows, the deviations are caused by the violation of the assumption of the analytical model that the magnetization is aligned with the applied field. While the deviations are particularly large in this system, one should expect such misalignment in other systems whenever the external field is not applied along an easy or hard axis.

Furthermore, measurements as a function of the in-plane angle of the applied field can provide insights regarding potentially anisotropic damping. For exchange bias systems, this topic is discussed in more detail in references [21, 23, 108].

Conductivity-Like Damping in Epitaxial Iron Films

The Gilbert damping parameter that enters the Landau–Lifshitz–Gilbert equation of motion is of paramount importance for many spintronic applications. It determines the critical current for switching in spin-transfer torque magnetic random access memory [109–111] and for auto-oscillations in spin-torque oscillators [112–114]. It also determines the mobility of domain walls [115–117] and the propagation of spinwaves [118, 119]. This has led to a renewed interest in theoretical predictions and experimental determination of the damping parameter in a wide variety of magnetic materials. While it may seem straightforward to determine the Gilbert damping parameter from field-swept FMR measurements based on Eq. (13), care has to be taken when trying to extract the intrinsic damping parameter of the material. There are a number of mechanisms that can lead to a Gilbert-like term in the LLG equation of motion, including spin-pumping and eddy current damping [120]. Furthermore, two-magnon scattering can contribute significantly to the measured FMR linewidth [56, 108, 121], in particular in thin films measured with the quasi-static field applied in the plane of the film.

For transition metals, relaxation caused by the spin-orbit coupling is expected to be the dominant contribution to the intrinsic damping [11, 16, 122]. This contribution to the damping can be quantified using the Kamberský torque correlation model [123–125], and its predicted conductivity-like increase of the damping parameter at low temperatures has been experimentally observed for cobalt and nickel already in 1974 [126]. However, in the same work, no significant increase was found for iron. Using state-of-the-art broadband ferromagnetic resonance spectroscopy, Khodadadi and coworkers have recently been able to confirm that Kamberský's torque correlation model accurately predicts the damping in high-quality epitaxial iron films [64]. They used broadband in-plane and out-of-plane measurements to confirm that two-magnon scattering does not contribute significantly to the in-plane linewidth measured in these films (see Fig. 11a). They investigated two sets of iron films grown on different substrates. Due to the smaller lattice mismatch, the films grown on (001)-oriented $MgAl_2O_4$ show a significantly better crystalline quality than the simultaneously deposited films on MgO. However, at room temperature

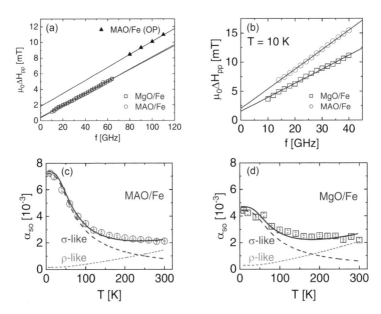

Fig. 11 (**a**) Frequency dependence of FMR linewidth ΔH_{PP} for MgAl$_2$O$_4$/Fe and MgO/Fe at room temperature. Linewidths measured under in-plane field are shown as open symbols, whereas those measured under out-of-plane (OP) field are shown as filled symbols. (**b**) Frequency dependence of FMR linewidth for MgAl$_2$O$_4$/Fe and MgO/Fe at $T = 10$ K. Temperature dependence of the spin-orbit-induced Gilbert damping parameter α_{so}, fit phenomenologically with the experimentally measured resistivity for (**c**) MgAl$_2$O$_4$/Fe and (**d**) MgO/Fe. The dashed and dotted curves in (**c**) and (**d**) indicate the conductivity-like and resistivity-like contributions, respectively; the solid curve represents the fit curve for the total spin-orbit-induced Gilbert damping parameter. For more details, see ref. [64]. Reprinted figure with permission from Khodadadi et al. [64]. Copyright (2020) by the American Physical Society

both films have the same damping parameter (see Fig. 11a). This changes drastically at low temperatures where the iron films grown on MgAl$_2$O$_4$ showed a significantly larger damping (see Fig. 11b). As shown in parts (c) and (d) of the same figure, this is caused by the increased conductivity-like contribution to the damping, as predicted by Kamberský's torque correlation model. These results show some of the new insights recent developments in the area of ferromagnetic resonance have enabled.

5 Summary

More than 100 years after its discovery, ferromagnetic resonance spectroscopy still remains a constantly evolving characterization method that can probe the magnetization dynamics over a very broad frequency range. In this chapter we have described the fundamental principles governing the ferromagnetic resonance condition and how they relate to the different experimental methods for its detection. The three recent examples illustrate how broadband ferromagnetic resonance can contribute to further our understanding of the underlying physical mechanisms influencing the magnetization dynamics. The ongoing development of new electrical detection schemes for ferromagnetic resonance is not only promising for obtaining a better and more unified understanding of the physical mechanisms that cause the rectification of microwave signals in magnetic materials but also in view of their potential application to measurements of nanomagnets and new spintronic devices.

Acknowledgments We would like to thank Prabandha Nakarmi, Anish Rai, Bhuwan Nepal, Arjun Sapkota, Upama Karki, and Marie Mewes for their help and constructive criticism during the preparation of the manuscript.

References

1. S. Von Sovskii, *Ferromagnetic Resonance: The Phenomenon of Resonant Absorption of a High-Frequency Magnetic Field in Ferromagnetic Substances* (Elsevier, Amsterdam, 2016). https://books.google.com/books?id=0OtPDAAAQBAJ
2. V. Arkad'Yev, Zh. Russk. Fiz.-Khim. Obshchestva, Otdel Fiz **44**, 165 (1912)
3. J. Dorfmann, Z. Phys. **17**, 98 (1923)
4. J. Griffiths, Nature **158**(4019), 670 (1946)
5. N. Mo, C. Patton, Rev. Sci. Instrum. **79**(4), 040901 (2008). https://doi.org/10.1063/1.2911921
6. C. Kittel, Phys. Rev. **71**, 270 (1947). https://doi.org/10.1103/PhysRev.71.270.2
7. C. Kittel, Phys. Rev. **73**, 155 (1948). https://doi.org/10.1103/PhysRev.73.155
8. J. Miltat, M. Donahue, *Numerical Micromagnetics: Finite Difference Methods* (American Cancer Society, Atlanta, 2007), chap. Numerical Micromagnetics: Finite Difference Methods. https://doi.org/10.1002/9780470022184.hmm202
9. T. Gilbert, Phys. Rev. **100**, 1243 (1955)
10. T. Gilbert, IEEE Trans. Magn. **40**(6), 3443 (2004)
11. K. Gilmore, Y.U. Idzerda, M.D. Stiles, Phys. Rev. Lett. **99**, 027204 (2007). https://doi.org/10.1103/PhysRevLett.99.027204
12. C. Mewes, T. Mewes, in *Handbook of Nanomagnetism*, ed. by R. Lukaszew (Pan Stanford Publishing, Singapore, 2015), pp. 71–96. https://books.google.com/books?id=I8Z5CgAAQBAJ
13. G. Rado, H. Suhl, *Magnetism: Magnetic Ions in Insulators, Their Interactions, Resonances, and Optical Properties*. Magnetism (Academic Press, London, 1963). https://books.google.com/books?id=PSRRAAAAMAAJ
14. J. Coey, *Magnetism and Magnetic Materials* (Cambridge University Press, Cambridge, 2010). https://doi.org/10.1017/CBO9780511845000
15. M. Farle, Rep. Progr. Phys. **61**(7), 755 (1998). https://doi.org/10.1088/0034-4885/61/7/001
16. B. Heinrich, *Ultrathin Magnetic Structures II* (Springer, Berlin, 1994), pp. 195–296
17. M. Hurben, C. Patton, J. Magn. Magn. Mater. **163**(1), 39 (1996). https://doi.org/10.1016/S0304-8853(96)00294-6
18. M.J. Hurben, C.E. Patton, J. Appl. Phys. **83**(8), 4344 (1998). https://doi.org/10.1063/1.367194
19. J. Smit, H. Beljers, Philips Res. Rep **10**, 113 (1955)
20. L. Baselgia, M. Warden, F. Waldner, S.L. Hutton, J.E. Drumheller, Y.Q. He, P.E. Wigen, M. Maryško, Phys. Rev. B **38**, 2237 (1988). https://doi.org/10.1103/PhysRevB.38.2237
21. A. Rai, M. Dunz, A. Sapkota, P. Zilske, J. Mohammadi, M. Meinert, C. Mewes, T. Mewes, J. Magn. Magn. Mater. **485**, 374 (2019). https://doi.org/10.1016/j.jmmm.2019.04.043

22. J. Mohammadi, G. Mankey, C. Mewes, T. Mewes, J. Appl. Phys. **125**(2), 023901 (2019). https://doi.org/10.1063/1.5052334
23. J. Mohammadi, J. Jones, S. Paul, B. Khodadadi, C. Mewes, T. Mewes, C. Kaiser, Phys. Rev. B **95**, 064414 (2017). https://doi.org/10.1103/PhysRevB.95.064414
24. J. Shaw, H. Nembach, T.J. Silva, C. Boone, J. Appl. Phys. **114**(24), 243906 (2013). https://doi.org/10.1063/1.4852415
25. H. Nembach, E. Jué, E. Evarts, J. Shaw, Phys. Rev. B **101**, 020409 (2020). https://doi.org/10.1103/PhysRevB.101.020409
26. C. Bilzer, Microwave susceptibility of thin ferromagnetic films: metrology and insight into magnetization dynamics. Ph.D., Université Paris Sud - Paris XI (2007). https://tel.archives-ouvertes.fr/tel-00202827
27. D. Stancil, A. Prabhakar, *Spin Waves: Theory and Applications* (Springer US, 2009). https://books.google.com/books?id=ehN6-ubvKwoC
28. H. Suhl, Phys. Rev. **97**, 555 (1955). https://doi.org/10.1103/PhysRev.97.555.2
29. Y.V. Goryunov, N.N. Garif'yanov, G.G. Khaliullin, I.A. Garifullin, L.R. Tagirov, F. Schreiber, T. Mühge, H. Zabel, Phys. Rev. B **52**, 13450 (1995). https://doi.org/10.1103/PhysRevB.52.13450
30. K. Lenz, H. Wende, W. Kuch, K. Baberschke, K. Nagy, A. Jánossy, Phys. Rev. B **73**, 144424 (2006). https://doi.org/10.1103/PhysRevB.73.144424
31. J. Lindner, I. Barsukov, C. Raeder, C. Hassel, O. Posth, R. Meckenstock, P. Landeros, D.L. Mills, Phys. Rev. B **80**, 224421 (2009). https://doi.org/10.1103/PhysRevB.80.224421
32. C. Patton, J. Appl. Phys. **39**(7), 3060 (1968). https://doi.org/10.1063/1.1656733
33. B. Kuanr, R. Camley, Z. Celinski, Appl. Phys. Lett. **87**(1), 012502 (2005). https://doi.org/10.1063/1.1968433
34. G. Caryotakis, Phys. Plasmas **5**(5), 1590 (1998). https://doi.org/10.1063/1.872826
35. B. Smith, M. Carpentier, *Microwave Engineering Handbook: Microwave Circuits, Antennas, and Propagation*. Electrical Engineering (Van Nostrand Reinhold, 1993). https://books.google.com/books?id=kKWONs44P94C
36. M. Shur, *GaAs Devices and Circuits*. Microdevices: Physics and Fabrication Technologies (Springer US, 1987)
37. C. Fay, R. Comstock, IEEE Trans. Microwave Theory Tech. **13**(1), 15 (1965)
38. R. Damon, J. Appl. Phys. **26**(10), 1281 (1955)
39. T. Itō, J. Res. Inst. Catal. Hokkaido Univ. **26**(2), 79 (1978)
40. T. Mewes, C.K.A. Mewes, E. Nazaretski, J. Kim, K.C. Fong, Y. Obukhov, D.V. Pelekhov, P.E. Wigen, P.C. Hammel, J. Appl. Phys. **102**(3), 033911 (2007). https://doi.org/10.1063/1.2761779
41. C. Patton, T. Kohane, Rev. Sci. Instrum. **43**(1), 76 (1972). https://doi.org/10.1063/1.1685449
42. Z. Frait, D. Fraitová, J. Magn. Magn. Mater. **15–18**, 1081 (1980). https://doi.org/10.1016/0304-8853(80)90895-1
43. R. Karim, K.D. McKinstry, J.R. Truedson, C.E. Patton, IEEE Trans. Magn. **28**(5), 3225 (1992)
44. S. Lebedev, C. Patton, M. Wittenauer, L. Saraf, R. Ramesh, J. Appl. Phys. **91**(7), 4426 (2002). https://doi.org/10.1063/1.1450057
45. J. Jackson, *Classical Electrodynamics*, 3rd edn. (Wiley, New Delhi, 2007). https://books.google.com/books?id=8qHCZjJHRUgC
46. H.E. Bussey, Proc. IEEE **55**(6), 1046 (1967)
47. J. Krebs, C. Vittoria, B. Jonker, G. Prinz, J. Magn. Magn. Mater. **54–57**, 811 (1986). https://doi.org/10.1016/0304-8853(86)90265-9
48. Z. Frait, D. Fraitová, G. Marchal, Phys. Status Solidi (b) **201**(1), 257 (1997). https://doi.org/10.1002/1521-3951(199705)201:1<257::AID-PSSB257>3.0.CO;2-P
49. S. Kalarickal, P. Krivosik, M. Wu, C. Patton, M. Schneider, P. Kabos, T. Silva, J. Nibarger, J. Appl. Phys. **99**(9), 093909 (2006). https://doi.org/10.1063/1.2197087
50. K.J. Kennewell, M. Kostylev, N. Ross, R. Magaraggia, R.L. Stamps, M. Ali, A.A. Stashkevich, D. Greig, B.J. Hickey, J. Appl. Phys. **108**(7), 073917 (2010). https://doi.org/10.1063/1.3488618

51. M. Kostylev, A.A. Stashkevich, A.O. Adeyeye, C. Shakespeare, N. Kostylev, N. Ross, K. Kennewell, R. Magaraggia, Y. Roussigné, R.L. Stamps, J. Appl. Phys. **108**(10), 103914 (2010). https://doi.org/10.1063/1.3493242
52. M. Kostylev, J. Appl. Phys. **119**(1), 013901 (2016). https://doi.org/10.1063/1.4939470
53. Y. Sugiyama, K. Kagawa, H. Ohmori, K. Hayashi, M. Hayakawa, K. Aso, IEEE Trans. Magn. **29**(6), 3867 (1993)
54. S. Zhang, J.B. Sokoloff, C. Vittoria, J. Appl. Phys. **81**(8), 5076 (1997). https://doi.org/10.1063/1.364511
55. C. Bilzer, T. Devolder, P. Crozat, C. Chappert, S. Cardoso, P.P. Freitas, J. Appl. Phys. **101**(7), 074505 (2007). https://doi.org/10.1063/1.2716995
56. H. Lee, Y.H.A. Wang, C.K.A. Mewes, W.H. Butler, T. Mewes, S. Maat, B. York, M.J. Carey, J.R. Childress, Appl. Phys. Lett. **95**(8), 082502 (2009). https://doi.org/10.1063/1.3207749
57. S. Schäfer, N. Pachauri, C.K.A. Mewes, T. Mewes, C. Kaiser, Q. Leng, M. Pakala, Appl. Phys. Lett. **100**(3), 032402 (2012). https://doi.org/10.1063/1.3678025
58. T. Kitazawa, Y. Hayashi, M. Suzuki, IEEE Trans. Microwave Theory Tech. **24**(9), 604 (1976)
59. B. Wadell, *Transmission Line Design Handbook*. Artech House Antennas and Propagation Library (Artech House, Norwood, 1991)
60. Y. Ding, T.J. Klemmer, T.M. Crawford, J. Appl. Phys. **96**(5), 2969 (2004). https://doi.org/10.1063/1.1774242
61. M.L. Schneider, A.B. Kos, T.J. Silva, Appl. Phys. Lett. **85**(2), 254 (2004). https://doi.org/10.1063/1.1769084
62. M.A.W. Schoen, J.M. Shaw, H.T. Nembach, M. Weiler, T.J. Silva, Phys. Rev. B **92**, 184417 (2015). https://doi.org/10.1103/PhysRevB.92.184417
63. S. Wu, K. Abe, T. Nakano, T. Mewes, C. Mewes, G. Mankey, T. Suzuki, Phys. Rev. B **99**, 144416 (2019). https://doi.org/10.1103/PhysRevB.99.144416
64. B. Khodadadi, A. Rai, A. Sapkota, A. Srivastava, B. Nepal, Y. Lim, D. Smith, C. Mewes, S. Budhathoki, A.J. Hauser, M. Gao, J.F. Li, D.D. Viehland, Z. Jiang, J. Heremans, P. Balachandran, T. Mewes, S. Emori, Phys. Rev. Lett. **124**, 157201 (2020). https://doi.org/10.1103/PhysRevLett.124.157201
65. A. Srivastava, K. Cole, A. Wadsworth, T. Burton, C. Mewes, T. Mewes, G.B. Thompson, R.D. Noebe, A.M. Leary, J. Magn. Magn. Mater. **500**, 166307 (2020). https://doi.org/10.1016/j.jmmm.2019.166307
66. K. Halbach, Phys. Rev. **119**, 1230 (1960). https://doi.org/10.1103/PhysRev.119.1230
67. G.A. Prinz, G.T. Rado, J.J. Krebs, J. Appl. Phys. **53**(3), 2087 (1982). https://doi.org/10.1063/1.330707
68. B. Heinrich, J.F. Cochran, M. Kowalewski, J. Kirschner, Z. Celinski, A.S. Arrott, K. Myrtle, Phys. Rev. B **44**, 9348 (1991). https://doi.org/10.1103/PhysRevB.44.9348
69. B. Heinrich, J. Cochran, Adv. Phys. **42**(5), 523 (1993). https://doi.org/10.1080/00018739300101524
70. A. Rai, A. Sapkota, A. Pokhrel, M. Li, M.D. Graf, C. Mewes, V. Sokalski, T. Mewes, Phys. Rev. B **99**, 184412 (2019)
71. S.M. Bhagat, L.L. Hirst, Phys. Rev. **151**, 401 (1966). https://doi.org/10.1103/PhysRev.151.401
72. B. Heinrich, D.J. Meredith, J.F. Cochran, J. Appl. Phys. **50**(B11), 7726 (1979). https://doi.org/10.1063/1.326802
73. R. Meckenstock, K. Harms, O. von Geisau, J. Pelzl, J. Magn. Magn. Mater. **148**(1), 139 (1995). https://doi.org/10.1016/0304-8853(95)00181-6
74. B. Khodadadi, J.B. Mohammadi, C. Mewes, T. Mewes, M. Manno, C. Leighton, C.W. Miller, Phys. Rev. B **96**, 054436 (2017). https://doi.org/10.1103/PhysRevB.96.054436
75. B. Khodadadi, J.B. Mohammadi, J.M. Jones, A. Srivastava, C. Mewes, T. Mewes, C. Kaiser, Phys. Rev. Appl. **8**, 014024 (2017). https://doi.org/10.1103/PhysRevApplied.8.014024
76. W.B. Weir, Proc. IEEE **62**(1), 33 (1974)
77. W. Barry, IEEE Trans. Microwave Theory Tech. **34**(1), 80 (1986)

78. G. Counil, J.V. Kim, T. Devolder, C. Chappert, K. Shigeto, Y. Otani, J. Appl. Phys. **95**(10), 5646 (2004). https://doi.org/10.1063/1.1697641
79. Y. Ding, T.J. Klemmer, T.M. Crawford, J. Appl. Phys. **96**(5), 2969 (2004). https://doi.org/10.1063/1.1774242
80. I. Neudecker, G. Woltersdorf, B. Heinrich, T. Okuno, G. Gubbiotti, C. Back, J. Magn. Magn. Mater. **307**(1), 148 (2006). https://doi.org/10.1016/j.jmmm.2006.03.060
81. J.F. Godsell, S. Kulkarni, T. O'Donnell, S. Roy, J. Appl. Phys. **107**(3), 033907 (2010). https://doi.org/10.1063/1.3276165
82. S. Tamaru, S. Tsunegi, H. Kubota, S. Yuasa, Rev. Sci. Instrum. **89**(5), 053901 (2018). https://doi.org/10.1063/1.5022762
83. V. Sharma, B.K. Kuanr, J. Alloys Compd. **748**, 591 (2018). https://doi.org/10.1016/j.jallcom.2018.03.086
84. T. Devolder, S. Couet, J. Swerts, S. Mertens, S. Rao, G.S. Kar, IEEE Magn. Lett. **10**, 1 (2019)
85. M. Tsoi, A. Jansen, J. Bass, W.C. Chiang, V. Tsoi, P. Wyder, Nature **406**(6791), 46 (2000)
86. S.I. Kiselev, J. Sankey, I. Krivorotov, N. Emley, R. Schoelkopf, R. Buhrman, D. Ralph, Nature **425**(6956), 380 (2003)
87. A. Tulapurkar, Y. Suzuki, A. Fukushima, H. Kubota, H. Maehara, K. Tsunekawa, D. Djayaprawira, N. Watanabe, S. Yuasa, Nature **438**(7066), 339 (2005)
88. Y.S. Gui, S. Holland, N. Mecking, C.M. Hu, Phys. Rev. Lett. **95**, 056807 (2005). https://doi.org/10.1103/PhysRevLett.95.056807
89. J.C. Sankey, P.M. Braganca, A.G.F. Garcia, I.N. Krivorotov, R.A. Buhrman, D.C. Ralph, Phys. Rev. Lett. **96**, 227601 (2006). https://doi.org/10.1103/PhysRevLett.96.227601
90. M. Harder, Z.X. Cao, Y.S. Gui, X.L. Fan, C.M. Hu, Phys. Rev. B **84**, 054423 (2011). https://doi.org/10.1103/PhysRevB.84.054423
91. Y. Wang, R. Ramaswamy, H. Yang, J. Phys. D: Appl. Phys. **51**(27), 273002 (2018)
92. Y. Tserkovnyak, A. Brataas, G.E.W. Bauer, Phys. Rev. Lett. **88**, 117601 (2002). https://doi.org/10.1103/PhysRevLett.88.117601
93. M.V. Costache, M. Sladkov, S.M. Watts, C.H. van der Wal, B.J. van Wees, Phys. Rev. Lett. **97**, 216603 (2006). https://doi.org/10.1103/PhysRevLett.97.216603
94. Y.S. Gui, N. Mecking, X. Zhou, G. Williams, C.M. Hu, Phys. Rev. Lett. **98**, 107602 (2007). https://doi.org/10.1103/PhysRevLett.98.107602
95. M. Harder, Y. Gui, C.M. Hu, Phys. Rep. **661**, 1 (2016). Electrical detection of magnetization dynamics via spin rectification effects. https://doi.org/10.1016/j.physrep.2016.10.002
96. L. Liu, T. Moriyama, D.C. Ralph, R.A. Buhrman, Phys. Rev. Lett. **106**, 036601 (2011). https://doi.org/10.1103/PhysRevLett.106.036601
97. E. Saitoh, M. Ueda, H. Miyajima, G. Tatara, Appl. Phys. Lett. **88**(18), 182509 (2006). https://doi.org/10.1063/1.2199473
98. M. Obstbaum, M. Härtinger, H.G. Bauer, T. Meier, F. Swientek, C.H. Back, G. Woltersdorf, Phys. Rev. B **89**, 060407 (2014). https://doi.org/10.1103/PhysRevB.89.060407
99. M. Weiler, J.M. Shaw, H.T. Nembach, T.J. Silva, IEEE Magn. Lett. **5**, 1 (2014)
100. H.Y. Inoue, K. Harii, K. Ando, K. Sasage, E. Saitoh, J. Appl. Phys. **102**(8), 083915 (2007). https://doi.org/10.1063/1.2799068
101. C.F. Pai, L. Liu, Y. Li, H.W. Tseng, D.C. Ralph, R.A. Buhrman, Appl. Phys. Lett. **101**(12), 122404 (2012). https://doi.org/10.1063/1.4753947
102. J.M. Shaw, H.T. Nembach, M. Weiler, T.J. Silva, M. Schoen, J.Z. Sun, D.C. Worledge, IEEE Magn. Lett. **6**, 1 (2015)
103. J.B. Mohammadi, K. Cole, T. Mewes, C.K.A. Mewes, Phys. Rev. B **97**, 014434 (2018). https://doi.org/10.1103/PhysRevB.97.014434
104. Y.J. Tang, B. Roos, T. Mewes, S.O. Demokritov, B. Hillebrands, Y.J. Wang, Appl. Phys. Lett. **75**(5), 707 (1999). https://doi.org/10.1063/1.124489
105. Y.J. Tang, B.F.P. Roos, T. Mewes, A.R. Frank, M. Rickart, M. Bauer, S.O. Demokritov, B. Hillebrands, X. Zhou, B.Q. Liang, X. Chen, W.S. Zhan, Phys. Rev. B **62**, 8654 (2000). https://doi.org/10.1103/PhysRevB.62.8654
106. J.C. Scott, J. Appl. Phys. **57**(8), 3681 (1985). https://doi.org/10.1063/1.334987

107. H. Xi, K.R. Mountfield, R.M. White, J. Appl. Phys. **87**(9), 4367 (2000). https://doi.org/10. 1063/1.373080
108. T. Mewes, R.L. Stamps, H. Lee, E. Edwards, M. Bradford, C.K.A. Mewes, Z. Tadisina, S. Gupta, IEEE Magn. Lett. **1**, 3500204 (2010)
109. J. Slonczewski, J. Magn. Magn. Mater. **159**(1), L1 (1996). https://doi.org/10.1016/0304-8853(96)00062-5
110. E. Chen, D. Apalkov, Z. Diao, A. Driskill-Smith, D. Druist, D. Lottis, V. Nikitin, X. Tang, S. Watts, S. Wang, S.A. Wolf, A.W. Ghosh, J.W. Lu, S.J. Poon, M. Stan, W.H. Butler, S. Gupta, C.K.A. Mewes, T. Mewes, P.B. Visscher, IEEE Trans. Magn. **46**(6), 1873 (2010)
111. E. Chen, D. Apalkov, A. Driskill-Smith, A. Khvalkovskiy, D. Lottis, K. Moon, V. Nikitin, A. Ong, X. Tang, S. Watts, R. Kawakami, M. Krounbi, S.A. Wolf, S.J. Poon, J.W. Lu, A.W. Ghosh, M. Stan, W. Butler, T. Mewes, S. Gupta, C.K.A. Mewes, P.B. Visscher, R.A. Lukaszew, IEEE Trans. Magn. **48**(11), 3025 (2012)
112. D. Ralph, M. Stiles, J. Magn. Magn. Mater. **320**(7), 1190 (2008). https://doi.org/10.1016/j. jmmm.2007.12.019
113. A. Slavin, Nat. Nanotechnol. **4**(8), 479 (2009)
114. J.V. Kim, in *Solid State Physics, Solid State Physics*, vol. 63, ed. by R.E. Camley, R.L. Stamps (Academic Press, London, 2012), pp. 217–294. https://doi.org/10.1016/B978-0-12-397028-2.00004-7
115. A.P. Malozemoff, J.C. Slonczewski, Phys. Rev. Lett. **29**, 952 (1972). https://doi.org/10.1103/ PhysRevLett.29.952
116. A. Mougin, M. Cormier, J.P. Adam, P.J. Metaxas, J. Ferré, Europhys. Lett. **78**(5), 57007 (2007). https://doi.org/10.1209/0295-5075/78/57007
117. A. Hubert, R. Schäfer, *Magnetic Domains: The Analysis of Magnetic Microstructures* (Springer, Berlin, 2008). https://books.google.com/books?id=uRtqCQAAQBAJ
118. T.J. Silva, M.R. Pufall, P. Kabos, J. Appl. Phys. **91**(3), 1066 (2002). https://doi.org/10.1063/ 1.1421040
119. Z. Liu, F. Giesen, X. Zhu, R.D. Sydora, M.R. Freeman, Phys. Rev. Lett. **98**, 087201 (2007). https://doi.org/10.1103/PhysRevLett.98.087201
120. C. Mewes, T. Mewes, *Handbook of Nanomagnetism: Applications and Tools* (2015), pp. 71–95
121. B. Heinrich, J. Cochran, R. Hasegawa, J. Appl. Phys. **57**(8), 3690 (1985). https://doi.org/10. 1063/1.334991
122. V. Kamberský, Phys. Rev. B **76**(13), 134416 (2007)
123. B. Heinrich, D. Fraitová, V. Kamberský, Physica Status Solidi (b) **23**(2), 501 (1967)
124. V. Kamberský, Can. J. Phys. **48**(24), 2906 (1970)
125. V. Kamberský, Czech. J. Phys. B **26**(12), 1366 (1976)
126. S.M. Bhagat, P. Lubitz, Phys. Rev. B **10**, 179 (1974). https://doi.org/10.1103/PhysRevB.10. 179

Part VI
Applications to Current Magnetic Materials

Magnetic Characterization of Geologic Materials with First-Order Reversal Curves

Ramon Egli

Abstract Since their introduction in Earth sciences in 1990, first-order reversal curve (FORC) measurements have become an important tool for characterizing geologic materials. The visualization of irreversible magnetic processes, such as transitions between magnetic states in individual crystals, in a two-dimensional map provides a unique tool for discriminating complex mixtures of magnetic particle assemblages according to their composition and size, thereby reducing the intrinsic non-uniqueness of simpler magnetic characterization tools. The use of FORC measurements in Earth sciences has rapidly evolved, both technically (e.g., measurement resolution, data processing) and theoretically. For instance, numerical unmixing techniques have been recently implemented on FORC measurements (FORC-PCA), and micromagnetic calculations are now used to model the FORC signature of specific magnetic particle assemblages. The scope of this work is to provide, for the first time, a didactic and comprehensive review of FORC diagrams and their application in Earth sciences which includes (1) the theoretical foundations; (2) the connection with Preisach theory and with other magnetic characterization tools; (3) detailed instructions for optimal planning, execution, and processing of FORC measurements; and (4) an in-depth analysis of the fundamental FORC signatures commonly occurring in rocks and sediments.

Keywords Rock magnetism · Mineral magnetism · Environmental magnetism · Magnetic hysteresis · Remanent magnetization curves · Coercivity analysis · Vibrating sample magnetometer (VSM) · Preisach theory · First-order reversal curves (FORC) · Magnetite · Domain state · Magnetic unmixing · Magnetic components · VARIFORC

R. Egli (✉)
Zentralanstalt für Meteorologie und Geodynamik (ZAMG), Wien, Austria
e-mail: ramon.egli@zamg.ac.at

© Springer Nature Switzerland AG 2021
V. Franco, B. Dodrill (eds.), *Magnetic Measurement Techniques for Materials Characterization*, https://doi.org/10.1007/978-3-030-70443-8_17

1 Introduction

Since their introduction to the geoscience community in 1990, first-order reversal curve (FORC) measurements have become a standard tool for the characterization of magnetic minerals in geological material. Knowledge of these minerals is important for a variety of geoscience applications, spanning from the reconstruction of past variations of the Earth magnetic field (paleomagnetism), of geologic processes (tectonics, transport in the atmosphere and the oceans), and of the environment in which sedimentary deposits have been formed. Unlike most applications in material sciences, magnetic measurements of rocks and sediments present unique technical and interpretation challenges. On the technical side, the concentration of magnetic minerals can be extremely low, sometimes in the sub-ppm range, and the mixture of widely heterogeneous magnetic contributions requires a very precise control of the experimental conditions (e.g., timing, applied field). These requirements often reach the limits of currently available instrumentation. At the interpretation level, the intrinsic non-uniqueness of magnetic inversion problems is exacerbated when it comes to the reconstruction of the wide range of compositional and physical properties of natural magnetic particles—often far from the ideal or well-defined cases treated in material science—for which little or no additional information is available. For instance, concentrations are often too low for mineralogic characterizations, such as X-ray fluorescence, and direct observation, e.g., by electron microscopy, might be biased by magnetic extraction procedures, by the tendency of nanometric iron oxides to stick on much larger, electron-opaque minerals, and by the lack of sufficient statistics. In this context, the two-dimensional "mapping" of irreversible hysteresis processes obtained from FORC measurements has proved itself to be an unparallel tool for the identification of fingerprint signatures for certain groups of magnetic particles: for instance, ultrafine magnetite particles produced by magnetotactic bacteria or precipitated authigenically in soils and sediments bear a unique FORC signature that is clearly distinguishable from that of larger magnetite grains of lithogenic origin coexisting in the same materials. The identification of this signature requires high-resolution high-precision measurements that were initially not available, demonstrating how technical aspects of FORC measurements go in hand with processing and interpretation techniques. The special requirements of geologic material characterization with FORC measurements created its own line of research from the same underlying principles of material science applications: the Preisach theory.

This chapter deals with all main aspects of FORC measurements of geologic materials—from theoretical foundations to specific applications. A particular effort is dedicated to providing a consistent and self-contained theoretical background and to establish connections between this relatively new measurement technique and other much older but also much faster types of measurements. This link has been seldomly exploited until now—FORC measurements tend to be presented as a completely separated set of data—yet it offers the opportunity to combine the strengths of fast-but-ambiguous measurements that can be performed on a large number of samples with an in-depth-but-slow technique to be applied on selected samples for calibration and interpretation purposes.

The chapter is divided into three main sections in addition to this introduction. Section 2 deals with the theoretical foundations of FORC measurements, within the framework of the Preisach theory, and establishes the relations existing with other measurement protocols commonly used for the characterization of magnetic minerals in geologic materials. Section 3 describes the FORC measurement protocol and discusses important technical aspects and challenges of such measurements, as well as various data processing techniques used for optimal signal extraction and representation. Section 4 provides an in-depth description of the fundamental FORC signatures encountered in geological materials, classified according to the domain state of the magnetic carriers, and their relationship with other magnetic signatures. Each section is self-contained and can be read independently from the others, even if certain aspects of the FORC signatures described in Sect. 4 require specific technical (Sect. 3) and theoretical (Sect. 2) knowledge. A rigorous mathematical treatment is provided when necessary and is accompanied by a qualitative description that can be appreciated by application-oriented readers. All FORC diagrams have been obtained from measurements with the PMC MicroMag 2900/3900 series VSM/AGM or the new Lake Shore 8600 series VSM and processed with the VARIFORC software [1] (https://www.conrad-observatory.at/index.php/downloads-en/category/3-variforc).

A consistent terminology and mathematical notation are used throughout all sections, as far as permitted by common practice. For instance, the magnetic field (common mathematical notation, H; SI unit, A/m) is usually expressed by the equivalent magnetic flux density $B = \mu_0 H$ (SI unit: T), where μ_0 is the magnetic permeability of vacuum. Therefore, B is used instead of H whenever the SI unit is involved (e.g., axes labels in figures). Furthermore, the Preisach fields H_A and H_B (or B_A and B_B) are replaced by H_r and H, respectively, when used in the context of FORC measurements, since they better recall their role in the measurement protocol as the so-called reversal and measurement fields, respectively. FORC coordinates, on the other hand, are consistently expressed as H_c and H_u (or B_c and B_u), whereby a universal convention does not exist for H_u, which is often called a bias field (H_b) or an interaction field (H_i). However, the H_u notation is preferred since it does not suggest incorrect interpretations of the FORC coordinates. All mathematical symbols are listed in the table at the end of this chapter.

2 The Preisach Model of Hysteresis

The Preisach model of hysteresis originates from a landmark paper [2] by F. Preisach published in 1935, which had the purpose of explaining the behavior of pinned domain walls in iron cores. A similar model was independently developed by D. H. Everett and W. I. Whitton [3] to explain the hysteresis of gas adsorption.

The appearance of the Preisach model in different contexts clearly indicated that it represents a general description of hysteretic processes, not necessarily bonded to a physical meaning. As a result, the Preisach model was separated from its physical connotations through a mathematical formalism developed by M. Krasnoselskii and others [4]. The basic idea behind the mathematical model is that all hysteretic phenomena can be described by the superposition of elemental hysteresis loops, called *hysterons*, in analogy with other spectral decomposition methods. Hysteresis-related material properties are then represented by a distribution of the two hysteron parameters, called *Preisach function*. The practical value of the Preisach formalism is that the Preisach function of a magnetic system, which is obtained from specific measurement protocols, can be used to predict the evolution of the system magnetization for any sequence of applied fields: in rock magnetism, these would be the various types of magnetization and demagnetization protocols used to characterize the magnetic minerals contained in geologic materials. An extensive review of Preisach models has been given by I. Mayergoyz [5].

The first application of the Preisach model to multiparticle systems of relevance in rock magnetism was discussed by L. Néel [6], who identified the Preisach hysterons with individual magnetic particles. Néel's intuition is based on the classic Stoner-Wohlfarth model of single-domain particles [7], whose hysteresis loops resemble hysterons. The Preisach-Néel model of hysteresis is the only one where hysterons have a direct physical interpretation and has been therefore extensively used to study the stability of magnetic recording media, after devising a method for reconstructing the Preisach function from remanent hysteresis measurements [8–10]. Because of the similar interest of the paleomagnetic community in the remanent magnetization of fine magnetic particles, D. Dunlop tested the Preisach-Néel model on magnetite particle assemblages of different sizes, concluding that domain states played a more important role than magnetostatic interactions for larger grains [11, 12]. In the former Soviet Union, the Preisach theory have been applied to rock magnetism by Lolij Sholpo and coworkers [13].

A renewed interest in Preisach theory in Earth sciences was triggered by P. Hejda and T. Zelinka [14], who devised a fully automatized in-field measurement protocol, based on first-order reversal curves (FORC), for the reconstruction of the Preisach function. Despite not being true Preisach functions, FORC functions reconstructed from such measurements were soon used as an equivalent tool for modeling the magnetic hysteresis of geologic materials [15–17]. Preisach functions continue to represent a fundamental tool for understanding FORC measurements. Furthermore, estimates of the Preisach function based on remanent measurements might be more appropriated than FORC functions for the selective investigation of remanence-carrying minerals [18]. This chapter reviews the main features of the Preisach model and its relevance for understanding commonly used magnetic parameters for the characterization of rock and sediments.

The Classical Preisach Model

As illustrated in Fig. 1, the irreversibility of hysteresis is manifested at the most elemental level by the difference between the ascending and descending branches of minor loops running between two almost identical fields H_A and H_B. The minor loop opening can be imagined to originate from an elemental loop, called *hysteron*, with rectangular shape, unit magnetization, and switching fields H_A and H_B defined by the fields at which the two branches merge. In analogy with major hysteresis loops, the hysteron opening is expressed by the *coercive field* (or coercivity) $H_c = (H_A - H_B)/2$, defined as half the difference between the two zero-crossing points. While major hysteresis loops are symmetric about $H = 0$, hysterons can be offset horizontally by a so-called bias field $H_u = (H_A + H_B)/2$. Hysterons are thus entirely determined by coordinates (H_A, H_B) in the so-called *Preisach space*, or (H_c, H_u) in a coordinate system rotated by 45° with respect to (H_A, H_B), called the *FORC space*, which is often used to represent the FORC function. Hysterons possess only two magnetic states: positive, or $+1$, and negative, or -1. These states, as long as existing, do not depend on the applied field. Single-domain (SD) particles with uniaxial anisotropy [7] behave like hysterons if their easy axis is aligned with

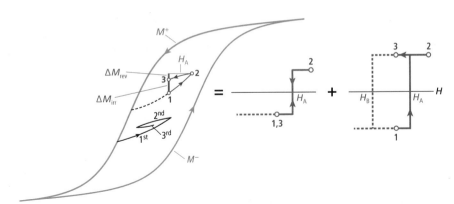

Fig. 1 Representation of reversible and irreversible magnetization changes with hysterons. M^+ and M^- are the descending and ascending branches of the major hysteresis loop. Every field sweep reversal produces a magnetization curve of higher order, from order 0 ($M^\pm$) to order 1 (first-order reversal curves) and higher. Magnetic irreversibility is shown for a minor hysteresis loop running between points 1, 2, and 3, where ΔM_{irr} and ΔM_{rev} are the irreversible and reversible magnetization changes occurring when sweeping the field from 1 to 2. Mathematically, ΔM_{irr} can be represented by a rectangular hysteresis loop, called hysteron, with two switching fields H_A and H_B: during the positive field sweep from 1 to 2, the negatively saturated hysteron is switched to positive saturation. This state is maintained during the final sweep to 3, if H_B is sufficiently small. Similarly, ΔM_{rev} is represented by a closed hysteron with $H_B = H_A$, for which the magnetization in 1 is recovered upon cycling the field to 2 and back

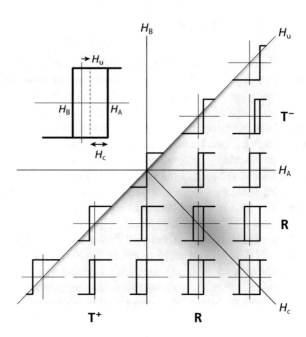

Fig. 2 The Preisach model. Hysterons with switching fields H_A and H_B (top left) are represented in the Preisach space $H_A \geq H_B$ (bottom right). Shading represents an example of Preisach function $p(H_A, H_B)$, whose amplitude is proportional to the relative number of hysterons with switching fields (H_A, H_B) required to reproduce the magnetic properties of a given magnetic material. The rotated coordinate system (H_c, H_u) represents the same hysterons through their coercive field H_c and bias field H_u. Hysterons along the $H_c = 0$ line are closed and represent reversible magnetization processes. In the absence of external fields, the Preisach space is divided into three regions: the memory rectangle **R**, where hysterons can have two states, and two transient triangles **T**, where hysterons are positively and negatively saturated, respectively

the direction of the applied field and if the magnetic moment component parallel to this direction is measured. In this case, H_u represents the stray field of nearby particles, as discussed later. In the framework of Preisach theory, hysterons must be understood as a mathematical object that provides the most compact description of hysteresis in terms of only two parameters.

A magnetic sample is fully characterized by the so-called *Preisach function* $p(H_A, H_B)$, defined as the probability density function of a large collection of hysterons (Fig. 2). This function is defined everywhere over the Preisach space defined by $H_A \geq H_B$. The $H_A = H_B$ diagonal, which corresponds to $H_c = 0$, represents completely closed hysterons. Superparamagnetic (SP) particles would be described by such hysterons. The other diagonal, given by $H_A = -H_B$ (or $H_u = 0$), is the place where hysterons are symmetric with respect to $H = 0$. Finally, the quadrant limited by $H_A > 0$ and $H_B < 0$, called the *memory region*, is occupied by hysterons that can have two so-called remanent states in $H = 0$ and thus a magnetic memory, while all hysterons outside this area, which occupy the *transient region*, are always positively ($H_A < 0$) or negatively ($H_B > 0$) magnetized if no external fields are applied. Measurement protocols based on remanence measurements support reconstructions of the Preisach function only inside the memory region because the remanent magnetic state of hysterons outside this region is fixed. The memory region is particularly relevant for paleomagnetic applications, since it is the only domain of the Preisach function that represents magnetic processes capable of recording the Earth's magnetic field. For practical reasons, it is sometimes more convenient to express the Preisach function in the transformed (H_c, H_u) coordinates used to represent FORC diagrams. In this case, the definition of the Preisach function as a probability density function with unit integral imposes $p^*(H_c, H_u) = 2p(H_A, H_B)$ for the newly defined function p^*.

The Preisach function is usually assumed to be symmetric about the $H_u = 0$ axis, because this ensures that the associated major hysteresis loop possesses the expected inversion symmetry $M^-(-H) = -M^+(H)$. However, the major loop symmetry does not force the Preisach function to be symmetric [19]. On the other hand, if hysterons of the Preisach function represent transitions between magnetic states, a symmetry about $H_u = 0$ follows from the fact that each state possesses an equivalent "anti-state," obtained by reversing the magnetization vector. In this case, each transition between states, and therefore each hysteron, is associated with a symmetric copy with opposed bias field.

Macroscopic samples with continuous hysteresis loops must, by definition, contain an infinite number of hysterons. It is therefore reasonable to assume that there are many hysterons, representing, for instance, similar processes occurring at several places in the sample, which share the same switching fields, thus contributing to the same point in Preisach space. The number of such identical hysterons is proportional to $p(H_A, H_B)$. Each hysteron can be in a positive or negative state, depending on the magnetic history of the sample and on the applied field, so that the total contribution of (H_A, H_B) hysterons to the sample's magnetization is proportional to their mean magnetic state $\gamma(H_A, H_B)$, with γ being comprised between -1 (all in a negative state), and $+1$ (all in a positive state), and $\gamma = 0$ denoting equal amounts of positive and negative states. As seen later, the magnetic randomization associated with $|\gamma| < 1$ can be obtained only with the help of thermal activations, for instance, by cooling a magnetic mineral from above its Curie temperature. It is important to recall that γ represents the *mean* state of hysterons with the same switching fields, not that of individual hysterons, which, by definition, is ± 1. Furthermore, hysterons can exist in both states only in fields $H_B < H < H_A$.

The bulk magnetization of the sample is the sum of all mean hysteron states, weighted by the Preisach function, and is mathematically represented by the integral

$$M = M_s \iint_{H_A > H_B} p(H_A, H_B)\, \gamma(H_A, H_B)\, \mathrm{d}H_A \mathrm{d}H_B \qquad (1)$$

over the Preisach space, with M_s being the *saturation magnetization*, that is, the maximum magnetization obtained with $\gamma = 1$ for all hysterons. Starting from an

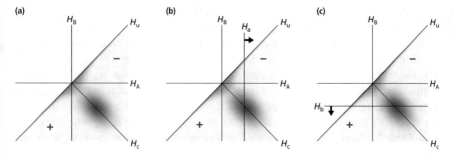

Fig. 3 The Preisach model of magnetization changes induced by external fields. The Preisach function is proportional to the darkness of the shaded area, while colors indicate the mean magnetic state γ of hysterons (gray, undefined; red, $+1$; blue, -1). (**a**) In a null external field, the transient regions are positively ($+$) and negatively ($-$) saturated, respectively, while the state of hysterons in the memory region depends on the past magnetic history. (**b**) A positive field sweep (arrow) sets all hysterons with "right-sided" switching fields H_A smaller than the actual field value H_a to positive saturation, regardless of their previous state (wiping property). (**c**) A negative field sweep (arrow) sets all hysterons with "left-sided" switching fields H_B larger than the actual field value H_b to negative saturation, regardless of their previous state (wiping property)

initial state fully specified by $\gamma(H_A, H_B)$ in a given initial field H_0, it is possible to calculate how the bulk magnetization evolves in a changing field. The field sweep direction is important, since magnetization changes are determined by the ascending hysteron branches if the field is increased and by the descending hysteron branches if the field is decreased. Increasing the field to $H_a > H_0$ brings all hysterons with $H_A < H_a$ to a positive state, resetting them to $\gamma = 1$. The cancellation of previous states is called *wiping out property*. In Preisach space, the positive field sweep is graphically represented by a vertical line at $H_A = H_a$ moving to the right, which leaves all hysterons to the left in a positive state (Fig. 3b). Positive saturation is eventually reached upon moving H_a beyond the maximum range of the Preisach function, in which case $\gamma = 1$ is obtained over the whole space. Similarly, decreasing the applied field to $H_b < H_0$ brings all hysterons with $H_B > H_b$ to a negative state, resetting them to $\gamma = -1$. In Preisach space, the negative field sweep is graphically represented by a horizontal line at $H_B = H_b$ moving to the bottom, which leaves all hysterons above in a negative state (Fig. 3c). Negative saturation is eventually reached upon moving H_b beyond the maximum range of the Preisach function, in which case $\gamma = -1$ over the whole space.

All magnetic experiments performed in a constant environment (temperature, pressure, etc.) consist of a series of field sweeps as described above, which can be represented in Preisach space. The starting point is usually a fully demagnetized sample in a null field (the Earth's magnetic field is negligibly small in most cases),

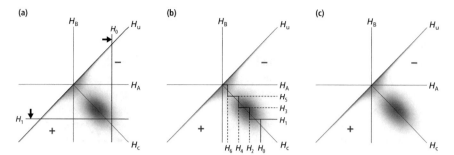

Fig. 4 The Preisach model of AF demagnetization (same color convention as in Fig. 3). Each transition from a local field minimum to a local field maximum and vice versa corresponds to a positive and a negative field sweep, respectively. The AF demagnetization is thus a sequence of field sweep to alternating positive and negative fields $H_0, H_1, H_2, \ldots$ of monotonically decreasing amplitude. (**a**) First two steps: positive sweep to H_0 and negative sweep to H_1. (**b**) Sequence of seven steps until the field has fully decayed. (**c**) Final AF-demagnetized state for the limit case of an infinite number of steps

with $\gamma = 0$ in the memory region and $\gamma = \pm 1$ in the remaining Preisach space. In geologic materials, this state is obtained when individual magnetic particles are magnetized randomly. Experiments starting with this state cannot be replicated, unless the sample is thermally demagnetized, because the application of external fields will always set some regions of the memory quadrant to $\gamma \neq 0$, even if the bulk remanent magnetization is zeroed. A well-known example is the alternating field (AF) demagnetization, obtained by exposing the sample to a series of field sweeps with decreasing amplitude, starting from a positive initial field H_0. The $0 \rightarrow H_0$ field sweep sets $\gamma = +1$ for all hysterons with $H_A < H_0$ (Fig. 4a). This is followed by a negative field sweep to $H_1 = -H_0 + \Delta H$, which sets all hysterons with $H_B > H_1$ to $\gamma = -1$ (Fig. 4b). The sequence of horizontal and vertical sweeps of the Preisach space creates a stepped line along $H_u = 0$, which separates two regions with $\gamma = -1$ and $\gamma = +1$, respectively (Fig. 4b). In the limit of $\Delta H \rightarrow 0$, the stepped line collapses onto the $H_u = 0$ axis (Fig. 4c) and defines a fully deterministic AF-demagnetized state. As seen later, the results of certain experiments, such as the acquisition of an isothermal remanent magnetization, depend on whether the initial state is truly random or AF demagnetized.

Modifications of the Classical Preisach Model

The classical Preisach model has some important limitations that prevent a correct representation of certain hysteresis processes. Countless modifications of the original model have been proposed to overcome this limitation, at the cost of a significant increase in complexity. Here, only the simplest modifications are discussed in general terms.

Regular Preisach functions fail at reproducing the reversible component of hysteresis, because the magnetization obtained from Eq. (1) remains constant once all hysterons have been switched to negative or positive saturation. In contrast, real hysteresis loops approach saturation asymptotically, and the magnetization continues to increase even after the two branches have completely merged. Two solutions have been proposed to overcome this problem. One of them has been originally proposed in the context of FORC analyses and consists in adding a so-called *reversible ridge* of the form $f_{rev}(H_u)\,\delta(H_u)$ along the H_u axis, where δ is the Dirac impulse and f_{rev} the first derivative of the reversible component of the hysteresis loop [20]. The second solution is formally equivalent to the introduction of a reversible ridge but avoids the difficulties of dealing with the Dirac impulse. It is based on the partition of the integration region in Eq. (1) into the memory rectangle $H_A > H$ and $H_B < H$, where H is the applied field, and the remaining triangular regions of the Preisach space. Since $\gamma = \pm 1$ is a fixed value over the triangular regions, the corresponding integrals are functions $f^{\pm}(H)$ representing the reversible component of hysteresis [5]. Because the memory rectangle moves with H, the modified Eq. (1) is called the moving Preisach model.

Even with these modifications in place, reversible magnetization changes can only be partially accounted, assuming a fixed contribution of hysterons with $H_c = 0$. The minimal example of a Stoner-Wohlfarth particle illustrates this limitation: two distinct functions are required to represent the curvature of the ascending and descending branches, so that the actual reversible magnetization changes occurring in assemblages of such particles depend on which of the two SD states is occupied for every combination of H_c and H_u values. In practice, f_{rev} becomes a function of the magnetic states in the memory region. A similar effect, this time over the entire Preisach space, needs to be taken into consideration when describing systems affected by a so-called mean internal field $H_i = DM$ proportional to the bulk magnetization M, where D is a proportionality constant. In this case, the total field $H + DM(H)$ transforms a hysteron with intrinsic switching fields $h_{A,B}$ into one with apparent switching fields $H_{A,B} = h_{A,B} - DM$. The resulting Preisach function depends on the magnetic states γ of the hysterons and is no longer a fixed distribution [21]. A similar problem arises with the local interaction field in assemblages of densely packed particles, which also depends on the bulk magnetization [22, 23]. Specific examples will be discussed in relation to the FORC function in section "Mean-Field Interactions".

Another important modification of the original Preisach model deals with thermal relaxation, in order to explain the time dependence of hysteresis measurements of so-called *viscous* magnetic systems. Thermal relaxation reduces the field required for a magnetic transition by overcoming the energy barrier separating two magnetic states before this barrier naturally disappears during a field sweep. The effective switching fields of hysterons are then given by $H_A = h_A - H_f$ and $H_B = h_B + H_f$, where H_f is the reduction of hysteron coercivity due to thermal relaxation, also known as *fluctuation field* [24]. Hysterons become completely closed, or superparamagnetic, when $H_f = (h_A - h_B)/2$, in which case they contribute only to the reversible component of hysteresis. In the other cases,

$$\frac{H_f}{H_c} \approx \left[\frac{k_B T}{\Delta E_0} \ln \frac{H_f}{\tau_0 \alpha} \right]^{1/q}, \tag{2}$$

is a function of the field sweep rate $\alpha = |\mathrm{d}H/\mathrm{d}t|$ [25], the absolute temperature T, and intrinsic parameters such as the energy barrier ΔE_0 for switching in a null field, the so-called attempt time $\tau_0 \approx 0.1 - 1$ ns [26], and the exponent q of the power law governing the field dependence of the energy barrier, with $q = 1.5$ for randomly oriented Stoner-Wohlfarth particles [27]. The fluctuation field causes a shrinking of the memory region, whose new boundaries are given by $H_A > H + H_f$ and $H_B < H - H_f$. These boundaries are no longer straight [28–30], because ΔE_0 is itself a function of H_c, with $\Delta E_0 \propto \mu_0 m H_c$ for Stoner-Wohlfarth particles with magnetic moment m.

Preisach Models of Selected Measurement Protocols

Measurement protocols and magnetic parameters of increasing complexity have been introduced over the years to characterize magnetic minerals in geologic materials [31–33], with FORC diagrams representing one of the last achievements. Magnetic characterization tools can be divided into three main categories according to their dimensionality:

- 0D: bulk parameters, associated binary or ternary diagrams (e.g., after Day et al. [34] and King et al. [35]), and principal component analyses [36]
- 1D: magnetization curves [37], associated shape comparisons [38], coercivity distributions [39], and numerical unmixing of curves/distributions [40–43]
- 2D: Preisach/FORC measurements [16, 18], associated isolation of specific signatures [1], numerical unmixing [44], and modified FORC protocols [45].

Even the most sophisticated characterization techniques, such as FORC, do not always overcome the intrinsic non-uniqueness of magnetic measurements, so that a combination of different techniques is sometimes required. In these cases, Preisach models provide a useful comparison tool, since reconstructed Preisach/FORC functions can be used to predict the results of lower-dimensional measurements, such as bulk parameters and magnetization curves [30], and compare them with measured ones. In the following, Preisach models of measurement protocols commonly used for the characterization of geologic materials are discussed.

Low-Field Susceptibility

The low-field susceptibility χ_{lf} is defined as the derivative dM/dH of the magnetization change induced by a small external field with respect to an initial remanent state. In geologic materials, χ_{lf} is the sum of contributions associated with dia-, para-, and antiferromagnetic minerals, as well as the initial susceptibility χ_i of ferrimagnetic minerals. Measurements are typically performed in a low amplitude alternating field $H_0 \cos\omega t$, with H_0 being of the order of 400 to 750 A/m, or 0.5–1 mT [46]. The resulting *Rayleigh hysteresis loop* is characterized by a small nonlinearity and opening that increase with increasing field amplitude (Fig. 5). The AF susceptibility is defined by the first Fourier coefficients of $M(H_0 \cos\omega t)/H_0$, which, for a Rayleigh loop with $\Delta M = M(H_0)$ and $2\delta M = M^+(0) - M^-(0)$, are given by $\chi' = \Delta M/H_0$ (in-phase or real component) and $\chi'' = 4\delta M/3\pi H_0$ (quadrature or imaginary component), respectively. The well-known Rayleigh law for the initial magnetization, $\Delta M = \chi_i H + \alpha_R H^2$, where α_R is the Rayleigh constant [47], yields the field-dependent in-phase susceptibility $\chi' = \chi_i + \alpha_R H_0$ and the corresponding quadrature term $\chi'' = 4\alpha_R H_0/3\pi$ [48].

This result can be understood using the classical Preisach model by considering that δM is the integral of the Preisach function over the memory region limited by $H_A < H_0$ and $H_B > -H_0$. Because of the limited extension of the memory region near the origin, the Preisach function can be approximated by the Taylor series $p(H_A, H_B) = p_0 + p_1 H_A + p_2 H_B + \dots$. The first term of the integral yields

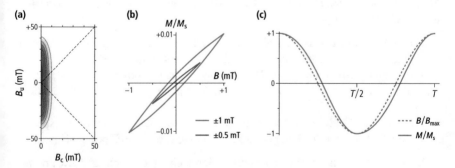

Fig. 5 Preisach model for Rayleigh hysteresis and low-field susceptibility measurements. (**a**) Simulated Preisach function of MD particles. The dashed lines represent the left limit of the memory region. (**b**) Rayleigh hysteresis loops generated with the Preisach function in (**a**). (**c**) Susceptibility measurement in an alternating field with period T and 1 mT amplitude, corresponding to the ±1 mT loop in (**b**). Notice the time lag of the magnetization with respect to the applied field for all points between minima and maxima

$\delta M \approx M_s p_0 H_0^2$, and therefore $\alpha_R = M_s p_0$, which means that the Rayleigh constant is determined mainly by the amplitude of the Preisach function near the origin. Stable SD particles have a finite coercivity and do not contribute to the Preisach function at the origin, so that $\alpha_R = 0$, in agreement with experiments. The Preisach function of multidomain (MD) particles, on the other hand, is concentrated near $H_c = 0$, yielding positive values of α_R and therefore a field-dependent χ_{lf}. The field dependence of χ_{lf} is commonly observed in titanomagnetite-bearing basaltic rocks [48] and in MD pyrrhotite crystals [49], albeit with deviations from the ideal Rayleigh law at higher fields. These deviations are expected to occur when the Preisach function is no longer approximated by a constant over the memory range defined by H_0.

IRM Acquisition Curves

The so-called *isothermal remanent magnetization* (IRM) is the remanent magnetization acquired by previously demagnetized material upon applying and then removing a so-called acquisition field. The expression "isothermal" originates from the fact that the IRM is acquired without changing the temperature. The IRM protocol results from the combination of two magnetization curves: the so-called *initial curve* $M_i(H)$, which is the in-field magnetization acquired while sweeping the applied field from zero to a given final value H_a, followed by a first-order curve departing from $M_i(H_a)$, while the applied field is swept back to zero (Fig. 6b). The resulting remanent magnetization M_{irm} is a function comprised between 0 for $H_a = 0$ and the so-called saturation remanence M_{rs} for H_a sufficiently large to saturate the sample. The sequence of IRM values acquired in increasingly large fields is called *IRM acquisition curve* (Fig. 6c). IRM acquisition curves are commonly used to characterize the coercivity of remanence-carrying minerals, e.g., through the median acquisition field [50], which is the H_a value for which $M_{irm} = M_{rs}/2$. The first derivative of $M_{irm}(H_a)$ defines the *IRM coercivity distribution* $f_{irm}(H_a)$, which is just one of the possible means of estimating a coercivity distribution (section "Coercivity Distributions"). IRM acquisition curves are widely used for coercivity analysis and detection of magnetic components [40].

In classical Preisach theory, IRM is represented by the switching of hysteron states in a vertical stripe of the memory region given by $0 < H_A < H_a$ from their original value dictated by the demagnetized state to $\gamma = +1$ (Fig. 6a). The shape of the IRM acquisition curve is sensitive to the initial state [30]: in the case of AF demagnetization, all hysterons below the H_c axis are already in a $\gamma = +1$ state, so that the acquired remanence depends only on the hysterons above this axis. If, on the other hand, the initial state is fully randomized, as with thermal demagnetization, all hysterons with $H_A < H_a$ contribute to IRM acquisition, which means that M_{irm} from a fully random initial state is always larger than M_{irm} acquired from the AF state (Fig. 6c). This difference is most pronounced for Preisach functions with a wide dispersion along H_u and is therefore particularly evident in samples containing MD

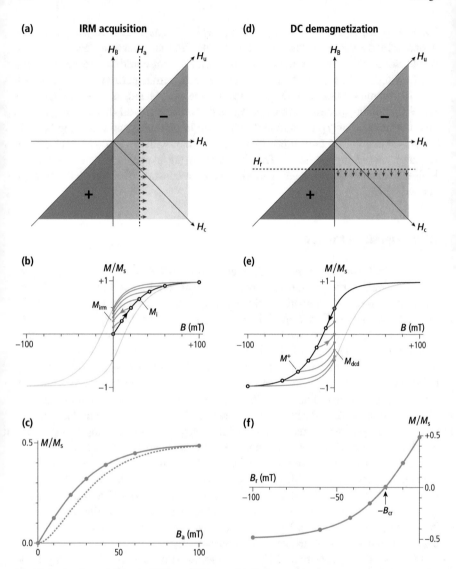

Fig. 6 Preisach model for the IRM acquisition and DC demagnetization protocols. (**a**) Representation of IRM acquisition Preisach space. The initial state is given by randomized hysteron states in the remanence region (gray). A positive field H_a switches all hysterons with $H_A \leq H_a$ to positive saturation. (**b**) Stepwise IRM acquisition protocol (black, initial curve; gray curves, field removal; gray dots, corresponding IRM steps). (**c**) IRM acquisition curve (solid line) with the IRM steps of (**b**) (dots). The dashed line is the IRM acquisition curve from an AF-demagnetized state for the same system. (**d**) Same as (**a**) for DC demagnetization. The initial state is a positive saturation IRM (all hysterons with $H_B < 0$ positively saturated). A negative field H_r switches all hysterons with $H_B \geq H_r$ to negative saturation. (**e**) Stepwise DCD protocol (black, descending branch of major hysteresis; gray curves, field removal; gray dots, corresponding DCD steps). (**f**) DCD curve (solid line) with the DCD steps of (**e**) (dots)

magnetite particles [51]. The only case where M_{irm} is independent of the initial state is when the Preisach function is concentrated along the H_c axis, forming a sharp ridge known as the *central ridge*, as it is the case for non-interacting uniaxial SD particles (section "Uniaxial SD Particles: Stoner-Wohlfarth Model"). The dependence of IRM acquisition curves on the initial state is particularly important when these curves are used for coercivity analysis, because coercivity components, especially the one covering the low-coercivity range, will be affected.

DC Demagnetization Curves

Direct-current demagnetization (DCD) of an IRM is obtained by applying and subsequently removing a so-called demagnetizing or *reversal field* H_r in the opposite direction. The resulting remanent magnetization, M_{dcd}, will be smaller than the original IRM and becomes negative if the reversal field amplitude is sufficiently large. DCD curves are measured starting from an IRM that is as close as possible to M_{rs} by applying a positive saturating field. Next, a negative reversal field H_r is applied. If the initial IRM was saturated, the field sweep to H_r occurs along the descending branch of the major hysteresis loop (Fig. 6e). During the subsequent removal of H_r, the magnetization follows a first-order reversal curve (FORC) departing from the hysteresis loop and ending with the remanent magnetization M_{dcd} at $H = 0$. The DCD curve is obtained by applying negative reversal fields of increasing amplitude, until negative saturation is reached (Fig. 6f). M_{dcd} thus begins with a positive value close to M_{rs} and ends with a negative value close to $-M_{\mathrm{rs}}$. The reversal field $H_r = -H_{\mathrm{cr}}$ for which $M(H_r) = 0$ is obtained defines the *coercivity of remanence* H_{cr} and is one of the parameters required to calculate the coercivity ratio H_{cr}/H_c used in the Day diagram [34]. The full DCD protocol described above is often shortened by imparting the initial IRM only once, followed by the sequence of reversal fields and remanence measurements. The shortened protocol is not exactly equivalent in the case of measurements affected by thermal relaxation, because the repeated application of negative fields to the same initial IRM causes a larger viscous decay than the case where a fresh IRM is acquired at every step.

The Preisach representation of the DCD demagnetization process is similar to that of an IRM, except for the fact that changes in the memory range occur in a horizontal band above $H_B = H_r$ and that the initial magnetization is represented by positive hysteron states $\gamma = +1$ (Fig. 6d). It is easy to show that, in the framework of classical Preisach models, the so-called Wohlfarth relation $M_{\mathrm{irm}}(H) = M_{\mathrm{rs}} - M_{\mathrm{dcd}}(-H)$ holds between the DCD curve starting from M_{rs} and the IRM acquisition curve obtained from a fully randomized initial state [52]. This is because hysterons in symmetric positions about the H_c axis, which contribute equally to the Preisach function, undergo a $\gamma = 0 \rightarrow 1$ transition in the case of IRM acquisition and a two times larger transition $\gamma = +1 \rightarrow -1$ in the case of DCD demagnetization. In practice, significant deviations from the Wohlfarth relation are commonly observed in MD and interacting SD particles, which are best represented with a so-called *Henkel plot* [53] of M_{irm} vs. M_{dcd}. A concave curve is obtained in these cases, instead of the straight line of the Wohlfarth relation. In addition to the use of IRM acquisition curves obtained from the AF demagnetization state instead of a fully random state [54], other causes for curved Henkel plots are related to magnetization processes that cannot be correctly described by the classical Preisach theory. One arises from thermal relaxation: the viscous decay of the initial M_{rs} while the reversal field is applied, or, in the case of the shortened protocol, during the entire measurement sequence, adds to the intrinsic initial magnetization decrease. This does not occur with IRM acquisition because a freshly acquired remanent magnetization is measured at each step. The other reason for curved Henkel plots is that correct modeling of strongly interacting SD particles [55, 56], MD particles, and other systems producing mean internal fields of significant magnitude require the use of moving Preisach models [57] (section "Modifications of the Classical Preisach Model"). In this case, the Preisach function is affected by the bulk magnetization, whose initial value is zero for IRM acquisition and equal to the saturation remanence in the case of DCD curves. Internal fields $H_i = DM$ characterized by $D < 0$ have a demagnetizing effect that help remove the initial IRM, yielding concave curves in Henkel plots, while the opposite occurs with $D > 0$. In MD particles, the demagnetizing factor D changes with the nucleation and annihilation of domain walls [58], being thus itself a function of the magnetization.

AF Demagnetization of IRM

AF demagnetization curves play an important role in paleo- and rock magnetism, owing to the fact that these are the only magnetization curves that can be obtained from the natural remanent magnetization (NRM) and from other weak-field remanent magnetizations. Stepwise AF demagnetization of an initial remanent magnetization is performed like the full AF demagnetization (section "The Classical Preisach Model"), except for the choice of the peak amplitude H_{af} of the AF field, which is stepwise increased after each remanence measurement until the sample becomes fully demagnetized. The AF peak field required to eliminate 50% of the initial remanent magnetization is called the *median demagnetizing field*. The original reason for measuring AF demagnetization curves of IRM was to compare them with the AF demagnetization of the NRM, in what is known as the Lowrie-Fuller test [59, 60], in order to gain some insights on the type of remanence carriers responsible for the Earth magnetic field recording.

In classical Preisach theory, AF demagnetization of an IRM proceeds similar to the DC demagnetization, except for the fact that only hysterons of the memory region characterized by $H_B > -H_{af}$ and $H_u > 0$ are switched to $\gamma = -1$, instead of all those with $H_B > H_r$ (Fig. 7a). This difference with the DCD curve is minor for small H_{af} values, so that, in a similar way, AF demagnetization proceeds faster than IRM acquisition. IRM acquisition and AF demagnetization are equivalent only if the Preisach function forms a narrow ridge along the H_c axis, as with non-interacting uniaxial SD particles (section "Uniaxial SD Particles: Stoner-Wohlfarth Model"). The shape of IRM acquisition and AF demagnetization curves can be

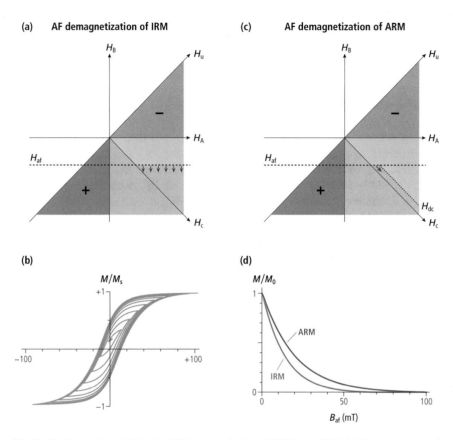

Fig. 7 The Preisach model for the AF demagnetization of IRM and ARM. (**a**) Representation of the AF demagnetization of IRM in Preisach space. The initial state is a positive saturation IRM (hysterons with $H_B < 0$ are positively saturated). An AF field with negative peak amplitude H_{af} sets hysterons with $H_u > 0$ to negative saturation. (**b**) ARM acquisition through the application of a biased alternating field with decreasing amplitude. The end point at $H = 0$ is the final ARM. (**c**) Same as (**a**) for the AF demagnetization of an ARM. The initial state is an ARM acquired in a DC bias field H_{dc} (hysterons with $H_u > H_{dc}$ are negatively saturated; those with $H_u < H_{dc}$ are positively saturated). In this case, AF demagnetization changes the state of hysterons with $0 < H_u < H_{dc}$ to negative saturation. (**d**) AF demagnetization curves of ARM and IRM for the SD-like hysteresis loop shown in (**b**)

compared either by transforming one into the other assuming that they represent identical coercivity distributions, in which case $M_{af}(H) = M_{irm}(0) - M_{irm}(H)$, or by plotting the normalized curves on a common field axis in what is known as a *Cisowski plot* [38]. In the latter case, differences between the two curves are tracked by deviations of the crossing point of the normalized curves from the ideal value $R = 0.5$ [61–66]. Additional differences between IRM acquisition and AF demagnetization curves arise, as with DC demagnetization, from thermal relaxations (the same magnetization is exposed repeatedly to the demagnetizing fields) and from magnetization-dependent changes of the Preisach function, due to mean internal field or to magnetic systems possessing more than one pair of magnetic states (section "The Pseudo-Single-Domain Magnetic Behavior"), such as so-called pseudo-single-domain particles [37].

ARM Acquisition and Demagnetization

The *anhysteretic remanent magnetization*(ARM) is the magnetization acquired in a slowly decaying alternating field possessing a small DC bias H_{dc} (Fig. 7b). The remanent magnetization M_a acquired after the AC field has fully decayed and the DC field is removed is a sigmoidal function of H_{dc}, with $M_a = \chi_a H_{dc}$ for $H_{dc} \rightarrow 0$, where χ_a is the so-called *ARM susceptibility*, and $M_a = M_{rs}$ for $H_{dc} \rightarrow \infty$. ARM shares important similarities with the thermoremanent magnetization (TRM) acquired by cooling rocks in the Earth's magnetic field [67, 68] and is therefore used as a TRM substitute in absolute paleointensity determination protocols where sample heating must be avoided [69, 70] and as a normalizer in relative paleointensity reconstructions [71]. TRM and ARM are both very sensitive to the energy barriers of irreversible magnetization processes [72, 73], the domain state of particles [74–77], and magnetostatic interactions [78, 79].

The classical Preisach model of ARM acquisition is similar to that of AF demagnetization, with the only difference that the partition line between $\gamma = +1$ and $\gamma = -1$ states is moved away from the H_c axis by H_{dc} (Fig. 7c). The slight asymmetry resulting from this offset produces a net magnetization corresponding to half of the integral of the Preisach function comprised between the diagonals given by $H_u = 0$ and $H_u = H_{dc}$, respectively. For small bias fields, this integral is equal to the product between H_{dc} and the integral of $p(H_c, H_u = 0)$ with respect to H_c, explaining the initial proportionality between M_a and H_{dc}. Preisach functions that are more concentrated along the H_c axis produce larger ARMs, due to the higher amplitudes of $p(H_c, H_u = 0)$. This effect is quantified by normalizing χ_a with M_s, so that χ_a/M_s is a measure of the H_u width of the Preisach function. Because of large reversible contributions from para- and superparamagnetic minerals in geologic materials, an IRM acquired in a field with the same amplitude as the AF peak field of ARM is chosen as normalizer instead of M_s, yielding the parameter χ_a/M_{irm} known as *ARM ratio*. Preisach models predict that the ARM ratio of SD particles is entirely controlled by magnetostatic interactions, which are solely responsible for the spread of $p(H_c, H_u)$ in the H_u direction. This model diverges for the limit case of non-interacting SD particles [80, 81], yielding an infinite value of χ_a.

The finite ARM susceptibility of non-interacting SD particles is explained by thermal relaxation effects near the minima and maxima of the AF field [82]. The slowly decaying series of minima and maxima make a SD particle switch instantaneously at every half-cycle, until the field amplitude has decayed below the critical field at which the particle possesses only one stable magnetic state. From this point on, switching continues through thermal activations by overcoming an energy barrier, which grows as the field amplitude decays further. Because of the growing energy barrier, the switching probability decreases, until the particle's moment becomes definitively frozen in its end state. In the case of an unbiased AF field, the end state is positive or negative with equal probability, and the net magnetization of many SD particles is zero. A small positive DC bias increases the probability of a positive final state to an extent that depends on how quickly the energy barrier grows during the AF decay, which in turn depends on how large the energy barrier is in a zero field. The resulting ARM ratio is a function of particle size, with typical values ranging between 1 and 5 mm/A for SD magnetite [73]. This result is easily integrated within the Preisach model after replacing the sharp partition at $H_u = H_{dc}$ between hysterons with positive and negative states with a gradual transition given by the sigmoidal shape of $M_a(H_{dc})/M_{rs}$ [83]. In this case, any Preisach function, also one representing non-interacting SD particles, yields a finite χ_a.

A modification of the Lowrie-Fuller test, based on the comparison between normalized AF demagnetization curves of IRM and ARM, has been often used as a domain state indicator; however, its usefulness is hampered by the dependence of the ARM acquisition process in MD particles on the density of domain wall pinning sites [77]. Nowadays, the main interest in AF demagnetization curves resides in their use as a normalizer of relative paleointensity [71] and for magnetofossil detection [84]. The Preisach model provides a straightforward explanation of differences between the shape of such curves and the corresponding median destructive fields through the abovementioned dependence of the ARM susceptibility on the H_u width of the Preisach function. Separable Preisach functions of the form $p = f(H_c)g(H_u)$, which are characterized by identical normalized profiles of p along H_u, yield AF demagnetization curves with shape differences that are determined only by the intrinsic coercivity dependence of χ_a/M_{rs}, for instance, that of non-interacting SD particles. This is because the H_u width of the Preisach function, and therefore the associated reducing effect on χ_a, is the same for all H_c values. Non-separable Preisach functions, on the other hand, modify the intrinsic coercivity dependence of χ_a/M_{rs} because the shape of H_u profiles depends on H_c. For instance, interacting SD particle assemblages (section "Weak Magnetostatic Interactions Between SD Particles") are characterized by a Preisach function whose H_u width is a decreasing function of H_c [85]: this makes low-coercivity hysterons less efficient in acquiring an ARM than high-coercivity ones. As a result, high-coercivity particles contribute proportionally more to the ARM than to the IRM, giving the AF demagnetization curve of ARM a higher resistance to AF demagnetization (Fig. 7d).

Coercivity Distributions

Magnetization curves are widely used in magnetic analyses of geologic materials for the identification of magnetic components with specific coercivity distributions. The identification of such coercivity components is performed either by fitting the magnetization curves, or their first derivatives, with a linear combination of functions believed to represent such components [40, 41] or by other methods based on principal component analysis [42, 86]. In both cases, the strong overlap between the coercivity ranges of individual magnetic components makes the numerical unmixing of magnetization curves very sensitive to the actual or assumed shape of the coercivity distributions of individual components [30, 41, 87]. The situation is further complicated by the fact that the shape of magnetization curves depends on the measurement protocol, as seen in the examples of the previous sections. Preisach models and the associated FORC diagrams provide a valuable tool for comparing magnetization curves and understanding the nature of magnetic components. Moreover, certain types of magnetization curves, such as the DC demagnetization (section "DC Demagnetization Curves"), coincide with a subset of the FORC measurement protocol and can be used to bridge the detailed information provided by FORC measurements with more conventional magnetic parameters.

All magnetization curves, if measured over a sufficiently wide field range, represent the progressive transition of a certain type of magnetization from an initial state $M_{ini}(H_{ini})$ to a final state $M_{fin}(H_{fin})$, with $H_{ini} = 0$ in most cases. For comparison purposes, these curves can be rescaled to share the same initial and final states, conventionally assumed to be 0 and 1, using

$$M^*(H) = \frac{M(H) - M_{ini}}{M_{fin} - M_{ini}}. \tag{3}$$

For instance, the Henkel plot, which is a tool for comparing IRM acquisition and DCD curves (section "DC Demagnetization Curves"), is obtained by rescaling M_{irm} and M_{dcd} with Eq. (3). The empirical coercivity distribution associated with each magnetization curve is defined by the first derivative of Eq. (3) with respect to H. These distributions are generally not identical, complicating the comparison of coercivity analysis results obtained from different measurement protocols. Coercivity distribution differences can be explained by four main factors: (1) coercivity-dependent intrinsic magnetic properties; (2) thermal relaxation effects; (3) magnetic memory, that is, the dependence of the measured magnetization on the sequence of previously applied fields; and (4) effects of the bulk magnetization on individual particles (e.g., interaction fields). Mechanism (3) is accounted by the classic Preisach model and can be investigated with FORC diagrams, while mechanism (4) requires a moving Preisach model, which, in limited cases, can also be deduced from FORC measurements.

The effects of mechanism (3) on coercivity distributions obtained from the magnetization curves described in sections "IRM Acquisition Curves," "DC Demagnetization Curves," "AF Demagnetization of IRM," "ARM Acquisition and Demagnetization" are exemplified with four different types of Preisach functions (Fig. 8). The first example is representative for non-interacting SD particle assemblages (section "Uniaxial SD Particles: Stoner-Wohlfarth Model"): in this case, all coercivity distributions coincide with the intrinsic distribution of hysteron coercivities (Fig. 8a). The second example is based on a separable Preisach function $p = f(H_c)g(H_u)$ commonly used in Preisach-Néel models of interacting particles [15, 86]. In this case, IRM acquisition from a thermally demagnetized state and AF demagneti-

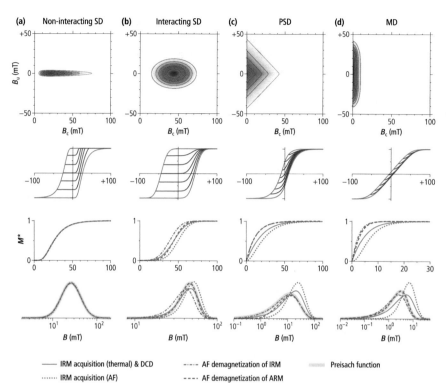

Fig. 8 Preisach functions and corresponding coercivity distributions for representative models of geologic materials. First row: Preisach functions plotted in (B_c, B_u) coordinates. Second row: corresponding hysteresis loop (blue) and FORCs (red), with no reversible component added. Third row: magnetization curves corresponding to commonly used measurement protocols, rescaled to be increasing functions comprised between 0 and 1. The protocols include IRM acquisition from the thermally demagnetized state and the AF-demagnetized state, DCD curves, and AF demagnetization curves of IRM and ARM. The last row shows the coercivity distributions corresponding to the first derivative of the normalized magnetization curves shown above. The marginal distribution of the Preisach function with respect to B_c, which represents the intrinsic coercivity distribution, is shown for comparison. (**a**) Non-interacting SD particles (central ridge along $B_u = 0$, typical example: magnetofossils). (**b**) Strongly interacting SD particles (oval contours, typical example: authigenic greigite). (**c**) Idealized PSD particles (triangular contours). (**d**) Idealized MD particles (vertical ridge along $B_c = 0$)

zation of ARM yield coercivity distributions with similar shape, while the AF demagnetization of IRM and the IRM acquisition from the AF-demagnetized state produce coercivity distributions that are shifted toward lower and higher fields, respectively (Fig. 8b). The Preisach function of the third example is characterized by triangular contour lines typical of so-called pseudo-single-domain (PSD) magnetite particles (this term is used in paleomagnetism to denote particles with coexisting SD and MD characteristics; see section "Multistate Systems: The PSD FORC Signature"). In this case, the AF demagnetization curves of ARM and IRM, on the one hand, and the IRM acquisition curve from the AF-demagnetized state, on the other, produce coercivity distributions with lowest and largest median fields, respectively, while IRM acquisition from a thermally demagnetized state yields intermediate results (Fig. 8c). The last example is based on a Preisach function featuring a vertical ridge along $H_c = 0$, which is typical of MD magnetite particles (section "Multistate Systems: The PSD FORC Signature"). The differences between coercivity distributions are qualitatively similar to those of the PSD example, but more pronounced (Fig. 8d).

In all four examples of Fig. 8, which are representative for the variety of Preisach functions encountered in geologic materials, the intrinsic coercivity distribution $f(H_c)$, defined as the integral of $p(H_c, H_u)$ over H_u, is best approximated by the coercivity distribution obtained from the AF demagnetization of ARM. Therefore, this type of magnetization curve is the most insensitive toward the H_u dependence of the Preisach function. This explains, at least in part, why coercivity analysis results obtained from the AF demagnetization of ARM are more consistent than those based on IRM curves [84].

3   The FORC Protocol

The FORC measurement protocol is one of the possible protocols that can be used to reconstruct the Preisach function. Its success in comparison to earlier protocols based on magnetic remanence measurements [11] is due to the rapidity of in-field measurements obtained with automated vibrating sample magnetometers (VSM). Preisach maps based on remanence measurements maintain certain advantages over FORC diagrams when stable remanence carriers need to be characterized in geologic materials that contain large amounts of other magnetization sources [18]. Standard FORC measurements in the geoscience community are based on the protocol introduced by Pike et al. [15]. Modifications of this protocol have been recently proposed to reduce the total measurement time [89] and to discriminate, at least partially, reversible and irreversible magnetization processes that contribute to the FORC function derived from the standard protocol [45]. These modifications can be easily understood using the knowledge that underlies the definition of the standard protocol, which is the main subject of the following discussion.

Despite the high degree of standardization of the FORC protocol, planning, execution, and processing of FORC measurements are far from being a trivial task, given the extreme magnetic variety of geologic materials and their low magnetization intensity. The resolution of minute magnetization changes over small field steps and the field control accuracy required by FORC measurements are extremely demanding even for most modern VSMs designed with this applications in mind, such as the 8600 VSM series from Lake Shore Cryotronics. Special data processing algorithms designed to provide optimal tradeoff between the opposed needs of field resolution and noise suppression [1, 90] add more complexity. All these factors need to be carefully considered, in order to avoid artifacts or results that lack the required field coverage, resolution, and signal-to-noise ratio. Knowledge of the FORC theory, the expected sample properties, instrumental limitations, and processing principles is therefore essential for a successful implementation of this powerful magnetic characterization technique. The state of the art of this knowledge is discussed in the following sections.

Definitions

As discussed in section "The Classical Preisach Model", hysteresis processes can be represented by elemental hysteresis elements, called hysterons, with a two-dimensional distribution of switching fields along the ascending and descending branches, called Preisach function. The Preisach function describes mathematically all irreversible magnetization changes occurring during sequences of zero- and first-order field sweeps— and thus various types of magnetizations and magnetization curves used for the characterization of magnetic minerals at constant temperature. The inverse task consists in reconstructing the Preisach function from a set of ascending and descending magnetization curves measured at constant temperature. These measurements must produce irreversible magnetization changes at any point over the domain of the Preisach function or over a selected region, such as the memory rectangle. Everett [91] demonstrated that such measurements can be represented as a sum of elemental sequences, consisting of a descending sweep to a given field H_b, followed by an ascending sweep to a field $H_a \geq H_b$. The elemental sequence sets all hysterons comprised within the triangular region given by $H_a \leq H_A$ and $H_b \geq H_B$ to positive saturation (dashed triangle in Fig. 9a), regardless of previous states. The total contribution of these hysterons to the Preisach function p is the Everett function

$$\xi(H_a, H_b) = \int_{H_b}^{H_a} \int_{H_b}^{H_a} p(H_A, H_B)\, dH_A dH_B,\tag{4}$$

defined as the integral of p over the above defined triangle. Changing one of the boundaries of the Everett function by a slight increase of H_a or H_b will affect the magnetic state of hysterons with $H_A = H_a$ or $H_B = H_b$, respectively, so that the derivative of ξ with respect to the changed limit is proportional to the contribution of hysterons with the corresponding switching field. The two switching fields of a hysteron are thus simultaneously defined by a small change of H_a and of H_b (hatched square in Fig. 9b). Accordingly, the Preisach function is obtained from the mixed derivative

$$p(H_A, H_B) = \frac{\partial^2 \xi}{\partial H_a \partial H_b}\bigg|_{H_a = H_A, H_b = H_B}\tag{5}$$

of the Everett function with respect to its boundaries.

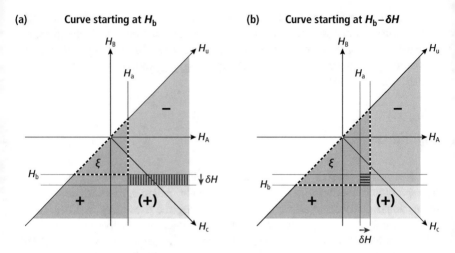

Fig. 9 The Preisach model of the Everett and the FORC functions. (**a**) Representation of the hysteron states obtained by sweeping the applied field to the reversal field H_b and then to $H_a > H_b$, which yield the magnetization $M(H_b, H_a)$ of the curve starting at H_b. Positive and negative saturation regions are labeled with "+" and "−," respectively; in parenthesis is the positive saturation of the initial state of FORC measurements. The Everett function ξ is the magnetization corresponding to the integral of the Preisach function over the area enclosed by the dashed triangle. The derivative of $M(H_b, H_a)$ with respect to H_b is equivalent to the difference between the curves starting at H_b and $H_b - \delta H$, respectively. This difference is produced by the change of hysteron states from positive to negative saturation over the hatched area. (**b**) The mixed derivative of $M(H_b, H_a)$ is determined by the decrease of the hatched area in (a) as H_a is increased to $H_a + \delta H$. This change corresponds to the magnetization of hysterons with Preisach coordinates located within the hatched square and is proportional the mixed derivative of ξ

The rather abstract pathway to the reconstruction of the Preisach function through the Everett function is best understood by turning to a practical implementation based on the FORC protocol first introduced by Hejda and Zelinka in 1990 [14]. In its most elemental form, the FORC protocol for generating a single value of the Everett function consists of three steps:

(1) Bring the specimen to positive saturation by applying a positive saturating field H_s.
(2) Perform a negative field sweep from H_s to a so-called *reversal field* H_r (which corresponds to H_b in the Everett formalism).
(3) Perform a positive field sweep from H_r to a field $H \geq H_r$ (which corresponds to H_a in the Everett formalism) in which the specimen magnetization is measured.

The initial saturation step is required to set all hysterons, and thus the whole specimen, into a well-defined, reproducible initial state, in this case positive saturation. Other resetting options are possible, such as AF demagnetization; however, they are more time-consuming. The measured magnetization, $M(H_r, H)$, is proportional to the Everett function. Negatively saturated hysterons with switching fields $H_A = H$ and $H_B = H_r$ are switched to positive saturation if the boundaries of the Everett function are changed by increasing $H = H_a$ and decreasing $H_r = H_b$, respectively. Therefore, the reconstructed Preisach function

$$\rho(H_r, H) = -\frac{1}{2} \frac{\partial^2 M}{\partial H_r \, \partial H} \tag{6}$$

is proportional the negative mixed derivative of M, with the factor ½ accounting for the fact that the magnetic state changes by 2 (from $\gamma = -1$ to $\gamma = +1$), instead of 1. Equation (6) is the definition of the *FORC function* [19]. The meaning of this definition can be simply understood by considering the FORC measurement of a specimen containing a single hysteron with switching fields H_A and H_B. All measurements obtained with a reversal field larger than H_B will yield the same positive saturation value and do not contribute to ρ, since the mixed derivative of a constant is zero. All measurements obtained with a reversal field smaller than H_B coincide with the lower branch of the hysteron, which is a step function changing from negative to positive saturation at H_A. The mixed derivative is still zero, because there is no dependency of M on H_r. The only two curves yielding a non-zero contribution to ρ are those starting immediately before and after $H_r = H_B$, respectively. The difference between these two curves, and thus the derivative of M with respect to H_r, is a step function, and differentiation of this function with respect to H yields a Dirac impulse centered on H_A. The resulting FORC function, $\rho = \delta(H_A, H_B)$, is zero everywhere, except for the Preisach coordinates associated with the single hysteron being measured.

The FORC function is usually plotted in the transformed coordinate system given by $H_c = (H - H_r)/2$ and $H_u = (H + H_r)/2$, which represent the coercivity and bias field of individual hysterons, respectively (section "The Classical Preisach Model"). Despite being formally identical to the Preisach function, reconstructions obtained with the FORC protocol, or any other sequence of real measurements, are affected by the limitation of the scalar Preisach model discussed in section "The Classical Preisach Model", namely, that magnetization-dependent changes of the Preisach function cannot be taken into account. Examples of this limitation include the dependence of reversible magnetic moment rotation on the state of Stoner-Wohlfarth particles [7] and the effects of interaction fields in dense particle aggregates [23, 24]. Therefore, both the Everett and the FORC function might depend on the measured magnetization and its past history, instead of being invariant functions of H_r and H. This lack of invariance allows the asymmetry of the FORC protocol [14] to be transmitted to ρ, so that $\rho(H_c, H_u) \neq \rho(H_c, -H_u)$. Practical consequences depend on the type of magnetic materials being investigated, with most evident asymmetries being associated with systems affected by strong negative interactions, such as in perpendicular recording media [92]. FORC functions of natural magnetic mineral assemblages are far more symmetric, due to the random cancellation of opposite effects in highly disordered magnetic systems: in this case, most evident residual asymmetries are observed with SD particles [16, 93, 94].

Stoleriu and Stancu [95] proposed an improved reconstruction of the Preisach function based on FORC measurements performed with the above-described protocol in conjunction with the complementary protocol based on descending FORCs (i.e., first-order curves starting from the ascending branch of the major loop). The sum of the corresponding FORC functions would yield a symmetric distribution that provides a better description of magnetic hysteresis. In the context of geologic material characterization, the FORC function is used as a diagnostic tool, for instance, to determine the domain state of magnetic particles, rather than for hysteresis reconstruction. Features associated with the FORC asymmetry, such as the negative amplitudes caused by reversible magnetic moment rotation in SD particles [93, 94], would be partially or totally removed by the symmetrization process, thus eliminating important diagnostic features.

Selected Properties of the FORC Function

Some useful relations exist between the FORC function and bulk magnetic parameters or magnetization curves, which facilitate the comparison of FORC measurements with simpler, rapidly measured, and widely used magnetic parameters. The theoretical background of these relations is derived in this section, with specific examples discussed in Sect. 4.

The first relation involves the irreversible component of the major hysteresis loop. For the descending branch, M_{irr} is defined by the initial part of the measured curves through

$$dM_{irr}(H_r) = M(H_r, H_r) - M(H_r - dH, H_r) , \tag{7}$$

where dM_{irr} is the irreversible magnetization change corresponding to the infinitesimal loop obtained through the $H_r \rightarrow H_r - dH \rightarrow H_r$ field sweep (Fig. 10a). It can be easily demonstrated that Eq. (7) is equivalent to the integration, in Preisach space, of the FORC function over a horizontal stripe with Preisach coordinates $H_A > H_r$ and $H_r - dH < H_B < H_r$, so that

$$\frac{dM_{irr}}{dH_r} = 2 \int_{H_r}^{\infty} \rho(H_A, H_r) dH_A \tag{8}$$

is the derivative of the irreversible component of the descending branch of the major hysteresis loop. In analogy with the definition of coercivity distributions from other magnetization curves, $f_{irr}(H) = \frac{1}{2} dM_{irr}/dH$ is the coercivity distribution associated with irreversible magnetization processes occurring during the measurement of the major hysteresis loop (Fig. 10b). Unlike the commonly used coercivity distributions discussed in section "Coercivity Distributions", f_{irr} is defined for positive and negative fields, as is the hysteresis loop. A single hysteron with no bias field contributes to f_{irr} with a Dirac impulse at $H = -H_c$. Because $H_c > 0$ and since other coercivity distributions are defined for positive fields, a better definition is given by $f_{irr}(H) = -\frac{1}{2} dM_{irr}(-H)/dH$, which is formally equivalent to the coercivity

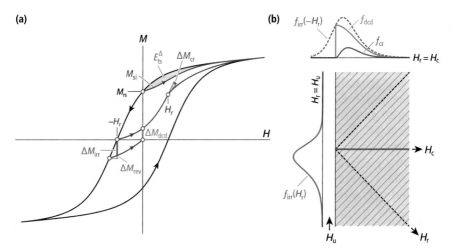

Fig. 10 Relevant FORC elements. (**a**) Measurement space with major hysteresis loop. M_{rs}—saturation remanence. M_{si}—saturation initial curve (the FORC starting at $H_r = 0$). E_{ts}^Δ—energy dissipated by transient irreversible processes in positive fields. ΔM_{irr}—irreversible magnetization change along the descending branch of the major hysteresis loop, defined by two FORCs starting at H_r and $H_r - \Delta H$, respectively. ΔM_{rev}—reversible magnetization change associated with ΔM_{irr}. ΔM_{dcd}—irreversible change of the remanent magnetization associated with ΔM_{irr}. ΔM_{cr}—contribution to the central ridge magnetization from processes with switching field $-H_r$, associated with a sudden change of the curve slope. (**b**) FORC space with associated coercivity distributions. Diagonal gray lines: traces of measured curves. Dashed black lines: boundary between the memory region (right) and the transient regions (above and below). Red line: central ridge along $H_u = 0$. The curve plotted on the left side of the FORC space is the coercivity distribution f_{irr} of irreversible hysteresis (M_{irr}), obtained by integrating the FORC function along the diagonal traces (gray) of the FORC space. The curves plotted above the FORC space (solid lines) are the coercivity distribution f_{dcd} of the DC demagnetization curve (M_{dcd}), obtained by integrating the FORC function along the diagonal traces within the memory region, and the central ridge coercivity distribution f_{cr} corresponding to FORC contributions along the central ridge. $f_{irr}(-H_r)$ (dashed line) is shown for comparison

distribution derived from the ascending branch of the major hysteresis loop [96] (Fig. 10b).

The integral of f_{irr} coincides, by definition, with the integral M_{forc} of the FORC function over the whole FORC space and with the half sum of all irreversible magnetization changes occurring along the major hysteresis loop. In the absence of reversible magnetization processes, M_{forc} is identical to the saturation magnetization M_s. Reversible magnetization processes reduce M_{forc} to a value between zero, when the hysteresis loop is fully closed, and M_s, when $M_{rs} = M_s$, where M_{rs} is the saturation remanent magnetization. FORC functions that are either symmetric about the H_c axis or fully contained within the memory region are characterized by $M_{forc} = M_{rs}$. Differences between M_{forc} and M_{rs} thus depend on asymmetric contributions of the FORC function outside the memory region. For instance, assemblages of Stoner-Wohlfarth particles are characterized $M_{forc} < M_{rs}$, because of the negative amplitude of the FORC function below the $H_u = -H_c$ diagonal [93].

The reconstruction of M_{irr} via Eq. (8) enables the determination of the reversible component $M_{rev} = M^+ - M_{irr}$ of the descending branch of the major hysteresis loop. This component can be added to the FORC function in the form of a ridge along $H_r = H$ in Preisach space or $H_c = 0$ in FORC space, so that $M_{forc} = M_s$ in all cases [20]. In practice, this operation is rarely performed, because it causes severe numerical processing problems.

Another interesting relation exists between the FORC function and DC demagnetization (DCD) curves. A DCD curve (section "DC Demagnetization Curves") is based on irreversible changes of the remanent magnetization obtained by applying increasingly large negative fields. The DCD measurement protocol is thus a subset of the FORC protocol based on the remanence measurement $M(H_r, 0)$ contained in each curve. By analogy with M_{irr}, the DCD curve is defined through

$$dM_{dcd}(H_r) = M(H_r, 0) - M(H_r - dH, 0), \tag{9}$$

for $H_r \leq 0$, where dM_{dcd} is the counterpart of dM_{irr} for the remanent magnetization (Fig. 10). The associated coercivity distribution is defined by $f_{dcd}(H) = -\frac{1}{2} dM_{dcd}(-H)/dH$. In terms of classical Preisach theory, $f_{dcd}(H_r)$ coincides with the integral of $\rho(H_r, H_A)$ over $H_A \geq 0$. Only the memory region of the Preisach space is sampled by f_{dcd} (Fig. 10b), and the integral of f_{dcd} coincides with the integral of the FORC function over the memory region, which is in turn equal to M_{rs}. Finally, it should be noted that Eqs. (8) and (9) are identical at $H_r = 0$, which means that $f_{irr}(0) = f_{dcd}(0)$.

The *transient irreversible magnetization* $M_{ts} = M_{forc} - M_{rs}$ coincides, by definition, with the integral of the Preisach function over the transient regions of Preisach space, that is, the two triangular regions surrounding the memory region [97] (Fig. 10b). A simple estimate of M_{ts} that does not require the measurement of a whole set of FORCs is obtained by measuring the so-called saturation initial curve M_{si} [97], which is the FORC starting at $H_r = 0$ (Fig. 10a). If the Preisach function is symmetric with respect to $H_u = 0$, the total energy dissipated by transient irreversible processes [98]

$$E_{ts}^\Delta = 2\mu_0 \int_0^\infty \left(M^+ - M_{si}\right) dH \tag{10}$$

is a proxy of M_{ts}/M_{forc}, after normalizing E_{ts}^Δ by the energy dissipated by the whole hysteresis loop (i.e., the integral of $M^+ - M^-$). Certain systems, such as non-interacting or weakly interacting SD particles, and slightly larger particles hosting a single magnetic vortex, are characterized by $E_{ts} = M_{ts} = 0$, which means that M^+ and M^- are completely reversible in positive and negative fields, respectively. MD particles, on the other hand, carry little remanence, since most of the irreversible magnetization processes (e.g., domain wall pinning/unpinning) occur in the transient region of the Preisach space, close to $H_c = 0$ [98].

Systems characterized by $M_{ts} = 0$ might possess an additional property captured by FORC measurements, which is the magnetic memory of field reversals. In this case, irreversible magnetization changes that have been triggered at H_r after sweeping from positive saturation are reverted when the opposite field $H = -H_r$ is reached during a FORC measurement. This is the typical case of uniaxial SD particles [7]: all particles with switching field $H_{sw} < -H_r$, which have been switched to negative saturation before starting a FORC at H_r, are switched back to positive saturation while measuring in $H \leq -H_r$, but not beyond this limit. This produces a sudden decrease of the slope of $M(H_r, H)$ at $H = -H_r$ (Fig. 10a), which contributes to the FORC function in the form of a sharp ridge along $H_u = 0$ called the *central ridge* [93, 94].

FORC Measurements

The FORC Protocol

The practical implementation of the FORC protocol described in section "Selected Properties of the FORC Function" is built on the minimal set of field sweeps required to measure a FORC, with additional steps dealing with calibration and measurement timing issues. All protocols currently used for the characterization of geologic samples are derived from Pike et al. [15] and are designed to cover a predefined rectangular region of the FORC space given by $H_c^{min} \leq H_c \leq H_c^{max}$ and $H_u^{min} \leq H_u \leq H_u^{max}$, whereby $H_c^{min} = 0$ is always used in practice and assumed in the following. In Preisach space, each FORC is represented by a horizontal line beginning at $H_A = H_r$ and ending at $H_A = H_{max}$, where $H_{max} = \min\left(2H_u^{max} - H_r, 2H_c^{max} + H_r\right)$ is chosen so, that $H_c \leq H_c^{max}$ and $H_u \leq H_u^{max}$. Reversal fields range from $H_u^{min} - H_c^{max}$ to H_u^{max} in regular steps of δH, and measurements of each FORC are performed with the same regular steps, so that the Preisach space is sampled by a square grid of points with resolution δH. For this to occur, it is necessary that the FORC space limits are expressed as multiples of δH. In FORC space, each curve is represented by a diagonal line, due to the rotation of the coordinate system (Fig. 11b). With this definition of the FORC space, measurements are performed with the following sequence:

(1) Apply a saturation field H_{sat} during a time t_{sat}.
(2) Sweep the applied field from H_{sat} to a calibration field H_{cal}, wait for a time t_{cal}, and perform a calibration measurement $M_{cal} = M(H_{cal})$.
(3) Sweep the applied field from H_{cal} to the first reversal field $H_{r,1}$ and wait for a time t_r.
(4) Measure the magnetization at regular steps δH while the field is slowly increased from H_r to the maximum field $H_{max}(H_{r,1})$, with each measurement taking a time t_m.
(5) Repeat steps (1–4) for the second reversal field value $H_{r,2} = H_{r,1} - \delta H$ and so on, until H_r^{min} is reached.

Non-measuring field sweeps are usually performed at the maximum rate $\alpha_0 = dH/dt$ enabled by the magnetometer, to save time. Measuring field sweeps

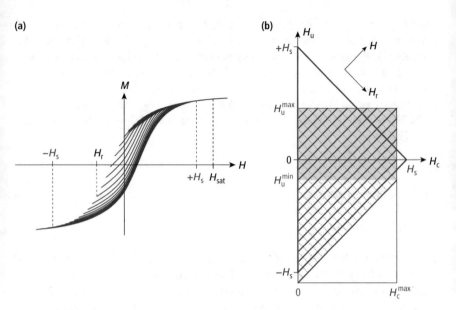

Fig. 11 Relation between hysteresis (**a**) and FORC space (**b**). Each curve in (**a**) contains measurement points that plot along an ascending diagonal in (**b**), and all curves together define a trapezoidal measurement space. The maximum extension of the FORC function is a triangular region determined by the field H_s at which the major loop becomes completely closed (red triangle in (**b**)). H_{sat} is the saturation field used in the FORC protocol. Conventional FORC protocols ensure that the rectangular area defined by $0 \leq H_c \leq H_c^{max}$ and $H_u^{min} \leq H_u \leq H_u^{max}$ (shaded rectangle) is covered by measurements. Modern FORC processing tools enable to plot the FORC function over the whole measurement range (upright rectangle). In this case, the choice of the FORC range parameters H_c^{max}, H_u^{min}, and H_u^{max} depends essentially on H_s and the dominant domain state of the ferrimagnetic minerals being measured

are determined by the field step and by the averaging time t_m of measurements in two different modes. In the continuous mode, measurements are taken one after the other, at intervals of δH, while the field is continuously swept at a constant rate $\alpha_m = \delta H$. In the discrete or point-by-point mode, the applied field is set to the nominal value established by the FORC protocol and allowed to stabilize for a time t_h, before a measurement is taken. The effective field sweeping rate is in this case given by $\alpha_m = \delta H/(t_0 + t_h + t_m)$, where t_0 is the time required to increase the applied field by δH. The point-by-point measuring mode is more accurate and less noisy because the field is stable, but more time-consuming.

Measurements taken at the calibration field are not directly necessary for the calculation of the FORC function; they are used to correct for a drift that might accumulate during the long time required to measure a typical set of FORCs (several hours to >1 day). The calibration field can be any value between the maximum field $H_u^{max} + H_c^{max}$ applied during FORC measurements and H_{sat}. In practice, it is chosen to lie just above $H_u^{max} + H_c^{max}$. Of all pauses defined by the FORC protocol, the only relevant one for cases where the hysteresis is time-dependent is t_r, the time spent at the reversal field. Its effect on the FORC function is discussed later in section "Thermally Activated SD Particles". The pause at the calibration field is used to enable full-field stabilization before a calibration measurement is taken and depends essentially on the magnitude of H_{sat}; 0.5–1 s are usually sufficient for this purpose. The pause at saturation should be kept small, since excessive time spent in a large field might activate high-coercivity viscous magnetizations, whose decay during measurements will have a negative impact on the calculation of the FORC function.

Choosing the FORC Protocol Parameters

The choice of adequate FORC space limits and measurement resolution is essential for obtaining useful results. This choice should be guided by preliminary measurements, which, at a minimum, include a hysteresis loop, possibly a major one, and by considerations about the type of magnetic particles that should be investigated. Ideally, the FORC space should encompass the field range of all irreversible processes, with measurements completely filling the area enclosed by the major loop. In this case, the field H_s at which the major loop becomes completely closed (not to be confused with the saturation field H_{sat} of the FORC protocol) sets H_c^{max} and H_u^{max}, depending on the expected shape of the FORC function. FORC functions of natural magnetic particles can be considered symmetric about the H_c axis in a first approximation; in this case, full coverage of a symmetric rectangular domain is obtained with $H_u^{min} = -H_u^{max} - H_c^{max}$. This choice does not take the trapezoidal form of the actual measurement space into account and the fact that the vertical extension of the FORC function decreases when moving to the upper limit of the H_c range, which means that the lower-right corner of the FORC space does not contain significant contributions. A more efficient choice of H_u^{min} takes advantage of these features, seeking full coverage of the vertical range only at $H_c = 0$. This coverage is obtained by setting $H_u^{min} = H_c^{max} - H_u^{max}$. Furthermore, coverage of the FORC function along the H_c axis near H_c^{max} imposes $H_u^{min} < 0$. Taking a safe margin of $\sim 0.1 H_u^{max}$ around the H_c axis, these conditions are satisfied by setting $H_u^{min} = -\max(H_u^{max}, H_c^{max} + 0.1 \ H_u^{max})$. FORC processing software such as VARIFORC [1] can deal with partially filled FORC spaces by setting the unmeasured domains to zero. Correct choices of H_u^{min} with the abovementioned criteria can save up to 50% measurement time by cutting the number of unnecessary identical FORCs that start in the negative saturation range of the major hysteresis loop (Fig. 11).

   With these considerations in mind, the number of independent FORC space parameters reduces to two: H_c^{max} and H_u^{max}. The choice of H_c^{max} and H_u^{max} is dictated by the shape of the FORC function. In practice, there are three main shape categories to consider (Fig. 8), which are dictated by the dominant domain state of magnetic particles:

(1) Non-interacting and weakly interacting SD: the FORC function is spread horizontally, with oval- or teardrop-shaped contours (Fig. 8a, b). Typical

examples include sediments, soils, and rapidly cooled rocks [99–101]. In this case, $H_c^{max} \approx H_s$ and $H_u^{max} \approx 0.5 H_c^{max}$ provide an adequate coverage.

(2) Strongly interacting SD and PSD: the FORC function is characterized either by oval contours [17] for the strongly interacting SD endmember, by triangular contours of the type shown in Fig. 8c for the idealized PSD endmember [16, 102–104], or by a blending of the two endmembers [16, 105, 106]. Typical examples include sediments dominated by detrital or aeolian inputs and a wide range of rocks. In these cases, full coverage of the FORC function is obtained with $H_c^{max} \approx H_b^{max} \approx H_s$.

(3) MD: the FORC function is characterized by a dominant vertical spread along $H_c \approx 0$ [18, 107, 108] (Fig. 8d). Typical examples include titanomagnetite-bearing slowly cooled rocks. Full coverage of this signature is obtained with $H_b^{max} \approx H_s$ and $H_c^{max} \approx 2 H_c$, where H_c is the coercive field of the major loop.

In all cases, H_s is an essential parameter for determining the correct choice of the FORC space. Practically all geologic samples contain high-coercivity minerals, essentially hematite (α-Fe$_2$O$_3$) with various degrees of Ti substitution, goethite (α-FeOOH), pyrrhotite (Fe$_{1-x}$S), or epsilon-maghemite (ε-Fe$_2$O$_3$), whose hysteresis loop does not close within the typical field range covered by electromagnets (1.5–3 T). Goethite, pyrrhotite, and fine-grained hematite do not saturate even in large fields generated by superconducting magnets [109, 110]. Therefore, the choice of H_s is based on practical considerations, rather than on its original definition. The small field step size required for a correct resolution of the FORC function in the low-coercivity range, which is mostly of the order 0.5–2 mT [1, 94], limits the H_c range of measurements obtainable in a reasonable amount of time to 0.1–0.2 T, unless a much coarser resolution is used to investigate only high-coercivity minerals at large fields. Few studies focusing on the pure mineral endmembers suggest that fine-grained hematite and goethite tend to have a non-interacting SD signature characterized by a central ridge and few, if any, contributions in a narrow region around it [111–113], sometimes accompanied by a vertical ridge along $H_c = 0$, which is a typical thermal relaxation feature of particle sizes close to the lower SD stability limit [114]. Large hematite-ilmenite crystals are characterized by a narrower coercivity distribution peaking around 0.2–0.3 T [115], which exceeds the typical H_c range of low-coercivity minerals such as magnetite or greigite. Due to their weak spontaneous magnetization, high-coercivity minerals are often overshadowed by low-coercivity ones. Furthermore, FORC measurements tend to be noisier at higher fields. In these situations, more effective methods based on the reconstruction of the Preisach function with remanent magnetization measurements should be used instead [18].

Once the FORC range has been chosen, the saturation field H_{sat} can be fixed to any value larger than the maximum applied field $H_c^{max} + H_u^{max}$ and does not need to coincide with the maximum available field. The best choice of H_{sat} is one that exceeds $H_c^{max} + H_u^{max}$ by the estimated magnitude of thermal fluctuation fields (section "Thermally Activated SD Particles"), so that the repeated exposure to positive fields close to $H_c^{max} + H_u^{max}$ during FORC measurements does not magnetize viscous particles beyond the initial state set by H_{sat}. Excessively large H_{sat} values lead to the progressive magnetization of high-coercivity viscous components, whose decay contaminates FORC measurements. In most cases, a value of H_{sat} that exceeds $H_{\text{c}}^{\text{max}} + H_{\text{u}}^{\text{max}}$ by 50% is a safe choice that guarantees that all FORCs start from the same initial state while avoiding high-coercivity viscous contributions.

Some of the pauses defined by the FORC protocol are useful for controlling measurement timing, especially in cases where the hysteresis is time-dependent, while the choice of other pauses is dictated only by technical aspects. The pauses at saturation (t_{sat}) and before calibration (t_{cal}) do not affect the FORC function. However, pauses, especially in the case of t_{sat}, must be avoided since they tend to magnetize viscous components and make measurements unnecessarily long. Therefore, t_{sat} and t_{cal} are best chosen to coincide with the time required to stabilize the field after a fast ramp. The same reasoning applies to the pause at reversal field (t_{r}); however, in this case, longer pauses can be chosen to enhance time-dependent hysteresis phenomena [114, 116, 117]. Values of t_{r} below the minimum time required for the full stabilization of the applied field have a detrimental effect on the quality of FORC measurements (see section "Field Control, Measurement Sensitivity, and Resolution"). The field stabilization time depends on the amplitude of the preceding field sweep and the sought precision; it can be checked on the real-time field readout of the controlling software, if available. About 1–2 s are required for FORC measurements up to 0.2 T with the MPC MicroMag 2900/3900 series VSM/AGM or the Lake Shore 8600 series VSM.

Field Control, Measurement Sensitivity, and Resolution

FORC measurements of geological materials are extremely demanding in terms of field control, measurement sensitivity, and resolution. Maximum deviations of the applied field from nominal values should not exceed the small field step size required to resolve critical FORC features, which is of the order of 0.5–1 mT. Resistive and superconducting magnets display drastically different performances in this respect. The field generated by the superconducting magnet of the Quantum Design MPMS3 SQUID magnetometer is controlled indirectly through the net current, and differences of up to 6 mT between nominal and effective field can be produced by flux trapping. These differences are instrument-specific and history-dependent, so that a posteriori corrections are possible only in the case of hysteresis loop measurements [118]. This limits the use of superconducting magnets to FORC measurements of high-coercivity materials that do not need small field steps. The resistive magnet of a VSM, on the other hand, is controlled by a Hall probe feedback and can attain, in principle, the same precision of Hall probe measurements. Hall probes possess a small, temperature-dependent DC offset δH_{hall}, which vertically offsets the FORC diagram. The Hall probe offset of the new Lake Shore 8600 VSM does not exceed 10 μT over the duration of longest FORC measurements (~66 h) after initial zeroing and when room temperature variations are limited to $\pm 2\,^\circ$C. Furthermore, the magnetic field in electromagnet-based magnetometers can

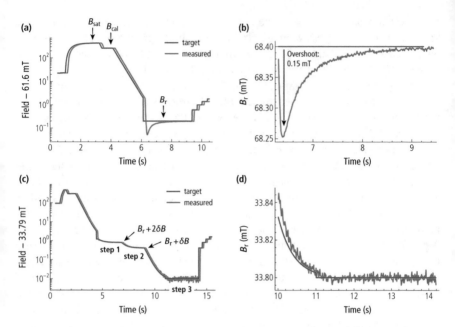

Fig. 12 Field control characteristics of the Lake Shore model 8600 VSM in proximity of the reversal field. **(a)** Target field and actual field measured by the built-in Hall sensor during the measurement of a single FORC with nominal reversal field $B_r = 68.4$ mT. A logarithmic scale is used to highlight small differences between actual and target fields at B_r. Measurement preparation include a saturation step at B_{sat}; a calibration measurement at B_{cal}; a fast, quasi-exponential field ramp to a field slightly larger than B_r; a final approach to B_r during the pause time set by the user (2 s in this example); and measurements while the field is increased in small steps (first four steps with $\delta B = 0.5$ mT in this example). Notice the B_r overshoot at ~6.4 s. **(b)** Detail of (a) showing the ~0.15 mT overshoot during the final approach to B_r. **(c)** Same as (a) for a modified protocol used by Wagner et al. [119] for an overshoot-free approach to B_r. The modification consists in adding two intermediate steps with target fields $B_r + 2\delta B$ and $B_r + \delta B$, respectively. **(d)** Detail of (c) showing the final approach to B_r

be swept at rates of up to 1 T/s, while the field sweep rate of superconducting magnet-based magnetometers (e.g., SQUID) is usually limited to ~20 mT/s, so that FORC measurements take much longer in the latter case.

The large inductivity of electromagnets introduces dynamic field control effects that tend to produce an overshoot whenever the field sweep direction is reversed. This can be a problem for field reversals at the beginning of each FORC, since the effective reversal field is lowered by the peak overshoot. Field overshooting with the Lake Shore 8600 VSM is largely suppressed by the built-in "RampOpti-mization.Overshoot" field control option: in this case, the field is rapidly ramped until a certain fraction of the target field is reached and then stabilized (Fig. 12a, b). The residual overshoot, as recorded by the built-in Hall probe at 100 Hz rate, is ≤0.2 mT for the typical ±100 mT field range of regular FORC measurements. Special applications of high-resolution FORC diagrams, such as the determination

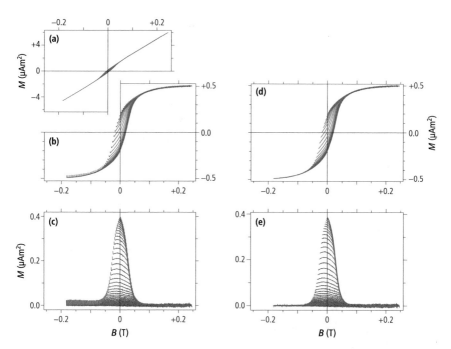

Fig. 13 High-resolution FORC measurements of a ferromanganese crust from the Pacific Ocean, which is representative for materials with low ferrimagnetic content and a large paramagnetic background. (**a**) Drift-corrected measurements obtained with a PMC 3900 VSM (stack of 16 measurement series, $\delta B = 0.5$ mT). (**b**) Same as (**a**), after subtracting the paramagnetic contribution. Notice the anomalous measurements at the beginning of each curve. Every eighth curve is shown for clarity. (**c**) Same as (**b**), after subtracting the lower branch of the hysteresis envelope reconstructed from measurements. Notice that the full scale of this plot is one order of magnitude smaller than that of the original measurements in (**a**). Anomalous first point measurements, especially at large negative fields, and measurement noise are clearly visible. (**d, e**) Same as (**b, c**), after replacing the first measurement of each curve with the extrapolated value from second-order polynomial regression of the next 6 points

of thermal relaxation effects from the H_u offset of the central ridge [119], require a modified protocol to eliminate any residual overshoot, e.g., by adding 1–2 target fields $H_r + k\,\delta H$, where k is a positive integer and δH the field step size of FORC measurements, before the nominal reversal field H_r is reached (Fig. 12c, d).

Field control problems near H_r produce a vertical ridge along $H_c = 0$ in the reconstructed FORC function, which might be confused with a magnetic viscosity signature. The two effects can be distinguished by the fact that the H_u range of the viscous ridge is limited by the saturation field H_s beyond which the major hysteresis is completely closed [116], while artifacts persist outside this range (Fig. 13b, c). In the case of evident artifacts, the first measurement of each FORC is best removed or replaced with an extrapolated value obtained from a second-order polynomial fit of the next points (Fig. 13d, e). This option is available with the VARIFORC processing software [1].

Another common problem often encountered with geologic samples is the low concentration of ferrimagnetic minerals in absolute terms, as well as relative to the induced magnetization of other minerals. The problem is best analyzed by considering a specimen with saturation magnetic moment m_s, saturation remanent moment m_{rs}, and a magnetic moment $m_p = \chi_{hf} H$ induced by dia-, para-, and antiferromagnetic minerals, where χ_{hf} is the high-field susceptibility in magnetic moment units. The measurement range required by FORC measurements with maximum field amplitude H_{max} is determined by the maximum magnetic moment $m_{max} = m_s + \chi_{hf} H_{max}$. On the other hand, the typical magnetic moment difference δm that needs to be resolved is smaller than the difference between two consecutive curves starting at reversal fields close to $-H_c$, where H_c is the coercive field. In the example of Fig. 13, which is representative for materials with low ferrimagnetic content and a strongly paramagnetic matrix, $m_{max} = 5$ μAm^2, $m_s = 0.5$ μAm^2, $\delta H = 0.5$ mT, and $\delta m = 5$ nAm^2, which is only 0.1% of m_{max}. For comparison, the empirical noise floor for this set of measurements is 1.6 nAm^2. Since only a small fraction of δm contributes to the FORC function, the signal-to-noise ratio (SNR) of the unsmoothed FORC diagram is generally <1.

Large values of χ_{hf} have detrimental effects on the quality of FORC measurements, because small changes of the associated magnetic moment, due, for instance, to the temperature dependence of χ_{hf} or of the instrument response, might reach a significant fraction of the total moment of ferrimagnetic minerals. For instance, in the above example, a room temperature variation of only 0.2 °C during the time required to measure a single curve produces a 0.03% change of m_{max}, that is, 3.3 nAm^2, which is of the same order of magnitude of δm. Similarly, a small lateral displacement of the sample holder from the centered position, due, for instance, to air flow, produces a change of the output signal, which, in the above example, is dominated by m_{max}. Using the quadratic output signal response of a similar magnetometer [120] and assuming that the specimen is perfectly centered, the expected maximum magnetic moment change produced by a lateral displacement of 0.1 mm is of the order 0.04% or 2 nAm^2 for the above example.

Drift and Stacking

Long-term stability is essential for obtaining high-quality FORC measurements. The main factors affecting the measurement stability are of a mechanical and thermal nature. In the case of VSM measurements, mechanical aspects include specimen resistance to vibrations and the stability of the measurement position. Powdered specimens are prone to become loose when vibrated for a long time, especially if they lack a fine matrix, and should be firmly pressed or glued. The magnetometer response near the centered measurement position is usually described by three parabolic functions of the horizontal and vertical specimen coordinates, respectively [120]. Accurate centering ensures that the effects of small displacements from the center position, e.g., by air flow or an unstable ground, are of second order; nevertheless, mechanical disturbances should be minimized.

Temperature variations affect the magnetic properties of the specimen, especially if most of the induced magnetization is carried by paramagnetic minerals, and should be avoided. Direct sun exposure and equilibration of the specimen temperature in the colder environment of water-cooled resistive magnets can be a major source of rapid initial drifts, noticeable even during the measurement of a hysteresis loop [121]. Preliminary measurements, such as hysteresis and DC demagnetization, which are anyways required to choose suitable FORC protocol parameters (section "Choosing the FORC Protocol Parameters"), provide enough buffer time for specimen temperature equilibration.

Calibration measurements performed before each FORC are used for stability monitoring and for correcting slow drifts. The calibration measurements $m_{c,i} = m_c(t_i)$ form a time series with increasing time intervals $t_{i+1} - t_i$ proportional to the number of measured points in the i-th FORC. A suitable interpolation $\tilde{m}_c(t)$ of this sequence (e.g., piecewise linear) is used by FORC processing tools to produce a drift-corrected version

$$\tilde{m}_{i,j} = \frac{\tilde{m}_c(t_{i,j})}{\tilde{m}_c(0)} m_{i,j} \tag{11}$$

of the j-th measurement in the i-th FORC, performed at the time $t_{i,j}$. The time stamps of calibration measurements are either stored in the measurement file or calculated from the FORC protocol. Calibration measurements possess their own noise, which is added to that of measurements by the drift correction procedure, unless the calibration measurement averaging time is chosen to be much larger than that of FORC measurements (section "Choosing the FORC Protocol Parameters"). Alternatively, a smoothed interpolation of calibration measurements can be used to suppress measurement noise. An essential prerequisite for the drift correction to be effective is that temporal changes of m_c must be slow and continuous, so that interpolated values are representative for the real state of the system at any time. Furthermore, if the origin of drift is thermal, the correction factor in Eq. (11) should always be very close to 1, because large temperature changes affect material properties that cannot be corrected for, such as coercivity. Typical long-term stabilities of the order of 0.2% over one day can be obtained if ambient temperature variations do not exceed $\pm 2\,^\circ$C (Fig. 14a, b).

Transient disturbances with time constants equal or shorter than the time required to measure individual curves cannot be corrected (Fig. 14c, d). The correlated error of the affected curves adds a characteristic diagonal striping to the reconstructed FORC diagram. Episodic curves affected by this problem can be identified and eliminated during FORC processing, e.g., with VARIFORC, without quality losses, as long as the measurement resolution provides a sufficient degree of redundancy among consecutive curves. Larger measurement averaging times will increase the time required to measure individual curves, and thus the time interval between consecutive calibration measurements, reducing the resolution of calibration measurements and the possibility of removing transient disturbances. The resulting measurement errors are strongly correlated (Fig. 14a), with detrimental effects on

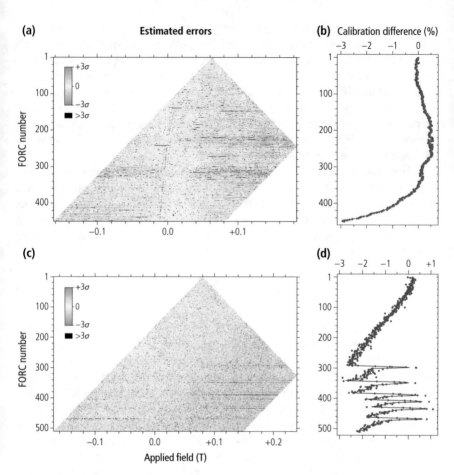

Fig. 14 FORC measurement errors and drift. (**a**) Typical error pattern for the PMC 2900 AGM magnetometer (magnetic moment range: 2 µAm^2), estimated using the VARIFORC software. Each pixel row in the plot represents a single curve, with color-coded errors comprised between ±3 standard deviations. Larger errors are treated as outliers and marked with black pixels. Horizontal striping denotes time-correlated errors caused by small fluctuations of the mechanical resonance controlling the response function. Also notice the mean error increase at large field amplitudes. (**b**) Calibration measurements for the same set of data. Abrupt variations of calibration measurements, e.g., around curves 240 and 320, cause larger correlated errors that cannot be completely removed. (**c**) Same as (**a**) for a Lake Shore 8600 VSM magnetometer (magnetic moment range: 3 µAm^2, step-by-step measurements). Contrary to the AGM example, VSM measurement errors are uncorrelated in time, as long as environmental conditions are stable, e.g., for curves 1–280. The mean error increases with increasing amplitude of the applied field, even if the field is stabilized before each measurement. (**d**) Calibration measurements for the same set of data. The regular trend of curves 1–280 is due to a magnetization decrease associated with a progressive ambient temperature increase, until air conditioning was turned on. From that point on, the sawtooth temperature variations produced by the on-off air conditioning cycle caused abrupt magnetization changes that cannot be completely compensated by drift correction, introducing detrimental correlated noise

the quality of FORC results. In these cases, it is better to stack repeated FORC measurements obtained with a shorter averaging time. Individual measurement files produced with the same measurement protocol can be merged with VARIFORC to produce a single FORC set available for further processing [1].

FORC Processing

FORC processing consists in estimating the FORC function defined by Eq. (6) from discrete measurements using suitable numerical methods. Because of the mixed derivative definition, measurement errors are strongly enhanced, which, for most geological materials, means that direct finite difference implementations of Eq. (6) yield FORC functions that are completely dominated by noise. Useful processing algorithms must include some methods to reduce noise, which result in smoothing. More noise suppression requires stronger smoothing, which might introduce significant processing artifacts while removing important details of the FORC function. Therefore, a balance must be found between the opposite requirements of noise suppression and detail preservation. To make things more complicated, the optimal balance between these requirements depends on local features of the FORC function. Several processing tools with increasing complexity have been developed since the first implementation by Pike et al. [15], and processing techniques are still evolving. However, even the most refined solutions cannot compensate for bad experimental design, such as insufficient measurement resolution and poor measurement stability. Careful measurement planning, as explained in section "FORC Measurements", is thus an essential prerequisite for obtaining reliable results.

In the following, the main processing techniques and their implementation in existing software are described.

Preprocessing

Preprocessing steps include checks of the measurement protocol and data integrity, unit conversion or normalization, drift correction (section "Drift and Stacking"), identification of bad measurements (outliers, unstable curves, first and last points of each curve in the case of field control issues) and their removal or substitution with extrapolated values, and stacking of repeated measurement sequences into a single dataset. VARIFORC [1] offers advanced preprocessing options, including diagnostic tools for the identification of specific measurement problems (e.g., field control issues), measurement error visualization (Fig. 14), and representation of the preprocessed data after paramagnetic correction (Fig. 13). The inspection of preprocessed data is essential for eliminating measurement artifacts right away or to recognize their effects in the finally produced FORC diagram and, where possible, to improve the measurement protocol.

Theoretical Principles of FORC Function Estimation

A main distinction can be made between non-local and local estimates of the FORC function. As suggested by the name, local estimates are based on a local regression of measurements with a given analytical function, which is then used to calculate the mixed derivative. The function used for this purpose is a second-order polynomial [15].

$$P_2(H_r, H) = a_1 + a_2 H + a_3 H_r + a_4 H^2 + a_5 H_r^2 + a_6 H_r H \qquad (12)$$

whose mixed derivative is simply given by a_6. The FORC function estimate at a point $(H_r, H) = (H_u - H_c, H_u + H_c)$ of the measurement space is thus given by

$$\hat{\rho}(H_r, H) = -\frac{a_6}{2}, \qquad (13)$$

where a_6 is the coefficient of Eq. (12) obtained from least-squares regression of a non-empty region U around (H_r, H), which must contain at least six measurement points. Besides practical reasons related to computational efficiency, the choice of a second-order polynomial is motivated by the fact that it coincides with a truncated Taylor series of the continuous magnetization function $\xi(H_r, H)$ sampled by the FORC measurements M (section "Definitions"). This means that P_2 converges to ξ for $U \to 0$, yielding an unbiased estimate of the FORC function, only if ξ is continuously differentiable. There are few exceptions to the continuous differentiability of ξ that will be discussed later. These exceptions are limited to one-dimensional subspaces of the measurement domain, so that the convergence of P_2 is almost granted over the entire FORC space. A simple strategy for achieving the maximum noise suppression with Eqs. (12) and (13) consists in choosing the largest U for which regression residuals $r = P_2 - M$ remain randomly distributed around zero. The maximum size of U depends on the local properties of ξ, whereby a strong dependence on H_r or H, for instance, over the steep flanks of the hysteresis loop, limits U along the corresponding direction. Other model functions can in principle be used instead of P_2; however, they all require more stringent prerequisites for obtaining unbiased estimates. For instance, a third-order polynomial regression requires ξ to have continuous second derivatives, a condition that is violated by simple models for hysteresis, such as the Stoner-Wohlfarth model [7] for identical, randomly oriented SD particles.

The simplest implementation of the polynomial regression method takes advantage of the regular measurement grid created by the conventional protocol [15] and identifies U with a $(2S + 1) \times (2S + 1)$ array of points around the chosen measurement point (Fig. 15), where the positive integer S is the *smoothing factor*. The calculation of a_6 is efficiently performed with a matrix inversion method. Because the measurement grid is rotated by 45° with respect to the FORC coordinates, edges of the FORC space require a different handling, because of the incompleteness of the rectangular region. Early solutions limited the calculation

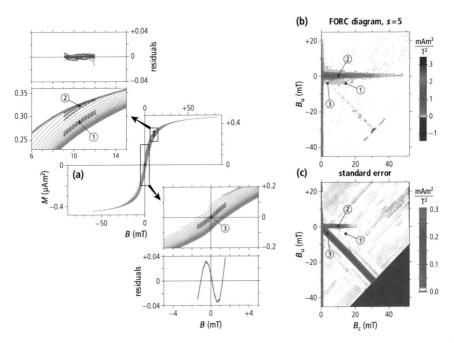

Fig. 15 Typical problems encountered with polynomial regression over different regions in FORC space. (**a**) High-resolution ($\delta B \approx 0.29$ mT) FORC measurements of equidimensional magnetosomes dispersed in kaolin. Three measurement regions with different processing requirements are highlighted, with expanded plots showing a regression example for each region, based on a smoothing factor of 5 (only three curves are highlighted for clarity). In region 1, curves are almost identical, up to a constant offset, and the concave or convex shape is well reproduced by a second-order polynomial. Residuals (plotted above) coincide with measurement errors. The mixed derivative of these curves, and therefore the FORC function amplitude, is small, due to their similarity. In region 2, curves are highly dissimilar and not continuously derivable. Therefore, they cannot be exactly reproduced by a second-order polynomial, so that residuals contain an additional misfit contribution. The mixed derivative in this region is large. In region 3, curves are almost identical, up to a constant offset, as for region 1, but they are characterized by a sigmoidal shape that cannot be reproduced by a second-order polynomial. Residuals are dominated by the misfit, but the contribution of this region to the FORC function is small, due to the similar shape of all curves. (**b**) FORC diagram obtained from the measurements in (**a**), with points marking the three selected regions. Regions 1, 2, and 3 correspond to a low-amplitude background, the central ridge, and the $B_c = -B_u$ diagonal, respectively. Even if the FORC diagram does not appear to be dominated by measurement noise or by processing artifacts, essential details of the FORC functions are concealed. (**c**) Estimated standard error of the FORC function: the three regions highlighted in (**a**) are clearly recognizable. Striping along ascending diagonals is due to correlated measurement errors (each stripe corresponds to a single curve)

of $\hat{\rho}$ to points where U is completely filled by measurements or added fictive measurements in place of the missing points, effectively extrapolating the FORC function [17, 18, 94, 122].

The field resolution of the estimated FORC function, defined as the size of the smallest resolvable detail, is given by $\delta H^* = \delta H(S + 0.5)$, where δH is the size of field steps. The use of large smoothing factors for noise suppression decreases resolution, and this must be compensated by a corresponding decrease of δH, which, however, is limited by the quadratic measurement time increase. Early measurements used ~100 curves [15, 16], which cover the range of typical FORC measurements of geologic materials in steps of ~3 mT. The typical smoothing factor required for processing weak sedimentary samples is 5, so that this combination of measuring and processing parameters yields $\delta H^* = 16.5$ mT. For comparison, the field resolution required to capture essential sedimentary signatures should not exceed ~5 mT, so that the minimum number of FORCs needs to be increased to ~330. The corresponding measurement time increases from a couple of hours to one day.

Non-local methods for estimating the FORC function are based on Fourier theory [123]. In Fourier space, the FORC function is obtained from the well-known identity for the Fourier transform of the derivatives, that is

$$\rho^*(k_r, k) = (2\pi i)^2 k_r k \, \xi^*(k_r, k) \tag{14}$$

where k_r and k are the wavenumbers associated with H_r and H, respectively, $i = \sqrt{-1}$, and

$$f^*(k_x, k_y) = \iint_\Omega f(x, y) \, e^{-2\pi i(k_x x + k_y y)} dx dy \tag{15}$$

is the Fourier transform of a two-dimensional function defined over Ω. If the measurements are distributed on a squared grid, ξ^* is replaced by the fast Fourier transform (FFT) of the measurement values, and the resulting estimate $\hat{\rho}^*$ of the FORC function in Fourier space is transformed back to the measurement space via inverse Fourier transform, so that

$$\hat{\rho}(H_r, H) = \int_{-\infty}^{+\infty} \int_{-\infty}^{+\infty} k_r k \, M^*(k_r, k) e^{-2\pi i(k_r H_r + k H)} dk_r \, dk. \tag{16}$$

This method is affected by the same noise enhancement problem as the local regression algorithm. The Fourier transform of noisy measurements contains high-frequency contributions that are amplified by the multiplication of M^* with $k_r k$. Therefore, M^* is multiplied by a low-pass filter W^* with $W^*(0, 0) = 1$ and $W^*(k_r, k) \to 0$ as $k_r, k \to \infty$. In the time domain, this operation is equivalent to the convolution $\hat{\rho} = \rho * W$ of ρ with the weight function W corresponding to the inverse Fourier transform of the filter. A simple implementation is obtained with the Gaussian filter $W^* = \exp[-\frac{1}{2}\pi^2 S^2(k_r^2 + k^2)]$, which defines the weight function

$$W(H_r, H) = \frac{1}{\pi S} \exp\left[-2\frac{H_r^2 + H^2}{S^2}\right] \tag{17}$$

in measurement space. The filter parameter S roughly corresponds to the smoothing factor used for polynomial regression [123] and can take any positive value. The Fourier method is extremely fast because convolution is a simple product in Fourier space. Its main limitation is the generation of numerical artifacts, known as Gibbs phenomenon, at the boundary of the measurement space, especially along $H_c = 0$. The Gibbs phenomenon is caused by the truncation of the Everett function at $H_r = H$, which requires infinitely large wavenumbers to be correctly reproduced. The suppression of large wavenumbers by low-pass filtering transforms the sharp boundary of the FORC space along $H_c = 0$ into an oscillating function. Nevertheless, the rapidity of the Fourier method can be exploited for test purposes when a rapid evaluation of FORC measurements is needed.

A formal equivalency between Fourier and polynomial regression methods is provided by the so-called Savitzky-Golay filter [124], which describes the regression of a grid of regularly spaced data in terms of convolution. In the case of a $(2S + 1) \times (2S + 1)$ grid to be fitted with P_2, the vector of regression coefficients is given by $\mathbf{a} = \mathbf{B} \cdot \mathbf{M}$, where $\mathbf{M}$ is the vector of measurements, obtained by joining all rows of the grid, and $\mathbf{B} = (\mathbf{X}^T \mathbf{X})^{-1} \mathbf{X}^T$, with $\mathbf{X}$ being a matrix where each row vector of the form $(1, H_r, H_r^2, H, H^2, H_r H)$ is constructed using the (H_r, H) coordinates of the corresponding measurement in $\mathbf{M}$. In the case of regularly spaced measurements, the matrix $\mathbf{B}$ is the same for all measurement points and needs to be calculated only once, obtaining a significant reduction of the total computation time [125]. This approach is very sensitive to violations of the assumption that measurements are placed exactly on a regular grid, which is practically always the case, so that regridding with a suitable interpolation method is required before calculating the FORC function.

Improved Regression Methods

As discussed in section "Theoretical Principles of FORC Function Estimation", the assumption of a regular measurement grid requires some form of extrapolation to obtain the FORC function near $H_c = 0$ (Fig. 16a). This region of the FORC space, which corresponds to the initial part of the measured curves, plays an important role in the characterization of geologic materials while being, at the same time, affected by the largest deviations from an ideal grid. This is particularly true for the first point in each curve, due to technical challenges of a precise control of the reversal field. Another problem with rectangular arrays of points used for regression is that they do not provide a continuous representation of the FORC function, because the regression results change abruptly when moving from one center point to the next. These disadvantages are overcome by weighting schemes that provide a continuous transition between selected measurements with respect to any point of the measurement space.

A universal weighting scheme for the solution of general problems in one and more dimensions, called Locally Weighted Regression and Smoothing Scatterplots (LOESS), was developed by Cleveland and Devlin [126, 127] and is implemented

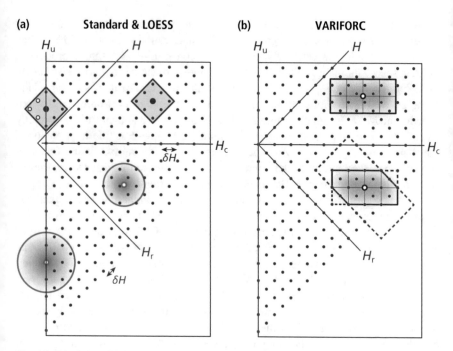

Fig. 16 Principles of FORC processing. The (H_c, H_u) FORC space is indicated in red and the (H_r, H) measurement space in blue. Measurements performed with a constant field step δH are indicated by dots. (**a**) Standard processing (e.g., FORCIT) is based on polynomial regression of a $(2S + 1) \times (2S + 1)$ array of measurement points (blue squares, corresponding to $S = 1$). Missing points at the left edge of the FORC space are replaced by fictive measurements. LOESS regression (e.g., FORCinel) is based on a fixed number of measurements closest to a reference point, weighted by a radial function of the distance to this point (circles, with shading proportional to the weight factors). The number of measurements for a given smoothing factor s loosely corresponds to the $(2S + 1)^2$ measurements considered by standard processing. The LOESS algorithm can deal with incomplete arrays (left), missing data, and irregularly spaced measurements for any point of the FORC space. Both methods produce the same smoothing over all directions. (**b**) VARIFORC processing is based on the weighted regression of measurements contained in a $(2s_c + 1) \times (2s_u + 1)$ rectangle centered on a reference point (upper example with $s_c = 2$ and $s_b = 1$, with shading proportional to the weight factors). Like the LOESS algorithm, it can deal with incomplete arrays, missing data, and irregularly spaced measurements for any point of the FORC space. Additional limitations of the regression region along the measurement coordinates are obtained by intersecting the regression rectangle (red) with a $(2s_r + 1) \times (2s + 1)$ rectangle with sides parallel to the H_r and H axes (blue). In this manner, VARIFORC can produce anisotropic smoothing along the four principal directions given by the measurement and the FORC coordinates

in the FORCinel software [128]. The LOESS algorithm requires the definition of a weight function $W(r)$ of normalized distance $r \geq 0$, which is positive, nonincreasing, and zero if $r \geq 1$, and a fixed number n of measurements that is considered for regression. The tricube function $W = (1 - r^3)^3$ is often used for this purpose. Regression of any set of regularly or irregularly spaced measurements $y_i = y(\mathbf{x}_i)$ with respect to a reference point $\mathbf{x}_0$ begins with the selection of the n measurement points that are closest to $\mathbf{x}_0$, with $r_{\max}$ being the maximum distance of these points to $\mathbf{x}_0$ (Fig. 16a). Contrary to the rectangular selection scheme discussed in section "Theoretical Principles of FORC Function Estimation", $r_{\max}$ depends on the distribution of points around $\mathbf{x}_0$, enabling the algorithm to overcome gaps and handle boundary regions.

Regression of the n selected measurements $y_{i,0} = y(\mathbf{x}_{i,0})$ with respect to $\mathbf{x}_0$, using a model function $f(\mathbf{x}, \mathbf{p})$ controlled by the parameter vector $\mathbf{p}$, is performed by minimizing the sum of squared residuals

$$R = \sum_{i=1}^{n} W\left(\frac{|\mathbf{x}_{i,0} - \mathbf{x}_0|}{r_{\max}}\right)\left[y_{i,0} - f\left(\mathbf{x}_{i,0}, \mathbf{p}\right)\right]^2. \tag{18}$$

In order to make the smoothing effectivity of the LOESS algorithm comparable with that of a squared array of $(2S + 1) \times (2S + 1)$ points, the number of selected measurements can be set to $(2S + 1)^2$. The FORCinel algorithm uses a slightly different choice given by [128].

$$n = 1 + \text{rnd}\left[\frac{n_{\text{tot}}}{n_{\text{obs}}}(2s + 1)^2\right], \tag{19}$$

where $\text{rnd}(x)$ is the rounding of x to the closest integer; n_{tot} is the total number of grid points within $r_{\max}$ from the selected point, including non-existing ones (e.g., those with $H < H_r$); and n_{obs} is the total number of observables, i.e., the measurements lying within $r_{\max}$ from the selected point that are effectively used for regression, which, for instance, might not include outliers.

The VARIFORC software [1] implements the LOESS algorithm differently, maintaining a predefined geometry for the data selection region U. This region is limited by four smoothing factors s_r, s, s_c, and s_u along four principal directions defined by the measurement coordinates H_r and H and by the FORC coordinates H_c and H_u, respectively. The default choice $s_r = s = \infty$ produces a homogeneous smoothing using measurements comprised within an upright rectangle of size $2(s_c + 1)\delta H \times 2(s_u + 1)\delta H$, centered on the regression point. Additional limits along the measurement coordinates are introduced with the other two smoothing factors, s_r and s, by intersecting, in FORC space, the original upright rectangle with another rectangle of size $\sqrt{8}\,(s_r + 1)\delta H \times \sqrt{8}\,(s + 1)\delta H$, counterclockwise rotated by $45°$ and centered on the regression point (Fig. 16b). These additional limits are designed to deal with critical regions of the measurement space where the measured magnetization is a strong function of H_r or H (e.g., case 3 in Fig. 15). Next, a one-dimensional weight function is defined as

$$w(x, d) = \begin{cases} 1, & |x| \leq d - 2 \\ 1 - \frac{1}{2}(|x| - d + 2)^2, & d - 2 < |x| \leq d - 1 \\ \frac{1}{2}(|x| - d)^2, & d - 1 < |x| d \\ 0, & \text{else} \end{cases} \tag{20}$$

with d being the half-size of the limiting rectangle in one of the four directions defined above, normalized by δH. This function produces a continuous transition from 1 to 0 within $2\delta H$ from the rectangle limit, using piecewise second-order polynomials. The total weight function is then obtained from the product of the one-dimensional weight functions of Eq. (20) along the four principal directions. This implementation of the LOESS algorithm ensures the best possible utilization of all measurements within a predefined region U, under the condition of delivering a continuous derivable FORC function estimate.

Overfitting, Underfitting, and Error Estimates

One of the main problems of FORC processing is the choice of a smooth- ing factor that guarantees the maximum level of noise suppression while not introducing smoothing artifacts. As seen in section "Theoretical Principles of FORC Function Estimation", second-order polynomial regression is guaranteed to converge to a continuous derivable FORC function for sufficiently small values of $\delta H^* = \delta H(S + 0.5)$. In this case, a small smoothing factor ensures that, for any point estimate of the FORC function with measurements from the selection area $U(S)$, regression residuals are completely random, that is, their amplitude and sign do not show any spatial trend within U. Regression residuals remain random upon increasing the smoothing factor from $S = 1$ to a critical value S_{opt}, but the mean residual amplitude will generally increase, as regressions of fewer points tend to fit random patterns produced by measurement errors, until a plateau is reached when the number of measurements included in U becomes statistically meaningful (Fig. 17). Only over this plateau, residuals coincide with the measurement errors.

If S is increased beyond S_{opt}, the Taylor approximation of $M(H_r, H)$ begins to fail, as it is no longer able to reproduce magnetization variations over the whole area of U. Accordingly, the amplitude of residuals begins to exceed the plateau given by measurement errors, because of an additional contribution related to the model misfit ΔM. The model misfit is spatially correlated, so that residuals reproduce the spatial pattern of ΔM, instead of being random. The model misfit tends to increase quadratically with S, so that a plot of the mean residual amplitude $\overline{R}$ vs. S defines a sigmoidal line with an inflection point at S_{opt} (top example in Fig. 17). FORC function estimates obtained with $S < S_{opt}$ are said to be *overfitted* because measurement errors are partially reproduced by the fit. On the other hand, estimates obtained with $S > S_{opt}$ are *underfitted*, because measurements are not completely reproduced by the fit. Underfitting introduces a systematic bias of the fitted values, which might lead to smoothing artifacts (signal removal). Between these two extremes, an *optimal fit* is obtained with $S \approx S_{opt}$. If the FORC function contains features that cannot be resolved with the given field step size of measurements, underfitting errors begin to grow with increasing S before measurement errors are significantly suppressed. In this case, $\overline{R}(S)$ will not possess an inflection point, and a definition of S_{opt} for the whole FORC domain is not possible (bottom example in Fig. 17), as discussed in section "Variable Smoothing Protocols".

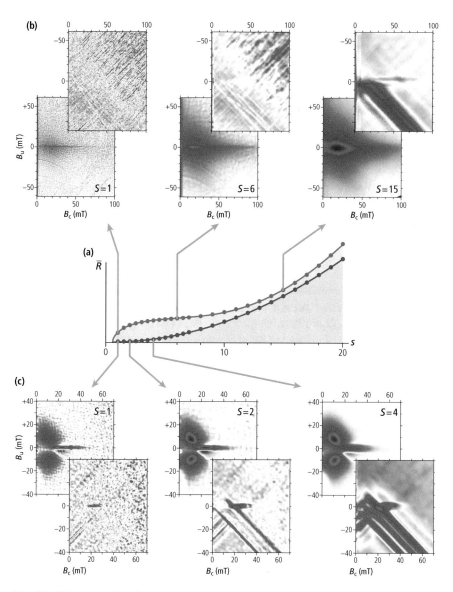

Fig. 17 Global smoothing factor optimization for a smooth FORC function (volcanic ash, top, data from Ludwig et al. [103]) and for a FORC function containing localized features (small magnetosome clusters, bottom, data from Katzmann et al. [129]). (**a**) Average regression residuals $\overline{R}$ for the two examples, as a function of the smoothing factor S. The inflection point of $\overline{R}(S)$ defines a global optimum at $S = 6$ for the volcanic ash example; no inflection point and no global optimum exist for the other example. (**b**) FORC diagrams (background) and corresponding maps of the standard deviation of regression residuals (foreground, white corresponds to $\overline{R}$, blue to lower values, and red to higher values) for the volcanic ash example. Large residuals along ascending diagonals correspond to measurement errors, which tend to increase with the applied field, while large residuals parallel to the descending diagonal correspond to regression errors at large values of S. Regression errors begin to appear at $S = 6$, but they are still much smaller than measurement errors. (**c**) Same as (**b**) for the magnetosome clusters. Regression errors are already dominant at $S = 2$

Analysis of the residuals obtained from a systematic variation of the smoothing factor provides an objective means to establish the best possible smoothing factor with the available measurements [128]. The existence of an inflection point in $\overline{R}(S)$ requires the magnetization function $\xi(H_r, H)$ to be sampled in a redundant manner, that is, in steps that are much smaller than required for sampling all features for ξ, especially if measurement errors are large. If this condition is not met, because ξ is not continuously differentiable or because δH is too large, overfitting artifacts are already introduced with the smallest possible smoothing factor, $S = 1$.

Optimally fitted FORC estimates $\hat{\rho}$ can be treated with standard statistical tools for calculating the standard error $\delta\hat{\rho}$ and the corresponding confidence interval, which is very useful for evaluating which regions of the FORC diagram contain information that can be considered significant [130]. The coefficients of a weighted second-order polynomial regressions of n measurements are given in vector form by

$$\hat{\mathbf{a}} = \left(\mathbf{X}^T \mathbf{W} \, \mathbf{X}\right)^{-1} \mathbf{X}^T \mathbf{W} \, \mathbf{y} \tag{21}$$

where the elements of $\hat{\mathbf{a}}$ correspond to Eq. (12); $\mathbf{X}$ is a matrix constructed with the $\mathbf{x} = (H_r, H)$ measurement coordinates, with one row vector of the form $(1, H_r, H_r^2, H, H^2, H_r H)$ for each measurement; $\mathbf{W} = \text{diag}\,[W(\mathbf{x}_1), \ldots, W(\mathbf{x}_n)]$ is the diagonal matrix of regression weights; and $\mathbf{y}$ is the vector of measurements. The mean squared error of the regression is

$$\varepsilon^2 = \frac{\mathbf{y}^T \, [\mathbf{W} - \mathbf{WB}] \, \mathbf{y}}{n - 6} \tag{22}$$

with $\mathbf{B} = \mathbf{X}(\mathbf{X}^T \, \mathbf{W} \mathbf{X})^{-1} \, \mathbf{X}^T \, \mathbf{W}$, while the mean squared error of the regression parameters is given by the diagonal elements of the covariance matrix

$$\hat{\mathbf{\Sigma}} = \left(\mathbf{X}^T \mathbf{W} \, \mathbf{X}\right)^{-1} \varepsilon^2. \tag{23}$$

A better estimate of the parameter errors proposed by Heslop and Roberts [130], which is robust against heteroscedastic noise (i.e., measurement errors that do not have a constant variance), is given by the so-called HC3 estimator [131].

$$\hat{\mathbf{\Sigma}}_{\text{HC3}} = \left(\mathbf{X}^T \mathbf{W} \, \mathbf{X}\right)^{-1} \mathbf{X}^T \mathbf{W} \, \text{diag}\left[\frac{[\mathbf{y} - \mathbf{By}]_i}{1 - B_{ii}} W(\mathbf{x}_i)\right] \, \mathbf{X}\left(\mathbf{X}^T \mathbf{W} \, \mathbf{X}\right)^{-1}. \tag{24}$$

The HC3 estimator performs well with t-tests used to assess regression coefficients [132]. Therefore, significant regions of the FORC diagrams can be tested against the null hypothesis that $\hat{\rho} = \hat{a}_6 = 0$ at a significance level $1 - \alpha$ (e.g., 95%, which means $\alpha = 0.05$) by checking that the associated probability

Table 1 SNR threshold of the estimated FORC function for passing the significance test $\hat{\rho} \neq 0$ at selected confidence levels $1 - \alpha$, for different smoothing factors s

s	n	90% ($\alpha = 0.2$)	95% ($\alpha = 0.05$)	98% ($\alpha = 0.02$)	99% ($\alpha = 0.01$)
1	9	2.35	3.18	4.54	5.84
2	25	1.73	2.09	2.54	2.86
3	49	1.68	2.02	2.42	2.70
4	81	1.67	2.00	2.38	2.64
5	121	1.66	1.98	2.36	2.62
7	225	1.65	1.97	2.34	2.60
10	441	1.65	1.97	2.33	2.59

$$p = 2\left[1 - t_{\alpha, n-6}\left(\frac{|\hat{a}_6|}{\sqrt{[\hat{\boldsymbol{\Sigma}}_{\mathrm{HC3}}]_{66}}}\right)\right] \tag{25}$$

is $< \alpha$, where $t_{\alpha, \nu}$ is the Student's t-distribution with ν degrees of freedom at the confidence level α. The argument of the t-distribution in Eq. (25) is the signal-to-noise ratio (SNR) of $\hat{a}_6$ and thus of the FORC function estimate, so that the level of significance depends essentially on this parameter and on the number of measurements used for regression. The SNR threshold for passing the significance test at a 95–98% confidence level converges to ~2 for smoothing factors >2 (Table 1): therefore, a SNR plot can be used to spot significant regions of the FORC diagram even without performing the significance test. Significant regions tend to be contained within SNR = 2 contour lines, which in turn follow the trend of regular contour lines of the FORC diagram. In practice, the significance contour follows the lowest-level contour line that maintains a certain continuity, without being split into separated "islands". A good practice for the representation of FORC diagrams consists in drawing the significance contour explicitly [130] or drawing regular contour lines only above the significance threshold. Significant regions of FORC functions that contain negative amplitudes will be split along the $\hat{\rho} = 0$ contours, since the associated SNR value near $\hat{\rho} = 0$ is arbitrarily small and does not pass the significance test.

Variable Smoothing Protocols

As discussed in section "Overfitting, Underfitting, and Error Estimates", there is an objective criterion for choosing the smoothing factor that provides the best compromise between the opposed needs of noise suppression and signal preservation. The application of this criterion, however, is not simple since the smoothing threshold above which artifacts start to emerge depends strongly on the local characteristics of the magnetization function $\xi(H_{\mathrm{r}}, H)$. These characteristics can be extremely heterogeneous, with regions along $H_{\mathrm{c}} = 0$, $H_{\mathrm{u}} = 0$, and $H = 0$ lines requiring an almost infinite resolution, while strong smoothing can be applied over the high-

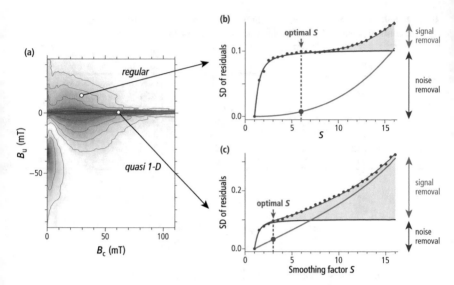

Fig. 18 Principles of local smoothing factor optimization. (**a**) FORC diagram of a magnetofossil-bearing pelagic carbonate (data from Ludwig et al. [103]). (**b**) Standard deviation of regression residuals (dots) as a function of the smoothing factor for a region of the FORC diagram where the FORC function is regular (i.e., continuously derivable). The gray and red lines represent the contributions of measurement noise and signal removal, respectively, to the total residuals. The optimal smoothing factor in this case corresponds to the plateau where maximum noise removal has been attained without altering the original signal. (**c**) Same as (**b**) for the central ridge, a quasi-one-dimensional region where the FORC function is not continuously derivable. In this case, signal removal occurs for any smoothing factor >1. Therefore, the optimal smoothing factor, if existing, is much smaller than over regular regions

coercivity range. As a result, the optimal smoothing factor can change locally by more than one order of magnitude in FORC diagrams of some geologic materials, such as magnetofossil-bearing sediments (Fig. 18). This example shows that optimal FORC processing of noisy measurements should be based on a local optimization of the smoothing factor, rather than a global one. Furthermore, strongly anisotropic smoothing is required to process quasi-unidimensional features such as the vertical ridge along $H_c = 0$ and the central ridge along $H_u \approx 0$, which are produced by viscous and non-interacting SD particle assemblages, respectively [1]. Anisotropic smoothing is obtained by replacing the squared region with dimensions defined by the classical smoothing factor with an upright or 45°-rotated rectangular region with dimensions defined by two smoothing factor values that can be interpreted as the principal values of a smoothing tensor.

The local optimization of anisotropic smoothing is a computationally demanding procedure that cannot be fully implemented in practice [90]. It is therefore replaced by a simplified but still widely applicable algorithm called VARIFORC [1]. This procedure is best understood by first considering the construction of a universal algorithm. Because the properties of the FORC function are not known a priori, the smoothing factor optimization must proceed iteratively from the first step, which consists in obtaining an estimate $\hat{\rho}_0$ of the FORC function based on a single isotropic smoothing factor that provides a statistically significant number of regression residuals for each point in the FORC diagram (e.g., $s = 7$ for >200 residuals). Depending on the characteristics of ξ, there will be two types of regression results: those that are clearly underfitted and those that are overfitted or correctly fitted (see section "Overfitting, Underfitting, and Error Estimates"). The two types of results are distinguishable by the statistical properties of their residuals. Overfitted or correctly fitted cases are characterized by uncorrelated residuals with no identifiable patterns, while residuals of underfitted cases display a significant correlation. The two cases can be discriminated by statistical tests for detecting non-randomness, such as the Wald-Wolfowitz run test [133]. In the next step, smoothing factors are updated according to the two cases: if residuals do not pass the non-randomness test, the regression region U is increased isotropically, e.g., by doubling the smoothing factor. If residuals do pass the non-randomness test, the size of U is decreased along one of the principal directions parallel to the H_c, H_u, H_r, and H axes for which the largest non-randomness is detected. This procedure is repeated until changes of U are no longer required. The practical implementation of this algorithm is limited by the random outcomes of the non-randomness test over regions of the FORC space where the size of U must be limited to few measurements along at least one direction. Therefore, individual point optimization must be replaced by regional optimization.

Egli [1] identified five universal features of the FORC function that guide local optimization of the smoothing factor. The first feature is related to the logarithmic dependence of magnetization curves in large fields, where *relative* field differences, rather than absolute ones, produce similar magnetization changes. Accordingly, the optimal horizontal and vertical sizes of regression regions should be proportional to H_c and $|H_u|$, respectively. Considering that a minimum degree of smoothing is required at the origin of the FORC space, the horizontal and vertical sizes of regression rectangles are defined by horizontal and vertical smoothing factors of the form

$$s_c = s_c^0 + \lambda_c \frac{H_c}{\delta H}, \quad s_u = s_u^0 + \lambda_u \frac{|H_u|}{\delta H}, \tag{26}$$

where s_c^0 and s_u^0 are the minimum smoothing factors at the origin of the FORC diagram and λ_c, λ_u are the rates of smoothing factor increase along the corresponding directions, which must be comprised between 0 and 1. In practice, $s_c^0 = s_u^0$ and $\lambda_c = \lambda_u$ are used, reducing the number of parameters to be chosen, but this limitation is not mandatory in VARIFORC. The special case $s_c^0 = s_u^0 = S$ and $\lambda_c = \lambda_u = 0$ corresponds to the classical processing with a constant smoothing factor and provides a good starting point for fixing S. The starting smoothing factor should provide a good SNR ratio and no evident underfit signs over a triangular region of the FORC diagram limited by the coercivity of remanence H_{cr} of the specimen (Fig. 19a). In practice, contour lines should be slightly wiggly, but not to the point of losing continuity, and the estimated error should not display evident

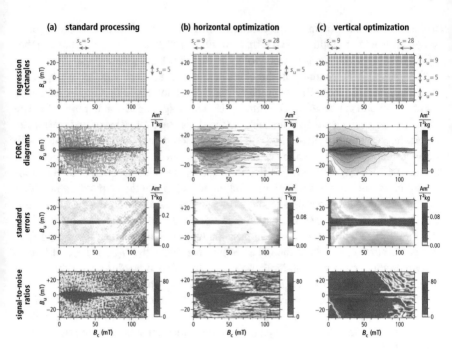

Fig. 19 VARIFORC smoothing factor optimization for a magnetofossil-bearing pelagic carbonate (data from Ludwig et al. [103]). (**a**) Standard processing with a uniform smoothing factor $s_c = s_u = 5$. From top to bottom: regression rectangles, FORC diagram, estimated error standard deviation, and estimated SNR (white corresponds to SNR = 3; blue regions do not pass the significance test). The central ridge is highly significant, due to its large amplitude, while the remaining features are barely significant and affected by large errors, as seen by the irregular contour lines. (**b**) Same as (**a**) with $s_c = 5 + 0.1 B_c/\delta B$ increasing linearly with B_c. The SNR has improved with respect to (**a**), but low-amplitude regions are still noisy. (**c**) Same as (**b**) with $s_u = 5 + 0.1 \, |B_u| \, /\delta B$ increasing linearly with $|B_u|$. The FORC function is sufficiently denoised over its whole domain. Notice the large regression errors over the central ridge in all three cases, because the FORC function is not continuously derivable along $B_u = 0$

patterns, except for those produced by correlated measurement errors (i.e., diagonal striping) and by localized features of the magnetization function requiring special handling later on.

At this stage, the outer regions of FORC space might be completely dominated by noise. This problem is corrected by increasing $\lambda_c = \lambda_u = \lambda$ until the estimated FORC function becomes significant over its whole domain (Fig. 19b, c). Typical values of λ are comprised between ~0.07 and ~0.2, whereby the upper limit, above which smoothing artifacts are introduced, depends on the measurement resolution. For instance, $\lambda = 0.14$ increases the smoothing factor by ~14 at the edges of typical FORC spaces ($\delta H \approx 1$ mT and $H_{c,\,max} \approx H_{u,\,max} \approx 100$ mT). Localized smoothing artifacts associated with diagonal bands where the error of the estimated FORC function raises well above the average are removed in the final processing stages and should not affect the choice of λ.

The use of variable smoothing factors defined by Eq. (26) depends, in the simplest case where the same criteria are applied along H_c and H_u, on two parameters: the initial smoothing factor S that is used near the origin and the rate λ of increase. A proper choice of these parameters should give a SNR of at least 2 over the triangular region with vertices $(0, \pm H_s)$ and $(H_s, 0)$, where H_s is the field amplitude at which the major hysteresis loop becomes (almost) closed. This goal might be impossible to achieve if measurements are excessively noisy or if field steps are not small enough. Furthermore, the magnetization function ξ might not be continuously derivable, or it might require unrealistically small field steps for the derivatives to be correctly sampled over specific regions of the FORC space, leading to undersampling artifacts. These artifacts might be removed or properly handled using local limitations of the smoothing factor, as explained in the following sections. If this is not possible, measurements should be repeated with higher resolution.

The Central Ridge

Points located along the $H_u \approx 0$ line in a FORC diagram have a special meaning for certain magnetic systems, which will be explained in section "Uniaxial SD Particles: Stoner-Wohlfarth Model". In Preisach theory, these points represent unbiased hysterons with switching fields $H_B = -H_A$. In the case of regular Preisach functions (i.e., continuously derivable), the contribution of these hysterons, as that of any 1D selection of the Preisach space, is infinitesimal. Certain magnetic systems, such as assemblages of non-interacting SD particles, are totally or in part represented by unbiased hysterons with finite contributions to the Preisach (and FORC) function that are entirely localized along $H_u \approx 0$. These contributions are correctly represented by an infinitely thin ridge called the *central ridge* [94]. The central ridge is mathematically defined with the help of the Dirac impulse δ as

$$\rho_{cr}(H_c, H_u) = M_{cr} f_{cr}(H_c) \, \delta(H_u), \tag{27}$$

where f_{cr} is the central ridge coercivity distribution and M_{cr} the total magnetization associated with the ridge, obtained by integrating ρ_{cr} over H_c and H_u. When looking at individual FORCs, the presence of the central ridge is signaled by a corner point, or discontinuity in the first derivative, at $H = -H_r$ (Fig. 10). The magnetization function ξ is therefore not continuously derivable at $H = -H_r$, where the Taylor series is undefined. This causes polynomial regression to fail for any finite field step size and any smoothing factor. Real central ridges have a finite width caused by thermal relaxation, so that δ is replaced by a function g_{cr} of finite width [85], which might be slightly offset with respect to $H_u = 0$ [117]. The typical width of g_{cr} is of the order of 0.2 mT or less for most sediments featuring a central ridge, requiring field steps as low as 30 μT to be correctly characterized with a smoothing factor of 3. Therefore, central ridges produced by non-interacting SD particles can be considered infinitely thin for all practical purposes. Because the central

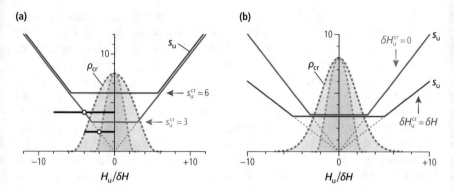

Fig. 20 VARIFORC smoothing factor limitation over the central ridge. (a) Vertical smoothing factor s_u along vertical profiles of the central ridge for two different limiting values s_u^{cr} (solid lines) and resulting central ridge profiles (thick dashed lines with shading), in the case of an ideal central ridge centered at $H_u^{cr} = 0$ with no intrinsic thickness (i.e., $\delta H_u^{cr} = 0$). The smoothing factor is constant over the region occupied by the processed central ridge and increases linearly outside this region until reaching the regular value. The linear increase rate is the largest possible for regression rectangles (circles with horizontal bars) that do not cross the intrinsic ridge, in this case at $H_u = 0$. (b) Same as (a) for $s_u^{cr} = 3$ and two selected values of the intrinsic thickness δH_u^{cr}. The linear increase rate of the smoothing factor is reduced with respect to (a) to maintain $s_u = s_u^{cr}$ over the full thickness of the processed ridge

ridge conveys important information about the origin of these SD particles, proper handling of the underfitting problem is required [94], along with the possibility for separating this feature from regular FORC contributions.

In practice, optimal processing of the central ridge requires the lowest possible degree of smoothing along H_u and regular smoothing along H_c. This is achieved by introducing a limitation of the smoothing factor at $H_u \approx 0$. VARIFORC implements this limitation using continuous vertical profiles of s_u given by

$$s_u^* = \min\left[s_u, s_u^{cr} \max\left[1, \frac{|H_u - H_u^{cr}|}{(s_u^{cr} + 1)\delta H + \delta H_u^{cr}}\right]\right], \tag{28}$$

where s_u is the unlimited smoothing factor (Eq. 26); s_u^{cr} is the smoothing factor over the central ridge; H_u^{cr} is the assumed vertical offset of the central ridge, due, for instance, to thermal relaxation effects; and δH_u^{cr} is the assumed intrinsic half-thickness, due to thermal relaxation effects or small magnetostatic interactions. The intrinsic half-thickness should not be confused with the apparent thickness resulting from processing. In practice, Eq. (28) is designed to provide a constant smoothing factor $s_u^* = s_u^{cr}$ over the H_u range occupied by the processed central ridge. Outside of this range, s_u^* increases linearly until the unlimited value is reached (Fig. 20). The rate of increase is as large as permitted by the condition that measurements belonging to the intrinsic range of the central ridge are not used for the regression of regular regions of the FORC space.

Fig. 21 Vertical profiles of the ideal central ridge processed with the smoothing factors indicated by numbers (dots) and corresponding best fits with Eq. (29) (lines)

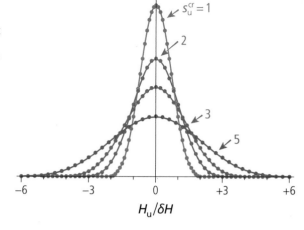

The estimated central ridge is described by the empirical approximation

$$\hat{\rho}_{cr}(H_c, H_u) \approx M_{cr} f_{cr}(H_c) \frac{1}{(3 + 2.8 s_u^{cr}) \delta H} \mathcal{B}\left(\frac{1}{2} + \frac{H_u}{(3 + 2.8 s_u^{cr}) \delta H}\right) * g_{cr}(H_u),$$
(29)

where $\mathcal{B}$ is the Beta distribution with shape parameters $\alpha = \beta = 9.7$ and the asterisk denotes the convolution with the intrinsic vertical profile g_{cr} of the central ridge. In the case of an ideal central ridge, $g_{cr} = \delta$ and vertical profiles of $\hat{\rho}_{cr}$ coincide with $\mathcal{B}$, with a full width to half-maximum (FWHM) of $(0.84 + 0.77 s_u^{cr}) \delta H$ (Fig. 21). The amplitude of $\hat{\rho}_{cr}$ is inversely proportional to the FWHM, so that low-resolution measurements and large smoothing factors tend to broaden $\hat{\rho}_{cr}$ and decrease its amplitude, to the point that it is no longer distinguishable from regular contributions to the FORC function. For example, the central ridge amplitude obtained from high-resolution measurements ($\delta H = 0.5$ mT) processed with $s_u^{cr} = 2$ is ~10 times larger than that of earlier measurements ($\delta H = 5$ mT) processed with a typical smoothing factor of 5.

The central ridge conveys specific information about the magnetic minerals associated with this FORC signature, for instance, through the coercivity distribution f_{cr} that can be extracted from the FORC diagram. The extraction of f_{cr} exploits the quasi-unidimensional nature of the central ridge, which, in the case of sufficient resolution, can be separated from regular FORC contributions [94]. For this purpose, vertical profiles of the FORC function are fitted with two empirical model functions $\hat{g}_{cr}(H_u)$ and $\hat{\gamma}_{cr}(H_u)$, representing the central ridge and the regular contributions, respectively. As seen with Eq. (29), the shape of $\hat{g}_{cr}$ depends on the intrinsic width of the central ridge and the smoothing factor s_u^{cr} used to process the FORC data near $H_u = 0$. A combination of one or two generic bell-shaped functions, such as the sech function used in VARIFORC [1], is sufficient to reproduce generic central ridge profiles, including skewed ones. In the limit case of infinite resolution, the

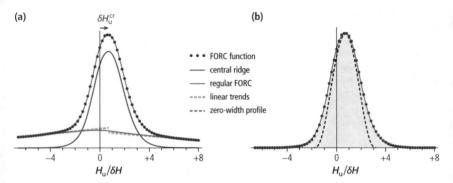

Fig. 22 Central ridge modeling. (**a**) Vertical profile of the central region of the FORC diagram of a magnetofossil-rich lake sediment. The original measurements in steps of $\delta B = 0.5$ mT have been regridded on a regular 0.25 mT mesh (dots) using the VARIFORC smoothing parameters $s_u^0 = 5$, $\lambda_u = 0.15$, $s_u^{cr} = 2$, $B_u^{cr} = +0.4$ mT, and $\delta B_u^{cr} = 0.4$ mT. The profile was fitted with the linear combination of two model functions representing the central ridge and regular FORC contributions, respectively, using Eqs. (30) and (31). Notice the central ridge offset of $+0.4$ mT, which was the reason for choosing $B_u^{cr} = +0.4$ mT. (**b**) Same as (**a**), after subtracting the regular FORC contributions to obtain the central ridge profile. The dashed line represents the result that would be obtained for a central ridge with zero intrinsic width. The shaded area is the integral of the profile used to calculate the central ridge distribution

dependence of the regular part of the FORC function on H_u is negligible over the infinitesimal interval occupied by the central ridge. Uniaxial, non-interacting SD particles, on the other hand, contribute to the lower quadrant with a function that is proportional to the derivative of the difference $M^+(H) - M^-(H)$ between the descending and ascending branches of the hysteresis loop of individual particles, which, in the absence of thermal relaxation effects, diverges as $|H_u|^{-1/2}$ near $H_u = 0$ [93]. This divergence is associated with the reversible rotation of SD magnetic moments near the switching field (section "Uniaxial SD Particles: Stoner-Wohlfarth Model"). Thermal relaxation and smoothing transform the background under the central ridge into a piecewise linear function constructed using the beta distribution in Eq. (29) as smoothing kernel (Fig. 22).

The model functions implemented in VARIFORC are given by

$$\hat{g}_{cr}(x) = q A \operatorname{sech}\left((x - x_1)/B_1\right) + (1 - q) A \operatorname{sech}\left((x - x_2)/B_2\right) \tag{30}$$

and

$$\hat{\gamma}_{cr}(x) = (a_2 - a_1)\left[q\operatorname{sechI}\frac{x - x_1}{C_1} + (1 - q)\operatorname{sechI}\frac{x - x_2}{C_2}\right] + a_1 + c_1 x -$$
$$(c_1 + c_2)\left[q\operatorname{sechII}\frac{x - x_1}{C_1} + (1 - q)\operatorname{sechII}\frac{x - x_2}{C_2}\right] - \tag{31}$$
$$(c_1 + c_2)\left[qx_1\operatorname{sechI}\frac{x - x_1}{C_1} + (1 - q)x_2\operatorname{sechI}\frac{x - x_2}{C_2}\right]$$

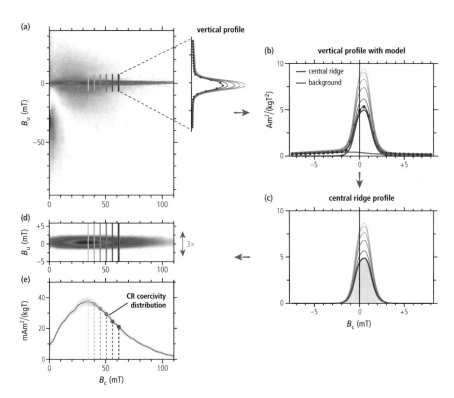

Fig. 23 Principles of central ridge post-processing. (**a**) FORC diagram of a magnetofossil-rich sediment from the Equatorial Pacific Ocean (data from Ludwig et al. [103]), with a sequence of vertical profiles across the central ridge. (**b**) Vertical profiles are modeled as the sum of central ridge and regular FORC contributions, as in Fig. 22. (**c**) Subtraction of the regular FORC contribution from each profile yields the isolated central ridge profiles. (**d**) The profiles in (**c**) can be used to reconstruct the FORC function of the isolated central ridge. The vertical scale has been exaggerated by a factor 3 for clarity. (**e**) Profile integrals in (**c**) (shaded) are used to reconstruct the central ridge coercivity distributions f_{cr}

where $x_{1,2}$, $A > 0$, $C_{1,2}$, $a_{1,2}$, $c_{1,2}$, and $0 \leq q < 1$ are model parameters and sechI and sechII are the first and second primitives of sech function, respectively, defined by integration from $-\infty$ to x. Other choices are possible, e.g., by replacing the sech function with similar bell-shaped functions. Vertical profiles of the estimated FORC function $\hat{\rho}$ are then fitted with $\hat{g}_{cr} + \hat{\gamma}_{cr}$ over a range of H_u values centered on the vertical position of the central ridge. This range includes a small portion of the regular part of the FORC function on each side (Fig. 22a). The estimated central ridge contribution, $\hat{\rho}_{cr}$, is then obtained from the subtraction of $\hat{\gamma}_{cr}$ from each profile (Fig. 22b). In the case of FORC measurements with proper resolution, the regular part of the FORC function is well approximated by the linear trends of $\hat{\gamma}_{cr}$ outside the range occupied by the central ridge, so that $\hat{\rho}_{cr}$ is relatively insensitive to the type of model functions used to fit vertical profiles.

Finally, the integral of $\hat{\rho}_{cr}$ over H_u yields an estimate of the central ridge coercivity distribution f_{cr} that is independent of the processing parameters (Fig. 23). Horizontal profiles of the central ridge are often used as a substitute of f_{cr}, due to simpler processing [113, 134, 135]; however, these profiles have a higher noise content [99], depend on the smoothing factor and on variations of the vertical offset, and do not possess the correct unit to represent a coercivity distribution, thus preventing quantitative comparisons with other magnetic parameters. The importance of a proper central ridge processing is illustrated with a recent example based on the study of sediments containing a mixture of conventional and so-called giant magnetofossils, a characteristic signature of a global warming event that occurred during the Paleocene-Eocene transition [119] (Fig. 24). In this example, the small vertical offset of the central ridge is a function of H_c that depends on different types of SD particles contributing to specific coercivity ranges, i.e., conventional magnetofossils with a minimum offset $B_u^{cr} \approx 0.4$ mT around $B_c = 50$ mT and needle-shaped giant magnetofossils with a maximum offset of ~0.7 mT at $B_c = 180$ mT. Because of these offset variations, there is not a single horizontal profile of the central ridge that correctly reproduces the shape of f_{cr}.

The Vertical Ridge

Thermal relaxation processes produce a time-dependent decrease of the magnetization in applied fields close to the reversal field H_r, which overlaps with the magnetization increase produced by the positive field sweep [114, 116, 117]. This viscous magnetization decay adds a positive contribution to the FORC function, which takes its maximum value along $H_c = 0$, and decreases in a pseudo-exponential manner at larger values of H_c. The horizontal extension of this feature corresponds to the mean fluctuation field of viscous magnetic particles (see section "Viscous SD Particles and the Vertical Ridge"). Fine particle systems with stable magnetizations are characterized by small fluctuation fields, so that thermal relaxation effects take the form of a narrow vertical ridge along $H_c = 0$. This ridge should not be confused with the effect of measurement artifacts occurring over the same range (e.g., Fig. 13). In the case of viscous non-interacting SD particles, the vertical ridge takes the form

$$\rho_{vr}(H_c, H_u < 0) = f_v(-H_u)\ \frac{e^{-t_r/\tau}}{H_f^2}\ \exp\left(-2\frac{H_c}{H_f}\right), \tag{32}$$

where f_v is the coercivity distribution of viscous particles, t_r is the pause at reversal (section "The FORC Protocol"), τ is a characteristic decay time, and H_f is an effective fluctuation field (section "Viscous SD Particles and the Vertical Ridge"). For sufficiently small values of H_f, the exponential function in Eq. (32) decreases to zero within the first one or two field steps, so that ρ_{vr} can be treated effectively as a ridge that requires a smoothing factor limitation along $H_c = 0$. In analogy with the central ridge, this limitation is implemented in VARIFORC [1] using horizontal profiles of the smoothing factor which are given by

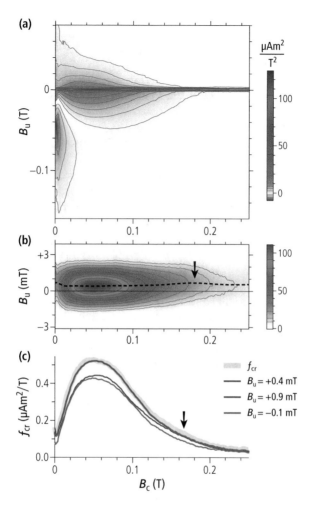

Fig. 24 Example of central ridge analysis with VARIFORC. (**a**) High-resolution ($\delta B = 0.4$ mT) FORC diagram of a sediment sample from Wilson Lake, New Jersey, deposited during the Paleocene-Eocene Thermal Maximum (data from Wagner et al. [119]). Quantile contours (section "FORC Data Rendering") correspond to 10% increments of the total FORC magnetization, from 10% to 90%. (**b**) Isolated central ridge (15× vertical exaggeration, quantile contours at 7, 12, 20, 30, 40, 50, 60, 70, 80, and 90% of the total magnetization). The dashed line represents the B_c-dependent vertical offset of the central ridge crest. The offset peak at ~0.18 T (arrow) is associated with needle-shaped giant magnetofossils. (**c**) Central ridge coercivity distribution f_{cr} (thick gray line), and three horizontal profiles of the central ridge close to the crest, rescaled to match the high-coercivity end of f_{cr}. Notice the differences around ~0.17 T, caused by variations of the vertical offset over the coercivity range of needle-shaped giant magnetofossils

$$s_c^* = \min\left[s_c, s_c^{vr} \max\left[1, \frac{H_c}{\left(s_c^{vr} + 1\right)\delta H + \delta H_c^0}\right]\right],\tag{33}$$

where s_c is the unlimited smoothing factor (Eq. 26), s_c^{vr} is the smoothing factor over the vertical ridge, and δH_c^0 is the assumed intrinsic thickness of the vertical ridge.

Processing FORC Data Associated with Thin Hysteresis Loops

The remanence ratio, or squareness M_{rs}/M_s, is a measure for the vertical opening of ferrimagnetic hysteresis loops. Thin hysteresis loops are characterized by $M_{rs}/M_s \ll 0.5$, which means that most magnetization changes measured with FORC are reversible and do not contribute to the FORC function. Yet, polynomial regression must reproduce the large reversible component very accurately to avoid processing artifacts. This can be a problem if the reversible component $M_{rev} = (M^+ + M^-)/2$ of the hysteresis loop contain steep gradients or, more generally, magnetization changes that are not well reproduced by a second-order polynomial. A typical example of interest for geologic materials is that of large contributions from superparamagnetic (SP) particles with sizes just below the stable SD stability range (Fig. 25a).

The reversible magnetization of SP particles with energy barrier $\Delta E = \beta_0 k_B T$, where β_0 is the Boltzmann factor (see below), is given by [24].

$$M_{rev}(H) = M_s \mathcal{L}\left(3\frac{\chi_{sp}H}{M_s}\right),\tag{34}$$

where M_s is their saturation magnetization, $\mathcal{L}(x) = \coth x - x^{-1}$ the Langevin function, and $\chi_{sp} = \mu_0 M_s m/3k_B T$ the superparamagnetic susceptibility of particles with magnetic moment m at the absolute temperature T. The susceptibility increases with m, reaching a maximum just before particles become blocked, that is, when $\beta_0 = \mu_0 m H_K/2k_B T \approx \ln(t_m/\tau_0)$, with t_m being the measurement time and H_K the so-called microcoercivity of the particles [31]. In this case, Eq. (34) becomes $M_{rev} = M_s \mathcal{L}(2\beta_0 H/H_K)$ with $\beta_0 \approx 20$ for $t_m \approx 1$ s. Natural magnetite SP particles with irregular shape, such as those forming in soils, are characterized by $H_K \approx 40$ mT [84], and M_{rev} starts to saturate in fields as low as 5 mT. This is also the typical resolution of processed FORC data (e.g., $\delta H = 1$ mT and $S = 5$), so that significant misfits of the polynomial regression occur in proximity of $H = 0$ (Fig. 25b). Such misfits can produce smoothing artifacts along the $H_u = -H_c$ diagonal in FORC space (Fig. 26a) or leave a noisy trace.

Mixtures of almost blocked SP particles with other ferrimagnetic components yield hysteresis loops that are constricted, or "wasp waisted" around $H = 0$, due to the large slope of the reversible magnetization [136]. Other magnetic systems can also produce constricted loops, for instance, particles hosting a single vortex as the only stable magnetic configuration in a null field [137]. FORC processing problems

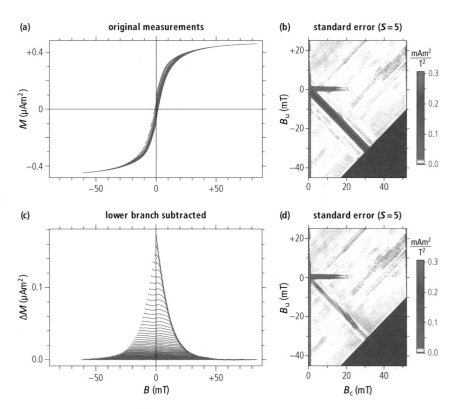

Fig. 25 FORC processing example for thin hysteresis with a large reversible gradient near $B = 0$. (a) High-resolution ($\delta B \approx 0.29$ mT) FORC measurements of equidimensional magnetosomes dispersed in kaolin. (b) Estimated standard error of the FORC function obtained with a homogeneous smoothing factor $S = 5$. Large underfitting errors occur along $B_u = 0$ (central ridge), $B_c = 0$ (vertical ridge associated with thermal relaxation), and the $B_c = -B_u$ diagonal (large reversible gradient near $B = 0$ due to superparamagnetic particles). (c) Same as (a), after subtracting the lower branch of the hysteresis loop reconstructed from FORC measurements. The resulting curves are almost completely flat, until SD particles are switched at $B = -B_r$, contributing to the central ridge. The highly expanded view of the area enclosed by the hysteresis loop highlights the small viscous magnetization decay between the first and the second point of each curve. (d) Same as (b), for the curves in (c). The error along $B_c = -B_u$ has been drastically reduced and reflects now the small but sharp magnetization changes occurring near $B = 0$, which are due to a positive magnetostatic coupling between superparamagnetic and stable SD particles within magnetosome chain fragments

with materials characterized by thin (constricted) loops of this type, or any other reversible magnetization with strong gradients, can be avoided by exploiting the natural insensitivity of the FORC function toward reversible magnetizations that do not depend on the reversal field. This means that any arbitrary magnetization curve can be subtracted from FORC measurements without altering the intrinsic properties of the FORC function, provided that the same curve is used for all measurements. If the purpose of this operation is to remove regression artifacts caused by M_{rev}, an

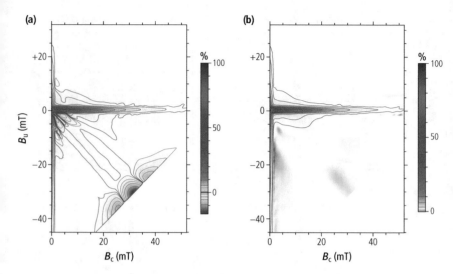

Fig. 26 FORC diagrams of equidimensional magnetosomes dispersed in kaolin obtained from the measurements shown in Fig. 25. (**a**) Original measurements processed with VARIFORC parameters $s_c^0 = s_u^0 = 15$, $\lambda_c = \lambda_u = 0.1$, and $s_c^{vr} = s_u^{cr} = 4$. Instead of highlighting low-amplitude features of the FORC function outside the area occupied by the central ridge and the vertical ridge, large smoothing factors produce artifacts along the $B_c = -B_u$ diagonal, due to the impossibility to fit the sigmoidal contribution of superparamagnetic particles near $B = 0$. Quantile contours (section "FORC Data Rendering") are drawn at 5, 10, 20, 30, 40, 50, 60, 70, 80, and 90% of the total magnetization. (**b**) Same as (**a**) but using lower-branch-subtracted measurements. Subtraction of the lower hysteresis branch removes the superparamagnetic contribution and the related artifacts along the $B_c = -B_u$ diagonal. The very-low-amplitude negative features in the lower quadrant are significant and do not represent processing artifacts

estimate $\hat{M}_{rev}$ of M_{rev} can be used, e.g., from an independent measurement of the hysteresis loop or using an analytical approximation. Such accessory measurements are not always available, so that $\hat{M}_{rev}$ is conveniently constructed using the FORC measurements.

The VARIFORC software identifies $\hat{M}_{rev}$ with a reconstruction of the lower branch $\hat{M}^-$ of the hysteresis loop [1]. Subtraction of $\hat{M}^-$ transforms FORC measurements into a set of curves with positive magnetization. This method is known as *lower branch subtraction*. For the common case where FORCs are fully contained by the hysteresis loop, the upper envelope of the lower-branch-subtracted curves coincides with the doubled even component $(M^+ - M^-)/2$ of the hysteresis loop (Fig. 25c). Extra smoothing can be applied to $\hat{M}^-$ in order to reduce the noise added to the original measurements by this operation, as long as the reversible magnetization features to be removed are preserved. This operation effectively removes the polynomial misfits caused by strong gradients of the reversible magnetization (Fig. 26b) with two beneficial effects: the suppression of processing artifacts and the avoidance of biased error estimates. A positive side effect of the lower branch subtraction is the better separation of individual curves, due to the highly expanded view of the area enclosed by the loop.

Smoothing Factor Limitations Along the Measurement Coordinates

As discussed in section "Processing FORC Data Associated with Thin Hysteresis Loops", strong reversible magnetization gradients introduce smoothing artifacts that can be removed by subtracting a curve containing such gradients, such as the lower branch of the hysteresis loop, from all measurements. However, processing problems can persist if such gradients are associated with equally sharp irreversible processes. For instance, magnetostatic interactions might couple the magnetizations of SP and SD particles inside clusters. A similar effect is produced by the exchange coupling between high- and very-low-coercivity components. In these cases, M_{rev} depends also on the magnetic state of the ferrimagnetic component, so that subtraction of a single estimate of M_{rev} from all measurements will leave traces of the original magnetization gradient around $H = 0$, which contribute to the FORC function along the $H_u = -H_c$ diagonal. A paradigmatic example is given by chondrules in some chondritic meteorites, where tetrataenite, a magnetically hard FeNi alloy, interacts with a much softer component, creating a positive-negative ridge doublet around the $H_u = -H_c$ diagonal [138] (Fig. 27). In such cases, it is desirable to limit the size of the regression rectangles across the diagonal, that is, along the measurement field H. This limitation is implemented in VARIFORC in a similar manner as for horizontal and vertical ridges [1], using the smoothing factor

$$ s = s^0 \max\left[1, \frac{|H - H^0|}{(s^0 + 1)\delta H + \delta H^0} \right] \tag{35} $$

where s^0 is the limiting smoothing factor over the diagonal region, H^0 the center coordinate of this region (e.g., $H^0 = 0$ for the $H_u = -H_c$ diagonal or $H^0 = 50$ mT in Fig. 27b), and δH^0 its width. This definition yields $s = s^0$ over the region of interest, with a linear increase to the regular smoothing factor outside of it. The default value $s^0 = \infty$ is used if no limitation of the smoothing factor along H is needed.

A similar limitation of the regression rectangles can be constructed along the other measurement coordinate, upon defining the smoothing factor s_r along H_r with Eq. (35), and after replacing the equation parameters with corresponding parameters s_r^0, H_r^0, and δH_r^0 for the limiting smoothing factor, the center coordinate, and the width of the critical H_r range, respectively. The two limiting smoothing factors s^0 and s_r^0 are used to construct a rectangle rotated by $45°$ with sides of length $2\sqrt{2}s\,\delta H$ and $2\sqrt{2}s_r\delta H$, respectively, which is intersected with the upright rectangle that defines the regression region (Fig. 16b). The choice of the side lengths is such that the upper-right and lower-left corners of a $2S\,\delta H \times 2S\,\delta H$ regression square are cut only if $s < S$ and the other two corners only if $s_r < S$. In the case of sufficiently small $s = s_r$ values, the regression region becomes a $45°$-rotated square, as in the original protocol of Pike et al. [15].

The combination of s_c, s_u, s, and s_r smoothing factors can be used to produce any type of regression region with limitations along the FORC and the measurement coordinates. For instance, in the case of the chondrule measurements by Acton et al. [138], variable smoothing factors $s_c = s_c^0 + \lambda H_c/\delta H$ and $s_u = s_u^0 + \lambda|H_u|/\delta H$

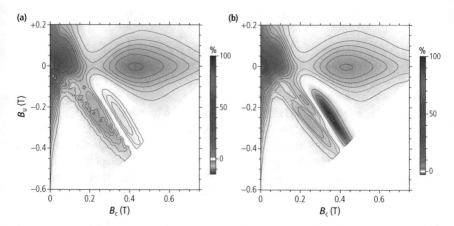

Fig. 27 FORC diagrams of a chondrule from the Bjurbole meteorite (data from Acton et al. [138]) obtained with two different sets of processing parameters. Quantile contours (section "FORC Data Rendering") from 10 to 90% of the total magnetization, in 10% steps. (**a**) Constant smoothing factor $S = 3$. The localized negative peak at $B_c = 0$, just below the central maximum, is a smoothing artifact caused by the steep magnetization gradient near $B = 0$. (**b**) VARIFORC smoothing with $s_c^0 = s_u^0 = 3$, $\lambda_c = \lambda_u = 0.05$, and limitation of the smoothing factor along the descending diagonal using $B^0 = 50$ mT, $\delta B^0 = 70$ mT, and $s^0 = 2$

are used to process a FORC function with amplitudes that decay progressively from a central maximum near the origin. No limitations are required along H_c and H_u, because of the lack of horizontal and vertical ridges, but the interaction features along the $H_u = -H_c$ diagonal require a limitation of the smoothing factor along H (Fig. 27b).

Processing FORC Data Associated with Highly Squared Hysteresis Loops

Highly squared hysteresis loops are characterized by strong magnetization gradients in proximity of the positive and negative coercive fields, which are caused by a narrow coercivity distribution concentrated around the mean switching field $\overline{H}_{sw} \approx H_c$. Squared hysteresis is a desirable property of permanent magnets, because the application of external fields of amplitude smaller than $\overline{H}_{sw}$ will produce only minimal magnetization changes [139]. In geological materials, squared hysteresis is produced by SD particles with a narrow coercivity distribution, the limiting case being that of identical randomly oriented Stoner-Wohlfarth particles (Fig. 28a). Magnetotactic bacteria represent an important natural realization of such uniaxial SD particles (Fig. 28b).

The FORC function of materials with highly squared loops is concentrated within a small region around $(H_c, H_u) = (H_{sw}, 0)$, which is best sampled by limiting FORC measurements to $-H_r \approx H \approx H_{sw}$. This requires a modification of the standard measurement protocol; however, the main reason for not doing so is the presence of

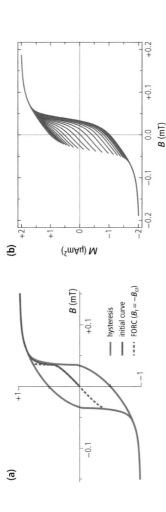

Fig. 28 Examples of highly squared hysteresis loops. (**a**) Identical, randomly oriented, isolated Stoner-Wohlfarth particles with a microcoercivity of 70 mT. Notice the infinite slope of the two branches at the corner points near the coercive field. The initial curve and the FORC starting at $-B_{\mathrm{cr}}$ are also shown. (**b**) FORC measurements of intact cultured cells of the magnetotactic bacterium *Magnetovibrio blakemorei* strain MV-1, which contain linear chains of prismatic magnetite crystals behaving like non-interacting uniaxial SD particles (data from Wang et al. [140])

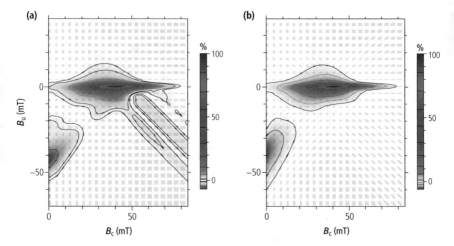

Fig. 29 FORC processing example for the magnetotactic bacterium measurements shown in Fig. 28 (data from Wang et al. [140]). The field step is $\delta B \approx 0.6$ mT. (**a**) VARIFORC smoothing with $s_c^0 = s_u^0 = 9$, $\lambda_c = \lambda_u = 0.1$, and $s_c^{cr} = 3$. Regression rectangles (not to scale) are superimposed for reference. Quantile contours (section "FORC Data Rendering") are drawn at 5 and 10–90% of the total magnetization in steps of 10%. Notice the smoothing artifacts along the diagonals departing from the central maximum in the lower quadrant, caused by the steep magnetization gradients near the coercive field. (**b**) Same as (**a**), with additional limitations $s_c^0 = s^0 = 9$ of the smoothing factor along the diagonals departing from the central maximum. The diagonal location and width of the limiting regions were determined automatically by VARIFORC using the maximum and the FWHM of the first derivative of the envelope hysteresis loop. The smoothing artifacts of (**a**) are suppressed

other magnetic signatures of the same particles (e.g., the negative region near $H_c = 0$ caused by reversible magnetic moment rotation) or magnetic components with normal hysteresis properties. In case of highly squared hysteresis, a limitation of the size of regression rectangles is required over two regions with high magnetization gradients, which, in FORC space, are given by $H_r \approx -\overline{H}_{sw}$ and $H \leq \overline{H}_{sw}$ (because $\partial M/\partial H_r$ is large before all curves merge above $+\overline{H}_{sw}$) and by $H \approx \overline{H}_{sw}$ and $H_r < -\overline{H}_{sw}$ (because $\partial M/\partial H$ is large for curves starting below $-\overline{H}_{sw}$). The two regions are represented by two diagonals extending over the lower half of the FORC space, which merge at the central maximum of the FORC function, located at $(H_c, H_u) = (\overline{H}_{sw}, 0)$ (Fig. 29a). If lower branch subtraction is used to process the measurements, the limiting region is formed by the same two diagonals, extending to the left of the merging point instead of below.

The location of the critical diagonal regions in FORC space requires two parameters: the mean position $\overline{H}_{sw}$ and the typical width $\delta\overline{H}_{sw}$ of the high-gradient segments of the hysteresis loop. The VARIFORC processing software [1] identifies $\overline{H}_{sw}$ with the field amplitude for which the first derivative of the envelope hysteresis obtained from FORC measurements is largest in the absolute value and $\delta\overline{H}_{sw}$ with the decrement of $\overline{H}_{sw}$ required to reduce the derivative to the half of its maximum amplitude. Appropriated smoothing factor limits $s_r^0 = s^0$ (section "Smoothing Factor Limitations Along the Measurement Coordinates") are entered by the user. Subtle smoothing artifacts extending over low-amplitude regions of the FORC function, such as those of Fig. 29, can be effectively identified by comparing FORC diagrams processed with and without smoothing factor limitations: only features that maintain their signature in both cases are real.

Demagnetizing Fields and FORC Deshearing

Specimens made of strongly magnetic materials possess an internal, so-called *demagnetizing field*, given by

$$H_i = -\mathbf{D} \cdot \mathbf{M}, \tag{36}$$

where $\mathbf{M}$ is the bulk specimen magnetization vector and $\mathbf{D}$ the so-called *demagnetizing tensor*, which depends on the specimen shape. The demagnetizing tensor is unitless in the SI unit system, since field and magnetization share the same unit, A/m. In this case, the sum of the eigenvalues of $\mathbf{D}$ is equal to 1. The internal field, and thus $\mathbf{D}$, is calculated from the distribution of surface magnetic charges $\sigma_m = \mathbf{M} \cdot \mathbf{n}$ produced at places where $\mathbf{M}$ is not parallel to the local surface element, represented by the normal vector $\mathbf{n}$. In general, H_i depends on the position inside the specimen, so that different parts experience different internal fields (Fig. 30). The only exception is represented by ellipsoidal shapes, for which $\mathbf{D}$ is homogeneous, with principal axes coinciding with those of the ellipsoid [141]. In the case of a sphere, the three eigenvalues of $\mathbf{D}$ are exactly 1/3. The demagnetizing tensor of more practical shapes, such as disks, cylinders, and cuboids, is usually approximated by an average value that is representative for a large fraction of the specimen volume [142, 143].

The dependence of the internal field on the specimen magnetization creates an undesirable coupling between the two quantities, such that the magnetic properties appear to change while measurements are performed. This situation is described by so-called moving Preisach models (section "Modifications of the Classical Preisach Model"), which also require using vectors instead of scalars in all cases where the external field is not applied along one of the principal axes. Complications caused by the demagnetizing field can be removed by cutting specimens in shapes that ease the calculation of the demagnetizing tensor and measuring them along a principal axis. In this case, Eq. (36) is written in scalar form as $H_i = -DM$, where the demagnetizing factor D coincides with one of the eigenvalues of $\mathbf{D}$. The intrinsic FORC properties of the material are then retrieved by substituting the external fields H_r and H applied during measurements with the total internal fields $H_r' = H_r - DM(H_r, H_r)$ and $H' = H + DM(H_r, H)$, respectively. This operation, called *deshearing*, greatly simplifies the interpretation of FORC diagrams of strongly magnetic specimens by removing fictive contributions [139]. Deshearing

Fig. 30 Homogeneously
magnetized disk with
magnetization **M** (shaded) in
an external field **H** and stray
field produced by the
magnetization (field lines).
The total field experienced by
the material inside the disk is
the sum of **H** and the internal
part of the stray field

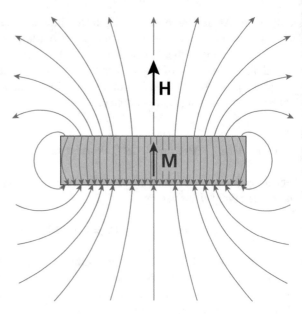

requires either resampling the corrected FORC measurements on a regular field grid and processing them like standard data or using regression methods that deal with irregularly spaced measurement points, such as FORCinel [128] and VARIFORC [1]. The latter implements deshearing by default.

An example of FORC deshearing is shown in Fig. 31 for a polycrystalline iron thin film where the internal field arises from a positive mean interaction field between crystals [144], rather than the negligible in-plane demagnetizing factor of a thin film. In this case, deshearing almost completely removes a complex FORC signature composed of two pairs of parallel diagonal ridges with positive and negative amplitudes, leaving a central maximum that represents a SD-like switching of the individual crystalline domains. Natural rocks are rarely magnetic enough for the demagnetizing factor to play a significant role, except for certain meteoritic materials and, on Earth, for basalts, whose saturation magnetizations can reach 20 kA/m [145]. In this case, the associated demagnetizing fields of the order of few mT are comparable with the coercive field and can significantly distort the intrinsic hysteresis properties of individual magnetic carriers. Large internal fields, however, exist inside ferrimagnetic particles. In most cases, these fields control the intrinsic magnetic properties of the particles through the associated magnetostatic energy term, so that deshearing these internal fields is meaningless. A notable exception is represented by large magnetite crystals, where the magnetic interaction between sub-volumes possessing "hard" and "soft" magnetic properties is described by an internal field of the form given by Eq. (36). In this case, FORC deshearing can be used to disentangle the mixed magnetic signature determined by the two interacting phases (section "Multistate Systems: The PSD FORC Signature").

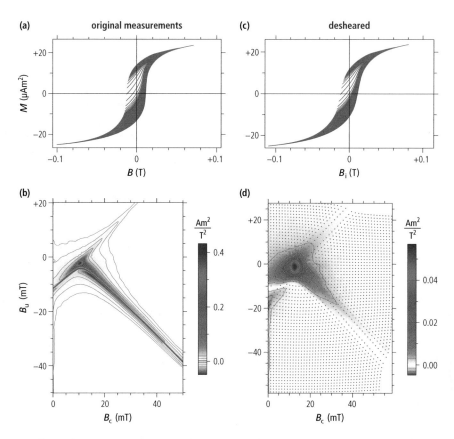

Fig. 31 Example of FORC deshearing for a polycrystalline iron thin film on a silicon substrate (data for sample B6 in Cao et al. [144]). (**a**) Original in-plane FORC measurements. Notice the extremely steep flanks of the hysteresis envelope near the coercive field. (**b**) Corresponding FORC diagram, processed with VARIFORC using the smoothing factor limitations for squared loops. The negative-positive pair of diagonal ridges departing from the central maximum is a typical feature of positive (magnetizing) mean-field interactions. (**c**) Desheared FORC measurements obtained assuming an internal field $B_i = \alpha M/M_s$ where M_s is the saturation magnetization and $\alpha = +12.8$ mT. (**d**) FORC diagram corresponding to the desheared measurements in (**c**) with superimposed grid of desheared measurement points (every second point). Deshearing enlarges the size of field steps along the diagonals affected by mean-field interactions, thereby reducing the amplitude of the FORC ridges. The value of α has been chosen to minimize the amplitude of the diagonal ridges. Failure to remove them completely is probably due to the fact that a distribution of α-values is needed to describe the interactions between single crystalline domains in different parts of the thin film. Nevertheless, the desheared FORC function has a much simpler structure and the central maximum has migrated toward $B_u = 0$

FORC Data Rendering

The correct representation of FORC data, especially those obtained from geologic materials, can be a challenging task, due to the extremely heterogeneous local properties of the FORC function, which typically include large low-amplitude regions and quasi one-dimensional high-amplitude ridges. Important details can be missed if plots are realized with unspecific color scales (Fig. 32a). For this reason, special nonlinear color scales have been implemented by some FORC processing tools, such as VARIFORC [1]. These scales are anchored to zero and represent negative/positive amplitudes of the FORC function with two different color gradients, usually blue-white and white-yellow-red or white-green-yellow-red. These gradients are used to discriminate negative low-amplitude regions, which are important diagnostic features of SD particles and mean-field interactions, from positive ones. Nonlinear color scales can be tuned to visualize the whole range of the FORC function (Fig. 32b), yet the perception of the relative contribution of different features remain difficult. For instance, the amplitude of the quasi-one-dimensional central ridge is entirely controlled by the measurement resolution and by the smoothing parameters (section "The Central Ridge"). Because high-resolution measurements are required to identify such features, their amplitude tends to be much larger than the rest of the FORC function, conveying the wrong impression that other signatures are unimportant. For instance, the amplitude of the central ridge in Fig. 32 is ~5 times larger than the remaining parts of the FORC function, yet its relative contribution to the total irreversible magnetization of this sample is only ~40%.

An objective representation of different parts of the FORC function can be obtained with contour lines. Equally spaced contour levels tend to concentrate around high-amplitude features (Fig. 32b); however, contour levels can be chosen to represent the relative contribution of all regions with lower amplitude to the total irreversible magnetization M_{forc} associated with the FORC function. A so-called quantile contour [146] with probability level $0 < p < 1$ is defined as the FORC function amplitude $\hat{\rho}_p$ for which the integral over all regions characterized by $\hat{\rho} \geq \hat{\rho}_p$ is equal to $(1 - \hat{\rho}_p) M_{forc}$. Quantile contours drawn at equally spaced p-levels divide the FORC space into regions with identical interquantile ranges and therefore identical contributions in terms of magnetization (Fig. 32c). Quantile contour levels are naturally more closely spaced at low p-values, capturing low-amplitude parts of the FORC function that contribute significantly to the total magnetization, because of their extension. Since quantiles are bounded to probability density functions, which are, by definition, positive, VARIFORC calculates quantile contour levels by considering only positive values of the FORC function; contours over negative regions are then drawn using $\hat{\rho}_{-p} = -\hat{\rho}_p$. Quantile contours are very effective at rendering any type of FORC function and are used throughout this work to represent FORC diagrams obtained from geologic materials.

4 FORC Diagrams of Natural Particle Assemblages

The number of publications dealing with the magnetic characterization of geologic materials with FORC diagrams has grown exponentially in recent years, along

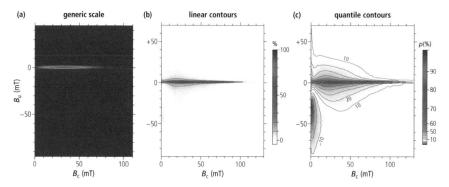

Fig. 32 High-resolution FORC diagrams of a magnetofossil-rich sediment from Lake Baldeggersee (Switzerland). (**a**) Representation with a color scale based on spectral colors (e.g., MATLAB "Jet" or Mathematica "Rainbow"). Only the central ridge is identifiable in this plot. (**b**) Representation with a color scale anchored to zero amplitudes (VARIFORC "Iridescent") and 9 equally spaced contours at 10, 20, 30, 40, 50, 60, 70, 80, and 90% of the full amplitude. The color scale is based on homogeneous transition between six reference colors. Notice the faint traces of negative amplitudes in the lower quadrant, near $B_c = 0$. Contours are concentrated on the central ridge, with the region enclosed by the lowest contour level contributing to ~40% of the total FORC magnetization. The remaining ~60% is only faintly rendered by the color scale and not represented by contours. (**c**) Same as (**b**), with the color scale automatically adapted to the distribution of FORC amplitudes, and equally spaced quantile contours with levels at -30, -20, -10, 10, 20, 30, 40, 50, 60, 70, 80, and 90% of the total magnetization. FORC regions between consecutive contours contribute equally in terms of magnetization. Unlike (**b**), all relevant features of the FORC function are captured

with the number of different FORC signatures attributed to specific types of magnetic particles. One of the main factors controlling these signatures is the domain state of the particles, with the usual distinction between single-domain (SD), multidomain (MD), and the vaguely defined intermediate pseudo-single-domain (PSD) regimes. A significant modeling effort has been undertaken to understand the fundamental characteristics of these regimes and, most recently, of the role played by vortex magnetic states in determining the FORC signature of PSD particles. The combination of forward modeling techniques, such as micromagnetic simulations, and inversion techniques, such as numerical unmixing of large collections of samples, led to the isolation of well-defined FORC signatures produced by particle assemblages with specific characteristics, mainly those made by one of the most widespread magnetic minerals in nature, magnetite. This section describes a collection of relevant FORC signatures from the forward modeling perspective (e.g., domain states, magnetostatic interactions, thermal relaxation) and from the inverse modeling perspective (different types of particle assemblages with similar properties). Special attention is dedicated to the correspondence between FORC measurements and other magnetic characterization tools, such as coercivity analysis, since a combination of different techniques is often required to overcome interpretation ambiguities.

Single-Domain Particles

The ideal magnetization pattern of SD particles consists in a single, homogeneously magnetized domain. The particle magnetization is thus entirely described by the magnetic moment vector $\mathbf{m} = V\mu_s \mathbf{u}$, where V is the particle's volume, μ_s its spontaneous magnetization, and $\mathbf{u}$ a unit vector parallel to the magnetization direction. The equilibrium moment of the particle in a quasi-static field $\mathbf{H}$ is a local minimum of the free magnetic energy

$$E = VK(\mathbf{u}) - \mu_0 V\mu_s \mathbf{H} \cdot \mathbf{u} \tag{37}$$

where K is the *magnetic anisotropy energy* per unit of volume, and the second term is the *Zeeman energy*, the energy density of a magnetized volume in a magnetic field. The magnetic anisotropy energy is minimal along preferential magnetization directions, which depend on the orientation of crystal axes and on the particle shape.

Uniaxial SD Particles: Stoner-Wohlfarth Model

The simplest case of magnetic anisotropy energy is that of uniaxial anisotropy. In this case, particles possess a single, so-called easy axis, along which K is minimal (Fig. 33). The simplest representation of uniaxial anisotropy is given by

$$K = \frac{1}{2}\mu_0\mu_s H_K \sin^2\theta, \tag{38}$$

where H_K is the so-called microscopic coercivity, or *microcoercivity* [31], and θ the angle between $\mathbf{u}$ and the easy axis. In this case, $K = 0$ when the magnetization is parallel to the easy axis ($\theta = 0$ or $\theta = 180°$) and $K > 0$ for all other orientations. SD particles with this type of anisotropy were modeled for the first time by Stoner and Wohlfarth in 1948 [7] and are called *Stoner-Wohlfarth (S-W) particles*. In the case of magnetite, uniaxial anisotropy originates from particle elongation, with the easy axis being parallel to the longest particle axis. If the particle shape is described by a prolate rotation ellipsoid,

$$H_K = \mu_s (D_1 - D_3) = \frac{\mu_s}{2}(1 - 3D_3), \tag{39}$$

where D_1 and D_3 are the demagnetizing factors along the short and long ellipsoid axes, respectively.

S-W particles possess only one or two stable magnetization states, depending on whether the amplitude of the external field is smaller or larger than a critical field

$$H_n(\varphi) = H_K \frac{\sqrt{1 - \tan^{2/3}\varphi + \tan^{4/3}\varphi}}{1 + \tan^{2/3}\varphi}, \tag{40}$$

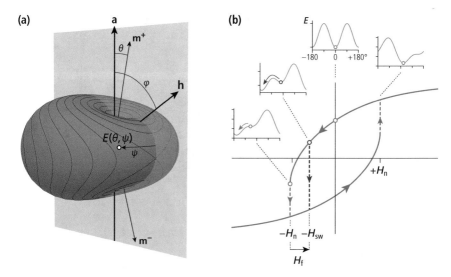

Fig. 33 The Stoner-Wohlfarth model of SD particles with uniaxial anisotropy. (a) Spherical plot of the free magnetic energy E of a particle, as a function of the polar and azimuthal angles θ and ψ of the magnetic moment vector **m** with respect to the easy axis **a** (orange surface, the distance from the origin is proportional to $E(\theta, \psi)$). In the absence of external fields, **m** at equilibrium is parallel or antiparallel to **a**. An external field parallel to the unit vector **h** deviates **m** in the plane defined by **a** and **h** (gray). If the external field amplitude is smaller than a critical value H_n, $E(\theta, \psi)$ has two local minima defining the equilibrium magnetic moments $\mathbf{m}^+$ and $\mathbf{m}^-$, which correspond to the descending and ascending branches of the hysteresis loop. (b) Hysteresis loop of the particle shown in (a), obtained from the component of $\mathbf{m}^\pm$ parallel to the field direction **h**. $E(\theta, 0)$ (insets) has two local minima if $|H| < H_n$. One of the two minima becomes a saddle point at $|H| = H_n$, causing a sudden transition to the only remaining minimum through a magnetization jump (dashed). This transition can occur earlier, in a switching field $|H_{sw}| < H_n$, if the energy barrier separating the two minima is overcome by thermal fluctuations

which depends on the angle φ between easy axis and applied field [7]. These magnetization states correspond to magnetic moment vectors $\mathbf{m}^\pm$ lying in the plane defined by the easy axis **a** and the external field **H**, which minimize the total magnetic free energy E of the particle (Fig. 33a). At $H = H_n$, one of the two minima of E converts to a saddle point, leaving only one minimum in larger fields. In the absence of thermal relaxation effects, the hysteresis loop $m(H) = m_0 \, \mathbf{u}(H) \cdot \mathbf{H}$ of individual S-W particles with magnetic moment $m_0 = V\mu_0$ is characterized by two curved branches that merge at the switching fields $H_{sw} = \pm H_n$ through a magnetization jump (Fig. 33b). These loops are similar to symmetric Preisach hysterons (section "The Classical Preisach Model"), except for the curvature of the two branches, which is due to the reversible rotation of the magnetic moment vector toward the external field direction (Fig. 34). The loops become increasingly closed when φ goes from 0 to 90°, because $m(H = 0) = m_0\cos\varphi$. The magnetization jumps $\pm\Delta m(\varphi)$ at $H_{sw} = \pm H_n$ follow a similar trend, with $\Delta m(0) = 2m_0$ and $\Delta m = 0$ at $\varphi \approx 77°$. At larger angles, the jumps reverse sign, reaching $\Delta m \approx -0.1m_0$ at

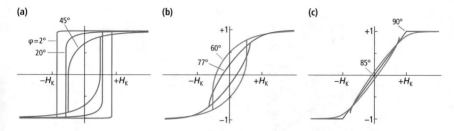

Fig. 34 Hysteresis loops of individual S-W particles with microcoercivity H_K at selected angles φ between easy axis and applied field. **(a)** $\varphi = 2, 20$, and $45°$, **(b)** $\varphi = 60$ and $77°$ (in the latter case the magnetization jump is zero), **(c)** $\varphi = 85°$ (negative magnetization jump) and $\varphi = 90°$ (closed hysteresis)

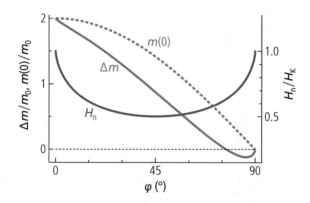

Fig. 35 Remanent magnetization $m(0)$, magnetization jump Δm at switching, and switching field H_n for S-W particles with angle φ between easy axis and applied field, in the absence of thermal fluctuations

$\varphi \approx 85°$, and disappear at $\varphi = 90°$. The switching field is smallest for $\varphi = 45°$ and largest for $\varphi = 0$ or $90°$, with $H_n(90° - \varphi) = H_n(\varphi)$ (Fig. 35).

At a macroscopic level, an assemblage of identical, aligned S-W particles that do not interact with each other behave exactly as the sum of individual particles, yielding the hysteresis loops shown in Fig. 34. All magnetic states that can be accessed by a sequence of applied fields coincide with one of the two branches $m^\pm$ of hysteresis. In the case of FORC measurements, all curves starting at reversal fields $H_r > -H_n$ coincide with the upper branch m^+, and all FORCs starting at reversal fields $H_r \leq -H_n$ coincide with the lower branch. Therefore, non-zero contributions to the FORC function are localized along the trace in FORC space of the curve beginning at $H_r = -H_n$: this is the ascending diagonal with start point $(H_c, H_u) = (0, -H_n)$ and end point $(H_c, H_u) = (H_n, 0)$ (Fig. 36a) and the only place where the derivative of measurements with respect to H_r is not zero. The FORC function is thus given by

$$\rho(H_r, H) = -\delta(H_r + H_n) \frac{\partial}{\partial H} \frac{m^+(H) - m^-(H)}{2} + \delta(H_r + H_n)\delta(H - H_n) \frac{\Delta m}{2},$$

(41)

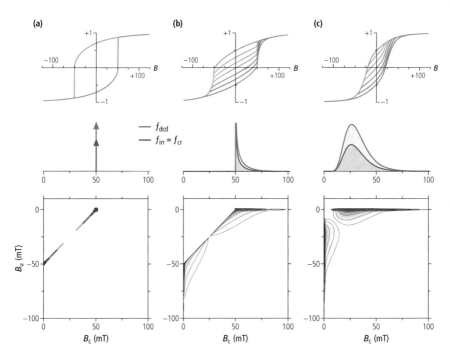

Fig. 36 FORC signature of non-interacting S-W particles with increasing heterogeneity. First row, descending hysteresis branch and selected FORCs; second row, coercivity distributions derived from FORC measurements (DC demagnetization, irreversible component of ascending branch, and central ridge); third row, FORC diagrams (quantile contours at 1, 2.5, 5, 10, and 50% of the total magnetization). (**a**) Identical, aligned S-W particles with $B_K = 100$ mT and $\varphi = 45°$. All FORCs coincide with one of the two major hysteresis branches, depending on whether the reversal field is smaller or larger than $-B_{sw} = -50$ mT. (**b**) Identical, randomly oriented S-W particles with $B_K = 100$ mT. (**c**) Randomly oriented S-W particles with lognormal microcoercivity distribution (logarithmic mean, 54 mT; logarithmic standard deviation, 0.35)

where δ is the Dirac impulse. The first term on the right-hand side of Eq. (41) is a diagonal ridge whose amplitude is proportional to the derivative of the difference between the two hysteresis branches, which is an odd function of H that diverges in proximity of $\pm H_n$ (Fig. 36a). This function depends on the curvature of the hysteresis loop and reflects the reversible magnetic moment rotation toward the applied field direction. The second term on the right-hand side of Eq. (41) produces a peak at $(H_c, H_u) = (H_n, 0)$, whose amplitude is proportional to the magnetization jump Δm through which m^- merges with m^+ at the positive switching field $+H_n$. The FORC function is zero over the whole upper quadrant, because all FORCs coincide with m^+ in applied fields larger than $+H_n$. The peak at $H_u = 0$ is the most elemental contribution to the central ridge by SD particles. The coercivity distribution associated with this type of FORC signature is a Dirac impulse centered at the switching field H_n with amplitude proportional to Δm in the case of f_{irr} and f_{cr} and to m_0 in the case of f_{dcd} (see section "Selected Properties of the FORC function" for a definition of these coercivity distributions). The relative contributions of the hysteresis curvature and the magnetization jumps to the FORC signature depend on the angle φ between easy axis and applied field: if $\varphi = 0$, the hysteresis is rectangular, like Preisach hysterons, yielding an isolated peak at $H_c = H_n$. At larger angles, the peak amplitude decreases, while the diagonal ridge caused by the hysteresis curvature increases. Finally, both contributions tend to zero as the hysteresis becomes completely closed at $\varphi = 90°$.

The FORC signature of aligned S-W particles is highly unrealistic, being everywhere singular, but it represents the fundamental unit upon which natural SD signatures are built. A first step toward realistic SD FORC models considers random particle orientations inside a non-magnetic matrix. The corresponding magnetic properties are obtained from the weighted integration of the magnetization of aligned particles, with weights proportional to the probability $\sin\varphi \, d\varphi$ of finding a particle whose easy axis forms an angle between φ and $\varphi + d\varphi$ with the field direction. Because each angle defines a different switching field $H_n(\varphi)$ comprised between $H_K/2$ and H_K, the resulting FORC diagram is given by the superposition of diagonal contributions similar to those of aligned particles (Fig. 36b). These contributions are concentrated near the diagonal line corresponding to $H_n = H_K/2$, due to the fact that most particles have switching fields close to this value, which is the local minimum of $H_n(\varphi)$ at $\varphi = 45°$. The peaks at $H_u = 0$ generated by individual angles overlap and form a central ridge that starts at $H_c = H_K/2$ and extends with declining amplitudes to $H_c = H_K$. Small negative amplitudes near $H_c = H_K$ are produced by the negative magnetization jump in the hysteresis of particles whose easy axis form an angle of $>77°$ with the applied field (Fig. 34c). The field range of all coercivity distributions is comprised between $H_K/2$ and H_K, and all distributions diverge at $H_K/2$. Slight differences between the shape of f_{irr} and f_{cr}, which originate from in-field measurements, and that of f_{dcd}, which is based on remanence measurements, are due to the distinct angular dependences of Δm (in-field) and $m(0)$ (remanence), respectively (Fig. 35). Despite the singularity at $H_K/2$, the integral of these distributions is finite and corresponds to the saturation remanence $M_{rs} = M_s/2$ in the case of f_{dcd}, and the irreversible fraction

$$\frac{M_{irr}}{M_s} = \frac{1}{2m_0} \int_0^{\pi/2} \Delta m(\varphi) \sin \varphi \, d\varphi \approx 0.274 \qquad (42)$$

of all magnetization changes occurring along a hysteresis branch, in the case of f_{irr} and f_{cr}. Given the similar shape of all coercivity distributions, Eq. (42) yields $f_{irr} = f_{cr} \approx \overline{S} f_{dcd}$ with $\overline{S} \approx 0.544$ [94]. The FORC function of identical, randomly oriented S-W particles is singular along the ascending diagonal starting at $(H_c, H_u) = (H_K/2, 0)$ and along $H_u = 0$, which is the location of the so-called central ridge.

The final step toward realistic models of SD particle assemblages consists in considering randomly oriented S-W particles with a distribution of shapes and thus of microcoercivities. Their magnetic properties are obtained by weighted integration of the magnetization of identical particles (e.g., Fig. 36b), with weights proportional to the microcoercivity distribution $f_K(H_K)$. Figure 36c shows the result obtained with a lognormal distribution corresponding to a typical coercivity distribution of natural magnetite particles [39]. Because f_K is usually a much wider distribution than the coercivity distribution of identical S-W particles, the resulting coercivity distributions are proportional to a rescaled version of f_K with median $H_{sw} \approx 0.53 H_K$ close to the switching field of particles with $\varphi \approx 45°$ [94]. The central ridge is now distributed along the range of coercivities determined by f_K, and it is surrounded, in the lower quadrant, by positive contributions associated with the reversible rotation of magnetic moments toward the applied field direction. Corresponding negative contributions peaking along $H_c = 0$ make the non-central ridge signature antisymmetric about the $H_u = -H_c$ diagonal. The pair of negative and positive contributions in the lower quadrant represents, together with the central ridge, the characteristic signature of uniaxial, non-interacting SD particles, as typically seen in magnetofossil-bearing sediments [94, 103].

Thermally Activated SD Particles

The S-W model of uniaxial SD particles assumes that switching between positive and negative states occurs exactly at $\pm H_n$, when one of the two local energy minima of Eq. (37) becomes a saddle point, enabling the sudden transition to the magnetic state corresponding to the other energy minimum. Random perturbations of the magnetization caused by the thermal energy $k_B T$, where k_B is the Boltzmann constant and T the absolute temperature, can trigger the transition between two magnetic states before the occupied local energy minimum disappears, by overcoming an *energy barrier* E_b. This barrier is defined as the smallest energy increase along all possible paths connecting two local energy minima. In the case of S-W particles, such paths lie in the plane defined by the easy axis and the applied field (Fig. 33), so that E_b is simply the difference between a local maximum and a local minimum of the free energy $E(\theta, \psi = 0)$. Thermally activated switching through the energy barrier E_b is a Poisson process with rate $\nu = \tau_0^{-1} \exp(-E_b/k_B T)$, where $\tau_0 \approx 0.1$–1 ns is the correlation time of thermal perturbations [26, 147, 148]. The energy barrier is a function of the applied field, with $E_b \to 0$ when $|H|$ approaches H_n from below (Fig. 37a). In proximity of H_n, the energy barrier is a power function of the so-called fluctuation field $H_f = H_n - |H|$, taking the form

$$\beta = \frac{E_b}{2k_B T} = \begin{cases} \beta_0 \frac{8\sqrt{2}}{9}\sqrt{1-(H_n/H_K)^2}\left(\frac{H_f}{H_n}\right)^{3/2}, & 0 < \varphi < \pi/2 \\ \beta_0\left(\frac{H_f}{H_n}\right)^2, & \varphi = 0, \pi/2 \end{cases}, \quad (43)$$

where β is the so-called Boltzmann factor and $\beta_0 = \mu_0 m H_K / 2k_B T$ the Boltzmann factor in zero field [27, 73]. The commonly assumed quadratic dependence on H_f is valid only for particles with $\varphi = 0$ and $90°$, which, however, represent a negligible contribution to the remanent magnetization of randomly oriented S-W particles. Magnetically stable SD particles are characterized by a relaxation time, defined by

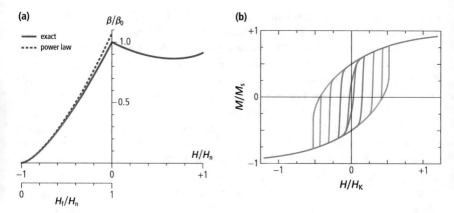

Fig. 37 Hysteresis of thermally activated uniaxial SD particles. (**a**) Energy barrier for switching from a positive to a negative state in an external field H at an angle $\varphi = 60°$ to the easy axis (solid line). The $H \to -H_\mathrm{n}$ limit is well approximated by a power law with exponent 3/2 (dashed line). (**b**) Thermally activated hysteresis loops for a large number of aligned S-W particles with $\varphi = 60°$ and Boltzmann factors $\beta_0 = 20, 30, 60, 200$, and ∞ (from inside outward). Other model parameters are $\tau_0 = 0.1$ ns and a field sweep rate of $2.5 H_K$/second

the inverse Poisson rate $\tau = \tau_0 e^{\beta_0}$, which is much larger than the measurement time, e.g., $\beta_0 = \ln(\tau/\tau_0) \approx 30$ for $\tau = 3$ h.

The hysteresis loop of many identical, aligned SD particles with uniaxial anisotropy depends on the occupation probabilities p^+ and $p^- = 1 - p^+$ of the m^+ and m^- states described in section "Uniaxial SD Particles: Stoner-Wohlfarth Model", which are governed by the kinetic equation

$$\frac{\mathrm{d}p^+}{\mathrm{d}t} = -v^+ p^+ + v^- \left(1 - p^+\right), \tag{44}$$

with $v^\pm$ being the field-dependent Poisson rates for the $m^+ \to m^-$ and $m^- \to m^+$ transitions, respectively [116, 149, 150]. The resulting hysteresis loop is then given by

$$M^\pm = p^+ m^+ + \left(1 - p^+\right) m^- \tag{45}$$

(Fig. 37b). Thermal activations transform the discrete magnetization jump at H_n into a continuous transition between positive and negative states. This regularization of the hysteresis loop is particularly evident for values of β_0 close to the lower stability range of SD particles. On the other hand, in the case of stable SD particles with $\beta_0 \gg \ln(\tau/\tau_0)$, where τ is the typical time required to measure a hysteresis loop, $p^\pm$ increases abruptly before H_n is reached. The width of the transition from $p^\pm = 0$ to 1 for stable SD particles made of magnetite (e.g., a $34 \times 34 \times 48$ nm prism: $H_K = 80$ mT and $\beta_0 = 500$) is comparable or smaller than the typical resolution of hysteresis or FORC measurements (Fig. 38), so that the S-W hysteresis model with

Fig. 38 Switching probability p for a uniaxial SD particle with $\varphi = 45°$, $B_K = 80$ mT, and selected values of β_0, as a function of $B_n - |B|$. Solid and dashed lines are calculation based on field sweep rates of 5 and 0.5 mT/s, respectively

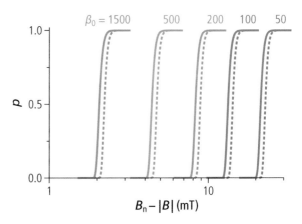

two discrete magnetization jumps is still a good approximation of real SD particles if H_n is replaced by $H_{sw} = H_n - H_f$, where H_f is the effective fluctuation field. In analogy with the thermal blocking process of SD particles in a cooling rock [151, 152], H_f is found by equating the particle relaxation time with a characteristic time of the field change. If the energy barrier is described by Eq. (43), H_f is the solution of

$$\beta(H_f) = -\ln\left[\frac{3}{2}\beta(H_f)\frac{\tau_0\alpha}{H_f}\right],\tag{46}$$

where α is the field sweep rate.

FORC measurements of thermally activated S-W particles can be modeled in a similar manner as the hysteresis loop by solving the kinetic Eq. (44) under consideration of the FORC protocol timing [116]. This includes different field sweep rates from positive saturation to the reversal field, and during measurements, as well as a pause at the reversal field (section "The FORC Protocol"). The FORC signature or thermally activated, stable SD particles is similar to that of the S-W model. The central ridge possesses a small intrinsic width [85], as well as a small vertical offset due to the timing asymmetry of the FORC protocol [117], with more time being spent at negative reversal fields than in positive fields (Fig. 39). This means that H_f is larger for $m^+ \rightarrow m^-$ transitions occurring at $H_r < 0$ than for the $m^- \rightarrow m^+$ transitions occurring during FORC measurements in positive fields. Accordingly, the central ridge occurs at $H = -H_r + H_f^+ - H_f^-$, where H_f^- is the effective fluctuation field during measurements and H_f^+ the effective fluctuation field for the negative field sweep to H. Figure 38 is a representative example for the fluctuation field differences occurring for stable SD magnetite particles subjected to field sweep rates that differ by one order of magnitude. The central ridge offset $H_u^{cr} = H_f^+ - H_f^-$ of stable SD magnetite particles is of the order of $0.01H_K$, or ~0.5 mT, for typical magnetofossil-bearing sediments. The dependence of H_u^{cr} on H_c can reveal important details on the nature of different SD components. For example, the presence of SD magnetite needles in magnetofossil-bearing sediments from the Paleocene-Eocene Thermal Maximum is revealed by a distinctive peak

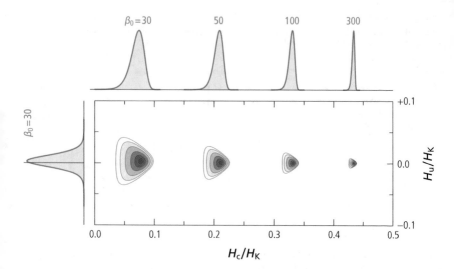

Fig. 39 Central ridges produced by four groups of identical uniaxial SD particles with $\beta_0 = 30$, 50, 100, and 300, respectively, whose easy axes form an angle of 60° with the applied field. Other model parameters are $2 H_K$/second for the field sweep rate from positive saturation to reversal field, with 0.2 s pause at the reversal field, 0.1 H_K/second for the field sweep rate during measurements, and $\tau_0 = 0.1$ ns. The contribution of each group has been normalized by its peak value for comparison purposes (contours at 1, 3, 10, 30, 60, and 90% of the peak value). Corresponding marginal distributions along H_c, obtained by integrating the central ridge over H_u, are plotted above the FORC diagram. The marginal distribution along H_u, obtained by integrating the central ridge over H_c, is plotted on the left for $\beta_0 = 30$. Notice the slight offset of toward positive H_u values

of H_u^{cr} at $H_c \approx 170$ mT (Fig. 24), well beyond the typical vertical offset related to conventional fossil magnetosome chains [119], due to the smaller mean needle volume.

Thermally activated switching at $|H_{sw}| < H_n$ increases the amplitude of the magnetization jumps responsible for the central ridge and eliminates the singularity of reversible contributions to the FORC function along $H_c = 0$ and $H_u = 0$, since the slope of the hysteresis loop branches just before switching is no longer infinite. In most cases, the intrinsic thickness of the central ridge remains equal to or smaller than the measurement resolution, so that this signature is preserved (Fig. 40). On the other hand, the regularization of reversible contributions eliminates the vertical ridge along $H_c = 0$, yielding positive and negative contributions in the lower quadrant, which are similar to those observed in sediments and rocks whose magnetic signature is dominated by SD particles [94, 103] (Fig. 41).

The regularization of reversible contributions to the FORC function facilitates the numerical isolation of the central ridge with the method described in section "The Central Ridge", because the small region of FORC space immediately above and below the central ridge can be approximated by a broken linear function (Fig. 22). The total central ridge magnetization M_{cr} and the irreversible component M_{irr} of the hysteresis of thermally activated SD particles with uniaxial anisotropy are

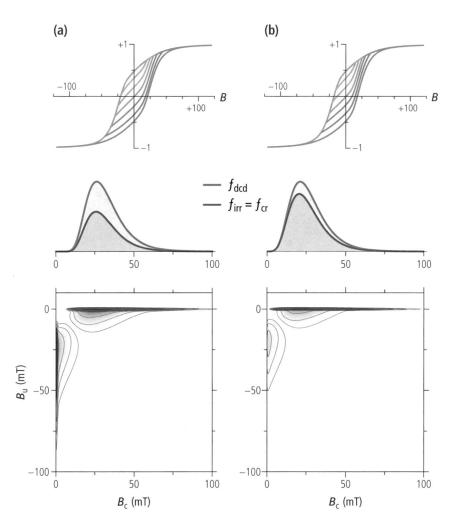

Fig. 40 Effect of thermal fluctuations on the FORC signature of non-interacting uniaxial SD particles. In this example, the particles have a lognormal microcoercivity distribution (logarithmic mean, 54 mT; logarithmic standard deviation, 0.35). First row, descending hysteresis branch and selected FORCs; second row, coercivity distributions derived from FORC measurements (DC demagnetization, irreversible component of descending branch, and central ridge); third row, FORC diagrams (quantile contours at 1, 2.5, 5, 10, and 50% of the total magnetization). (**a**) No thermal fluctuations (same as Fig. 36c). (**b**) With thermal fluctuations causing a decrease of the switching field by $B_f = 5$ mT for all particles. Notice the reduction of reversible contributions to the FORC diagram and the elimination of the negative vertical ridge

comprised between $0.274M_s$ for the $\beta_0 \rightarrow \infty$ limit (Eq. 42) and $0.5M_s$ for the quasi-rectangular hysteresis loops obtained when $|H_{sw}| \gg H_n$. Natural materials often contain additional contributions from interacting SD or non-SD particles, which are distributed over the FORC space and overlap with the reversible contributions of SD

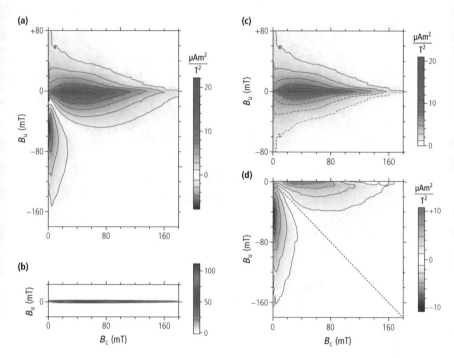

Fig. 41 Reconstruction of the FORC signature of reversible magnetic moment rotation in uniaxial SD particles. **(a)** High-resolution ($\delta B = 0.4$ mT) FORC diagram of a sediment sample from Wilson Lake, New Jersey, deposited during the Paleocene-Eocene Thermal Maximum (see Fig. 24) after subtraction of the central ridge **(b)** with VARIFORC (quantile contours every 10%). Notice the asymmetry between positive contributions in the upper and lower quadrant, respectively. Because the FORC function of uniaxial SD particles is zero in the upper quadrant, positive contributions above $B_u = 0$ must be attributed to other magnetic carriers. **(b)** Isolated central ridge that has been subtracted from the bulk FORC function to obtain the diagram in **(a)**. **(c)** Reconstruction of the FORC function produced by non-SD magnetic carriers, assuming it to be symmetric with respect to $B_u = 0$. Dashed contours represent the extrapolated part obtained by reflection of the upper quadrant about $B_u = 0$. **(d)** FORC signature of reversible magnetic moment rotation in uniaxial SD particles obtained by subtracting the non-SD contribution in **(c)** from the FORC function in **(a)**. The result is almost perfectly symmetric about the $B_c = -B_u$ diagonal, as expected for uniaxial SD particles

particles. In this case, the central ridge is the only signature that can be rigorously separated from remaining contributions without using additional assumptions about the nature of the magnetic particles contributing to the observed signal.

Viscous SD Particles and the Vertical Ridge

Another FORC signature of thermally activated SD particle assemblages, in the form of a vertical ridge along $H_c = 0$, is produced by relaxation effects near the reversal field [114]. During field sweeping from positive saturation to a negative reversal field H_r at a rate α_r, particles with increasingly large switching fields are set

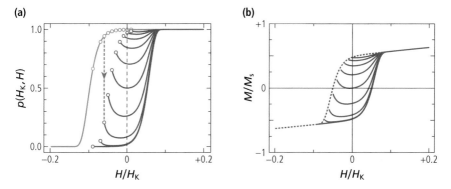

Fig. 42 Simulated FORC measurements for identical, thermally activated, uniaxial SD particles with normalized zero-field energy barrier $\beta_0 = 25$ and an angle $\varphi = 60°$ between easy axis and applied field. The assumed field sweep rates to the reversal field and during FORC measurements are $\alpha_r = 2H_K/\text{sec}$ ond and $\alpha_m = 0.1H_K/\text{sec}$, respectively, with $\tau_0 = 0.1$ ns. (a) Occupation probability for the descending hysteresis branch (gray line) and FORCs (red lines) starting between $H_r/H_K = -0.13$ and -0.05 in steps of 0.01 (circles). The gap between the descending branch and the FORCs is due to the viscous magnetization decay during the pause $t_r = 0.2$ s at the reversal field (dashed arrow). (b) Simulated FORC measurements obtained from (a). The hysteresis loop obtained with a constant field sweep rate equal to α_m is shown for comparison (dashed line). Notice the small gap between the descending branch of the hysteresis loop and the FORCs, due to the faster field sweep rate used to reach H_r in typical measurement protocols

to negative saturation. Once the reversal field has been reached, particles continue to switch during the pause time t_r, as well as during the first FORC measurement steps, when the applied field is still very close to H_r, due to thermal activations (Fig. 42a). The result of this time-dependent process, also known as *magnetic viscosity*, is a continuous decrease of the magnetization below the initial value obtained when H_r has been just reached. The initial decrease occurring during the measurements of the first few points produces concave curves (Fig. 42b) whose exact shape depends on the distribution of energy barriers and the timing of the FORC protocol [114, 116, 117].

A rough approximation that helps in understanding the origin of the vertical ridge and its dependence on the FORC protocol is given by

$$M(H_r, H) \approx M_0(H_r, H) + M_V \exp\left(-\frac{t_r}{\tau} - \frac{H - H_r}{H_f}\right), \qquad (47)$$

where M_0 is the non-viscous magnetization, M_V a viscous magnetization that decays over time, τ the time constant of viscous decay, $H_f(\beta_0, \alpha_m)$ an effective fluctuation field that depends on the energy barrier of the particles and the field sweep rate α_m during measurements, and t_r the pause time at reversal. The mixed derivative of the viscous term in Eq. (47) is an exponential function of the same form, so that, using $H - H_r = 2H_c$, the corresponding viscous contribution to the FORC function is given by

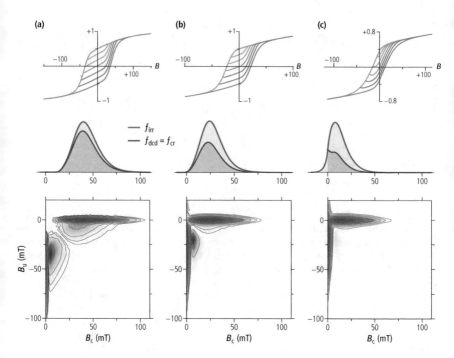

Fig. 43 Simulated FORC measurements of randomly oriented, thermally activated, uniaxial SD particles with lognormal-distributed microcoercivities (logarithmic mean, 125 mT; logarithmic standard deviation, 0.2) and normalized zero-field energy barriers $\beta_0 = 72.5$ (**a**), 36.2 (**b**), and 21.7 (**c**). First row, descending hysteresis branch and selected FORCs; second row, coercivity distributions derived from FORC measurements (DC demagnetization, irreversible component of descending branch, and central ridge); third row, FORC diagrams (quantile contours at 0.5, 1, 2, 5, 15, 40, 60, and 80% of the total magnetization). Data from Lanci and Kent [116]

$$\rho_{vr}(H_c, H_u) = M_V \frac{e^{-t_r/\tau}}{H_f^2} \exp\left(-2\frac{H_c}{H_f}\right). \tag{48}$$

The viscous magnetization M_V is proportional to the number of particles that are switched by the reversal field H_r, which is in turn proportional to the switching field distribution at $H_{sw} = -H_r$. Using the central ridge distribution $f_{cr}(-H_r)$, Eq. (32) is obtained. The most important aspects of these results are that the maximum amplitude of the vertical ridge occurs at $H_c = 0$ and that the ridge amplitude is reduced by the pause time t_r, because of the viscous decay occurring before each curve is measured.

Numerical simulations of particles with increasingly small energy barriers show how the amplitude of the vertical ridge increases, while reversible contributions decrease, due to the increasing squareness of thermally activated loops (Fig. 43). Furthermore, the central ridge moves to lower coercivities (Fig. 43b), until significant contributions from particles that decay spontaneously in a null field start to appear at $H_c = 0$ (Fig. 43c). At this point, thermal activations are sufficiently large to switch some particles even before the field is reversed, so that the vertical ridge extends to positive fields. As the energy barrier is further decreased, more particles become SP, no longer contributing to the FORC diagram, and the few that still have an open hysteresis loop are characterized by $(H_c, H_u) \to 0$. This effect is clearly demonstrated by measurements on growing magnetotactic bacteria [153, 154].

Weak Magnetostatic Interactions Between SD Particles

In discussing the FORC signature of uniaxial SD particles, it has been so far assumed that they are completely isolated from each other. The lack of magnetic interactions ensures that the local field acting on each particle coincides with the external field applied during measurements. A good approximation of this condition is obtained with particles that are well dispersed in a non-magnetic matrix, with nearest-neighbor distances that are sufficiently large to make the external dipole field associated with each particle negligibly small in comparison to the switching field. If this condition is not met, each particle is subjected to an additional internal field given by the sum of the dipole fields produced by all other particles, also known as *interaction field*. Because of the random nature of particle positions and orientations in geologic materials, the interaction field can be treated as a stochastic variable $H_i = \overline{H}_i + H_{int}$ with a mean and a random component [155]. The mean component $\overline{H}_i$ often coincides with the demagnetizing field, in which case it is proportional to the specimen magnetization, through a proportionality constant depending on the specimen shape (section "Demagnetizing Fields and FORC Deshearing"). The random component H_{int} is controlled mainly by the position and magnetic moment orientation of nearest-neighbor particles. In the case of randomly distributed SD particles with random magnetic moment orientations, the random component parallel to any given measurement direction is described by the symmetric, bell-shaped probability density function [85]

$$W(H_{int}; \eta, \mu_s) = \frac{e^{\alpha/\beta}}{\pi\beta\sqrt{1+H_{int}^2/\alpha^2}} K_1\left(\frac{\alpha}{\beta}\sqrt{1+H_{int}^2/\alpha^2}\right)$$

$$\alpha = \frac{\pi\eta\mu_s(1+\varepsilon)}{12}\left[\frac{\ln(2+\sqrt{3})}{2\sqrt{3}}+1\right], \quad \beta = \frac{\eta\mu_s^2}{54\alpha(1+\delta)} \tag{49}$$

$$\varepsilon = 0.359\arctan^{0.892}(12.15\eta), \quad \delta = 3.2145\sqrt{\eta}\,e^{-7.07\eta},$$

where K_1 is the modified Bessel function of the second kind; $0 \le \eta < 1$ the packing fraction, defined as the total volume occupied by the particles, divided by the total volume of the specimen; and μ_s the spontaneous magnetization of the particles. The width of W increases with the packing fraction, and the shape is comprised between the limits of a Lorentz function for $\eta \ll 0.05$ and a Gauss function for $\eta \gg 0.05$ (Fig. 44a). For $\eta \le 0.1$, the half-width-to-half-maximum (HWHM) of W, which is a measure of the typical magnitude of H_{int}, is $\sim 0.363\,\eta\,\mu_s$ (Fig. 44b).

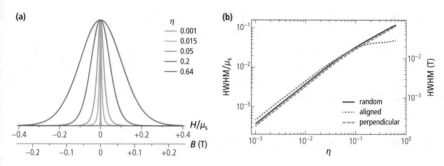

Fig. 44 Random component of the interaction field produced by randomly distributed SD particles. (**a**) Interaction field distributions for selected values of the packing fraction η. The upper scale is the interaction field normalized by the spontaneous magnetization μ_s of the particles; the lower scale is the unnormalized interaction field for magnetite particles ($\mu_s = 480$ kA/m). (**b**) Half-width-to-half-maximum (HWHM) of the interaction field distribution as a function of the packing fraction for particles with randomly oriented moments, moments aligned with the field direction, and random moments perpendicular to the field direction. The limits for $\eta \to 0$ are 0.363 η(random), 0.455 η (aligned), and 0.319 η(perpendicular). Notice the onset of a weaker dependence on η above $\eta \approx 0.1$, especially in the case of aligned moments. Data from Egli [85]

In the Preisach-Néel model of interacting SD particles [6], H_{int} is assumed to be independent from the bulk magnetization, so that the ensemble of interaction fields acting on all particles, and thus the distribution of bias fields given by the Preisach function, does not change during FORC measurements. The resulting FORC function would contain a broadened central ridge of the form

$$\rho_{cr}(H_c, H_u) = f_{cr}(H_c) \; W(H_u), \tag{50}$$

with vertical profiles proportional to the interaction field distribution W (e.g., Fig. 8b). In reality, the interaction field distribution depends on the magnetic moment orientations and thus on the bulk magnetization [23]. In extreme cases, the magnetic coupling between particles is sufficiently strong to promote a collective behavior, as in chains of closely spaced SD particles [156] or in small clusters of densely packed particles [129, 157]. The corresponding FORC signatures depend strongly on the type of relation between bulk magnetization and H_{int}, which is in turn controlled by the spatial arrangement of particles. Distribution anisotropy, which is the directional dependence of interparticle spacing, plays a fundamental role: for instance, linear arrangements, such as magnetosome chains, yield a collective behavior that is equivalent to that of an isolated SD particle with uniaxial anisotropy [94, 156], which is very different from that of SD particles confined in a plane [92].

Therefore, a general description of interacting SD particles is possible only in the case of weak magnetostatic interactions.

Weak magnetostatic interactions are defined by the preservation of individual particle switching events: the change of H_{int} caused by the switching of one particle will not induce other switching events. This condition is fulfilled if the typical maximum amplitude of H_{int}, given, for instance, by the 90% quantile of W, does not exceed the typical minimum switching field H_{sw}^{min} of the SD particles, given, for instance, by the 10% quantile of the coercivity distribution. The maximum packing fraction of weakly interacting SD particles corresponding to the above definition is $\eta \approx 0.4 H_{sw}^{min}/\mu_s$. In the case of magnetite particles with $H_{sw}^{min} = 20$ mT, this limit would be ~1.3%, which corresponds to typical interparticle distances equal to ~3 diameters. In the weakly interacting regime, W depends only marginally on the bulk magnetization, with limit cases given by aligned magnetic moments (when a strong field is applied) and magnetic moments perpendicular to the field direction (when most particles are about to be switched) (Fig. 44b). Therefore, W is still described by Eq. (49) with slightly different parameters [85].

The statistical independence of the interaction field from the bulk magnetization does not mean that H_{int} does not change locally during FORC measurements; however, it enables a general description of the evolution of H_{int}. Two limit cases explain this situation when considering a random set of weakly interacting SD particles with a distribution $f(H_c)$ of switching fields. The first case describes the behavior of a low-coercivity particle with H_c close to the lower end of $f(H_c)$. During the course of FORC measurements, there will be a reversal field H_r that is just sufficient to switch this particle to negative saturation, when $H_r = -H_{int} - H_c$. The same particle will be switched back to positive saturation during the measurement of the curve starting at H_r, when the applied field H reaches $-H_{int} + H_c$. Because H_c is small, most particles retain the same magnetic polarity during the field sweep from H_r to H, so that changes of H_{int} are caused only by small reversible magnetic moment rotations (Fig. 45a, b). Therefore, the irreversible magnetic behavior of this particle is effectively described by a hysteron with constant bias field $H_u = H_{int}$ and contributes to the central ridge at a point with FORC coordinate (H_c, H_{int}). A large number of similar low-coercivity particles record different realizations of the random variable H_{int}, reproducing the interaction field distribution along H_u.

Consider now the opposite case of a high-coercivity particle with H_c close to the upper end of $f(H_c)$. Because of its large H_c, this particle will be switched to positive or negative saturation only after most of the other particles have been already switched. During the course of FORC measurements, the high-coercivity particle is switched to negative saturation when the reversal field $H_r = -H_{int}^- - H_c$ has been reached, where H_{int}^- is the interaction field produced when all particles are in a negative state. The same high-coercivity particle is switched back to positive saturation during the measurement of the curve starting at H_r, when the applied field H reaches $-H_{int}^+ + H_c$, where H_{int}^+ is the interaction field produced when all particles are in a positive state. Because the magnetic moment of most particles is rotated by $180°$ when the field is swept from H_r to H, $H_{int}^- \approx -H_{int}^+$ (Fig. 45c, d) and the irreversible magnetic behavior of the high-coercivity particle is effectively

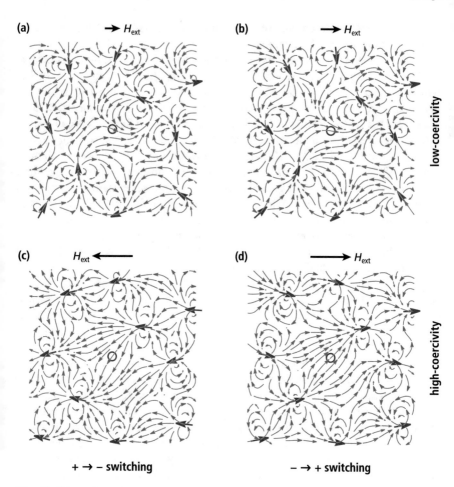

Fig. 45 Schematic representation of the local interaction field H_{int} (blue arrows) when the selected particle (red circle) is switched to negative (left plots) and positive (right plots) saturation. Red arrows are the magnetic moments of neighbor particles and H_{ext} is the external field. (**a, b**) Only a small change of H_{ext} is required to cycle a low-coercivity particle between positive and negative saturation: in the limit case of a vanishingly small coercivity, the magnetic moments of most particles and therefore H_{int} remain unchanged. (**c, d**) A high-coercivity particle is switched after the other particles, so that negative and positive switching occur in antiparallel interaction fields with similar amplitudes

described by a hysteron with switching fields $H_B = H_{int}^+ - H_c$ and $H_A = -H_{int}^+ + H_c$. This hysteron contributes to the central ridge at a point with FORC coordinates $\left(H_c - H_{int}^+, 0\right)$. Because positive and negative values of H_{int}^+ are equally likely, high-coercivity particles with intrinsic coercivity close to H_c will contribute to the central ridge at $(H_c, 0)$, as if the interaction field would not exist.

With the above examples in mind, particles with intermediate coercivities are exposed to an interaction field that can be divided into a fixed component generated by particles with higher coercivity and a switching component generated by particles

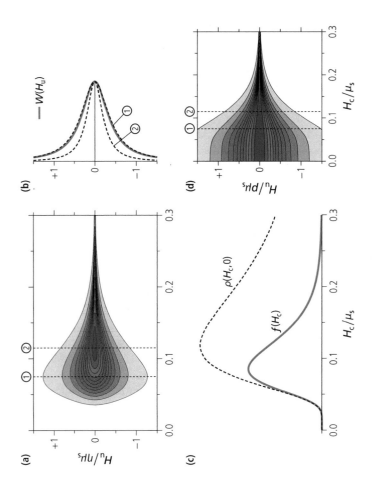

Fig. 46 (continued)

with lower coercivity. Vertical profiles of the central ridge are affected only by the fixed interaction field component produced by the fraction $F(H_c)$ of non-switching particles given by the integral of the switching field distribution $f(H_{sw})$ over $H_{sw} > H_c$ and are thus proportional to $W(H_u; \eta F, \mu_s)$ with W given by Eq. (49). Because the shape of W is fixed at the low packing fractions required for weak interactions, the central ridge is finally given by [85]

$$\rho_{cr}(H_c, H_u) = M_{rs}\overline{S}\frac{f(H_c)}{F(H_c)} W\left(\frac{H_u}{F(H_c)}; \eta, \mu_s\right), \tag{51}$$

where M_{rs} is the saturation remanence and $\overline{S}$ is the mean amplitude of normalized magnetization jumps in single-particle hysteresis loops (section "Uniaxial SD Particles: Stoner-Wohlfarth Model"). Because of the limits $F(0) = 1$ and $F(\infty) = 0$, vertical profiles of the FORC function possess the full width of the interaction field distribution at $H_c \to 0$ and null width at the high-coercivity end. This dependence of the vertical spread of ρ_{cr} on H_c gives the FORC function typical teardrop-shaped contour lines (Fig. 46), as seen in diluted SD particle assemblages [15]. Unlike in the Preisach-Néel model, the coercivity and interaction field distributions cannot be identified with horizontal and vertical profiles of the FORC function running through the central maximum. A correct estimate of the coercivity distribution is obtained from the marginal distribution corresponding to the integral of $\rho(H_c, H_u)$ with respect to H_u [122], while the interaction field distribution is represented by vertical profiles of ρ at H_c values close to the left limit of the coercivity range (Fig. 46a–c). Vertical profiles are best evaluated by plotting a normalized version of the FORC function, where each column of values is divided by the function value at $H_u = 0$ (Fig. 46d). In this plot, contour lines give a direct measure of the narrowing of vertical profiles as a function of H_c, with the left limit approaching the full width of the interaction field distribution.

Real assemblages of interacting SD particles often display intermediate properties between those of the model illustrated above and the Preisach-Néel model, as in the example shown in Fig. 47, which shows FORC measurements of extracted magnetosomes dispersed in a non-magnetic matrix. In the upper quadrant of the FORC diagram, which is unaffected by reversible contributions associated with in-

---

Fig. 46 FORC signature of a fully random assemblage of weakly interacting S-W particles. (a) FORC function for a lognormal distribution of microcoercivities with a median of $0.2\mu_s$ (~120 mT in the case of magnetite) and a logarithmic standard deviation of 0.4. Notice the different normalizations for H_c and H_u, with μ_s being the spontaneous magnetization of the particles and η the packing fraction. The corresponding full H_u range in the case of a 1% magnetite dispersion is ± 9 mT. (b) Vertical profiles (dashed lines) of the FORC function at $H_c = 0.075\mu_s$ and $0.115\mu_s$ (vertical dashed lines in (a)) and the interaction field distribution W. Profile 1 almost coincides with W, while profile 2, which runs through the central maximum of the FORC function, is narrower. (c) Horizontal profile of the FORC function at $H_u = 0$ (dashed line) and switching field distribution $f(H_c)$. (d) FORC function, normalized by its value at $H_u = 0$: contour lines show the decreasing width of vertical profiles as a function of H_c. Data from Egli [85]

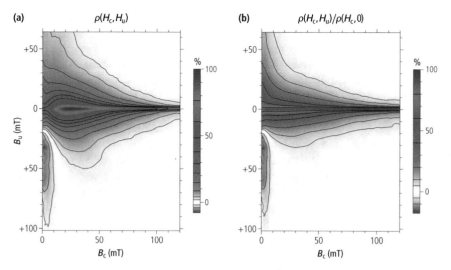

Fig. 47 FORC signature of magnetostatic interactions for a sample containing prismatic magnetosomes extracted from cultured magnetotactic bacteria, strain MV-1 (data from Wang et al. [140]). (a) FORC diagram with quantile contours at 5, 10, 20, 30, 40, 50, 60, 70, 80, and 90% of the total magnetization. Notice the teardrop shape of the contours, which is typical for random assemblages of interacting particles, and the negative region produced by the reversible rotation of magnetic moments in the applied field. (b) Same as (a), after normalizing each column of values with the maximum FORC amplitude (at $H_u \approx 0$) in the same column. Contours at 5, 10, 20, 40, 60, and 80% of the maximum amplitude. These contours show the narrowing of vertical profiles of the FORC function at higher coercivities

field magnetic moment rotation, the width of vertical profiles decreases towards the high coercivity end, but not to the same extent predicted by Eq. (47). Instead, there is a rapid decrease from a HWHM value of ~15 mT at $H_c = 0$ to ~8 mT at $H_c = 10$ mT, followed by a much slower decrease at higher fields. The 0–10 mT coercivity range is likely associated with clusters of strongly interacting immature magnetosomes, while the remaining range is covered by weakly interacting mature crystals or chain fragments. The weaker dependence on H_c can be explained by assuming a certain degree of heterogeneity in the sample: if similar structures, such as chain fragments with higher coercivities, form separated clusters, interactions within individual clusters with different coercivities will produce a superposition of magnetostatic interaction signatures, which tend to homogenize the H_c-dependent vertical spread of the central ridge.

Mean-Field Interactions

As seen in section "Weak Magnetostatic Interactions Between SD Particles", magnetostatic interaction fields can be represented by the sum of a random component with zero expectation and a deterministic component, called *mean interaction field*. The mean interaction field has two sources: the demagnetizing field (section "Demagnetizing Fields and FORC Deshearing") and distribution anisotropy [83, 155, 158–160]. Distribution anisotropy arises when the mean distance between particles is direction dependent. In both cases, the mean interaction field $\overline{H}_i = \alpha M$ is proportional to the bulk magnetization through a positive or negative proportionality coefficient α.

So-called *magnetizing interactions* are characterized by $\alpha > 0$ and tend to stabilize the existing magnetization. Magnetizing interactions decrease the coercive field and make the hysteresis loop more squared, because the internal field amplifies external field variations. The central ridge is offset toward the lower quadrant (Fig. 48a). This offset, which is most pronounced at the lower coercivity end and disappears at the higher coercivity end, is explainable by the fact that low-coercivity particles are switched back to positive saturation in a positive mean field caused by the positive bulk magnetization. The central ridge offset declines at higher coercivities in the same manner as the vertical spread caused by random interactions, because high-coercivity particles are switched to negative saturation and back to positive saturation in internal fields of similar amplitude but opposite sign (section "Weak Magnetostatic Interactions Between SD Particles"). A boomerang-shaped pair of positive and negative lobes appears below the central ridge, extending along 45° diagonals that departs from the central maximum (Fig. 48a; see also Fig. 31b for another example). If a random interaction component is also present, vertical smearing of the central ridge and the boomerang-shaped lobes yields a positive central maximum and a minor, negative minimum located just below, respectively [161].

Demagnetizing interactions are characterized by $\alpha < 0$ and tend to destabilize or destroy the existing magnetization. The effects of demagnetizing interactions on hysteresis and on the central ridge are opposed to those of demagnetizing interactions. The hysteresis loop is stretched along the field axis, and the apparent coercivity increases, because the internal field acts as a partial shield against the external field. The central ridge possesses a positive vertical offset at the low-coercivity end (Fig. 48b). This offset is explainable by the fact that low-coercivity particles are switched back to positive saturation when bulk mean magnetization is positive, so that a negative internal field needs to be overcome. The remaining parts of the FORC diagram are not simply complementary to the magnetizing interactions case: instead, positive contributions to the FORC function appear below the low-coercivity end of the central ridge along a sort of wishbone structure [92], along with a small negative peak below the high-coercivity end of the central ridge.

In both cases, FORC contributions below the central ridge originate from the differential compression or expansion of reversible magnetization changes before the curves merge with the upper branch of the major hysteresis loop. Curves starting at different reversal fields are affected by different internal fields, proportionally to the difference in bulk magnetization, and these internal field differences produce different biases of single-particle hysteresis loops.

Strong mean-field interactions are often encountered in patterned structures [155], such as thin films [144] and particle arrays [92] used in magnetic memory

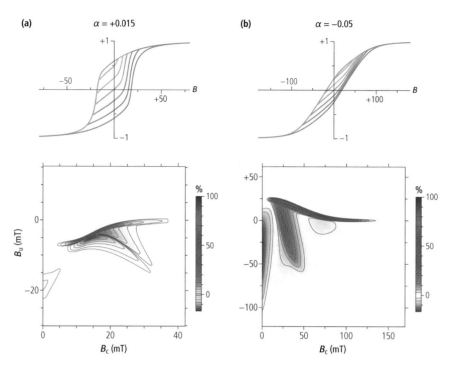

Fig. 48 Effect of mean-field interactions on the magnetic properties of the uniaxial SD particles shown in Fig. 40. The mean internal field is given by $\mu_0 \overline{H}_i = \alpha M / M_s$ with (**a**) $\alpha = +0.015$ for magnetizing interactions and (**b**) $\alpha = -0.05$ for demagnetizing interactions. Top: major hysteresis loop and simulated FORC measurements for selected reversal fields. Bottom: corresponding FORC diagrams with quantile contours at 5, 10, 20, 30, 40, 50, 60, 70, 80, and 90% of the total magnetization

devices, because of their distribution anisotropy. The dispersion of SD particles in natural materials, on the other hand, is much more random and lacks the long-range distribution anisotropy of engineered materials. Magnetic grains embedded in a host mineral that imparts a long-range order represent an exception. Examples include exsolution of magnetic phases during rock cooling [162, 163] and magnetic particle growth by iron diffusion along dislocations and cracks [164]. The random arrangement of host crystals containing such inclusions causes the superposition of signatures associated with different degrees of magnetizing and demagnetizing interactions, whose effect is not identical to that of random interactions. While positive and negative vertical offsets of the central ridge produce the H_c-dependent vertical spread of random interactions (section "Weak Magnetostatic Interactions Between SD Particles"), the mean-field FORC signatures shown in Fig. 48 combine into complex patterns that break the symmetry between upper and lower quadrants. Strong mean-field interactions produce collective particle behaviors with different characteristics, to the point that the resulting magnetic properties are no longer identifiable with those of interacting SD particles. Notable examples include magnetosome chains, which are distinguishable from isolated, uniaxial SD particles only on the basis of their coercivity distribution [94, 165], and small dense clusters, which share the same signature of larger particles hosting a single magnetic vortex [129]. These examples can be considered as extreme cases of the moving Preisach model [22], where small magnetization changes triggers avalanche processes that tend to reinforce (regenerative effects) or cancel (degenerative effects) the original magnetization change [9].

SD Particles with Multiaxial Anisotropy

SD particles with uniaxial anisotropy possess a maximum of two stable magnetization states determined by a single axis of preferred magnetization, called the easy axis. In the absence of external fields, the magnetic moment is parallel or antiparallel to this axis. Uniaxial anisotropy might be intrinsic to the crystal structure, as in cobalt, or it might be produced by particle elongation. Multiaxial anisotropy, on the other hand, is characterized by the presence of more than one preferred magnetization axis related to the crystal structure, to complex particle shapes [166], or both.

Above the Verwey transition temperature (125 K), magnetite possesses a cubic magnetocrystalline anisotropy with easy axes parallel to the four $\langle 111 \rangle$ crystal axes (Fig. 49). The associated anisotropy energy per unit of volume is given by

$$E = K_1 \left(\alpha_1^2 \alpha_2^2 + \alpha_2^2 \alpha_3^2 + \alpha_3^2 \alpha_1^2 \right) + K_2 \alpha_1^2 \alpha_2^2 \alpha_3^2 \tag{52}$$

where $\alpha_k = \mathbf{c}_k \cdot \mathbf{u}$ is the angle between the magnetization direction $\mathbf{u}$ and the k-th $\langle 100 \rangle$ axis $\mathbf{c}_k$ of the crystal and K_1, K_2 are the cubic anisotropy constants ($K_1 = -13.5$ kJ/m^3 and $K_2 = -2.8$ kJ/m^3 at room temperature [167, 168]). Magnetostriction, which is the magnetocrystalline anisotropy change caused by magnetization-induced crystal deformation, is negligible for the size range of SD magnetite particles [169], so that the magnetic properties are entirely determined by the particle shape and by the crystal axes orientations relative to shape. Partial cancelling effects of the two anisotropy sources in nearly equidimensional, often irregularly shaped SD magnetite particles [170, 171] produce complex configurations with several easy axes or planes. This situation is best illustrated by treating the two anisotropy sources separately at first.

Irregular particle shapes can be approximated, to first order, by triaxial ellipsoids with axes $a \geq b \geq c$, for which it is possible to calculate a triaxial demagnetizing tensor $\mathbf{D}$ [141] with corresponding shape anisotropy energy

$$E = \frac{\mu_0}{2} \mathbf{M} \cdot \mathbf{D} \cdot \mathbf{M} \tag{53}$$

per unit of volume, where $\mathbf{M}$ is the magnetization vector. The uniaxial shape anisotropy of S-W particles is a particular case corresponding to shapes that can

Fig. 49 Spherical plot of the cubic magnetocrystalline anisotropy energy of magnetite. The energy is minimal along the four ⟨111⟩ axes and maximal along the three ⟨100⟩ axes

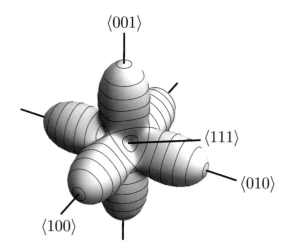

$\langle 001 \rangle$

$\langle 111 \rangle$

$\langle 010 \rangle$

$\langle 100 \rangle$

be modeled by a prolate rotation ellipsoid ($a > b = c$). In this case, the easy axis is parallel to the longest axis of the ellipsoid. Oblate ellipsoids with rotation symmetry ($a = b > c$) yield an easy plane, instead of an easy axis, in which the magnetic moment is unconstrained. Such particles do not possess a stable magnetization since infinitesimal fields can rotate the magnetic moment within the easy plane. The corresponding hysteresis loop is completely closed and discontinuous at $H = 0$. Triaxial ellipsoids possess an easy axis parallel to a, but the lack of rotation symmetry creates a preferred plane through the intermediate axis, in which the magnetic moment is rotated. Particles with pronounced uniaxiality, produced by preferential growth along one crystallographic direction, are well approximated by the prolate rotation ellipsoid of the Stoner-Wohlfarth model (section "Uniaxial SD Particles: Stoner-Wohlfarth Model") because dimensional differences between the minor axes and the contribution of magnetocrystalline anisotropy are negligible for elongations in excess of ~1.2.

On the other hand, the anisotropy of particles formed by a uniform growing mechanism is dictated by random differences between the principal axes. As a result, these particles are representable by a quasi-spherical ellipsoid whose principal axes lengths are random realizations of a single random variate representing the particle size. A large collection of such particles contains cases with arbitrarily small anisotropy, whose abundance is inversely related to the width of the size distribution. Accordingly, more equidimensional particles are obtained in case of well-constrained axes lengths. Because of the lack of a minimum uniaxial anisotropy threshold, particles can have a vanishing coercivity even without thermal relaxation. The corresponding FORC diagram shares the same properties of S-W particles—a central ridge and positive-negative contributions in the lower quadrant, antisymmetric about the $H_u = -H_c$ diagonal—but they are characterized by an exponential-like coercivity distribution peaking at $H_c = 0$ (Fig. 50a). As a consequence of the lack of a minimum coercivity, the hysteresis loop and the FORC measurements have discontinuous first derivatives at $H = 0$. The simulation

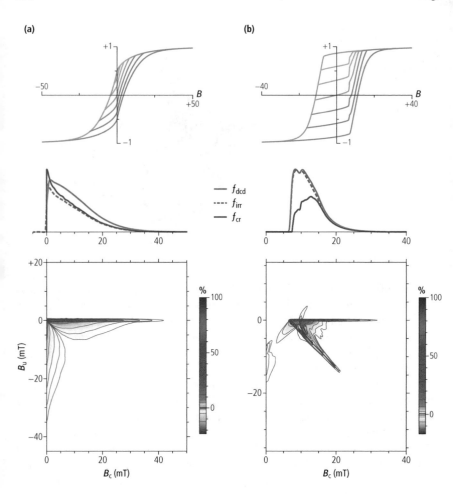

Fig. 50 Simulated magnetic properties of non-S-W SD particles with no thermal relaxation. Top: major hysteresis loop and simulated FORC measurements for selected reversal fields. Middle: three coercivity distributions obtained from FORC data subsets. Bottom: FORC diagrams with quantile contours at 2.5, 5, 10, 20, 30, 40, 50, 60, and 80% of the total magnetization. (a) One million randomly oriented, quasi-equidimensional magnetite-like particles ($\mu_s = 480$ kA/m) with no magnetocrystalline anisotropy and shape anisotropy corresponding to triaxial ellipsoid shapes with principal axes of length $a_i = a_0(1 + \delta a)$, where a_0 is an arbitrary diameter and δa the random realization of a Gaussian random variate with zero expectation and a standard deviation of 0.08. (b) Randomly oriented, perfectly equidimensional magnetite particles ($\mu_s = 480$ kA/m) with cubic magnetocrystalline anisotropy ($K_1 = -13.5$kJ/m^3, $K_2 = -2.8$kJ/m^3)

results of Fig. 50a are remarkably similar to the FORC diagram of dispersed equidimensional magnetosomes (Fig. 26); however, isolated magnetosomes are affected by thermal relaxation, as seen from the absence of the signature of reversible magnetic moment rotation in the lower quadrant.

Consider now magnetite particles with cubic magnetocrystalline anisotropy only. In this case, particles have up to four pairs of stable magnetization states, depending on the applied field, instead of one pair as with triaxial shape anisotropy. In a null external field, these states correspond to the magnetic moment being parallel or antiparallel to one of the four $\langle 111 \rangle$ crystal axes. If the crystals are randomly oriented, the maximum angle θ of the closest $\langle 111 \rangle$ axis to any given direction is comprised in a spherical triangle with sides limited by $\theta < \pi/4$ and $\theta < \arctan(\mathrm{sech}\theta)$, instead of $\theta < \pi/2$, as for uniaxial anisotropy [172]. This means that the saturation remanence is built by magnetic moments with a lesser angular scatter relative to the direction of the applied field, resulting in a remanence ratio $M_{\mathrm{rs}}/M_{\mathrm{s}} \approx 0.832$ instead of 0.5. The corresponding hysteresis loop is more squared than that of S-W particles (Fig. 50b). The maximum switching field is $B_{\mathrm{c}}^{\max} = 4|K_1|/3M_{\mathrm{s}} \approx 37.5$ mT [173], which corresponds to a coercivity of ~11 mT in the case of randomly oriented crystals. The additional easy axes introduce intermediary transitions between positive and negative saturation, which are marked by magnetization jumps occurring within a narrow field range around $+6.5$ mT (Fig. 50b). The difference in jump amplitude and position between consecutive curves produces a positive-negative doublet of diagonal ridges in the lower quadrant of the FORC diagram, which merge with the central ridge at $(B_{\mathrm{c}}, B_{\mathrm{u}}) \approx (7, 0)$ mT. After jumping to an intermediate easy axis, the magnetic moment vector rotates further until the final jump to the easy axis closest to the positive field direction occurs, contributing to the central ridge. The spaces between the ridges merging at $(B_{\mathrm{c}}, B_{\mathrm{u}}) \approx (7, 0)$ mT, and left of the diagonal ridge contain positive and negative contributions from reversible magnetic moment rotation.

A slightly less complex signature is obtained in the case of particles with a distribution of cubic anisotropy constants, due, for instance, to cation substitution [174]. In this case, the central ridges of particles with different anisotropy constants add constructively, while replicas of the diagonal ridge doublet departing at different values of H_{c} tend to be smoothed out, leaving a negative contribution below the low-coercivity end of the central ridge. Even so, the FORC signature of cubic anisotropy is easily overshadowed by shape anisotropy. Numerical simulations of nearly equidimensional magnetite particles with mixed anisotropy and random reciprocal orientations of shape and crystalline axes, which are representative of irregular shapes with no preferential growth directions, display a characteristic FORC signature that is not a simple superposition of the two cases discussed above (Fig. 51). This distinct signature is caused by complex energy landscapes created by the sum of the energy contributions in Eqs. (52) and (53).

The central ridge f_{cr} does not feature a minimum coercivity threshold, as it might be expected from magnetocrystalline anisotropy, nor does it peak at $H_{\mathrm{c}} = 0$ as it would be expected from the lack of a minimum uniaxial anisotropy. Instead, a Gaussian-like coercivity distribution is obtained, which extends between $H_{\mathrm{c}} = 0$, where $f_{\mathrm{cr}}(0) = 0$, and a maximum coercivity value determined by the maximum ratio a/c between the longest and the shortest ellipsoid axes. The lack of zero-field switching events imposed by $f_{\mathrm{cr}}(0) = 0$ eliminates the slope discontinuity of magnetization curves at $H = 0$, which characterizes SD particles with the same shape but no magnetocrystalline anisotropy. The characteristics of the other two coercivity distributions are affected by irreversible transitions between magnetic

Fig. 51 Simulated magnetic properties of nearly equidimensional magnetite particles with no thermal relaxation and coexisting shape and magnetocrystalline anisotropies with same characteristics as in Fig. 50. Top: major hysteresis loop and simulated FORC measurements for selected reversal fields. Middle: three coercivity distributions obtained from FORC data subsets. Bottom: FORC diagrams with quantile contours at 2.5, 5, 10, 20, 30, 40, 50, 60, and 80% of the total magnetization. The dashed lines represent the left limit of the memory region

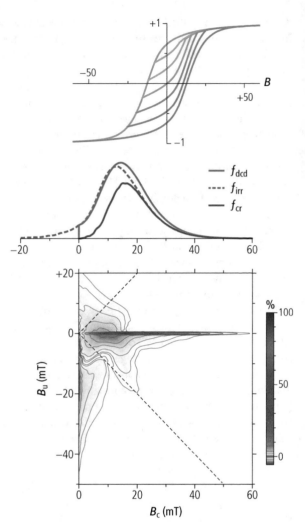

states occurring already before the field direction is reversed, as seen by the fact that f_{irr}, the coercivity distribution associated with the irreversible component of the ascending hysteresis branch (section "Selected Properties of the FORC function"), extends to negative fields. Interestingly, such transitions require the presence of both magnetocrystalline and shape anisotropy since they are not observed from the separated simulations of Fig. 50. These transitions can be explained by the existence of magnetic states with no zero-field stability, caused by the complex energy landscape resulting from the superposition of two anisotropy sources with different symmetry. As far as the saturation remanence is concerned, part of it is unstable and can be switched in arbitrarily small fields, as seen from the fact that $f_{dcd} > 0$.

Magnetic states without stable remanence are associated with positive amplitudes of the FORC function outside of the memory region, especially in the upper quadrant (Fig. 51). These states represent the extension of a continuous contribution located around the low-coercivity end of the central ridge. In the upper quadrant, the FORC function is sharply terminated at $H_c \approx 18$ mT, while a wider distribution of H_c values is covered in the lower quadrant, up to ~40 mT. The extension of the central ridge beyond this limit is caused by few particles whose uniaxial shape anisotropy is large enough to suppress other magnetic states.

Thermal activations are expected to eliminate magnetic states with smallest energy barriers, such as those associated with the flat energy landscape resulting from certain superpositions of shape and magnetocrystalline anisotropy in non-elongated, irregularly shaped SD particles. The resulting FORC diagram should contain less non-central ridge contributions; however, their preferential occurrence around the low-coercivity tail of the central ridge should not change. Indeed, remarkably similar FORC signatures have been found in ferromanganese nodules from the Western Pacific Ocean containing equidimensional (e.g., octahedral-like) magnetite crystals with a median size of ~60 nm [175]. The central ridge peak of these nodules at <20 mT suggests that these SD crystals are not arranged in linear chains, as expected for preserved magnetofossil structures. Non-central ridge contributions qualitatively similar to those of Fig. 51 might also be confused with those of magnetic particles hosting a single vortex (section "Two-State Magnetic Systems"), so that attention must be paid to details, such as the peak position of the central ridge, the shape of vertical profiles across the central ridge, and the field ranges covered by other FORC features.

Selected Natural Examples

Sediments and rapidly cooled rocks might contain abundant SD particles; however, grain sizes are rarely so well constrained to be entirely within the relatively narrow SD stability range [176–179], so that SD FORC signatures get unavoidably mixed with the contribution of particles with other domain states. The selection of cases with dominant SD contributions is therefore pivotal for understanding the signature of natural SD particle assemblages with different origins. Physical [180–182] and chemical [103, 183] separation techniques can be used to obtain SD endmembers that can serve as reference for numerical unmixing techniques, such as coercivity analysis and FORC-PCA.

Lithogenic SD magnetite is rarely so well dispersed to display the theoretical features of S-W particles discussed in previous sections. A notable exception is represented by some rapidly cooled glasses, such as the Tiva Canyon Tuff on Yucca Mountain (Nevada), which contains Ti-substituted magnetite needles with average lengths increasing from 15 nm (mostly SP) at the base of the sequence to >50 nm at the top (stable SD) [184]. The corresponding FORC diagrams [117] feature a central ridge and, depending on the degree of magnetic viscosity, the thermal relaxation features discussed in section "Viscous SD Particles and the Vertical Ridge". In most cases, the central ridge represents a minor or negligible contribution to the total magnetization, even when contributions from larger PSD particles are practically absent, like for the silicate-hosted magnetic minerals [180] shown in Fig. 52a. In this case, contour lines are slightly teardrop-shaped, as expected for a heterogeneous assemblage of weakly interacting SD particles (section "Weak Magnetostatic Interactions Between SD Particles"), where individual silicate grains might host ferrimagnetic inclusions with different mean coercivities and concentrations. Even fine-grained lithogenic inputs, such as those isolated from a pelagic carbonate through selective dissolution of magnetofossils [103], contain a distribution of grain sizes that is sufficiently wide to include non-SD particles (Fig. 52b), as seen from traces of a typical PSD signature (section "Multistate Systems: The PSD FORC Signature") outlined by low-amplitude contour lines. In this example, contour lines over the upper quadrant show a continuous transition from a teardrop shape for large amplitudes near the central ridge, which might be attributed to magnetostatic interactions, to divergent, triangular-shaped low-amplitude contours which are typical of PSD particles. Because of the lack of a precise control of the aspect ratio, lithogenic SD magnetite particles display a wide range of coercivities, from $H_c = 0$ to >100 mT. Corresponding coercivity components are therefore characterized by dispersion parameters of 0.35–0.5, well above the 0.1–0.3 range of coercivity components associated with biogenic or authigenic SD magnetite [84].

Natural high-coercivity minerals, such as pyrrhotite (Fe_7S_8, $\mu_s \approx 80$ kA/m), hematite (α-Fe_2O_3, $\mu_s \approx 2.5$ kA/m), goethite (α-FeOOH, $\mu_s \approx 2$ kA/m), and epsilon-maghemite (ε-Fe_3O_4, $\mu_s \approx 80$ kA/m), possess a much smaller spontaneous magnetization than magnetite, so that magnetostatic interactions are proportionally reduced. FORC measurements of SD pyrrhotite, hematite, and goethite are generally dominated by a central ridge that might extend to fields >0.5 T [111, 113], well above the theoretical upper limit of 0.3 T for magnetite, if not excessively lowered by thermal relaxations. Sediments often contain variable amounts of secondary hematite or goethite particles, which contribute to a low-amplitude central ridge that begins to emerge near the high-field limit of regular FORC measurements (usually 0.12–0.15 T) and continues to much higher fields (Fig. 52c). These high-coercivity SD contributions are not easily detectable with FORC measurements for three main reasons: (1) the ~200 times smaller spontaneous magnetization of the associated minerals and their wider coercivity distributions result in much lower amplitudes of the FORC function, (2) high-field FORC measurements need to be performed at a lower resolution to keep the total measurement time within reasonable limits, and (3) measurement noise tends to increase at large fields. Therefore, high-coercivity minerals remain mostly undetected in conventional FORC diagrams. Other experimental protocols are better suited to the characterization of these minerals: one of them is based on logarithmically spaced remanence measurements of the Preisach function and has been used to resolve the central ridge of exsolved hematite-ilmenite systems [18].

Secondary magnetite particles formed in sedimentary environments are gaining increasing attention because of their importance as paleoenvironmental indicators [32] and as magnetic carriers in continuous records of past variations of the Earth magnetic field [186, 187]. Secondary magnetite might precipitate inorganically

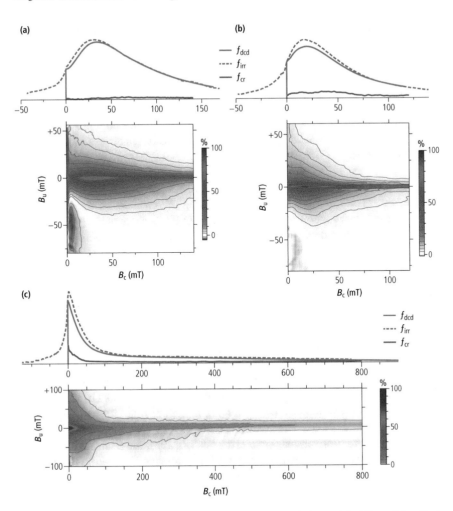

Fig. 52 FORC diagram examples of lithogenic fractions containing mainly SD particles. Each panel shows three coercivity distributions obtained from FORC measurement subsets (top) and the FORC diagram (bottom). (**a**) Silicate-hosted magnetic minerals separated from bulk sediment from the Northern Pacific Ocean, ~100 km offshore from central Japan, using a Frantz isodynamic magnetic separator (data from Chang et al. [180]). Contours are drawn at 5% and 10–90% quantiles in steps of 10%. (**b**) Pelagic carbonate sediment from the Equatorial Pacific, from which free-standing iron oxide crystals <1 μm in size (mainly magnetofossils) have been selectively dissolved with a citrate-bicarbonate-dithionite treatment (data from Ludwig et al. [103]). Magnetic particles surviving the chemical treatment are larger than the upper SD size limit and/or embedded into silicate minerals. Contours are drawn at 10–90% quantiles in steps of 10%. (**c**) Fluvial sediment containing a large fraction of high-coercivity minerals, probably a mixture of hematite and goethite (data from Scheidt et al. [185]). Contours are drawn at 20–80% quantiles in steps of 20%

[188, 189], often through redox reactions promoted by dissimilatory metal-reducing bacteria [190], or it is produced intracellularly by magnetotactic bacteria [191]. The tendency of SD magnetite synthesized in aqueous solution to form clumps [170], the occurrence of magnetite particle clusters around dissimilatory metal-reducing bacteria [190] and on the surface of primary minerals being reduced [192], and the once-believed widespread reductive dissolution of magnetosomes produced by magnetotactic bacteria [193, 194] led originally to the common belief that secondary magnetite would display the classical Preisach signature of magnetostatic interactions. Early FORC measurements, which were based on relatively large field steps of the order of 2–5 mT, apparently supported the interacting nature of sedimentary SD signatures [16], because processing-related smoothing effects (see section "The Central Ridge") were masking the presence of a central ridge. Only after the introduction of high-resolution FORC measurements [94] and a re-evaluation of existing data [106] did it become evident that many sediments contain a central ridge with a negligibly small intrinsic width. This ridge is produced only by SD particles that are well dispersed in the sediment matrix, either in isolated form or as magnetosome chains representing the fossil remains of magnetotactic bacteria. The minimum distance between isolated particles or magnetosome chains imposed by the sharpness of the central ridge is of the order ~9 times their length [103].

FORC measurements of the bulk material cannot discriminate between primary and secondary SD magnetite crystals; however, measurements of the same material before and after selective dissolution of ultrafine iron oxides reveal that the central ridge is mostly associated with particles directly dispersed in the sediment matrix, rather than lithogenic magnetite inside insoluble silicate minerals [103]. While the bacterial origin of SD magnetite contributing to central ridge in many freshwater and marine sediments has been widely confirmed by the direct observation of abundant magnetofossils in magnetic extracts [106, 195], the contribution of other formation pathways, in particular dissimilatory iron reduction [196], is still uncertain. Soils provide a useful comparison term, because the concentration of viable magnetotactic bacteria in pedogenic horizons is too low to explain the observed SD magnetite concentrations [197]. Early FORC measurements of fossilized soils (paleosols) from North America [101] hinted to the presence of a central ridge with properties similar to models of non-interacting viscous particles [87, 116]. Modern high-resolution measurements (Fig. 53a) confirm these findings: the central ridge coercivity distribution f_{cr} peaks at $H_c < 20$ mT, as expected for the irregularly shaped, nearly equidimensional magnetite particles extracted from soils [199], and is compatible with the pedogenic coercivity component "P" of fossil and modern soils obtained with numerical unmixing methods [84, 198]. The FORC signature remaining after subtraction of the central ridge is typical of larger lithogenic magnetite particles and coincides with that of the parent material on which the soil is formed [101].

The central ridge of typical magnetofossil-rich sediments (Fig. 53b) peaks at 30–40 mT, and its coercivity distribution contains two narrow coercivity components [94, 103], originally called "biogenic soft" (BS) and "biogenic hard" (BH), which can be associated to chains of equidimensional and elongated magnetosomes,

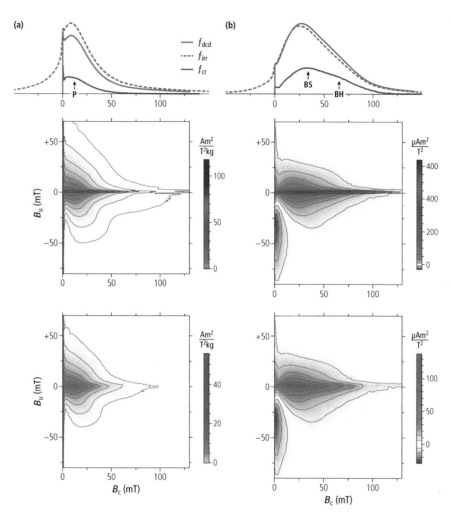

Fig. 53 FORC examples of sediments containing significant amounts of secondary SD magnetite particles. (**a**) Well-developed Paleosol S1 from the Chinese Loess Plateau (Spassov et al. [198]) (specimen courtesy of S. Spassov). (**b**) Magnetofossil-rich sediment from Lake Baldeggersee [84]. Top row: three coercivity distributions obtained from high-resolution FORC measurement subsets. Arrows point to the coercivity distribution peaks of the magnetic components "pedogenic magnetite P" [84, 198], "biogenic soft BS," and "biogenic hard BH" [84] contributing to the central ridge. Middle row: FORC diagrams. Bottom row: FORC diagram after removing the central ridge. Same contour levels (at 10–90% quantiles of the total magnetization in steps of 10%), same color scale, and same field scales have been used for ease of comparison

respectively [84, 200–202]. Intact magnetosome chains are expected to possess a minimum level of uniaxial anisotropy, owing to the chain geometry [135, 202], which, together with the strict magnetosome size control within the stable SD size range [177, 178], excludes contributions at or close to $H_c = 0$. In this case, the

0–10 mT coercivity range must be controlled by isolated, nearly equidimensional magnetite particles similar to those found in soils. Unlike the case of soils, the non-central ridge signature of magnetofossil-rich sediments is not exclusively associated with lithogenic particles, as it contains contributions from partially collapsed and multi-stranded magnetosome chains [135, 202] (section "Two-State Magnetic Systems"). The two examples shown in Fig. 53 illustrate the importance of high-resolution FORC measurements and proper processing for a clear identification of the central ridge and its quantitative analysis.

The Pseudo-Single-Domain Magnetic Behavior

The magnetic properties of ferrimagnetic particles are strongly affected by their size, with well-defined limits represented by single-domain (SD) and multidomain (MD) crystals. Certain minerals, such as magnetite, possess a size range with intermediate properties, which has been called pseudo-single domain (PSD). This term has been originally introduced by Stacey [203, 204] to describe ~0.2–3 μm magnetite particles that are too large to be SD, yet they hold a relatively intense and very stable thermoremanent magnetization, in contrast to MD particles [31]. Because of the limited grain size range of stable SD magnetite (~30–100 nm), PSD particles represent the most common recorders of the Earth magnetic field in rocks.

When applied to magnetic properties, the term "PSD" empirically describes a continuous transition between SD and MD endmembers. This transition is captured, for instance, by the hysteresis shape parameters M_{rs}/M_s and H_{cr}/H_c used in the so-called Day diagram [34, 205] (Fig. 54). Rectangular hysteresis loops typical of SD particles have elevated remanence ratios ($M_{rs}/M_s \approx 0.5$ in the case of uniaxial anisotropy and random orientations) and coercivity ratios (H_{cr}/H_c) close to 1. MD particles, on the other hand, are characterized by almost closed hysteresis loops with $M_{rs}/M_s \ll 0.5$ and $H_{cr}/H_c \gg 1$. Bulk samples containing PSD particles in the ~0.2–3 μm size range feature intermediate properties, usually limited by $0.02 < M_{rs}/M_s < 0.5$ and $2 < H_{cr}/H_c < 5$. These limits form the so-called PSD range of the Day diagram (Fig. 54). The PSD range can also be populated by strongly interacting SD particles [211], as well as binary or ternary mixtures of SP, SD, and MD particles [212, 213]. It is therefore not surprising that most geologic materials plot within the PSD range, with lower and upper M_{rs}/M_s limits given by theoretical SD-MD and SD-SP mixing trends, respectively [205]. A similar ambiguity exists with any other domain state indicator based on bulk magnetic parameters [33, 35], so that the term "PSD behavior" instead of "PSD particles" is more appropriate for describing bulk magnetic measurements that do not unambiguously indicate a SD or a MD signature.

Despite the indefiniteness of PSD-like magnetic properties, there is a well-defined third magnetic state for magnetite particles with intermediate size ranges, which explains the PSD behavior originally described by Stacey. From the micro-magnetic point of view, ferrimagnetic materials can be classified with the help of two fundamental parameters: the *magnetic hardness*

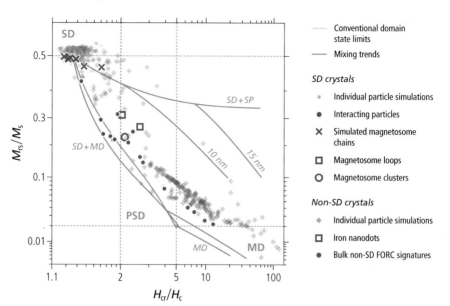

Fig. 54 Diagram after Day et al. [34], expressing the shape of hysteresis loops in terms of coercivity ratio H_{cr}/H_c and the remanence ratio M_{rs}/M_s (also known as squareness). Dashed lines represent conventional boundaries for SD, PSD, and MD magnetite particles, and solid lines represent mixing or grain size trends after Dunlop [205]. A pseudo-logarithmic scale has been used to equalize the size of the regions occupied by the three domain states. Filled dots represent measurements of bulk samples containing (1) SD particles with increasing degrees of magnetostatic interactions (data from Moskowitz et al. [206] and Li et al. [207]) and (2) magnetite particles with a PSD or MD FORC signature (data from sample V1 in Ludwig et al. [103]; core R41 in Lascu et al. [208]; sample Har7 426.5 in Church et al. [18]). Small diamonds represent micromagnetic simulations of individual silicate-hosted magnetite crystals whose exact shape has been determined using focused ion beam nanotomography (data from Nikolaisen et al. [209]). The domain states of these particles (SD or PSD) have been determined according to the presence or the lack of magnetic vortices in simulated hysteresis loops. Large symbols represent the bulk properties of particle assemblages with well-defined properties, which include, in the case of SD particles, micromagnetic simulations of magnetosome chains [201], measurements of closed loops of equidimensional magnetosomes produced by the $\Delta mamJ \Delta limJ \Delta$islet mutant of *M. magneticum* [201], and measurements of small clusters of equidimensional magnetosomes produced by the $\Delta mamJ$mutant of *M. gryphiswaldense* [129] and, in the case of single-vortex particles, measurements of metallic iron nanodots [210]

$$\kappa = \sqrt{\frac{|K|}{\mu_0 \mu_s^2}} \tag{54}$$

defined as the dimensionless ratio of magnetocrystalline anisotropy (K) to dipole ($\mu_0\mu_s^2$) energy and the *exchange length*

$$l_{ex} = \sqrt{\frac{A}{\mu_0 \mu_s^2}},\qquad(55)$$

which is based on the ratio of exchange constant A to the dipole energy [214, 215]. Magnetically soft materials, such as magnetite ($\kappa = 0.2$), are characterized by $\kappa < 1$. Relevant dimensions of magnetic structures can be expressed by these two quantities. Of particular importance here are the Bloch wall width $\delta_w = \pi l_{ex}/\kappa$, which is the typical width of a domain wall, and the maximum size $R_{sd} \approx 36 \kappa l_{ex}$ of a SD particle. If $R_{sd} > \delta_w$, increasing the size of a particle produces a direct transition from a homogeneously magnetized SD state to a MD configuration as soon as there is enough volume to accommodate a domain wall. If, on the contrary, $R_{sd} < \delta_w$, particles whose sizes are comprised between R_{sd} and δ_w are too large to be homogeneously magnetized, but too small to accommodate a domain wall. Instead, they will host a so-called magnetic vortex [216], in which the magnetization vectors form a circular pattern around a so-called vortex tube with axial magnetization (Fig. 55). The vortex configuration reduces the stray field energy of the particle by minimizing the area over which magnetization vectors are not parallel to the surface, compatibly with an increase of the exchange energy in the vortex tube, where the magnetization is strongly inhomogeneous. The formation of magnetic vortices is hindered by magnetocrystalline anisotropy, which tend to keep the magnetization direction parallel to one of the easy axes. Because the reduction in stray field energy is proportional to μ_s and the increase in magnetocrystalline anisotropy energy is proportional to $|K|$, the existence of vortices depends on κ, the ratio between these two quantities, with the $R_{sd} > \delta_w$ condition yielding $\kappa < \sqrt{\pi/36} \approx 0.3$. Magnetite particles ($\kappa = 0.2$) feature magnetic vortices over a range of sizes above R_{sd}. Sometimes, a few magnetic vortices coexist in particles with irregular shapes, especially elongated ones [209]. As the particle size increases, the volume outside vortex tubes tends to split into regions where the magnetization direction is sub-parallel to one magnetocrystalline easy axis. These regions can be effectively regarded as magnetic proto-domains separated by transition planes or proto-domain walls, where the magnetization vectors rotate from one easy axis to the other. These proto-domain structures progressively evolve into classic domains and domain walls, respectively, as the crystal size is further increased [216].

In analogy with SD particles, the orientation of the vortex tube is stabilized by magnetocrystalline anisotropy and by the particle shape. Therefore, the vortex tube, which is the only part contributing significantly to the total magnetic moment, behaves essentially like a SD particle, granting the magnetization stability sought in paleomagnetism [217]. On the other hand, the application of external fields favors the transition to a SD state, which might or might not continue to exist metastably once the field is removed. Depending on the stability of the SD state in zero field, particles can exhibit an elevated remanent magnetization typical of truly SD particles, expressed by M_{rs}/M_s values close to 0.5, or a much lower remanent magnetization arising from the vortex tubes, with M_{rs}/M_s values close to those of MD particles. Micromagnetic models of magnetite particles with real shapes

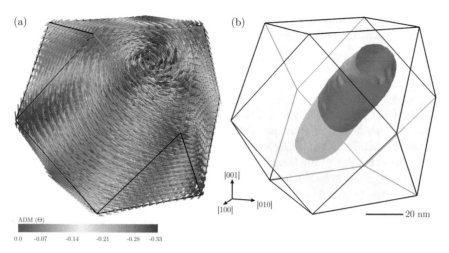

ADM (Θ)

0.0   -0.07   -0.14   -0.21   -0.28   -0.33

Fig. 55 Vortex magnetization in a 100 nm magnetite cuboctahedron. **(a)** Vectors are color-coded according to the angle formed with the crystallographic axes (red, parallel to the nearest ⟨100⟩ hard axis for the highest magnetocrystalline anisotropy energy density; blue, parallel to the nearest ⟨111⟩ easy axis for the lowest magnetocrystalline anisotropy energy density). The vortex core intersects the [111] plane highlighted by the yellow triangle. Notice how the vectors form small angles to the particle's faces, minimizing the stray field energy, except in proximity of the vortex core. **(b)** Helicity isosurface of the vortex core. The helicity $h = \mathbf{M} \cdot (\nabla \times \mathbf{M})$ is defined as the curl of the magnetization, projected onto itself. Far from the vortex core, vectors lie in planes perpendicular to the direction $\nabla \times \mathbf{M}$ of the vortex core, so that $h = 0$. The out-of-plane vector components near the vortex core give a non-zero helicity. Figure reproduced from Nagy et al. [216]

obtained from ion beam nanotomography reveal that individual crystals hosting one or two magnetic vortices are characterized by hysteresis loops that span all three domain states defined by the Day diagram (Fig. 54). Particles with metastable SD-like remanent magnetizations are characterized by H_{cr}/H_c values not exceeding ~3: above this threshold, most individual particles form a trend line defined by the inverse relationship $M_{rs}/M_s \approx 0.45 (H_{cr}/H_c)^{-1}$ between remanence and coercivity ratios. This trend is parallel to that of MD particles, which are characterized by ~8 times smaller remanence ratios.

Micromagnetic models show that ensembles of vortex particles can display SD or PSD bulk magnetic properties depending on their size and shape distribution and on the type of magnetization being considered. On the other hand, vortex particles possess a characteristic FORC signature, which will be discussed in section "Two-State Magnetic Systems", which is fundamentally distinct from that of SD and MD particles. These observations led to the recent proposal to replace the "PSD behavior" term with "vortex state", which would have the advantage of maintaining the same universality while referring to a physical description of the magnetization [104]. This proposal has its own caveats, because, as shown in section "Two-State Magnetic Systems", the magnetic signature of true vortex state particles can be indistinguishable from that of specific magnetic structures made of interacting SD particles. In these cases, the common feature captured by FORC diagrams is the existence of magnetic systems characterized by two or more pairs of states with different magnetic moments and stabilities in small and large fields, respectively.

Two-State Magnetic Systems

Uniaxial SD particles possess a single pair of stable magnetic states with nearly homogeneous magnetization: parallel and antiparallel to the easy axis (section "Uniaxial SD Particles: Stoner-Wohlfarth Model"). The corresponding FORC signature is relatively simple and dominated by the central ridge, which marks the abrupt transitions between these two states as the field is increased. Additional pairs of magnetic states are introduced by multiaxial anisotropy (section "SD Particles with Multiaxial Anisotropy"). In symmetric cases, such as that of cubic magnetocrystalline anisotropy, all states possess the same magnetic moment and the same intrinsic stability in external fields. SD particles with symmetric multiaxial anisotropy can therefore be considered *one-state systems*, if magnetic configurations with identical geometry are considered expressions of the same state. The overlap of anisotropy sources with different symmetries, such as cubic magnetocrystalline and triaxial shape, creates the next level of complexity, resulting in the FORC signature of Fig. 51, which bears some similarities with that of PSD particles. PSD signatures thus appear to be generated by magnetic systems featuring *unequal* magnetic states.

The simplest example of unequal states is given by *two-state systems* possessing two types of states with different magnetic moments and field stabilities. Single-vortex (SV) particles such as the one shown in Fig. 55 are typical examples of a two-state system, featuring one or more pairs of high-moment states with a nearly homogeneous magnetization and one or more pairs of low-moment states with a vortex configuration. The high-moment, SD-like states tend to be more stable in large external fields, because they lower the Zeeman energy, while the low-moment states are more stable in a null field, because they minimize the stray field energy. Magnetic states occur always in pairs characterized by exactly antiparallel magnetization vectors.

If a two-state system is subjected to a varying external field, transitions will occur between states according to their field stability. Figure 56 shows three examples of two-state systems with a very similar FORC signature, consisting of a central ridge with one or two small negative lobes immediately above and below it and two large positive lobes placed symmetrically above and below its lower-coercivity end. The color scale rendering of the positive lobes depends on the maximum amplitude of the central ridge, which, as discussed in section "The Central Ridge", is controlled by the measurement resolution and by the smoothing factor used for processing. Quantile contour lines, which express the relative amount of magnetization carried by larger FORC amplitudes (section "FORC Data Rendering"), are similar in all three examples and thus describe the same type of signature.

The first example is a micromagnetic simulation of randomly oriented, double-stranded chains of equidimensional SD particles [201] (Fig. 56a). Double-stranded chains represent one of the possible structures of fossil remains of magnetotactic

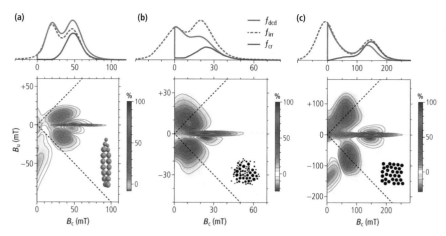

Fig. 56 Magnetic properties of three magnetic systems possessing two pairs of stable magnetic states with high and low magnetic moment, respectively. The top row shows the coercivity distributions derived from FORC data subsets; the bottom row shows the FORC diagrams, with dashed lines indicating the limits of the memory region. (**a**) Numerical model of randomly oriented, double-stranded chains of equidimensional magnetosomes formed from the fold collapse of single-stranded chains (an example is shown in the inset, data from Amor et al. [201]). Contour levels correspond to 5, 10, 20, 35, 50, 70, and 80% quantiles of the total magnetization. (**b**) Sample of a $\Delta mamJ$ mutant of *M. gryphiswaldense* forming magnetosome clusters (an example is shown in the inset, data from Katzmann et al. [129]) instead of linear chains (data from Katzmann et al. [129]). Contour levels correspond to 5, 10, 20, 35, 60, and 80% quantiles of the total magnetization. (**c**) Metallic iron nanodots with 67 nm diameter, deposited on a planar substrate (inset, data from Dumas et al. [210]) and measured in the plane. Contour levels correspond to 5, 10, 20, 40, 60, and 80% quantiles of the total magnetization. In (a-b) individual particles are SD, and the existence of low- and high-moment states depends on the magnetic moment arrangement. In (c) individual particles possess SD-like and vortex states

bacteria in sediments. In null or weak fields, the strong uniaxial anisotropy of these double-stranded chains forces the magnetic moments of all particles to be parallel or antiparallel to the chain axis with two possible combinations: in the more stable low-moment state, the magnetic moment of one strand is antiparallel to that of the other strand, and the magnetic flux lines form a closed loop [201]. The net magnetic moment is proportional to the difference between the number of particles in the two strands, with equal strands yielding a null moment. The high-moment state is obtained when the magnetic moments of the two strands are parallel or sub-parallel to each other, a condition that is always satisfied in a sufficiently large external field. The stability of the high-moment state in zero field depends on the exact chain geometry. In the example of Fig. 56a, the high-moment state is stable in zero field, yielding SD-like hysteresis characteristics (corresponding to the rightmost cross of the "simulated magnetosome chain" series in Fig. 54). SD-like hysteresis characteristics can also be deduced from the fact that all irreversible components of the FORC function—in this example everything except the negative amplitudes near $H_c = 0$ in the lower quadrant—are comprised within the memory region (right of the dashed lines in Fig. 56a). In particular, the lack of large FORC contributions outside the memory region means that the ascending branch of the hysteresis loop remains reversible until the field becomes positive, so that the associated coercivity distribution f_{irr} is nearly zero for negative arguments.

The example of Fig. 56b is based again on a system of interacting SD particles. This time the SD particles, which are produced by a magnetotactic bacterium mutant whose genes responsible for a linear chain arrangement of the magnetosomes have been inactivated, are arranged into small, mostly elongated clusters of 10–30 particles [129]. The magnetic configurations of these structures have not been modeled; however, the measured FORC signature is similar to that of the previous example, except for the two positive lobes above and below the central ridge extending across the memory region (dashed lines) instead of being fully contained in it. This means that most high-moment states decay into a low-moment state before the field is zeroed, lowering M_{rs}/M_s to values typical for the PSD range ("magnetosome clusters" in Fig. 54). In this example, low-moment states must be associated with the magnetic moments of individual particles forming some flux closure patterns similar to those observed in other interacting SD particle systems [218]. Finally, the example of Fig. 56c shows the signature of SV iron nanoparticles. The particles are closely spaced, like in the other examples; however, the low-moment state is associated with magnetic vortices within individual crystals, like in Fig. 55, rather than flux closure patterns. Similar preparations with slightly smaller crystals feature the classic FORC signature of SD particles [210], thus excluding that the FORC signature of Fig. 56c is caused by flux-closure pattern like in the previous two examples. The positive lobes above and below the central ridge extend across the memory region, lowering M_{rs}/M_s in a similar manner as for the magnetosome clusters. Unlike the other two examples of Fig. 56, there is a clear asymmetry between the two positive lobes, which is caused by reversible vortex movements [137].

Other types of two-state magnetic systems have been described in the literature, including 0.75–1.5 μm magnetite crystals with monoclinic twinning below the Verwey transition temperature [219] and toroidal iron oxide nanoparticles [220]. In the former case, the strong uniaxial magnetocrystalline anisotropy of monoclinic magnetite at low temperature prevents the formation of true vortex states ($\kappa \approx 0.94$), and the observed FORC signatures, similar to the examples in Fig. 56, are attributable to the interaction between twin domains in the same particle. The only common characteristic of all examples discussed so far, which include interacting SD particles, true single vortex particles, and peculiar crystal shapes, is the existence of two types of magnetic states with small and large magnetic moments, respectively, rather than a specific magnetization geometry. Because of the inherent multiplicity of magnetic configurations leading to two-state or multiple-state systems, identification of the corresponding magnetic signatures with "vortex particles" or "vortex behavior" requires additional information, e.g., from electron microscopy.

The FORC signature of two-state magnetic systems is pivotal for understanding the magnetic properties of a wider class of magnetic systems (e.g., single particles or particle aggregates) possessing two or more than two types of magnetic states, which collectively contribute to the observed PSD signatures. Because the above mentioned two-state systems share similar FORC signatures, the numerical simulation of two-stranded magnetosome chains (Fig. 56a) will be used as a reference example for further discussion, because it gives access to individual magnetic states contributing to the bulk magnetization. Hysteresis loops are constricted (or "wasp-waisted") near $M = 0$ (Fig. 57a), a feature that is shared by all systems with bimodal coercivity distributions [136, 221] such as those shown in Fig. 56. The constricted loop shape is determined by the clustering of individual magnetic moment values close to positive/negative saturation (high-moment states, HMS) and near $M = 0$ (low-moment states, LMS). Accordingly, the first steep section of a major hysteresis branch corresponds to HMS-to-LMS transitions, followed by a less steep section associated with reversible LMS magnetization changes and a second steep section due to LMS-to-HMS transitions.

These transitions are best viewed using the simulated FORC measurements of a single two-stranded chain example taken randomly from the original set of 10^5 chains used to calculate the bulk properties (Fig. 57b). All FORCs in this chain example can be divided into three groups of overlapping curves (labeled as A, B, and C), depending on their starting (reversal) field. Group A includes all curves M_A beginning with the positive HMS and coincides with the descending branch of the major hysteresis loop. Because the FORC function definition is based on the first derivative of these curves with respect to the reversal field H_r, and thus on differences between consecutive curves, the identical curves of group A do not produce any contribution in the FORC diagram. As H_r is decreased in successive steps, a critical point in the descending branch is reached, where the positive HMS state is no longer stable, so that a sudden transition occurs to the positive LMS state. Curves starting with this state form group B and share identical magnetizations M_B. As the applied field is increased during the measurement of M_B, the positive LMS state evolves reversibly, until a critical positive field is reached, above which it is no longer stable. At this point, a sudden transition to the HMS state occurs, producing a magnetization jump labeled with 1 in Fig. 57b.

The difference $M_A - M_B$ between the last curve of group A and the first curve of group B contributes to the FORC diagram along the diagonal trajectory starting at the reversal field that caused the HMS-to-LMS transition (A-B in Fig. 57c). This contribution is given by the first derivative of $M_A - M_B$ with respect to the applied field. It contains small contributions from reversible magnetization changes of the two states, as well as a sharp peak in correspondence with the magnetization jump 1 of M_B caused by the denucleation of the LMS state. The FORC coordinates of this peak, which contributes to the upper positive lobe, depend on the reversal field in which the positive LMS nucleates and the measurement field in which the same LMS denucleates. The two fields have different amplitudes, owing to the different stabilities of the two states, resulting in a peak with positive H_u coordinate in the FORC diagram. In practice, M_A and M_B can be considered as the two branches of a minor hysteresis loop with a positive field offset coinciding with H_u. All the successive curves of group B do not contribute to the FORC diagram, since the corresponding magnetizations are identical.

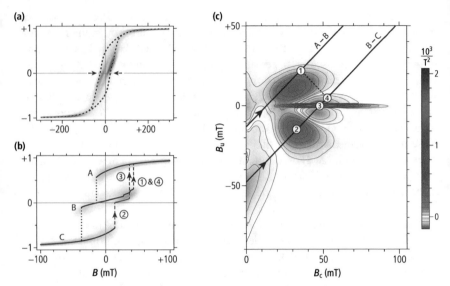

Fig. 57 FORC signature of two-state systems, exemplified by a numerical model of 10^5 randomly oriented, double-stranded chains of equidimensional magnetosomes (data from Amor et al. [201]). **(a)** Hysteresis loop (dashed line) and probability of individual double-stranded chains to possess the combination of applied field and magnetic moment given by the plot coordinates (blue shading). Darker regions highlight the pair of high-moment states (close to ± 1) and the pair of low-moment states (close to the flanks of the hysteresis loop). Arrows point to the constriction of the hysteresis loop resulting from these two pairs of states. **(b)** Simulated FORC measurements for a single double-stranded chain example taken from the collection of 10^5 simulations (lines). The chain axis forms an angle of ~38.5° with the applied field direction. All curves collapse onto three groups with identical magnetic moments, labeled as A, B, and C. The blue shading represents the magnetic state probability as in **(a)**, this time only for chain axes at angles comprised between 28.5 and 48.5°. Dotted lines indicate the transitions between low- and high-moment states along the descending branch of the hysteresis loop. Dashed lines indicate transitions along FORCs, with circled numbers referring to their contribution to the FORC diagram in **(c)**. **(c)** FORC diagram obtained from the numerical simulations. The sketched diagonal lines show the trajectories of the last curve in group A and the last curve in group B of the example chain shown in **(b)**. Magnetization jumps at the locations indicated by circled numbers represent pointwise contributions of the example chain in **(b)** to the FORC function

As H_r is further decreased, a critical point in the descending branch is reached, where the positive LMS state is no longer stable, so that a sudden transition occurs to the negative HMS state. Curves starting after this point belong to group C and coincide with the lower branch of the major hysteresis loop. Two transitions occur during the measurement of M_C: when the negative LMS is nucleated (magnetization jump 2 in Fig. 57c) and when this state becomes unstable and is replaced by the positive HMS (magnetization jump 3). The difference $M_B - M_C$ between the last curve of group B and the first curve of group C gives the second family of contributions to the FORC diagram along the diagonal trajectory starting at the reversal field that caused the LMS-to-HMS transition (B-C in Fig. 57c). Besides minor contributions from reversible magnetization changes along the two curves, the derivative of $M_B - M_C$ with respect to the applied field features two sharp positive peaks, corresponding to the magnetization jumps 2 and 3 of M_C, and a sharp negative peak corresponding to magnetization jump 1 of M_B (jump 4 in Fig. 57c). Magnetization jump 2 contributes to the lower positive lobe of the FORC diagram, owing to its negative H_u coordinate. Both lobes reflect the different field amplitudes required to nucleate and denucleate a LMS and thus represent the unique signature of this state.

Magnetization jump 3 is characterized by $H_u = 0$ and therefore contributes to the central ridge. The reason for H_u being exactly zero originates from inversion symmetry: a positive state of an isolated magnetic system can be transformed into a negative state with same properties by changing the sign of the magnetization vectors and of the applied field. In this case, the same physical process produces jump 3 and the transition from which group C originates namely, the denucleation of a LMS state. The corresponding fields are thus equal in amplitude but opposite in sign, since the positive LMS state is involved in one case and the symmetrically negative LMS state in the other. Finally, the magnetization jump 1 of M_B contributes with a peak on the B–C trajectory at the same applied field of its contribution to the A–B trajectory (dashed line in Fig. 57c). The contributions of jump 1 to the two trajectories have opposite signs, due to the fact that M_B appears with opposite signs in the corresponding magnetization differences.

Jumps 1 and 3 correspond to the denucleation of opposite LMS states, which are characterized by similar low magnetic moment components along the applied field. Therefore, the stabilities of these states in a positive field are similar, which means that the two jumps occur in similar fields. Whether the field required for jump 1 is larger than that of jump 2 or vice versa depends on micromagnetic details, so that, in most cases, the negative peak labeled as 4 in Fig. 57c is located indifferently above or below the central ridge, but close to it, generating two nearly symmetric negative lobes. Contrary to the central ridge, whose position at $H_u = 0$ is forced by the inversion symmetry, the positive and negative lobes are diffused features, due to the contribution of individual peaks 1, 2, and 4 at FORC coordinates determined by the LMS nucleation/denucleation fields of many different configurations. The remaining FORC diagram features, such as the negative contributions in the lower quadrant, close to $H_c = 0$, reminiscent of SD particles, are caused by reversible changes of the LMS and HMS states as the field is swept during measurements [104]. It is important to remark that all FORC features of Fig. 57c are explained by the mere existence of a LMS and a HMS state, without the need of a detailed micromagnetic knowledge.

Rock and Paleomagnetic Significance of Two-State Systems

The bimodal coercivity distribution and associated FORC signature of two-state systems are rarely observed in geologic materials, because of the overlap with other signatures, yet it is essential for understanding the characteristics of coercivity distributions obtained from different types of magnetization processes and their implications for numerical unmixing techniques such as coercivity analysis [40, 41, 86], hysteresis-PCA [43] and FORC-PCA [44].

The bimodal characteristics of the coercivity distributions f_{irr} and f_{dcd} of two-state systems (Fig. 56) are explainable with the contribution of the two positive lobes and the central ridge to a low- and the high-coercivity component, respectively. The peak position of the low-coercivity contribution depends essentially on the reversal field of the A–B trajectory, which is the nucleation field H_n of a LMS state. If the HMS state is particularly stable, as in Fig. 56a, the mean H_n is negative, and the positive lobes of the FORC diagram are located within the memory rectangle (dashed lines in Fig. 56). This yields a low-coercivity contribution to f_{irr} and f_{dcd} that peaks at the positive field $-H_n$. In the case of less stable HMS states (Fig. 56c), LMS nucleate, on average, in positive H_n values, which locate the positive lobes of the FORC diagram located outside the memory rectangle. The low-coercivity contribution to f_{irr} peaks at the negative field $-H_n$, and f_{dcd} contains only the positive tail of this contribution. In practice, f_{dcd} is similar to a truncated version of f_{irr} with largest values near $H_c = 0$. This means that a significant fraction of the saturation remanent magnetization can be remagnetized in arbitrarily small fields, especially in the case of systems with less stable HMS, such as in vortex particles. This instability is due to the fact that large ensembles will always contain cases where the LMS nucleation field is very close to zero.

The poor M_{rs} stability of some two-state systems, which is not necessarily caused by thermal relaxation as in the case of viscous SD particles, is apparently at odds with the commonly observed paleomagnetic stability of rocks containing PSD particles [204], if one does not consider that the nature of the thermoremanent (TRM) magnetization acquired by those rocks upon cooling in the Earth magnetic field is completely different from a saturation remanence. TRM and its room temperature analogue, the anhysteretic remanent magnetization (ARM), favor LMS states over HMS ones because they tend to possess lower magnetic energies [217, 222] in zero field (and in the small Earth magnetic field). As seen with the central ridge of two-state systems, larger magnetic fields are required to denucleate LMS states, which also means that larger alternating field (AF) peak values are required to demagnetize a TRM or and ARM than an IRM. This enhanced resistance against AF demagnetization is exploited by the so-called Lowrie-Fuller test [59], in which the shape of AF demagnetization curves is compared to assess the stability of a natural remanent magnetization. The selectivity of the central ridge toward the denucleation of SD states in SD particles and LMS states in larger particles, which are those preferentially contributing to the TRM, is at the basis of a FORC method for estimating the ancient magnetic field recorded by heat-sensitive rocks (especially extraterrestrial ones), designed to avoid the laboratory heating steps of conventional absolute paleointensity methods [29].

With the above considerations in mind, a main distinction can be made between (1) magnetization curves and/or coercivity distributions that capture all magnetic configurations of two-state systems and (2) those that are selective only to LMS states. The first group includes IRM acquisition curves, backfield or DCD demag-

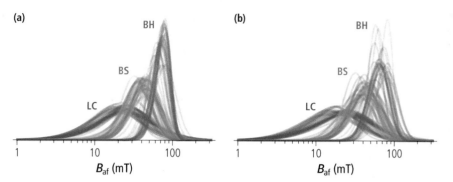

Fig. 58 Coercivity components obtained from the analysis of detailed AF demagnetization curves of ARM (**a**) and IRM (**b**) for different types of sedimentary samples (data from Egli [84]). LC—low-coercivity component (lake and marine sediments, Chinese loess, and paleosols) attributable to various sources, including lithogenic PSD particles, pedogenic magnetite, and other authigenic magnetite particles. BS—component attributed to chains of equidimensional magnetofossils ("biogenic soft," lake and marine sediments). BH—component attributed to chains of elongated magnetofossils ("biogenic hard," lake and marine sediments). All distributions are normalized to yield a unit magnetization, so that the peak value is a measure of their width. Notice how the components obtained from IRM curves are slightly wider (especially LC and BH) and less well constrained, despite the superior measurement quality associated with larger IRM moments

netization curves (f_{dcd}), irreversible magnetization changes along the hysteresis loop (f_{irr}), and the AF demagnetization of IRM. The second group includes the central ridge of FORC diagrams (f_{cr}) and AF demagnetization curves of weak-field magnetizations (ARM and TRM). The first group is preferentially used for coercivity analysis, because measurements are generally faster, but is also more problematic with respect to solution stability and interpretation, since it contains more coercivity components than the second group. A systematic comparison between AF demagnetization curves of ARM and IRM for magnetofossil-rich sediments shows that the coercivity components derived from IRM curves are wider, and therefore more overlapped, and centered on lower fields [84]. Most importantly, the properties of magnetofossil coercivity components derived from IRM curves are less consistent than their ARM counterparts (Fig. 58), as expected from the fact that magnetofossil magnetic states selected by the ARM magnetization process (SD in the case of single-stranded chains and LMS in the case of double- or multistranded chains) have similar mean denucleation fields [201].

Finally, two-state systems can introduce significant differences between coercivity analysis results obtained with parametric methods on the one hand, which are based on model curves, and PCA analyses on the other hand. These differences are either intrinsic, or they arise from problems [223] related to the solution stability of parametric methods and to incorrect processing (e.g., type and number of model functions, number of endmembers). Intrinsic differences are caused by specific assumptions underlying the two methods. Parametric methods assume that the coercivity distribution of a magnetic component, defined as an assemblage of particles with a certain distribution of physical and chemical properties derived from a common production process, is representable by a unimodal, quasi-lognormal function [41, 87]. In the case of PCA analyses, any combination of coercivity components that co-vary in a similar manner within a group of samples (e.g., from a sediment core) can define a PCA endmember, so that features produced by a single magnetic component can get distributed among endmembers if this component does not possess strictly invariant properties. This "coercivity leakage" problem is particularly pronounced in the case of magnetic components with two-state properties, because they tend to possess bimodal coercivity distributions with low- and high-coercivity proportions that are very sensitive to physical parameters such as grain size in the case of vortex particles [224] or particle arrangement in the case of SD particle systems [158, 160, 201] (Fig. 56). Small variations of these parameters cause a decoupling of low- and high-coercivity contributions in PCA analysis and possible grouping of these contributions with those of other magnetic components. In this case, coercivity components and endmembers no longer represent discrete groups of magnetic particles.

A paradigmatic example is given by FORC-PCA analyses of magnetofossil-bearing sediments [208, 225, 226]. Variable proportions of one-state single-stranded chains and two-state multistranded or collapsed chains can split the magnetofossil signature into two endmembers that do not correspond to specific chain geometries: for instance, the central ridge of both types of chains can contribute to one end-member, and the FORC lobes of two-state chains to another endmember, possibly together with the PSD signature of lithogenic particles, if these co-vary with the magnetofossil content, as it is the case for certain pelagic environments [227]. This might explain why some FORC endmembers interpreted as magnetofossil contain the signature of two-state chains (e.g., EM2 in Figs. 8 and 10 of Lascu et al. [208]), while others do not (e.g., EM2 in Fig. 11 of [225] or EM3 in Fig. 4 of [226]). In the latter case, two-state chains are either absent, or their non-central ridge contribution is contained in a different endmember.

The abovementioned ambiguities of numerical unmixing methods, which are inherent to the non-uniqueness of all magnetic characterization methods, need particular attention in the case of magnetic systems possessing few magnetic states, in particular two-state ones, because of the discreteness of these states and their sensitivity to physical parameters that are easily affected by the environment.

Multistate Systems: The PSD FORC Signature

FORC signatures of increasing complexity are obtained for systems with more than two stable magnetization states. Such systems can be very different at a microscopic level; common examples occurring in geologic materials include isolated particles hosting one or more magnetic vortices that interact with surface irregularities and crystal defects [209], particles containing few magnetic domains with strong shape-induced imbalance [228], skeletal titanomagnetites [221], and small clusters of interacting SD particles [129, 157]. Multistate systems exhibit a PSD behavior if

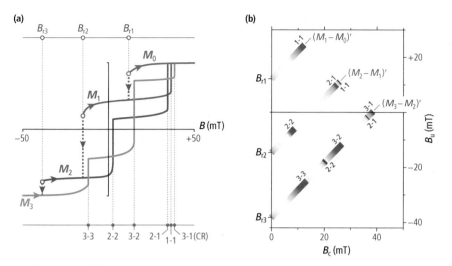

Fig. 59 Micromagnetic FORC simulation of a system featuring three pairs of stable magnetic states (in this case a linear chain of 7 slightly elongated SD magnetite particles with long axes perpendicular to the chain axis, which forms an angle of 75° with the applied field, data from Egli and Winkelhofer [96]). (**a**) Simulated measurements. M_0 to M_3 are the magnetizations of four groups of fully overlapped curves starting at reversal fields $\geq B_{r1}$, $\geq B_{r2}$, $\geq B_{r3}$, and $<B_{r3}$, respectively. (**b**) FORC diagram corresponding to the curves in (a). Non-zero contributions are localized along diagonals starting at fields B_{r1} to B_{r3} and are given by the first derivative of the difference between consecutive curve groups. Number pairs k-i along the diagonal indicate positive (violet) and negative (blue) contributions by the i-th magnetization jump of M_k, with curve and jump numbering as in (a). Color shading along the diagonals indicate the amplitude of the FORC function (zero in white, positive in yellow and red, negative in blue)

consecutive state transitions during a field sweep involve a major reorganization of the magnetization, as opposed to MD particles, where such transitions coincide with microscopic domain wall movements (Barkhausen jumps). The universality of the PSD behavior enables us to discuss all relevant aspects of the corresponding FORC signature using a single example, which is taken from a micromagnetic simulation of a collapsed chain of magnetosomes previously used for a similar purpose [96] (Fig. 59).

The major hysteresis loop of a typical PSD system consists of reversible sections separated by a certain number N of irreversible transitions between stable states (dashed lines with $N = 3$ in Fig. 59a). These transitions divide the FORCs departing from the descending branch of the hysteresis loop into $N + 1$ groups of fully overlapped curves, whose magnetizations are denoted by M_0, M_1, $\ldots$, M_N, respectively. The first group, M_0, includes all curves starting before the first transition on the descending branch, that is, in fields $\geq B_{r1}$. The second group, M_1, includes curves starting at reversal fields comprised between B_{r1} and the field B_{r2} of the second transition on the descending branch, and so on, until the last group M_N is reached, which includes all curves starting after the last transition on the descending branch. This last group coincides with the ascending branch of the major hysteresis loop. Overlapping curves belonging to the same group are characterized by $\partial M/\partial B_r = 0$, since M is independent of B_r, and do therefore not contribute to the FORC function. The only non-zero contributions to the FORC function originate from the difference between curves starting immediately before and immediately after an irreversible transition on the descending branch. In this case, the k-th transition yields $\partial M/\partial B_r = (M_{k-1} - M_k)\,\delta(B_r - B_{r,k})$, where δ is the Dirac impulse, and the FORC function can be written as

$$\rho(B_r, B) = \frac{1}{2}\delta\left(B_r - B_{r,k}\right)\ \frac{\partial(M_k - M_{k-1})}{\partial B}. \tag{56}$$

In practice, non-zero FORC contributions are localized along diagonals starting at $B_c = 0$ and $B_u = B_{r,k}$ on the left edge of the FORC space (Fig. 59b). These diagonals contain two types of contributions: reversible ones, which are given by the different slopes of M_{k-1} and M_k, and irreversible ones, which are concentrated at places where M_{k-1} or M_k undergo a transition between states. Transitions between states are almost always associated with magnetization jumps: these jumps are labeled with number pairs in Fig. 59, so that k-$i$denotes the i-th jump on M_k, with counting starting from the jump occurring at the largest field. Every jump contributes with a sharp peak to the FORC diagram (Fig. 59b), and these peaks are preceded by tails caused by reversible magnetization changes that usually anticipate a state transition. Peaks produced by magnetization jumps occurring on all M_k except the first and the last one appear twice in the FORC diagram: one time along the diagonal starting at $B_u = B_{r,k}$, with positive sign for a positive jump (such as those of Fig. 59), and one time along the diagonal starting at $B_u = B_{r,k+1}$, this time with the opposite sign. M_0 is fully reversible by definition, so it does not contribute, while each magnetization jump along M_N produces only one peak with the same sign. The first magnetization jump of M_N (i.e., the one occurring at the largest field) has a special importance, because the corresponding peak N-1, which is always located at $B_u = 0$ in the FORC diagram, contributes to the central ridge. This property is called *local reversal symmetry* [229], and is explained by the inversion symmetry of the major hysteresis loop: M_{N-1} starts at $B_{r,N}$, which is the field in which the last irreversible transition occurs on the descending branch, while M_N, which coincides with the ascending branch, has its last transition in the opposite field $-B_{r,N}$. The Preisach coordinates of the peak produced by this transition are $B_r = B_{r,N}$ and $B = -B_{r,N}$, which converts to the FORC coordinates $B_c = -B_{r,N}$ and $B_u = 0$.

As seen in Fig. 59b, peaks and reversible contributions are contained in a triangular domain with lower limit given by the last diagonal trace starting at $B_u = B_{r,N}$ and upper limit given by the final transition fields at which each curve merges with the major hysteresis loop. These final transition fields are nearly equal in amplitude to $B_{r,N}$, the last transition field of the descending branch, so that the triangular domain of the FORC function is inferiorly limited by the $B_r \geq B_{r,N}$ diagonal and superiorly limited by the $B \lesssim -B_{r,N}$ diagonal. The two limits merge at the FORC coordinate $(B_c, B_u) = (-B_{r,N}, 0)$ on the central ridge. FORC contributions inside this triangular domain occur at places dictated by the transition fields between intermediate states, which can be distributed anywhere between $\pm B_{r,N}$. Some of these states are nucleated within FORCs but not along the major hysteresis loop, as, for instance, the state associated with the portion of M_2 in Fig. 59a which is comprised between jumps 2-2 and 2-1. Furthermore, and contrary to what is generally observed with bulk measurements [230], FORCs are not necessarily enclosed by the major hysteresis loop, as seen by the sections of M_1 and M_2 lying below M_3 in Fig. 59a.

The next step for generating a bulk PSD signature consists in adding the contributions of many systems like the one of Fig. 59. An example is shown in Fig. 60 with the superposition of micromagnetic FORC simulations of five magnetite particles, replicated for different orientations with respect to the field direction, whose shape has been taken from focused ion beam-scanning electron microscopy (FIB-SEM) of an obsidian fragment [231]. The remanent state of these particles includes SD, SV, and multi-vortex configurations. Each micromagnetic simulation features a series of negative and positive peaks that is limited in FORC space by a triangular domain whose size is determined by the field amplitude $B_s = B_{r,N}$ at which the two branches of the major hysteresis loop merge. Because the saturation range is entered at this point, B_s is effectively the saturation field. As seen in Fig. 60, individual peaks tend to cluster over regions where the measured bulk FORC function has larger amplitudes. The location of these regions depends on the type of magnetic systems being analyzed, but, as a general trend, they tend to expand and fill the whole triangular domain spanned by B_s as the system heterogeneity increases. For instance, multi-vortex magnetite grains contribute mostly to the memory region of the FORC diagram, averaging out to create a broad central peak that fills the gap between the positive lobes of SV particles [231]. A simple maximum entropy model of the PSD FORC signature assumes that the contribution of peaks generated by systems with the same saturation field B_s is statistically the same everywhere inside the triangular domain spanned by B_s, except for the central ridge peaks, which are forced to occur at $(B_c, B_u) = (B_s, 0)$ if not biased by internal fields. The corresponding FORC function consists of the superposition of the central ridge to a triangular domain with constant amplitude given by $(M_{irr} - M_{cr})/B_s^2$, where M_{irr} is the bulk irreversible magnetization of all systems with saturation field B_s and M_{cr} their central ridge magnetization. The integral $M_{irr} - M_{cr}$ of this region, added to the central ridge contribution M_{cr}, yields M_{irr}, as expected from the definition of irreversible magnetization.

The relative contribution M_{cr}/M_{irr} of the central ridge to the total irreversible magnetization is proportional to the mean amplitude of the corresponding magnetization jumps (e.g., 3-1 in Fig. 59a). Assuming that magnetization jumps along the major hysteresis loop have the same amplitude on average, $M_{cr} = M_{irr}/N$ is obtained, where N is the mean number of jumps in individual systems ($N = 1$ for SD particles, $N = 2$ for two-state systems, $N = 3$ in Fig. 59, etc.). In real PSD systems, the central ridge is usually visible only at higher coercivities (e.g., Fig. 60a), and it is typically characterized by a relatively large vertical spread, as well as a small offset toward

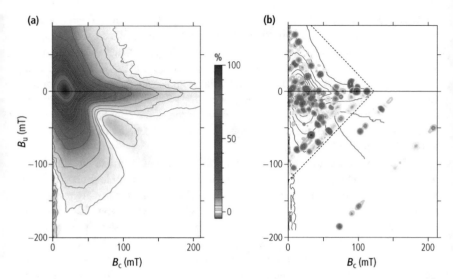

Fig. 60 Continuous limit of PSD FORC signatures. (**a**) FORC diagram of a bulk obsidian fragment containing irregularly shaped magnetite crystals with volume cubic root sizes comprised between ~40 and ~400 nm (data from Lascu et al. [231], quantile contours at 2, 5, and 10–90% of the total magnetization in steps of 10%). Notice the negative region in the lower quadrant and the upward offset of the central maximum. (**b**) Contour lines of the FORC diagram in (a) overlaid with micromagnetic simulations of five representative crystals, replicated for different orientations with respect to the field direction, with exact crystal shapes obtained from focused ion beam-scanning electron microscopy (FIB-SEM) of the obsidian fragment. The dashed line represents the triangular domain containing the peaks produced by most simulations

the lower quadrant. Contrary to the case of interacting SD particles (section "Weak Magnetostatic Interactions Between SD Particles"), vertical profiles of the central ridge tend to expand at larger values of B_c. The origin of the internal fields causing these apparent interaction signatures is discussed later.

The continuum limit of the processes described above yields a FORC function of the type:

$$\rho(H_c - H_u) = \left(1 - N^{-1}\right) \frac{M_{irr}}{F_s^2(0)} F_s(H_c) F_s(|H_u|) + \frac{m_{irr}}{2N} f_s(H_c) g(H_u) \qquad (57)$$

where f_s is the distribution of saturation fields H_s in individual systems,

$$F_s(H) = \int_H^\infty f_s(x) \frac{dx}{x^2} \qquad (58)$$

the contribution of all systems with $H_s \geq H$ to the triangular domain delimited by a given field H, and g a function representing the vertical profile of the central ridge. The two terms on the right-hand side of Eq. (57) represent FORC contributions within the triangular domain limited by H_s and of the central ridge, respectively.

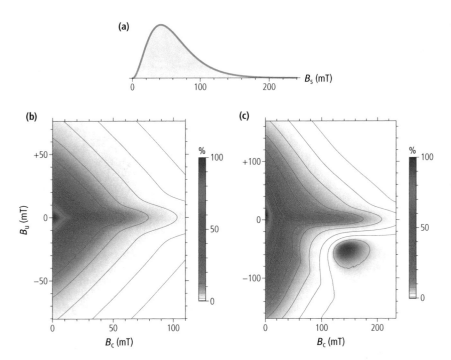

Fig. 61 Synthetic example of PSD FORC signature obtained with Eqs. (57) and (58). (**a**) Distribution of saturation fields assumed for the calculations. (**b**) FORC function according to Eqs. (57) and 58) with $N = 9$, the saturation field distribution in (**a**), and vertical central ridge profiles given by a Gaussian function with standard deviation $\sigma = 1 + 0.06 B_c$. Contours are drawn at 0.1, 1, 5, 15, 30, 50, 70, and 90% quantiles of the total magnetization. (**c**) FORC function for a material with the intrinsic FORC properties shown in (**b**), after accounting for the effects of a mean internal field $B_i = -\alpha M/M_s$ with $\alpha = 100$ mT. In the case of magnetite composition ($M_s = 480$ kA/m), B_i is equivalent to $H_i = -\alpha M$ with $\alpha = 0.167$. Internal field calculations include a reversible magnetization proportional to $\mathcal{L}(B/B_0)$, where $\mathcal{L}$ is the Langevin function and $B_0 = 30$ mT, with the amplitude required to obtain $M_{rs}/M_s = 0.11$. Same contour lines as in (**b**). Notice the expansion of the FORC function domain with respect to (**b**), due to the partial screening of the external field by B_i, the appearance of a negative region below the central ridge, and the upward shift of the function maximum

These contributions are clearly recognizable in the example shown in Fig. 61b, obtained with Eqs. (57) and (58) and $N = 9$, in the form of triangular contour lines with the central ridge protruding over the high-coercivity range. The FORC function of this example, as any other generated with Eqs. (57) and (58), is symmetric about $H_u = 0$, unlike real PSD signatures, which feature a wide region with very small negative amplitudes below the central ridge, comprised between the diagonal $H_u = -H_c$ and $H_u = 0$ (e.g., Fig. 60a). The overlap between this region and positive, low-coercivity amplitudes in the lower quadrant creates a characteristic indentation of the contour lines, which is seen in the FORC signature of most PSD systems made of non-SD particles, but also with strongly interacting SD

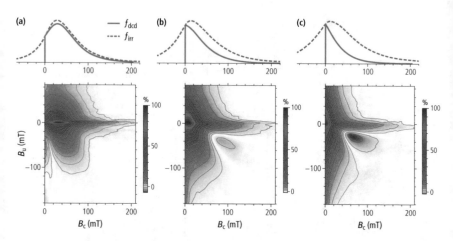

Fig. 62 PSD signatures of sediment material obtained from a piston core collected on the western margin of the North Atlantic (core R41, data from Lascu et al. [208]). The three examples correspond to siliciclastic fractions of increasing sizes obtained with gravity settling: **(a)** < 4 μm ($M_{rs}/M_s = 0.249$, $H_{cr}/H_c = 2.346$), **(b)** 10–20 μm ($M_{rs}/M_s = 0.0734$, $H_{cr}/H_c = 3.945$), and **(c)** 40–63 μm ($M_{rs}/M_s = 0.0250$, $H_{cr}/H_c = 9.727$). The top row shows the coercivity distributions obtained from FORC data representing the irreversible component of the hysteresis loop (f_{irr}) and the DCD demagnetization curve (f_{dcd}). The bottom row shows the FORC diagrams, with contour levels corresponding to magnetization quantiles of 2, 5, and 10–90% in steps of 10%. The color scales of **(b)** and **(c)** are identical, in order to highlight the larger negative amplitudes in **(c)**

particles (e.g., Fig. 3 in Carvallo et al. [17]). In both cases, this feature, which also includes an upward shift of the central maximum of the FORC function, can be phenomenologically explained by the presence of demagnetizing interactions (section "Mean-Field Interactions"). Shearing the FORC model of Fig. 61b by a mean internal field $B_i = -\alpha M/M_s$, with $\alpha \approx 0.1$ (Fig. 61c) introduces a signature that is remarkably similar to that of natural samples, especially those containing coarse magnetite fractions. The grain size dependence of this signature is clearly visible in siliciclastic fractions obtained from sediment cores dominated by coarse lithogenic minerals [208] (Fig. 62).

Despite the common occurrence of signatures typical of demagnetizing interactions in FORC diagrams of PSD and strongly interacting SD particles, the origin of the associated internal fields is fundamentally different, as suggested by a comparison with Preisach maps based on remanent magnetization measurements. Preisach maps are nearly symmetric about $H_u = 0$ in the case of interacting SD particles [17], while they maintain, at least in part, the same asymmetry of FORC diagram in the case of non-SD particles [18]. The bulk saturation remanent magnetization of most samples with PSD signature is too small to produce a significant demagnetizing field, unless the magnetic particles are assumed to be tightly clustered, and local magnetic interactions are drastically reduced by the non-dipole nature of the stray field produced by PSD particles in their remanent state [232]. On the other hand, the identification of $B_i = -\alpha M/M_s$ with the demagnetizing field of individual particles, instead of the bulk sample implies that $\alpha = \mu_0 D \mu_s$, yielding the correct range of α values required to generate the observed FORC signatures: for instance, $\alpha \approx 0.2$ for equidimensional magnetite ($D = 1/3$, $\mu_s = 480$ kA/m). Therefore, demagnetizing interaction signatures in FORC diagrams of PSD and MD particles can be attributed to the partial screening effect [233] of reversible magnetization changes on sites where a new magnetic state is nucleated. This screening effect also explains the increase of the vertical range of the FORC function, since the internal field values at H_r and H add constructively near $H_c \approx 0$, while they tend to cancel out along $H_u \approx 0$.

In conclusion, PSD FORC signatures contains variable proportions of three fundamental features: (1) a central ridge with pronounced vertical spread and slight vertical offset, which includes a central maximum located at $H_c \geq 0$ and $H_u \geq 0$, (2) a low-amplitude negative region located below the high-coercivity end of the central ridge, and (3) positive contributions above and below the central ridge that smoothly merge with the low-coercivity range of the central ridge, yielding triangular contour lines, especially in the upper quadrant. The central ridge's vertical spread and offset, the positive vertical offset of the central maximum, and the negative regions are compatible with magnetic screening effects inside individual particles.

The typical grain size trend of PSD signatures generated by non-SD particles is shown in Fig. 62. The signature of small PSD particles (Fig. 62a) is similar to that of two-state systems discussed in section "Rock and Paleomagnetic Significance of Two-State Systems": it is dominated by a central ridge whose low-coercivity end is surrounded by two positive lobes in the upper and lower quadrants, respectively. Unlike the selected two-state examples of Fig. 56, which are generated by very homogenous systems, the lobes of natural PSD signatures tend to merge with the central ridge. The wide grain size distribution of natural assemblages of small PSD particles likely includes a minor contribution from truly SD particles, as inferred from the negative amplitudes in the lower quadrant near $H_c = 0$. The bulk hysteresis parameters are similar to those of two-state vortex particles, within the classic PSD range of the Day plot, but above the SD + MD mixing line (Fig. 54). The signature of demagnetizing interactions is very subdued, being recognizable only from the asymmetry of low-amplitude contour lines. At larger particle sizes (Fig. 62b), the sharp central ridge along $H_u \approx 0$ becomes blurred and identifiable only at higher coercivities. The central peak of the FORC function moves toward $H_c = 0$, while the lobes are transformed into a region with triangular contour lines and a wide extension along H_u. Any trace of SD signatures disappears, while the signature of demagnetizing interactions becomes evident. The bulk hysteresis parameters are still within the classic PSD range of the Day diagram (Fig. 54), but they mark the beginning of a trend determined by the inverse proportionality between remanence and coercivity ratios, which is typical of MD particles [205] and caused by the screening effect of the internal demagnetizing field. As particle size increases further (Fig. 62c), the demagnetizing interaction signature becomes more evident, while the central peak of the FORC function has reached $H_c = 0$. The bulk hysteresis parameters have a mixed character, with the remanence ratio being still in the PSD range, while the coercivity ratio is already MD (Fig. 54). This trend continues at even larger sizes with typical MD hysteresis properties and a FORC function that contines to contain residual PSD signatures [18] (section "Multidomain Particles").

Reverting to the original problem of the PSD signature ambiguity, two transitions appear to affect the hysteresis and FORC properties of soft ferrimagnetic minerals such as magnetite. The first transition is sharp and occurs between systems featuring only one type of magnetic states (SD) and system with two or more types of magnetic states (SD-like and single-particle vortices or supervortices). The corresponding FORC signatures are sufficiently well defined to discriminate, at least qualitatively, between one- and multistate systems (Fig. 62a). Individual multistate systems, and also very homogeneous bulk samples containing such systems, have very scattered hysteresis properties limited only by the SD-MD and SD-SP mixing lines, respectively. Caution should therefore be used when interpreting hysteresis parameters as the result of mixing SD, MD, and SP endmembers. The scatter of bulk hysteresis parameters is caused by the fact that the stability of HMS and LMS is very sensitive to physical parameters such as particle size and shape. The second transition occurs when a given H_{cr}/H_c threshold, which appears to be comprised between 2.5 and 3, is exceeded (Fig. 54). Above this threshold, bulk hysteresis parameters are controlled by the inverse proportionality between M_{rs}/M_s and H_{cr}/H_c typical of MD crystals, while the further evolution of FORC properties appears to be controlled by an increasingly effective screening of the external field (Figs. 61 and 62b, c). This screening effect is caused by the internal demagnetizing field $H_i = -D_{eff} M$, where D_{eff} is an effective demagnetizing factor that depends on particle shape and magnetization configuration. A distribution of D_{eff} values explains the vertical broadening of the central ridge, to the point that it is no longer distinguishable from other low-coercivity contributions, while the negative sign of H_i is responsible for the demagnetizing mean-field signatures visible in the FORC diagram.

The simultaneous onset of the MD trend for the bulk hysteresis parameters and the appearance of negative mean-field signatures in the FORC diagram (i.e., the region of negative FORC amplitudes below the high-coercivity range of the central ridge) can be identified with the upper limit for particles or systems with true PSD behavior. As far as the FORC diagram is concerned, a proper identification of negative mean-field signatures relies on negative, low-amplitude regions of the FORC function located between $H_u = -H_c$ and $H_u = 0$. Optimized FORC protocols with sufficiently wide measurement ranges, sufficient quality for resolving low-amplitude details of the FORC function, and proper processing that avoids smoothing artifacts while providing sufficient noise suppression (section "FORC Processing"), are required for a proper identification of the mean-field signature.

In micromagnetic terms, the grain size trend characterized by the sudden transition from SD to two-state signatures and the subsequent progressive transition toward MD properties can be understood by the appearance of the vortex state and the subsequent progressive division on the volume that surrounds the vortex tube into domains where the magnetization vector is parallel to one of the magnetocrystalline easy axes, separated by transition regions that can be effectively regarded as domain wall precursors [216]. The MD transition appears to be more gradual for micromagnetic simulations of cuboctahedral crystals than spherical ones [216], which means that complex surfaces, such as those of many natural PSD particles [231], might extend this transition over a wide range of grain sizes. However, as discussed in section "Multidomain Particles", the coexistence of PSD and MD signatures over a wide grain size range appears to be related to the existence of high-stress regions with PSD behavior, partially screened by the reversible magnetization of the remaining particle volume.

It is finally worth discussing the PSD-MD trend in the case of coercivity distributions. A sharp central ridge exists only before the onset of negative mean-field signatures, and in this case its contribution to the total irreversible magnetization is marginal, except for specimens containing only two-state systems such as those of Fig. 56. Because magnetic state transitions along the major hysteresis loop occur in positive and negative fields, f_{irr} is a broad distribution, and $f_{dcd}(0) > 0$ (Fig. 62a). The two distributions are very similar in the positive field range, and this property is shared with most two-state systems (Fig. 56), so that f_{dcd} can be regarded as a truncated version of f_{irr} that reflects FORC contributions comprised within the memory region. As negative mean-field signatures start to appear at the PSD-MD transition (Fig. 62b, c), the maximum of f_{irr} begins to move toward $H = 0$, while f_{dcd} approaches a one-sided exponential-like function as it loses contributions at higher coercivities. This loss of contributions inside the memory region, which reflects the decrease of M_{rs}/M_s, is due to the increasing vertical spread of the FORC function. The large vertical spread of the FORC function near $H_c = 0$ contributes to f_{irr}, but not to f_{dcd}. This trend continues until f_{irr} becomes symmetric about $H = 0$ and f_{dcd} decays to zero in a very narrow range of fields close to $H = 0$, approaching the ideal signature of MD particles (see section "Multidomain Particles").

Multidomain Particles

The last FORC signature that needs to be discussed for a complete description of geologic materials is that of multidomain crystals. As suggested by the name, a multidomain crystal is divided into discrete magnetic domains with homogeneous magnetization separated by *domain walls*, which are thin planar regions where the magnetization vector rotates between the directions of adjacent domains. The typical thickness of these regions is given by the so-called Bloch wall width $\delta_w = \pi l_{ex}/\kappa$, where l_{ex} is the exchange length (Eq. 55) and κ the magnetic hardness (Eq. 54); this width is of the order of 100 nm for magnetite [234–236]. The number and size of magnetic domains and the orientation of their magnetization are controlled by energy minimization principles. Relevant terms include the domain wall energy, the magnetostatic energy, the Zeeman energy, and the magnetocrystalline energy [237]. The domain wall energy per unit of wall surface, $E_w \approx \pi \sin(\theta_w/2) \sqrt{2AK}$, where θ_w is the angle between the magnetization vector of the adjacent domains, accounts for the energy required to overcome the exchange coupling, expressed by the exchange energy constant A, and to rotate the magnetization away from the preferred axes of magnetocrystalline anisotropy, expressed by the anisotropy constant K. The magnetostatic energy accounts for the energy of domain magnetizations in the stray field produced by the intersection of the magnetization vector with the crystal surface. In analogy with SD particles, the magnetostatic energy per unit of volume is given by $E_d = -\mu_0 \mu_s^2 D_{eff}$, where μ_s is the spontaneous magnetization and D_{eff} an effective demagnetizing factor (see section "Demagnetizing Fields and FORC Deshearing"), which depends on the shape of the crystal and the number of domains. A simple expression for D_{eff} valid for a large number n of domains is $D_{eff} \approx D_{SD}/n$, where D_{SD} is the demagnetizing factor of the same crystal when it contains only one magnetic domain [58]. In practice, the decrease of D_{eff} with the number of domains can be understood by the fact that the stray field produced at the surface of the crystal becomes more localized as the number of domains increases. Finally, the expressions for the Zeeman and magnetocrystalline energies are the same as for SD particles (section "Uniaxial SD Particles: Stoner-Wohlfarth Model").

In the absence of external fields, the number of domains is given by a balance between the total domain wall energy, which is proportional to the number n of domain walls, and the magnetostatic energy, which is inversely proportional to n. Minimization of these two energies gives the optimal number of domains, which turns out to be proportional to the square root of the particle size [31]. The application of an external field has three main effects on the domain structure: at small fields, domain walls move so that favorably magnetized domains (i.e., when the magnetization vector forms a $< 90°$ angle to the external field) grow at the expense of unfavorably magnetized ones, decreasing the Zeeman energy (Fig. 63a). If the field is applied at an angle to the magnetocrystalline easy axes, the domain magnetization rotates away from the original easy axis direction, but so to maintain the domain wall angles θ_w. For magnetite crystals, this rotation is completed when the external field reaches ~20 mT. Finally, unfavorably magnetized domains begin to denucleate (disappear) in larger external fields, until the entire crystal contains a single magnetic domain, thus entering the saturation range. The interplay between these three processes, which requires complex micromagnetic models to be analyzed in a realistic manner, determines the sigmoidal shape of MD hysteresis loops. The essential magnetic signatures of MD particles, on the other hand, are captured by much simpler conceptual models, in which rotation of the magnetization vectors and domain (de)nucleation are neglected.

The minimalistic model used to understand the FORC signature of MD particles considers a cuboidal crystal that contains only two domains separated by a single Bloch wall (Fig. 63b). Furthermore, it is assumed that the external field is parallel to one of the magnetocrystalline easy axes, so that the magnetization vectors of the two domains are parallel and antiparallel to the field, respectively. The relevant energy terms in this case are E_d, the Zeeman energy, and E_w. The sum of these terms yields the total energy as a function of the external field H and the domain wall position $0 \leq x \leq L$, where L is the size of the crystal:

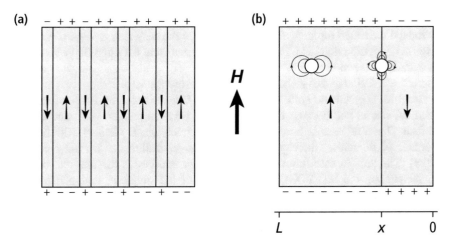

Fig. 63 1D model of a MD cuboid. (**a**) Magnetic domains with alternate magnetization directions (arrows) separated by Bloch walls, in an external field H. Surface magnetic charges indicated with "+" and "−" produce the stray field caused by the intersection of the magnetization vectors with the surface, in the absence of closure domains. (**b**) Magnetic configuration equivalent to (**a**), consisting of a single Bloch wall with position x comprised between 0 and the cuboid length L. Also shown are two defects, represented by spherical voids, and their stray field, when fully contained in a magnetic domain (left) or crossed by a domain wall (right)

$$E_{tot}(x, H) = \frac{1}{2}\mu_0\mu_s^2 AD_{eff}(1 - 2x/L)^2 + \mu_0\mu_s A(L - 2x/L) H + AE_w(x),$$

(59)

with A being the domain wall area [107, 108]. The first term on the right-hand side of Eq. (59), the magnetostatic energy, is a quadratic function of the domain wall position, which takes a minimum at $x = L/2$, that is, when the domain wall is placed exactly in the middle. In this case, the equal and opposed magnetic moments of the two domains cancel out, yielding a zero net magnetic moment. The second term in Eq. (59) is a linear function of the domain wall position, with coefficient proportional to the external field H. This term moves the local minimum, and so the domain wall position x, so that the favorably magnetized domain (in this case the left one) grows at the expense of the other domain, until saturation is reached at $x = 0$. Finally, A and E_w do not depend on the wall position in a defect-free cuboidal crystal, so that the last term on the right-hand side of Eq. (59) is a constant. In this case, Eq. (59) has a single local minimum, which is also the global minimum, at $1 - 2x/L = -H/(\mu_s D_{eff})$ for any given external field. The total magnetic moment of the crystal is then simply given by $\mu_s A(2x - L)$ for $0 \le x \le L$. This result yields a closed hysteresis loop with the form of a saturating linear function:

$$M(H) = \begin{cases} \chi H, & |H| < \chi^{-1}\mu_s \\ \mu_s & \text{else} \end{cases},$$

(60)

where $\chi = D_{\text{eff}}^{-1}$ is the magnetic susceptibility of the crystal. This type of hysteresis is indeed measured on large single crystals when the easy axis is aligned with the external field direction [31]. Full loop closure means that the saturation remanence is exactly zero and the FORC function is zero everywhere. Ideal MD particles are therefore invisible to remanence and FORC measurements.

Real MD crystals do hold a small remanent magnetization by virtue of unsymmetrical shapes and because of the interaction between domain walls and crystal defects. The first case is only relevant for relatively small particles with an odd number of domains, whereby the resulting magnetization has a PSD character [228]. The second case, known as *domain wall pinning*, is the one relevant for understanding the MD FORC signature. One of the simplest pinning models is the inclusion theory developed by Kersten [238], and later refined by Néel [239], which considers a crystal containing non-magnetic inclusions with minimum sizes comparable to the domain wall thickness [240]. These inclusions, which can be effectively identified with voids, create an internal stray field due to the virtual surface charges forming where the magnetization vectors intersect the inclusion surface. In the case of a spherical inclusion fully contained within a magnetic domain, this stray field is equivalent to that produced by a magnetic dipole with dipole moment $\mu_s V$, where V is the volume of the inclusion. This field generates a magnetostatic energy $E_{\text{inc}} = \mu_0 V \mu_s^2 D_{\text{inc}}$, where D_{inc} is the shape-dependent demagnetizing factor of the inclusion (e.g., $D_{\text{inc}} = 1/3$ for a sphere). If a domain wall intersects the inclusion, the non-homogeneous wall magnetization produces a more complex surface charge distribution that decreases the equivalent magnetic moment of the stray field and therefore also E_{inc}. Similar effects also occur with other types of defects, such as dislocation, through the magnetostriction energy. If domain walls move in a crystal containing many sparse defects, the additional energy terms E_{inc}, which depend on the wall position, can be assimilated to the total wall energy E_w. Because defects lower E_w, domain walls tend to be immobilized by them, until the additional energy created by wall bending and by the non-ideal position becomes sufficiently large to suddenly detach the wall from its pinning site(s). This sudden detachment creates a microscopic step in the magnetization curve known as *Barkhausen jump*. In practice, domain wall pinning is modeled by adding a random function of the wall position to Eq. (59). Néel [239] used a sawtooth function of random amplitude. In later refinements, the domain wall displacement has been treated in the framework of random walk theories [241, 242].

For the purpose of understanding the fundamental characteristics of MD hysteresis, random fluctuations of the domain wall energy can be represented by a sinusoidal function of the form $A\varepsilon \sin(\pi x/d)$, where ε is fluctuation energy amplitude per unit of domain wall area and d a fluctuation dimension of the order of the size of inclusions. Both parameters might vary with the amplitude of the applied field, especially when saturation is approached. The total energy landscape obtained by adding a fluctuation term to Eq. (59) now contains a series of local minima in addition to the global one, depending on the amplitude of the external field (Fig. 64a). These local minima cause the domain wall to lag the global minimum by an

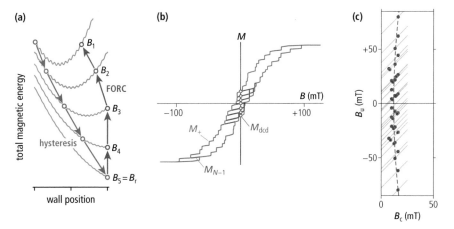

Fig. 64 Conceptual model of MD hysteresis. (**a**) Total magnetic energy (gray curves) as a function of wall position for the two-domain particle of Fig. 63b, when immersed in external fields B_1 to B_5. Curve undulation represents fluctuations of the domain wall pinning energy, modeled with a sinusoidal function. During a field sweep from positive saturation, the wall position is determined by the first local minimum of the energy function from the left (e.g., blue dot on the B_1 curve). As the field is decreased to the point that the occupied minimum disappears, the wall jumps to the next local minimum (next blue dot on the B_2 curve), and so on. The sequence of magnetization values corresponding to the wall positions indicated by the blue dots builds the descending branch of the hysteresis loop (arrows). The field sweep is stopped at B_5 and reversed, so that $B_5 = B_r$ is the reversal field of a FORC measurement. The field is now increased back to B_1. The local minimum occupied in B_r remains stable until a critical value is reached (in this example B_3). From this point on, the magnetization increases again through a series of jumps to the rightmost stable minimum (orange dots), building a FORC curve (arrows) that, from B_3 onward, coincides with the ascending branch of the major hysteresis loop. (**b**) Magnetic hysteresis (blue) and selected FORCs (orange) corresponding to the energy model shown in (**a**). Dots indicate remanent magnetizations contributing to the DC demagnetization curve M_{dcd}. Only FORCs contributing to intermediate remanent magnetization and the FORC starting just after the last magnetization jump on the descending branch are shown for clarity. (**c**) FORC function corresponding to (**b**). Non-zero contributions, in the form of isolated positive and negative peaks indicated by orange and blue dots, respectively, occur along the diagonal traces (gray lines) of FORCs starting immediately after a magnetization jump on the descending branch of the major hysteresis loop. The dashed line indicates the average horizontal position of the peaks

amount that depends on ε and d as the field is swept during a hysteresis or FORC measurement. This lag is quantified by the field $H_c = \pi \varepsilon / \mu_0 \, \mu_s \, d$ that needs to be added to the Zeeman energy term in Eq. (59) to eliminate the local minima caused by the fluctuation term. This field is, by definition, equal to the coercive field. The saturation remanence $M_{rs} = \chi H_c$ that arises from domain wall lags is obtained from Eq. (60) using the slope χ of the hysteresis loop, which remains unchanged [31]. This MD hysteresis model can now be used to calculate FORCs. During the preparatory field sweep from positive saturation, the domain wall in Fig. 63b moves from right to left occupying always the rightmost local minimum. A Barkhausen jump occurs every time an occupied energy minimum disappears, and the domain wall position jumps to the next local minimum (Fig. 64a). When the field sweep direction is changed at the reversal field H_r, the occupied minimum remains stable until the field has increased by $2H_c$. During this time, the domain wall does not move irreversibly, and the magnetization remains nearly constant. Once the field has increased by $2H_c$, the ascending branch of the major hysteresis loop is reached, and a series of Barkhausen jumps brings the particle to positive saturation along this branch. In practice, given the stochastic nature of the pinning process, few jumps can occur before the ascending branch of the hysteresis loop is reached, as in the example of Fig. 64b. Nonetheless, the general trend for all FORC curves is to merge with the ascending branch at $H \approx H_r + 2H_c$ [107].

The FORC measurements of a single MD crystal with coercivity H_{c0} can therefore be modeled by a series of ramp functions:

$$M\,(H_r, H) = \chi\,R\,(H - H_r - 2H_{c0}) \tag{61}$$

with $R(x) = \theta(x)x$, where θ is the Heaviside unit step function. The corresponding FORC function, obtained from the mixed derivative of Eq. (61) using $R'(x) = \theta(x)$ and $\theta'(x) = \delta(x)$, where δ is the Dirac impulse, and using the FORC coordinates $H_c = (H - H_r)/2$ and $H_u = (H + H_r)/2$, is

$$\rho(H_c, H_u) = \frac{1}{4}\chi\delta(H_c - H_{c0})\,\Pi(H_u/2H_s)\,, \tag{62}$$

with $H_s = \chi^{-1}\,\mu_s$ being the saturation field and $\Pi(x) = 1$ for $|x| \leq 1/2$ and $\Pi(x) = 0$ otherwise being the unit rectangle function. In practice, Eq. (62) describes a vertical ridge placed at $H_c = H_{c0}$ and extending vertically between $H_u = \pm H_s$. Using $H_s \approx \mu_s\,D_{\text{eff}}$ with $D_{\text{eff}} \approx 0.33$ for equidimensional particles, one can see that the vertical extension of the FORC function is of the order 200 mT for magnetite, which is much larger than that of SD and PSD particles. The integration of the FORC function in Eq. (62) over the FORC space yields a total irreversible magnetization $M_{\text{irr}} = \mu_s$, as expected from the fact that reversible magnetization changes, for instance, by rotation of the magnetization when the field direction does not coincide with an easy axis, have not been considered.

The vertical ridge of Eq. (62) is in reality composed of a large number of discrete positive and negative peaks corresponding to individual Barkhausen jumps occurring in each FORC before merging with the ascending branch of the major hysteresis loop (Fig. 64c). These jumps are similar to those of the PSD FORC model discussed in section "Multistate Systems: The PSD FORC Signature", but have a markedly different distribution in FORC space, due to the fact that only a small initial portion of each curve is distinct from the other curves. The final step toward a general model for the MD FORC signature consists in considering a large assemblage of MD particles possessing a distribution $f(H_{c0})$ of coercivities and a distribution $g(H_s)$ of saturation fields due to a range of shape-dependent D_{eff} values. Integration of Eq. (62) over these distributions yields

$$\rho(H_c, H_u) = \frac{\mu_s}{4}f(H_c)\,G(H_u) \tag{63}$$

with

$$G(H_u) = \int_{H_u}^{\infty} \frac{1}{H_s} g(H_s) \, dH_s \qquad (64)$$

In the case of nearly equidimensional particles, the range of variation of $g(H_s)$ is relatively limited, and $G(H_u)$ is a nearly flat function for $|H_u|$ values smaller than the lower limit of the H_s range, which then decays to zero for larger arguments. If $f(H_{c0})$ is a function peaking at $H_{c0} = 0$, Eq. (63) represents a vertical ridge extending along the left limit of the FORC space. This is the signature commonly observed for large MD particles, whereby the vertical extension is often truncated by the limited H_u range chosen with the FORC protocol [107]. Examples covering the whole vertical range, such as those of Fig. 65, are rare and mostly affected by artifacts related to the first measurement of each curve at large fields (section "Field Control, Measurement Sensitivity, and Resolution", Fig. 13).

The simple MD model illustrated above ignores possible variations of the pinning energy function (e.g., ε) and of the demagnetizing factor D_{eff} as saturation is approached. A general increase of D_{eff} close to saturation is expected because of the elimination of closure domains [58]. This decreases the slope of the hysteresis loop, determining the more gradual saturation approach observed in real MD particle assemblages. Because H_s controls the extension of the vertical ridge in the FORC function of individual MD particles, the ridge amplitude distribution G in Eq. (63) shows a more gradual decay at large values of $|H_u|$ [108]. As far as the pinning energy is concerned, the field required to unpin a Bloch domain wall is an increasing function of the angle between the domain magnetization direction and the field direction. This dependence is irrelevant for magnetite at room temperature, because the small cubic magnetocrystalline anisotropy is easily overcome, and a full alignment of the magnetization is obtained in field amplitudes of ~20 mT or less. On the other hand, materials with strong uniaxial magnetocrystalline anisotropy, such as magnetite below the Verwey transition temperature, will maintain the domain magnetization parallel to the easy axis until the saturation range is reached [243]. In the case of polycrystalline grains, domain walls will move more easily inside regions with easy axes sub-parallel to the external field, so that magnetization changes close to the saturation region will be controlled by domain wall displacements in regions with easy axes nearly perpendicular to the field. These regions are characterized by larger H_{c0} values, so that the horizontal FORC coordinate of the vertical ridge becomes a function of $|H| = |H_c + H_u| \approx |H_u|$ [108]. An example is shown in Fig. 65c for titanomagnetite at 50 K: in this case, the increase of H_{c0} with $|H_u|$ gives the central ridge a characteristic crescent shape.

FORC diagrams of MD particle assemblages often contain traces of the PSD signatures discussed in section "Multistate Systems: The PSD FORC Signature", right of the vertical ridge. These PSD signatures include of a vertically scattered and slightly offset central ridge overlying a negative region, and can be observed also in samples that do not contain PSD-sized particles (Fig. 65a, b). The PSD signature of MD crystals is eliminated by thermal annealing [107]. Thermal annealing is

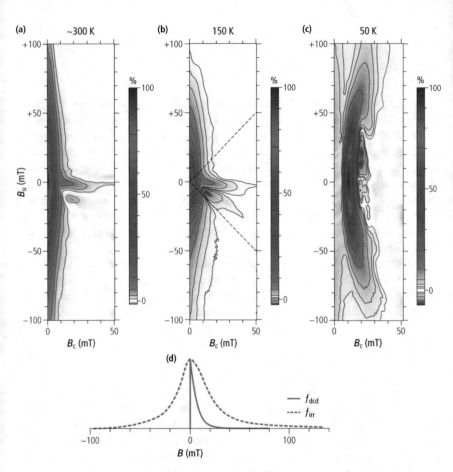

Fig. 65 FORC diagrams at selected temperatures (**a-c**) and coercivity distributions at 150 K (**d**) of a polycrystalline titanomagnetite specimen ($Fe_{3-x}Ti_xO_4$ with $x = 0.4$). Data from Church et al. [108], after correction of first point measurement artifacts with VARIFORC. Contours are drawn at 2, 5, and 10–90% quantiles of the total magnetization in steps of 10%. The dashed lines in (b) highlight the limits of the memory region, which controls f_{dcd} in (d)

expected to redistribute pinning sites homogeneously throughout the grain volume in a manner that is consistent with the stochastic pinning model that yields the MD FORC function discussed above. Concentrated stress in the unannealed state, on the other hand, might block domain walls over a large portion of the grain volume, which then behaves as a sort of vortex particle that is partially screened from the external field by the reversible magnetization of the stress-free regions. Growth of the relative amplitude of PSD signatures in MD magnetite-bearing ore after manual powdering [244] supports this interpretation. The coexistence of low- and high-stress regions in unannealed grains has a significant effect on the shape of MD coercivity distributions: f_{irr}, which is sensitive to all irreversible magnetization changes (e.g., Barkhausen jumps) along the major hysteresis loop, between negative and positive saturation, tends to be perfectly symmetric about $H = 0$ in the ideal case of homogeneously distributed pinning sites. The generally stronger contributions in positive fields are ascribable to the presence of high-stress regions. On the other hand, the DC demagnetization curve and the associated coercivity distribution f_{dcd} depend only on FORC contributions located within the memory region. Only a small fraction of ideal MD signature consisting in a vertical ridge located near $H_c = 0$ intersects the memory region, while the opposite is true for the PSD signature (Fig. 65b). Therefore, f_{dcd} is mainly controlled by the PSD-like magnetization of high-stress regions, especially at higher coercivities. Even if f_{dcd} might extend to relatively large fields (e.g., 200 mT in Fig. 62c), the PSD-like remanent magnetization is not an ideal paleomagnetic carrier, because of its coupling with the magnetically unstable low-stress regions.

5   Conclusions and Outlook

As illustrated in this review, FORC measurements provide a powerful magnetic characterization tool for geologic materials. The affinity of FORC measurements with Preisach theory has two main advantages. First, many "ordinary" magnetic parameters used for the characterization of rocks and sediments can be understood as 0D (bulk magnetizations) or 1D (coercivity distributions) projections of the FORC function, which means that a quantitative link can be established between time-consuming FORC measurements, which are typically limited to few examples, and faster measurements more suited to the systematic investigation of a large number of samples. Second, specific magnetization processes produce easily distinguishable FORC features. Some features, such as vertical and horizontal ridges, are highly localized in one dimension and can be separated, sometimes quantitatively, from other diffuse contributions. One of the most typical examples is the central ridge, which enables the identification of non-interacting SD particles even if the magnetization is dominated by other magnetic carriers. More diffuse features of the FORC function can be identified through specific combinations of positive and negative contributions, such as the lobes associated with transitions from high- to low-moment states in two-state systems or the signatures of mean-field interactions. These features define the FORC fingerprints of specific natural magnetic particle assemblages.

Unlike most engineered materials, natural magnetic particle assemblages are characterized by broad and highly overlapped distributions of magnetic properties. The high degree of non-uniqueness in the interpretation of magnetic measurements obtained from geologic materials has been traditionally overcome—at least in part—by multiparameter approaches and numerical unmixing techniques. FORC diagrams do not solve the non-uniqueness problem either, but they possess a much higher discrimination power. For instance, a binary mixture of SD and MD particles, which might be totally undistinguishable from particles with intermediary sizes based on bulk parameters (e.g., H_{cr}/H_c vs. M_{rs}/M_s in the Day diagram), would be clearly identifiable in a FORC diagram. Moreover, identical coercivity distributions can be split into the contribution of different magnetic components according to the domain states identifiable with FORC measurements (e.g., coarse lithogenic vs. fine authigenic magnetite particles in sediments).

FORC diagrams, however, have their own limits: for instance, it is not possible to discriminate among different types of particles with the same magnetic behavior, such as the SD and non-SD examples of two-state magnetic systems discussed in section "Two-State Magnetic Systems", unless the exact shape of the FORC function of specific magnetic components is known a priori. This knowledge might be gained from forward (e.g., micromagnetic) models or from numerical unmixing techniques (e.g., FORC-PCA). Both techniques have their own caveats: forward models require a detailed knowledge of physical parameter distributions (grain size, geometry, arrangement in space), which is often not available, and numerical unmix-

Table of Symbols and Mathematical Notations

Symbol	Meaning	Reference
A	Exchange constant	Section "The Pseudo-Single-Domain Magnetic Behavior"
ARM	Anhysteretic remanent magnetization	Section "ARM Acquisition and Demagnetization"
$\mathcal{B}$	Beta distribution	Section "The Central Ridge"
D	Demagnetizing factor (or tensor)	Section "Demagnetizing Fields and FORC Deshearing"
D_{eff}	Effective demagnetizing factor	Sections "Multistate Systems: The PSD FORC Signature," "Multidomain Particles"
DCD	DC demagnetization	Section "DC Demagnetization Curves"
$E, E_{tot}, \Delta E$	Energy and energy difference	Sections "Modifications of the Classical Preisach Model," "Multidomain Particles"
E_b	Energy barrier	Section "Thermally Activated SD Particles"
E_d	Magnetostatic (demagnetizing) energy	Section "Multidomain Particles"
E_w	Domain wall energy	Section "Multidomain Particles"
f_{cr}	Central ridge coercivity distribution	Sections "Selected Properties of the FORC Function," "The Central Ridge"
f_{dcd}	Coercivity distribution defined by the DCD	Section "Selected Properties of the FORC Function"
f_{irr}	Coercivity distribution defined by M_{irr}	Section "Selected Properties of the FORC Function"
f_K	Microcoercivity distribution	Section "Uniaxial SD Particles: Stoner-Wohlfarth Model"
f_{rev}	Derivative of the reversible hysteresis component	Section "Modifications of the Classical Preisach Model"
f_s	Distribution of saturation fields	Section "Multistate Systems: The PSD FORC Signature"

(continued)

Symbol	Meaning	Reference
f_v	Coercivity distribution of viscous particles	Section "The Vertical Ridge"
FWHM	Full width to half-maximum	—
H	Applied field (during a magnetic measurement)	Section "The Classical Preisach Model"
H^0	Center position of the smoothing factor limitation along H	Section "Smoothing Factor Limitations Along the Measurement Coordinates"
H_A, H_B	Switching fields of a hysteron	Section "The Classical Preisach Model"
H_a	Acquisition field	Section "IRM Acquisition Curves"
H_{af}	Peak alternating field amplitude	Section "AF Demagnetization of IRM"
H_c	Coercive field (coercivity) and FORC coordinate	Section "The Classical Preisach Model"
H_c^{max}	Maximum H_c (FORC protocol)	Section "The FORC Protocol"
H_{cal}	Calibration field (FORC protocol)	Section "The FORC Protocol"
H_{dc}	DC field applied during ARM acquisition	Section "ARM Acquisition and Demagnetization"
H_f	Fluctuation field	Sections "The Vertical Ridge," "Thermally Activated SD Particles"
H_i	Random interaction field	Section "Demagnetizing Fields and FORC Deshearing"
$\overline{H}_i$	Mean interaction field	Sections "Weak Magnetostatic Interactions Between SD Particles" "Mean-Field Interactions"
H_{int}	Random interaction field	Section "Weak Magnetostatic Interactions Between SD Particles"
H_K	Microcoercivity	Sections "Processing FORC Data Associated with Thin Hysteresis Loops," "Uniaxial SD Particles: Stoner-Wohlfarth Model"
H_n	Critical (nucleation) field	Section "Uniaxial SD Particles: Stoner-Wohlfarth Model"
H_r	Reversal field	Section "DC Demagnetization Curves"
H_s	Saturation field	Section "Definitions"
H_{sat}	Saturation field (FORC protocol)	Section "The FORC Protocol"
H_{sw}	Switching field	Section "Uniaxial SD Particles: Stoner-Wohlfarth Model"
H_u	Bias field and FORC coordinate	Section "The Classical Preisach Model"
H_u^{cr}	Vertical position of the central ridge	Sections "The Central Ridge," "Thermally Activated SD Particles"
H_u^{min}, H_u^{max}	Lower and upper limit of H_u (FORC protocol)	Section "The FORC Protocol"

(continued)

Symbol	Meaning	Reference
δH	Field step (FORC protocol)	Section "The FORC Protocol"
δH^0	Width of the smoothing factor limitation along H	Section "Smoothing Factor Limitations Along the Measurement Coordinates"
δH^*	Field resolution of the FORC diagram	Section "Theoretical Principles of FORC Function Estimation"
δH_c^{vr}	Intrinsic thickness of the vertical ridge	Section "The Vertical Ridge"
δH_u^{cr}	Intrinsic thickness of the central ridge	Section "The Central Ridge"
HMS	High-moment state	Section "Two-State Magnetic Systems"
HWHM	Half-width to half-maximum	—
IRM	Isothermal remanent magnetization	Section "IRM Acquisition Curves"
K	Anisotropy energy constant	Section "Single-Domain Particles"
K_1, K_2	Cubic anisotropy energy constants	Section "SD Particles with Multiaxial Anisotropy"
k_B	Boltzmann constant	—
$\mathcal{L}$	Langevin function	Section "Processing FORC Data Associated with Thin Hysteresis Loops"
l_{ex}	Exchange length	Section "The Pseudo-Single-Domain Magnetic Behavior"
LMS	Low-moment state	Section "Two-State Magnetic Systems"
m	Magnetic moment	—
M	Bulk magnetization	Section "The Classical Preisach Model"
M^+, m^+	Descending branch of the major hysteresis loop	Sections "The Classical Preisach Model," "Uniaxial SD Particles: Stoner-Wohlfarth Model"
M^-, m^-	Ascending branch of the major hysteresis loop	Sections "The Classical Preisach Model," "Uniaxial SD Particles: Stoner-Wohlfarth Model"
M^*	Rescaled magnetization curve	Section "Coercivity Distributions"
M_a	Anhysteretic remanent magnetization	Section "ARM Acquisition and Demagnetization"
M_{af}	AF demagnetization curve	Section "AF Demagnetization of IRM"
M_{cal}	Calibration measurement (FORC protocol)	Section "The FORC Protocol"
M_{cr}	Central ridge magnetization	Section "Thermally Activated SD Particles"
M_i	Initial magnetization curve	Section "IRM Acquisition Curves"
M_{dcd}	DC demagnetization curve	Section "DC Demagnetization Curves"
M_{forc}	Total integral of the FORC function	Section "Selected Properties of the FORC Function"
M_{irm}	IRM acquisition curve	Section "IRM Acquisition Curves"
M_{irr}	Irreversible component of the major hysteresis branch	Section "Selected Properties of the FORC Function"
M_{rev}	Reversible component of the major hysteresis branch	Section "Selected Properties of the FORC Function"

(continued)

Symbol	Meaning	Reference
M_{rs}	Saturation remanence	Section "IRM Acquisition Curves"
M_s	Saturation magnetization	Section "The Classical Preisach Model"
M_{si}	Saturation initial curve	Section "Selected Properties of the FORC Function"
M_{ts}	Transient irreversible magnetization	Section "Selected Properties of the FORC Function"
MD	Multidomain (domain state)	—
p	Preisach function	Section "The Classical Preisach Model", Fig. 2
$p^{\pm}$	Occupation probability of magnetic states	Section "Thermally Activated SD Particles"
PSD	Pseudo-single domain (domain state)	—
q	Power law exponent	Section "Modifications of the Classical Preisach Model"
R_{sd}	Maximum SD grain size	Section "The Pseudo-Single-Domain Magnetic Behavior"
S	Smoothing factor (isotropic)	Section "Theoretical Principles of FORC Function Estimation"
$\overline{S}$	Mean normalized magnetization jump amplitude	Section "Weak Magnetostatic Interactions Between SD Particles"
s_c, s_u, s_r, s	Smoothing factors (VARIFORC)	Section "Improved Regression Methods"
s_c^0, s_u^0	Smoothing factors at the origin (VARIFORC)	Section "Variable Smoothing Protocols"
s_c^{vr}	Limiting smoothing factor for the vertical ridge	Section "The Vertical Ridge"
s_u^{cr}	Limiting smoothing factor for the central ridge	Section "The Central Ridge"
s^0	Limiting smoothing factor along H	Section "Smoothing Factor Limitations Along the Measurement Coordinates"
SD	Single domain (domain state)	—
SV	Single vortex (domain state)	—
S-W	Stoner-Wohlfarth particles	Section "Uniaxial SD Particles: Stoner-Wohlfarth Model"
t	Time	—
t_{cal}	Waiting time before calibration (FORC protocol)	Section "The FORC Protocol"
t_m	Measurement time (FORC protocol)	Section "The FORC Protocol"
t_r	Waiting time at reversal field (FORC protocol)	Section "The FORC Protocol"
t_{sat}	Duration of H_{sat} (FORC protocol)	Section "The FORC Protocol"
T	Absolute temperature or period	—

(continued)

Symbol	Meaning	Reference
V	Particle volume	Section "Single-Domain Particles"
W	Distribution of interaction fields	Section "Weak Magnetostatic Interactions Between SD Particles"
α	Field sweep rate	Section "Modifications of the Classical Preisach Model"
α_m	Field sweep rate of measuring sweeps (FORC)	Section "The FORC Protocol"
α_R	Rayleigh constant	Section "Low-Field Susceptibility"
α_0	Field sweep rate of non-measuring sweeps (FORC)	Section "The FORC Protocol"
$\alpha_1, \alpha_2, \alpha_3$	Angles to $\langle 100 \rangle$ axes	Section "SD Particles with Multiaxial Anisotropy"
β	Boltzmann factor	Section "Thermally Activated SD Particles"
β_0	Boltzmann factor in zero field	Sections "Processing FORC Data Associated with Thin Hysteresis Loops," "Thermally Activated SD Particles"
$\hat{\gamma}_{cr}$	Background function under the central ridge	Section "The Central Ridge"
γ	Mean magnetic state of hysterons	Section "The Classical Preisach Model"
δ	Dirac impulse	—
δ_w	Bloch wall width	Section "The Pseudo-Single-Domain Magnetic Behavior"
η	Packing fraction of particles	Section "Weak Magnetostatic Interactions Between SD Particles"
κ	Magnetic hardness	Section "The Pseudo-Single-Domain Magnetic Behavior"
λ_c, λ_u	Smoothing factor increasing rates (VARIFORC)	Section "Variable Smoothing Protocols"
μ_s	Spontaneous magnetization	Section "Single-Domain Particles"
μ_0	Vacuum permeability	—
$v^{\pm}$	Transition rates	Section "Thermally Activated SD Particles"
ξ	Everett function, FORC magnetization function	Sections "Definitions," "Theoretical Principles of FORC Function Estimation"
ρ	FORC function	Section "Definitions"
$\hat{\rho}$	Estimated FORC function	Section "Theoretical Principles of FORC Function Estimation"
$\rho_{cr}, \hat{\rho}_{cr}$	Central ridge and its estimate	Section "The Central Ridge"
ρ_{vr}	Vertical ridge	Section "The Vertical Ridge"
τ	Viscous decay time	Section "Thermally Activated SD Particles"
τ_0	Characteristic time of thermal fluctuations	Section "Modifications of the Classical Preisach Model"
χ_a	ARM susceptibility	Section "ARM Acquisition and Demagnetization"

(continued)

Symbol	Meaning	Reference
χ_i	Initial susceptibility (ferrimagnetic materials)	Section "Low-Field Susceptibility"
χ_{lf}	Low-field susceptibility	Section "Low-Field Susceptibility"
χ_{hf}	High-field susceptibility	Section "Field Control, Measurement Sensitivity, and Resolution"
χ_{sp}	Superparamagnetic susceptibility	Section "Processing FORC Data Associated with Thin Hysteresis Loops"

ing does not necessarily lead to the isolation of physically meaningful endmembers. Therefore, further advances in the magnetic characterization of geologic materials must be based on a mixed approach that combines fast magnetic characterization techniques with FORC diagrams, forward modeling, and numerical unmixing.

References

1. R. Egli, VARIFORC: an optimized protocol for calculating non-regular first-order reversal curve (FORC) diagrams. Glob. Planet. Chang. **110**, 302–320 (2013)
2. F. Preisach, Über die magnetische Nachwirkung. Z. Phys. **94**, 277–302 (1935)
3. D.H. Everett, W.I. Whitton, A general approach to hysteresis. Trans. Faraday Soc. **48**, 749–757 (1952)
4. M. Krasnoselskii, A. Pokrovskii, *Systems with Hysteresis* (Nauka, Moskow, 1983)
5. I. Mayergoyz, *Mathematical Models of Hysteresis and Their Applications* (Elsevier, 2003)
6. L. Néel, Sur les effets d'un couplage entre grains ferromagnétiques. C.R. Acad. Sci. **246**, 2313–2319 (1958)
7. E.C. Stoner, E.P. Wohlfarth, A mechanism of magnetic hysteresis in heterogeneous alloys. Philos. Trans. Royal Soc. Lond. A **240**, 599–642 (1948)
8. J.D. Woodward, E. DellaTorre, Particle interaction in magnetic recording tapes. J. Appl. Phys. **31**, 56–62 (1960)
9. G. Bate, Statistical stability of the Preisach diagram for particles of γ-Fe_2O_3. J. Appl. Phys. **33**, 2263–2269 (1962)
10. E.D. Daniel, I. Levine, Experimental and theoretical investigation of the magnetic properties of iron oxide recording tape. J. Acoust. Soc. Am. **32**, 1–15 (1960)
11. D. Dunlop, Preisach diagrams and remanent properties of interacting monodomain grains. Philos. Mag. **19**, 369–378 (1969)
12. D.J. Dunlop, M.F. Wescott-Lewis, M.E. Bailey, Preisach diagrams and anhysteresis: do they measure interactions? Phys. Earth Planet. Inter. **65**, 62–77 (1990)
13. V.I. Belokon, V.V. Kochegura, L.E. Sholpo, *Methods of paleomagnetic studies of rocks (in Russian)* (Nedra, Leningrad, 1973), 248 p
14. P. Hejda, T. Zelinka, Modeling of hysteresis processes in magnetic rock samples using the Preisach diagram. Phys. Earth Planet. Sci. Lett. **63**, 32–40 (1990)
15. C.R. Pike, A.P. Roberts, K.L. Verosub, Characterizing interactions in fine magnetic particle systems using first order reversal curves. J. Appl. Phys. **85**, 6660–6667 (1999)
16. A.P. Roberts, C.R. Pike, K.L. Verosub, First-order reversal curve diagrams: a new tool for characterizing the magnetic properties of natural samples. J. Geophys. Res. **105**, 28461–28475 (2000)

17. C. Carvallo, D.J. Dunlop, Ö. Özdemir, Experimental comparison of FORC and remanent Preisach diagrams. Geophys. J. Int. **162**, 747–754 (2005)

18. N.S. Church, K. Fabian, S.A. McEnroe, Nonlinear Preisach maps: Detecting and characterizing separate remanent magnetic fractions in complex natural samples. J. Geophys. Res. Solid Earth **121**, 8373–8395 (2016)

19. M.W. Gutowski, On the symmetry of a Preisach map. Phys. Stat. Sol. B **243**, 343–346 (2006)

20. C.R. Pike, First-order reversal-curve diagrams and reversible magnetization. Phys. Rev. B **68**, 104424 (2003)

21. V. Basso, G. Bertotti, Description of magnetic interactions and Henkel plots by the Preisach hysteresis model. IEEE Trans. Magn. **30**, 64–72 (1994)

22. M. Cerchez, L. Stoleriu, A. Stancu, Interaction effects in high density magnetic particulate media. Physica B **343**, 48–52 (2004)

23. A. Muxworthy, W. Williams, Magnetostatic interaction fields in first-order-reversal-curve diagrams. J. Appl. Phys. **97**, 063905 (2005)

24. L. Néel, Some theoretical aspects of rock magnetism. Adv. Phys. **4**, 191–243 (1955)

25. V. Basso, C. Beatrice, M. LoBue, P. Tiberto, G. Bertotti, Connection between hysteresis and thermal relaxation in magnetic materials. Phys. Rev. B **61**, 1278–1285 (2000)

26. B.M. Moskowitz, R.B. Frankel, S.A. Walton, P.E. Dickson, K.K.W. Wong, T. Douglas, S. Mann, Determination of the preexponential frequency factor for superparamagnetic maghemite particles in magnetoferritin. J. Geophys. Res. **102**, 22671–22680 (1997)

27. R.H. Victora, Predicted time dependence of the switching field for magnetic materials. Phys. Rev. Lett. **63**, 457–460 (1989)

28. A. Stancu, L. Spinu, Temperature- and time-dependent Preisach model for a Stoner-Wohlfarth particle system. IEEE Trans. Magn. **34**, 3867–3875 (1998)

29. A.R. Muxworthy, D. Heslop, A Preisach method for estimating absolute paleofield intensity under the constraint of using only isothermal measurements: 1.Theoretical framework. J. Geophys. Res. **116**, B04102 (2011)

30. D. Heslop, G. McIntosh, M.J. Dekkers, Using time- and temperature-dependent Preisach models to investigate the limitations of modelling isothermal remanent magnetization acquisition curves with cumulative log Gaussian functions. Geophys. J. Int. **157**, 55–63 (2004)

31. D.J. Dunlop, Ö. Özdemir, *Rock Magnetism* (Cambridge University Press, Cambridge, 1997)

32. M.E. Evans, F. Heller, *Environmental Magnetism: Principles and Applications of Enviromagnetics* (Academic Press, Amsterdam, 2003)

33. A.P. Roberts, P. Hu, R.J. Harrison, D. Heslop, A. Muxworthy, H. Oda, T. Sato, L. Tauxe, X. Zhao, Domain state diagnosis in rock magnetism: evaluation of potential alternatives to the Day diagram. J. Geophys. Res. Solid Earth **124**, 5286–5314 (2019)

34. R. Day, M. Fuller, V.A. Schmidt, Hysteresis properties of titanomagnetites: grain size and composition dependence. Phys. Earth Planet. Inter. **13**, 260–267 (1977)

35. J. King, S.K. Banerjee, J. Marvin, Ö. Özdemir, A comparison of different magnetic methods for determining the relative grain size of magnetite in natural materials: some results from lake sediments. Earth Planet. Sci. Lett. **59**, 404–419 (1982)

36. S.J. Watkins, B.A. Maher, Magnetic characterisation of present-day deep-sea sediments and sources in the North Atlantic. Earth Planet. Sci. Lett. **214**, 379–394 (2003)

37. S.L. Halgedahl, Revisiting the Lowrie-Fuller test: Alternating field demagnetization characteristics of single-domain through multidomain glass-ceramic magnetite. Earth Planet. Sci. Lett. **160**, 257–271 (1998)

38. S. Cisowski, Interacting vs. non-interacting single domain behavior in natural and synthetic samples. Phys. Earth Planet. Inter. **26**, 56–62 (1981)

39. D.J. Robertson, D.E. France, Discrimination of remanence-carrying minerals in mixtures, using isothermal remanent magnetisation acquisition curves. Phys. Earth Planet. Inter. **82**, 223–234 (1994)

40. P.P. Kruiver, M.J. Dekkers, D.J. Heslop, Quantification of magnetic coercivity components by the analysis of acquisition curves of isothermal remanent magnetisation. Earth Planet. Sci. Lett. **189**, 269–276 (2001)

41. R. Egli, Analysis of the field dependence of remanent magnetization curves. J. Geophys. Res. **108**, 2081 (2003)
42. Z. Gong, M.J. Dekkers, D. Heslop, T.A.T. Mullender, End-member modelling of isothermal remanent magnetization (IRM) acquisition curves: a novel approach to diagnose remagnetization. Geophys. J. Int. **178**, 693–701 (2009)
43. D. Heslop, A.P. Roberts, A method for unmixing magnetic hysteresis loops. J. Geophys. Res. **117**, B03103 (2012)
44. R.J. Harrison, J. Muraszko, D. Heslop, I. Lascu, A.R. Muxworthy, A.P. Roberts, An improved algorithm for unmixing first-order reversal curve diagrams using principal component analysis. Geochem. Geophys. Geosyst. **19**, 1595–1610 (2018)
45. X. Zhao, A.P. Roberts, D. Heslop, G. Paterson, Y. Li, J. Li, Magnetic domain state diagnosis using hysteresis reversal curves. J. Geophys. Res. Solid Earth **122**, 4767–4789 (2017)
46. F. Hrouda, Low-field variation of magnetic susceptibility and its effect on the anisotropy of magnetic susceptibility of rocks. Geophys. J. Int. **150**, 715–723 (2002)
47. L., R.S. Rayleigh Sec, Notes on electricity and magnetism—III. On the behaviour of iron and steel under the operation of feeble magnetic forces. Philos. Mag. Ser. **5**(23), 225–245
48. M. Jackson, B. Moskowitz, J. Rosenbaum, C. Kissel, Field-dependence of AC susceptibility in titanomagnetites. Earth Planet. Sci. Lett. **157**, 129–139 (1998)
49. H.-U. Worm, D. Clark, M.J. Dekkers, Magnetic susceptibility of pyrrhotite: grain size, field and frequency dependence. Geophys. J. Int., 127–137 (1993)
50. P. Dankers, Relationship between median destructive field and remanent coercive forces for dispersed natural magnetite, titanomagnetite and hematite. Geophys. J. R. Astron. Soc. **64**, 447–461 (1981)
51. E. McClelland, V.P. Shcherbakov, Metastability of domain state in multidomain magnetite: Consequences for remanence acquisition. J. Geophys. Res. **100**, 3841–3857 (1995)
52. E.P. Wohlfarth, Relations between different modes of acquisition of the remanent magnetization of ferromagnetic particles. J. Appl. Phys. **29**, 595–596 (1958)
53. O. Henkel, Remanenzverhalten und Wechselwirkungen in hartmagnetischen Teilchenkollektiven. Phys. Status Solidi **7**, 919–929 (1964)
54. P.D. Mitchler, E. Dan Dahlberg, E. Engle, R.M. Roshko, Henkel plots in a thermally demagnetized scalar Preisach model. IEEE Trans. Magn. **31**, 2499–2503 (1995)
55. R. Proksch, B. Moskowitz, Interactions between single domain particles. J. Appl. Phys. **75**, 5894–5896 (1994)
56. P. Hejda, E. Petrovský, T. Zelinka, The Preisach diagram, Wohlfarth's remanence formula and magnetic interactions. IEEE Trans. Magn. **30**, 896–898 (1994)
57. V. Basso, M. Lo Bue, G. Bertotti, Interpretation of hysteresis curves and Henkel plots by the Preisach model (invited). J. Appl. Phys. **75**, 5677–5682 (1994)
58. D.J. Dunlop, On the demagnetizing energy and demagnetizing factor of a multidomain ferromagnetic cube. Geophys. Res. Lett. **10**, 79–82 (1983)
59. W. Lowrie, M. Fuller, On the alternating field demagnetization characteristics of multidomain thermoremanent magnetization in magnetite. J. Geophys. Res. **76**, 6339–6349 (1971)
60. M. Fuller, T. Kidane, J. Ali, AF demagnetization characteristics of NRM, compared with anhysteretic and saturation isothermal remanence: an aid in the interpretation of NRM. Phys. Chem. Earth **27**, 1169–1177 (2002)
61. D.F. McNeill, Biogenic magnetite from surface Holocene carbonate sediments, Great Bahama Bank. J. Geophys. Res. **95**, 4363–4371 (1990)
62. J.L. Kirschvink, A.T. Maine, H. Vali, Paleomagnetic evidence of a low-temperature origin of carbonate in the Martian meteorite ALH84001. Science **275**, 1629–1633 (1997)
63. J.C. Larrasoaña, J.M. Parés, E.L. Pueyo, Stable Eocene magnetization carried by magnetite and iron sulphides in marine marls (Pamplona-Arguis formation, southern Pyrenees, Northern Spain). Stud. Geophys. Geodaet. **47**, 237–254 (2003)
64. R.I.F. Trinidade, M.S. D'Agrella-Filho, M. Babinski, E. Font, B.B. Brito Neves, Paleomagnetism and geochronology of the Bebedouro cap carbonate: evidence for continental-scale Cambrian remagnetization in the São Francisco craton, Brazil. Precambrian Res. **128**, 83–103 (2004)

65. E. Font, R.I.F. Trinidade, A. Nédélec, Remagnetization in bituminous limestones of the Neoproterozoic Araras Group (Amazon craton): Hydrocarbon maturation, burial diagenesis, or both? J. Geophys. Res. **111**, B06204 (2006)

66. E. Font, C. Veiga-Pires, M. Pozo, C. Carvallo, A.C. de Siqueira Neto, P. Camps, S. Fabre, J. Mirão, Magnetic fingerprint of southern Portuguese speleothems and implications for paleomagnetism and environmental magnetism. J. Geophys. Res. Solid Earth **119**, 7993–8020 (2014)

67. S. Levi, R.T. Merrill, A comparison of ARM and TRM in magnetite. Earth Planet. Sci. Lett. **32**, 171–184 (1976)

68. Y. Yu, D.J. Dunlop, Ö. Özdemir, Are ARM and TRM analogs? Thellier analysis of ARM and pseudo-Thellier analysis of TRM. Earth Planet. Sci. Lett. **205**, 325–336 (2003)

69. G.A. Paterson, D. Heslop, Y. Pan, The pseudo-Thellier paleointensity method: new calibration and uncertainty estimates. Gephys. J. Int. **207**, 1596–1608 (2016)

70. G.A. Lerner, A.V. Smirnov, L.V. Surovitckii, E.J. Piispa, Nonheating methods for absolute paleointensity determination: Comparison and calibration using synthetic and natural magnetite-bearing samples. J. Geophys. Res. Solid Earth **122**, 1614–1633 (2017)

71. J.-P. Valet, L. Meynadier, A comparison of different techniques for relative paleointensity. Geophys. Res. Lett. **25**, 89–92 (1998)

72. S.L. Halghedahl, R. Day, M. Fuller, The effect of cooling rate on the intensity of weak-field TRM in single-domain magnetite. J. Geophys. Res. **85**, 3690–3698

73. R. Egli, W. Lowrie, Anhysteretic remanent magnetization of fine magnetic particles. J. Geophys. Res. **107**, 2209 (2002)

74. D.J. Dunlop, Thermoremanent magnetization of nonuniformly magnetized grains. J. Geophys. Res. **103**, 30561–30574 (1998)

75. D.J. Dunlop, A.J. Newell, R.J. Enkin, Transdomain thermoremanent magnetization. J. Geophys. Res. **99**, 19741–19755 (1994)

76. K. Fabian, Statistical theory of weak field thermoremanent magnetization in multidomain particle ensembles. Geophys. J. Int. **155**, 479–488 (2003)

77. S. Xu, D. Dunlop, Toward a better understanding of the Lowrie-Fuller test. J. Geophys. Res. **100**, 22533–22542 (1995)

78. N. Sugiura, ARM, TRM and magnetic interactions: concentration dependence. Earth Planet. Sci. Lett. **42**, 451–455 (1979)

79. V.P. Shcherbakov, N.K. Sycheva, B.E. Lamash, Monte Carlo modelling of TRM and CRM acquisition and comparison of their properties in an ensemble of interacting SD grains. Geophys. Res. Lett. **23**, 2827–2830 (1996)

80. E.P. Wohlfarth, A review of the problem of fine-particle interactions with special reference to magnetic recording. J. Appl. Phys. **35**, 783–790 (1964)

81. E. Kneller, Magnetic-interaction effects in fine-particle assemblies and in thin films. J. Appl. Phys. **39**, 945–955 (1968)

82. W.F. Jaep, Anhysteretic magnetization of an assembly of single-domain particles. J. Appl. Phys. **40**, 1297–1298 (1969)

83. R. Egli, Theoretical considerations on the anhysteretic remanent magnetization of interacting particles with uniaxial anisotropy. J. Geophys. Res. **111**, B12S18 (2006)

84. R. Egli, Characterization of individual rock magnetic components by analysis of remanence curves, 1. Unmixing natural sediments. Stud. Geophys. Geodaetica **48**, 391–446 (2004)

85. R. Egli, Theoretical aspects of dipolar interactions and their appearance in first-order reversal curves of thermally activated single-domain particles. J. Geophys. Res. **111**, B12S17 (2006)

86. D. Heslop, M. Dillon, Unmixing magnetic remanence curve without a priori knowledge. Geophys. J. Int. **170**, 556–566 (2007)

87. R. Egli, Characterization of individual rock magnetic components by analysis of remanence curves. 2. Fundamental properties of coercivity distributions. Phys. Chem. Earth **29**, 851–867 (2004)

88. M. Pardavi-Horvath, E. Della Torre, F. Vajda, A variable variance Preisach model. IEEE Trans. Magn. **29**, 3793–3795 (1993)

89. X. Zhao, D. Heslop, A.P. Roberts, A protocol for variable-resolution first-order reversal curve measurements. Geochem. Geophys. Geosyst. **16**, 1364–1377 (2015)

90. D. Heslop, A. Roberts, H. Oda, X. Zhao, R.J. Harrison, A.R. Muxworthy, P.X. Hu, T. Sato, An automatic model selection-based machine learning framework to estimate FORC distributions. J. Geophys. Res. Solid Earth **125**, e2020JB020418 (2020)

91. D. Everett, H. A general approach to hysteresis—Part 4. An alternative formulation of the domain model. Trans. Faraay Soc. **51**, 1551–1557 (1955)

92. C.R. Pike, C.A. Ross, R.T. Scalettar, G. Zimanyi, First-order reversal curve diagram analysis of a perpendicular nickel nanopillar array. Phys. Rev. B **71**, 134407 (2005)

93. A.J. Newell, A high-precision model of first-order reversal curve (FORC) functions for single-domain ferromagnets with uniaxial anisotropy. Geochem. Geophys. Geosys. **6**, Q05010 (2005)

94. R. Egli, A.P. Chen, M. Winklhofer, K.P. Kodama, C.S. Horng, Detection of noninteracting single domain particles using first-order reversal curve diagrams. Geochem. Geophys. Geosys. **11**, Q01Z11 (2010)

95. L. Stoleriu, A. Stancu, Using experimental FORC distribution as input for a Preisach-type model. IEEE Trans. Magn. **42**, 3159–3161 (2006)

96. R. Egli, M. Winklhofer, Recent developments on processing and interpretation aspects of first-order reversal curves (FORC). Sci. Proc. Kazan Federal Univ. **156**, 14–53 (2014)

97. K. Fabian, T. von Dobeneck, Isothermal magnetization of samples with stable Preisach function: A survey of hysteresis, remanence, and rock magnetic parameters. J. Geophys. Res. **102**, 17659–17677 (1997)

98. K. Fabian, Some additional parameters to estimate domain state from isothermal magnetization measurements. Earth Planet. Sci. Lett. **213**, 337–345 (2003)

99. D. Heslop, A.P. Roberts, L. Chang, Characterizing magnetofossils from first-order reversal curve (FORC) central ridge signatures. Geochem. Geophys. Geosyst. **15**, 2170–2179 (2014)

100. A.P. Roberts, D. Heslop, X. ZHao, C.R. Pike, Understanding fine magnetic particle systems through use of first-order reversal curve diagrams. Rev. Geophys. **52**, 557–602 (2014)

101. C.E. Geiss, R. Egli, W. Zanner, Direct estimates of pedogenic magnetite as a tool to reconstruct past climates from buried soils. J. Geophys. Res. **113**, B11102 (2008)

102. A.R. Muxworthy, D.J. Dunlop, First-order reversal curve (FORC) diagrams for pseudo-single-domain magnetites at high temperature. Earth Planet. Sci. Lett. **203**, 369–382 (2002)

103. P. Ludwig, R. Egli, S. Bishop, V. Chernenko, T. Frederichs, G. Rugel, S. Merchel, M.J. Orgeira, Characterization of primary and secondary magnetite in marine sediment by combining chemical and magnetic unmixing techniques. Glob. Planet. Chang. **110**, 321–339 (2013)

104. A.P. Roberts, T.P. Almeida, N.S. Church, R.J. Harrison, D. Heslop, Y. Li, J. Li, A.R. Muxworthy, W. Williams, X. Zhao, Resolving the origin of pseudo-single domain magnetic behavior. J. Geophys. Res. **122**, 9534–9558 (2017)

105. C. Carvallo, Ö. Özdemir, D.J. Dunlop, First-order reversal curve (FORC) diagrams of elongated single-domain grains at high and low temperatures. J. Geophys. Res. **109**, B04105 (2004)

106. A.P. Roberts, L. Chang, D. Heslop, F. Florindo, J.C. Larrasoaña, Searching for single domain magnetite in the "pseudo-single-domain" sedimentary haystack: Implications of biogenic magnetite preservation for sediment magnetism and relative paleointensity determination. J. Geophys. Res. **117**, B08104 (2012)

107. C.R. Pike, A.P. Roberts, M.J. Dekkers, K.L. Versoub, An investigation of multi-domain hysteresis mechanisms using FORC diagrams. Phys. Earth Planet. Inter. **126**, 11–25 (2001)

108. N. Church, J.M. Feinberg, R. Harrison, Low-temperature domain wall pinning in titano-magnetite: quantitative modeling of multidomain first-order reversal curve diagrams and AC susceptibility. Geochem. Geophys. Geosyst. **12**, Q07Z27 (2011)

109. M.J. Dekkers, Magnetic properties of natural pyrrhotite Part I: Behaviour of initial susceptibility and saturation-magnetization-related rock-magnetic parameters in a grain-size dependent framework. Phys. Earth Planet. Inter. **52**, 376–393 (1988)

110. P. Rochette, P.-E. Mathé, L. Esteban, H. Rakoto, J.-L. Bouchez, Q. Liu, J. Torrent, Non-saturation of the defect moment of goethite and fine-grained hematite up to 57 Teslas. Geophys. Res. Lett. **32**, L22309 (2005)

111. A.P. Roberts, Q. Liu, C.J. Rowan, L. Chang, C. Carvallo, J. Torrent, C.-S. Horng, Characterization of hematite (α-Fe_2O_3), goethite (α-FeOOH), greigite (Fe_3S_4), and pyrrhotite (Fe_7S_8) using first-order reversal curve diagrams. J. Geophys. Res. **111**, B12S35 (2006)

112. M. Ahmadzadeh, C. Romero, J. McCloy, Magnetic analysis of commercial hematite, magnetite, and their mixtures. AIP Adv. **8**, 056807 (2018)

113. P. Liu, A.M. Hirt, D. Schüler, R. Uebe, P. Zhu, T. Liu, H. Zhang, Numerical unmixing of weakly and strongly magnetic minerals: examples with synthetic mixtures of magnetite and hematite. Geophys. J. Int. **217**, 280–287 (2019)

114. C.R. Pike, A.P. Roberts, K.L. Verosub, First-order reversal curve diagrams and thermal relaxation effects in magnetic particles. Geophys. J. Int. **145**, 721–730 (2001)

115. S.J. Brownlee, J.M. Feinberg, T. Kasama, R.J. Harrison, G.R. Scott, P.R. Renne, Magnetic properties of ilmenite-hematite single crystals from Ecstall pluton near Prince Rupert, British Columbia. Geochem. Geophys. Geosyst. **12**, Q07Z29 (2011)

116. L. Lanci, D.V. Kent, Forward modeling of thermally activated single-domain magnetic particles applied to first-order reversal curves. J. Geophys. Res. **123** (2018). https://doi.org/10.1002/2018JB015463

117. T.A. Berndt, L. Chang, S. Wang, S. Badejo, Time-asymmetric FORC diagrams: a new protocol for visualizing thermal fluctuations and distinguishing magnetic mineral mixtures. Geochem. Geophys. Geosyst. **19**, 1595–1610 (2018)

118. Spassov, S., Egli, R. Method for correcting remanent fields in magnetic property measurement systems without field compensation. COST Action TD1402 RADIOMAG, 2016

119. C.L. Wagner, R. Egli, I. Lascu, P.C. Lippert, K.J.T. Livi, H.B. Sears, In-situ identification of giant, needle-shaped magnetofossils in Paleocene-Eocene Thermal Maximum sediments. Proc. Natl. Acad. Sci. U. S. A. **118** (2021). https://doi.org/10.1073/2018169118

120. S. Foner, Versatile and sensitive vibrating-sample magnetometer. Rev. Sci. Instrum. **30**, 548–557 (1959)

121. M. Jackson, P. Solheid, On the quantitative analysis and evaluation of magnetic hysteresis data. Geochem. Geophys. Geosyst. **11**, Q04Z15 (2010)

122. M. Winklhofer, G.T. Zimanyi, Extracting the intrinsic switching field distribution in perpendicular media: a comparative analysis. J. Appl. Phys., 08E710 (2006)

123. T.A. Berndt, L. Chang, Waiting for Forcot: accelerating FORC processing $100\times$ using a Fast-Fourier-Transform algorithm. Geochem. Geophys. Geosyst. **20**, 6223 (2019)

124. A. Savitzky, M.J.E. Golay, Smoothing and differentiation of data by simplified least-squares procedures. Anal. Chem. **36**, 1627–1639 (1964)

125. D. Heslop, A.R. Muxworthy, Aspects of calculating first-order reversal curve distributions. J. Magn. Magn. Mater. **288**, 155–167 (2005)

126. W.S. Cleveland, Robust locally weighted regression and smoothing scatterplots. J. Am. Stat. Assoc. **74**, 829–836 (1979)

127. W.S. Cleveland, S.J. Devlin, Locally weighted regression: an approach to regression analysis by local fitting. J. Am. Stat. Assoc. **83**, 596–610 (1988)

128. R.J. Harrison, J.M. Feinberg, FORCinel: an improved algorithm for calculating first-order reversal curve distributions using locally weighted regression smoothing. Geochem. Geophys. Geosyst. **9**, Q05016 (2008)

129. E. Katzmann, M. Eibauer, W. Lin, Y. Pan, J.M. Plitzko, D. Schüler, Analysis of magnetosome chains in magnetotactic bacteria by magnetic measurements and automated image analysis of electron micrograph. Appl. Environ. Microbiol. **79**, 7755–7762 (2013)

130. D. Heslop, A.P. Roberts, Estimation of significance levels and confidence intervals for first-order reversal curve distributions. Geochem. Geophy. Geosyst. **13**, Q12Z40 (2012)

131. J.G. MacKinnon, H. White, Some heteroskedasticity-consistent covariance matrix estimators with improved finite sample properties. J. Econ. **29**(29), 305–325 (1985)

132. J.S. Long, L.H. Erwin, Using heteroscedasticity consistent standard errors in the linear regression model. Ann. Stat. **54**, 217–224 (2000)
133. J.V. Bradley, *Distribution-Free Statistical Tests* (Prentice-Hall, 1968), p. 388
134. J. Li, K. Ge, Y. Pan, W. Williams, Q. Liu, H. Qin, A strong angular dependence of magnetic properties of magnetosome chains: Implications for rock magnetism and paleomagnetism. Geochem. Geophys. Geosyst. **14**, 3887–3907 (2013)
135. L. Chang, R.J. Harrison, T.A. Berndt, Micromagnetic simulations of magnetofossils with realistic size and shape distributions: Linking magnetic proxies with nanoscale observations and implications for magnetofossil identification. Earth Planet. Sci. Lett. **527**, 115790 (2019)
136. L. Tauxe, T.A.T. Mullender, T. Pick, Potbellies, wasp-waists, and superparamagnetism in magnetic hysteresis. J. Geophys. Res. **101**, 571–583 (1996)
137. C. Pike, A. Fernandez, An investigation of mangetic reversal in submicron-scale Co dots using first order reversal curve diagrams. J. Appl. Phys. **85**, 6668–6676 (1999)
138. G. Acton, Q.Z. Yin, K.L. Verosub, L. Jovane, A. Roth, B. Jacobsen, D.S. Ebel, Micromagnetic coercivity distributions and interactions in chondrules with implications for paleointensities of the early solar system. J. Geophys. Res. **112**, B03S90 (2007)
139. T. Schrefl, T. Shoji, M. Winklhofer, H. Oezelt, M. Yano, G. Zimanyi, First order reversal curve studies of permanent magnets. J. Appl. Phys. **111**, 07A728 (2012)
140. H. Wang, D.V. Kent, M.J. Jackson, Evidence for abundant isolated magnetic nanoparticles at the Paleocene-Eocene boundary. Proc. Natl. Acad. Sci. U. S. A. **110**, 425–430 (2013)
141. J.A. Osborn, Demagnetizing factors of the general ellipsoid. Phys. Rev. **67**, 351–357 (1945)
142. R.I. Joseph, Ballistic demagnetizing factor in uniformly magnetized cylinders. J. Appl. Phys. **37**, 4639–4643 (1966)
143. A. Aharoni, Demagnetizing factors for rectangular ferromagnetic prisms. J. Appl. Phys. **83**, 3432–3434 (1998)
144. Y. Cao, K. Xu, W. Jiang, T. Droubay, P. Ramuhalli, D. Edwards, B.R. Johnson, J. McCloy, Hysteresis in single and polycrystalline iron thin films: Major and minor loops, first order reversal curves, and Preisach modeling. J. Magn. Magn. Mater. **395**, 361–375 (2015)
145. M. Kono, Magnetic properties of DSDP Leg 55 basalts. Deep Sea Drilling Project Rep. Public. **40**, 723–736 (2007)
146. J.P. McDermott, D.K. Lin, Quantile contours and multivariate density estimation for massive datasets via sequential convex hull peeling. IIE Trans. **39**, 581–591 (2007)
147. L. Néel, Théorie du traînage magnétique des ferromagnétiques en grains fins avec applications aux terres cuites. Ann. Géophys. **5**, 99–136 (1949)
148. W.F. Brown, Thermal fluctuation of a single domain particle. Phys. Rev. **130**, 1677–1686 (1963)
149. G. Bertotti, *Hysteresis in Magnetism.* (Academic Press, 1998) 978-0-12-093270-2
150. J.J. Lu, H.L. Huang, I. Kilk, Field orientations and sweep rate effects on magnetic switching of Stoner-Wohlfarth particles. J. Appl. Phys. **76**, 1726–1732 (1994)
151. D. York, A formula describing both magnetic and isotopic blocking temperatures. Earth Planet. Sci. Lett. **39**, 89–93 (1978)
152. D.J. Dunlop, The rock magnetism of fine particles. Phys. Earth Planet. Inter. **26**, 1–26 (1981)
153. C. Carvallo, S. Hickey, D. Faivre, N. Menguy, Formation of magnetite in Magnetospirillum gryphiswaldense studied with FORC diagrams. Earth Planets Space **61**, 143–150 (2009)
154. J. Li, Y. Pan, G. Chen, Q. Liu, L. Tian, W. Lin, Magnetite magnetosome and fragmental chain formation of Magnetospirillum magneticum AMB-1: transmission electron microscopy and magnetic observations. Geophys. J. Int. **177**, 33–42 (2009)
155. Y.G. Velázquez, A.L. Guerrero, J.B. Martinez, E. Araujo, M.R. Tabasum, B. Nysten, L. Piraux, A. Encinas, Relation of the average interaction field with the coercive and interaction field distributions in First order reversal curve diagrams of nanowire arrays. Sci. Rep. **10**, 21396 (2020)
156. I.S. Jacobs, C.P. Bean, An approach to elongated fine-particle magnets. Phys. Rev. **100**, 1060–1067 (1955)

157. M.A. Valdez-Grijalva, L. Nagy, A.R. Muxworthy, W. Williams, A.P. Roberts, D. Heslop, Micromagnetic simulations of first-order reversal curve (FORC) diagrams of framboidal greigite. Geophys. J. Int. **222**, 1126–1134 (2020)
158. A. Stephenson, Distribution anisotropy: Two simple models for magnetic lineation and foliation. Phys. Earth Planet. Inter. **82**, 49–53 (1994)
159. J. Lam, Magnetic hysteresis of a rectangular lattice of interacting single-domain ferromagnetic spheres. J. Appl. Phys. **72**, 5792–5798 (1992)
160. A.R. Muxworthy, W. Williams, Distribution anisotropy: the influence of magnetic interactions on the anisotropy of magnetic remanence, in *Magnetic fabric: Methods and applications*, ed. by F. Lüneburg, C. M. Aubourg, C. Jackson, M. Martín-Hernández, (Geol. Soc., London, 2004), pp. 37–41
161. A. Stancu, C. Pike, L. Stoleriu, P. Postolache, D. Cimpoesu, Micromagnetic and Preisach analysis of the first order reversal curves (FORC) diagram. J. Appl. Phys. **93**, 6620–6622 (2003)
162. R.J. Harrison, R.E. Dunin-Borkowski, T. Kasama, E.T. Simpson, J.M. Feinberg, Properties of rocks and minerals - magnetic properties of rocks and minerals. Treat. Geophys. **2**, 579–630 (2007)
163. M.E. Evans, D. Krása, W. Williams, M. Winklhofer, Magnetostatic interactions in a natural magnetite-ulvöspinel system. J. Geophys. Res. **111**, B12S16 (2006)
164. F. Tang, R.J.M. Taylor, J.F. Einsle, J.F. Borlina, R.R. Fu, B.P. Weiss, H.M. Williams, W. WIlliams, L. Nagy, P.A. Midgley, E.A. Lima, E.A. Bell, T.M. Harrison, E.W. Alexander, R.J. Harrison, Secondary magnetite in ancient zircon precludes analysis of a Hadean geodynamo. Proc. Natl. Acad. Sci. U. S. A. **116**, 407–412 (2019)
165. M. Charilaou, M. Winklhofer, A.U. Gehring, Simulation of ferromagnetic resonance spectra of linear chains of magnetite. J. Appl. Phys. **109**, 093903 (2011)
166. W. Wernsdorfer, K. Hasselbach, A. Benoit, G. Cernicchiaro, D. Mailly, B. Barbara, L. Thomas, Measurement of the dynamics of the magnetization reversal in individual single-domain Co particles. J. Magn. Magn. Mater., 38–44 (1995)
167. Z. Kakol, J.M. Honig, Influence of deviations from ideal stoichiometry on the anisotropy parameters of magnetite $Fe_{3(1-\delta)}O_4$. Phys. Rev. B **40**, 9090–9097 (1989)
168. R. Aragón, Cubic magnetic anisotropy of nonstoichiometric magnetite. Phys. Rev. B **46**, 5334–5338 (1992)
169. K. Fabian, F. Heider, How to include magnetostriction in micromagnetic models of titanomagnetite grains. Geophys. Res. Lett. **23**, 2839–2842 (1996)
170. B.A. Maher, Magnetic properties of some synthetic sub-micron magnetites. Geophys. J. Int. **94**, 83–96 (1988)
171. D. Faivre, P. Agrinier, N. Mengui, P. Zuddas, K. Pachana, A. Gloter, J.-Y. Laval, F. Guyot, Mineralogical and isotopic properties of inorganic nanocrystalline magnetites. Geochim. Cosmochim. Acta **68**, 4395–4403 (2004)
172. C.E. Johnson, W.F. Brown, Theoretical magnetization curves for particles with cubic anisotropy. J. Appl. Phys. **32**, 243S–244S (1961)
173. N.A. Usov, S.E. Peschany, Theoretical hysteresis loops for single-domain particles with cubic anisotropy. J. Magn. Magn. Mater. **174**, 247–260 (1997)
174. M.A. Valdez-Grijalva, A.R. Muxworthy, First-order reversal curve (FORC) diagrams of nanomagnets with cubic magnetocrystalline anisotropy: a micromagnetic approach. J. Magn. Magn. Mater. **471**, 359–364 (2019)
175. X.D. Jiang, X. Zhao, Y.M. Chou, Q.S. Liu, A.P. Roberts, J.B. Ren, X.M. Sun, J.H. Li, X. Tang, X.Y. Zhao, C.C. Wang, Characterization and quantification of magnetofossils within abyssal manganese nodules from the western Pacific Ocean and implications for nodule formation. Geochem. Geophys. Geosyst. **21**, e2019GC008811
176. R.F. Butler, S.K. Banerjee, Theoretical single-domain grain size range in magnetite and titanomagnetite. J. Geophys. Res. **197, 80**, 4049–4058
177. A.R. Muxworthy, W. Williams, Critical single-domain/multidomain grain sizes in noninteracting and interacting elongated magnetite particles: Implications for magnetosomes. J. Geophys. Res. **111**, B12S12 (2006)

178. A.J. Newell, Transition to superparamagnetism in chains of magnetosome crystals. Geochem. Geophys. Geosyst. **10**, Q11Z08 (2009)
179. A.R. Muxworthy, W. Williams, A.P. Roberts, M. Winklhofer, L. Chang, M. Pósfai, Critical single domain sizes in chains of interacting greigite particles: Implications for magnetosome crystals. Geochem. Geophys. Geosyst. **14**, 5430–5441 (2013)
180. L. Chang, A.P. Roberts, D. Heslop, A. Hayashida, J. Li, X. Zhao, W. Tian, Q. Huang, Widespread occurrence of silicate-hosted magnetic mineral inclusion in marine sediments and their contribution to paleomagnetic recording. J. Geophys. Res. Solid Earth **121**, 8415–8431
181. C. Franke, T. Frederichs, M.J. Dekkers, Efficiency of heavy liquid separation to concentrate magnetic particles. Geophys. J. Int. **170**, 1053–1066 (2007)
182. C. Shen, J. Beer, P.W. Kubik, M. Suter, M. Borkovec, T.S. Liu, Grain size distribution, 10-Be content, and magnetic susceptibility of micrometer-nanometer loess materials. Nucl. Instrum. Methods Phys. Res. B, 613–617 (2004)
183. N.J. Vidic, J.D. TenPas, K.L. Verosub, M.J. Singer, Separation of pedogenic and lithogenic components of magnetic susceptibility in the Chinese loess/paleosol sequence as determined by the CBD procedure and a mixing analysis. Geophys. J. Int. **142**, 551–562 (2000)
184. M. Jackson, B. Carter-Stiglitz, R. Egli, P. Solheid, Characterizing the superparamagnetic grain distribution $f(V, H_k)$ by thermal fluctuation tomography. J. Geophys. Res. **111**, B12S07 (2006)
185. S. Scheidt, R. Egli, T. Frederichs, U. Hambach, A mineral magnetic characterization of the Plio-Pleistocene fluvial infill of the Heidelberg Basin. Geophys. J. Int. **210**, 743–764 (2017)
186. T. Ouyang, D. Heslop, A.P. Roberts, C. Tian, Z. Zhu, Y. Qiu, X. Peng, Variable remanence acquisition efficiency in sediments containing biogenic and detrital magnetites: Implications for relative paleointensity signal recording. Geochem. Geophys. Geosyst. **15**, 2780–2796 (2014)
187. L. Chen, D. Heslop, A.P. Roberts, L. Chang, X. Zhao, H.V. McGregor, G. Marino, L. Rodriguez-Sanz, E.J. Rohling, H. Pälike, Remanence acquisition efficiency in biogenic and detrital magnetite and recording of geomagnetic paleointensity. Geochem. Geophys. Geosys. **18**, 1435–1450 (2017)
188. J.A. Dearing, K.L. Hay, S.M.J. Baban, A.S. Huddleston, E.M.H. Wellington, P.J. Loveland, Magnetic susceptibility of soil: an evaluation of conflicting theories using a national data set. Geophys. J. Int. **127**, 728–734 (1996)
189. J. Torrent, V. Barrón, Q. Liu, Magnetic enhancement is linked to and precedes hematite formation in aerobic soils. Geophys. Res. Lett. **33**, L02401 (2006)
190. D. Fortin, S. Langley, Formation and occurrence of biogenic iron-rich minerals. Earth-Sci. Rev. **72**, 1–19 (2005)
191. D. Faivre, D. Schüler, Magnetotactic bacteria and magnetosomes. Chem. Rev. **108**, 4875–4898 (2008)
192. C.M. Hansel, S.G. Benner, J. Neiss, A. Dohnalkova, R.K. Kukkadapu, S. Fendorf, Secondary mineralization pathways induced by dissimilatory iron reduction of ferrihydrite under advective flow. Geochim. Cosmochim. Acta **67**, 2977–2992
193. R. Karlin, Magnetite diagenesis in marine sediments from the Oregon Continental Margin. J. Geophys. Res. **95**, 4405–4419 (1990)
194. B.W. Leslie, S.P. Lund, D.E. Hammond, Rock magnetic evidence for the disso¬lution and authigenic growth of magnetic minerals within anoxic marine sediments of the California continental borderland. J. Geophys. Res. **95**, 4437–4452 (1990)
195. R. Kopp, J.L. Kirschvink, The identification and biogeochemical interpretation of fossil magnetotactic bacteria. Earth Sci. Rev. **86**, 42–61 (2008)
196. Z. Gibbs-Eggar, B. Jude, J. Dominik, J.-L. Loizeau, F. Oldfield, Possible evidence for dissimilatory bacterial magnetite dominating the magnetic properties of recent lake sediments. Earth Planet. Sci. Lett. **168**, 1–6 (1999)
197. J.A. Dearing, J.A. Hannam, A.S. Anderson, E.M.H. Wellington, Magnetic, geochemical and DNA properties of highly magnetic soils in England. Geophys. J. Int. **144**, 183–196 (2001)

198. S. Spassov, F. Heller, R. Kretzschmar, M.E. Evans, L.P. Yue, D.K. Nourgaliev, Detrital and pedogenic magnetic mineral phases in the loess/paleosol sequence at Lingtai (Central Chinese Loess Plateau). Phys. Earth Planet. Inter. **140**, 255–275 (2003)

199. M.W. Hounslow, B.A. Maher, Quantitative extraction and analysis of carriers of magnetization in sediments. Geophys. J. Int. **124**, 57–74 (1996)

200. A.P. Chen, V.M. Berounsky, M.K. Chan, M.G. Blackford, C. Cady, B.M. Moskowitz, P. Kraal, E.A. Lima, R.E. Kopp, G.R. Lumpkin, B.P. Weiss, P. Hesse, N.G.F. Vella, Magnetic properties of uncultivated magnetotactic bacteria and their contribution to a stratified estuary iron cycle. Nat. Commun. **5**, 4797 (2014)

201. M. Amor, R. Egli, J. Wan, J. Carlut, C. Gatel, I.M. Anderssen, E. Snoeck, A. Komeili, Magnetic flux closure in magnetotactic bacteria producing multiple chains of magnetite and the signature of magnetofossils. https://doi.org/10.21203/rs.3.rs-187824/v1 (2021) (submitted for publication)

202. R. Egli, Characterization of individual rock magnetic components by analysis of remanence curves. 3. Bacterial magnetite and natural processes in lakes. Phys. Chem. Earth **29**, 869–884 (2004)

203. F.D. Stacey, Theory of the magnetic properties of igneous rocks in alternating fields. Philos. Mag. **6**, 1241–1260 (1961)

204. —. A generalized theory of thermoremanence, covering the transition from single-domain to multi-domain magnetic grains. Philos. Mag. 1962, 7, 1887–1900

205. D.J. Dunlop, Theory and application of the Day plot (M_{rs}/M_s versus H_{cr}/H_c) 1. Theoretical curves and testes using titanomagnetite data. J. Geophys. Res. **107**, 2056 (2002)

206. B.M. Moskowitz, R.B. Frankel, D.A. Bazylinski, Rock magnetic criteria for the detection of biogenic magnetite. Earth Planet. Sci. Lett. **120**, 283–300 (1993)

207. J. Li, W. Wu, Q. Liu, Y. Pan, Magnetic anisotropy, magnetostatic interactions and identification of magnetofossils. Geochem. Geophys. Geosyst. **13**, Q10Z51 (2012)

208. I. Lascu, R.J. Harrison, Y. Li, J.R. Muraszko, J.E.T. Channell, A.M. Piotrowski, D.A. Hodell, Magnetic unmixing of first-order reversal curve diagrams using principal component analysis. Geochem. Geophys. Geosyst. **16**, 2900–2915 (2015)

209. Nikolaisen, E. S., Harrison, R., Fabian, K., McEnroe, S. A. Hysteresis of natural magnetite ensembles: Micromagnetic of silicate-hosted magnetite inclusions based on focused-ion-beam nanotomography. Geochem. Geophys. Geosyst. **21**, e2020GC009389 (2020)

210. R.K. Dumas, C.-P. Li, I.V. Roshchin, I.K. Schuller, K. Liu, Magnetic fingerprints of sub-100 nm Fe dots. Phys. Rev. B **75**, 134405 (2007)

211. A.M. Muxworthy, W. Williams, D. Virdee, Effect of magnetostatic interactions on the hysteresis parameters of single-domain and pseudo-single-domain grains. J. Geophys. Res. **108**, 2517 (2003)

212. D.J. Dunlop, Theory and application of the Day plot (M_{rs}/M_s versus H_{cr}/H_c) 2. Application to data for rocks, sediments, and soils. J. Geophys. Res. **2002**, 107 (2057)

213. A.P. Roberts, L. Tauxe, D. Heslop, X. Zhao, Z. Jiang, A critical appraisal of the "Day" diagram. J. Geophys. Res. Solid Earth **123**, 2618–2644 (2018)

214. J.M.D. Coey, *Magnetism and Magnetic Materials* (Cambridge University Press, 2010)

215. W. Rave, K. Fabian, A. Hubert, Magnetic states of small cubic particles with uniaxial anisotropy. J. Magn. Magn. Mater. **190**, 332–348 (1998)

216. L. Nagy, W. Williams, L. Tauxe, A.R. Muxworthy, From nano to micro: evolution of magnetic domain structures in multidomain magnetite. Geochem. Geophys. Geosyst. **20**, 2907–2918 (2019)

217. L. Nagy, W. Williams, A.R. Muxworthy, K. Fabian, T.P. Almeida, P.O. Conbhui, V.P. Shcherbakov, Stability of equidimensional pseudo-single-domain magnetite over billion-year timescales. Proc. Natl. Acad. Sci. U. S. A. **114**, 10356–10360 (2017)

218. R.J. Harrison, R.E. Dunin-Borkowski, A. Putnis, Direct imaging of nanoscale magnetic interactions in minerals. Proc. Natl. Acad. Sci. U. S. A. **99**, 16556–16561 (2002)

219. A.V. Smirnov, Low-temperature magnetic properties of magnetite using first-order reversal curve analysis: Implications for the pseudo-single-domain state. Geochem. Geophys. Geosyst. **7**, S11011 (2006)
220. G.R. Lewis, J.C. Loudon, R. Tovey, Y.-H. Chen, A.P. Roberts, R.J. Harrison, P.A. Midgley, E. Ringe, Magnetic vortex states in toroidal iron oxide nanoparticles: combining micromagnetics with tomography. Nano Lett. **20**, 7405–7412 (2020)
221. L. Tauxe, H.N. Bertram, C. Seberino, Physical interpretation of hysteresis loops: micromagnetic modeling of fine particle magnetite. Geochem. Geophys. Geosyst. **3**, 1055 (2002)
222. M. Winklhofer, K. Fabian, F. Heider, Magnetic blocking temperatures of magnetite calculated with a three-dimensional micromagnetic model. J. Geophys. Res. **102**, 22695–22709 (1997)
223. K. He, X. Zhao, Y. Pan, X. Zhao, H. Quin, T. Zhang, Benchmarking component analysis of remanent magnetization curves with a synthetic mixture series: insight into the reliability of unmixing natural samples. J. Geophys. Res. Solid Earth **125**, e2020JB020105 (2020)
224. A.J. Newell, R.T. Merrill, Nucleation and stability of ferromagnetic states. J. Geophys. Res. **105**, 19377–19391 (2000)
225. J.E.T. Channell, R.J. Harrison, I. Lascu, I.N. McCave, F.D. Hibbert, W.E.N. Austin, Magnetic record of deglaciation using FORC-PCA, sortable-silt grain size, and magnetic excursion at 26 ka, from the Rockall Trough (NE Atlantic). Geochem. Geophys. Geosyst. **17**, 1823–1841 (2016)
226. T. Yamazaki, W. Fu, T. Shimono, Y. Usui, Unmixing biogenic and terrigenous magnetic mineral components in red clay of the Pacific Ocean using principal component analyses of first-order reversal curve diagrams and paleoenvironmental implications. Earth Plantes Space **72**, 120 (2020)
227. A.P. Roberts, F. Florindo, G. Villa, L. Chang, L. Jovane, S.M. Bohaty, J.C. Larrasoaña, D. Heslop, J.D. Fitz Gerald, Magnetotactic bacterial abundance in pelagic marine environments is limited by organic carbon flux and availability of dissolved iron. Earth Planet. Sci. Lett. **310**, 441–452 (2011)
228. K. Fabian, A. Hubert, Shape-induced pseudo-single-domain remanence. Geophys. J. Int. **138**, 717–726 (1999)
229. H.G. Katzgraber, F. Pázmándi, C.R. Pike, K. Liu, R.T. Scalettar, K.L. Verosub, G.T. Zimányi, Reversal-field memory in the hysteresis of spin glasses. Phys. Rev. Lett. **89**, 257202 (2002)
230. M. Pohlit, P. Eibisch, M. Akbari, F. Porrati, M. Huth, J. Müller, First order reversal curves (FORC) analysis of individual magnetic nanostructures using micro-Hall magnetometry. Reivew of Scientific Instruments **87**, 113907 (2016)
231. I. Lascu, J.F. Einsle, M.R. Ball, R.J. Harrison, The vortex state in geologic materials: a micromagnetic perspective. J. Geophys. Res. Solid Earth **123**, 7285–7304 (2018)
232. D. Cortés-Ortuño, K. Fabian, L. V. De Groot, Single particle multipole expansions from micromagnetic tomography, Geochem. Geophys. Geosys.., e2021GC009663 (in press, 2021)
233. T. Moon, R.T. Merrill, Magnetic screening in multidomain material. J. Geomag. Geoelec. **38**, 883–894 (1986)
234. S. Xu, D.J. Dunlop, Micromagnetic modeling of Bloch walls with Néel caps in magnetite. Geophys. Res. Lett. **23**, 2819–2822 (1996)
235. T.G. Pokhil, B.M. Moskowitz, Magnetic domains and domain walls in pseudo-single-domain magnetite studies with magnetic force microscopy. J. Geophys. Res. **102**, 22681–22694 (1997)
236. R.B. Proksch, S. Foss, E.D. Dahlberg, High resolution magnetic force microscopy of domain wall fine structures. IEEE Trans. Magn. **30**, 4467–4472 (1994)
237. C. Kittel, Theory of the structure of ferromagnetic domains in films and small particles. Phys. Rev. **70**, 965–971 (1946)
238. M. Kersten, Zur Theorie der Koerzitivktraft. Z. Phys. **124**, 714–741 (1948)
239. L. Néel, Bases d'une nouvelle théorie générale du champ coercitif. Ann Univ. Grenoble. **22**, 299–343 (1946)
240. C. Kittel, Physical theory of ferromagnetic domains. Rev. Mod. Phys. **21**, 541–583 (1949)

241. G. Bertotti, V. Basso, A. Magni, Stochastic dynamics in quenched-in disorder and hysteresis. J. Appl. Phys. **85**, 4355–4357 (1999)
242. A. Magni, C. Beatrice, G. Durin, G. Bertotti, Stochastic model for magnetic hysteresis. J. Appl. Phys. **86**, 3252–3261 (1999)
243. A. Kosterov, Low-temperature magnetization and AC susceptibility of magnetite: effect of thermomagnetic history. Geophys. J. Int. **154**, 58–71 (2003)
244. B. Reznik, A. Kontny, F.R. Schilling, Magnetic fatigue: Effect of seismic-related loading on magnetic and structural behavior of magnetite. Earth Space Sci Open Arch (2020)

Characterization of Magnetic Nanostructures with the First-Order Reversal Curves (FORC) Diagram Technique

Alexandru Stancu

Abstract The first-order reversal curves (FORC) diagram method has become the state-of-the-art tool in magnetic characterization of materials with hysteresis. This book chapter is intended as a guide in the understanding the FORC diagrams features observed in a variety of magnetic materials. Our goal is to explain the typical FORC features and give the proper tools for a quantitative approach to the FORC diagram users. We discuss very simple magnetic systems made from wires with rectangular hysteresis loops (hysterons). In noninteracting systems of wires with a distribution of coercivities, the major hysteresis loop is sufficient to find this distribution. FORC tool is dedicated essentially to find information about the interactions. This is why we have focused the discussion on this topic. We show several simple examples of nanostructures in which we can observe the fundamental relation between the interactions and the shape of the FORC diagram. These examples give insight on the main typical FORC features detected in many magnetic systems: T-shaped, wishbone, and boomerang FORC diagrams. These results are certainly important as the systems producing these diagrams are really simple and this fact facilitates the evidence of the causal links.

Keywords FORC diagram · Preisach model · Hysteresis · Magnetic nanostructures · Nanomagnetism · Magnetic wires · Bit patterned media

1 Introduction

The magnetic characterization of materials based on the first-order reversal curves (FORC) and on the diagram that can be derived from these experimental data is becoming the preferred tool to perform complex analyses of materials showing hysteretic behavior. As we shall present in details in this chapter, FORC diagram

A. Stancu (✉)
Faculty of Physics, Alexandru Ioan Cuza University of Iasi, Iasi, Romania
e-mail: alstancu@uaic.ro

© Springer Nature Switzerland AG 2021
V. Franco, B. Dodrill (eds.), *Magnetic Measurement Techniques for Materials Characterization*, https://doi.org/10.1007/978-3-030-70443-8_18

provides a unique opportunity to find valuable physical information on complex magnetic structures. We have also to mention that the area covered by the FORC technique is not restricted only to ferromagnetic hysteresis. A variety of other physical hysteretic processes and materials have benefitted from this analysis: ferro-electric materials, spin transition temperature, light-induced and pressure hysteresis, magneto-caloric phenomena, etc. [1–5]. Partially, the discussion presented here could be extended in these cases, as well. The classical measurement of the major hysteresis loop (MHL) has a limited value especially when one analyses a system of magnetic elements in interaction as it doesn't provide an evaluation of the intensity of the interactions between these elements. It can be shown that a system with narrow distribution of coercivities but with strong interparticle interactions could have virtually the same major hysteresis loop with a system with weak interactions but with a wider distribution of coercive fields. The solution for this problem was the design of more intricate experiments that could qualitatively reveal the type of interactions between the magnetic components of the studied sample (Henkel plot [6], deltaM plot [6–8], etc.). The FORC diagram method, although initially was also seen as a qualitative tool for magnetic characterization, has a tremendous potential as a quantitative method as well. This actually became in recent years the point of maximum interest in the magnetics scientific community. We have to mention that various numerical algorithms used to produce the FORC distribution and its contour plot, named FORC diagram, are not discussed in this chapter. There are a number of excellent articles discussing this problem in detail [9–13], and a few numerical tools are available to obtain the FORC diagrams from experimental data. The essential problem we have, as soon as we have obtained an appropriate FORC diagram for a magnetic system, is to interpret it correctly and to extract as much physical information from it as possible. In order to explain the ability of the FORC diagram technique to provide useful information about a hysteretic system, we should undoubtedly start with the concept of *hysteron*—as an elementary brick of hysteresis. The idea of a rectangular hysteresis loop as a building block for measured hysteresis loops originated in the study of ferromagnetic systems. Weiss and de Freudenreich [14], more than 100 years ago, were the first to discuss this notion and to introduce concepts that many are associating now with the better known classical Preisach model (CPM) presented by Preisach in 1935 [15]. The real original contribution of Preisach in his famous paper was to introduce a geometrical representation for the rectangular hysterons. As each hysteron is characterized by two switching fields (up-down and down-up), the straightforward idea is to use them as coordinates of a plane (known now as the Preisach plane—see Fig. 1). In a 3D image of this representation, one should add the third axis—perpendicular to the Preisach plane—representing the magnetic moment of the hysteron. In this way, one can imagine a real sample represented as a collection of hysterons. This is the central idea of the Preisach model and of the FORC-type technique.

One question one can ask is how realistic this representation is, at least in ferromagnetism. If one applies field along the easy axis of a uniaxial single-domain ferromagnetic particle, the system has only two stable equilibrium states possible (along or against the field direction). In this case, one obtains a rectangular hysteresis

Fig. 1 Preisach plane and
rectangular hysteron

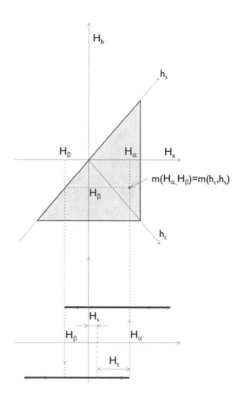

loop, as previously described as a hysteron, the building block of the Preisach model. One should mention that the switching fields up-down and down-up have the same absolute value (with the notations in Fig. 1 $H_\alpha = -H_\beta$, which represents points on the second bisector of the Preisach plane) for one isolated (noninteracting) element. The idea of asymmetry of the hysteresis loop due to *the interaction field* created by the neighboring magnetic elements within the sample is specific to the Preisach model. In the CPM, asymmetrical hysterons are allowed, and the degree of asymmetry is a measure of the interaction field. Different elements of the magnetic system could show different interaction fields, and, as a result, the hysterons are distributed not only on the second bisector of the Preisach plane but on the entire plane. Actually, there are some constraints: (i) the switching fields have limit values on both up-down and down-up cases; and (ii) one allows only the physically correct loops with the down-up switching field bigger than the up-down switching field. These constraints allow to restrict our analysis on the Preisach plane only to a triangular area, shown as a hashed area in Fig. 1. Outside that area, we should not have any hysteron associated to the Preisach plane.

One should also clarify the different representations of the Preisach plane. The representation used in Fig. 1 is based on the use of the switching fields as coordinates for the hysterons (the (H_a, H_b) Preisach plane). We can also introduce the rotated set of axes, (h_c, h_s), where the (O, h_c) axis is along the hysterons' coercivities and the perpendicular axis, (O, h_s), measures the asymmetry of the hysterons along the field axis (the shift of the hysterons along this axis). As we have mentioned, this shift can be related to the interaction field. Summarizing, the moment of one hysteron associated to the point (H_α, H_β) has the coordinates

$$\begin{cases} h_c = \left(H_\alpha - H_\beta\right) \frac{\sqrt{2}}{2} = \frac{(H_\alpha - H_\beta)}{2}\sqrt{2} = H_c\sqrt{2} \\ h_s = \left(H_\alpha + H_\beta\right) \frac{\sqrt{2}}{2} = \frac{(H_\alpha + H_\beta)}{2}\sqrt{2} = H_s\sqrt{2} \end{cases} \tag{1}$$

where H_c is the coercive field of the hysteron and $H_s = -H_i$ is the value of the shift field (see Fig. 1). The interaction field, H_i, is in the opposite direction to the shift field. When the hysteron is shifted due to the interactions along the positive direction of the applied field, H, it means that one needs a stronger field to switch the hysteron from down to up than from up to down. This is compatible with a negative interaction field. The Preisach plane could be also represented in the (H_c, H_s) coordinate system with the advantage of showing directly the coercive and interaction fields. The (h_c, h_s) representation has only a theoretical value and it shows the correct link with the Preisach (H_α, H_β) plane.

2 Classical Preisach Model

The idea that any ferromagnetic system can be represented as a superposition of a certain distribution of rectangular hysterons is fundamental in the CPM. In order to become a practical tool for describing the hysteretic behavior of a given ferromagnetic sample, one has to develop methods to find the distribution of hysterons associated to the sample, named the Preisach distribution [16]. If a unique distribution is found, then one should develop algorithms to calculate the magnetic response of the sample under various sequences of applied fields. In this way, one can show the power of prediction of CPM, when the field sequence is different from the one used in the process of finding the Preisach distribution of the sample. The quality of the predictions could be then tested experimentally. This short description of the model reveals that one can discuss separately two problems: (i) the process of evaluation of the Preisach distribution, called identification, and (ii) the method used to evaluate the magnetic moment of the sample with a known Preisach distribution after a sequence of applied fields. In order to understand the identification techniques, including the one based on the FORC measurement, we should first comprehend how the CPM works when the distribution is known.

We can start this analysis by asking the question: is it possible to calculate the total magnetic moment of the sample when we know the Preisach distribution and the applied field? We know that the distribution of hysterons will contain hysterons with various coercivities and interaction fields. If the applied field is known, the important thing is to compare this field with the characteristic switching fields of each component hysteron. If the applied field is bigger than the down-up switching

Fig. 2 (**a**) The status of the hysterons in the Preisach plane when a field H is applied to the sample. (**b**) Preisach plane in positive saturation state (H_s is a field sufficient to saturate the sample). (**c**) Preisach plane when, after positive saturation, the field decreased to a value $H < H_s$

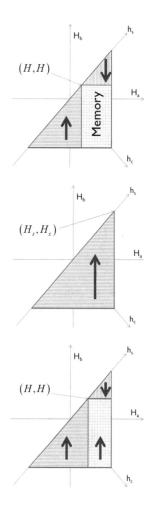

field, H_α, then the hysteron will certainly be in the "up" position. If the applied field is smaller than the up-down switching field, H_β, the hysteron will be in the "down" position. However, when the applied field is between the two switching fields we don't know, a priori, the status of the hysteron. It depends on the previous position of the moment (up or down), and that position will not be changed by the applied field. All the hysterons with this property will be responsible for the memory of the system. It is rather easy to identify the memory region in the Preisach plane for a given applied field (see Fig. 2(a)—region "Memory"). This region depends on the applied field. When the field is larger than the highest down-up switching field of the magnetic entities in the sample, then all the particles will be saturated in the positive (up) direction. This is an example of memory deletion with a saturation field. Of course, the discussion is valid also for the negative saturation.

If the applied field is changed, one observes that the (up, down, memory) regions also change. If the field is increasing, the triangular "up" region increases its surface

Fig. 3 The triangular region in the Preisach plane corresponding to the Everett integral calculated with (2)

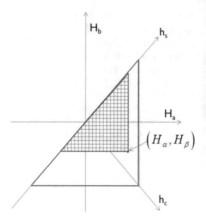

at the expense of the other two regions. If the field is decreasing, the triangular "down" region increases its surface. This observation is of paramount importance in order to understand how the magnetic moment can actually be calculated. First, we know that the "memory" region moment cannot be calculated unless we know the previous state of the hysterons before the field was applied. Consequently, the Preisach model will give results only if one starts from a state in which we know the status of all the hysterons. The positive and negative saturations are such states. If initially one saturates the sample in the "up" position (positive saturation), then all the hysterons will be in the "up" position. From this state we are able to evaluate the value of the magnetic moment for a lower value of the applied field. The difference between the positive sample saturation and this state is due to the triangular region that switched from "up" to "down" position (see Fig. 2b, c). From this example one can observe that the magnetic moment variation between two states can be calculated using the contribution of triangular regions in the Preisach plane.

The moment of such a triangular zone is given by the Everett integral [17]:

$$E\left(H_\alpha, H_\beta\right) = \int_{H_\beta}^{H_\alpha} \left[\int_{H_\beta}^{H_a} P\left(H_a, H_b\right) dH_b \right] dH_a \tag{2}$$

where $P(H_a, H_b)$ is the Preisach distribution value in the point of (H_a, H_b) in the Preisach plane and $E(H_\alpha, H_\beta)$ is the total moment of the hashed triangular region (see Fig. 3). For example, the saturation magnetic moment of the sample is given by $m_s = E(H_s, -H_s)$ where H_s is the applied field sufficient to saturate the sample. The magnetic moment on the descending branch of the major hysteresis loop (MHL) is given by the difference between the saturation moment and twice the value of the Everett integral $E(H_s, H)$ (because the triangular zone switched from $+1$ (up) to -1 (down)):

$$m_{MHL} = m_s - 2E\left(H_s, H\right). \tag{3}$$

Fig. 4 (**a**) FORC with the starting point on the descendant branch of the MHL. (**b**) Preisach plane diagram in the state A. (**c**) Preisach plane diagram in the state B

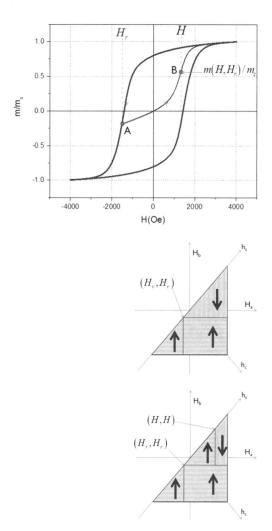

For the FORCs starting on the descending branch of MHL, one should first saturate the sample in the positive direction and then decrease the field along the MHL. At a certain field, named the reversal field H_r (point A in Fig. 4a), the field variation is stopped for a short time, and, after that, the field is increased in order to obtain again the positive saturation. The FORC is the curve measured between point A and the positive saturation. At a point B (see Fig. 4a), the applied field is H, but the measured value of the moment is a function of two variables: (H, H_r). The explanation of this fact in CPM is given by the two diagrams of the Preisach plane in the A and B states (see Fig. 4b, c). The moment of the sample at point A can be calculated as

$$m_A = m_s - 2E\,(H_s, H_r) \tag{4}$$

The moment on the FORC at point B differs from the moment m_A with the contribution of a triangular zone that can be calculated with an Everett integral, and is given by:

$$m_B = m_A + 2E(H, H_r).$$ (5)

Using the definition of the Everett integral (2) the moment at point B is given by:

$$m_B \equiv m_{FORC}(H, H_r) = m_s - 2 \int_{H_r}^{H_s} \left[\int_{H_r}^{H_a} P(H_a, H_b) \, dH_b \right] dH_a$$

$$+ 2 \int_{H_r}^{H} \left[\int_{H_r}^{H_a} P(H_a, H_b) \, dH_b \right] dH_a.$$ (6)

The identification technique of the Preisach distribution based on the FORC data was proposed by Mayergoyz in 1985 [18] and was based on the observation that the second order mixed derivative of the moment on the FORC, m_{FORC}, expressed as in (6), is given by:

$$\frac{\partial^2 m_{FORC}(H, H_r)}{\partial H \partial H_r} = -2P(H, H_r)$$ (7)

From this relation one obtains the well-known expression for the Preisach (and for the FORC) distributions:

$$P(H, H_r) = -\frac{1}{2} \frac{\partial^2 m_{FORC}(H, H_r)}{\partial H \partial H_r}.$$ (8)

Mayergoyz proposed this technique only in magnetic systems correctly described by CPM. He also enunciated the necessary and sufficient conditions for a system to be represented by a CPM. This fundamental theoretical result is known as the *Identification Theorem* and states that a CPM system should have two properties: wiping-out and congruency. The first property essentially affirms that a CPM system, after a complete minor loop, should return to the same magnetic state. The second one states that the minor loops within any given field limits should be congruent. Unfortunately, there is virtually no ferromagnetic system obeying with sufficient accuracy these conditions. This explains why many years after Mayergoyz's article there were no reports of use of the FORC technique. A real breakthrough was produced by Pike et al. from University of California, Davis. In the Ref. [19], the authors show the main problems with the FORC identification technique for CPM systems and link these problems with a number of physical causes. The main result of this analysis was the decision to use the FORC technique for any ferromagnetic system without requiring any condition (like the ones from the Identification Theorem). In this way, one can produce, for any ferromagnetic sample, an experimental FORC distribution, which is not the Preisach distribution. The expectation is that this FORC distribution have features specific to each magnetic phase from the sample. This point of view is at the origin of the qualitative use of the FORC distributions, which was coined as *magnetic fingerprinting* [20]. The FORC distributions and especially their contour plot representations, known as FORC diagrams (looking similar to fingerprints, etc.), became an instant success, and an increased number of laboratories became interested in implementing this experimental technique.

The qualitative FORC studies are specific to a number of domains, especially those related to geology [21–24] but also to other research fields. Systematic studies have shown that typical features that are observed in many FORC distributions can't be actually linked to some magnetic phases in the composition of materials. It was understood from the initial studies that these are in fact evidence of specific aspects concerning the geometry of the samples. A relevant moment in the evolution of understanding the significance of the FORC diagrams, opening the way from qualitative to quantitative interpretation of the experimental results, was the reanalysis of the relation between the Preisach model and the FORC distribution [25]. It was pointed out that indeed the very restrictive conditions of CPM can't be fulfilled by the ferromagnetic systems. However, a number of modified Preisach-type models were developed in order to relax these conditions. Moving Preisach model [26], Variable Variance Preisach Model [27] and Preisach Model for Patterned Media [28] could be mentioned in this context. All these Preisach-type models are able to simulate FORC data, and, when the mixed second order derivative is calculated with Mayergoyz's algorithm, one can produce most of the features observed in experimental results for real magnetic samples. This, in principle, opens the opportunity to find, from the FORC data, the Preisach distribution and a number of supplementary parameters specific to each modified Preisach model. In this way, a quantitative approach for the interpretation of the FORC diagrams becomes achievable.

3 Magnetic Ensembles of Real Magnetic Hysterons

One enduring effort in the Preisach community was to find real magnetic systems containing magnetic entities in interaction, each isolated element having a rectangular hysteresis loop and with interaction fields along the applied field direction. Such magnetic systems were named at a meeting at the 2000 INTERMAG Conference— *standard samples*. Also, at the same meeting, the idea of using data generated by micromagnetic models was assessed as *standard problems* for the Preisach-type models to evaluate the correlation between the Preisach models parameters and the physical micromagnetic data [29].

A similar approach is advisable for the quantitative FORC analysis. The first condition is to use magnetic bricks with rectangular hysteresis loop. As we already mentioned, a uniaxial single-domain ferromagnetic particle has a rectangular loop when the external field is applied along the easy axis of the particle. The other condition which the systems should obey is that the interaction fields should be also acting along the same direction. These two concurrent conditions restrict drastically our choice for a magnetic structure. If one also adds a condition to be able to calculate easily the interactions, we can imagine a system of quasi-identical single-domain particles in a 2D network with the easy axes perpendicular to a plane. It is interesting to observe that similar magnetic systems are intensely studied due to their potential application in various fields from recording media to tuned absorbents of electromagnetic waves: arrays of magnetic wires produced by electrochemical deposition in specially designed matrices, which controls the geometrical characteristics of the arrays (for details on magnetic wire systems we recommend [30]). So, instead of a single-domain particle, which will have a very small magnetic moment, in this case we consider a long magnetic wire with a significant magnetic moment. The possible problem would be the fact that the wire is not a single-domain element. However, the system has two clear stable states of the magnetic moments in the wire, along the axis of the cylinder. The reversal process is evidently not a coherent rotation of the moment of the wire. In this case one has a process of nucleation of a domain wall at the end of the wire followed by a rapid propagation of the wall toward the other end of the wire. The process is really rapid (near 1 ns) compared to the typical duration of measurements made in Vibrating Sample Magnetometers (VSMs) [31]. This means that actually the isolated wire will have a rectangular hysteresis loop. It is a near ideal physical hysteron as the moment is big enough to be easily measured in a classical VSM.

The next logical question is about the interaction field in such an array of magnetic wires. Without losing the generality of the problem, one can assume a rectangular structure of identical cylindrical wires. The axes of symmetry of the wires are all parallel among them and all perpendicular to a plane. We shall call that configuration a 2D-perpendicular array of magnetic wires [32–34]. If one discusses the interactions in magnetic wire structures, a legitimate question is how one can calculate this field taking into account that the field created by the other wires is different at different points of each wire. For wires much longer than the diameter of the circular base, it can be shown that the interaction field can only be accurately calculated if a really large number of wires are taken into account [35]. Moreover, it is appropriate to use the interaction field in the center of the wire. In this way a second physical condition desirable from the Preisach model point of view is fulfilled, that is, the interaction field is also along the easy axis of the magnetic particles.

To evaluate the behavior of a 2D-perpendicular array, as the one described before, in any sequence of applied field, one uses a very simple model in which each wire is represented as a rectangular hysteron. The most complicated problem in this case is to calculate with sufficient accuracy the interaction field in each wire. If the cylindrical wire is saturated, the uncompensated magnetic charges are present only on the circular bases of the cylinder. Each wire will create a magnetostatic field which can be calculated as the superposition of the fields created by two discs: one charged positively and one negatively. It was also shown that the approximation of the disc uniformly charged with a monopolar magnetic charge has sufficient accuracy in the calculation of the interaction field.

To discuss the typical results obtained with simple Ising-Preisach model (at zero Kelvin) as described in [32–34] and with similar physical data, we shall analyze an array (of 10^4 wires) with the following characteristics: $L = 6\,\mu\text{m}$, $r = 40$ nm, $M_s = 485.0\,\frac{\text{emu}}{\text{cm}^3} = 4.85 \times 10^5\,\frac{\text{A}}{\text{m}}$. Each wire is represented in the model by a hysteron with an intrinsic coercivity. In these simulations we have randomly associated coercivities with a Gaussian distribution with a mean value of $100\,Oe$ and with a dispersion $\sigma = (50/3)\,Oe$. Consequently, we have wires with intrinsic coercivities between $50\,Oe$ and $150\,Oe$ in our array. The switching condition could be simply triggered by the switching fields of each wire. Another modeling solution for the evaluation of the switching could be the use of a model for the energy barrier between the two stable states of the wire (up/down) and to use a Metropolis Monte Carlo (MMC) technique [36] to calculate the switching. At zero Kelvin, there should be no difference between the two algorithms. The MMC method, however, offers the prospect for the model to evaluate the magnetic behavior at different temperatures.

4   Results of Simulations

The following results have been obtained with an Ising-Preisach model for a rectangular array of magnetic wires with the characteristics mentioned in the previous section. The FORC diagrams and the data for other figures were produced by doFORC package [13]. One can observe that the interactions between the wires give a "T-shaped" FORC diagram. First, we should present a few quantitative aspects of the diagram. In Fig. 5(b) we have marked on the diagram two lines: (i) one at $+337$ Oe and (ii) one at -537 Oe. These values were calculated starting from the value of -387 Oe, which is the demagnetizing field at the positive saturation in

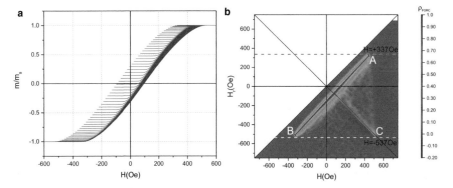

Fig. 5 FORCs and diagram for simulation of 2D-perpendicular array. The wires have coercivities between 50 Oe and 150 Oe. (**a**) The FORCs; (**b**) the FORC diagram

the middle of the sample. We have assumed that the least coercive wire is the first to switch at $(387 - 50)\,\text{Oe} = +337\,\text{Oe}$ and the last to switch is the most coercive. When this most coercive wire will switch, all the other wires have been already switched, so the field created by these wires will act on the positive direction and to be able to switch this last wire one needs $-(387 + 150)\,\text{Oe} = -537\,\text{Oe}$.

When we discuss the effect of interactions for the 2D-perpendicular array of magnetic wires we have to put emphasis on at least two elements:

1. The effect of magnetostatic interactions between the wires in the array is considerably different from the one described in the classical Preisach model. This is unexpected, as we have seen that each wire can be realistically assimilated with a Preisach rectangular hysteron.
2. The interaction field has different effects on the wires depending on their coercivity.

To make that clearer, we remember that in the classical Preisach model, the interactions have the effect of shifting the rectangular loop of each hysteron with a constant (interaction) field independent of the magnetic state of the system. The interaction field induces the asymmetry of the rectangular hysterons. The positive and negative switching fields may have different absolute values. The interaction fields of the entire ensemble have a characteristic distribution. Scientists have analyzed this hypothesis many years ago and they realized that this distribution cannot, realistically, be independent of the magnetic state. The concept of *statistically stable Preisach distribution* was used to solve this problem, as discussed in [37–38]. However, some magnetization processes, like the remanent anhysteretic magnetization, as shown by Wohlfarth in [38], is really hard to understand even in this context. The most significant contribution to this problem was to introduce a phenomenological dependence of the interaction field distribution on the magnetic state. Della Torre has developed the *moving Preisach model*, based on the hypothesis that there is a linear dependence of the interaction field on the magnetic moment of the sample [26]. Many of the problems observed in the classical version of the Preisach model were solved in the moving Preisach model (e.g., the problem of the remanent anhysteretic susceptibility [39]). Essentially, in the moving Preisach model we can understand the interaction field as a stable distribution (as seen in the classical Preisach model) moving with a value dependent on the total magnetic moment of the sample. A new parameter, α, the moving constant—which is controlling how much the stable distribution is moving—is in this way introduced in the Preisach model. However, most of the researchers using the Preisach model have conserved the image of interactions producing a shift of the rectangular hysteresis loop which is symmetrical if measured for the isolated wire (or other magnetic fundamental brick of the studied system). In this way, interactions cannot change the intrinsic coercivity of the wire. Nonetheless, we have seen in our simulation that starting with wires with intrinsic coercivities between 50 Oe and 150 Oe, we have detected in the FORC distribution symmetric hysterons with a coercivity of 537 Oe, much larger than the most coercive wire in the system (more than three times higher!). This is not a trivial fact, and a clear explanation is needed. As we have shown in [32], this increase of the *apparent* coercivity is also the effect of the interaction of the intrinsically most coercive wire with its neighbors. The switch of the wire with the smallest coercivity (the first to switch) will be favored by the interactions. The switch will happen near positive saturation (at positive applied fields) and the switch back up is obtained on the FORCs also near the positive saturation. As the neighboring moments will be up at both switches, the interaction field will be virtually constant, and the effect will be the shift of the rectangular hysteresis loop of this hysteron of low coercivity along the field axis toward positive fields. The coercivity of the hysterons with small coercivity will be conserved by the interactions. They will behave almost like ideal hysterons in the CPM. The case of the higher coercivity hysterons is different. The interaction field created by all the neighboring wires, which have been switched already down, is actually stabilizing the highest coercivity wire up state. A high value of the negative external field will be necessary to switch this wire (in absolute value the sum of the coercive and interaction fields—see Fig. 5(b)). To switch it back on the FORCs, it is easy to imagine that this wire will again be the last to be switched back up. But before that, all the neighbors will have been already in the up position because they have smaller coercivities. Again, the interaction field will stabilize the down position of the most coercive wire. Actually, it will be again necessary to apply a field equal to the absolute value of the one required to switch down the same wire on the major hysteresis loop. This explains two facts: (i) the absolute values of both switching fields are higher than the intrinsic coercive field of the most coercive wire; and (ii) the absolute values of these switching fields will be virtually equal. This process is at the origin of the presence of a high coercivity hysteron in point C on the FORC diagram from Fig. 5(b). One can make this analysis more profound by observing that point C is the unique representation on the FORC diagram of the wire with the highest coercivity. The question is whether all the other wires will have a unique representation in the FORC diagram/distribution.

To understand the conditions for such biunivocal relation between the wires and points in the Preisach plane, we have to reexamine the essence of the FORC technique. We know that, at some reversal field, one wire will be switched from up to down on the descending branch of the major hysteresis loop. Let's analyze what happens with this switched wire on the FORC. We know that in the process of restoring the positive saturation on this FORC, this wire will certainly switch back to the up position at a certain value of the applied field. We also know that the FORC distribution is found from a second order mixed derivative of the magnetic moment measured on the FORC. The derivative with respect to the applied field is also known as magnetic susceptibility, noted with $\chi(H, H_r)$. So, the FORC distribution will be related to the derivative of the magnetic susceptibility with respect to the reversal field:

$$\chi(H, H_r) = \frac{\partial m_{\text{FORC}}(H, H_r)}{\partial H}$$

$$\rho(H, H_r) = -\frac{1}{2}\frac{\partial^2 m_{\text{FORC}}(H, H_r)}{\partial H \partial H_r} = -\frac{1}{2}\frac{\partial \chi(H, H_r)}{\partial H_r}$$

One may understand the FORC technique as follows:

1. One hysteron in the sample, initially saturated in the up position, will switch down at some reversal field. Taking into account the fact that we use a finite field step, we can trace the switching up-down with a resolution related to this field step.
2. At one value of the reversal field, the hysteron didn't switch down and on the next FORC it was switched.
3. On the first FORC, the wire will stay up during the entire time we measure this curve. However, on the next FORC, as the wire will start in the down position, it will switch back up at a certain value of the applied field, a fact that will be observed in the magnetic susceptibility of the sample for that applied field. The derivative of the magnetic susceptibility on the two successive FORCs as a function of the reversal field will evidence a difference in susceptibility due to the switch down-up observed only on the second FORC and not present in the first one.

In this way we track both switching fields and can locate the moment contribution of the wire in a point in the Preisach plane. This discussion could give the impression that the corresponding wire—image on the FORC diagram—is unique. There is however another condition to be fulfilled by the switching process of the mentioned wire on the next FORCs, characterized by more negative values of the reversal field. The switch down-up of this wire should ideally happen at the same value of the applied field as it was observed on the first FORC started by the wire in the down position. To check this property, we have made a simulation in which we calculated the switching fields on each FORC for the wire with the smallest coercivity. In Fig. 6 we show the result of the simulation for two cases: (a) the noninteracting array and (b) the interacting array. We see that in the noninteracting array, the switching behaves as we think it should: the wire switches on all the FORCs at the same value of the field. This assures that the only image of this wire in the FORC diagram will be recorded at the first FORC started with the wire in the down position. All the following FORCs will not produce images as any such two successive FORCs will see the same increased value of the susceptibility due to the wire switch up at the same value of the field, regardless of the reversal field. The derivative of the susceptibility with regard to the reversal field will not give a contribution due to this wire. Actually, the FORCs will track gradually all the wires in the order of their switch up-down field.

However, when interactions are important, we see in Fig. 6(b) that the switch down-up is produced at a different value of the applied field on different FORCs with the exception of very low values (negative) of the reversal field that have already saturated the entire sample in the negative direction.

This behavior is at the origin of multiple images of the wire switch (down-up) in different positions for different values of the reversal field. In [33] we have called this fact—multiplicity of the images for one magnetic entity. What makes this unpleasant fact even more disagreeable is that the multiplicity is dependent on the intrinsic coercive field of the wire. The most susceptible wires to produce more

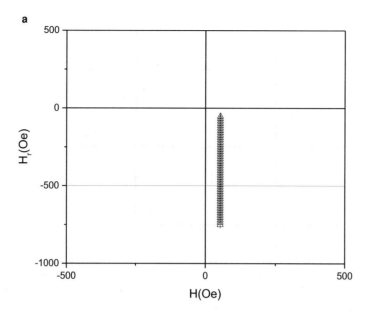

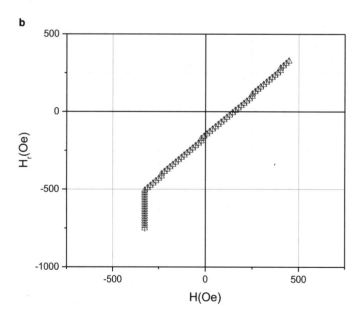

Fig. 6 (**a**) Switching field of the lowest coercivity wire when the reversal field is changing (noninteracting case). (**b**) Switching field of the lowest coercivity wire when the reversal field is changing (strong interacting system)

images are the lower coercive wires. The most coercive wire has only one image. This is true for the magnetic systems dominated by demagnetizing-type interactions.

As we need a system with either magnetizing or demagnetizing effects, we consider a 1D chain of magnetic wires. We shall study the FORCs of two such magnetic chains: the 1D-perpendicular chain (where the wires are all perpendicular on one straight line, Fig. 7a) and the 1D-longitudinal chain where all the wires are along one straight line (Fig. 7b). We have simulated two systems of 1601 wires with the same diameter and material as in the 2D-perpendicular system discussed previously. For the 1D-perpendicular system, we took wires with 3 μm length and 350 nm distance between the wires. For the 1D-longitudinal system, we considered wires of 0.6 μm and with a distance between the centers of the wires of 0.8 μm. We have simulated the FORCs measured with the field applied along the length of the wires in both cases. We mention that the 1D-perpendicular is a system with demagnetizing interactions, as the (interaction) field created by the saturated wires in one direction is in the opposite direction. The maximum value of the demagnetizing field in the central wire of the sample is evaluated as 16.5 Oe. On the FORC diagram obtained for this system, we also have represented the region (50, 150) Oe for the intrinsic coercivities of the isolated wires. The effect of the interaction field is similar to the one discussed for the 2D-perpendicular sample. Near the positive saturation, the interaction field will assist the switching of the wires, and we have represented the up-down switching field for the wire with the lowest coercivity (50 Oe). The line for this first expected switch is traced at $(16.5 - 50)$ Oe $= -33.5$ Oe. The last expected switch is near the negative saturation for the wire with the highest coercivity at $-(150 + 16.5)$ Oe $= -166.5$ Oe. The first observation is that the FORC diagram is bigger than the region of intrinsic coercivities and its shape is typical for many magnetic samples. Many researchers in the field are calling this a *wishbone shape* FORC diagram, and usually this shape is signaling demagnetizing-type interactions between the magnetic elements within the sample. A simple technique is recommended for this type of diagrams. One can recalculate the set of FORC data by eliminating the mean field (demag field) and recalculate the FORC diagram [40]. Various parameters can be used to find the optimum value for the mean field correction, as shown in [41, 42]. This technique could offer a very good approximation of the intrinsic coercivities of the magnetic elements in a sample (e.g., the pillars in a Bit Patterned Medium [43]) and is of interest for many applications. As an example of this technique, please see the corrected experimental FORC diagrams presented in the Fig. 6 in Ref. [40].

The case of 1D-longitudinal chain of magnetic wires shows a quite different FORC diagram shape. It is often named *boomerang shape* FORC diagram and usually indicates the presence of magnetization-type interactions between the constituting elements. In the chain of wires, we have indeed magnetizing interactions. For example, near positive saturation each wire experiences an interaction field created by the neighbor wires in the direction of the magnetic moment, stabilizing this position. The switch of the wires is produced by higher fields (in absolute value). We show, in a similar manner, the expected values for the switch up-down on the major hysteresis loop of the wires with the lowest and the highest coercivities. The

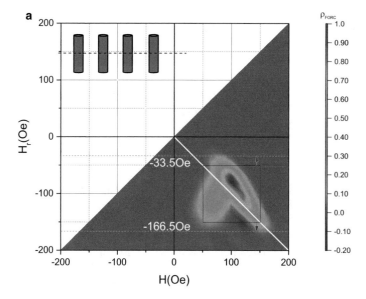

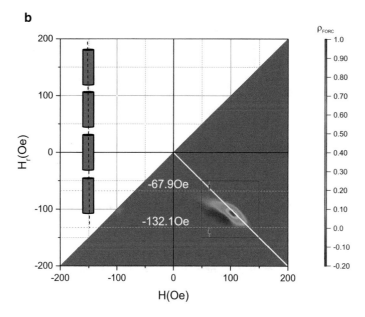

Fig. 7 (**a**) FORC diagram for a 1D-perpendicular chain of magnetic wires. (**b**) FORC diagram for a 1D-longitudinal chain of magnetic wires

maximum value of the interaction field in the central wire at the positive saturation is 17.9 Oe, and, consequently, the lowest coercivity wire will be switched at an applied field of $(-50 - 17.9)$ Oe $= 67.9$ Oe, and the most coercive will be switched at $(-150 + 17.9)$ Oe $= -132.1$ Oe. We see that the last wire to be switched down will experience an interaction field created by the neighbor wires which have been already switched. In this case the interaction field is assisting the switch of the last wire. We see that the entire FORC diagram for this sample is smaller than the rectangular area where the intrinsic switching fields reside. We also observe that the center of the FORC distribution is shifted on the opposite side of the second bisector of the Preisach plane (plotted with a white line in the FORC diagrams representations) compared with 1D-perpendicular case.

Increasing magnetizing interactions, the two limit lines within which all the switching occur become closer and closer. So, they tend to a situation in which all the wires inside the chain are switching in a very narrow field range. Gradually, with increased interactions, the FORC diagram collapses toward one point on the second bisector of the Preisach plane. As the field range of the switching is smaller, it is more difficult to measure/simulate FORCs inside the major hysteresis loop. This is a general problem when we try to use the FORC technique for systems showing really abrupt switching. Nonuniform field steps were suggested [40] to solve this problem. Essentially, it was suggested to use smaller field steps for the reversal field in the region of the switching on the major hysteresis loop. However, even if this technique proved useful in certain cases, its efficiency is certainly limited. This can be easily understood if we focus our attention on the region of the Preisach plane corresponding to the narrow switching field range. This region remains narrow in the Preisach plane regardless of the number of subintervals in which we are partitioning it. So, by increasing the number of FORCs measured in this region, the result will be an increase of details just within this small region on the Preisach plane and, actually, does not increase too much the quality of the full diagram. If we take this analysis further, we could increase more the coupling between the wires (including exchange interaction) and, by simulating these cases, we shall observe that the FORC diagram of the chain collapses toward a point in the second bisector of the Preisach plane. The chain is acting like a single hysteron. A detailed analysis shows that the switching process of the chain in this case is triggered by the lowest coercivity wire in the system and is followed by the switching of the other wires in an avalanche-type process. The general observation is that the demagnetizing-type interactions expand the region covered by the FORC diagram compared to the one for noninteracting wires, while magnetizing-type interactions produce a diagram in a smaller region compared to the noninteracting wires. If these magnetizing interactions are sufficiently strong, the entire diagram could collapse, and finally the whole magnetic system is represented by one hysteron only. In terms of multiplicity of images in the FORC distribution for one magnetic element (wire), we can say that demagnetizing interactions are related to multiplicities higher than one, and the systems with magnetizing interactions are expected to have subunitary multiplicity, meaning that more than one wire will be represented at the same point in the Preisach plane. The limit case is when the entire system is represented with the total magnetic moment of the sample at only one point. In this case we obviously lose the ability to find information concerning the individual wires. This shows the limitation of the FORC diagram in the classical form (which assumes that the measurement is made quasi-statically). Essentially, the conclusion of this analysis is that in case of intense interactions—in this case of ferromagnetic type— more than one magnetic internal brick, which could be identified separately with other types of measurements (like microscopy), can have a strongly correlated magnetic behavior in external magnetic fields and consequently seen in magnetic measurements as hysterons. In this case these hysterons could be the representatives of these strongly interacting entities. At the limit, if the entire sample has only one representative hysteron, we lose by quasi-static magnetic measurements the ability to see the magnetic representation of the internal components by the FORC diagram technique. For the moment we have shown this in magnetizing type interactions. These could include exchange ferromagnetic interactions, as well, but we have seen that in some cases magnetostatic interactions could be sufficient to produce the same result. The increase of field rate in the FORC measurement was proved to be one way to transform a collapsed diagram into one which offers information concerning the magnetic structure of the sample. We have named this approach as *dynamic FORCs* [44]. We are not providing here details of this technique which seems to give an interesting solution of this problem.

In order to observe the effect of stronger interactions in 1D-perpendicular magnetic chains, we have to take into account wires with smaller length and with smaller distance between them compared with the previous examples.

At the limit one can consider each magnetic element (pillar) as a punctiform dipole. To evidence more clearly the effect of interaction fields separately from their coercive fields, we have taken dipoles of an identical coercivity of 100 Oe for all the magnetic elements and a distance between the dipoles and their magnetic moments in such a way that the interaction field created by one dipole in the closest neighbor to be about 195.3 Oe. The interaction field in the next neighbor is in this case $195.3/8 = 24.4$ Oe (due to the dependence of the field created by a punctiform dipole with distance). The first observation is that if we only take into account the closest eight neighbors and we consider all in an up position, we get an interaction field of $2 * 195.3 * (1 + 1/8 + 1/27 + 1/64) = 460.0$ Oe, and if we take all the dipoles into the calculation of the interaction field, we obtain 469.6 Oe. This is drastically different from the case of long magnetic wires in which we discussed that, to accurately obtain the true interaction field, we have to take into account a large number of wires. Here, only a few neighbors are producing almost the entire interaction field. In Table 1 we have presented a number of configurations of the closest neighbors of one dipole in order to evaluate the main steps in the interaction field created by the switch of a certain neighboring dipole. When we look at the hysteresis loop and FORCs (Fig. 8(a)), we see that the major hysteresis loop has three distinct regions which are also visible in the FORC diagram. On the major loop plot, we have evidenced two lines at $\pm \exp(-2)$ which can be theoretically proven as remanence values for this type of strongly interacting chain of dipoles (a detailed discussion of this problem in a system of strongly interacting identical

Table 1 (Analytic evaluation of the interaction field in a given chain of dipoles in 1D-perpendicular configuration)

No.	Dipoles configuration	Magnetic field in the central dipole	Comments
1.	⇑ ⇕ ⇑	390.6 Oe	Maximum interaction field created by the nearest two neighbors
2.	⇑ ⇑ ⇕ ⇑ ⇑	439.4 Oe	Maximum interaction field created by the nearest four neighbors
3.	⇑ ⇑ ⇑ ⇕ ⇑ ⇑ ⇑	453.9 Oe	Maximum interaction field created by the nearest six neighbors
4.	⇑ ⇑ ⇑ ⇑ ⇕ ⇑ ⇑ ⇑ ⇑	460.0 Oe	Maximum interaction field created by the nearest eight neighbors (Line 1 in Fig. 8(b))
5.	⇑ ⇑ ⇑ ⇑ ⇕ ⇑ ⇑ ⇑ ⇓ ⇓ ⇑ ⇑ ⇑ ⇕ ⇑ ⇑ ⇑ ⇑	453.9 Oe	From eight neighbors seven are up and one is down
6.	⇑ ⇑ ⇑ ⇑ ⇕ ⇑ ⇑ ⇓ ⇑ ⇑ ⇓ ⇑ ⇑ ⇕ ⇑ ⇑ ⇑ ⇑	445.5 Oe	From eight neighbors seven are up and one is down
7.	⇑ ⇑ ⇑ ⇑ ⇕ ⇑ ⇓ ⇑ ⇑ ⇑ ⇑ ⇓ ⇑ ⇕ ⇑ ⇑ ⇑ ⇑	411.2 Oe	From eight neighbors seven are up and one is down (Line 2 in Fig. 8(b))
8.	⇑ ⇓ ⇑ ⇑ ⇕ ⇑ ⇓ ⇑ ⇑ ⇑ ⇑ ⇓ ⇑ ⇕ ⇑ ⇑ ⇓ ⇑	396.7 Oe	From eight neighbors six are up and two are down
9.	⇑ ⇓ ⇑ ⇑ ⇕ ⇑ ⇑ ⇓ ⇓ ⇓ ⇑ ⇓ ⇑ ⇕ ⇑ ⇑ ⇓ ⇑	390.6 Oe	From eight neighbors five are up and three are down
10.	⇓ ⇓ ⇑ ⇑ ⇕ ⇑ ⇓ ⇓ ⇓	370.0 Oe	From eight neighbors two are up and six are down (Line 3 in Fig. 8(b))
11.	⇓ ⇓ ⇓ ⇑ ⇕ ⇑ ⇓ ⇓ ⇓	321.2 Oe	From eight neighbors two are up and six are down (Line 4 in Fig. 8(b))
12.	⇑ ⇑ ⇑ ⇑ ⇕ ⇓ ⇑ ⇑ ⇑ ⇑ ⇑ ⇑ ⇓ ⇕ ⇑ ⇑ ⇑ ⇑	69.4 Oe	From eight neighbors seven are up and one is down (Line 5 in Fig. 8(b))
13.	⇑ ⇑ ⇓ ⇑ ⇕ ⇓ ⇑ ⇑ ⇑ ⇑ ⇑ ⇑ ⇓ ⇕ ⇑ ⇓ ⇑ ⇑	20.6 Oe	From eight neighbors six are up and two are down (Line 6 in Fig. 8(b))
14.	⇑ ⇑ ⇓ ⇑ ⇕ ⇓ ⇑ ⇓ ⇑ ⇑ ⇓ ⇑ ⇓ ⇕ ⇑ ⇓ ⇑ ⇑	6.10 Oe	From eight neighbors five are up and three are down

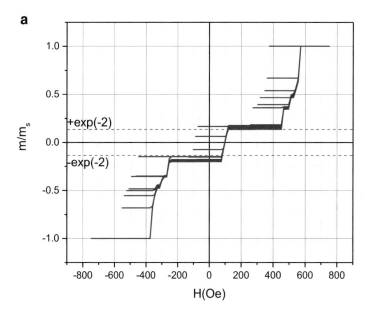

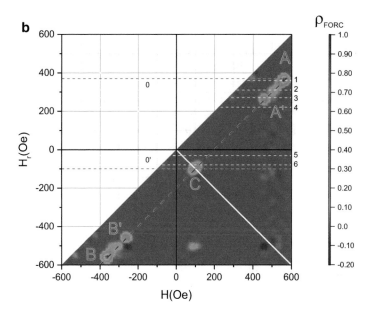

Fig. 8 (**a**) FORCs 1D-perpendicular—dipole system. (**b**) FORC diagram for the 1D-perpendicular—dipole system

dipoles can be found in [45]). In Fig. 8(b) we have noted a central peak (C) and several peaks, like (A-A′, B-B′). The central peak shows small interactions and is centered at 100 Oe. To understand the structure of the A-A′ peaks and the extension of the central peak, we have plotted a number of lines corresponding to possible changes in the structure of the eight nearest neighbor dipoles, as detailed in the table. We can easily see that the FORC diagram accurately records the continuous change in the status of the neighboring dipoles. At the initial stages of the switch on the major hysteresis loop, one can expect that the switching dipoles will be in the field created by most of the other dipoles in an up position (line 0 in Fig. 8(b)). Then, gradually, the most probable vicinity for the switching dipoles will contain one neighboring dipole down, then two, and so on. The central peak corresponds to the loop between the positive and negative remanence, dominated by long chains of dipoles with alternating up-down orientations of their magnetic moments.

A short conclusion of the discussion concerning the FORC diagrams of several magnetic systems is that in order to make quantitative analysis on these diagrams, one always needs two elements: (i) a proper Preisach image of the system based on the hypothesis of rectangular hysterons and (ii) a second physical model sufficiently sophisticated to be able to describe the typical features observed in the FORC experiment.

5 Summary

I think that most of the readers of this chapter are certainly interested in the use of the FORC diagram as a quantitative tool for complex hysteretic systems. We have shown that this technique unquestionably can be used in this capacity. However, we have to avoid the error made by some researchers to use the FORC diagram directly as a Preisach distribution. The "quantitative" part of the FORC analysis should be based on a physical model able to simulate the main features observed in the experimental diagrams. Our discussion made around a few magnetic systems like the 1D chains (longitudinal and perpendicular) and 2D-perpendicular of magnetic wires show that the essence of the magnetic interactions in the magnetic systems can be incorporated in such simple structures. In fact, the value of a physical model used in a quantitative FORC diagram study is greater when the systems used in simulations are simpler. However, it is not a bad idea to use more complex physical simulation tools like the micromagnetic simulations. In some cases, this solution is not computationally feasible due to the time and memory constraints in large magnetic structures simulations.

References

1. A. Stancu, D. Ricinschi, L. Mitoseriu, P. Postolache, M. Okuyama, First-order reversal curves diagrams for the characterization of ferroelectric switching. Appl. Phys. Lett. **83**, 3767 (2003)

2. C. Enachescu, R. Tanasa, A. Stancu, E. Codjovi, J. Linares, F. Varret, FORC method applied to the thermal hysteresis of spin transition solids: first approach of static and kinetic properties. Physica B: Condens. Matter **343**, 15–19 (2004)
3. R. Tanasa, C. Enachescu, A. Stancu, J. Linares, E. Codjovi, F. Varret, et al., First-order reversal curve analysis of spin-transition thermal hysteresis in terms of physical-parameter distributions and their correlations. Phys. Rev. B **71**, 014431 (2005)
4. C. Enachescu, R. Tanasa, A. Stancu, F. Varret, J. Linares, E. Codjovi, First-order reversal curves analysis of rate-dependent hysteresis: the example of light-induced thermal hysteresis in a spin-crossover solid. Phys. Rev. B **72**, 054413 (2005)
5. V. Franco, F. Béron, K.R. Pirota, M. Knobel, M.A. Willard, Characterization of the magnetic interactions of multiphase magnetocaloric materials using first-order reversal curve analysis. J. Appl. Phys. **117**, 17C124 (2015)
6. O. Henkel, Remanenzverhalten und Wechselwirkungen in hartmagnetischen Teilchenkollektiven. Phys. Status Solidi **7**, 919 (1964)
7. G.W.D. Spratt, P.R. Bissell, R.W. Chantrell, E.P. Wohlfarth, Static and dynamic experimental studies of particulate recording media. J. Magn. Magn. Mater. **75**, 309–318 (1988)
8. A. Stancu, P.R. Bissell, R.W. Chantrell, Interparticle interactions in magnetic recording media as obtained from high-order measurements by a Preisach model. J. Appl. Phys. **87**, 8645–8652 (2000)
9. A.P. Roberts, D. Heslop, X. Zhao, C.R. Pike, Understanding fine magnetic particle systems through use of first-order reversal curve diagrams. Rev. Geophys. **52**, 557–602 (2014)
10. R. Egli, VARIFORC: an optimized protocol for calculating non-regular first-order reversal curve (FORC) diagrams. Glob. Planet. Chang. **110**, 302–320 (2013)
11. R.J. Harrison, J.M. Feinberg, FORCinel: An improved algorithm for calculating first-order reversal curve distributions using locally weighted regression smoothing. Geochem. Geophys. Geosyst. **9**, Q05016 (2008)
12. J.B. Abugri, P.B. Visscher, S. Gupta, P.J. Chen, R.D. Shull, FORC+ analysis of perpendicular magnetic tunnel junctions. J. Appl. Phys. **124**, 043901 (2018)
13. D. Cimpoesu, I. Dumitru, A. Stancu, doFORC tool for calculating first-order reversal curve diagrams of noisy scattered data. J. Appl. Phys. **125**, 023906 (2019)
14. P. Weiss, J. de Freudenreich, Etude de l'aimantation initiale en fonction de la température. Arch. Sci. Phys. Nat. **42**, 449–470 (1916)
15. F. Preisach, Uber die magnetische Nachwirkung. Z. Phys. **94**, 277–302 (1935)
16. A. Stancu, C. Enachescu, R. Tanasa, J. Linares, E. Codjovi, F. Varret, FORC experimental method for physical characterization of spin crossover solids, in *Frontiers in Condensed Matter Physics Research*, ed. by J. V. Chang, (Nova Science Publishers, 2006), pp. 59–110
17. D.H. Everett, A general approach to hysteresis. Part 4. An alternative formulation of the domain model. Trans. Faraday Soc. **51**, 1551–1557 (1955)
18. I.D. Mayergoyz, Hysteresis models from the mathematical and control theory points of view. J. Appl. Phys. **57**, 3803 (1985)
19. C.R. Pike, A.P. Roberts, K.L. Verosub, Characterizing interactions in fine magnetic particle systems using first order reversal curves. J. Appl. Phys. **85**, 6660–6667 (1999)
20. H.G. Katzgraber, G. Friedman, G.T. Zimányi, Fingerprinting hysteresis. Phys. B Condens. Matter **343**, 10–14 (2004)
21. C.R. Pike, A.P. Roberts, M.J. Dekkers, K.L. Verosub, An investigation of multi-domain hysteresis mechanisms using FORC diagrams. Phys. Earth Planet. Inter. **126**, 11–25 (2001)
22. A.R. Muxworthy, D.J. Dunlop, First-order reversal curve (FORC) diagrams for pseudo-single-domain magnetites at high temperature. Earth Planet. Sci. Lett. **203**, 369–382 (2002)
23. C. Carvallo, A. Muxworthy, Low-temperature first-order reversal curve (FORC) diagrams for synthetic and natural samples. Geochem. Geophys. Geosyst. **7**, Q09003 (2006). https://doi.org/10.1029/2006GC001299
24. F. Wehland, A. Stancu, P. Rochette, M.J. Dekkers, E. Appel, Experimental evaluation of magnetic interaction in pyrrhotite bearing samples. Phys. Earth Planet. Inter. **153**, 181–190 (2005)

25. A. Stancu, C. Pike, L. Stoleriu, P. Postolache, D. Cimpoesu, Micromagnetic and Preisach analysis of the First Order Reversal Curves (FORC) diagram. J. Appl. Phys. **93**, 6620 (2003)
26. E. Della Torre, Effect of interaction on the magnetization of single domain particles. IEEE Trans. Audio Acoust. **AU-14**, 86–92 (1966)
27. E. Martha Pardavi-Horvath, F. Della Torre, A. Vajda, Variable variance Preisach model. IEEE Trans. Magn. **29**, 3793–3795 (1993)
28. A. Stancu, L. Stoleriu, P. Postolache, R. Tanasa, New Preisach model for structured particulate ferromagnetic media. J. Magn. Magn. Mater. **290-291**, 490–493 (2005)
29. A. Stancu, L. Stoleriu, M. Cerchez, P. Postolache, D. Cimpoesu, L. Spinu, Standard problems for phenomenological Preisach-type models. Physica B-Condens. Matter **306**, 91–95 (2001)
30. M. Vazquez (ed.), *Magnetic Nano- and Microwires* (Woodhead Publishing, Cambridge, UK, 2015)
31. W. Wernsdorfer, D. Mailly, K. Hasselbach, A. Benoit, J. Meier, J.-P. Ansermet, B. Barbara, Nucleation of magnetization reversal in individual nanosized nickel wires. Phys. Rev. Lett. **77**, 1873–1876 (1996)
32. C.-I. Dobrotă, A. Stancu, What does a first-order reversal curve diagram really mean? A study case: array of ferromagnetic nanowires. J. Appl. Phys. **113**, 043928 (2013)
33. C.-I. Dobrotă, A. Stancu, Tracking the individual magnetic wires' switchings in ferromagnetic nanowire arrays using the first-order reversal curves (FORC) diagram method. Phys. B Condens. Matter **457**, 280–286 (2015)
34. M. Nica, A. Stancu, FORC diagram study of magnetostatic interactions in 2D longitudinal arrays of magnetic wires. Phys. B Condens. Matter **475**, 73–79 (2015)
35. L. Clime, P. Ciureanu, A. Yelon, Magnetostatic interactions in dense nanowire arrays. J. Magn. Magn. Mater. **297**, 60–70 (2006)
36. N. Metropolis, A.W. Rosenbluth, M.N. Rosenbluth, A.H. Teller, E. Teller, Equation of state calculations by fast computing machines. J. Chem. Phys. **21**, 1087 (1953)
37. G. Bate, Statistical Stability of the Preisach Diagram for Particles of γ-Fe_2O_3. J. Appl. Phys. **33**, 2263 (1962)
38. E.P. Wohlfarth, A review of the problem of fine-particle interactions with special reference to magnetic recording. J. Appl. Phys. **35**, 783 (1964)
39. C. Papusoi, A. Stancu, Anhysteretic remanent susceptibility and the moving Preisach model. IEEE Trans. Magn. **29**, 77–81 (Jan 1993)
40. P. Postolache, M. Cerchez, L. Stoleriu, A. Stancu, Experimental evaluation of the Preisach distribution for magnetic recording media. IEEE Trans. Magn. **39**, 2531–2533 (2003)
41. R. Tanasa, A. Stancu, Statistical characterization of the FORC diagram. IEEE Trans. Magn. **42**, 3246–3248 (2006)
42. A.S. Samardak, A.V. Ognev, A.Y. Samardak, E.V. Stebliy, E.B. Modin, L.A. Chebotkevich, et al., Variation of magnetic anisotropy and temperature-dependent FORC probing of compositionally tuned Co–Ni alloy nanowires. J. Alloys Compd. **732**, 683–693 (2018)
43. T.R. Albrecht, Bit-patterned magnetic recording: theory, media fabrication, and recording performance. IEEE Trans. Magn. **51**, 0800342 (2015)
44. D. Cimpoesu, I. Dumitru, A. Stancu, Kinetic effects observed in dynamic first-order reversal curves of magnetic wires: experiment and theoretical description. J. Appl. Phys. **120**, 173902 (2016)
45. R. Tanasa, A. Stancu, Deterministic and non-deterministic switching in chains of magnetic hysterons. J. Phys. Condens. Matter **23**, 426002 (2011)

FORC Diagrams in Magnetic Thin Films

Dustin Gilbert

Abstract The first-order reversal curve (FORC) technique allows the use of bulk measurements to extract nanoscale features from hysteretic systems and separate intrinsic and interaction effects. Magnetic thin films in particular possess an extremely diverse and anisotropic energy landscape, which is complicated by interactions that exist on several length scales and the effects of nonmagnetic features such as pinning defects and nucleation sites. The diversity of these energies makes magnetic thin films, heterostructures, and patterned nanostructures highly tunable and relevant to applications but also exceedingly challenging to evaluate and design. The FORC technique is a unique tool for performing these evaluations; however, the technique is often considered challenging due to a lack of accessible material on the origins of the FORC features. In response to this, this chapter presents some common FORC diagrams found in thin films and evaluates their origins using mechanistic approaches. That is, by considering specific magnetic switching events within the reversal of a magnetic system and how the magnetism evolves during the FORC measurement, key features in the FORC distribution will be developed. Establishing an understanding of the FORC distribution using this mechanistic approach expands the ability to effectively utilize the FORC technique to evaluate these complex systems.

Keywords Hysteresis · Thin films · Heterostructures · Domain growth · Granular films

The first-order reversal curve (FORC) technique has been developed as an approach to extract microscopic details of a hysteretic system through macroscopic measurements [1–5]. While this approach can be used to investigate virtually any hysteretic system, magnetism is the by far the most common. In magnetic thin films and thin film heterostructures, the reversal behavior is exceedingly complex due to the

D. Gilbert (✉)
University of Tennessee, Knoxville, TN, USA
e-mail: dagilbert@utk.edu

mixing of the many magnetic energy terms [3]. These energies come together to define the magnetic behavior (reversal and field response), and in doing so, the underlying physics can be obscured. In one clear example, we consider the impact of defects in thin films of high-anisotropy material. In these systems, small defects— including potentially a single displaced atom—cause low-anisotropy regions, which then reverse prematurely. From this defect site, a magnetic domain wall, typically with a width of ≈5 nm, can rapidly propagate across macroscopic length scales, reversing the whole sample at a much lower magnetic field than theoretical values [6]. Alternatively, for the case of strong magnetostatic interactions, a film may only reverse until it is demagnetized [7].

The FORC technique is well suited to evaluate the magnetic reversal behavior within all magnetic systems, including thin film and thin film heterostructures. These measurements can typically extract qualitative details, such as the reversal process, and generate an understanding of the magnitude of the interactions [8] and distribution of intrinsic coercivities or defect densities. For some unique cases, the FORC technique can quantitatively determine these details [9].

This chapter focuses on FORC measurements performed on thin films and thin film heterostructures. This class of materials represents some of the most exciting and technologically high-impact systems, including hard drives [10], magnetic random access memory (MRAM), logic and spintronic technologies [11], topological insulator heterostructures, and most magnetic sensors. Thin films can possess extremely strong, albeit short-range, interactions in the exchange energy, which generates a new reversal mechanism—namely, through domain growth—that is not accessible in small patterned elements [12] or nanoparticles [13]. The domain growth reversal mechanism presents a unique challenge to the FORC technique, in that the reversal process cannot be generally treated as a mean-field process. This emphasizes another challenge in thin films, namely, the crucial role of defects. Defects in thin films can act as low-anisotropy or high-anisotropy sites, responsible for nucleating or pinning magnetic domains [14].

This chapter will present a perspective on FORC starting with thin films comprised of isolated grains, then to grains with dipolar coupling, then to continuous, single-phase films with in-plane and out-of-plane anisotropy, and finally to multilayer films. This progression builds on each subsequent section to generate a comprehensive picture of the FORC diagram with respect to thin film magnetism.

1   The FORC Measurement Sequence

A simplistic, overarching narrative of the FORC measurements can be presented such as the following: a hysteretic system is initially prepared in a well-known (typically saturated, single-phase) state; the system is then transformed into a mixed phase state; the evolution of the system from this mixed phase state is measured and will clearly depend both on the intrinsic qualities and the interactions within the system; and by measuring this evolution in a systematic series of mixed states and comparing the differences, these factors (the intrinsic and interaction properties) can be separated.

In a more technical explanation, the FORC technique measures a series of minor hysteresis loops. For magnetism, FORC measurements start in the (typically positive) saturated state. From positive saturation, the magnetic field is decreased to a reversal field, H_R. From H_R, the applied magnetic field, H, is increased and the magnetization measured as the system is brought back to positive saturation, measuring a single FORC branch. This sequence is repeated for H_R values between the initial reversal or nucleation field and negative saturation; measurements beyond this range nominally capture no signal but will improve the boundaries of the dataset. As the FORC measurement proceeds, H_R becomes more negative, and as it does the magnetization decreases. If the system is treated as a collection of hysteretic elements, the magnetization decreasing corresponds to some of the elements switching from positive saturation to negative, undergoing a "down-switching" event. Proceeding to the next H_R (at a more negative magnetic field), the magnetization is even lower, indicating new down-switching events. Thus, the H_R parameter probes the down-switching events. Then, from each H_R, H is increased back toward positive saturation, measuring a unique FORC branch, with a magnetization $M(H, H_R)$. As H becomes more positive, the magnetization will increase, indicating the complementary up-switching events; the H-axis thus probes up-switching events.

From a mechanistic perspective, at each H_R the system is prepared in a new state that is evolved out of the state at the previous H_R. This configuration will be determined by the value of H_R, the intrinsic qualities of the magnetic material, and the interactions in the system; the intrinsic qualities do not change with H_R, while the interactions do. Along each FORC branch the evolution of the magnetization will be different depending on these same qualities. By analyzing the differences between the FORC branches by applying a derivative, the changes in the evolution are mapped out. Thus, the progression is changes in H_R lead to changes in the magnetic configuration, which in turn changes how the state evolves along the FORC branch. With this in mind, the FORC distribution is determined by calculating the derivative dM/dH, where M is the magnetization, to capture the up-switching events on each branch and then a second derivative with respect to H_R to capture the changes in these up-switching events between adjacent FORC branches: $\rho = -d^2M/(dHdH_R)$. For convention, the FORC distribution is sometimes presented with a normalizing pre-factor of $1/(2M_S)$, where M_S is the saturation magnetization.

A critical detail which must be kept in mind throughout all the sections of this chapter is that the FORC technique has no ability to associate a specific reversal event to a specific region of the sample. This means that a feature in the FORC distribution likely is the result of the "down-switching" of one region and the "up-switching" of a different region. For this reason, the term "hysteron" is established as a fundamental unit of hysteresis, being comprised of up-and down-switching events without any regard for the physical origins of the reversal. The FORC distribution is a map of the hysteron density and *does not* map the coercivity or interactions of distinct elements within the sample except in special cases— typically at the boundaries of the FORC distribution. With this in mind, the text will refer to the "lowest (highest) coercivity" element, meaning the magnetically softest (hardest) regions, which would reverse first (last) in a system with no interactions.

Data will also be presented in an alternative coordinate space, termed the bias and coercivity axes (H_B, H_C), respectively. As noted above, the hysteron is defined by the down-switching and up-switching field, represented by its H_R and H coordinates, respectively. As with any hysteresis loop, the coercivity is defined as the half difference between the up- and down-switching fields, and the bias is their average, $H_B = (H + H_R)/2$, $H_C = (H - H_R)/2$. While bias in magnetism is frequently associated with "exchange bias," which arises due to coupling between a ferromagnet and an antiferromagnet [15], since we are measuring minor loops, the bias here can be caused by many things, including interactions between ferromagnetic regions or the cross-correlation of the up- and down-switching events of two separate regions in the definition of the hysteron. Therefore, the bias axis *can* identify interactions, as in the former case but may also only represent the intrinsic qualities of two different regions of the sample and therefore would not represent any interactions.

With the hysteron model muddling the magnetic signals, it initially seems that deconvoluting the FORC distribution may be impossible. However, the FORC distribution is well-known to encode the qualitative "fingerprints" of the reversal mechanisms. Furthermore, by following physical reasoning, some details of the FORC features can still be extracted quantitatively [9, 16]. The goal of the next sections is to lay the foundations of this understanding and identify key points to aid in the interpretation of the FORC diagram.

2 FORC Measurements of Isolated Granular Films

The first section focuses on a system of isolated grains on a thin film [17]. This might be experimentally realized as a film comprised of islands, magnetic defects, or dilute particles or nanostructures on a surface [16, 17]. At positive saturation, all of the magnetic elements are co-aligned in the positive direction. Following the FORC measurement procedure, the field is reduced to H_R. The first down-switching event can be attributed to the region with the lowest intrinsic anisotropy, since there are no interactions. Without interactions, the down-switching will occur when the reversal field equals the negative coercivity, $H_R = -H_C{}^1$. Along the subsequent FORC branch, e.g., $H_R < -H_C{}^1$, this region is already down-switched and will always up-switch at $H = +H_C{}^1$, since there are no interactions, as shown in Fig. 1. There are no up-switching events on the FORC branch starting at $H_R > -H_C{}^1$ (thus $dM/dH = 0$ everywhere), and now at $H_R = -H_C{}^1$ there is an up-switching event

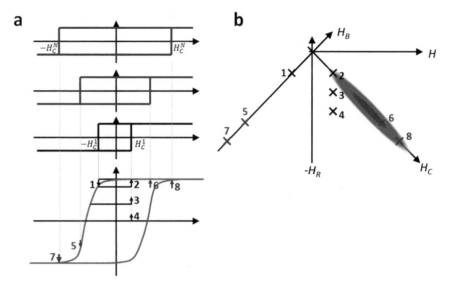

Fig. 1 (**a**) Illustrative diagram showing the reversal of three different regions of a sample of isolated regions and how those individual loops evolve throughout the major hysteresis loop. (**b**) Locations of the switching events identified in panel a, shown in the FORC distribution

at $H = +H_C{}^1$. Assuming the up-switching occurs as a step reversal, $dM/dH = 0$ everywhere else except at the up-switching field of this region, at which point $dM/dH = \Delta m/\Delta H$ where m is the magnetization of the reversed region and ΔH is the field step in H. Calculating the FORC distribution, it is zero everywhere except at $H_R = -H_C{}^1$, and $H = +H_C{}^1$, where $\rho = -((\Delta m/\Delta H) - 0)/(\Delta H \Delta H_R)$. Since convention has H_R measured from positive saturation to negative, $\Delta H_R < 0$ and $\rho = \Delta m/(\Delta H|\Delta H_R|)$. Integrating this feature clearly recovers the magnetization of the reversed element. Calculating this feature in (H_C, H_B) coordinate space, the feature appears at $(H_C = H_C{}^1, H_B = 0)$.

Proceeding with the FORC measurement, subsequent FORC branches will reverse new regions. However, these reversals will not change the up-switching field of the first element; the up-switching field of this first region is still at $H = +H_C{}^1$, independent of H_R. Thus, on every FORC branch with $H_R < -H_C{}^1$, at $H = +H_C{}^1$ there will be a feature with magnitude Δm. The derivative dM/dH at $H = +H_C{}^1$ will always be the same, so the H_R derivative will always be zero. Due to this ever-unchanging up-switching field, each region appears only once in the FORC distribution. The reversals of the other elements will occur at their respective H_R and H values and, following the same calculations as above, will appear at an H_C coordinate associated with their intrinsic coercivity and $H_B = 0$. Therefore, in the special case of noninteracting elements, the (H_C, H_B) coordinate space indeed maps out the intrinsic material properties. The key signature of this type of reversal is a feature, extended in H_C, with a small, symmetric distribution in H_B.

3 FORC Measurements of Interacting Granular Films

This section introduces long-range interactions to the isolated granular film [18]. The original FORC distribution from the noninteracting case is shown in greyscale along the H_C axis. These interactions will be treated as mean-field-like, such that $H_{\text{Interaction}} = \alpha(M/M_S)$; non-nearest neighbor interactions manifest by segregating the FORC distribution [9, 19, 20]. Considering the lowest-coercivity region again and demagnetizing interactions (e.g., $\alpha < 0$), then from positive saturation, the interaction field experienced by every region of the sample is simply $-|\alpha|$, making the effective field experienced by the sample $H - |\alpha|(M(H, H_R)/M_S)$. Since every region experiences the same interaction field, the reversal order will be the same as the isolated case; the lowest-coercivity region will down-switch at $H_R - |\alpha| = -H_C{}^1$ or from the measurement perspective $H_R = -H_C{}^1 + |\alpha|$. Beginning a FORC branch at $H_R = -H_C{}^1 + |\alpha|$, and disallowing any self-contribution to the mean-field term, the interactions experienced by this region didn't change before versus after the down-switching event, so the subsequent up-switching event is also be biased by $-|\alpha|$, occurring at $H - |\alpha| = +H_C{}^1$, e.g., $H = +H_C{}^1 + |\alpha|$. Transforming this FORC feature into (H_C, H_B) coordinates, the feature now appears at $(H_C = H_C{}^1, H_B = |\alpha|)$. Thus, both the coercivity of the lowest-coercivity region and the mean-field interactions can be determined by this one point in the FORC distribution.

Since the proposed interactions are mean-field-like, the reversal order is only determined by the intrinsic coercivity. Accordingly, the up-switching field for the lowest-coercivity particle will constantly shift, occurring at $H = H_C{}^1 + |\alpha|(M(H_R)/M_S)$. It is easy to visualize that at the system's coercive field, $M = 0$; thus the up-switching event should occur at $H = H_C{}^1$ but will be associated with the down-switching event, $H_R = H_C{}^{n/2}$. As shown in Fig. 2, this first up-switching event on each branch will never have a matching event on the previous H_R branch and thus will always appear in the FORC distribution, in contrast to the noninteracting case. This emphasizes the mixed quality of the hysteron: new regions down-switching are responsible for the changes at H_R, but the same low-coercivity region is responsible for the up-switching on every branch. Approaching negative saturation, the H_R values are probing the down-switching field of the highest coercivity regions. The down-switching field of these regions is also biased by the interaction field, occurring at $H_R - |\alpha|(-M_S/M_S) = -H_C{}^N$, e.g., $H_R = -H_C{}^N - |\alpha|$. Progressing along the FORC branch from this H_R value, the FORC feature from the first switching event, arising from the lowest-coercivity region, will occur at $H = H_C{}^1 - |\alpha|$. Calculating the coercivity and bias field coordinates: $(H_C = ((H_C{}^1 + H_C{}^N)/2), H_B = ((H_C{}^1 - H_C{}^N)/2) - |\alpha|)$. Based on the endpoints of the leading FORC feature, occurring at $(H_C = H_C{}^1, H_B = |\alpha|)$ and $(H_C = ((H_C{}^1 + H_C{}^N)/2), H_B = ((H_C{}^1 - H_C{}^N)/2) - |\alpha|)$, the asymmetry in the H_B axis quantitatively maps the span of the intrinsic coercivity distribution, while the H_C coordinate identifies the average of the coercivity maxima.

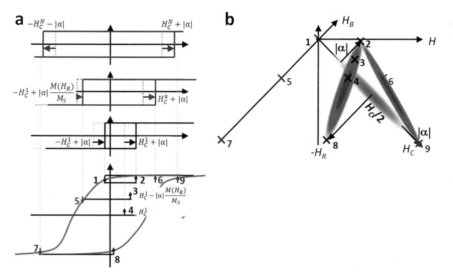

Fig. 2 (**a**) Illustrative diagram showing the reversal of three different regions of a sample with net demagnetizing interactions and how those individual loops evolve throughout the major hysteresis loop. (**b**) Locations of the switching events identified in panel a, shown in the FORC distribution. $|\alpha|$ indicates the length of the black arrow along the H_C axis, highlighting the displacement of feature 9

A third point that can be quantitatively identified is the up-switching field of the highest coercivity region. The previous paragraph identified the down-switching event of this region at $H_R = -H_C^N - |\alpha|$. Along the FORC branch, this highest coercivity element will also be the last to up-switch before positive saturation, occurring at $H = H_C^N + |\alpha|$. Calculating the bias and coercivity axis coordinates, this feature will appear at ($H_C = H_C^N + |\alpha|$, $H_B = 0$), e.g., on the H_C axis.

Based on these three points, a pair of lines can be constructed between the extremum points: ($H_C = H_C^1$, $H_B = |\alpha|$), ($H_C = ((H_C^1 + H_C^N)/2)$, $H_B = ((H_C^1 - H_C^N)/2) - |\alpha|$), and ($H_C = H_C^N + |\alpha|$, $H_B = 0$), e.g., points 2, 8, and 9. The generalized shape of the FORC distribution illustrated in Fig. 2a is called a "wishbone structure" and can be developed from these three points. First, the angle from the H_B axis of the line between points (2) and (8) can be calculated to be $\mathrm{Arctan}[-(H_C^1 - H_C^N)/(4|a|-(H_C^1 - H_C^N))]$, which goes to zero for large $|\alpha|$ relative to ($H_C^1 - H_C^N$), and 45° as $|\alpha|$ becomes small. This means that the leading feature becomes parallel to the H_B axis as the interaction strength becomes large compared to the coercivity distribution.

Calculating the angle between points 8-2-9 yield an equation:

$$\mathrm{Arctan}\left[\frac{\left(2\,|\alpha| - \left(H_C^1 - H_C^N\right)\right)^2}{\left(2\,|\alpha| - \left(H_C^1 - H_C^N\right)\right)^2 + 2\,|\alpha|\left(H_C^1 - H_C^N\right)}\right]$$

While this equation has many variable dependencies, both the numerator and denominator are always positive. This means that the angle between points 8-2-9 will always be acute! Exploring the extremum values, for large $|\alpha|$ or small ΔH_C, the angle tends toward $45°$, and for small $|\alpha|$, the angle also tends toward $45°$. Interestingly, this equation has a zero at $|\alpha| = \Delta H_C/2$, which corresponds to the case that the leading edge and the backbone meet, e.g., points 8 and 9 coincide, and so below this value, the FORC distribution will only be tilted away from the H_C axis.

While we identified three crucial points at the end of each of the arms of the wishbone, the whole wishbone can be described based on the magnetization at H_R and the intrinsic distribution. Specifically, the last up-switching event on each FORC branch can be tabulated as occurring at $H = H_C{}^k + |\alpha|$, where $H_C{}^k$ is the intrinsic coercivity of the last region which down-switched at H_R; since this is a mean-field system, the reversal order is determined only by the intrinsic distribution; the last up-switching event on every branch has the largest coercivity and is the last to down-switch at H_R. The down-switching field for this region is also calculated as $H_R = -H_C{}^k + |\alpha|(M(H_R)/M_S)$. Using the calculated switching fields and the known reversal sequence, the backbone of the feature can be calculated as $H_C = H_C{}^k + (|\alpha|/2)(1 - (M(H_R)/M_S))$ and $H_B = (|\alpha|/2)(1 + (M(H_R)/M_S))$. Furthermore, the first up-switching event along each branch can be calculated to occur at $H = H_C{}^1 + |\alpha|(M(H_R)/M_S)$, since it is always the same lowest-coercivity region that is up-switching. Transforming these coordinates into (H_C, H_B) coordinates reveals that the leading feature occurs at $H_C = (H_C{}^1 + H_C{}^k)/2$ and $H_B = (H_C{}^1 - H_C{}^k)/2 + |\alpha|(M(H_R)/M_S)$. These calculations agree with the previously reported endpoints but also show that the backbone feature always occurs at positive values of H_B that are dependent only on the magnetization at the reversal field and the mean-field coefficient.

The coordinates can more generally be calculated for every point in the FORC distribution, using the mean-field term; however, most switching events are matched on adjacent branches and so do not appear in the distribution. Nevertheless, these calculated points illustrate that the switching fields are shifted in the FORC parameter spaces only through a simple additive term which is only dependent on the magnetization and the mean-field coefficient. As a result, for a system with mean-field interactions, the interactions and the intrinsic properties can be decoupled by transforming the independent variables, H and H_R, using the dependent variable M, in an approach called "de-shearing" [17, 19, 21]. Specifically, the magnetic field experienced by any region at any point in the FORC measurement is $H = H_{\text{Applied}} + |\alpha|(M(H,H_R)/M_S)$; applying the transform $H \rightarrow H - |\alpha|(M(H_R)/M_S)$, and $H_R \rightarrow H_R - |\alpha|(M(H_R)/M_S)$, the mean-field interactions are removed. For the wishbone feature, this results in the FORC distribution losing its leading edge and collapsing to the H_C axis, recovering the noninteracting case, e.g., the greyscale feature. The resultant distribution should be nearly symmetric in $\pm H_B$, indicating no residual interactions. The challenges of applying this technique include that (1) it corrects for only mean-field interactions,

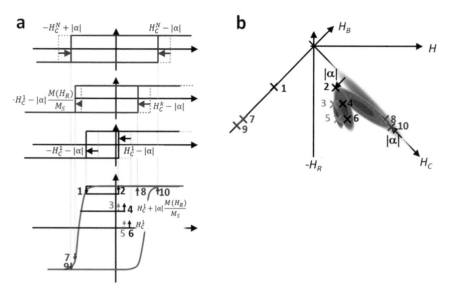

Fig. 3 (**a**) Illustrative diagram showing the reversal of three different regions of a sample with net magnetizing interactions and how those individual loops evolve throughout the major hysteresis loop. Feature 3 and 5 identifies the up-switching events 2 and 4, which is absent on the subsequent FORC branches. (**b**) Locations of the switching events identified in panel a, shown in the FORC distribution. $|\alpha|$ indicates the length of the black arrow along the H_C axis, highlighting the displacement of feature 10

(2) the transformation only applies to the demagnetizing case, and (3) the transformed data may have significant variation in field step size, which can present challenges in calculating the derivative.

Applying the same mechanistic considerations to a system with magnetizing interactions works reasonably well while the interactions are weak compared to the intrinsic coercivity distribution. Specifically, for mean-field magnetizing interactions, the field experienced by the sample will now be $H_R + |\alpha|(M(H,H_R)/M_S)$ and a more negative magnetic field is required to cause the first down-switching event, at $H_R = -H_C{}^1 - |\alpha|$. Along the subsequent FORC branch, the up-switching event will be biased by the same interaction field, occurring at $H = H_C{}^1 - |\alpha|$. Transforming these coordinates into (H_C, H_B) representation, the FORC feature appears at $(H_C = H_C{}^1, H_B = -|\alpha|)$, as shown in Fig. 3, thus the symmetry with the demagnetizing case is apparent. Proceeding to the next, more negative H_R, the next smallest coercivity particle will reverse, and along the subsequent FORC branch, the first up-switching event will occur at $H = H_C{}^1 - |\alpha|(M(H_R)/M_S)$, which is more positive than $H_C{}^1 - |\alpha|$. Accordingly, at $H = H_C{}^1 - |\alpha|$ there is an up-switching event on the $H_R = -H_C{}^1 - |\alpha|$ branch, but not on either of the neighboring branches. In Fig. 3 this is seen in feature 4, for example. Calculating the FORC feature indicates that there will be a positive/negative feature pair at this coordinate;

since these features are from the same magnetic region, they will have equal and opposite magnitude. As $M(H_R)$ continues to decrease with H_R, the first up-switching event continues to move to more positive values of H, generating a series of positive/negative feature pairs. Finally, the region with the largest coercivity reverses at $H_R = -H_C^N + |\alpha|$. Along the subsequent FORC branch, the lowest-coercivity particle up-switches at $H = H_C^1 + |\alpha|$, generating a FORC feature at $(H_C = (H_C^1 + H_C^N)/2, H_B = ((H_C^1 - H_C^N)/2) + |\alpha|)$. Also along this FORC branch, the highest coercivity region will be the last to up-switch, occurring at $H = H_C^N - |\alpha|$, generating a FORC feature at $(H_C = H_C^N - |\alpha|, H_B = 0)$.

Again, the magnetizing case has a leading edge—although now it is a positive negative feature pair—which encodes the interaction strength, and an endpoint on the H_C-axis. The backbone of the feature resulting from the last up-switching event can again be calculated as occurring at $(H_R = -H_C^k - |\alpha|(M(H_R)/M_S)$, $H = H_C^k - |\alpha|)$ or $H_C = H_C^k - (|\alpha|/2)(1 - (M(H_R)/M_S))$ and $H_B = -(|\alpha|/2)(1 + (M(H_R)/M_S))$. The magnetizing case appears to be the mirror symmetry of the demagnetizing case, with the backbone always at H_B, e.g., below the H_C coordinate axis.

This description presents a simplified version of the magnetizing case, and the breakdown of that simplification provides a convenient segue to the continuous film case by considering the magnetizing case with interactions which are comparable to the intrinsic coercivity distribution. Returning to the lowest-coercivity element, which down-switches at $H_R = -H_C^1 - |\alpha|$, the interaction field experienced by the rest of the sample decreases by $|\alpha|(M^1/M_S)$ after the switching event, where M^1 is the magnetization of this first region. For sufficiently large M^1 or $|\alpha|$ or $H_C^2 \approx H_C^1$, where H_C^2 is the intrinsic switching field of the next lowest-coercivity region, the reduced field can switch the next lowest-coercivity region, starting a cascade event. Physically, this implies that the reversal of the first element reduces the overall system stability, which then precipitates the reversal of the second element; this in turn could precipitate further reversals. The same arguments can be made along the subsequent FORC branch. Thus the FORC diagram will capture one reversal event for this collection of reversals, effectively concealing their intrinsic coercivity distribution. This becomes exceedingly problematic for large $|\alpha|$ since it can hide many reversal events under the feature of a single triggering event. This becomes particularly prevalent for the case of non-mean-field interactions, which can be much stronger. For such a case, the reversal of one region can nucleate an avalanche event that will grow from that initial location. This is more commonly known as domain nucleation and propagation or growth. In many continuous thin films, the exchange interaction dominates the local energy environment and is much larger than the other terms that contribute to the intrinsic anisotropy distribution. As a result, continuous films frequently reverse by domain propagation, in direct analogy to the magnetizing case presented above. An avalanche reversal or a domain nucleation/propagation can grow until negative saturation is reached without something to stop it, such as a pinning site, or a counterbalancing interaction.

4 FORC Measurements of Continuous Films with In-Plane Anisotropy

The reversal behavior of continuous films with in-plane anisotropy nominally occurs similar to the magnetizing case discussed above. Coming from positive saturation, the lowest-coercivity region overcomes its intrinsic coercivity and interactions with neighboring elements and down-switches. For the case of a continuous film, the interactions are different in that the exchange energy is typically much larger than the magnetic field, making $|\alpha|$ exceedingly large and suggesting a large, negative H_R is required to achieve switching. Two details undermine this consideration: (1) continuous films suffer from defect sites which can reverse prematurely, and (2) in the absence of a strong anisotropy, large areas of magnetic moments, e.g., domains, can reverse together reducing the interaction cost to only the domain perimeter. Since the Zeeman energy grows with the area of a domain, while the exchange cost only grows with the domain perimeter, this latter effect would suggest domains would grow continuously. However, there are typically magnetic forces, including pinning defects, and magnetostatic energy (not for planar systems with in-plane anisotropy) that stop the domain propagation. As a consequence of these effects, the reversal of continuous thin films does not typically capture a quantitative intrinsic coercivity or interaction.

The FORC diagram of continuous thin films with in-plane anisotropy, shown in Fig. 4, tends to be dominated by the defect density and the distribution of the magnetocrystalline anisotropy for both nucleation and domain growth, or if these are both very low, the film can undergo coherent rotation. Specifically, measuring the FORC distribution of a continuous film with very few defects, the magnetic reversal can be initiated at one of the rare defect sites, or a sample edge, in-which the magnetic moment is canted away from the positively saturated magnetic field; the canting could be caused by magnetostatic energies from a canted facet, or a region with magnetocrystalline misalignment, or an interstitial or vacancy defect. Coming from positive saturation, these regions will reverse at a magnetic field, H_{Nuc}, that will depend on the details of the nucleating defect. In a film with in-plane anisotropy the reversal typically begins at $H_R < 0$, since there are few interactions which are acting to demagnetize the sample; reversing at $H > 0$ requires a magnetic energy to counter the Zeeman energy. Once nucleated, the reversed domain has a beneficial Zeeman energy which scales with the volume of the domain and an exchange-driven domain wall energy cost which scales with the perimeter area. For films, the volume scales as the area of the domain times the film thickness, while the perimeter area is the perimeter length times the film thickness. The one-dimensional advantage of the Zeeman relative to the domain wall energy means that the energy of the system is continuously decreased as the domain grows. Thus, in a system with few defects, the domains will rapidly propagate until the domain wall encounters a feature which pins the expansion. This feature may be another defect or a high-anisotropy region or even a nonmagnetic inclusion. Using the nucleation/propagation field as the first H_R which shows a change in magnetism, the subsequent FORC branch is measured as

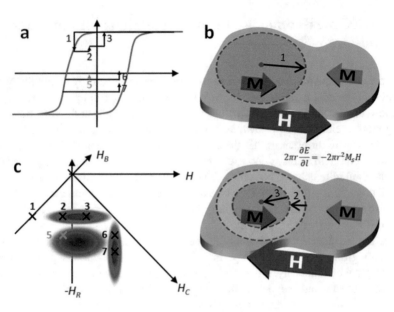

Fig. 4 (**a**) Illustrative diagram showing the reversal events within the system, with 1 identifying the initial domain nucleation and propagation and then 2, 3 indicating the domain contraction returning to saturation. Approaching negative saturation, the system foregoes the domain contraction, 5, and must re-nucleate domains, identified at features 6 and 7. The domain growth and contraction are shown in (**b**) and the resultant FORC diagram in (**c**)

the field is increased toward positive saturation. While $H < 0$ the magnetic domain's configuration typically changes very little due to the same arguments as above— a contraction of a negatively aligned domain would incur a Zeeman energy cost which scales with the domain volume. Approaching $H = 0$, the domain wall energy becomes comparable then larger than the Zeeman energy, and it becomes beneficial for the domain to contract. The contraction will begin in regions in which the domain is not pinned, manifesting a feature in the FORC diagram with an H_R value corresponding to the nucleating defect, and an H coordinate corresponding to the contraction. The contraction field for an in-plane film can be calculated by setting the Zeeman energy equal to the domain wall energy: $-H M_S \pi r^2 t = 2 \pi r t (dE/dl)$ or $H = -2 (dE/dl)/(M_S r)$ describing a circular domain with radius r, and a domain wall energy per unit length of (dE/dl); this equation does not take into account the pinning sites and hence only identifies the field at which the domain begins to contract. As the field continues to become more positive, the domain continues to contract, maintaining a balance between the decreasing Zeeman energy and the domain wall energy and the pinning sites. In the FORC diagram, the feature will spread along the H coordinate at this fixed value of H_R. During the contraction, the domain wall can de-pin from the defect sites which may cause local propagations of the domain, similar to Barkhausen jumps [22]. Once $H > 0$ the Zeeman energy and the domain wall energy both benefit from the collapse of the domain, and as such the contraction accelerates with respect to the field change. Together, the initial domain nucleation/propagation and then subsequent contraction cause a feature located at a single H_R coordinate, H_{Nuc}, with a spread along the H-axis. If the field along the FORC branch achieves $H = -H_{Nuc}$, then the initial nucleation site will again become active, nucleating a positively oriented domain, which expands to complete the reversal. As a result, the up-switching field is always less than or equal to the negative of the initial nucleation field $H \leq -H_{Nuc} = H_R$, and thus this initial feature is located at $H_B \leq 0$.

Continuing to more negative values of H_R, the domain wall expands, crossing more pinning sites. One could consider these defects as "stronger" due to the fact that a larger magnetic field is required to overcome the preceding pinning sites. One could also consider that the domain wall is now trapped between pinning sites, with some resisting further growth and others resisting contraction. The pinning sites that resist contraction will increase the magnetic field necessary to achieve up-switching along FORC branches measured from these H_R, resulting in a reduced dM/dH initially along each FORC branch which causes a negative feature in the FORC distribution. Similar to the magnetizing case discussed above, these suppressed switching events must occur along the FORC branch—since the sample does eventually achieve positive saturation—and so must therefore form a positive/negative feature pair. However, the constraint continues to exist that once $H = -H_{Nuc}$, a new positive domain will be formed and propagate outward. Since $-H_{Nuc} \leq -H_R$, the FORC feature is once more constrained to $H_B < 0$. Furthermore, for the case of few pinning sites, re-nucleation will become the dominant reversal mechanism and so will generate a FORC feature at $H = -H_{Nuc}$.

Putting these features together, for the case of strong pinning in which re-nucleation plays a dominant role in the switching, the FORC feature forms two arms, located at $H_R = -H_{Nuc}$ and $H = -H_{Nuc}$, both located at $H_B < 0$, with a negative feature located between the arms, as shown in Fig. 4. While this feature has been reported, in practice, it is much more common to have a bowed shape, rather than arms at a strict $90°$, or even a feature simply located along the H_C axis. The distinction arises from the ability to propagate domains, with more pinning sites behaving more as a system of isolated regions.

5   FORC Measurements of Continuous Films with Out-of-Plane Anisotropy

Continuous films with out-of-plane anisotropy possesses the same energy terms as the in-plane case, that is, the Zeeman energy balanced against the domain wall energy and pinning sites, but also has a demagnetizing magnetostatic energy discussed above. This new energy term causes a distinctly different reversal by constantly motivating the system toward a demagnetized state. Following the same considerations as the continuous in-plane film, the magnetic field is reduced from positive saturation and nucleates from a defect site. The nucleation site is influenced by both the magnetic field and the magnetostatic interaction, which is $4 \pi M_S$ for a saturated thin film. Once an H_R is achieved which facilitates a nucleation event, the domain will rapidly propagate, reducing the magnetostatic energy, in exchange for increasing the domain wall energy. Unlike the in-plane case, the magnetic field from the magnetostatic interaction can easily be larger than the applied field, and so the Zeeman energy can play a role by helping the domain growth (if reversal occurs at $H_R < 0$) or opposes the expansion (if reversal occurs at $H_R > 0$). Similar to the in-plane case, the magnetostatic energy scales with the magnetization, which in turn scales with the domain volume, while the domain wall energy scales with the perimeter length, and as such the domains are able to rapidly propagate up to the system's coercive field; the magnetostatic energy increases beyond the coercive field, opposing further domain growth. Measuring a FORC branch from the nucleation field $H_R = H_{Nuc}$, as H is increased the domain, may remain initially static due to the combined influences of pinning, the magnetostatic energy (which opposes returning to saturation), and the Zeeman energy (assuming $H_R < 0$). Once $H > 0$, the Zeeman energy works against the magnetostatic energy to move the system back to positive saturation. Recalling that both the Zeeman and magnetostatic energies scale with volume, an increase in H (corresponding to the Zeeman energy) increases the domain size, which increases the magnetostatic interactions, resulting in a balancing act. In effect, the Zeeman energy, which prefers the saturated state, is balanced against the magnetostatic energy, which prefers the demagnetized state. This is in contrast to the in-plane case, during which the Zeeman and domain wall energies both favor the saturated state along the FORC branch, with the pinning sites only acting to stop saturation. As a result of the balancing act in films with an out-of-plane anisotropy, the progression from H_R to saturation tends to be much longer in H than the in-plane case, which has no such balancing mechanism, as shown in Fig. 5. Saturation is achieved once the Zeeman energy overcomes the total magnetostatic field $4\pi M_S$. However, some works have shown the presence of residual bubbles which exist beyond the apparent saturation point [7]. These features are the result of the extreme energies required to collapse small domains with continuous domain walls, an effective topological protection [23, 24]. Other features, such as pinning defects, will stagger the field progression. The resulting FORC feature therefore will span in H potentially from the initial nucleation field to at least $H = 4\pi M_S$, all at a single value of $H_R = H_{Nuc}$.

The initial nucleation event does not typically reach HC since the magnetostatic energy becomes smaller as M approaches zero and eventually becomes small compared to the pinning strength of defects. Along subsequent FORCs, as H_R becomes more negative, the negatively oriented domains grow following a similar balancing mechanism as the positive domains under increasing H, that is, as H_R is decreased, reducing the magnetization, the magnetostatic energy increases, resisting the change. Increasing H along the subsequent FORC branch, between $H_R < H < 0$, the Zeeman energy decreases, allowing the domains to tend toward a demagnetized state, and for $H > 0$ the positively oriented domains grow as the system approaches saturation. In both cases, and for each field step, the balance between the Zeeman

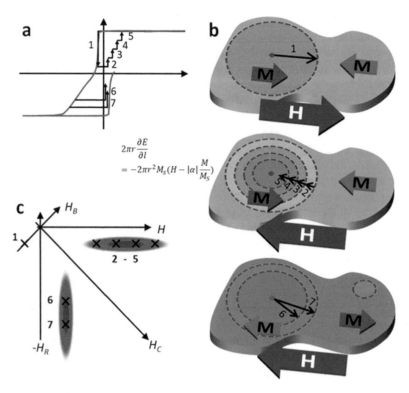

Fig. 5 (**a**) Illustrative diagram showing the reversal events within the system, with 1 identifying the initial domain nucleation and propagation and then 2–5 indicating the domain contraction returning to saturation. Approaching negative saturation, the system foregoes the domain contraction and must re-nucleate domains, identified at features 6 and 7. The domain growth and contraction are shown in (**b**) and the resultant FORC diagram in (**c**)

and magnetostatic energies is maintained, encouraging the domains to resize only slightly between field steps. Since the rebalancing occurs at each field step and is independent of HR, the resultant feature will be very subtle and appear at $H = H_R$, e.g., $H_C = 0$.

Approaching negative saturation, the positively oriented domains, which present both a Zeeman energy and domain wall energy cost, annihilate to form larger, negatively oriented domains. Along the subsequent FORC branch, the positive domain which had previously grown under increasing H, but is now annihilated, will not initially contribute to the magnetization. This manifests in the FORC branch as a decrease in dM/dH, which in turn causes a negative feature in the FORC distribution. Thus, the annihilation of domains can be identified in this system by the presence of negative features in the FORC diagram, with H_R indicating the annihilation field and H identifying the former growth. Continuing along the FORC branch, the larger, negatively oriented domain can re-nucleate a positive domain, rather than undergoing the much longer contraction process. The re-nucleation field is determined by the intrinsic qualities of the defect sites, as before, and the interaction fields from the local environment. Since the magnetostatic energy continues to increase as H_R becomes more negative, the nucleation field should become offset to $-H$ to compensate. However, experiments show that the feature tends to be nearly vertical, at $H = -H_{\text{Nuc}}$, indicating that the nucleation field may be more determined by the local interactions, which will be independent of H_R.

Work by Davies et al. [7] showed in this system that, even past the apparent saturation field, the FORC distribution continues to change. Using X-ray microscopy, the changes were associated with residual bubble domains in the film. Approaching the nucleation/propagation field, these bubbles would act as nucleation sites, with rapid propagation occurring before the nucleation field. As the small bubble domains are saturated at large H_R, the nucleation field becomes defined only by the defects. As a result, the vertical feature may become "sharper" on the $+H$ side. The $-H_R$ end of the vertical FORC feature occurs when the last bubble domain is saturated out, at an H_R which overcomes the crush field of the small bubble domains; this field is much larger than the "apparent" saturation from the major hysteresis loop, demonstrating the ability of FORC to uncover nanoscale magnetic phenomena using macroscopic probes.

Both the domain growth mechanisms present a FORC diagram with an initial feature along the H-axis and later features along the H_R-axis. Previous works have shown that the alignment of these features is related to the interactions being demagnetizing (as is the case for samples with out-of-plane anisotropy) versus magnetizing (for the in-plane film) [25]. This work used the Henkel (ΔM) method as a secondary approach to identify the nature of the interaction [26].

In the above section, the origins of four FORC diagrams were discussed in the context of thin films, highlighting the roles of interactions and intrinsic behavior in defining the shape of the FORC features but also the role of defects and domain growth. For films which cannot sustain domain growth, the local reversal mechanisms manifest features oriented along the (H_C, H_B) axes, with tilting and stretching that encodes the distributions in the interactions and intrinsic coercivities. By comparison, continuous films with out-of-plane anisotropy generate FORC features with symmetries along the (H, H_R) axes, in which the alignment of the features encodes the interactions. The domain growth samples can be further complicated by the presence of significant pinning defects, which prevent domain growth and return the film to the more granular case.

6 FORC Measurements of Heterostructured Films

Heterostructured films are particularly relevant to thin films since they typically rely on interfacial exchange coupling. Since the direct exchange, and more exotic Ruderman-Kittel-Kasuya-Yosida (RKKY) or Dzyaloshinskii-Moriya interaction (DMI) exchange, all disappear by $\approx$20 nm length scales, it is appropriate to discuss them in the context of thin films.

Exchange Springs

Exchange springs are bilayer films in which there are magnetically hard (large-coercivity) and soft (low-coercivity) layers, which are directly exchange coupled at their interface [27, 28]. This is typically performed for two reasons: (1) the magnetically hard layer anchors the system to large magnetic fields, while the soft layer provides a large saturation magnetization – resulting in a large energy product for the system, or (2) the soft layer reduces the reversal field of a magnetically hard layer – to improve writability in, e.g., hard drive materials [29]. The mechanics behind exchange springs are that the exchange interaction, which tends to be >10 T over a few unit cells, couples the hard and soft layers at the interface, preventing the latter from reversing. Away from the interface (e.g., beyond the exchange length), the soft layer undergoes reversible rotation in response to the magnetic field. At a large negative magnetic fields, the effective domain wall in the soft layer can be injected into the hard layer, inducing domain reversal prematurely from the hard/soft interface [30].

The FORC measurement for an exchange spring system is prepared with a film consisting of two or more layers, with the requirement of one having a large intrinsic anisotropy and one a lower anisotropy, identified as magnetically hard and soft, respectively [27, 31], as shown in Fig. 6. The system is positively saturated such that both layers possess a parallel alignment along a magnetic easy axis.

Fig. 6 (**a**) Illustrative diagram showing the reversal of three different regions of a sample of isolated regions and how those individual loops evolve throughout the major hysteresis loop, with 1 and 2 identifying the reversible "winding," 3 indicating the reversal of the hard layer. The reversal of the hard layer removes the reversal events which are identified at 4 and 5 and now instead occur at 6 and 7. Thus 1, 4, and 6 are all from the same region of the sample. The reversal events are shown illustratively in panel (**b**) and the resultant FORC diagram in (**c**)

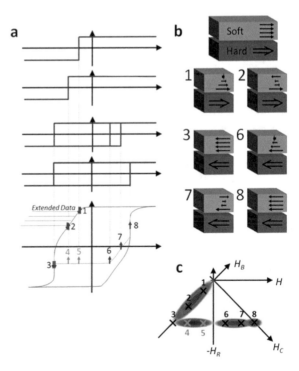

Following the FORC measurement procedure, at $H_R > 0$ the sample tends to remain positively saturated due to the hard layer. At $H_R < 0$ the soft layer begins to rotate in response to the applied field magnetic field, typically called "winding." This action incurs a significant exchange energy cost with the interfacial moments oriented parallel to the hard layer, while moments away from the interface rotate to follow the field. Increasing H from a wound state, the exchange energy is directly traded with the Zeeman energy, unwinding the soft layer; in many ways this is similar to the continuous film with out-of-plane anisotropy in that there is a balance that is continuously maintained between the Zeeman energy, and in this case the exchange. The winding and unwinding of the soft layer are typically considered to be reversible. In order for the FORC distribution to capture reversible features, the data must be processed with a so-called data extension [16, 32, 33]. Specifically, reversible features appear in the first step of the FORC branch, between H_R and $H_R + \Delta H$, where ΔH is the field step in H. For this value of H, there is no data on the antecedent FORC branch; thus the change in the derivative can only be calculated in the forward direction. By only calculating the derivative in the forward direction, changes in the magnetization in this first step are lost or will become negative features in subsequent FORCs. By defining the magnetization as a constant $M(H < H_R) \equiv M(H_R)$, the data in this immeasurable region can be calculated. Recognizing that this definition is a constant value of M, its first-order derivative (dM/dH) is zero everywhere; thus the derivative in H_R is also zero everywhere except at the boundary, e.g., $H = H_R$ or $H_C = 0$. This feature at $H = H_R$ is the reversible feature we seek to capture.

Processing the FORC diagram for an exchange biased system initially generates a reversible feature, located at $H_C = 0$, representing the winding of the soft magnetic layer. Progressing to more negative values of H_R further winds the soft layer, resulting in continued growth of the reversible feature. At some value of H_R, the magnetically hard layer will begin to reverse; this field will be called H_R^{hard}. As the hard layer reverses, the coupling causes the preferred orientation of the soft layer to switch also. As a result, the unwinding of the soft region no longer occurs at the same values of H that it had in the preceding FORC branches. In effect, the dM/dH of the FORC branch decreases in the unwinding region, at $H_R \leq H_R^{hard}$. Even though the reversible winding occurs only at the boundary of the dataset $(H = -H_R)$ and appears only if the dataset is extended, the decrease in dM/dH manifests a feature in the FORC diagram, away from the boundary. Similar to the case of the domain reversal FORC, this feature will manifest at a single value of $H_R = H_R^{hard}$ and span in H across the entire region in which the unwinding had previously occurred. Furthermore, since this feature is caused by a decrease in dM/dH, it will appear as a negative feature in the FORC diagram.

Proceeding along the FORC diagram starting at $H_R = H_R^{hard}$, the winding of the soft layer will now occur at positive fields (since the hard layer is anchoring the soft layer in the negative direction). This feature will also appear at $H_R = H_R^{hard}$ and by symmetry will span the same range in H as the initial, reversible winding feature.

Necessarily, since the integral of the FORC feature captures the total magnetization, the positive and negative feature pairs from the winding will be equal and opposite in magnitude. Interestingly, this means, without extending the dataset, the reversible feature appears in the FORC diagram twice, but is not captured by the integral of the FORC diagram, consistent with the common understanding that reversible features are not captured in the FORC diagram by standard calculation techniques.

7   Dynamic FORC Measurements

Dynamic FORC measurements are a recent consideration [34], leveraging what had previously been an issue in the FORC measurement procedure. Specifically, the FORC measurement sequence proceeds by saturating the sample, then proceeding to an H_R, and then measuring M while increasing H. If the measurement is not paused at H_R, then thermally activated reversal can occur during the first several steps of the FORC diagram [35]. In this region, the magnetization is decreasing despite increasing H, causing a negative feature in the FORC diagram. This feature is typically considered an artifact that the sample was measured too fast, but it does encodes temperature-dependent behavior which is dependent on both the interactions and intrinsic qualities. Recent works [34] have proposed that interesting dynamic data can be extracted by performing high-speed FORC measurements. Measuring the FORC distribution versus the sweep rate as well will encode details such as the domain propagation rate, pinning potential, and the effects of thermally activated propagation. While the details of this technique are still being experimentally established, it is a promising new direction in FORC measurements.

8   FORC Measurements Beyond Magnetometry

While FORC measurements are traditionally performed using magnetometry, the FORC technique can be performed on virtually any hysteretic system which has a well-defined saturated state and intermediate states that are stable on the timeframe of the measurement. Some works have performed FORC measurements with probes other than magnetization [36–40]. The most notable of these approaches has been the use of electrical transport, e.g., resistance measurements. These measurements can be susceptible to magnetization through the anomalous Hall effect or magnetoresistance but can also be used to probe metal-insulator transitions. Using basic analysis of the electrical pathways, resistance-based FORC measurements are able to locally probe regions within the sample and thus are becoming more common in thin films.

Magnetoresistance

FORC measurements performed using magnetoresistance effects [36] present a more complex, but also more detailed, understanding of the magnetization. One cause of the increased complexity is that the magnetoresistance measurements are not sensitive to the bulk magnetization but rather the local magnetic environment along the pathway that the current proceeds. This means that, in a heterogeneous film, the magnetoresistance probe will be scaled against the parallel circuit current division within the system, and so some regions can be invisible. This is immediately apparent if one considers a thin film heterostructure with a magnetic insulating layer, for example, $Fe/Y_3Fe_5O_{12}$ (YIG).

Despite its challenges, the magnetoresistance provides a highly localized probe of the system which cannot be captured by the bulk measurement. In a clear example of this, we can consider the magnetoresistance measurement of a giant magnetoresistance (GMR) heterostructure, $[Co/Cu]_N$ or $[Fe/Cr]_N$. The magnetometry measurement will, at best, capture a series of stepped reversals as each of the magnetic layers switches between parallel and antiparallel configurations and less than that if the layers reverse by a domain growth mechanism. However, probing the reversal with the FORC technique gives a localized understanding of the ordering. Specifically, the GMR is large and positive when the alignment of the magnetization in adjacent layers is antiparallel. As a result, the GMR FORC encodes the degree of magnetic correlation across the nonmagnetic spacer layer. Using this technique previously, it was demonstrated that the field cycling performed during the FORC measurement can generate a ground state with maximal disorder, which is not accessible along the major loop [36]. This detail is invisible to the magnetometry FORC measurement which captures only the overall magnetization and is blind to local magnetic correlation.

Anomalous Hall FORCs [41] are also dependent on the path of the electrons through the system and so again are preferentially sensitive to the low-resistance regions. This can be useful in probing specific regions within the sample.

Metal-Insulator Transitions

Some works have been performed which seek to probe nonmagnetic transitions using the FORC technique. One such effort has been the investigation of metal-insulator transitions (MIT) [37]. Here, the temperature is used as the independent variable (e.g., a saturating temperature, reversal temperature, and applied temperature), while the resistance is measured. Similar to the magnetoresistance and Hall measurements discussed above, the signal will be augmented by the electrical transport pathways. In an MIT system, the large changes in the resistance will tend to be masked by percolative pathways within the system. In other words, the resistance will change very little initially since the electrons can follow a resistance path that circumvents the high-resistance regions. For this reason, the FORC distribution of an MIT will be very different than if it was measured by an area-averaging optical technique, for example. Thus, considering what the FORC distribution would describe in an MIT FORC, the data would encode the tendency to form or break the percolative networks which shunt the probing current. Thus, the MIT FORCs illustrate that choosing different probes presents opportunities to investigate different details of a system, depending on the metrics of interest.

9 Final Thoughts

The FORC technique is an incredibly useful tool for evaluating hysteretic systems. Few techniques can perform a bulk measurement and capture the minute intrinsic qualities and interactions within a system. With the FORC technique becoming more accessible, including new processing software [42, 43] and being included as an easy-to-use sequence within major commercial equipment, a strong foundation in the FORC technique becomes increasingly crucial. This chapter was written to provide a mechanistic understanding of the FORC technique, focusing on measurements common in magnetic thin films. The hope is that, by understanding the physical origin of the FORC features, and how these come together to form the specific structures within the FORC diagram, such a foundation is provided. Furthermore, understanding the origin of the FORC features allows new insights, including quantitative evaluation, and allows the FORC technique to be generalized and used in new and novel ways including the evaluation of a wider variety of systems using generalized "interaction" and "coercivity" metrics.

References

1. C.R. Pike, C.A. Ross, R.T. Scalettar, G. Zimanyi, Phys. Rev. B **71**, 134407 (2005)
2. A.P. Roberts, C.R. Pike, K.L. Verosub, J. Geophys. Res. Solid Earth **105**, 28461 (2000)
3. A. Stancu, C. Pike, L. Stoleriu, P. Postolache, D. Cimpoesu, J. Appl. Phys. **93**, 6620 (2003)
4. I. Mayergoyz, *Mathematical Models of Hysteresis* (Springer, 1991), p. 1
5. M. Almasi-Kashi, A. Ramazani, S. Izadi, E. Jafari-Khamse, Phys. Scr. **90**, 085803 (2015)
6. J. Fischbacher, A. Kovacs, H. Oezelt, M. Gusenbauer, T. Schrefl, L. Exl, D. Givord, N.M. Dempsey, G. Zimanyi, M. Winklhofer, et al., Appl. Phys. Lett. **111**, 072404 (2017)
7. J.E. Davies, O. Hellwig, E.E. Fullerton, G. Denbeaux, J.B. Kortright, K. Liu, Phys. Rev. B **70**, 224434 (2004)
8. J.E. Davies, D.A. Gilbert, S.M. Mohseni, R.K. Dumas, J. Åkerman, K. Liu, Appl. Phys. Lett. **103**, 022409 (2013)
9. D.A. Gilbert, G.T. Zimanyi, R.K. Dumas, M. Winklhofer, A. Gomez, N. Eibagi, J.L. Vicent, K. Liu, Sci. Rep. **4**, 4204 (2014)
10. D. Weller, A. Moser, L. Folks, M.E. Best, L. Wen, M.F. Toney, M. Schwickert, J. Thiele, M.F. Doerner, IEEE Trans. Magn. **36**, 10 (2000)
11. S.S.P. Parkin, M. Hayashi, L. Thomas, Science **320**, 190 (2008)

12. J.M. Shaw, S.E. Russek, T. Thomson, M.J. Donahue, B.D. Terris, O. Hellwig, E. Dobisz, M.L. Schneider, Phys. Rev. B **78**, 024414 (2008)
13. E. Bonet, W. Wernsdorfer, B. Barbara, A. Benoît, D. Mailly, A. Thiaville, Phys. Rev. Lett. **83**, 4188 (1999)
14. A.R. Ali, G. Said, Physica B+C **112**, 241 (1982)
15. J. Nogués, I.K. Schuller, J. Magn. Magn. Mater. **192**, 203 (1999)
16. D.A. Gilbert, J.-W. Liao, L.-W. Wang, J.W. Lau, T.J. Klemmer, J.-U. Thiele, C.-H. Lai, K. Liu, APL Mater. **2**, 086106 (2014)
17. C.R. Pike, A.P. Roberts, K.L. Verosub, J. Appl. Phys. **85**, 6660 (1999)
18. B.F. Valcu, D.A. Gilbert, K. Liu, IEEE Trans. Magn. **47**, 2988 (2011)
19. S. Ruta, O. Hovorka, P.-W. Huang, K. Wang, G. Ju, R. Chantrell, Sci. Rep. **7**, 45218 (2017)
20. C.-I. Dobrotă, A. Stancu, Phys. B Condens. Matter **457**, 280 (2015)
21. P. Sergelius, J.G. Fernandez, S. Martens, M. Zocher, T. Böhnert, V.V. Martinez, V.M. de la Prida, D. Görlitz, K. Nielsch, J. Phys. D. Appl. Phys. **49**, 145005 (2016)
22. S. Zapperi, P. Cizeau, G. Durin, H.E. Stanley, Phys. Rev. B **58**, 6353 (1998)
23. W. Jiang, P. Upadhyaya, W. Zhang, G. Yu, M.B. Jungfleisch, F.Y. Fradin, J.E. Pearson, Y. Tserkovnyak, K.L. Wang, O. Heinonen, et al., Science **349**, 283 (2015)
24. N. Nagaosa, Y. Tokura, Nat. Nanotechnol. **8**, 899 (2013)
25. D.A. Gilbert, J.-W. Liao, B.J. Kirby, M. Winklhofer, C.-H. Lai, K. Liu, Sci. Rep. **6**, 32842 (2016)
26. P.E. Kelly, K.O. Grady, P.I. Mayo, R.W. Chantrell, IEEE Trans. Magn. **25**, 3881 (1989)
27. E.F. Kneller, R. Hawig, IEEE Trans. Magn. **27**, 3588 (1991)
28. E.E. Fullerton, J.S. Jiang, M. Grimsditch, C.H. Sowers, S.D. Bader, Phys. Rev. B **58**, 12193 (1998)
29. D. Suess, T. Schrefl, S. Fähler, M. Kirschner, G. Hrkac, F. Dorfbauer, J. Fidler, Appl. Phys. Lett. **87**, 012504 (2005)
30. D. Suess, J. Lee, J. Fidler, T. Schrefl, J. Magn. Magn. Mater. **321**, 545 (2009)
31. J.E. Davies, O. Hellwig, E.E. Fullerton, J.S. Jiang, S.D. Bader, G.T. Zimányi, K. Liu, Appl. Phys. Lett. **86**, 262503 (2005)
32. C.R. Pike, Phys. Rev. B **68**, 104424 (2003)
33. M. Winklhofer, R.K. Dumas, K. Liu, J. Appl. Phys. **103**, 07C518 (2008)
34. D. Cimpoesu, I. Dumitru, A. Stancu, J. Appl. Phys. **120**, 173902 (2016)
35. C.R. Pike, A.P. Roberts, K.L. Verosub, Geophys. J. Int. **145**, 721 (2001)
36. R.K. Dumas, P.K. Greene, D.A. Gilbert, L. Ye, C. Zha, J. Åkerman, K. Liu, Phys. Rev. B **90**, 104410 (2014)
37. J.G. Ramírez, A. Sharoni, Y. Dubi, M.E. Gómez, I.K. Schuller, Phys. Rev. B **79**, 235110 (2009)
38. D.A. Gilbert, E.C. Burks, S.V. Ushakov, P. Abellan, I. Arslan, T.E. Felter, A. Navrotsky, K. Liu, Chem. Mater. **29**, 9814 (2017)
39. A. Stancu, D. Ricinschi, L. Mitoseriu, P. Postolache, M. Okuyama, Appl. Phys. Lett. **83**, 3767 (2003)
40. M.K. Frampton, J. Crocker, D.A. Gilbert, N. Curro, K. Liu, J.A. Schneeloch, G.D. Gu, R.J. Zieve, Phys. Rev. B **95**, 214402 (2017)
41. J. W. Lau, D. A. Gilbert, V. Provenzano, K. B. Stritch, J. Liao, C.-H. Lai, K. Liu (eds), 59th Annual Conference on Magnetism and Magnetic Materials, Honolulu, HI, USA, (2014)
42. R.J. Harrison, J.M. Feinberg, Geochem. Geophys. Geosyst. **9** (2008)
43. R. Egli, Glob. Planet. Change **110**, 302 (2013)

First-Order Reversal Curve (FORC) Measurements for Decoding Mixtures of Magnetic Nanowires

Mohammad Reza Zamani Kouhpanji and Bethanie J. H. Stadler

Abstract The progression of nanotechnology has resulted in the exploitation of very complex magnetic nanostructures with anomalous magnetic responses, which cannot be fully understood using hysteresis loop measurements. First-order reversal curve (FORC) measurements are among the most powerful tools for characterization of complex bulk, micro-, and nano-systems that are broadly used in the field of magnetism. The main advantage of FORC is that it scans the whole area of hysteresis loops in a two-dimensional fashion, leading to detailed information regarding the intrinsic magnetic nanostructures and the magnetic interactions. In this review, we specifically focus on the current state of FORC measurements and their use in decoding magnetic nanoparticles assemblies.

Keywords FORC measurement · Magnetic nanowires · Quantitative signatures · Fast decoding

1 Introduction

Mayergoyz [1–3] proposed the original FORC measurement as an identification technique for the classical Preisach model [4], which describes magnetic hysteresis loops as a superposition of a large number of independent relays, called hysterons. Hysterons represent the switching of a single magnetic element with rectangular hysteresis loops, such as those of isolated magnetic nanoparticles acting like Stoner-Wohlfarth particles. Traditionally, FORC measurements start by applying a large magnetic field to ensure the positive saturation of a sample. Next, the applied field (H) is reduced to a predefined field, known as a reversal field (H_r), and the magnetization is then measured, while H is retuned to positive saturation; see Fig. 1. This process is repeated with decreasing H_r until negative saturation, leading to a

M. R. Z. Kouhpanji · B. J. H. Stadler (✉)
Department of Electrical and Computer Engineering, University of Minnesota Twin Cities, Minneapolis, MN, USA
e-mail: stadler@umn.edu

© Springer Nature Switzerland AG 2021   651
V. Franco, B. Dodrill (eds.), *Magnetic Measurement Techniques for Materials Characterization*, https://doi.org/10.1007/978-3-030-70443-8_20

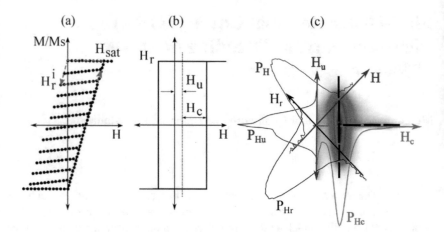

Fig. 1 Traditional first-order reversal curves (FORC). (**a**) Traditional FORC data showing measurement points, (**b**) a single hysteron, and (**c**) traditional FORC parameter heat-map and its projection on different field axes. The figure is adopted from [5, 6]

set of magnetization curves, $M(H, H_r)$. The FORC distribution (ρ) is defined as the second derivative of the magnetization with respect to these fields:

$$\rho = -\frac{1}{2}\frac{\partial^2 M(H, H_r)}{\partial H \partial H_r} \tag{1}$$

In FORC analysis, ρ is plotted as a contour (also sometimes as heat-map) onto two-dimensional axes representing the coercive field (x-axis, $H_c = \frac{1}{2}(H - H_r)$), the interaction field (y-axis, $H_u = \frac{1}{2}(H + H_r)$). Once the heat-maps are created, one can project out the entire heat-map on the aforementioned axes, by taking an integral over the orthogonal axis, to quantify the coercivity or interaction field distributions. Figure 1 schematically illustrates the collection and analysis of FORC data.

This traditional data collection and analysis of FORC has been broadly used to determine coercivity and interaction distributions of magnetic elements in samples from geology [7, 8] and engineered nanostructures, including spherical nanoparticles [9], elongated nanoparticles [10–15], thin films [16, 17], patterned dots [18], magnetic tunnel junctions [19], and patterned recording media [20, 21]. This measurement also has been used to demonstrate the magnetic response of single components, multiphases such as intermetallic alloys, and multicomponents such as magnetic/nonmagnetic segmented nanowires. Alongside the experimental studies, several theoretical studies, based on mean-field models, also have been carried out to simulate and/or interpret the information in experimental FORC data [18, 22–25]. Despite this wide range of interesting studies, FORC has three main drawbacks when it comes to decoding magnetic nanoparticle mixtures. These drawbacks are (1) low reliability for quantitative decoding, (2) extremely slow measurements, and (3) slow data analysis that could induce artifacts by taking derivatives, integrals, and over-smoothing. In the following section, we discuss each drawback individually and we review solutions to overcome them.

2 Quantitative Decoding

According to Eq. (1), the Preisach distribution predicts the probability of finding a hysteron that switches in the interval between $[H, H_r]$ and $[H + \Delta H, H_r + \Delta H_r]$. Traditional FORC thus indicates the presence of different magnetic components, but it does not precisely quantify the amount of these components [22]. Note that here quantification is defined as determining the amount of each component or phase present in a sample, not the quantification of the intrinsic properties of those components or phases.

Traditional FORC uses projections of the FORC parameter onto the H_c and H_u axes to determine the distribution of coercivity and interactions within a sample. Intuitively, the coercivity and interaction distributions are not appropriate for quantifying the amount of each magnetic subcomponent present in a sample. For example, unlike thin films, collections of nanoparticles with distinct coercivities do not lead to distinct features in hysteresis curves. This fact is frustrated by artifacts that often appear in nanoparticle FORC heat-maps, such as the horizontal "T" formed by the vertical and horizontal dashed lines in Fig. 1. This feature leads to an anomalous spike in the H_u projection and also erroneously emphasizes the low coercivity features, according to Dobrotă and Stancu [20]. Also, some magnetic nanostructures contain both interacting and noninteracting magnetic subcomponents, so a measured interaction field distribution would only capture the interacting subcomponents regardless of the amount of the noninteracting subcomponents present.

It has been recently shown that two features in the FORC protocol can be used to overcome this limitation, namely, the irreversible switching field (ISF) and the backfield remanence magnetization (BRM) [5, 26, 27]. ISF determines the required field to induce irreversible switching of moments. Therefore, the ISF can be understood by the moment remaining after a sample is subjected to a reversal field (H_r) and then a reduced field equal to the previously applied H_r (see the blue line in Fig. 2a). Interestingly, these measurements only require the first few points of each reversal curve, which are substantially lower than the number of points used in FORC and similar to those in a standard (fast) hysteresis loop, as compared in Fig. 2b, c. BRM is the moment remaining after the applying each H_r and then reducing the field to zero, which involves even fewer points as shown in Fig. 2d. The reduced number of points will be discussed more in the speed and analysis sections below.

Figure 3 gives an example of the accuracy of determining the volume fraction of two types of Co nanowires using the ISF distribution, BRM distribution, and the backfield remanence coercivity (BRC) which is the derivative of the BRM. When the ratio of volumes for two types of nanowires in a combination is less

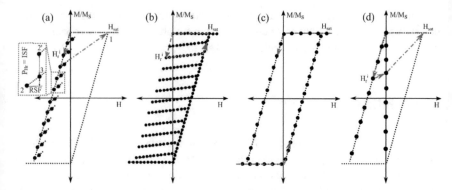

Fig. 2 A schematic showing fast measurements that provide quantitative analysis of magnetic components or phases in a complex magnetic sample: (**a**) data needed for the projection method (first few points of each FORC), (**b**) traditional FORC data collection, (**c**) hysteresis loop measurement, and (**d**) magnetization at zero field. Note: ISF stands for the irreversible switching field and RSF stands for the reversible switching field [6, 26]. In all subfigures, the green arrows indicate the measurement direction

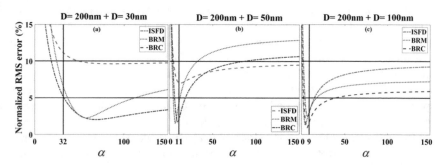

Fig. 3 Using the distributions in ISF, BRM, and BRC, mixtures of two types of nanowires could be quantitatively decoded where α is the volume ratio between the two nanowire types. The measured mixtures contain (**a**) 200 nm and 30 nm diameter Ni nanowires, (**b**) 200 nm and 50 nm Ni nanowires, (**c**) 200 nm and 100 nm Ni nanowires. BRM showed the best match where the minimum in error is the calculated volume ratio compared to the real χ (vertical line). Although (**a**) is not as good a match, there is $32\times$ less volume of 30 nm which is a very small number that was still detectable [25]

than 1–10, then that ratio can be accurately determined using these fast techniques. The main shortcoming of this approach is that the magnetic response of each component/phase must be known initially to calibrate each magnetic response within the whole sample. This shortcoming limits this method's application to the discovery of the unknown magnetic nanostructures, unless accompanied by other measurements. However, this approach is a great asset in expanding the application of magnetic nanoparticles for multiplexing/demultiplexing magnetic biolabels or encoding/decoding the magnetic nanobarcodes, where initial knowledge regarding the magnetic properties of the nanoparticles is readily available [27, 28]. Next we'll look at other methods that have been proposed to hasten FORC measurements and give the rationale for our fast approach.

3 Measurement Speed

One of the main disadvantages of the traditional FORC measurement is that it requires many data points and is therefore an extremely slow data acquisition method. Long measurement times are not efficient for either laboratory development or industry quality control. Practically, increasing the number of reversal curves increases the accuracy of capturing complex details, which can push single sample measurement times to several hours or even days. For example, if a sample consists of two components, one with a very narrow and one with a broad coercivity distribution, the number of FORCs must be drastically increased to capture the narrow coercivity distribution. Unfortunately, long measurements are associated with field drift that can change the magnetization curves.

To overcome the slow measurement drawback, it was proposed to scan only half of the hysteresis loop area [29]. This would be beneficial for single-phase magnetic nanostructures, such as an array of identical magnetic nanowires, where the interaction fields are negligible, that is, Stoner-Wohlfarth type samples. However, in bistable systems where coercivity and interaction fields are correlated, this method cannot be used due to incomplete data. Figure 4 shows the data collection for this method and compares it with the traditional FORC measurement [29]. Briefly, this

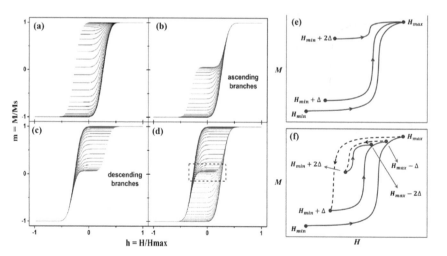

Fig. 4 A schematic of a shortened FORC measurement, (**a**) traditional FORC measurement, (**b**) new method measuring the ascending branches, (**c**) new method measuring the descending branches, (**d**) combined ascending and descending branches, (**e**) and (**f**) are the measurement procedure [29]

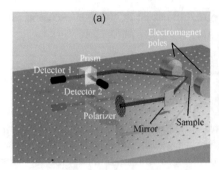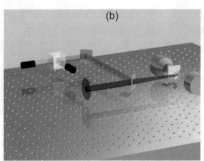

Fig. 5 Schematic of the magneto-optic Kerr effect (MOKE) setup combined with the traditional FORC setup to speed up the data collection, (**a**) demonstrates the setup for measuring the out of plane MOKE-FORC signal, (**b**) demonstrates the setup for measuring the in-plane MOKE-FORC signal [30]

method starts by sweeping the magnetic field from negative saturation ($H_{\min}$) to positive saturation ($H_{\max} = -H_{\min}$). Next, the field is reverted to $H_{\min} + \Delta H$ recording the magnetization of the descending branch. The ascending branch is swept to $H_{\max} - \Delta H$, and the magnetization is also recorded. This is different from the usual FORC method in which each ascending branch reaches $H_{\max}$. The magnetic field is inverted again to $H_{\min} + 2\Delta H$. The new ascending branch is measured up to $H_{\max} - 2\Delta H$. The procedure is repeated until the magnitude of the magnetic field diminishes to zero. Note that this modification is a creative way to half the required data points of FORC measurements.

Another solution was proposed by Grafe et al., where they suggested combining magneto-optic Kerr effect measurements (MOKE) with traditional FORC data collection to speed up the measurements [30]. This method applies the same magnetic field sequence as traditional FORC measurements, but it collects the MOKE signal instead of the magnetization. This approach does not have limitations on the number of phases as the previous approach; however, it can be done only on samples where the optical signal can readily access all of the magnetic components (e.g., thin films). Furthermore, because of the limited field of view (restricted by the laser spot), this method is useful only for very homogeneous magnetic nanostructures. For inhomogeneous thin films, this method does not result in fast measurements, as it will require several measurements at different spots and the data analysis will be an open question. In short, MOKE-FORC provides a good way to accelerate magnetization acquisition (at a small point) without reducing the data points of FORC. Figure 5 shows the proposed MOKE-FORC setups.

Irreversible switching field was introduced in the previous section, and it is schematically shown in Fig. 2. We often call measurements of ISF distributions *the projection method*, because it is mathematically equivalent to the projection of the FORC heat-map onto the H_r axis [6, 26, 27, 31, 32]. In traditional FORC analysis, many points per curve would be measured in order to plot the FORC parameter as

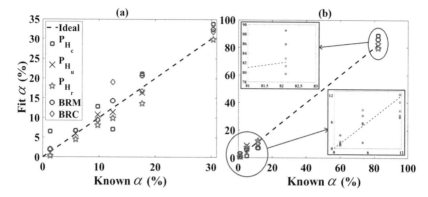

Fig. 6 FeCo nanowires with different diameters were decoded using projections onto three of the primary axes shown in Fig. 1 and using backfield remanent magnetization (BRM) and its derivative (BRC). Note that the projection onto the H_r axis is the mathematically equivalent to the irreversible switching field (ISF) distribution and can therefore be measured with very few points. (**a**) Combinations including two different types of MNWs. (**b**) One mixture of all four types of MNWs. It is interesting to see that most of these methods matched the known volume ratios (dashed line) with volume ratios from 0 to 80% of one barcode vs another. Please see [5] for more information

a heat-map. Projections are then the integration over all orthogonal points onto an axis. So, the projection on H_r is

$$\int_{H_r}^{\infty} \rho\,(H, H_r)\,\mathrm{d}H = -\frac{1}{2} \left. \frac{\partial M\,(H, H_r)}{\partial H_r} \right|_{H=\infty} + \frac{1}{2} \left. \frac{\partial M\,(H, H_r)}{\partial H_r} \right|_{H=H_r}$$

$$= 0 + \frac{1}{2} \left. \frac{\partial M\,(H, H_r)}{\partial H_r} \right|_{H=H_r} \qquad (2)$$

Note that the first term is zero because the magnetization does not change with H_r at large values of applied field (H) due to saturation. As a result, the authors proposed to measure only a few data points on each reversal curve; see Fig. 2a to determine ISF distribution, which is also the projection of a FORC heat-map onto the reversal field (H_r) axis. Not only does this approach significantly accelerate the measurement, but it also accelerates the data analysis because it only requires one derivative (rather than two derivatives followed by integrals) over limited data points to fully and reliably determine the ISF, which can be used to approximate distributions in coercivities and interaction fields. The processing of this data into a projection on the reversal field axis will be discussed in the next section. Interestingly the ISF distribution was better for decoding FeCo nanowires, Fig. 6, than it was for Ni nanowires, Fig. 3. It should be mentioned that the BRM measurement can also be used for fast measurement because it requires the same number of data points similar to hysteresis loops, and it provides excellent decoding for mixtures of Ni nanoparticles as shown in Fig. 3. Together, these two fast methods make a fast, powerful decoding tool. Note the BRC is the derivative of the BRM.

4   Data Processing

The traditional FORC data processing requires two field derivatives followed by integrals for data analysis. Taking two derivatives amplifies noise and also eliminates features that depend on either the initial magnetization state or the spontaneous magnetization. For example, taking the derivative with respect to the applied field (H) omits the magnetization features that are only a function of the reversal field (H_r). Even though during the projection onto the H_c (or H_u) axis, one takes the integral over all H_u (or H_c), the features that have been erased by the derivatives are not recovered. In addition, smoothing is required for data processing which can induce spurious features [33–36] while concealing real features. In this context, plenty of software has been proposed to process the data while suppressing noise. For example, the FORCinel software program uses locally weighted regression smoothing (LOESS) [36]. FORCinel is able to implement non-integer values for the smoothing factor, enabling finer control over the degree of smoothing. It also is able to apply a variable smoothing factor, which was first demonstrated by Egli [37], for preserving central ridge in FORC heat-maps induced by single-domain noninteracting nanowires. As a result, it can extrapolate across the data and eliminate the overlaying data caused by the instrumentation instability. Another example, Cimpoesu et al. introduced the DoFORC software for processing the noise scattered data [38]. DoFORC is able to perform smoothing on the measured FORCs prior to data processing and also during the data processing of the partial derivatives. Consequently, it provides flexibility to compute residuals induced by different smoothing to characterize the difference between the predicted and observed values, leading to a better approximation of the smoothing factor. This generalizes the cross-validation to measure the predictive performance, providing three degrees of freedom to understand the amounts of smoothing being performed by different smoothing methods once the same value of the smoothing factor is applied. Figure 7 renders the fitting quality of the DoFORC for processing the data with four different noise levels.

Later on, in the same year, Grob et al. reported on processing the data in Fourier space [39]. In short, they took benefit from the diversity of Fourier space to not only accelerate the computations but also move away from the conventional smoothing factor toward real field resolution [39]. By comparing the baseline resolution of the processed data in Fourier space and the resolution of the processed data in the field space, the authors provided an empirical equation that converts the smoothing factor from one space to the other to have a quantitative comparison between the procedures. Figure 8 shows a comparative analysis of the proposed algorithm and the traditional algorithm, known as the FORCinel algorithm. As can be seen from Fig. 8c, d, both methods produce identical FORC heat-maps. However, as

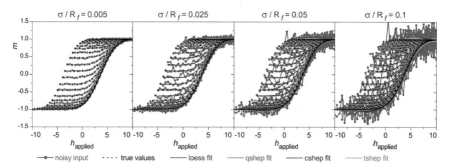

Fig. 7 Analyzing the fitting quality of DoFORC software in terms of the noise variance to the range of the testing function (σ/R_f) that employs four smoothing algorithms, (1) locally weighted regression smoothing (LOESS), (2) modified quadratic polynomial Shepard method (qshep), (3) cubic polynomial Shepard method (cshep), and cosine series Shepard method (tshep). As the noise increases (larger σ/R_f), all four methods overlap [38]

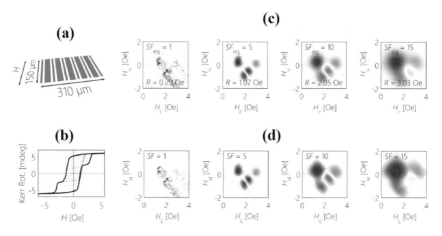

Fig. 8 Comparison of FORC processing with different smoothing factors and algorithms, (**a**) a scheme of the samples, (**b**) the FORC data, (**c**) processed data using the Fourier space, and (**d**) processed data using the traditional algorithm in the field space [39]

the smoothing factor increases, the methods do not yield similar heat-maps. This indicates that the fitting kernel is much larger than the peak's width, which the authors claimed is not real.

More recently, Berndt et al. proposed to utilize fast Fourier transform (FFT) rather than implementing the traditional derivatives [40]. This approach, named FORCOT, also predicts the optimal smoothing factor directly from the noise spectrum in Fourier space. It does not require recalculation for different smoothing factors before reaching the minimum mean square root as employed in the FORCinel software. The authors treated the FORC smoothing as a pure image processing problem rather than considering a collection of partial hysteresis curves, viewing the mixed second derivative as pixels of a very noisy image. They then

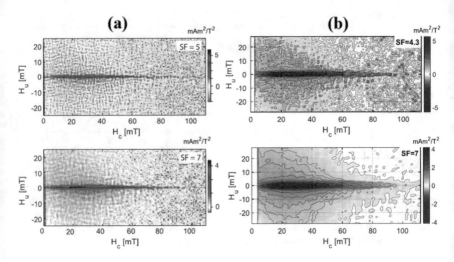

Fig. 9 The results of the traditional FORC algorithm (FORCinel), (**a**) and the FORCOT, (**b**) for two different smoothing factors [36]

aimed to filter the noise from this image to a degree that is visually optimal in the sense that real-signal features should visually appear more prominent than the noise. Figure 9 shows an example comparing the results of the traditional FORC analysis using the FORCinel software and the FORCOT software. Since the FORCOT software employs the Fourier space for processing rather than the field space, the edge artifacts, which are due to the elimination of the first point at the reversal point, do not impact the FORC heat-maps. These edge artifacts are also known as zero-field coercivity artifacts that had been addressed and resolved in the FORC+ software [34]. Note that FORC+ is a software tool that does not use any smoothing factor for analyzing the data, and it is capable of decomposing the irreversible and reversible magnetization.

As mentioned, the purpose of using smoothing is to suppress the measurement noise so it is not amplified during the data processing of the traditional FORC measurements. The projection method (aka the irreversible switching field distribution as mentioned in the previous section) does not require the two derivatives, so it does not need any smoothing. Figure 10 depicts the projection method's data processing, where one can simply measure the crosses (two for each FORC) to calculate the irreversible switching field (ISF, the blue arrow in Fig. 10). For very noisy measurements, magnetically unstable samples or poor acquisition systems, one can either increase the time averaging or increase the number of data points (circles in Fig. 10) for a better signal to noise ratio. It has been shown that two or four data points provide similar results [26]. Note that one may fit lines (red and green lines in Fig. 9) to the data points prior for calculating the projection for achieving a cleaner signal. Once the ISF is determined, its features, the location of its peak, broadening, and tails where they go to zero are sufficient to calculate the coercivity and to some extent interaction field [31, 32]. In fact, ISF was shown in the previous

Fig. 10 Schematically illustrating the calculation of ISF from the projection method data. The upper (red) FORC begins at H_r^i and has four measured points (black dots). A straight red line is least-squares fit to these points. The lower FORC begins at H_r^{i+1} and its points are similarly fit to the green line [26]

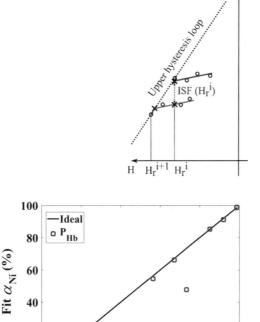

Fig. 11 The projection method uses ISF, also known as the projection on the reversal field axis to determine the quantitative volume ratios of Ni nanowires compared to the volume of Co nanowires in a mixture [6]

section to be sufficient in decoding FeCo nanowires with different diameters because their coercivities were quite different. Figure 11 shows that ISF can also be useful in decoding mixtures of Co and Ni nanowires for the same reason. In short, since the projection method requires only a limited number of data points and it directly determines the FORC heat-map projection on the H_r axis using one derivative, its processing is substantially faster and simpler than the traditional FORC analysis.

5 Summary and Future Outlooks

A brief description of the current state of FORC measurements and data processing has been presented. For the specific application of decoding magnetic nanowire barcodes, traditional FORC has three main drawbacks: qualitative (not quantitative) analysis, slow measurement speeds, and processing. To date, the literature has focused on resolving these drawbacks individually instead of addressing all drawbacks together. However, the projection method seems promising to overcome all these challenges when the application goal is the quantification of magnetic components or phases while increasing measurement speed by reducing the number of data points. The projection method requires a significantly smaller number of data points, and it bypasses the need for two derivatives and an integral in calculating the projection onto the reversal field axis, leading to a substantially faster data processing. In addition, the projection method does not require any smoothing, so it is free of the artifacts that are often found in projections on H_u and H_c axes. The projection method could be combined with other techniques, such as MOKE, to open a route for even faster FORC protocols.

References

1. I.D. Mayergoyz, The classical Preisach model of hysteresis and reversibility. J. Appl. Phys. **69**(8), 4602–4604 (1991)
2. I.D. Mayergoyz, Hysteresis models from the mathematical and control theory points of view. J. Appl. Phys. **57**(8), 3803–3805 (1985)
3. I.D. Mayergoyz, Mathematical models of hysteresis (Invited). IEEE Trans. Magn. **22**(5), 603–608 (1986)
4. F. Preisach, Uber die magnetische nachwirkung. Mitteilung aus dem Zentrallaboratorium des Wernerwerkes der Siemens Halske **277**, 277–302 (1935)
5. M.R. Zamani Kouhpanji, B.J.H. Stadler, Beyond the qualitative description of complex magnetic nanoparticle arrays using FORC measurement. Nano Express **1**(1), 010017 (2020)
6. M.R. Zamani Kouhpanji, B.J.H. Stadler, Projection method as a probe for multiplexing/demultiplexing of magnetically enriched biological tissues. RSC Adv. **10**(22), 13286–13292 (2020)
7. D. Heslop, M. Dillon, Unmixing magnetic remanence curves without a priori knowledge. Geophys. J. Int. **170**(2), 556–566 (2007)
8. A.P. Roberts, D. Heslop, X. Zhao, C.R. Pike, Understanding fine magnetic particle systems through use of first-order reversal curve diagrams. Am. Geophys. Union **52**, 557–602 (2014)
9. A. Mohtasebzadeh, L. Ye, T. Crawford, Magnetic nanoparticle arrays self-assembled on perpendicular magnetic recording media. Int. J. Mol. Sci. **16**(8), 19769–19779 (2015)
10. M.R. Zamani Kouhpanji, B. Stadler, Magnetic nanowires toward authentication. Part. Part. Syst. Charact. **12**, 2000227 (2020)
11. M.R. Zamani Kouhpanji, B. Stadler, Unlocking the decoding of unknown magnetic nanobarcode signatures. Nanoscale Adv. **3**, 584–592 (2021)
12. A. Ramazani, V. Asgari, A.H. Montazer, M.A. Kashi, Tuning magnetic fingerprints of FeNi nanowire arrays by varying length and diameter. Curr. Appl. Phys. **15**, 819–828 (2015)
13. K. Nielsch, R.B. Wehrspohn, J. Barthel, J. Kirschner, S.F. Fischer, T. Schweinb, High density hexagonal nickel nanowire array. J. Magn. Magn. Mater. **249**, 234–240 (2002)
14. J.G. Fernández, V.V. Martínez, A. Thomas, V.M. de la Prida Pidal, K. Nielsch, Two-step magnetization reversal FORC fingerprint of coupled bi-segmented Ni/Co magnetic nanowire arrays. Nano **8**(7), 1–15 (2018)
15. R. Lavin, J.C. Denardin, J. Escrig, D. Altbir, A. Cortés, H. Gómez, Magnetic characterization of nanowire arrays using first order reversal curves. IEEE Trans. Magn. **44**(11), 2808–2811 (2008)
16. R.K. Dumas, P.K. Greene, D.A. Gilbert, L. Ye, C. Zha, J. Akerman, Accessing different spin-disordered states using first-order reversal curves. Phys. Rev. B **90**(104410), 1–7 (2014)
17. D.A. Gilbert et al., Probing the A 1 to L 10 transformation in FeCuPt using the first order reversal curve method. APL Mater. **2**, 086106 (2014)

18. D.A. Gilbert et al., Quantitative decoding of interactions in tunable nanomagnet arrays using first order reversal curves. Sci. Rep. **4**(4204), 1–5 (2014)
19. J.B. Abugri, P.B. Visscher, S. Gupta, P.J. Chen, R.D. Shull, FORC+ analysis of perpendicular magnetic tunnel junctions. J. Appl. Phys. **124**(4), 043901 (2018)
20. S. Ruta, O. Hovorka, P.W. Huang, K. Wang, G. Ju, R. Chantrell, First order reversal curves and intrinsic parameter determination for magnetic materials: limitations of hysteron-based approaches in correlated systems. Sci. Rep. **7**, 1–12 (2017)
21. B.F. Valcu, D.A. Gilbert, K. Liu, S. Technology, Fingerprinting inhomogeneities in recording media using the first-order reversal curve method. IEEE Trans. Magn. **47**(10), 2988–2991 (2011)
22. C.-I. Dobrotă, A. Stancu, Tracking the individual magnetic wires' switchings in ferromagnetic nanowire arrays using the first-order reversal curves (FORC) diagram method. Phys. B Condens. Matter **457**, 280–286 (2015)
23. C.I. Dobrotă, A. Stancu, Mean field model for ferromagnetic nanowire arrays based on a mechanical analogy. J. Phys. Condens. Matter **25**, 3 (2013)
24. C.I. Dobrotă, A. Stancu, PKP simulation of size effect on interaction field distribution in highly ordered ferromagnetic nanowire arrays. Phys. B Condens. Matter **407**(24), 4676–4685 (2012)
25. L. Stoleriu, A. Stancu, M. Cerchez, Micromagnetic analysis of the physical basis of vector Preisach-type models, in *Magnetic storage systems beyond 2000*, (Springer Netherlands, Dordrecht, 2001), pp. 369–372
26. M.R. Zamani Kouhpanji, A. Ghoreyshi, P.B. Visscher, B.J.H. Stadler, Facile decoding of quantitative signatures from magnetic nanowire arrays. Sci. Rep. **10**(1), 15482 (2020)
27. M.R. Zamani Kouhpanji, J. Um, B.J.H. Stadler, Demultiplexing of magnetic nanowires with overlapping signatures for tagged biological species. ACS Appl. Nano Mater. **3**(3), 3080–3087 (2020)
28. M.R. Zamani Kouhpanji, B.J.H. Stadler, A guideline for effectively synthesizing and characterizing magnetic nanoparticles for advancing nanobiotechnology: a review. Sensors **20**(9), 2554 (2020)
29. E. De Biasi, Faster modified protocol for first order reversal curve measurements. J. Magn. Magn. Mater. **439**, 259–268 (2017)
30. J. Gräfe, M. Schmidt, P. Audehm, G. Schütz, E. Goering, Application of magneto-optical Kerr effect to first-order reversal curve measurements. Rev. Sci. Instrum. **85**(2) (2014)
31. M.R. Zamani Kouhpanji, B.J.H. Stadler, Assessing the reliability and validity ranges of magnetic characterization methods. ArXiv **1–9** (2020)
32. M.R. Zamani Kouhpanji, P.B. Visscher, B.J.H. Stadler, Underlying magnetization responses of magnetic nanoparticles in assemblies. arXiv **1**(612), 1–7 (2020)
33. C.I. Dobrotă, A. Stancu, What does a first-order reversal curve diagram really mean? A study case: Array of ferromagnetic nanowires. J. Appl. Phys. **113**(4) (2013)
34. P.B. Visscher, Avoiding the zero-coercivity anomaly in first order reversal curves: FORC+. AIP Adv. **9**(3), 035117 (2019)
35. C. Pike, A. Fernandez, An investigation of magnetic reversal in submicron-scale Co dots using first order reversal curve diagrams. J. Appl. Phys. **85**(9), 6668–6676 (1999)
36. R.J. Harrison, J.M. Feinberg, FORCinel: an improved algorithm for calculating first-order reversal curve distributions using locally weighted regression smoothing. Geochem. Geophys. Geosyst. **9**(5) (2008)
37. R. Egli, VARIFORC: an optimized protocol for calculating non-regular first-order reversal curve (FORC) diagrams. Glob. Planet. Change **110**, 302–320 (2013)
38. D. Cimpoesu, I. Dumitru, A. Stancu, DoFORC tool for calculating first-order reversal curve diagrams of noisy scattered data. J. Appl. Phys. **125**(2) (2019)
39. F. Groß et al., gFORC: a graphics processing unit accelerated first-order reversal-curve calculator. J. Appl. Phys. **126**(16), 163901 (2019)
40. T.A. Berndt, L. Chang, Waiting for Forcot: accelerating FORC processing 100× using a fast-fourier-transform algorithm. Geochem. Geophys. Geosyst. **20**(12), 6223–6233 (2019)

Soft Magnetic Materials

Michael E. McHenry, Paul R. Ohodnicki, Seung-Ryul Moon, and Yuval Krimer

Abstract This chapter provides a review of soft magnetic materials (SMM's) beginning with a cursory discussion of fundamental and technical magnetic properties followed by a summary of technically important soft magnetic materials classes. This is followed by a review of measurement techniques which are important for materials characterization in emerging SMM's.

Keywords Soft magnetic materials (SMMs) · Alloys · FeCo alloys · Silicon steels · Permalloys · Ferrites · Amorphous magnetic ribbons (AMRs) · Metal amorphous nanocomposites (MANCs) · Magnetic anisotropy · Magnetostriction · AC power losses · B-H loops · First-order reversal curves · Eddy current losses · Magnetic domains

1 Introduction

Useful engineering magnetic materials have collective magnetism [1], resulting in magnetization (net *atomic dipole moment* per unit volume), **M**, in the absence of an applied magnetic field, **H**, over macroscopic volumes called *magnetic domains*. Collective magnetization in $H = 0$ is called the *spontaneous magnetization*. Atomic dipole moments arise chiefly from the spin (and orbital) angular momentum of d- or f- electrons in unfilled shells.

Among collective magnetism, *direct exchange coupling* leads to parallel aligned dipole moments in *ferromagnetism*. In ferromagnets the spontaneous magnetization

M. E. McHenry (✉) · Y. Krimer
Carnegie Mellon Univ., Pittsburgh, PA, USA
e-mail: mm7g@andrew.cmu.edu

P. R. Ohodnicki
University of Pittsburgh, Pittsburgh, PA, USA

S.-R. Moon
Gridbridge, Raleigh, North Carolina, USA

© Springer Nature Switzerland AG 2021
V. Franco, B. Dodrill (eds.), *Magnetic Measurement Techniques for Materials Characterization*, https://doi.org/10.1007/978-3-030-70443-8_21

665

has a maximum value, M_s, the ***saturation magnetization***, at $T = 0$ K, disappearing through a 2nd-order phase transition at the ***Curie temperature***, T_c. The saturation magnetization, gives rise to a ***saturation induction, $B_s = M_s$*** (SI units); $B_s = 4\pi M_s$ (cgs units) [2]. Indirect exchange (***superexchange***) mediated by overlap of magnetic orbitals with p-orbitals results in ***ferrimagnetism***. Dipole moments of different magnitude couple antiferromagnetically (antiparallel) yielding a partially compensated magnetization that disappears at a temperature, T_N, called the ***Neel temperature***. Due to partial cancellation of atomic moments, ferrites typically have smaller inductions.

Ferromagnetic (ferrimagnetic) materials are classified as ***soft*** or ***hard magnets***. In a hard magnet, magnetic hysteresis is useful to store a large metastable ***remanent magnetization***, in the absence of a field. Soft magnetic materials (SMMs) can be switched from one bistable magnetic state to another in small applied magnetic fields. This chapter illustrates synthesis → structure → properties → performance relationships in SMM's focusing on bulk materials for power applications and reviews techniques for measuring pertinent magnetic properties of SMMs.

Early SMM choices were motivated by attractive intrinsic properties, such as B_s and T_c. Alloy developments for applications involved optimizing extrinsic properties, such as ***remanent induction*** and ***coercivity*** by processing to achieve suitable microstructures. For SMMs, small power losses per cycle are desirable, and the prominent mechanism for these losses differs at increasing frequencies. Important applications of SMMs include (a) inductors and inductive components; (b) low- and high-frequency transformers; (c) Alternating current machines, motors and generators; (d) Magnetic lenses for charged particle beams and magnetic amplifiers.

Magnetostatics and ***magnetodynamics*** were developed in the 1800s. Michael Faraday observed that an applied voltage to a copper coil resulted in a voltage across a secondary coil wound on the same iron core. Faraday's law of induction (1831), $V = N \, d\Phi/dt$, defines an induced voltage in terms of the number of turns, N, in a primary exciting coil and the rate of change in the magnetic flux, Φ caused by an AC current. Faraday's law is one of Maxwell's equations [3] presented in the 1860s prior to a more compact expression using vector calculus [4]. Realization in devices fueled technologies including ***power transformers*** [5]. The 1884 demonstrations of open and closed core ***transformer***s were followed by their use to supply power to >1000 Edison lamps at the 1885 Budapest exhibition [6].

Desired technical properties for SMMs include:

(1) ***High Magnetic Permeability:*** Permeability, $\mu = \sim dB/dH \sim B/H = (1 + \chi)$, is a material property, the slope of flux density, B, vs. the applied field, H, which reduces to B/H in linear response theory. High μ materials can produce large flux changes for small fields.

(2) ***Low Hysteresis Loss:*** Hysteresis loss is the energy consumed in cycling a material between +H and − H. The power loss of an AC device includes a term equal to the frequency multiplied by the hysteretic loss per cycle. At high

frequencies AC eddy current losses depend on another material's property, the resistivity, ρ.

(3) **_Large Saturation and Remnant Magnetizations:_** A large saturation induction, B_s, is desirable because it represents the ultimate response of soft magnetic materials.

(4) **_High T_c:_** The ability to use SMMs at elevated temperatures depends on a material's T_c.

2 Soft Magnetic Materials Classes

Figure 1 shows DC permeability and saturation induction and figures of merit for SMMs used in power electronic applications. Lower induction permalloys are notable for their large magnetic permeabilities. Fe, Co, and Ni are the only elemental room temperature ferromagnets. In their cubic allotropic forms, all are good SMMs, because their high symmetry provides low values of **_cubic magnetocrystalline anisotropy_** and low **_magnetostriction_** coefficients.

FeCo Alloys: are SMMs important for applications needing high saturation inductions and high-temperature operation. The largest RT saturation induction ($B_s \sim 2.5$ T) occurs with minor substitutions to $Fe_{65}Co_{35}$ binary alloys. Dipole moments are understood by energy band theory and the Slater Pauling curve [8]. In $Fe_{1-x}Co_x$ alloys, a chemical phase transformation from an ordered CsCl-type structure to a disordered BCC phase causes embrittling which poses challenges to rolling thin sheet. Consideration of both alloy resistivity (important for determining eddy current losses) and alloy additions which influence mechanical properties is

Fig. 1 Figures of merit, saturation induction, and permeability, for selected SMMs. Adapted from Ref. [7]

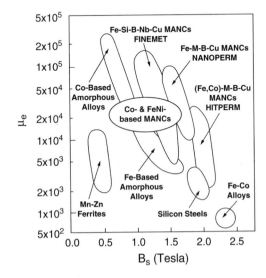

used to design alloys for rotor applications in electric motors. Magnetostriction considerations are also important in design of FeCo SMMs for applications. Notable commercial alloys include Permendur ($Fe_{1-x}Co_x$ $x \sim 0.5$), Supermendur, and Hiperco-50, a similar ($Fe_{1-x}Co_x$ $x \sim 0.5$) alloy with a 2% V addition to increase strength and electrical resistivity. Small Nb additions to Hiperco-50 result in grain refinement in Hiperco-50 HS that improves mechanical properties and reduces degradation in long term aging at elevated T's [9].

Silicon Steels: Si and Al provide the largest increase in electrical resistivity when added to Fe. Studies of Fe–Si date back to 1885 [10]. Alloys with 2–4.5 wt.% Si were the most important SMM by volume and market value by 1934 [11]. *Non-oriented (NO) steels* and *grain-oriented (GO) Si steels* are *electrical steels*. NO Si steels are a commodity in rotating machinery, GO silicon steels in transformers. GO Si steels are further subdivided into *regular grain-oriented (RGO) high-permeability grain-oriented (HGO)* materials [12] After 1951 Fe–Si transformer sheet was produced by cold rolling followed by recrystallization [13]. Polycrystalline NO materials were followed by GO materials with (110)/[001] (*Goss texture*). Although Si steels have B_s as high as 2.1 T, practically only $0.7B_s$ ($\sim$1.4 T) is achievable in low fields in Goss textured materials.

Reducing losses in transformer grade Si steels focuses on improving Goss texture in GO steel and reducing eddy current losses by decreasing thickness and surface treatment. SMM's core loss is partitioned between (a) magnetic hysteresis, (b) eddy currents, and (c) anomalous losses. Magnetic softness is rooted in (1) low magnetocrystalline anisotropy, (2) low magnetostriction coefficients, and (3) high-domain wall mobility. Low eddy current and domain wall losses are achieved with large electrical resistivities and thin laminates. Resistivity increases with Si accompanied by embrittlement approaching an Fe_3Si intermetallic composition. This poses challenges for thermomechanical processing and limits laminate thickness. Surface modification of Si steel sheet decreases domain size and reduce anomalous eddy current losses.

Processing developments for GO Si steel include promoting polycrystalline Goss recrystallization texture, controlling grain growth, and development of insulating layers to electrically isolate laminates and develop interfacial stress states to reduce magnetostrictive losses. Anomalous eddy current losses are reduced by forcing rotational magnetization processes through pinning magnetic domain walls by scratching or laser scribing. Powder products include Sendust $Fe_{1-x-y}Si_xAl_y$ with an optimum composition near the zero anisotropy zero magnetostrictive composition ($x = 0.1$ and $y = 0.05$). Sendust's attractive soft magnetic properties are challenged by its brittleness and therefore it is chiefly used in powder form.

Permalloys: Fe–Ni alloys are important crystalline high-permeability SMMs. Ni-rich alloys are called *permalloys*. Important permalloys are (1) 78% Ni permalloy, which has a zero magnetostriction coefficient, is sold under the names supermalloy, Mu-metal and Hi-mu 80; (2) 65% Ni permalloy with near zero magnetocrystalline anisotropy and exhibiting a strong response to field annealing and (3) 50% Ni permalloy having high B_s and a strong response to field annealing to produce square loop materials. Magnetic properties and response to field annealing to induce anisotropy in permalloys depend on whether the material has a disordered fcc or ordered $L1_2$ (Ni_3Fe) intermetallic structure. Permalloy crystallographic texture is developed by ***thermomechanical processing*** taking advantage of strong properties variation with direction. These fcc-derivative alloys have [111] easy magnetization directions and strong anisotropy in magnetostriction. Rolling to reduce alloy strip thicknesses can benefit high-frequency magnetic properties limited by eddy current losses.

An important set of alloys in the Ni_xFe_{1-x} family are the ***Invar alloys*** with ~36 at.% Ni alloys. The ***Invar effect (Invar anomaly)*** occurs because of the magnetoelastic effects near T_c. Magnetoelastic effects (magnetostriction) give rise to spontaneous changes in the lattice parameters as a function of temperature. Like the ***magnetocaloric effect***, these are largest where the T-dependence of the magnetization is largest, i.e., near T_c. In Invar alloys the magnetostrictive volume change can be tuned to precisely cancel the thermal expansion coefficient near room temperature, making the material dimensions T-independent [14].

Ferrites: are double oxides of Fe^{3+} and a divalent metal, M^{2+}. If M^{2+} is Fe^{2+} the ferrite is ***magnetite, Fe_3O_4***, the first magnetic mineral known to man. It was called ***lodestone*** and used in early compass needles. ***Cubic ferrites*** (except $CoFe_2O_4$) are preeminent SMMs adopting a spinel structure. ***Hexagonal ferrites*** are hard magnets with hexagonal, trigonal, or rhombohedral structures. Studies of ferrite properties date back to work of Snoek [15] and Verwey [16] of Phillips Research Labs who from 1933–1948 studied electron hopping conductivity mechanisms in ferrites from ***ferric, Fe^{2+}*** to ***ferrous, Fe^{3+}*** cations. The theory of ferrimagnetism was developed by Louis Neel in 1948 who explained the T-dependent magnetic response of these SMMs [17].

Application of ferrites involves considerations of induction, Neel temperature, magnetic anisotropy, magnetostriction coefficients, and electrical resistivity. Two technically important ferrite SMM classes are Mn–Zn- and Ni–Zn-ferrites [18]. Mn–Zn-ferrites have resistivities of 100–1000 Ω cm useful for frequencies up to ~1 MHz. Ni–Zn ferrites have resistivities of 10^6 Ω cm useful for frequencies up to ~1 GHz. The resistivity (conductivity) in ferrites is thermally activated, strongly depending on charged defect concentrations and cation disorder [19]. Spinel ferrite applications include inductor and transformer cores, switch mode power supplies, converters, microwave devices, filters, and electromagnetic interference (EMI) absorbers. Of technical interest for applications are high magnetic permeabilities and large resistivities (to limit eddy current losses). The magnetic permeability in ferrites is strongly influenced by microstructure, e.g., permeability is known to increase with grain size in Mn–Zn ferrites.

Amorphous Metal Ribbons (AMRs): have a frozen liquid or glassy structure [20]. In AMRs a supercooled liquid metal transforms into a ***metallic glass*** at lower temperatures, at cooling rates too fast to allow crystallization of thermodynamically stable crystalline phase(s). Typical AMR alloys are two or more components, the primary component largely responsible for engineering properties and additional components serving as ***glass formers***. Alloys that most easily cast as AMRs are near a ***eutectic composition*** in a binary or multicomponent phase diagram as these are compositions for which the liquid phase remains stable to the lowest temperature.

The choice of the primary component in AMRs produced for power magnetic applications is from among Fe, Co, and Ni (or combinations). Fe-based alloys are most common due to attractive magnetic properties and alloy economics. Common choices for glass formers, are **metalloids**, such as B and Si that yield deep eutectics [7, 21]. The first splat quenched ferromagnetic amorphous alloy [21] and the first $Fe_{80}P_{13}C_7$ AMR were both produced in the Caltech lab of Pol Duwez [22]. Hasegawa reviews early developments of commercial AMRs (Metglas) [23]. Chen and Polk's [24] studies of effects of Si and Al on glass forming ability in Fe led to commercial metallic glasses for magnetic applications [25]. The first commercial AMR, Metglas 2826, an $Fe_{40}Ni_{40}P_{14}B_6$ alloy was produced by Allied Signal in 1973. The Fe–B system was developed for AMR [26] and soon followed by Fe–B–Si [27, 28] and Fe–B–Si–C [29, 30]. These alloys exhibited significantly lower losses than Si steels and improved efficiencies fueled further AMR development. These include AMR materials produced commercially by Metglas, Vacuumschmelze, Hitachi Metals, and others. Commercially produced Co-based and NiFe-based AMRs are commodities [31].

Metal Amorphous Nanocomposites (MANCs) [32]: are produced by controlled crystallization of a multicomponent AMR precursor to yield a nanocrystalline phase embedded in an amorphous matrix. Thin AMR and MANCs are preferable in a variety of magnetic applications where **eddy current losses** limit high-frequency, high power density magnetic applications. The kinetics of nanocrystallization is controlled by solid-state nucleation and growth (N&G) kinetics. Progress of an isothermal phase transformation is described by plotting volume fraction (i.e., of primary crystallites) transformed, X (t,T), as a function of temperature, T, and time, t, in a TTT curve [33].

These materials have been termed **nanocrystalline materials** as commercial products, such as FINEMET@™. However, their superior properties derive from both the distinct properties of the amorphous and nanocrystalline phases, and therefore the term nanocomposite is more accurate. Chemical partitioning [34] of species in primary nanocrystallization and, in particular, large resistivities of the composites, due to the amorphous phase surrounding nanocrystals, is important to suppress eddy currents at higher frequencies. MANC resistivities are predicted in multiphase models that consider volume fractions and chemistries of the crystalline and amorphous phases, respectively [35]. The influence of chemical composition on resistivity in crystalline and amorphous phases is further understood in terms rigid band ideas [36].

MANC alloy development is analogous to crystalline and AMR alloys. The first MANCs were Fe–Si- and Fe-based FINEMET [37] and NANOPERM [38], mimicking Fe–Si crystalline alloys (electrical steels). There are now Co-based [39] and FeNi-based [40] MANCs. Co-based MANCs have magnetic anisotropies tunable over five orders of magnitude and a method for inducing anisotropy by strain annealing [41]. Materials with large **transverse strain-annealed/strain-induced anisotropy** allow tunable permeabilities and spatially varying the same in cores for transformers and inductors.

3   Measurement Techniques for Materials Characterization in Emerging SMMs

The physics of amorphous magnets [36] fuels magnetic applications of AMRs and MANCs enabled by low switching losses. Power losses are reported at particular levels of magnetic induction and frequency fit to Steinmetz relationships [42] originally used to parameterize losses in Si steels. It is typical today to fit power loss to a modified three-term *Steinmetz equation*:

$$P_{\mathrm{L}} = k_{\mathrm{hys}} f B^{\beta} + k_{\mathrm{eddy}} f^2 B^2 + k_{\mathrm{exc}} f^{1.5} B^{1.5} \tag{1}$$

where k_{hys}, k_{eddy}, and k_{exc} are used to partition power loss density for (i) hysteresis, (ii) classical eddy current, and (iii) excess loss components. f is the AC frequency (kHz), B is the peak flux density (T), and β is a power constant near 1. It is common to lump the last two terms as total eddy current losses with a power law exponent $1.5 < a < 2$. Static hysteresis loss dominates at low-f and eddy currents dominate at high-f.

The partitioning constant k_{eddy} is proportional to the square thickness of the material and inversely proportional to resistivity. Thus, high resistivity is desired for magnetic components operating at and typically are 2–4× larger in AMRs significantly reducing eddy current losses. Hysteretic losses are diminished in AMR, MANCs, and bulk amorphous materials as a result of random magnetic anisotropy [43]. Magnetic induction is sacrificed for glass forming ability (and resistivity) in AMRs as compared with crystalline materials (e.g., Si steels). MANCs offer a middle ground between the highest inductions achievable in crystalline materials and the high resistivity of amorphous materials. The ability to cast AMRs in thin cross sections significantly impacts eddy current losses to lower high-frequency losses that is also achieved in MANCs.

The *anomalous eddy current (AEC) losses*, 3rd term in the Steinmetz equation is associated with dynamic domain wall motion in a SMM. AEC contributions originate at a mesoscale due to local heterogeneities in magnetic domain structure [44]. Just as laser scribing is used in pinning magnetic domain walls in Si steels to limit dissipation at high-fs, inducing anisotropy in AMR and MANCs lowers losses at high-f [45]. In AMRs anisotropy is induced by transverse (to AMR length) field annealing, rolling [46], and recently strain annealing [47].

Measurements of B-H loops and associated losses in SMMs at application relevant excitation frequencies can be challenging. High-performing SMMs have extremely low coercivities and losses such that accurate characterization of B-H characteristics under quasi-static excitation conditions is difficult for conventional magnetometry techniques. Demagnetizing fields can also result in significant distortions in the measurements of magnetic anisotropies and associated B-H loops in cases where sample geometries are not toroidal. AC permeametry techniques are thus used to characterize the B-H characteristics of SMMs which are typically

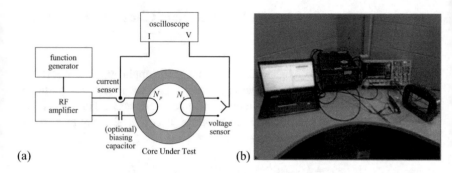

Fig. 2 Arbitrary waveform core loss test system (CLTS) (**a**) conceptual setup (**b**) actual

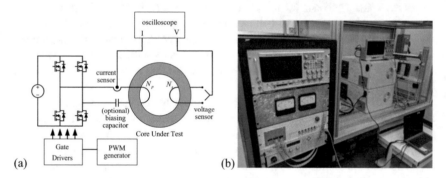

Fig. 3 Square waveform core loss test system (CLTS) (**a**) conceptual setup (**b**) actual

constructed in a toroidal configuration to minimize demagnetizing fields (Figs. 2 and 3).

For AC permeametry measurements, a time-varying excitation waveform is applied to a toroidal magnetic core using a primary winding, and the induced voltage measured on a secondary winding is used to estimate the B-H loop and the associated core losses as a function of frequency. Excitation waveforms are often sinusoidal, although arbitrary and square waveform excitation can also be utilized to characterize SMMs. Non-sinusoidal excitation is particularly valuable for application relevant waveforms with examples of an experimental setup for arbitrary and square-wave excitations shown in Figs. 2 and 3, respectively. Figure 4 illustrates three different excitation voltage waveforms and corresponding flux density waveforms.

In the arbitrary waveform core loss testing system (CLTS) illustrated, a function generator can produce any arbitrary small signal, and the small signal is amplified and applied to a core under test (CUT) using a linear amplifier. The arbitrary waveform CLTS is advantageous in that any waveforms can be easily applied to characterize a CUT; however, the linear amplifier has limited electrical capabilities, such as ±75 V & ±6A peak ratings as well as a 400 V/μs slew rate in the case of the experimental setup illustrated. Therefore, a full core characterization may not

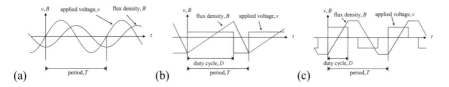

Fig. 4 Excitation voltage waveforms and corresponding flux density waveforms (**a**) sinusoidal excitation with sinusoidal flux, (**b**) asymmetrical excitation with sawtooth flux, and (**c**) symmetrical excitation with trapezoidal flux

be possible in some cases where true application relevant excitation conditions are desired, such as in low permeability cores, high frequency, and/or large sized cores. In cases where higher power and voltage excitation are required, a dedicated square waveform CLTS can be utilized to perform various square waveform measurements with different duty cycles, as shown in Fig. 4(b) and (c). To achieve such excitation waveforms, an H-bridge-based excitation circuit can be utilized with wide-bandgap-based switching devices such as commercially available 1200 V SiC MOSFET devices enabling the extension of core characterization to conditions which are relevant for larger scale components in end-use applications such as solid-state transformers.

To perform the measurements of dynamic B-H loops, two windings are placed around the core under test. The amplifier or square-wave CLTS excites the primary winding with an applied voltage (v), and the current (i) of the primary winding is measured, in which the current information is converted to the magnetic field strengths H as

$$H(t) = \frac{N_p \cdot i(t)}{l_m}, \tag{2}$$

where N_p is the number of turns in the primary winding. A DC-biasing capacitor is typically inserted in series with the primary winding to provide zero average voltage applied to the primary winding which avoids undesired DC-bias in the B-H loop and associated core loss measurements. During measurements, the secondary winding is open, and the voltage across the secondary winding (v_s) is measured, in which the voltage information is integrated to derive the flux density B as

$$B(t) = \frac{1}{N_s \cdot A_e} \int_0^T v_s(\tau) \, d\tau, \tag{3}$$

where N_s is the number of turns in the secondary winding and T the excitation waveform period.

In Fig. 4(a), the excitation voltage is sinusoidal, and its flux waveform is also a sinusoidal shape. In Fig. 4(b), the excitation voltage is a two-level square waveform with asymmetrical duty cycle between high-level and low-level voltages, and its flux waveform is a sawtooth shape referred as an asymmetrical waveform. Its duty

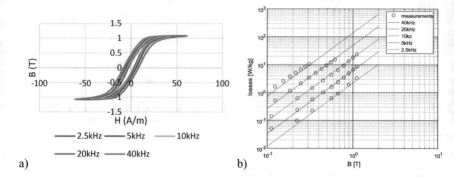

Fig. 5 (**a**) B-H loops of an Fe-based MANC core with increasing excitation frequency from 2.5 kHz to 40 kHz. (**b**) Losses as a function of peak induction and frequency with fits to the Steinmetz Equation

cycle is defined as the ratio between the applied high voltage time and the period, and the duty cycle can range from 0% to 100%. Furthermore, the average excitation voltage is adjusted to be zero via the DC-biasing capacitor, and, thus, the average flux is also zero. In Fig. 4(c), the excitation voltage is a three-level square voltage with symmetrical duty cycle between high-level and low-level voltages, and its flux waveform is a trapezoidal shape referred as symmetrical waveforms. Its duty cycle is defined as the ratio between the applied high-level voltage time and the period, and the duty cycle can range from 0% to 50%. At 50% duty cycles, the asymmetrical and symmetrical waveforms are identical.

Examples of measured B-H loops and losses as a function of frequency for a sinusoidal excitation of an Fe-based MANC core are illustrated in Fig. 5(a). Fits to the traditional Steinmetz expressions are presented in Fig. 5(b) showing an increase in B-H loop area and core loss with increasing frequency, as expected for increased classical and anomalous eddy current losses. Losses will also vary as a function of excitation waveform, which can be captured through two approaches: (1) empirical loss maps such as effective Steinmetz coefficient fitting to empirical data and (2) by correcting the losses predicted from sinusoidal Steinmetz coefficients through advanced loss modeling such as modified Steinmetz equations (MSE), generalized Steinmetz equations (GSE), and many others [48, 49].

For magnetic component design, nonlinear B-H characteristics are critical, but it is often enough to have detailed information on core saturation behavior without explicitly including core losses (though losses can be included in efficiency estimates with traditional loss modeling techniques). In such cases, the anhysteretic approximation can be employed for which the B-H loop is effectively replaced by the average of the positive and negative branch to produce a curve that does not contain any hysteresis but does include information about core saturation characteristics. An estimation of the anhysteretic response is illustrated in Fig. 6(a) along with a corresponding plot of permeability parameterized as a function of applied magnetizing field intensity H in Fig. 6(b). It can be noted that all

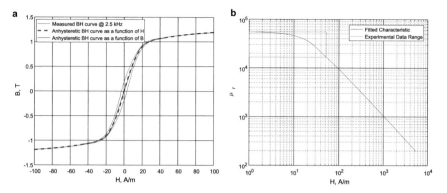

Fig. 6 (**a**) B-H loops of an Fe-based MANC core at 2.5 kHz along with an anhysteretic fit to the B-H loop. (**b**) Measured and extrapolated permeability as a function of magnetizing field intensity for the corresponding anhysteretic magnetizing curve

of the information presented can be estimated and parameterized based upon the measured B-H loop using the techniques described above. In some cases, parameterization of the permeability as a function of induction, B, may be preferred for the purpose of magnetic component designs [50, 51]. Estimation of anhysteretic characteristic is performed using a genetic optimization program, which can be found at the website: (https://engineering.purdue.edu/ECE/Research/Areas/PEDS/go_system_engineering_toolbox).

To provide magnetic component designers with the necessary tools for optimized designs, accurate core characterization is required for a range of materials and excitation conditions. In addition, parameterization of important parameters such as B-H loops, absolute permeability, differential permeability, and core losses using pragmatic methods such as loss modeling approaches can allow for faster and more widespread adoption by magnetics designers. In order to address this need, detailed data sheets have been recently developed for a number of commercial and custom magnetic cores by the National Energy Technology Laboratory and published in the form of technical reports. Examples of the latest published data sheets can be found in Refs. [52–57] (Fig. 7).

In MANCs, magnetic anisotropy controls **magnetic domain structure** and methods to tune properties [58, 59]. Important length scales include those for **mesoscale eddy currents** coupling to domain wall motion to cause AEC losses. These require evolution in magnetic measurements for characterization. Efforts to spatially and temporally image magnetic domains as a function of frequency and T can be facilitated by **magneto-optic Kerr effect (MOKE) tools** [60, 61] and **magnetic first-order reversal curves (FORC)** [62]. Magnetic measurements can be coupled to observations of **nanoheterogeneity** by high-resolution TEM or **atom probe field ion microscopy (APFIM)**. Figure 8(ii) shows APFIM of an Fe–Si–B MANC: (a) Fe, Si and B distributions, (b) interfaces, and (c) partitioning between Fe_2B and α-Fe phases [63]. Figure 8(iii) shows FORC distributions for

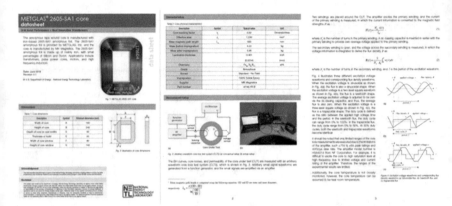

Fig. 7 Example data sheet prepared for Metglas 2605-SA1 AMR at NETL

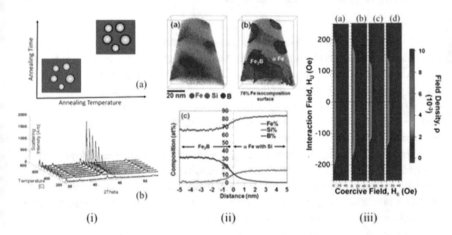

Fig. 8 MANC structural evolution and magnetic coupling (**i**) (**a**) $X(t,T)$ and (**b**) T-dependent XRD for crystallization of a $(Fe_{65}Co_{35})_{81+x}B_{12}Nb_{4-x}Si_2Cu_1$ MANC [36]; (**ii**) APFIM of an Fe–Si–B MANC showing (**a**) Fe, Si, and B distributions, (**b**) interfaces, and (**c**) partitioning between Fe_2B and α-Fe phases [63]; (**iii**) FORC for $(Fe_{65}Co_{35})_{81+x}B_{12}Nb_{4-x}Si_2Cu_1$ (**a**) $x = 2$, 370 °C (**b**) $x = 2$, 430 °C (**c**) $x = 4$, 370 °C (**d**) $x = 4$, 430 °C [64]

$(Fe_{65}Co_{35})_{81+x}B_{12}Nb_{4-x}Si_2Cu_1$ at different temperatures: (a) $x = 2$, 370 °C; (b) $x = 2$, 430 °C; (c) $x = 4$, 370 °C; and (d) $x = 4$, 430 °C [64].

MANCs present options to induce anisotropies since anisotropy can be introduced by applying one or more force variables in nanocrystallizing an AMR precursor. Annealing in an external magnetic field or self-field annealing [65] creates a preferred easy axis typically ascribed to ***atomic pair ordering***. In strain-annealed Fe-rich MANCs, with BCC nanocrystals, anisotropy results from lattice parameter distortions in nanocrystals from residual stress in the amorphous matrix [66]. Stacking faults improve mechanical properties and strain annealing response in Co-rich MANCs with close packed nanocrystals [67].

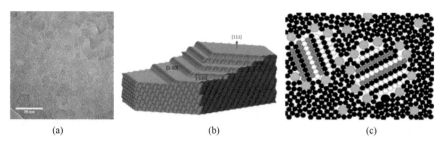

(a)   (b)   (c)

Fig. 9 (**a**) Co-based MANC alloy HRTEM after strain annealing showing stacking faults [54]; (**b**) model illustrating growth-induced stacking faults in Co nanocrystals; and (**c**) cartoon of a *Co-based MANC material* analog where structure allows further tuning of MANC properties

Both Co-based [67] and FeNi-based [40] MANCs exhibit strain-induced anisotropies, transverse in the 1st and longitudinal in the 2nd. Co-based MANC structures are shown in Fig. 9(a) in HRTEM image, proposed growth faulting occurring in strain annealing (Fig. 9(b)) and a cartoon for a MANC nanostructure where nanocrystal faulting in contributes to permeability tuning. This strain-induced anisotropy is attractive in power electronic systems exploiting high-f wide band gap semiconductors [68]. FeNi-based MANCs have recently been evaluated for applications in solid-state power transformers [69] and high-speed motor applications [70].

Dynamic magnetization processes in soft magnetic materials are strongly dependent on domain structure. At frequencies below approximately 10 MHz, eddy currents induced by wall motion are a significant source of core losses. Above 10 MHz, wall motion becomes damped but rotational processes differ depending on the orientation of the domain to the local driving field. The response to fields, in performance-related metrics (magnetic losses, plastic deformation), is nonlinear and complicated to describe in AMRs and MANCs. Magnetic properties are observable by measuring **global magnetization** and **local magnetic domain structure**. Studies of **magnetic domain structure** by **magneto-optic Kerr effect (MOKE) microscopy** [71] and monitoring intergranular coupling strength, (by **first-order reversal curves (FORC)**) offer new insights for understanding micromagnetics of MANCs.

Figure 10 shows magnetic domains imaged by MOKE in a field-annealed Co-based MANC. The image shows the domain structure of the upper portion of a 3 mm disc punched from a ribbon, where the red arrow indicates the long axis of the ribbon. Contrast in MOKE images can be enhanced by subtracting image intensity at the desired state from the same sample in a reference condition. The reference condition can be the sample at the remnant field, in saturation, or under a sinusoidal excitation that is faster than the image capturing period. In Fig. 10a, the reference condition was the latter, where the excitation frequency (10 Hz, background average over 3 s) was significantly faster than the image capture time. In this way, features of the sample not related to the domain structure, such as surface defects, can be minimized and the domain contrast maximized. The blue arrow shows the sensitive direction of the Kerr polarization, and the domain structure is a series of transverse

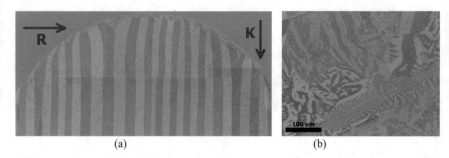

Fig. 10 (**a**) Magnetic domains imaged by MOKE in a 3 mm diameter field-annealed Co-based MANC disc and (**b**) a bulk crystalline non-oriented 3 wt% Si Steel sample

bar domains. The spacing of the domains is proportional to the induced anisotropy, where larger induced anisotropies lead to higher wall energies and consequently wider domain patterns. Near disc edges, the effects of.

Shape demagnetization are evident in the distortion of the bar domains. Figure 10b shows the surface structure of a non-oriented 3 wt% Si steel sample produced in a similar manner as Fig. 10a. Grain size in this material is $\sim$100 μm, many times larger than the MANC grain size. The complex domain structure of a given grain can continue through or abruptly change at grain boundaries depending on the orientation of the grains and the local field. Grain-oriented Si steels produce domain structures similar to the bar domains of Fig. 10a.

MOKE images provide information regarding the surface grain structure that can differ from the bulk. Care must be taken to avoid effects such as stress concentrations from improper polishing techniques that alter the domain structure in unintended ways. Aside from qualitative information regarding the domain structure, quantitative information can be obtained from measurements of the angles between magnetization in domains across a wall or in hysteresis loops produced from Kerr images during cyclic excitation. Additionally, MOKE data can be useful to set boundary conditions for micromagnetic calculations of bulk domain structures.

First-order reversal curves (FORCs) give information not possible to obtain from a hysteresis loop alone. These curves include the distribution of switching and interaction fields, and identification of multiple phases in composite or hybrid materials containing more than one phase. A FORC curve (Fig. 8 (iii)) is measured by saturating a sample in a field H_{sat}, decreasing the field to a reversal field H_a and then measuring moment versus field H_b as the field is swept back to H_{sat}. This process is repeated for many values of H_a, yielding a series of FORCs. The measured magnetization at each step as a function of H_a and H_b gives $M(H_a, H_b)$, which is plotted as a function of H_a and H_b in field space. The FORC distribution $\rho(H_a, H_b)$ is the mixed second derivative, i.e., $\rho(H_a, H_b) = -(1/2)\partial [2] M(H_a, H_b)/\partial H_a \partial H_b$. The FORC diagram is a 2D or 3D contour plot of $\rho(H_a, H_b)$. It is common to change the coordinates from (H_a, H_b) to $H_c = (H_b - H_a)/2$ and $H_u = (H_b + H_a)/2$. H_u

represents the distribution of interaction or reversal fields, and H_c represents the distribution of switching or coercive fields.

An emerging trend involves exploiting advanced magnetometry techniques to reveal a deeper understanding of the magnetization reversal mechanisms within a given magnetic material using first-order reversal curve (FORC) analysis. FORC techniques require a large number of B-H measurements. While FORC techniques are ideally suited for permanent magnet materials where static magnetic properties can be readily sampled as in vibrating sample magnetometry (VSM) and alternating gradient magnetometry (AGM) techniques, the application to soft magnetic materials is challenging due to the low DC and AC coercivities which typically require permeametry. Recent literature has begun to apply the technique of AC FORC to soft magnetic materials through custom measurement methods and experimental facilities. In most cases, quasi-static measurements are performed with a goal of minimizing the effects of eddy currents due to dynamic magnetization processes on the measured B-H characteristics [72]. However, there exists an opportunity to exploit such techniques for the purpose of new insights into dynamic magnetization processes moving into the future. Significant additional complexity of the magnetization process is expected, however, as the detailed AC FORC measurements will depend upon the shape of selected excitation waveforms as well as the excitation frequency.

4 Conclusion

This chapter has reviewed some of the current state of the art in soft magnetic materials (SMMs) discussing fundamental and technical magnetic properties and technically important soft magnetic materials classes. The significant proliferation of new soft magnetic materials has led to development of sophisticated new measurement techniques which allow for new understanding of both traditional and emerging SMMs.

References

1. M.E. McHenry, D.E. Laughlin, Chapter 19—Magnetic properties of metals and alloys, in *Physical Metallurgy*, 5th edn., (2015), pp. 1881–2008. https://doi.org/10.1016/B978-0-444-53770-6.00019-8
2. SI (mksa) units are the preferred standard units and have been recently been revised, see Goldfarb
3. J.C. Maxwell, *A Treatise on Electricity and Magnetism* (Dover, New York, 1954)
4. J.C. Maxwell, A dynamical theory of the electromagnetic field. Phil. Trans. Roy. Soc. Lond. **155**, 459–512 (1865). https://doi.org/10.1098/rstl.1865.0008
5. D.J. Allan, Power transformers—the second century. Power Eng. J. **5**(1), 5–14 (1991). https://doi.org/10.1049/pe:19910004

6. F.J. Uppenborn, *History of the Transformer* (E. & F. N. Spon, London, 1889), pp. 35–41
7. M.E. McHenry, M.A. Willard, D.E. Laughlin, Amorphous and nanocrystalline materials for applications as soft magnets. Prog. Mater. Sci. **44**, 291–433 (1999). https://doi.org/10.1016/S0079-6425(99)00002-X
8. J.M. MacLaren, T.C. Schulthess, W.H. Butler, R.A. Sutton, M.E. McHenry, Calculated exchange interactions and curie temperature of equiatomic B2 FeCo. J. Appl. Phys. **85**, 4833–4835 (1999). https://doi.org/10.1063/1.370036
9. R.T. Fingers, R.P. Carr, Z. Turgut, Effect of aging on magnetic properties of Hiperco@™ 27, Hiperco@™ 50, and Hiperco 50 HS@™ alloys. J. Appl. Phys. **91**, 7848 (2002)
10. J. Hopkinson, Magnetization of iron. Trans. Roy. Soc. (Lond.) A **176**, 455 (1885)
11. C.-W. Chen, *Magnetism and Metallurgy of Soft Magnetic Materials* (Dover, New York, 1986)
12. R. Boll, H. Warlimont, Applications of amorphous magnetic materials in electronics. IEEE **17**, 3053–3058 (1981)
13. R.M. Bozorth, *Ferromagnetism* (IEEE Press, Piscataway, N. J, 1993)
14. D. Gignoux, Magnetic properties of metallic systems 3, in *Chap. 5 in Materials Science and Technology, A Comprehensive Treatment. Electronic and Magnetic Properties of Metals and Ceramics*, ed. by K. H. J. Buschow, (VCH, Weinheim, 1995), pp. 367–453
15. J.L. Snoek, Magnetic and electrical properties of the binary systems $MOFe_2O_3$. Physica **3**, 463–483 (1936)
16. E.J.W. Verwey, Physical properties and cation arrangement of oxides with spinel structures. Part I: Cation arrangements in spinels. Nature **144**, 327 (1939)
17. L. Neel, Proprietes Magnetique des Ferrites: Ferrimagnetism et Antiferromagnetisme. Ann. Phys. **3**, 137 (1948)
18. A.M. Goldman, *Modern Ferrite Technology* (Springer, Pittsburgh, PA, 2006)
19. R.A. McCurrie, *Ferromagnetic Materials Structure and Properties* (Academic Press, San Diego, 1982)
20. P. Duwez, R.H. Willens, W. Klement, Continuous series of metastable solid solutions in silver–copper alloys. J. Appl. Phys. **31**, 1136–1137 (1960)
21. C.C. Tsuei, P. Duwez, J. Appl. Phys. **37**, 435 (1966)
22. P. Duwez, S.C.H. Lin, J. Appl. Phys. **38**, 4096–4097 (1967)
23. R. Hasegawa, Amorphous magnetic materials: a history. J. Mag. Mag. Mat. **100**, 1–12 (1991)
24. H.S. Chen, D.E. Polk, US Patent No. 3856513 (1974); US Reissue Patent RE 32925
25. R. Hasegawa, R.C. O'Handley, L.E. Tanner, R. Ray, S. Kavesh, Appl. Phys. Lett. **29**, 219 (1976)
26. R.C. O'Handley, C.P. Chou, N. DeCristofaro, J. Appl. Phys. **50**, 3603 (1979)
27. N. DeCristofaro, A. Datta, L.A. Davis, R. Hasegawa, in *Proc. 4th Intern. Conf. Rapidly Quenched Metals*, ed. by T. Masumoto, K. Suzuki, (The Japan Inst. Metal, 1982), p. 1031
28. N. DeCristofaro, A. Freilich, D. Nathasingh, US Patent No. 4219355 (1980)
29. M. Mitera, T. Masumoto, N.S. Kazama, J. Appl. Phys. **50**, 7609 (1979)
30. F.E. Luborsky, J. Walter, IEEE Trans. Magn. **MAG-16**, 572 (1980)
31. Commercial examples of magnetic AMRs include: Metglas 2605, an Fe-based metallic glass; Metglas 2714, a Co-based metallic glass and Metglas 2826, an FeNi-based metallic glass. Metglas 2605 has B and Si as glass formers. Alloy precursors are produced by melting high purity Fe, ferroboron and ferrosilicon
32. Hitachi Metals offers nanocrystalline FeSi-based alloys with Nb used as a growth inhibitor and Cu as a nucleation agent called FT3 grades. Hitachi has a 14 µm FINEMET commercially available: Vacuumschmelze licenses FINEMET from Hitachi and offers a variety of alloy grades under the Tradename Vitroperm. Vacuumschmelze's best advertised is 15 µm, Metglas advertises 16 µm
33. M.E. McHenry, F. Johnson, H. Okumura, T. Ohkubo, A. Hsiao, V.R.V. Ramanan, D.E. Laughlin, The kinetics of nanocrystallization and implications for properties in FINEMET, NANOPERM, and HITPERM nanocomposite magnetic materials. Scripta Mat. **48**, 881–887 (2003)

34. V. DeGeorge, S. Shen, P. Ohodnicki, M. Andio, M.E. McHenry, Multiphase resistivity model for magnetic nanocomposites developed for high frequency high power transformation. J. Elect. Mat. **43**(1), 96–108 (2014). https://doi.org/10.1007/s11664-013-2835-1

35. S. Shen, S.J. Kernion, P.R. Ohodnicki, M.E. McHenry, Two-current model of the composition dependence of resistivity in amorphous $(Fe_{100-x}Co_x)_{89-y}Zr_7B_4Cu_y$ alloys using a rigid-band assumption. J. Appl. Phys. **112**, 103705–103709 (2012). https://doi.org/10.1063/1.4765673

36. R.C. O'Handley, Physics of ferromagnetic amorphous alloys. J. Appl. Phys. **62**, R15–R49 (1987)

37. Y. Yoshizawa, S. Oguma, K. Yamauchi, New Fe-based soft magnetic alloys composed of ultrafine grain structure. J. Appl. Phys. **64**, 6044–6046 (1988)

38. A. Makino, K. Suzuki, A. Inoue, T. Masumoto, Mat. Trans. JIM **32**, 551 (1991)

39. P.R. Ohodnicki, Y.L. Qin, D.E. Laughlin, M.E. McHenry, M. Kodzuka, T. Ohkubo, K. Hono, M.A. Willard, Composition and non-equilibrium crystallization in partially devitrified Co-rich soft magnetic nanocomposite alloys. Acta Mat. **56** (2008)

40. N. Aronhime, V. DeGeorge, V. Keylin, P. Ohodnicki, M.E. McHenry, The effects of strain-annealing on tuning permeability and lowering losses in Fe–Ni based metal amorphous nanocomposites. J. Mater. (2017). https://doi.org/10.1007/s11837-017-2480-x

41. A. Leary, P.R. Ohodnicki, M.E. McHenry, Soft magnetic materials in high frequency, high power conversion applications. J. Metals **41**, 772–781 (2012). https://doi.org/10.1007/s11837-012-0350-0

42. C.P. Steinmetz, Am. Inst. Electr. Eng. Trans. **3**, 3 (1892)

43. G. Herzer, Nanocrystalline soft magnetic alloys, in *Ch. 3 of Handbook of Magnetic Materials*, ed. by K. H. J. Buschow, vol. 10, (Elsevier Science, Amsterdam, 1997), p. 415

44. G. Bertotti, IEEE Trans. Magn. **24**, 621 (1988)

45. R. Hasegawa, K. Takahashi, B. Francoeur, P. Couture, Magnetization kinetics in tension and field annealed Fe-based amorphous alloys. J. Appl. Phys. **113**, 17A312 (2013)

46. S.J. Kernion, M.J. Lucas, J. Horwath, Z. Turgut, E. Michel, V. Keylin, J. Huth, A.M. Leary, S. Shen, M.E. McHenry, J. Appl. Phys. **113**, 17A306–17A311 (2013)

47. S.J. Kernion, A. Leary, S. Shen, J. Luo, J. Grossman, V. Keylin, J.F. Huth, M.S. Lucas, P.R. Ohodnicki, M.E. McHenry, Giant induced magnetic anisotropy in strain-annealed co-based nanocomposite alloys. Appl. Phys. Lett. **101**, 102408–102412 (2012). https://doi.org/10.1063/1.4751253

48. M. Albach, T. Durbaum, A. Brockmeyer, Calculating core losses in transformers for arbitrary magnetizing currents a comparison of different approaches, in *PESC Record. 27th Annual IEEE Power Electronics Specialists Conference*, vol. 2, (1996), pp. 1463–1468

49. L. Jieli, T. Abdallah, C.R. Sullivan, Improved calculation of core loss with nonsinusoidal waveforms, in *Conference Record of the 2001 IEEE Industry Applications Conference. 36th IAS Annual Meeting (Cat. No.01CH37248)*, vol. 4, (2001), pp. 2203–2210

50. Scott D. Sudhoff, Magnetics and magnetic equivalent circuits. In: Power Magnetic Devices: A Multi-Objective Design Approach, Vol. 1, Wiley-IEEE Press, 2014, pp. 488

51. G.M. Shane, S.D. Sudhoff, Refinements in anhysteretic characterization and permeability modeling. IEEE Trans. Magn. **46**(11), 3834–3843 (2010)

52. National Energy Technology Laboratory. *Technical Reports Series (TRS)*. Available: https://netl.doe.gov/TRS

53. National Energy Technology Laboratory. *3% Silicon Steel Core Material (Grain-Oriented Electrical Steel) Datasheet*. Available: https://www.netl.doe.gov/sites/default/files/netl-file/Core-Loss-Datasheet%2D%2D-3-percent%5B1%5D.pdf

54. National Energy Technology Laboratory. *6.5% Silicon Steel Core Material (Non-Grain-Oriented Electrical Steel) datasheet*. Available: https://www.netl.doe.gov/sites/default/files/netl-file/Core-Loss-Datasheet%2D%2D-6_5-percent%5B1%5D.pdf

55. National Energy Technology Laboratory. *MnZn Ferrite Material (EPCOS N87) Datasheet*. Available: https://www.netl.doe.gov/sites/default/files/netl-file/Core-Loss-Datasheet%2D%2D-MnZn-Ferrite%2D%2D-N87%5B1%5D.pdf

56. National Energy Technology Laboratory. *METGLAS® 2605-SA1 core datasheet.* Available: https://www.netl.doe.gov/sites/default/files/netl-file/METGLAS-2605-SA1-Core-Datasheet_approved%5B1%5D.pdf

57. National Energy Technology Laboratory. *Nanocrystalline Material (FINEMET) Datasheet.* Available: https://www.netl.doe.gov/sites/default/files/netl-file/Core-Loss-Datasheet%2D%2D-Nano-crystalline%5B1%5D.pdf

58. J. Long, M.E. McHenry, D.E. Laughlin, C. Zheng, H. Kirmse, W. Neumann, Analysis of hysteretic behavior in a FeCoB-based nanocrystalline alloy by a Preisach distribution and electron holography. J. Appl. Phys. **103**, 07E710–07E712 (2008). https://doi.org/10.1063/1.2829395

59. C. Zheng, H. Kirmse, J. Long, D.E. Laughlin, M.E. McHenry, Magnetic domain observation in nanocrystalline FeCo base alloys by Lorentz microscopy and electron holography. Microsc. Microanal. **13**(Suppl. S03), 282–283 (2007). https://doi.org/10.1017/S143192760708141X

60. S. Flohrer, R. Schäfer, C. Polak, G. Herzer, Interplay of uniform and random anisotropy in nanocrystalline soft magnetic alloys. Acta Mater. **53**(10), 2937–2942 (2005)

61. S. Flohrer, R. Schafer, J. Mccord, S. Roth, L. Schultz, G. Herzer, Magnetization loss and domain refinement in nanocrystalline tape wound cores. Acta Mater. **54**, 3253–3259 (2006)

62. B. Doderill, P. Ohodnicki, M.E. McHenry, A. Leary, High-Temperature First-Order-Reversal-Curve (FORC) study of magnetic nanoparticle based nanocomposite materials. MRS Adv. **2**(49), 2669–2674 (2017)

63. S. Katakam, A. Devaraj, M. Bowden, S. Santhanakrishnan, C. Smith, R.V. Ramanujan, T. Suntharampillai, R. Banerjee, N.B. Dahotre, Laser assisted crystallization of ferromagnetic amorphous ribbons: a multimodal characterization and thermal model study. J. Appl. Phys. **184901**, 114 (2013)

64. S. Shen, P.R. Ohodnicki, S.J. Kernion, A. Leary, V. Keylin, J. Huth, M.E. McHenry, Nanocomposite alloy design for high frequency power conversion applications, in *Energy Technology 2012: Carbon Dioxide Management and Other Technologies*, (2012). TMS Warrendale, PA), pp. 265–274. https://doi.org/10.1002/9781118365038.ch32

65. K. Suzuki, G. Herzer, Magnetic field induced anisotropies and exchange softening in Fe-rich nanocrystalline soft magnetic alloys. Scripta Mater. **67**, 548–553 (2012)

66. M. Ohnuma, K. Hono, T. Yanai, H. Fukunaga, Y. Yoshizawa, Direct evidence for structural origin of stress-induced magnetic anisotropy in Fe–Si–B–Nb–Cu nanocrystalline alloys. Appl. Phys. Lett. **83**, 2859–2861 (2003)

67. A. Leary, V. Keylin, A. Devaraj, V. DeGeorge, P. Ohodnicki, M.E. McHenry, Stress induced anisotropy in Co-rich magnetic nanocomposites for inductive applications. J. Mat. Res. **31**(20), 3089–3107 (2016) Editor's Feature

68. P. Ohodnicki, A. Leary, M. E. McHenry, F. Faete, G. Kojima, A. Hefner, State-of-the-art of HF soft magnetics and HV/UHV silicon carbide semiconductors. In: PCIM Europe 2016, Conference for Power Electronics, Intelligent Motion, Renewable Energy and Energy Management (2016)

69. M. Nazmunnahar, S. Simizu, P.R. Ohodnicki, S. Bhattacharya, M.E. McHenry, Finite Element Analysis (FEA) Modeling of High Frequency Single-Phase Solid-State Transformers Enabled by Metal Amorphous Nanocomposites (MANC) and calculation of leakage inductance for different winding topologies. IEEE Trans. Mag. (2019). https://doi.org/10.1109/TMAG.2019.2904007

70. S. Simizu, P.R. Ohodnicki, M.E. McHenry, Metal Amorphous Nanocomposite (MANC) Soft Magnetic Material (SMM) enabled high power density, rare earth free rotational machines. IEEE Trans. Mag. **54**(5), 1–5 (2018). https://doi.org/10.1109/TMAG.2018.2794390

71. A. Huber, R. Schafer, *Magnetic Domains: The Analysis of Magnetic Microstructures* (Springer, 1998)

72. M. Rivas, P. Gorria, C. Munoz-Gomez, J.C. Martinez-Garcia, Quasistatic AC FORC measurements for soft magnetic materials and their differential interpretation. IEEE Trans. Magn. **53**(11), 1–6 (2017) Art no. 2003606

Permanent Magnet Materials

Satoshi Okamoto

Abstract The characterization of permanent magnets is important for industries and academics. Moreover, the magnetic properties of permanent magnets have many aspects from macroscopic and microscopic views. For these different standpoints, we must select proper magnetic measurements and analyses. In this chapter, various magnetic measurements and analyses are concisely explained to help the readers who have different backgrounds.

Keywords Permanent magnet · Nd–Fe–B · Magnetization curve · FORC · Magnetic viscosity · Magnetization reversal process · XMCD

1 Introduction

Permanent magnets have been important materials for motors and generators that convert electricity to power and vice versa. Their importance has increased because of the recent drastic shift of vehicle powertrain from fossil fuel to electricity. Therefore, the demand for high-performance permanent magnets has increased significantly [1–3]. To respond to this demand, how can we characterize permanent magnets? Which measurement method and analysis should be used? From the industrial standpoint, the most important parameter of permanent magnets is the energy product $(BH)_{max}$, which is defined as the maximum product of magnetic flux density B and magnetic field H. Thus, for this purpose, accurate measurement of the intrinsic magnetization curve of a permanent magnet is critical. On the other hand, the physical origin of coercivity and the magnetization reversal process of permanent magnets are the major concerns for academic researchers. These issues are not only of academic interests but also important to improve the properties of permanent magnets. For these research purposes, time- and temperature-dependent

S. Okamoto (✉)
IMRAM, Tohoku University, Sendai, Japan
e-mail: satoshi.okamoto.c1@tohoku.ac.jp

© Springer Nature Switzerland AG 2021
V. Franco, B. Dodrill (eds.), *Magnetic Measurement Techniques for Materials Characterization*, https://doi.org/10.1007/978-3-030-70443-8_22

magnetization curve measurements and/or high-spatial-resolution magnetic imaging techniques are required.

In this chapter, general notes for the magnetization curve measurements of permanent magnets are explained in Sect. 2. Then, after brief explanations on coercivity and magnetization reversal analyses for permanent magnets, some examples for these analyses are presented in Sects. 3 and 4.

2 General Notes for Magnetization Curve Measurements of Permanent Magnets

Magnetization curves of permanent magnets are measured by various methods, such as dc B-H tracer, pulse B-H tracer, and vibrating sample magnetometer (VSM) [4]. In a dc B-H tracer, a magnet sample is sandwiched by electromagnet poles, making a closed magnetic circuit. Consequently, the intrinsic magnetization curve of the magnet sample can be measured without any demagnetization correction treatments, because the demagnetization field does not exist inside the magnet sample placed in the closed magnetic circuit. This advantage of the dc B-H tracer that is attributed to the use of electromagnets becomes a disadvantage for the measurement of very high coercivity magnet samples because of the limitation of the magnetic field of 3 T or less. Moreover, the saturation of the magnetic pole at the high-field region causes a nonlinear signal change because of the mirror effect. On the other hand, a pulse B-H tracer can apply a much larger magnetic field; however, this method uses an open magnetic circuit. Therefore, the magnetization curve measured by this method is deformed from its intrinsic curve shape because of the shape-dependent demagnetization field. In addition to the demagnetization field, the application of a large pulse field may further deform the magnetization curve because of the eddy current effect when the magnet sample is metallic. A VSM has a very high sensitivity and is widely used for academic research. However, because a VSM has an open magnetic circuit like a pulse B-H tracer, the magnetization curve also deforms due to the demagnetization field.

Usually, a cuboid sample is used for the magnetization curve measurements because of easy sample shaping and easy sample mounting on the sample holder of the measurement apparatus. However, the demagnetization factor inside the cuboid sample is not uniform, making the demagnetization field correction difficult. Some textbooks state that a spheroid sample is effective for the demagnetization field correction because of the uniform demagnetization field inside the body. This, however, is valid only when the sample is uniformly magnetized, that is, in the fully saturated state. When permanent magnets are demagnetized, this assumption is not valid. During the demagnetization process of permanent magnets, large magnetic domains are observed in Nd–Fe–B sintered magnets [5]. Thus, the demagnetization field becomes nonuniform during the demagnetization process even for the spheroid samples. For anisotropic permanent magnets, a pillar-shaped

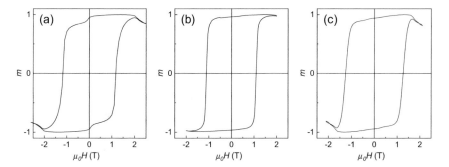

Fig. 1 Magnetization curves of pillar-shaped (**a**) Nd–Fe–B sintered, (**b**) Nd–Fe–B hot-deformed, and Sm_2Co_{17} magnets. (**a**) and (**b**) are measured at RT, and (**c**) is measured at 200 °C

sample that has the long axis along the magnetic easy axis can minimize this difficulty of demagnetization correction [6].

Finally, the surface-damaged layer of permanent magnets should be mentioned. Figure 1 shows the magnetization curves of pillar-shaped Nd–Fe–B sintered, Nd–Fe–B hot-deformed, and Sm_2Co_{17} magnets measured by VSM. The nonlinear decrease in magnetization in the high-field region is attributed to the mirror effect. For the Nd–Fe–B sintered magnet shown in Fig. 1(a), a kink on the magnetization curve at around zero magnetic field is obviously found, indicating the presence of the magnetic soft phase of the surface-damaged layer [7]. This surface-damaged layer is due to the mechanical polishing process of sample shaping. The thickness of the surface-damaged layer of Nd–Fe–B sintered magnets is estimated to be several tens of micrometers [4], corresponding to a thickness of approximately several grains. This surface-damaged layer thickness strongly depends on the permanent magnet material and its microstructure. In fact, Nd–Fe–B hot-deformed and Sm_2Co_{17} magnets, shown in Fig. 1(b, c), do not exhibit the kink at around zero magnetic field on their magnetization curves, indicating that the surface-damaged layer is negligibly thin in these magnets.

3 Coercivity Analysis of Permanent Magnets

Generally, the coercivity of permanent magnets is small compared with the anisotropy field, which is the theoretical upper limit of coercivity [8, 9]. This is called as "Brown's paradox" [10]. Because the coercivity is a very important parameter of permanent magnets, the coercivity mechanism has long been studied to solve Brown's paradox. The magnetization reversal process of permanent magnets is roughly categorized into two types, the nucleation and wall pinning types, as illustrated in Fig. 2(a, b), respectively. The former proceeds via avalanche-like domain wall propagation initiated by a nucleation of a small reversed domain. The

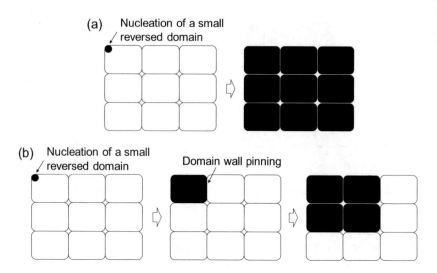

Fig. 2 Illustrations of magnetization reversal processes of (**a**) nucleation and (**b**) domain wall pinning

latter is also initiated by a nucleation of a small reversed domain. However, the domain does not expand until a certain critical field due to the domain wall pinning.

Two different approaches for these issues have been studied so far. One is the static analysis developed by Kromüller [8, 9]. The other is the thermal activation analysis by Givord [11–13]. In the following subsections, these models and examples of analyzed results are briefly explained.

Static Analysis

Kromüller assumed a one-dimensional model of a very thin soft magnetic layer sandwiched by a hard magnetic phase that mimics a defect layer or a grain boundary. He calculated the coercivity H_c for nucleation and/or pinning cases based on the one-dimensional micromagnetics theory [8, 9]. Consequently, he deduced the following simple equation:

$$H_c = \alpha H_k - N_{eff} M_s, \tag{1}$$

where H_k is the anisotropy field, M_s is the saturation magnetization, α is the reduction coefficient related to the soft-region magnetic anisotropy and/or easy axis orientation, and N_{eff} is the effective local demagnetization coefficient. Kromüller elaborately studied the form of α for various cases of nucleation/pinning, and as a result, he derived that α is given as a function of r_0/δ_B, where r_0 is the thickness of the soft magnetic phase and δ_B is the domain wall thickness of the hard magnetic phase. This equation has been widely accepted for many experimental researchers to analyze the coercivity empirically, and the values of α and N_{eff} have been evaluated from the plot of temperature-dependent H_c/M_s vs. H_k/M_s. Usually, the value of H_k in this analysis is treated as the literature data. Thus, the obtained α and N_{eff} are regarded as temperature-independent empirical parameters. However, as mentioned above, α is a temperature-dependent parameter, because α is the function of r_0/δ_B. Therefore, the naive empirical adoption of Eq. (1) may lead to values of α and N_{eff} that are quite different from the ones originally considered by Kromüller.

Thermal Activation Analysis

Kromüller's theory does not consider the thermal activation effect on the magnetization reversal. One might say that the thermal activation effect does not need to be considered for permanent magnets, because permanent magnets are bulk materials. However, the actual magnetization reversal is triggered by the nucleation of a very small reversed domain with nanometer scale. In this size range, thermal activation becomes crucial. In fact, when a permanent magnet is kept in a constant reverse magnetic field, the magnetization logarithmically decreases with time because of the thermal activation effect. This behavior is known as magnetic viscosity [14]. The time-dependent magnetization $M(t)$ in the magnetic viscosity is expressed as.

$$M(t) = M(0) - S \ln t, \qquad (2)$$

where S is the magnetic viscosity coefficient. Wohlfarth [15] and Gaunt [16] developed the theory of the magnetic viscosity and derived the following relation:

$$S = -k_B T \chi_{irr} / \left(\frac{dE}{dH} \right), \qquad (3)$$

where χ_{irr} is the irreversible magnetic susceptibility, $k_B T$ is the thermal energy, H is the applied field, and E is the energy barrier for the magnetization reversal. Generally, E is a function of H given as.

$$E(H) = E_0 (1 - H/H_0)^n, \qquad (4)$$

where E_0 is the barrier height at $H = 0$, H_0 is the critical field for the magnetization reversal without thermal activation effect, and n is a constant depending on the magnetization reversal process, i.e., $n = 2$ for coherent rotation [17] and $n = 1$ for a weak pinning case [16]. Thus, by substituting Eq. (4) into Eq. (3), we get the relation of

$$S/\chi_{\text{irr}} (\equiv H_{\text{f}}) = k_{\text{B}}T / \left[n\frac{E_0}{H_0}\left(1 - \frac{H}{H_0} \right)^{n-1} \right] \tag{5}$$

Thus, the defined H_{f} is called as the fluctuation field, and is obtained from the separately measured values of S and χ_{irr}. H_{f} can be also obtained only from the magnetic viscosity measurements as [18]

$$H_{\text{f}}(H) = \frac{\Delta H}{\Delta \ln (S/t)}. \tag{6}$$

Givord et al. found that the values of S and χ_{irr} of Nd–Fe–B sintered magnets exhibited the same trends against H [11], indicating that H_{f} is a constant irrespective of H. This fact leads to $n = 1$ in Eqs. (4) and (5). Considering the normal measurement condition of coercivity using VSM, that is, several seconds of data acquisition time for each data point, E corresponds to $25k_{\text{B}}T$. Moreover, assuming the effective magnetization reversal field $H = H_{\text{c}} + N_{\text{eff}}M_{\text{s}}$, Eq. (4) with $n = 1$ is transformed into.

$$H_{\text{c}} = \frac{E_0}{M_{\text{s}}v_{\text{act}}} - N_{\text{eff}}M_{\text{s}} - 25H_{\text{f}}, \tag{7}$$

where $v_{\text{act}} = k_{\text{B}}T/M_{\text{s}}H_{\text{f}}$ is the activation volume. Consequently, the follow equation is given by assuming $E_0 = \alpha\gamma_{\text{w}}v_{\text{act}}^{2/3}$ [12]:

$$H_{\text{c}} = \alpha\frac{\gamma_{\text{w}}}{M_{\text{s}}v_{\text{act}}^{1/3}} - N_{\text{eff}}M_{\text{s}} - 25H_{\text{f}}, \tag{8}$$

where γ_{w} is the domain wall energy. Obviously, Eq. (8) derived from the thermal activation model has the similar form of Eq. (1) proposed by Kromüller.

Note that Eqs. (7) and (8) are only valid for $n = 1$ in Eq. (4). Very recently, Okamoto et al. proposed a more general thermal activation analysis for permanent magnets based on the magnetic viscosity measurements [19]. Magnetic viscosity measurements starting from various magnetic fields just above H_{c} also give the time-dependent coercivity $H_{\text{c}}(t)$, with an example shown in Fig. 3. $H_{\text{c}}(t)$ for the magnet with the energy barrier of Eq. (4) was formulated by Sharrock [20] as.

$$H_{\text{c}}(t) = H_0 \left[1 - \left\{ \frac{k_{\text{B}}T}{E_0} \ln\left(\frac{f_0 t}{\ln 2} \right) \right\}^{1/n} \right], \tag{9}$$

where f_0 is the attempt frequency with the order of 10^9–10^{11} Hz. Here, there are three unknown parameters (H_0, E_0, and n), and these cannot be determined simultaneously only from the experimentally values of $H_{\text{c}}(t)$. However, these three parameters are determined from the analysis by combining $H_{\text{c}}(t)$ and H_{f} [19]. Figure 4 shows examples of the thus obtained values of H_0, E_0, and n of Nd–Fe–B hot-deformed magnets. HD and GBD denote the differently processed magnets

Fig. 3 Example of magnetic viscosity curves of a hot-deformed Nd–Fe–B magnet measured at 200 °C. The intersection with the line of $m = 0$ give the time-dependent coercivity $H_c(t)$

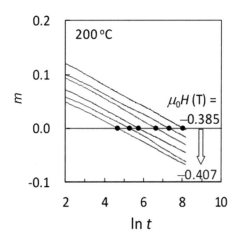

with quite different $\mu_0 H_c$ of 1.1 and 2.2 T, respectively, at ambient temperature. Moreover, these values of $\mu_0 H_c$ decrease significantly with temperature. The values of $\mu_0 H_0$ are well consistent with these different values of $\mu_0 H_c$ of HD and GBD magnets and their large temperature dependences. On the other hand, the values of n are almost 1 irrespective of magnets and temperature. The values of E_0 is also less dependent on magnets and temperature. These experimental results reflect the feature of the magnetization reversal process of Nd–Fe–B hot-deformed magnets, as discussed in the next section.

4 Magnetization Reversal Process of Permanent Magnets

Magnetic Imaging

Direct observation techniques of the magnetization reversal process of permanent magnets are magnetic imaging, such as magneto-optical Kerr (MOKE) and X-ray magnetic circular dichroism (XMCD) microscopies.

MOKE microscopy utilizes the change of optical polarization of the reflected light from the magnet surface. Therefore, the polished mirror surface is indispensable. This means that the MOKE signal reflects the magnetization state of the polished surface of the magnets. In fact, the coercivity obtained from the MOKE measurement is much smaller than that of the magnet measured by VSM [21], indicating that the polished surface layer loses the hard magnetic property due to the mechanical damage. Nevertheless, it is conceivable that the magnetic domain image by the MOKE microscopy reflects, to some extent, the magnetization state underneath the surface-damaged layer because of the strong magneto-static interaction. Figure 5 shows examples of MOKE images of a Nd–Fe–B sintered magnet. The domain wall displacements inside the grains are clearly observed.

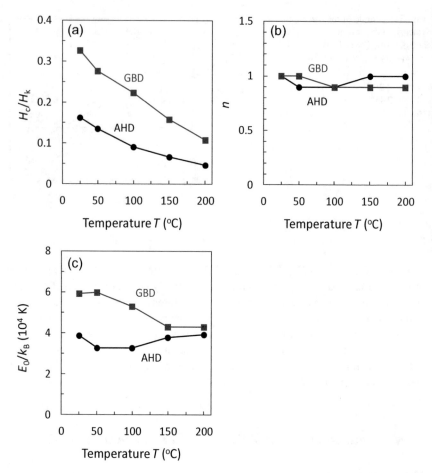

Fig. 4 Energy barrier parameters obtained of a hot-deformed Nd–Fe–B magnets from of thermal activation analysis. (**a**) H_0, (**b**) n, (**c**) E_0 are the parameters of Eq. (4). GBD and HD denote the differently processed magnets with different coercivities

XMCD microscopy is available at synchrotron radiation facilities. Unlike MOKE microscopy, XMCD microscopy does not need the mirror polished surface, because the XMCD signal is obtained from the difference in the X-ray absorption for different helicities. Very recently, Nakamura et al. established the sample fracturing technique inside the XMCD microscopy chamber, which is in ultrahigh vacuum atmosphere [22]. Because Nd–Fe–B sintered magnets favor to fracture at the thin Nd-rich grain boundary phase rather than the $Nd_2Fe_{14}B$ main phase grains, this technique makes it possible to observe the very fresh and non-damaged Nd–Fe–B magnet surface covered with the thin Nd-rich grain boundary phase. Using this technique, they successfully demonstrated that the magnetization curve of a Nd–Fe–B sintered magnet obtained by XMCD agrees very well with that by VSM,

Fig. 5 Example of a MOKE image of a Nd–Fe–B sintered magnet

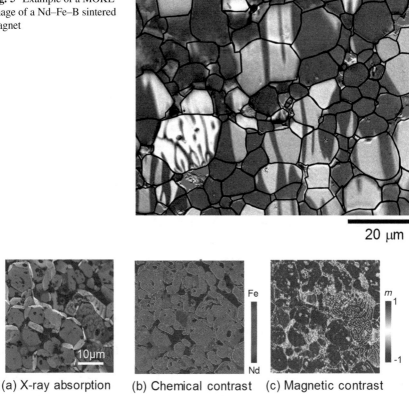

20 μm

(a) X-ray absorption   (b) Chemical contrast   (c) Magnetic contrast

Fig. 6 Examples of multiple images of a Nd–Fe–B sintered magnet using a MOKE microscopy. (**a**) X-ray absorption, (**b**) chemical contrast of Fe and Nd, (**c**) magnetic contrast images. Experiment was performed at BL25SU of SPring-8

indicating that the magnetization state of the fractured surface is almost identical to that of the bulk inside. Moreover, XMCD microscopy makes it possible to obtain multiple images of topographic, chemical, and magnetic contrast, as in the examples shown in Fig. 6. Since Nd–Fe–B sintered magnets consist of various phases other than the $Nd_2Fe_{14}B$ main phase, such as the metallic Nd-rich phase and oxidized Nd as triple junction and grain boundary phases, this multiple imaging of XMCD microscopy is very powerful to reveal the relation of the magnetization reversal process with the microstructure [23].

FORC Analysis

As explained in the preceding section, magnetic imaging is the direct method to visualize the magnetization reversal process of permanent magnets; however, the information is limited to the surfaces of the magnets. In contrast, first-order reversal curve (FORC) analysis is regarded as a method to visualize the magnetization reversal process of bulk magnets, even though it is not in a direct way. In this sense, magnetic imaging and FORC analysis are complementary. Originally, FORC analysis was established to evaluate the coercivity and interaction field dispersions based on the Preisach model [24] (please see Sect. 3.1). The Preisach model assumes a magnet consisting of a large number of hysteresis units known as hysterons. This assumption, however, is inappropriate for many permanent magnets, especially for Nd–Fe–B magnets. As seen in Figs. 5 and 6, because many multi-domain grains are observed in the Nd–Fe–B sintered magnets, the magnetization reversal in Nd–Fe–B sintered magnets cannot be described by the simple assembly of hysterons. In the FORC measurement, many reversal magnetization curves that start from H_r on the demagnetization curve are recorded with H, and then, the magnetization m on each reversal curve is given as a function of H_r and H. The FORC distribution ρ is defined as a second-order derivative of m with respect to H and H_r:

$$\rho (H, H_r) = -\frac{\partial}{\partial H_r} \left(\frac{\partial m}{\partial H} \right). \tag{10}$$

This equation represents the variation of magnetic susceptibility ($\partial m/\partial H$) on H_r, corresponding to the irreversible change in the magnetic susceptibility. Thus, the FORC diagram of permanent magnets can be regarded as the contour map of the irreversible magnetization reversal on the (H, H_r) plane rather than the understanding based on the Preisach model.

Figure 7 shows the FORC diagrams of pillar-shaped Nd–Fe–B sintered [25], Nd–Fe–B hot-deformed [26], and Sm_2Co_{17} magnets [27]. These magnet samples are the same as shown in Fig. 1. The magnetization curves of these magnets shown in Fig. 1 are similar rectangles; however, their FORC diagram patterns are quite different. The FORC diagram pattern of the Nd–Fe–B sintered magnet exhibits two spots at around the low-field and high-field regions, indicating that there are two regions of irreversible magnetization reversals. The high-field spot is easily assigned to the magnetization reversal at around the coercivity. On the other hand, the low-field spot evidences that the nontrivial amount of magnetization reversal occurs at very low-field region in the Nd–Fe–B sintered magnet. This low-field spot has been widely observed in various Nd–Fe–B sintered magnets [28]. According to the recent study on the XMCD microscopy observation under the field sequence for the low-field spot of the FORC diagram, the magnetization reversal for the low-field spot is assigned to the domain wall displacement inside the multi-domain grains [25]. In contrast, the FORC diagrams of the Nd–Fe–B hot-deformed and the Sm_2Co_{17} magnets exhibit the high-field spot only, which corresponds to the magnetization reversal at around the coercivity. These different FORC diagrams clearly reflect the different magnetization reversal process of these magnets. Whereas there are multi-domain grains in Nd–Fe–B sintered magnets in which the domain wall smoothly moves, the domain walls in Nd–Fe–B hot-deformed and the Sm_2Co_{17} magnets cannot move smoothly in the low-field region because of the very high density

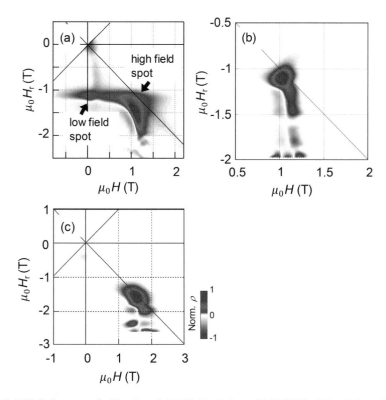

Fig. 7 FORC diagrams of pillar-shaped (**a**) Nd–Fe–B sintered, (**b**) Nd-Fe-B hot-deformed, and (**c**) Sm_2Co_{17} magnets. Samples are the same shown in Fig. 1. (**a**) and (**b**) are measured at RT, and (**c**) is measured at 200 °C

of pinning sites in these magnets. Consequently, the low-field spot in the FORC diagram of these two magnets disappears. Moreover, the positions of these high- and low-field spots and their width give more detailed information on the magnetization reversal process of permanent magnets [29].

5 Summary

Although the study on permanent magnets has very long history, there are still many issues to be studied. Recently, multilateral analyses become very important. The magnetic measurements and analyses explained in this chapter are one approach of them. Moreover, recent advancements of electron microscopy, computer science, synchrotron radiation, and so on are very dramatic. By combining with these cutting-edge technologies, it is highly expected that the magnetic measurements and analyses for permanent magnets will move into new stages. As explained in

Sect. 4.2, FORC analysis combined with XMCD microscopy is one example. This kind of evolution will not only deepen our understandings but also open new fields of studies on permanent magnets.

References

1. J.M.D. Coey, Hard magnetic materials: A perspective. IEEE Trans. Magn. **47**, 4671–4681 (2011). https://doi.org/10.1109/TMAG.2011.2166975
2. S. Sugimoto, Current status and recent topics of rare-earth permanent magnets. J. Phys. D. Appl. Phys. **44**, 064001 (2011). https://doi.org/10.1088/0022-3727/44/6/064001
3. S. Hirosawa, Current status of research and development toward permanent magnets free from critical elements. J. Magn. Soc. Jpn. **39**, 85–95 (2015). https://doi.org/10.3379/msjmag.1504R004
4. Yamamoto H (2019) Basics of permanent magnet and its magnetic measurement. NEODI-CONSUL (in Japanese)
5. J.D. Livingston, Magnetic domains in sintered Fe-Nd-B magnets. J. Appl. Phys. **57**, 4137–4139 (1985). https://doi.org/10.1063/1.334644
6. T. Yomogita et al., First-order reversal curve analysis in hot-deformed Nd–Fe–B magnets, in *Proc. 24th Inter. Workshop on Rare-Earth and Future Permanent Magnets and Their Applications (REPM)*, vol. 2016, (2016), pp. 649–653
7. T. Fukagawa, S. Hirosawa, Coercivity generation of surface $Nd_2Fe_{14}B$ grains and mechanism of fcc-phase formation at the $Nd/Nd_2Fe_{14}B$ interface in Nd-sputtered Nd–Fe–B sintered magnets. J. Appl. Phys. **104**, 013911 (2008). https://doi.org/10.1063/1.2952556
8. H. Kronmüller, Theory of nucleation fields in inhomogeneous ferromagnets. Phys. Stat. Sol. (b) **144**, 385–396 (1987). https://doi.org/10.1002/pssb.2221440134
9. H. Kronmüller, H.-D. Durst, Analysis of the magnetic hardening mechanism in RE-FeB permanent magnets. J. Magn. Magn. Mater. **74**, 291–302 (1988). https://doi.org/10.1016/0304-8853(88)90202-8
10. W.F. Brown, Virtues and weaknesses of the domain concept. Rev. Mod. Phys. **17**, 15–19 (1945). https://doi.org/10.1103/RevModPhys.17.159
11. D. Givord et al., Magnetic viscosity in different NdFeB magnets. J. Appl. Phys. **61**, 3454 (1987). https://doi.org/10.1063/1.338751
12. D. Givord et al., Experimental approach to coercivity analysis in hard magnetic materials. J. Magn. Magn. Mater. **83**, 183–188 (1990). https://doi.org/10.1016/0304-8853(90)90479-A
13. D. Givord, The physics of coercivity. J. Magn. Magn. Mater. **258–259**, 1–5 (2003). https://doi.org/10.1016/S0304-8853(02)00988-5
14. R. Street, J.C. Woolley, A study of magnetic viscosity. Proc. Phys. Soc. A **62**, 562–572 (1949)
15. E.P. Wohlfarth, The coefficient of magnetic viscosity. J. Phys. F **14**, L155–L159 (1984)
16. P. Gaunt, Magnetic viscosity and thermal activation energy. J. Appl. Phys. **59**, 4129–4132 (1986). https://doi.org/10.1063/1.336671
17. E.C. Stoner, E.P. Wohlfarth, A mechanism of magnetic hysteresis in heterogeneous alloys. Philos. Trans. R. Soc. Lond. Ser. A **240**, 599–642 (1948). https://doi.org/10.1098/rsta.1948.0007
18. M. El-Hilo et al., Fluctuation fields and reversal mechanisms in granular magnetic systems. J. Magn. Magn. Mater. **248**, 360–373 (2002). https://doi.org/10.1016/S0304-8853(02)00146-4
19. S. Okamoto, Temperature-dependent magnetization reversal process and coercivity mechanism in Nd–Fe–B hot-deformed magnets. J. Appl. Phys. **118**, 223903 (2015). https://doi.org/10.1063/1.4937274
20. M.P. Sharrock, Time dependence of switching fields in magnetic recording media. J. Appl. Phys. **76**, 6413–6418 (1994). https://doi.org/10.1063/1.358282

21. M. Takezawa, Analysis of the demagnetization process of Nd–Fe–B sintered magnets at elevated temperatures by magnetic domain observation using a Kerr microscope. J. Appl. Phys. **115**, 17A733 (2014). https://doi.org/10.1063/1.4866894
22. Okamoto S unpublished data
23. T. Nakamura et al., Direct observation of ferromagnetism in grain boundary phase of Nd–Fe–B sintered magnet using soft x-ray magnetic circular dichroism. Appl. Phys. Lett. **105**, 202404 (2014). https://doi.org/10.1063/1.4902329
24. D. Billington et al., Unmasking the interior magnetic domain structure and evolution in Nd–Fe–B sintered magnets through high-field magnetic imaging of the fractured surface. Phys. Rev. Mater. **2**, 104413 (2018). https://doi.org/10.1103/PhysRevMaterials.2.104413
25. I.D. Mayergoyz, Mathematical models of hysteresis. IEEE Trans. Magn. **MAG-22**, 603–608 (1986). https://doi.org/10.1109/TMAG.1986.1064347
26. K. Miyazawa et al., First-order reversal curve analysis of a Nd–Fe–B sintered magnet with soft X-ray magnetic circular dichroism microscopy. Acta Mater. **162**, 1–9 (2019). https://doi.org/10.1016/j.actamat.2018.09.053
27. T. Yomogita et al., Temperature and field direction dependences of first-order reversal curve (FORC) diagrams of hot-deformed Nd–Fe–B magnets. J. Magn. Magn. Mater. **447**, 110–115 (2018). https://doi.org/10.1016/j.jmmm.2017.09.072
28. T. Schrefl et al., First order reversal curve studies of permanent magnets. J. Appl. Phys. **111**, 07A728 (2012). https://doi.org/10.1063/1.3678434
29. S. Okamoto et al., Temperature dependent magnetization reversal process of a Ga-doped Nd–Fe–B sintered magnet based on first-order reversal curve analysis. Acta Mater. **178**, 90–98 (2019). https://doi.org/10.1016/j.actamat.2019.08.004

Magnetocaloric Characterization of Materials

Victorino Franco (iD)

Abstract Magnetocaloric characterization is most typically considered for materials that will be applied in thermomagnetic energy conversion devices, but it can also be of relevance for any material that exhibits a thermomagnetic phase transition. This chapter overviews the most usual methods for magnetocaloric characterization, including direct and indirect methods, and shows the appropriate choice of measurement protocols depending on the type of magnetocaloric material under study. In addition, it gives an overview on how magnetocaloric characterization can be used to study phase transitions, including the determination of critical exponents in cases that the conventional methods are not applicable, the quantitative determination of the order (first or second) of thermomagnetic phase transitions, and the emerging use of TFORC techniques for the study of these transitions.

Keywords Magnetocaloric effect · Magnetic refrigeration · Thermomagnetic phase transitions · Critical phenomena · Critical exponents · Universal scaling · Temperature first order reversal curves (TFORC)

1 Introduction

Climate at different points on Earth is rather different, ranging from extremely hot to extremely cold. However, we humans want to carry out our lifestyle with minimum variations regardless of where we are located. This implies the necessity of temperature control systems that allow us to live comfortably in challenging environments and refrigeration systems to keep our food in proper conditions, just to mention a few examples. By the time this book was being written, the general public became aware of how critical is the cold transportation and storage of medical products, as the first viable vaccine against a coronavirus that was severely altering our way of life required transportation at temperatures well below room

V. Franco (✉)
Department of Condensed Matter Physics, University of Seville, Seville, Spain
e-mail: vfranco@us.es

© Springer Nature Switzerland AG 2021
V. Franco, B. Dodrill (eds.), *Magnetic Measurement Techniques for Materials Characterization*, https://doi.org/10.1007/978-3-030-70443-8_23

temperature. This constituted a major challenge in order to be able to reach most of the population and is an example of how important the search for reliable and efficient refrigeration systems is for our society.

Among the different final uses of energy in residential and commercial sectors, refrigeration and air conditioning account for a relevant fraction of electricity use, with numbers varying from country to country due to their different climate. Recent data from EIA [1] estimate that 87% of US households are furnished with air conditioners and ~114 million units account for an annual energy demand of 186 billion kWh of electricity. There is no doubt that our way of life relies on our capability of cooling food and controlling the temperature of our living and working environments. If developing countries adopt similar trends in cooling habits, there could be a 50 times increase in the demand of air conditioners [2]. In the European Union, heating and cooling in buildings and industry account for half of the EU's energy consumption and in order to fulfil the EU's climate and energy goals, the heating and cooling sector must drastically reduce their energy consumption and decrease their use of fossil fuels [3].

In fact, magnetic refrigeration has increased in popularity in recent years [4]. This has been due to the discovery of magnetic materials that exhibit a giant change in their temperature when the applied magnetic field is changed adiabatically close to room temperature [5]. This has been accompanied by a noticeable increase in the development of magnetic refrigerator prototypes [4, 6]. A magnetocaloric material can be used in an analogous way as the gas that is compressed and expanded in a conventional refrigerator just by performing cycles of magnetization and demagnetization (Fig. 1). The environmental friendliness of magnetic refrigeration, which is energetically more efficient than the conventional refrigerator devices and does not require the use of ozone depleting or greenhouse effect related gases, is a strong driving force to advance in this research area and can help us find solutions to the current and future challenges that we face in making our life sustainable.

Since the discovery of the giant magnetocaloric effect in $Gd_5(Si,Ge)_4$ in 1997 [5], magnetic refrigeration is considered a "promising alternative" for temperature control. There have been numerous publications in scientific journals and major research agencies have funded projects on this topic. Industry has also been actively involved in this research, with several relevant prototypes that serve some niche applications. However, despite the broad variety of materials reported in the literature, this "promising alternative" has not materialized in the consumer market yet. The reason for this delay in reaching the market has different aspects, but probably the key limitations are related to the performance of the magnetocaloric materials. The aim of this chapter is to present what we consider the most relevant aspects for a correct characterization of magnetocaloric materials so that results can be appropriately compared when data are provided by different labs and measured different experimental conditions. In addition to this technological aspect of magnetocaloric materials, we will show some examples of applications of the magnetocaloric effect beyond refrigeration and temperature control: the study of phase transitions and critical phenomena, which can be performed quite efficiently via the properly characterized magnetocaloric response of the materials.

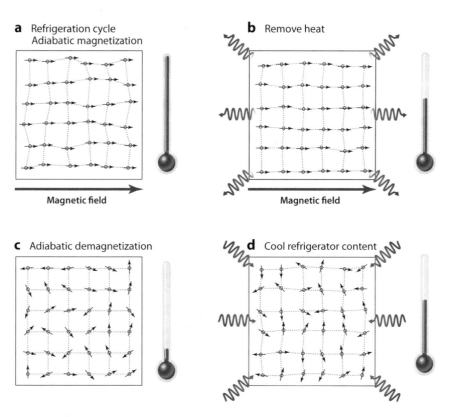

Fig. 1 The four stages of a magnetic refrigeration cycle: (**a**) adiabatic magnetization, (**b**) removal of heat, (**c**) adiabatic demagnetization, and (**d**) cool refrigerator contents. Reprinted from [7] with permission from Annual Reviews

2 Types of Magnetocaloric Materials

The first approach to classify magnetocaloric materials is according to their composition. Different families of alloys and compounds are formed, and properties are studied depending on doping of the starting material. However, the different families can have similar properties, making the classification more confusing [4, 7]. Alternatively, magnetocaloric materials can be classified according to the order of the phase transition that they undergo, either first- (FOPT) or second-order (SOPT) type phase transitions. Both types of transitions have their advantages and disadvantages, with the former having large magnetocaloric response, but usually at the expense of thermal hysteresis, rate-dependent behavior, and a modest cyclic performance. SOPT materials, on the other hand, have a more modest magnetocaloric response, but they do not suffer from thermal hysteresis. Therefore, one of the current goals of magnetocaloric materials research is to combine the best of both types of materials: large response without hysteresis and a trade-off between static and cyclic performance. A recent example of that trade-off is the study of high entropy alloys (HEA) for magnetocaloric applications. While conventional rare-earth-free HEA have a modest performance, a thoughtful design of the composition to incorporate a magneto-structural phase transition allows an increase of one order of magnitude in the magnetocaloric response [8].

There have been several attempts to develop series of alloys and compounds that bring us to this intermediate point where the transition changes from FOPT to SOPT, known as tricritical point or critical point of the second-order phase transition. La(Fe,Si)$_{13}$ [9, 10] and MnFePSi [11] alloy series are examples of a gradual change from FOPT to SOPT. For these types of studies, it is important to have a quantitative procedure to determine the order of the phase transition based on magnetic measurements [12], as some of the previously used methods are either non-general [13, 14] or qualitative [15–18].

3 Relevant Magnitudes for the Characterization of Magnetocaloric Materials

As indicated above, the magnetocaloric effect is connected to the temperature change of the sample due to a change in its entropy. Therefore, the characterization of a magnetocaloric material can be made by determining either the magnitude of the adiabatic temperature change, ΔT_{ad}, or the isothermal entropy change, ΔS_T. However, if the magnitudes of both ΔT_{ad} and ΔS_T are known and combined in some new related metrics, then a better estimate about the applicability of the material for magnetic refrigeration can be determined.

Isothermal Entropy Change

Arguably, the most used magnitude for the characterization of magnetocaloric materials is the isothermal entropy change induced by the application of a magnetic field. The reason for its extensive appearance in the literature is that it can be obtained by numerically processing properly measured magnetization curves as a function of temperature and magnetic field. By using Maxwell's relation, the temperature and field dependence of ΔS_T can be obtained as:

$$\Delta S_T (T, H_{max}) = \mu_0 \int_0^{H_{max}} \left(\frac{\partial M}{\partial T} \right)_H dH, \tag{1}$$

where H_{max} is the maximum field used for the calculations, and it is assumed that the magnetization process starts form 0 field. From this expression it is evident that, in order to exhibit a large isothermal entropy change, the material should have an abrupt change of magnetization as a function of temperature. Since the susceptibility of paramagnetic salts diverges when approaching 0 K, they are suitable as magnetocaloric materials in the cryogenic regime. In contrast, room temperature applications require a phase transition that causes this strong temperature dependence of magnetization near the desired working temperature of the device. These phase transitions can be purely magnetic, like the Curie temperature of Gd, magnetoelastic like for $La(Fe,Si)_{13}$, or magneto-structural like the case of $Gd_5(Si,Ge)_4$. In this latter case, a low temperature orthorhombic phase transforms into a monoclinic phase at temperatures close to room temperature, providing the required magnetization change.

Adiabatic Temperature Change

The adiabatic temperature change can be measured directly by applying a magnetic field to the sample in adiabatic conditions. It can also be related to other magnitudes, like the temperature and field dependence of the specific heat of the sample, c_H, and the temperature and field dependence of magnetization:

$$\Delta T_{ad} = \mu_0 \int_0^{H_{max}} \frac{T}{c_H} \left(\frac{\partial M}{\partial T} \right)_H dH. \tag{2}$$

In many publications, however, it is considered that c_H is field-independent, leading to an approximation that relates the adiabatic temperature change and the isothermal entropy change via the zero field specific heat:

$$\Delta T_{ad} \approx \mu_0 \frac{T}{c_p} \int_0^{H_{max}} \left(\frac{\partial M}{\partial T} \right)_H dH = \frac{T \Delta S_T}{c_p}. \tag{3}$$

This means that although a large ΔS_T is necessary to achieve a large ΔT_{ad}, it is not a sufficient condition to realize an excellent magnetocaloric material because heat capacity also plays a very relevant role.

Refrigerant Capacity and Its Variants

In a refrigeration device, the main goal is to transfer heat between the hot and cold reservoirs. This can be quantified by the refrigerant capacity, RC:

$$RC(\Delta H) = \int_{T_{cold}}^{T_{hot}} \Delta S_T(T, \Delta H) \, dT, \tag{4}$$

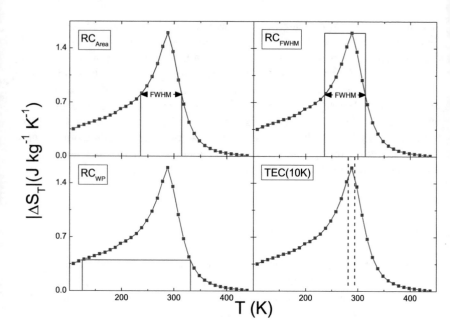

Fig. 2 Isothermal entropy change of a typical transition metal based amorphous alloy for which the three commonly accepted definitions of *RC* are applied, where *RC* corresponds to the area enclosed by the blue line. The difference among them is due to the approximations performed for the calculation of the integral in Eq. (4). In the case of *TEC* (10 K), dashed lines correspond to $T_{mid} \pm 5$ K

where ΔH corresponds to the change of magnetic field during the process, typically between 0 and H_{max}, and T_{cold} and T_{hot} are the temperatures of the two reservoirs. However, the characterization of magnetocaloric materials is generally not done in a refrigerator device in contact with two reservoirs. In order to extract the characteristics of the material, it has been broadly assumed that the temperatures of the reservoirs correspond to the full width at half maximum of the magnetic entropy change peak, δT_{FWHM}. While this is a rather useful simplification, it can lead to the overestimation of the suitability of magnetocaloric materials with a very low magnetic entropy change peak that is extended over a rather broad temperature interval. This broad peak would produce artificially large *RC* values that would not be suitable for applications. It is currently accepted that the usable temperature range of the material is restricted to the region where its ΔT_{ad} is above 1 K.

In the literature, different authors use different ways of calculating the integral in Eq. (4). An illustration of the most extended variants found in the literature is presented in Fig. 2. RC_{FWHM} corresponds to taking the full width at half maximum of the peak for the integration limits and approximating the integral as the peak value times that width ($RC_{FWHM} = \Delta S_T^{pk} \delta T_{FWHM}$). This magnitude is also known as relative cooling power (*RCP*) [19]. RC_{Area} corresponds to properly calculating the area under the peak within that δT_{FWHM} temperature range. The third definition, proposed by Wood and Potter [20], RC_{WP}, corresponds to the area of the largest rectangle which can be inscribed into the $\Delta S_T(T)$ curve. In all cases, when hysteretic losses are relevant for the studied material, the subtraction of these losses from the calculated RC would be necessary [21].

To overcome the limitations of RC, alternative figures of merit have been recently proposed. The coefficient of refrigerant performance, CRP compares the net work in a reversible cycle with the positive work on the refrigerant [22]:

$$CRP\,(H_{max}) = \frac{\Delta S_T \Delta T_{rev}}{\mu_0 \int_0^{H_{max}} M\,\mathrm{d}H},\qquad(5)$$

This expression is a ratio of the reversible refrigerant capacity divided by the magnetic work performed on the magnetic refrigerant, and, therefore, the field dependence of the magnetization curve at the transition temperature is needed. As only the reversible response is considered, hysteresis is eliminated from this definition.

The most recent addition to these figures of merit is the temperature averaged entropy change (TEC) [23]:

$$TEC\,(\Delta T_{lift}) = \frac{1}{\Delta T_{lift}} \max_{T_{mid}} \left\{ \int_{T_{mid}-\frac{\Delta T_{lift}}{2}}^{T_{mid}+\frac{\Delta T_{lift}}{2}} \Delta S_T(T)\,\mathrm{d}T \right\},\qquad(6)$$

where ΔT_{lift} corresponds to either the temperature span of a single layer in a multi-material refrigerator bed (typically around 3 K), or to the largest adiabatic temperature change obtained for a magnetocaloric material by applying the field using permanent magnets (typically around 10 K). T_{mid} is the center of that temperature span that maximizes TEC; depending on the shape of the magnetic entropy change curve of the material, T_{mid} might not be symmetrically positioned around the peak of entropy change. The advantage of TEC vs. CRP is that it only requires the measurement of isothermal entropy change curves, simplifying the initial screening of the potential performance of the materials.

Most papers in the literature do not indicate the value of TEC, making it difficult to calculate its value from the published curves, as it would require access to the data or digitizing the $\Delta S_T(T)$ curves from the publication. A simplification that allows estimating a value for TEC can be made, assuming that the midpoint of the TEC definition (Eq. 6) lies at the temperature of the peak:

$$TEC\,(\Delta T_{lift}) \approx \frac{\Delta S_T\left(T_{pk} - \frac{\Delta T_{lift}}{2}\right) + \Delta S_T\left(T_{pk}\right) + \Delta S_T\left(T_{pk} + \frac{\Delta T_{lift}}{2}\right)}{3}\qquad(7)$$

While this is just an approximation, it facilitates the calculation of this magnitude for previously published data using a limited number of points.

All these figures of merit will have their own limitations, and the evaluation of the potential applicability of any material should be put in the context of the operating conditions in which it would be used. There are additional requisites for a magnetocaloric material beyond its magnetocaloric response to be practically useful, like mechanical integrity, appropriate thermal conductivity, large electrical resistivity, machinability/formability to prepare it in the desired shape for a refrigerator bed, etc. Most of them are not considered in the search for new materials, at least not at the first stages. The ultimate check would be to implement the material into a real magnetic refrigerator and assess its eventual performance. However, this is not within reach for many laboratories working on magnetic materials.

4   Indirect Characterization Methods

The most extensively used way of characterizing magnetocaloric materials is by calculating ΔS_T or ΔT_{ad} from other experimentally measured magnitudes, either magnetization or heat capacity. The popularity of these indirect methods is due to the broad availability of commercial instruments that allow performing the measurements. This allows us to classify indirect characterization methods in those that either use magnetometry or calorimetry.

Magnetometry

A simple inspection into the literature of magnetocaloric materials leads us to realize that most papers use magnetometry to determine the isothermal entropy change. By measuring the temperature and field dependence of magnetization and applying Maxwell's relation (Eq. 1), it is relatively straightforward to determine ΔS_T as a function of temperature (Fig. 3). If the integration is performed up to different maximum fields within the experimental range, the field dependence of the isothermal entropy change can also be obtained.

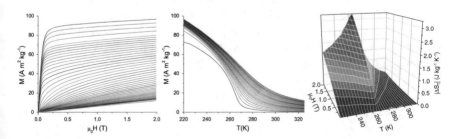

Fig. 3 Field and temperature dependence of magnetization of a $Gd_{50}Zn_{50}$ alloy and $|\Delta S_T|$ calculated from these data using Maxwell's relation (Eq. 1). Further details of the alloy can be found in [24]

We must consider, however, that in the case of FOPT materials, there exists a discontinuity of magnetization at the transition temperature, causing a divergency in the derivative of magnetization vs. temperature. In this case, the Clausius Clapeyron equation can be used instead. For some years, this led to the belief that the notable artifacts in the $\Delta S_T(T)$ curves that occurred near the FOPT were due to the inapplicability of Maxwell's relation. However, the ultimate reason for these spurious spikes that overestimate ΔS_T for FOPT is the coexistence of phases during the measurement [25, 26] and artifacts can be avoided if the measurement protocol incorporates a procedure to "erase the memory" of the material in between isothermal measurements. By following these discontinuous procedures, Maxwell's relation can still be used to determine the entropy change. If the memory of the material is not erased between magnetization curves, the spurious spikes would occur in the temperature region where the material exhibits thermal hysteresis.

In the case of measurement of $M(T)$ curves for different applied fields, the only precaution that has to be taken is that the initial state for each curve has to be reached always in the same way. For isothermal $M(H)$ measurements, the correct measurement protocol implies the following steps [27]:

1. Change temperature to reach the fully transformed condition of the sample (either in zero field or with the maximum applied field).
2. Keeping the same applied field (zero or maximum), reach the temperature at which the magnetization curve is going to be measured, T_{meas}.
3. Measure the isothermal magnetization curve with increasing (if steps 1 and 2 were performed in zero field) or decreasing field (in the case that those steps were performed with maximum field).
4. Repeat steps 1–3 for as many different T_{meas} as desired in the $\Delta S_T(T)$ curve.

Once the set of $M(T,H)$ data are collected, numerically processing the data using Eq. 1 requires some interpolation of the curves, to distribute the experimental points on a grid of suitable T and H values that allow correct calculation of the numerically approximated integrals and derivatives. Nowadays there are software packages that automatically make this data processing [28].

Magnetometry measurements are suitable for a large variety of samples and current devices are able to measure not only bulk materials but also powders, thin films, liquids, etc. This contributes to the versatility of this characterization procedure.

As numerical derivatives as a function of temperature are involved, it is important to find a trade-off between the resolution required for the $\Delta S_T(T)$ curves, which implies small temperature increments, and the level of noise in the curves emerging from the derivatives, which can be minimized with larger temperature increments.

An additional precaution that has to be taken during magnetometric measurements concerns the ramp rate of the applied field (for isothermal measurements) or of the temperature (for isofield measurements). In the case of isofield magnetization curves, ramping the temperature too fast would induce an apparent lag in the measurements that could be misinterpreted as hysteresis in the sample when heating and cooling it, leading to potentially erroneous conclusions about the order of the phase transition. The case of isothermal magnetization measurements is more subtle. Researchers usually tend to ramp field at a fast rate in order to save experimental time. However, we need to recall that the magnetocaloric effect consists in the temperature change of the sample when it is submitted to an adiabatic field change. Therefore, if the field is ramped too quickly, the temperature of the sample will not remain in isothermal conditions, distorting the measured results. The consideration of how fast is too fast depends on the characteristics of the sample and its coupling to the sample holder, especially in the temperature range close to the phase transition [29, 30].

This ramp rate limitation poses a difficulty for the dynamic characterization of the magnetocaloric response. As the isothermal entropy change is determined from quasistatic magnetic measurements, the rate-dependent response of the material when used in a refrigerator cannot be experimentally measured in this way. The cyclic response can be, nevertheless, predicted by calculating the overlap of the heating and cooling entropy change curves [27].

Demagnetizing Field

The influence of the demagnetizing field deserves a separate mention, as it will affect all experiments involving the application of an external field. Actually, all the equations above refer to the internal field experienced by the sample, which will only coincide with the applied magnetic field in the case of toroidal samples magnetized along the circumferential direction. In any other case, the internal field experienced by the sample is related to the applied field H_a via the demagnetizing field H_d:

$$H = H_a - H_d, \tag{8}$$

which in turn depends on the magnetization of the sample through the demagnetizing factor N:

$$H_d = N \cdot M. \tag{9}$$

For weakly magnetic samples, like paramagnets, the value of the demagnetizing field is irrelevant due to the small value of magnetization. However, the internal and applied fields are rather different in ferromagnets, producing a relevant modification of the $M(H)$ curves when compared to the $M(H_a)$ ones. Curves will not be affected for cases where $M = 0$, i.e., at the coercive field, or when the sample is fully saturated, provided that the internal field also produces saturation.

As a consequence, for samples measured up to large enough magnetic fields, the peak isothermal entropy change is not significantly affected, producing only minor changes in the peak entropy change [31]. However, the field dependence of the magnetocaloric response is notably affected, especially in the low-field regime, as evidenced both experimentally [32] and by numerical simulations [31]. In the case of field-driven FOPT materials, an additional constraint has to be taken into account: the internal magnetic field has to be large enough to drive the transition, which can significantly increase the requisite of applied field for not optimally designed sample shapes.

In order to minimize the influence of the demagnetizing field, samples should be prepared as thin plates magnetized in plane or as long wires magnetized along the wire axis. Numerical correction of the demagnetizing field effect would be possible in single-phase, single-piece materials [33, 34], although it is rather complicated in the case of composites and coexisting phases as N will no longer depend exclusively on the external shape of the sample but will be a combination of external shape and shape of the different phases [35, 36].

Calorimetry

The measurement of heat capacity as a function of temperature and magnetic field is an alternative procedure that leads to the indirect determination of the magnetocaloric response of materials. There are several relevant differences with respect to the magnetometric methods, as summarized in Table 1.

Although calorimeters are broadly available in laboratories dealing with Materials Science, most of these devices are limited to measuring the temperature dependence of heat capacity in the absence of a magnetic field. With these, the magnetocaloric response of a material cannot be calculated, although they can be helpful, in combination with magnetometric measurements, to estimate $\Delta T_{ad}(T)$ via Eq. (3).

Table 1 Comparison between direct and indirect characterization techniques

	Indirect		Direct
	Magnetometry	Calorimetry	
Magnitude	$\Delta S_T(T)$	$\Delta S_T(T)$ and $\Delta T_{ad}(T)$	$\Delta T_{ad}(T)$
Availability	In most magnetic labs	Less usual with magnetic field	Traditionally rare; increasing in recent years
Commercial devices	Many	Some but with specific measurement limitations	Very few
Sample size	Very small samples	Relatively small samples	Relatively large (depending on sensing technique)
Cyclic response	Via mathematical processing	Via mathematical processing	Experimentally
Dynamic response	Not possible	Not possible	Experimentally

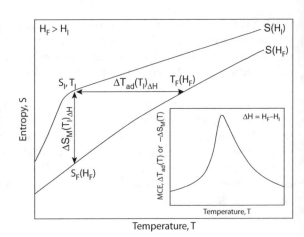

Fig. 4 Magnetocaloric effect as the adiabatic temperature rise, $\Delta T_{\text{ad}}(T_I)_{\Delta H}$, and the magnetic entropy change, $\Delta S_{\text{M}}(T_I)_{\Delta H}$ shown for a single initial temperature T_I and a given magnetic field change $\Delta H = H_F - H_I$ in a ferromagnetic material. The inset displays MCE typical for a ferromagnet as a function of temperature. Reprinted from [43] with the permission of AIP Publishing

Among those calorimeters that allow the application of a magnetic field [37–41], measurements can be made either using a heat pulse technique or by slowly scanning temperature. While for materials with SOPT, there is no relevant difference in the results, the abrupt transformations of FOPT materials are better measured in scanning mode. Additional precautions have to be taken when using relaxation calorimeters for FOPT as the usual time constant fitting approach is not applicable [42]. At the same time, the application of the heat pulse will partially transform the material, while the measured relaxation curve does not necessarily correspond to the reversal process. This should also be taken into account in the analysis of the raw calorimetric data.

The principle of calculating the magnetocaloric response relies on the determination of the total entropy of the material as a function of temperature for different magnetic fields [43]:

$$S_{\text{H}}(T) = \int_0^T \frac{c_{\text{H}}}{T} \, dT + S_{0,H} \tag{10}$$

where $S_{0,H}$ is the zero-temperature entropy term for that given value of magnetic field H. Once these curves are calculated, the isothermal entropy change caused by an increment of field from 0 to H_{max} is calculated by (Fig. 4):

$$\Delta S_{\text{T}}(T) = \left[S_{H_{\text{max}}}(T) - S_0(T) \right]_T, \tag{11}$$

where we can consider that the applied field does not affect the zero-temperature entropy.

To calculate the adiabatic temperature change, it is necessary to numerically invert the experimental $S_{\text{H}}(T)$ curves to obtain $T_{\text{H}}(S)$ curves. Following an analogous procedure to Eq. (11), ΔT_{ad} is calculated as (Fig. 4):

$$\Delta T_{\text{ad}}(T) = \left[T_{H_{\text{max}}}(S) - T_0(S) \right]_S. \tag{12}$$

This apparently simple procedure has a relevant limitation: the integration in Eq. (10) has to be made from 0 K up to the desired temperature. For materials with phase transitions in the cryogenic range, the initial temperature can reasonably approach 0 K without a large increase in the experimental time and associated cost. However, when the transition occurs close to room temperature, making calorimetric measurements down to very low temperatures is an experimental burden. It has become usual to make a linear extrapolation of the integrand down to 0 K, simplifying Eq. (10) into

$$S_H(T) \approx \frac{1}{2} c_H(T_{ini}) + \int_{T_{ini}}^{T} \frac{c_H}{T} dT, \tag{13}$$

where the zero entropy term has been disregarded and T_{ini} corresponds to the lowest experimental temperature used for the integration. While this seemed to be a reasonable approximation that would not produce large errors if T_{ini} is low enough, it has been recently shown [44] that it significantly affects the results, both at the peak of the magnetocaloric response and even at the high temperature tails (Figs. 5 and 6). In the case of ΔS_T, the error introduced by this linear extrapolation is a temperature independent offset. However, for ΔT_{ad} the extrapolation implies an error that increases with temperature.

Despite this apparently serious limitation of the calculation of ΔS_T and ΔT_{ad} from calorimetric measurements, it has also been proven that it is not necessary to reach very low temperatures to avoid this extrapolation error: there is an optimal initial temperature to perform the integration that provides the same result as the integration from 0 K. This optimal temperature can be identified experimentally and is the same for the ΔS_T and ΔT_{ad} curves. The procedure is equally valid for FOPT and SOPT [44, 45] and constitutes a way for reducing both experimental time and cost of measurements.

In the case of rather small samples, microelectromechanical systems (MEMS) nanocalorimeters can be used [46]. In them, a heater and a temperature sensor are deposited onto a SiN membrane, reducing the thermal mass of the sensor. This also allows ultrahigh heating rates and AC characterization. Although these nanocalorimeters are very useful to determine transition temperatures, quantitative measurements are challenging in the case of very small samples due to the difficulty of separating the contribution of the sample from the addenda used to attach the sample to the sensor.

5 Direct Characterization Methods

Following the definition of the magnetocaloric effect, direct measurement methods consist of recording the temperature change of the sample upon application/removal of a magnetic field in adiabatic conditions. The main advantage of this kind of procedure is that it can mimic the operating conditions of magnetocaloric devices,

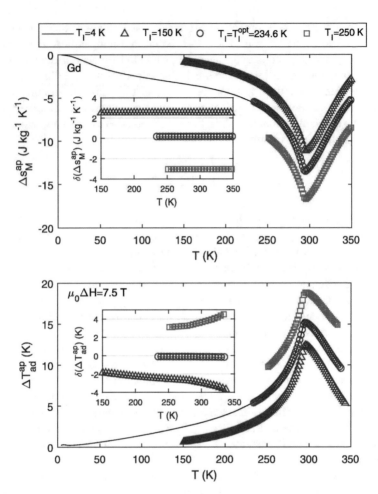

Fig. 5 Approximated magnetic entropy change, Δs_M^{ap}, (upper panel) and adiabatic temperature change, ΔT_{ad}^{ap}, (lower panel) vs. temperature curves at 7.5 T for different initial temperatures (T_{ini}) for pure Gd. The insets show the temperature dependence of the difference of each approximated curve with respect to the curve integrated from 4 K (δ). Reprinted from [45] with permission from Elsevier

being able to experimentally characterize dynamic and cyclic responses (in contrast with the restriction to a calculated cyclic response of the indirect methods [27]).

Despite its apparent simplicity, experimental implementation of these types of methods is not free from difficulties. The most important ones are the thermal mass of the temperature sensor, the adiabaticity of the experimental conditions and the characteristic thermal response time of the temperature sensor. In the case of small samples, the thermal mass of the sensor should be much less than the thermal mass of the sample. This is challenging in the case of ribbon samples and requires piling

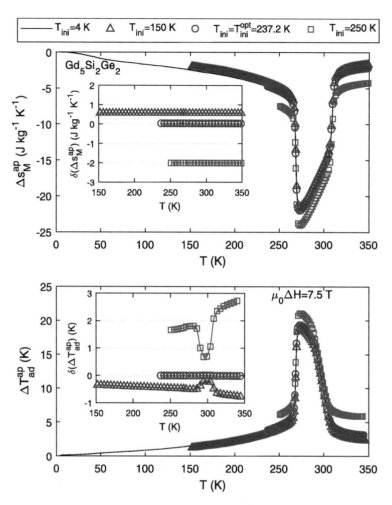

Fig. 6 Approximated magnetic entropy change, Δs_M^{ap}, (upper panel) and adiabatic temperature change, ΔT_{ad}^{ap}, (lower panel) vs. temperature curves at 7.5 T for different initial temperatures (T_{ini}) for $Gd_5Si_2Ge_2$. The insets show the temperature dependence of the difference of each approximated curve with respect to the curve integrated from 4 K (δ). Reprinted from [45] with permission from Elsevier

up several samples to ensure that the thermal mass of the sensor is comparatively small and has to be accompanied by recalibration [47].

There are different ways to apply the changing magnetic field. It can be either by actually changing the value of the applied magnetic field (with an electromagnet, superconducting magnet, pulse magnetizer or with a Halbach array) or by reciprocating the sample in and out of the magnetic field region. The temperature change of the sample can be made either with contact sensors (e.g., thermocouples, CERNOX™ resistance thermometers, etc.) or noncontact sensors (e.g., the thermoacoustic method, infrared thermometry, etc.). The reader is referred to [48] for an overview of the most traditional methods.

The few commercial measurement systems that exist nowadays usually consist of a Halbach permanent magnet field source (typically applying fields between 1.5 and 2 T), a ΔT measuring system that consists of a differential thermocouple, with one of the junctions in contact with the sample and the other at the sample holder, and a temperature control system that sets the base temperature of the sample and sample holder before the application of the field. Varying the ramp rate of the field, it is possible to determine the frequency dependence of the magnetocaloric response. Recording the temperature variation of the sample as the field is applied allows accurate determination of the field dependence of ΔT_{ad}, which does not have to necessarily correspond to a mean field behavior [49].

It is important to note that there is an intrinsic lag in the measurement system due to the thermal mass and time response of the sensor. Therefore, it is necessary to take this into account in order to compare the differences in AC performance of different materials [50, 51]. For cases in which the response might be small (due to the characteristics of the sample or because of the application of a small magnetic field), a lock-in technique can be implemented to enhance the signal-to-noise ratio [52].

Magnetocaloric measurements in pulsed magnetic fields imply additional challenges for the reliable determination of the sample temperature change because a 50 T magnetic field can be reached in typically 250 ms. One possible solution is to develop calorimeters by depositing a ~ 10 nm thick heater on one side of the sample and using bare-chip resistive thermometers [53]. The use of a highly effective impedance system to generate a constant AC current during the pulse and a four contact AC method with a digital lock-in system for the characterization of resistance of the thermometer allows for the determination of ΔT_{ad} of $Sr_3Cr_2O_8$ samples. In the case of metallic samples, their large conductivity makes the deposited film heater less appropriate. It is also possible to fabricate a resistive film thermometer (typically a 100 nm thick patterned Au_xGe_{1-x} alloy film) directly on the surface of the sample to measure its ΔT_{ad}. This minimizes the heat capacity of the sensor and maximizes the thermal conductance between it and the sample [54]. The electrical isolation between sample and sensor can be achieved by depositing an isolation layer (e.g., 150 nm thick CaF_2 film).

A possible approach to diminish the thermal mass of the addenda and decrease, therefore, the time constant of the measurement device, is to use noncontact temperature measurement techniques like infrared detectors [56, 57]. An original alternative approach relied on the use of the mirage effect, which is an optical beam deflection technique used in photothermal spectroscopy to measure thermal diffusivity and the thermo-optical absorption spectrum of materials [55]. The temperature change of the sample when submitted to the applied magnetic field produces a gradient of the refractive index in the surrounding medium, detected by the deflection of a laser beam passing through the layer of that medium beside the sample (Fig. 7).

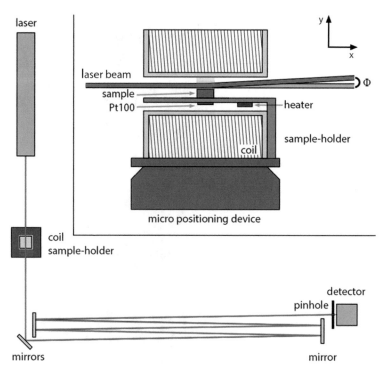

Fig. 7 Sketch of the setup for direct ΔT_{ad} measurements with pulsed magnetic fields. On the top: details of the sample holder inserted onto the coil and mounted above a micro positioning device. The magnetic pulsed field generates a temperature variation of the sample that deflects the laser beam, aligned close to its surface, of an angle Φ. Reprinted from [55] with permission from AIP Publishing

The broader availability of infrared cameras has also facilitated the use of infrared thermography for the determination of the magnetocaloric effect. This has the advantage of being able to determine the temperature change of several samples placed in the magnetic field, implementing a high-throughput measurement system [58, 59]. In the case of small signals, the use of lock-in thermography enables the determination of temperature changes as low as a few millikelvin and provides information about the magnitude and the phase of the response (Fig. 8) [58].

The protocol for measuring the magnetocaloric material, as indicated in the previous section, is rather important in order to avoid spurious results emerging from the coexistence of phases along a FOPT. In the case of ΔS_T, failing to erase the memory of the material produces a spike in the response that often causes researchers overestimate the characteristics of the material. On the other hand, ill-designed protocols for the measurement of ΔT_{ad} produce lower values than those intrinsic to the material [60]. It is therefore crucial to apply similar discontinuous protocols for the characterization of both magnitudes, although it is not usual to do it when determining ΔT_{ad}.

Fig. 8 dT_{mod} and φ images for the $Gd_{1-x}Y_x$ and Co samples under the conditions of (**a**) $H_{bias} = 275$ mT, $dH_{mod} = 24$ mT, and $f = 0.5$ Hz, (**b**) $H_{bias} = 1127$ mT, $dH_{mod} = 10$ mT, and $f = 5.0$ Hz, and (**c**) $H_{bias} = 51$ mT, $dH_{mod} = 2$ mT, and $f = 25.0$ Hz. Reprinted from [58] with permission from AIP Publishing

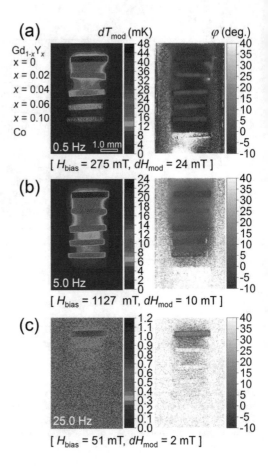

Although in this section we have restricted ourselves to direct measurement of ΔT_{ad}, it is worth mentioning that properly designed, usually custom made, calorimeters also allow for the direct determination of ΔS_T [41, 61–64].

In recent years, driven by the rich phenomenology of magnetocaloric materials with a FOPT, there is a trend to simultaneously characterize the different magnitudes that are modified during the phase transformation. A paradigmatic example is the in situ measurement of magnetization and volume magnetostriction in materials with magneto-volume effect like La(Fe,Si)$_{13}$ [65]. This kind of experimental device will certainly help the study of phase transitions, provide a better understanding of the different driving forces that cause the transformation, and will facilitate the optimization of magnetocaloric materials for technological applications.

6 Magnetocaloric Effect to Study Phase Transitions

While most of magnetocaloric effect characterization nowadays focuses on the properties of new (and not so new) materials for their potential use in magnetic refrigeration devices, there was a time when room temperature magnetic refrigeration was not yet considered. By then, magnetocaloric characterization was a useful tool to enhance the knowledge that we had about phase transitions [66, 67]. More recently, this fundamental aspect of magnetocaloric research has been revisited and new methods for determining critical exponents or to quantitatively determine the order of phase transitions have been proposed. We will briefly review some of these.

Critical Scaling in Second-Order Phase Transition Materials

It is well-known that second-order phase transitions follow scaling laws close to the critical region [68]. For SOPT magnetocaloric materials, like Gd, this implies that the magnetocaloric magnitudes follow power laws of the field of the type *magnitude* $\propto H^{\text{exponent}}$. Each different magnitude will scale with a different exponent, which is related to the critical exponents of the material [69–71] (Table 2). As a consequence, by studying the field dependence of MCE, we should be able to determine the critical exponents of the material. This determination was traditionally done using the conventional Kouvel-Fisher method [72]. Results obtained for simple cases, like a purely SOPT material formed by a single magnetic phase, using the traditional method and those obtained from the field dependence of MCE are in good agreement [70]. The main advantage of the MCE-based procedure is that it does not require an iterative approach, being more straightforward to use. However, there are cases for which the conventional methods are not applicable, like materials formed by different ferromagnetic phases or if several phase transitions coexist. It has been demonstrated that the scaling of MCE is capable of providing accurate values of the critical exponents even for these more complicated situations [73].

Table 2 Exponents controlling the field dependence of different magnitudes related to MCE (*magnitude* $\propto H^{\text{exponent}}$)

Magnitude	Exponent
$T_{pk} - T_C$ (not mean field)	$1/\Delta$
$T_{pk} - T_C$ (mean field)	0
$\Delta S_T(T = T_c)$	$1 + 1/\delta(1 - 1/\beta) = (1 - \alpha)/\Delta$
ΔS_T^{pk}	$1 + 1/\delta(1 - 1/\beta) = (1 - \alpha)/\Delta$
RC_{Area} or RC_{FWHM}	$1 + 1/\delta$

The critical exponents Δ, δ, β, and α fulfil $\Delta = \beta\delta$ and $1 - \alpha = \beta + \Delta - 1$, with β and δ characterizing the magnetization behavior along the coexistence curve ($H = 0$, $T < T_c$) and the critical isotherm ($T = T_c$), respectively

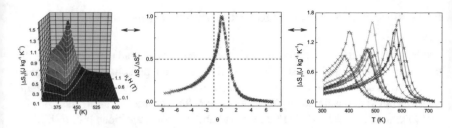

Fig. 9 Universal scaling (central panel) can be used to predict the field and temperature dependence of MCE (left panel) or the behavior of different alloys in a series of compositions (right panel)

These scaling laws are the basis that justifies the phenomenological construction of a universal curve for the isothermal entropy change, although that universal curve was initially proposed without requiring any prior knowledge of the critical exponents of the materials [15]. This phenomenological approach has practical applications (Fig. 9) like the extrapolation of experimental data to fields and temperatures less commonly accessible in laboratories [74], the prediction of MCE responses of different alloys from a series of materials with similar values of critical exponents [74], or even noise reduction of the experimental data and enhancement of data resolution recorded for low applied magnetic fields [69]. The scaling procedure can be performed in four simple steps:

1. Determining the temperature-dependent isothermal entropy change of the material for each magnetic field value.
2. Normalizing each isothermal entropy change curve with respect to its own peak.
3. Identifying the reference temperature (T_r) of each curve that corresponds to a specific value of the normalized curve (which is typically chosen between 0.5 and 0.7 for temperatures above the transition).
4. Rescaling the temperature axes of these curves as $\theta = (T - T_C)/(T_r - T_C)$.

This description of the procedure is based on the assumption that there is a single phase in the material and that the influence of the demagnetizing factor is negligible. Otherwise, instead of using a single reference temperature above the transition, two reference temperatures (one below and one above T_c) would be needed and the rescaling temperature would be:

$$\theta = \begin{cases} -(T - T_C)/(T_{r1} - T_C); & T \leq T_C \\ (T - T_C)/(T_{r2} - T_C); & T > T_C \end{cases} \tag{14}$$

Once put in the proper context of scaling relations [75], this phenomenological universal curve was further extended to the adiabatic temperature change of magnetocaloric materials [49]. The reader should be aware, however, that universal scaling is restricted to the critical region of the material, i.e., temperatures relatively close to the transition temperature and fields that are not too large [76] (the notions of "close" and "large" are material-dependent). The applicability limits for scaling in magnetocaloric materials [77] fit inside the usual experimental conditions of fields up to 10 T. This has been experimentally confirmed by numerous publications [78–89]. It is also worth mentioning that the demagnetizing field, the presence of various coexisting phases in the same sample, or the distribution of Curie temperature of the phases due to sample inhomogeneity might alter the field dependence of magnetocaloric magnitudes [31, 32, 90–92]. In the case of composites formed by magnetic phases with relatively close transition temperatures [24, 93], deviations from the universal curve can be used to deconvolute the magnetocaloric response of the individual phases and to quantify the relative fraction of phases present in the composite [93], providing results in good agreement with the determination of phase fractions performed by X-ray diffraction.

Another alternative application of the deviations from universal scaling is related to the study of skyrmions [94]. MnSi exhibits helimagnetic order and an additional skyrmion phase for a range of magnetic fields and temperatures. By studying the deviations from universal scaling of the magnetocaloric response, it was possible to determine that the maximum concentration of skyrmion vortices happens at 28.5 K and a magnetic field of 0.16 T.

Quantitative Determination of the Order of the Phase Transition

The most extended method to determine the order of a magnetic phase transition is the Banerjee criterion [13]. However, it is based on a mean field equation of state, and not all materials can be classified as mean field materials. Consequently, there were cases for which this criterion predicted a SOPT behavior, while calorimetric measurements showed a FOPT [95]. Universal scaling was proposed as an alternative to solve this discrepancy by using purely thermomagnetic data [14]. The success of the scaling approach is based on the fact that it is not restricted to any specific equation of state but only assumes that SOPT scale. Recent theoretical studies indicate that the scaling of MCE in the proximity of the critical region is valid for both SOPT materials and for those at the tricritical point [96].

Despite solving a controversy on the apparent discrepancies between calorimetry and thermomagnetic data for the determination of the order of the phase transition, the method based on scaling is qualitative, and, as such, different researchers might appreciate in a different way to which extent a set of curves collapse. It was necessary to be able to find a quantitative alternative.

The field dependence of the isothermal entropy change can be expressed as a power law of the field

$$\Delta S_T \propto H^n, \tag{15}$$

where the exponent n, which is field- and temperature-dependent, can be locally calculated as a logarithmic derivative:

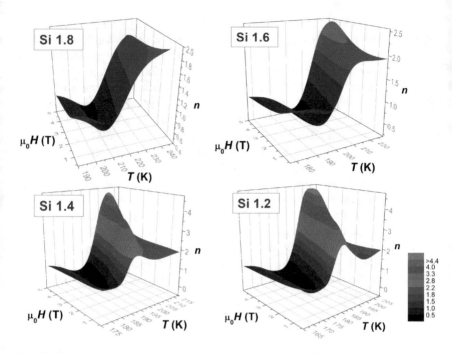

Fig. 10 Field and temperature dependence of the exponent n for $La_1Fe_{13-x}Si_x$ alloys. The values of $n > 2$ for Si 1.2, 1.4, and 1.6 clearly indicate their FOPT behavior, while Si 1.8 is a SOPT. Reprinted from [12] under CC BY 4.0 license

$$n = \frac{d \ln |\Delta S_T|}{d \ln H}. \tag{16}$$

In the case of second-order phase transition materials undergoing a ferro-paramagnetic phase transition, $n = 1$ for temperatures in the fully ferromagnetic range (well below the Curie temperature), $n = 2$ for temperatures well above T_c, and has a minimum at temperatures close to T_c. [69] However, for FOPT, it was recently discovered that the exponent n exhibits an overshoot above 2 in the temperature range of the phase transition, which can be used as a quantitative method for the determination of the order of the phase transition (Fig. 10) [12]. The applicability of this method has been demonstrated by using not only simulated data but also experimental results of $La(Fe,Si)_{13}$, Heusler alloys, cobaltites [97], Tb_3Ni single crystals with peculiar magnetic behavior [98], and even new high entropy alloys with a magnetostructural phase transition and unusually large magnetocaloric response [8].

The field dependence of the isothermal entropy change can also be used to determine the critical composition that makes the phase transition change from first to second order, as the critical exponents of that so-called tricritical point correspond to $n = 2/5$ [99]. For the series of $LaFe_{13-x}Si_x$, the tricritical point corresponds to a silicon concentration of $x \approx 1.65$, in agreement with the results shown in Fig. 10.

Study of Thermomagnetic Hysteresis in Magnetocaloric Materials by Using TFORC

FORC, which stands for first-order reversal curves, is a well-known methodology for the study of hysteresis in magnetic materials [100], with several chapters in this book showing its full potential in the case of M(H) hysteresis loops. However, good magnetocaloric materials have small field hysteresis, making the study of conventional FORC complicated and difficult to interpret [101]: while the FORC technique will give information of interaction fields in the magnetic field region for which the material is hysteretic, the magnetocaloric response is more sensitive to large fields. Therefore, absolute values of interaction fields obtained by these two techniques were (and should be) completely different.

Nevertheless, magnetocaloric materials with a FOPT also exhibit thermal hysteresis. This can be studied by replacing the H driving force in conventional FORC by temperature, becoming a TFORC technique [102]. While it is relatively easy to sweep the magnetic field in a reasonable time to obtain the required number of minor loops for FORC experiments, temperature variations are much slower, thus this long measuring time being one of the major limitations to broadly adopt this technique. However, the thermomagnetic phase transitions of magnetocaloric materials can be driven both by field and by temperature, which made us propose an effective temperature T^* that combines the thermal and magnetic energy provided to the material during the transformation. This allowed us to transform the technique into T*FORC, with a notable reduction of experimental time [103].

Another limitation that prevented a broad application of TFORC was the limited theoretical basis for the interpretation of the observed distributions. While FORC analysis has been carried out for decades and there is a clear understanding of the physical motivation of the observed "wishbones" and "boomerangs" in the distributions, TFORC in magnetocaloric materials started a short time ago, and there was a lack of theoretical developments to set the basis for the interpretation. Recently, the use of thermomagnetic models has allowed us to interpret some of the most usual figures that are observed in TFORC distributions (Fig. 11) [104]. It is expected that this particular technique will extend its use in the near future. While, due to the scope of this book, these TFORC and T*FORC techniques have been shown here for thermomagnetic hysteresis, they can be readily translated to other caloric materials like elastocaloric or electrocaloric by replacing the driving force of the measurement and analysis by the suitable one.

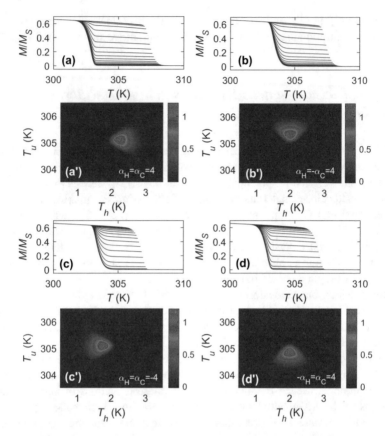

Fig. 11 Different TFORC distributions extracted from minor $M(T)$ loops showing how detailed features of a thermomagnetic phase transition, not easily distinguishable in the magnetization curves, correspond to different characteristic figures in the TFORC plane. (**a**) to (**d**) show the reversal magnetization curves, while (**a'**) to (**d'**) are the corresponding TFORC distributions. For details about the different parameters on the graphs, the reader is referred to [104]. Reprinted from [104] under CC BY 4.0 license

7 Conclusions

The characterization of magnetocaloric materials has two main applications: the search for new materials with optimized properties for thermomagnetic energy conversion, like magnetic refrigeration, and the study of phase transitions in materials that are either applicable in other areas, like spintronics and skyrmions, or they are simply important for fundamental science. One of the aims of this chapter is to convince the reader that it is not necessary to have a good magnetocaloric material in order to have relevant results from its magnetocaloric characterization.

Magnetocaloric characterization implies at least two driving forces: magnetic field and temperature, which complicates the measurement protocols. In addition,

FOPT involve the coexistence of phases, which can produce artifacts in the magnetocaloric measurements. To avoid these difficulties, measurement protocols must ensure that the sample is fully saturated, i.e., erasing the previous history of the material, using both driving forces. This is relevant for both indirect and direct magnetocaloric characterization.

In the recent years, direct magnetocaloric measurement techniques are evolving more into optical methods in order to facilitate high-throughput characterization of materials. AC techniques are also becoming more relevant, as the dynamic response of magnetocaloric materials is a key in the eventual implementation in a magnetic refrigerator. At the same time, detailed studies of thermomagnetic hysteresis are being performed with emerging techniques like TFORC and its variants like T*FORC.

Acknowledgments Work supported by AEI/FEDER-UE (grant MAT-2016-77265-R and PID2019-105720RB-I00), US/JUNTA/FEDER-UE (grant US-1260179), Consejería de Economía, Conocimiento, Empresas y Universidad de la Junta de Andalucía (grant P18-RT-746) and Army Research Laboratory under Cooperative Agreement Number W911NF-19-2-0212.

References

1. U.S. Energy Information Administration, http://www.eia.gov/outlooks/ieo
2. M. Sivak, Will AC put a chill on the global energy supply? Am. Sci. **100**, 330 (2013)
3. EUROPEAN COMMISSION https://ec.europa.eu/energy/sites/ener/files/documents/1_EN_ACT_part1_v14.pdf, 2016
4. V. Franco, J.S. Blázquez, J.J. Ipus, J.Y. Law, L.M. Moreno-Ramírez, A. Conde, Magnetocaloric effect: from materials research to refrigeration devices. Prog. Mater. Sci. **93**, 112 (2018)
5. V.K. Pecharsky, K.A. Gschneidner, Giant magnetocaloric effect in $Gd_5(Si_2Ge_2)$. Phys. Rev. Lett. **78**, 4494 (1997)
6. B.F. Yu, M. Liu, P.W. Egolf, A. Kitanovski, A review of magnetic refrigerator and heat pump prototypes built before the year 2010. Int. J. Refrig.-Rev. Int. Du Froid **33**, 1029 (2010)
7. V. Franco, J.S. Blázquez, B. Ingale, A. Conde, The magnetocaloric effect and magnetic refrigeration near room temperature: materials and models. Ann. Rev. Mater. Res. **42**, 305 (2012)
8. J.Y. Law, L.M. Moreno-Ramírez, Á. Díaz-García, A. Martín-Cid, S. Kobayashi, S. Kawaguchi, T. Nakamura, V. Franco, MnFeNiGeSi high-entropy alloy with large magnetocaloric effect. J. Alloys Compd. **855**, 157424 (2021)
9. O. Gutfleisch, A. Yan, K.H. Muller, Large magnetocaloric effect in melt-spun $LaFe_{13-x}Si_x$. J. Appl. Phys. **97**, 10M305 (2005)
10. J. Liu, J.D. Moore, K.P. Skokov, M. Krautz, K. Lowe, A. Barcza, M. Katter, O. Gutfleisch, Exploring $La(Fe,Si)_{(13)}$-based magnetic refrigerants towards application. Scripta Mater. **67**, 584 (2012)
11. M.F.J. Boeije, M. Maschek, X.F. Miao, N.V. Thang, N.H. van Dijk, E. Bruck, Mixed magnetism in magnetocaloric materials with first-order and second-order magnetoelastic transitions. J. Phys. D. Appl. Phys. **50**, 174002 (2017)
12. J.Y. Law, V. Franco, L.M. Moreno-Ramirez, A. Conde, D.Y. Karpenkov, I. Radulov, K.P. Skokov, O. Gutfleisch, A quantitative criterion for determining the order of magnetic phase transitions using the magnetocaloric effect. Nat. Commun. **9**, 2680 (2018)

13. B.K. Banerjee, On a generalised approach to first and second order magnetic transitions. Phys. Lett. **12**, 16 (1964)
14. C.M. Bonilla, J. Herrero-Albillos, F. Bartolome, L.M. Garcia, M. Parra-Borderias, V. Franco, Universal behavior for magnetic entropy change in magnetocaloric materials: an analysis on the nature of phase transitions. Phys. Rev. B **81**, 224424 (2010)
15. V. Franco, J.S. Blazquez, A. Conde, Field dependence of the magnetocaloric effect in materials with a second order phase transition: a master curve for the magnetic entropy change. Appl.Phys. Lett. **89**, 222512 (2006)
16. G.F. Wang, Z.R. Zhao, X.F. Zhang, L. Song, O. Tegus, Analysis of the first-order phase transition of $(Mn,Fe)_2(P,Si,Ge)$ using entropy change scaling. J. Phys. D. Appl. Phys. **46**, 295001 (2013)
17. E.M. Clements, R. Das, L. Li, P.J. Lampen-Kelley, M.H. Phan, V. Keppens, D. Mandrus, H. Srikanth, Critical behavior and macroscopic phase diagram of the monoaxial chiral helimagnet $Cr_{1/3}NbS_2$. Sci. Rep. **7**, 12 (2017)
18. B. Uthaman, P. Manju, S. Thomas, D.J. Nagar, K.G. Suresh, M.R. Varma, Observation of short range ferromagnetic interactions and magnetocaloric effect in cobalt substituted $Gd_5Si_2Ge_2$. Phys. Chem. Chem. Phys. **19**, 12282 (2017)
19. K.A. Gschneidner, V.K. Pecharsky, Magnetocaloric materials. Ann. Rev. Mater. Sci. **30**, 387 (2000)
20. M.E. Wood, W.H. Potter, General-analysis of magnetic refrigeration and its optimization using a new concept—maximization of refrigerant capacity. Cryogenics **25**, 667 (1985)
21. V. Provenzano, A.J. Shapiro, R.D. Shull, Reduction of hysteresis losses in the magnetic refrigerant $Gd_5Ge_2Si_2$ by the addition of iron. Nature **429**, 853 (2004)
22. E. Bruck, H. Yibole, L. Zhang, A universal metric for ferroic energy materials. Philos. Trans. Royal Soc. A: Math. Phys. Eng. Sci. **374**, 20150303 (2016)
23. L.D. Griffith, Y. Mudryk, J. Slaughter, V.K. Pecharsky, Material-based figure of merit for caloric materials. J. Appl. Phys. **123**, 034902 (2018)
24. J.Y. Law, L.M. Moreno-Ramírez, J.S. Blázquez, V. Franco, A. Conde, Gd+GdZn biphasic magnetic composites synthesized in a single preparation step: increasing refrigerant capacity without decreasing magnetic entropy change. J. Alloys Compd. **675**, 244 (2016)
25. L. Tocado, E. Palacios, R. Burriel, Entropy determinations and magnetocaloric parameters in systems with first-order transitions: study of MnAs. J. Appl. Phys. **105**, 093918 (2009)
26. A.M.G. Carvalho, A.A. Coelho, P.J. von Ranke, C.S. Alves, The isothermal variation of the entropy (ΔS_T) may be miscalculated from magnetization isotherms in some cases: MnAs and $Gd_5Ge_2Si_2$ compounds as examples. J. Alloys Compd. **509**, 3452 (2011)
27. B. Kaeswurm, V. Franco, K.P. Skokov, O. Gutfleisch, Assessment of the magnetocaloric effect in La,Pr(Fe,Si) under cycling. J. Magn. Magn. Mater. **406**, 259 (2016)
28. V. Franco, (Lake Shore Cryotronics, https://www.lakeshore.com/products/product-detail/8600-series-vsm/mce-analysis-software, 2014)
29. J.D. Moore, K. Morrison, K.G. Sandeman, M. Katter, L.F. Cohen, Reducing extrinsic hysteresis in first-order $La(Fe,Co,Si)_{(13)}$ magnetocaloric systems. Appl. Phys. Lett. **95**, 252504 (2009)
30. B.R. Hansen, C.R.H. Bahl, L.T. Kuhn, A. Smith, K.A. Gschneidner, V.K. Pecharsky, Consequences of the magnetocaloric effect on magnetometry measurements. J. Appl. Phys. **108**, 043923 (2010)
31. C. Romero-Muniz, J.J. Ipus, J.S. Blazquez, V. Franco, A. Conde, Influence of the demagnetizing factor on the magnetocaloric effect: Critical scaling and numerical simulations. Appl. Phys. Lett. **104**, 252405 (2014)
32. R. Caballero-Flores, V. Franco, A. Conde, L.F. Kiss, Influence of the demagnetizing field on the determination of the magnetocaloric effect from magnetization curves. J. Appl. Phys. **105**, 07A919 (2009)
33. D.X. Chen, E. Pardo, A. Sanchez, Demagnetizing factors for rectangular prisms. IEEE Trans. Magn. **41**, 2077 (2005)

34. D.X. Chen, J.A. Brug, R.B. Goldfarb, Demagnetizing factors for cylinders. IEEE Trans. Magn. **27**, 3601 (1991)
35. M. Le Floc'h, J.L. Mattei, P. Laurent, O. Minot, A.M. Konn, A physical model for heterogeneous magnetic materials. J. Magn. Magn. Mater. **2191**, 140–144 (1995)
36. R. Bjørk, Z. Zhou, The demagnetization factor for randomly packed spheroidal particles. J. Magn. Magn. Mater. **476**, 417 (2019)
37. V.K. Pecharsky, K.A. Gschneidner, Heat capacity near first order phase transitions and the magnetocaloric effect: An analysis of the errors, and a case study of $Gd_5(Si_2Ge_2)$ and Dy. J. Appl. Phys. **86**, 6315 (1999)
38. V.K. Pecharsky, J.O. Moorman, K.A. Gschneidner, A 3–350 K fast automatic small sample calorimeter. Rev. Sci. Instrum. **68**, 4196 (1997)
39. T. Plackowski, Y.X. Wang, A. Junod, Specific heat and magnetocaloric effect measurements using commercial heat-flow sensors. Rev. Sci. Instrum. **73**, 2755 (2002)
40. S. Jeppesen, S. Linderoth, N. Pryds, L.T. Kuhn, J.B. Jensen, Indirect measurement of the magnetocaloric effect using a novel differential scanning calorimeter with magnetic field. Rev. Sci. Instrum. **79**, 083901 (2008)
41. J. Marcos, F. Casanova, X. Batlle, A. Labarta, A. Planes, L. Manosa, A high-sensitivity differential scanning calorimeter with magnetic field for magnetostructural transitions. Rev. Sci. Instrum. **74**, 4768 (2003)
42. J.C. Lashley, M.F. Hundley, A. Migliori, J.L. Sarrao, P.G. Pagliuso, T.W. Darling, M. Jaime, J.C. Cooley, W.L. Hults, L. Morales, D.J. Thoma, J.L. Smith, J. Boerio-Goates, B.F. Woodfield, G.R. Stewart, R.A. Fisher, N.E. Phillips, Critical examination of heat capacity measurements made on a quantum design physical property measurement system. Cryogenics **43**, 369 (2003)
43. V.K. Pecharsky, K.A. Gschneidner, Magnetocaloric effect from indirect measurements: magnetization and heat capacity. J. Appl. Phys. **86**, 565 (1999)
44. L.M. Moreno-Ramírez, J.S. Blázquez, J.Y. Law, V. Franco, A. Conde, Optimal temperature range for determining magnetocaloric magnitudes from heat capacity. J. Phys. D. Appl. Phys. **49**, 495001 (2016)
45. L.M. Moreno-Ramírez, V. Franco, A. Conde, H. Neves Bez, Y. Mudryk, V.K. Pecharsky, Influence of the starting temperature of calorimetric measurements on the accuracy of determined magnetocaloric effect. J. Magn. Magn. Mater. **457**, 64 (2018)
46. Y. Miyoshi, K. Morrison, J.D. Moore, A.D. Caplin, L.F. Cohen, Heat capacity and latent heat measurements of CoMnSi using a microcalorimeter. Rev. Sci. Instrum. **79**, 074901 (2008)
47. J.Y. Law, V. Franco, R.V. Ramanujan, Direct magnetocaloric measurements of Fe-B-Cr-X (X = La, Ce) amorphous ribbons. J. Appl. Physi. **110**, 023907 (2011)
48. A.M. Tishin, Y.I. Spichkin, *The Magnetocaloric Effect and its Applications* (Institute of Physics Publishing, Bristol, 2003)
49. V. Franco, A. Conde, J.M. Romero-Enrique, Y.I. Spichkin, V.I. Zverev, A.M. Tishin, Field dependence of the adiabatic temperature change in second order phase transition materials: application to Gd. J. Appl. Phys. **106**, 103911 (2009)
50. C. Romero-Torralva, C. Mayer, V. Franco, A. Conde, Dynamic effects in the characterization of the magnetocaloric effect of LaFeSi-type alloys. In: *2015 IEEE Magnetics Conference (INTERMAG)* (2015), p. 1
51. T. Gottschall, K.P. Skokov, F. Scheibel, M. Acet, M.G. Zavareh, Y. Skourski, J. Wosnitza, M. Farle, O. Gutfleisch, Dynamical effects of the martensitic transition in magnetocaloric heusler alloys from direct delta T-ad measurements under different magnetic-field-sweep rates. Phys. Rev. Appl. **5**, 024013 (2016)
52. A.M. Aliev, A.B. Batdalov, V.S. Kalitka, Magnetocaloric properties of manganites in alternating magnetic fields. JETP Lett. **90**, 663 (2010)
53. Y. Kohama, C. Marcenat, T. Klein, M. Jaime, AC measurement of heat capacity and magnetocaloric effect for pulsed magnetic fields. Rev. Sci. Instrum. **81**, 104902 (2010)

54. T. Kihara, Y. Kohama, Y. Hashimoto, S. Katsumoto, M. Tokunaga, Adiabatic measurements of magneto-caloric effects in pulsed high magnetic fields up to 55 T. Rev. Sci. Instrum. **84**, 074901 (2013)

55. F. Cugini, G. Porcari, C. Viappiani, L. Caron, A.O. dos Santos, L.P. Cardoso, E.C. Passamani, J.R.C. Proveti, S. Gama, E. Bruck, M. Solzi, Millisecond direct measurement of the magnetocaloric effect of a Fe_2P-based compound by the mirage effect. Appl. Phys. Lett. **108**, 012407 (2016)

56. F. Cugini, G. Porcari, M. Solzi, Non-contact direct measurement of the magnetocaloric effect in thin samples. Rev. Sci. Instrum. **85**, 074902 (2014)

57. J. Döntgen, J. Rudolph, T. Gottschall, O. Gutfleisch, S. Salomon, A. Ludwig, D. Hägele, Temperature dependent low-field measurements of the magnetocaloric ΔT with sub-mK resolution in small volume and thin film samples. Appl. Phys. Lett. **106**, 032408 (2015)

58. Y. Hirayama, R. Iguchi, X.-F. Miao, K. Hono, K.-I. Uchida, High-throughput direct measurement of magnetocaloric effect based on lock-in thermography technique. Appl. Phys. Lett. **111**, 163901 (2017)

59. K. Wang, Y. Ouyang, Y. Shen, Y. Zhang, M. Zhang, J. Liu, High-throughput characterization of the adiabatic temperature change for magnetocaloric materials. J. Mater. Sci. **56**, 2332 (2020)

60. L.M. Moreno-Ramírez, A. Delgado-Matarín, J.Y. Law, V. Franco, A. Conde, A.K. Giri, Influence of thermal and magnetic history on direct ΔT_{ad} measurements of $Ni_{49+x}Mn_{36-x}In_{15}$ Heusler alloys. Metals **9**, 1144 (2019)

61. F. Casanova, A. Labarta, X. Batlle, F.J. Perez-Reche, E. Vives, L. Manosa, A. Planes, Direct observation of the magnetic-field-induced entropy change in $Gd_5(Si_xGe_{1-x})_4$ giant magnetocaloric alloys. Appl. Phys. Lett. **86**, 262504 (2005)

62. V. Basso, M. Kupferling, C.P. Sasso, L. Giudici, A Peltier cell calorimeter for the direct measurement of the isothermal entropy change in magnetic materials. Rev. Sci. Instrum. **79**, 063907 (2008)

63. K.K. Nielsen, H.N. Bez, L. von Moos, R. Bjork, D. Eriksen, C.R.H. Bahl, Direct measurements of the magnetic entropy change. Rev. Sci. Instrum. **86**, 103903 (2015)

64. V. Basso, C.P. Sasso, M. Kupferling, A Peltier cells differential calorimeter with kinetic correction for the measurement $c_p(H, T)$ and $\Delta s(H, T)$ s(H, T) of magnetocaloric materials. Rev. Sci. Instrum. **81**, 113904 (2010)

65. D.Y. Karpenkov, A.Y. Karpenkov, K.P. Skokov, I.A. Radulov, M. Zheleznyi, T. Faske, O. Gutfleisch, Pressure dependence of magnetic properties in $La(Fe,Si)_{13}$: multistimulus responsiveness of caloric effects by modeling and experiment. Phys. Rev. Appl. **13**, 034014 (2020)

66. K.P. Belov, L. Cherniko, S.A. Nikitin, V.V. Tikhonov, E. Talalaev, T. Kudryavt, V. Ivanovsk, Determination of exchange interaction of sublattices in gadolinium iron-Garnet on basis of magnetocaloric effect. Sov. Phys. JETP-USSR **34**, 588 (1972)

67. K.P. Belov, E.V. Talalayeva, L.A. Chernikova, T.I. Ivanova, V.I. Ivanovsky, G.V. Kazakov, Observation of spin reorientation based on measurements of magnetocaloric effect. Zhurnal Eksperimentalnoi I Teoreticheskoi Fiziki **72**, 586 (1977)

68. H.E. Stanley, Scaling, universality, and renormalization: three pillars of modern critical phenomena. Rev. Mod. Phys. **71**, S358 (1999)

69. V. Franco, A. Conde, Scaling laws for the magnetocaloric effect in second order phase transitions: from physics to applications for the characterization of materials. Int. J. Refrig.-Rev. Int. Du Froid **33**, 465 (2010)

70. V. Franco, A. Conde, V. Provenzano, R.D. Shull, Scaling analysis of the magnetocaloric effect in $Gd_5Si_2Ge_{1.9}X_{0.1}$ (X=Al, Cu, Ga, Mn, Fe, Co). J. Magn. Magn. Mater. **322**, 218 (2010)

71. V. Franco, A. Conde, V.K. Pecharsky, K.A. Gschneidner, Field dependence of the magnetocaloric effect in Gd and $(Er_{1-x}Dy_x)Al_2$: Does a universal curve exist? Europhy. Lett. (EPL) **79**, 47009 (2007)

72. J.S. Kouvel, M.E. Fisher, Detailed magnetic behavior of nickel near its curie point. Phys. Rev. A: Gen. Phys. **136**, 1626 (1964)

73. M. Sanchez-Perez, L.M. Moreno-Ramirez, V. Franco, A. Conde, M. Marsilius, G. Herzer, Influence of nanocrystallization on the magnetocaloric properties of Ni-based amorphous alloys: determination of critical exponents in multiphase systems. J. Alloys Compd. **686**, 717 (2016)

74. V. Franco, C.F. Conde, J.S. Blazquez, A. Conde, P. Svec, D. Janickovic, L.F. Kiss, A constant magnetocaloric response in FeMoCuB amorphous alloys with different Fe/B ratios. J. Appl. Phys. **101**, 093903 (2007)

75. V. Franco, A. Conde, J.M. Romero-Enrique, J.S. Blazquez, A universal curve for the magnetocaloric effect: an analysis based on scaling relations. J. Phys.-Condens. Matter **20**, 285207 (2008)

76. A. Smith, K.K. Nielsen, C.R.H. Bahl, Scaling and universality in magnetocaloric materials. Phys. Rev. B **90**, 104422 (2014)

77. C. Romero-Muñiz, R. Tamura, S. Tanaka, V. Franco, Applicability of scaling behavior and power laws in the analysis of the magnetocaloric effect in second-order phase transition materials. Phys. Rev. B **94**, 134401 (2016)

78. P. Alvarez, P. Gorria, J.L.S. Llamazares, M.J. Perez, V. Franco, M. Reiffers, J. Kovac, I. Puente-Orench, J.A. Blanco, Magneto-caloric effect in the pseudo-binary intermetallic YPrFe17 compound. Mater. Chem. and Phys. **131**, 18 (2011)

79. J.Y. Fan, L. Pi, L. Zhang, W. Tong, L.S. Ling, B. Hong, Y.G. Shi, W.C. Zhang, D. Lu, Y.H. Zhang, Investigation of critical behavior in $Pr_{0.55}Sr_{0.45}MnO_3$ by using the field dependence of magnetic entropy change. Appl. Phys. Lett. **98**, 072508 (2011)

80. R.W. Li, C.J. Zhang, Y.H. Zhang, Study of magnetic entropy and ESR in ferromagnet CuCr2Te4. J. Magn. Magn. Mater. **324**, 3133 (2012)

81. X.C. Zhong, J.X. Min, Z.G. Zheng, Z.W. Liu, D.C. Zeng, Critical behavior and magnetocaloric effect of $Gd_{65}Mn_{35-x}Ge_x$ (x=0, 5, and 10) melt-spun ribbons. J. Appl. Phys. **112**, 033903 (2012)

82. J.C. Debnath, P. Shamba, A.M. Strydom, J.L. Wang, S.X. Dou, Investigation of the critical behavior in $Mn_{0.94}Nb_{0.06}CoGe$ alloy by using the field dependence of magnetic entropy change. J. Appl. Phys. **113**, 093902 (2013)

83. R. Pelka, P. Konieczny, M. Fitta, M. Czapla, P.M. Zielinski, M. Balanda, T. Wasiutynski, Y. Miyazaki, A. Inaba, D. Pinkowicz, B. Sieklucka, Magnetic systems at criticality: different signatures of scaling. Acta Phys. Polonica A **124**, 977 (2013)

84. S.K. Giri, P. Dasgupta, A. Poddar, T.K. Nath, Large magnetocaloric effect and critical behavior in $Sm_{0.09}Ca_{0.91}MnO_3$ electron-doped nanomanganite. Epl **105**, 47007 (2014)

85. D.D. Belyea, M.S. Lucas, E. Michel, J. Horwath, C.W. Miller, Tunable magnetocaloric effect in transition metal alloys. Sci. Rep. **5**, 15755 (2015)

86. S. Mahjoub, M. Baazaoui, R. M'Nassri, N.C. Boudjada, M. Oumezzine, Critical behavior and the universal curve for magnetocaloric effect in $Pr_{0.6}Ca_{0.1}Sr_{0.3}Mn_{1-x}Fe_xO_3$ (x=0, 0.05 and 0.075) manganites. J. Alloys Compd. **633**, 207 (2015)

87. R.W. Li, P. Kumar, R. Mahendiran, Critical behavior in polycrystalline $La_{0.7}Sr_{0.3}CoO_3$ from bulk magnetization study. J. Alloys Compd. **659**, 203 (2016)

88. B. Sattibabu, A.K. Bhatnagar, K. Vinod, S. Rayaprol, A. Mani, V. Siruguri, D. Das, Studies on the magnetoelastic and magnetocaloric properties of $Yb_{1-x}Mg_xMnO_3$ using neutron diffraction and magnetization measurements. RSC Adv. **6**, 48636 (2016)

89. H. Han, L. Zhang, X.D. Zhu, H.F. Du, M. Ge, L.S. Ling, L. Pi, C.J. Zhang, Y.H. Zhang, Critical phenomenon in the itinerant ferromagnet $Cr_{11}Ge_{19}$ studied by scaling of the magnetic entropy change. J. Alloys Compd. **693**, 389 (2017)

90. V. Franco, R. Caballero-Flores, A. Conde, Q.Y. Dong, H.W. Zhang, The influence of a minority magnetic phase on the field dependence of the magnetocaloric effect. J. Magn. Magn. Mater. **321**, 1115 (2009)

91. J.Y. Law, V. Franco, R.V. Ramanujan, The magnetocaloric effect of partially crystalline Fe–B–Cr–Gd alloys. J. Appl. Phys. **111**, 113919 (2012)

92. L.M. Moreno-Ramirez, J.J. Ipus, V. Franco, J.S. Blazquez, A. Conde, Analysis of magnetocaloric effect of ball milled amorphous alloys: demagnetizing factor and Curie temperature distribution. J. Alloys Compd. **622**, 606 (2015)

93. Á. Díaz-García, J.Y. Law, P. Gębara, V. Franco, Phase deconvolution of multiphasic materials by the universal scaling of the magnetocaloric effect. JOM **72**, 2845 (2020)

94. M. Ge, L. Zhang, D. Menzel, H. Han, C.M. Jin, C.J. Zhang, L. Pi, Y.H. Zhang, Scaling investigation of the magnetic entropy change in helimagnet MnSi. J. Alloys Compd. **649**, 46 (2015)

95. J. Herrero-Albillos, F. Bartolome, L.M. Garcia, F. Casanova, A. Labarta, X. Batlle, Nature and entropy content of the ordering transitions in RCo_2. Phys. Rev. B **73**, 134410 (2006)

96. C. Romero-Muniz, V. Franco, A. Conde, Two different critical regimes enclosed in the Bean-Rodbell model and their implications for the field dependence and universal scaling of the magnetocaloric effect. Phys. Chem. Chem. Phys. **19**, 3582 (2017)

97. J.Y. Law, V. Franco, A. Conde, S.J. Skinner, S.S. Pramana, Modification of the order of the magnetic phase transition in cobaltites without changing their crystal space group. J. Alloys Compd. **777**, 1080 (2019)

98. A. Herrero, A. Oleaga, A.F. Gubkin, A. Salazar, N.V. Baranov, Peculiar magnetocaloric properties and critical behavior in antiferromagnetic Tb_3Ni with complex magnetic structure. J. Alloys Compd. **808**, 151720 (2019)

99. V. Franco, J.Y. Law, A. Conde, V. Brabander, D.Y. Karpenkov, I. Radulov, K. Skokov, O. Gutfleisch, Predicting the tricritical point composition of a series of LaFeSi magnetocaloric alloys via universal scaling. J. Phys. D. Appl. Phys. **50**, 414004 (2017)

100. C.R. Pike, A.P. Roberts, K.L. Verosub, Characterizing interactions in fine magnetic particle systems using first order reversal curves. J. Appl. Phys. **85**, 6660 (1999)

101. V. Franco, F. Béron, K.R. Pirota, M. Knobel, M.A. Willard, Characterization of the magnetic interactions of multiphase magnetocaloric materials using first-order reversal curve analysis. J. Appl. Phys. **117**, 17c124 (2015)

102. V. Franco, T. Gottschall, K.P. Skokov, O. Gutfleisch, First-Order Reversal Curve (FORC) analysis of magnetocaloric Heusler-type alloys. IEEE Magn. Lett. **7**, 6602904 (2016)

103. V. Franco, Temperature-FORC analysis of a magnetocaloric Heusler alloy using a unified driving force approach (T*FORC). J. Appl. Phys. **127**, 133902 (2020)

104. L.M. Moreno-Ramírez, V. Franco, Setting the basis for the interpretation of Temperature First Order Reversal Curve (TFORC) distributions of magnetocaloric materials. Metals **10**, 1039 (2020)

Magnetostrictive Materials

Alfredo García-Arribas

Abstract Ferromagnetic materials present an intrinsic coupling between their magnetic and elastic properties. Magnetostriction is the direct manifestation of this effect by which a material deforms upon magnetization. The inverse phenomenon, i.e., magnetoelasticity, produces a change in the magnetic state when the material is deformed. The inherent fundamental interest of this coupling and the obvious opportunity for applications has motivated an intense research activity on the subject that has not decayed over nearly 180 years. In this chapter, the basic concepts regarding magnetostriction and magnetoelastic phenomena are first briefly introduced, together with a description of the most performing materials. After that, the main part of the chapter is devoted to review the experimental methods usually employed to characterize the magnetostriction of materials, with a section dedicated to the particular case of materials in the form of thin films. To finalize, a concise overview of applications is presented, in which the principles and strategies that configure sensor and actuator devices are briefly discussed, including a short accounting of opportunities in energy harvesting.

Keywords Magnetostriction · Magnetoelasticity · Magnetostrictive materials · Magnetic measurements · Magnetostrictive actuators · Magnetoelastic sensors

A. García-Arribas (✉)
Departamento de Electricidad y Electrónica, Universidad del País Vasco, UPV/EHU, Leioa, Spain

BCMaterials, Basque Center for Materials, Applications and Nanostructures, UPV/EHU Science Park, Leioa, Spain
e-mail: alfredo.garcia@ehu.es

© Springer Nature Switzerland AG 2021
V. Franco, B. Dodrill (eds.), *Magnetic Measurement Techniques for Materials Characterization*, https://doi.org/10.1007/978-3-030-70443-8_24

1   Introduction

Basic Concepts and Definitions

Magnetostriction

When a magnetic material is magnetized by the effect of an applied field, its dimensions change, a phenomenon that is known as *magnetostriction*. The effect was discovered by J. P. Joule in 1842 who observed a slight increase in the length of an iron rod when it was magnetized along its length [1]. The magnetostriction is quantified as a strain, that is, a dimensionless relative change of length, and it is usually denoted by λ:

$$\lambda = \frac{\Delta l}{l}. \tag{1}$$

The largest value of the strain is reached when the specimen is magnetically saturated. This value λ_s is called *saturation magnetostriction coefficient*, being a characteristic parameter of the nature of the material. Magnetostriction is usually small, producing strains of some parts per million (ppm). To get an approximate idea, it can be compared to thermal expansion. In a typical material, the strain caused by magnetostriction is comparable with the deformation produced by one-degree temperature change. Despite its small magnitude, the magnetostrictive effect has important consequences, in the form of applications, as the generation of acoustic waves (first sonar devices developed during World War I were made from nickel), or in the form of undesired effects, as the vibrations that produce the rumbling noise of electrical transformers.

Magnetostriction can be positive or negative, that is, a magnetic field applied in a given direction can produce either an increase or a decrease in length, respectively. This is a reflection of its microscopic physical origin, which relies in the spin-orbit coupling and makes that the overall macroscopic strain depends on the crystal symmetry. At this point, we must recall that the field-induced phenomenon described so far, occurring upon magnetization, is strictly known as *linear magnetostriction*. The complete phenomenology includes the *volume magnetostriction* caused spontaneously by the magnetic order (below Curie temperature) and the *forced magnetostriction*, evidenced at large fields beyond the magnetic saturation (in the so-called paraprocess). The latter two are of limited interest for applications (except for the case of invar alloys which show compensation between thermal expansion and spontaneous magnetostriction that maintains their volume unchanged with temperature). We will limit ourselves to the most basic phenomena related to linear magnetostriction (which we will call in the following only magnetostriction).

The magnetostriction occurring during magnetization is mostly caused by the rotation of the magnetization, since magnetization reversal within domains doesn't produce magnetostrictive deformation. The value of the saturation magnetostriction in single crystals depends on the direction in which the field is applied with respect to the crystal axes. In polycrystalline samples, provided that the grain orientation is random, the saturation magnetostriction coefficient can be determined by averaging. In these cases, and in isotropic materials, the magnetostrictive strain is completely determined by the saturation magnetostriction coefficient λ_S. The deformation of the specimen in a given direction can be quantified as

$$\lambda = \frac{3}{2}\lambda_s \left(\cos^2\theta - \frac{1}{3} \right). \tag{2}$$

where θ is the angle that forms the magnetization with the direction in which the strain is measured.

Magnetoelastic Effects

The magnetostriction reveals the existence of a coupling between the elastic and magnetic properties of magnetic materials. The coupling is two-way and causes that, if a mechanical stress is applied to a magnetostrictive material, the magnetic configuration is concomitantly altered. For instance, in a material with a positive magnetostriction, that is, which elongates when magnetized, an applied tensile stress will increase the magnetization, and a compressive stress will decrease it, even if no magnetic field is applied. This *magnetoelastic* phenomenon is also known as *inverse magnetostriction* or *Villari effect*. There are other effects related to magnetostriction that couple torsion with magnetism. The *Wiedemann effect* describes the torsion generated by a helicoidal magnetic field: a magnetic wire gets twisted under the combination of a longitudinal magnetic field and the circular magnetic field generated by a current flowing through it. The inverse effect, known as *Matteucci effect*, accounts for the change in magnetization produced by an applied mechanical torque. All the above effects are often reunited in what are called *magnetoelastic effects*.

ΔE Effect

The magnetostriction and, in general, the magnetoelastic coupling have a relevant consequence in the way that magnetic materials respond to an applied stress: in a non-magnetostrictive material, the relation between the applied stress σ and the produced strain ϵ is given by the Young modulus $E = \sigma/\epsilon$. In a magnetostrictive material, an applied stress causes a rotation of the magnetization that produces an additional strain. If, for instance, $\sigma > 0$ (tensile stress), this additional strain is always positive independently of the sign of the magnetostriction coefficient since, if λ_s is positive, the magnetization will rotate toward the direction of the stress, and the sample gets longer, and, if λ_s is negative, the magnetization will rotate away for the direction of stress, and the sample gets longer as well. Therefore, in a magnetostrictive material, depending on the magnetic state of the material, i.e., on the values of the applied stress and magnetic field, the Young modulus can be significantly reduced. This is known as the ΔE *effect.*

Magnetoelastic Waves and Resonance

A strain perturbation can propagate as a sound wave through a material. If the material is magnetostrictive, the strain wave is intimately coupled to a magnetization one. These traveling magnetoelastic waves can be excited (and detected) either by mechanical or magnetic means. In the latter case, the waves can be generated in one end of the material by a coil fed with an alternating current, while the detection can be performed at the other end by the voltage induced in a search coil. If the wavelength of the magnetoelastic waves is congruent with the dimensions of the sample, stationary waves are developed that produces the *magnetoelastic resonance.* For instance, in a magnetostrictive sample of length L, with density ρ and Young's modulus E, allowed to oscillate freely (unclamped at both ends), the resonance frequency, corresponding to the first mode of oscillation, is given by

$$f_r = \frac{1}{2L}\sqrt{\frac{E}{\rho}}. \tag{3}$$

Interestingly, because of the ΔE effect described in the previous paragraph, the resonance frequency for a given sample can be tuned by modifying its magnetic state, that is, by applying a suitable bias field.

This extremely concise account of the phenomenology related to magnetostriction and magnetoelastic coupling summarizes the knowledge on the subject, which is quite old and well established at the moment. Much more detailed information can be found, for instance, in the early review of Lee [2]. The excellent books on general Magnetism as Cullity [3], Chikazumi [4], and Coey [5] contain more or less extensive content on magnetostriction. A very detailed treatise on all aspects of the subject is the book by Tremolet de Laicheserie [6].

Magnetostrictive Materials

All magnetically ordered materials experience magnetostriction. The materials relevant for applications are those in which the effect is either very large or nearly vanishing, together with those materials that combine a significant magnetostrictive effect with other interesting features as good mechanical properties.

Pure $3d$ metals in polycrystalline form present saturation magnetostriction coefficients of: $\lambda_s = -7 \times 10^{-6}$ (Fe), -62×10^{-6} (Co), and -34×10^{-6} (Ni) [7]. Their alloys display quite different values depending on composition [8]. For instance, in $Fe_{50}Ni_{50}$, $\lambda_s = 28 \times 10^{-6}$ [9], whereas $Fe_{19}Ni_{81}$ presents null magnetostriction which contributes to their exceptionally soft magnetic behavior. Electrical Fe–Si steel (~3% at. Si) has a magnetostriction of about 20 ppm [10] (responsible of transformer's humming). Iron-rich cobalt-aluminum alloys (*Alcofer*) present magnetostriction coefficients above 40 ppm [11].

Especially important for magnetoelastic applications are the amorphous alloys, also known as metallic glasses. Typically, they are composed of about 70–80 at.% transition metals with up to 30% metalloids (Si, B, P, etc.) and fabricated in the form of ribbons by rapid quenching from the melt, which preserves the topological and chemical disorder of the liquid in the solid state [12]. The lack of crystallinity confers them excellent magnetic and mechanical properties that can even be upgraded by suitable thermal treatments under an applied magnetic field and/or stress [13]. Commercial brands are Metglas (a company of Hitachi Metals America) and Vitrovac (from Vacuumschmelze, Germany). Just to give an example of the variety of existing materials, Metglas 2615 ($Fe_{80}P_{16}C_3B_1$) presents a quite large magnetostriction of $\lambda_s = 29 \times 10^{-6}$ and Vitrovac 6025 [$Co_{66}Fe_4(MoSiB)_{30}$] a very low one: $\lambda_s = 5 \times 10^{-7}$.

Piezomagnetic ferrites constitute another family of magnetic materials with good magnetoelastic properties. *Ferroxcube* and *Kearfoot* present saturation magnetostriction coefficients in the range of -26 to -30 ppm [14].

Rare Earth elements such as Td and Dy present extraordinarily large magnetostriction values at low temperatures and high magnetic fields [15]. Alloyed with Fe constitute *terfenol-D*, with composition $Tb_xDy_{1-x}Fe_2$ ($x \sim 0.3$), which can attain strains up to 2×10^3 ppm in relatively small fields (few kOe) [16, 17]. This material commands the so-called giant magnetostriction materials that includes *galfenol* $Fe_{1-x}Ga_x$ that reach values of 265 ppm for $x = 0.19$ [18] and *alfenol* (Fe_xAl_{1-x}), up to 184 ppm for $x = 0.185$ [19]. The search for good magnetostrictive materials has extended to other compositions [19] and remains active nowadays [20].

2 Magnetostriction Measurement Techniques

There are two main different strategies for determining the magnetostriction of materials. The first one consists in a direct measurement of the magnetostrictive strain. The second implies an indirect determination of the magnetostriction coefficient from the influence of the magnetoelastic coupling on the magnetic behavior of the material. Devices and procedures based on both types of approaches are briefly described below, but it is worth pointing out that only direct evaluation methods are capable of measuring the absolute value of the strain produced by the magnetostriction, which is certainly necessary for the assessment of the actual actuation capabilities of the materials. Thin film materials present unique measurement hurdles due to the influence of the substrate, and a brief subsection is dedicated to them at the end of this section.

Direct Measurements

Strain Gauges

The determination of stresses using strain gauges is a well-developed technique widely used in mechanical analysis of materials and structures. A strain gauge is a resistive sensor commonly composed by a metallic thin film meander deposited onto a plastic substrate (Fig. 1a), which is glued on the surface of the object whose deformation is to be measured. The deformation of the gauge conductor produces a change in its resistance R that is proportional to the strain ϵ: $\Delta R/R = K\epsilon$, being K the gauge factor. The gauge factor, that is, the sensitivity of the gauge, is determined by both the change in dimensions of the conductor and the resistivity change caused by the deformation [21]. Typically, metallic strain gauges are made of *constantan* (Cu–Ni), *Karman* (Cu–Ni–Al–Fe) or *isoelastic* (Fe–Ni–Cr) alloys, displaying a gauge factor of about $K = 2$. Therefore, for typical deformations, the relative resistance variation of the gauge is very small. Semiconductor gauges present a larger but nonlinear sensitivity ($K \sim 200$) and a lower measuring range.

This method was early used to determine magnetostriction on single crystals [22, 23]. The main limitation of using gauges is that the sample must be large enough (mostly thick enough, provided that its lateral dimensions are larger than the gauge) so the presence of the gauge doesn't hinder the magnetostrictive strain. Additionally, a serious source of error of this method is the sensitivity of the resistance of the gauge to temperature variations and, most important in magnetostriction measurements, the contribution of the intrinsic magnetoresistance of the material composing the gauges [24]. The two abovementioned shortcomings, i.e., the small magnitude of the gauge resistance variation due to the deformation and the undesired contributions of temperature and magnetoresistance, are largely mitigated by the use of a Wheatstone bridge (Fig. 1b). It consists in two confronted voltage dividers conceived to suppress the contribution of the common value of the resistance and produce an output proportional to resistance differences. Let's denote $x = K\epsilon$ the contribution of magnetostriction and y the relative resistance

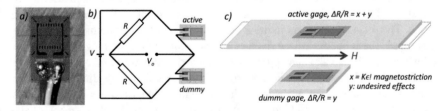

Fig. 1 (a) Photograph of a strain gauge. (b) Wheatstone bridge connection of the active and dummy gauges in consecutive branches. Common resistors R complete the bridge. The output $V_o \propto x$, is compensated from undesired effects. (c) Representation of an active strain gauge glued to a magnetostrictive specimen and a dummy gauge on a non-magnetostrictive sample, experiencing the same environment, for compensation

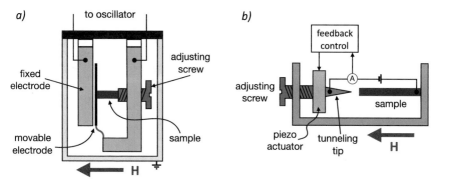

Fig. 2 Schematic representation of (**a**) capacitive dilatometer to measure magnetostriction (adapted from Refs. [27, 30]; (**b**) tunneling tip dilatometer (adapted form Ref. [31]

variation produced by the undesired effects. The gauge glued to the specimen under test would produce a resistance variation $\Delta R/R = x + y$. An additional "dummy" gage can be used, which is submitted to the same temperature and magnetic field than the "active" gauge but glued onto a non-magnetostrictive material, so $\Delta R/R = y$ (Fig. 1c). It is easy to confirm that, when the gauges are connected in consecutive branches of the Wheatstone bridge, the voltage output is proportional to the difference of the resistance variations between the two gauges ($V_o \propto x$), thus compensating the undesired effects.

The strain gauge method is routinely used to measure magnetostriction in materials and can be integrated in compact systems for high field and low temperature measurements [25].

Capacitive Bridges

Capacitive dilatometers are customarily used to measure thermal expansion [26]. Basically, the strain to be measured displaces one of the capacitor plates, producing a variation in capacitance that is detected through the variations in the resonance frequency in a *LC* circuit. The construction of the capacitance cell normally follows a three terminal configuration in which the grounded third terminal assures electrostatic shielding and provides thermal stability (Fig. 2a). Different, improved versions of capacitance bridges have been used to measure magnetostriction in a wide range of temperatures and materials [27–29]. Using a cantilever configuration, the method can be extended to measure magnetostriction in thin films [30].

Tunneling and Atomic Force Tip Measurements

The magnetostriction of a conductive material can be quite precisely determined by the tunneling current from a proximity tip, as described by Brizzolara [31]. Figure

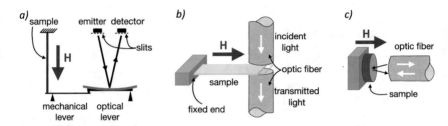

Fig. 3 (**a**) Simplified scheme of the optical lever used in Ref. [34]. (**b**) Fiber-optic method described in Ref. [35]. (**c**) Principle used in Ref. [36]

2b schematizes the procedure, where a feedback mechanism maintains constant the distance between the tip and the magnetostrictively strained sample, using a piezoelectric actuator. Strains of 10^{-9} can be discriminated.

In a much smaller scale, an atomic force microscope (AFM) has been used to observe the magnetostriction of individual galfenol-based nanowires 150 nm in diameter [32]. The sophisticated procedure involves welding both ends of the wire by electron beam deposition to an AFM reference grid. The magnetic field applied along the wire produces a magnetostrictive elongation, causing a bending that is measurable through the AFM and can be used to estimate the magnetostriction, with a value of 40 ppm in that particular case.

Although the above methods measure macroscopic magnetostriction, scanning probes are well suited for local measurements. For instance, AFM tips in contact mode have been used to determine the local magnetostriction response of Cobalt thin films [33].

Optical and Interferometric Methods

Different optical methods for a direct measurement of the magnetostrictive strain have been developed, with increasing degree of sophistication as the technology advanced, providing a considerable resolution in all cases. Early systems used optical levers, in which the deformation of the specimen produced a change in the reflection angle of an optical beam. For instance, McKeehan and Cioffi [34], using a clever setup combining a mechanical lever, an optical lever and a system of slits (Fig. 3a), were capable of determining strains of 2×10^{-9} in permalloy wires. The advent of fiber-optic dilatometers allowed for simplified measuring systems, benefiting from the inexistence of physical contact between the sample and the probe. Squire and Gibbs [35] proposed a system adapted for samples in the form of ribbons in which the sample acts as a shutter for the light transmitted through a small gap between the fibers (Fig. 3b). This is a very convenient method since it minimizes the load on the measured sample, which is attached by one end but free to move on the other, with no additional element attached to it. The method was probed capable of determining magnetostriction coefficients of 10^{-8}. Samata and co-workers [36] optimized a fiber-optic dilatometer working in reflection mode in which the magnetrostriction is determined by the ratio of the incident and reflected light (Fig. 3c), a configuration that was later improved using a single fiber and an annular photodiode for detection [37].

Interferometric measurement methods were introduced by Kwaaitaal [38] and later by Kakuno and Gondó [39], who adapted a Michelson-type interferometer, devised for vibration measurements, to the determination of magnetostrictive displacements with a resolution of 1 nm. Many other configurations have been proposed since then: Doppler interferometry [40], Mach-Zehnder [41], and even the more exotic Speckle interferometry [42].

It is to be noted that magnetostrictive-mediated fiber-optic strain sensing has become a very sensitive procedure to measure small magnetic fields. The principle was demonstrated in 1983 [43] and improved using fiber Bragg grating (FBG) strained by terfenol-D [44]. These techniques are reviewed in Ref. [45].

Indirect Measurements

The methods described here make use of the Villari effect, also called inverse magnetostriction, which account for the influence of stress on the magnetic behavior of materials. To best describe the principle of the methods, let's consider the simple situation of a magnetic sample with a uniaxial anisotropy K as represented in Fig. 4a. If a magnetic field H is applied perpendicularly to the easy axis, the equilibrium angle θ of the magnetization will be the one that minimizes the sum of the magnetostatic energy $E_m = - \mu_0 H M_s \cos \theta$ and the anisotropy energy $E_a = K \cos^2 \theta$, where M_s is the saturation magnetization. The net magnetization measured in the direction of the field will be $M = M_s \cos \theta$, giving.

$$M = \frac{1}{2} \frac{\mu_0 M_s^2}{K} H \ (\sigma = 0).\tag{4}$$

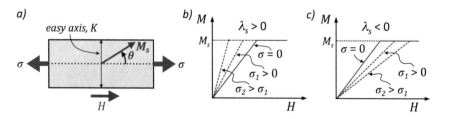

Fig. 4 (a) Magnetostrictive sample subjected to a tensile stress σ and a magnetic field H. The sample presents a uniaxial transverse anisotropy K. The equilibrium magnetization makes an angle θ. (b) Change in the magnetization curve with increasing tensile stresses for positive magnetostriction. (c) Same for negative magnetostriction

In this case, the magnetization curve is linear (see Fig. 4), presenting a constant susceptibility $\chi = M/H$ until the saturation is reached at $H = H^k$, with the anisotropy field defined as

$$H^k = \frac{2K}{\mu_0 M_s}. \tag{5}$$

If a stress σ is applied, a new magnetoelastic energy term must be considered $E_{me} = -\frac{3}{2}\lambda_s\sigma\cos^2\theta$, which can be readily evaluated as the work done by the stress σ along the deformation given by Eq. (1) when the magnetization is rotated to the angle θ. The energy minimization process yields

$$M = \frac{1}{2}\frac{\mu_0 M_s^2}{\left(K - \frac{3}{2}\lambda_s\sigma\right)}H \quad (\sigma \neq 0). \tag{6}$$

Comparing to Eq. (4), we see that the stress in a magnetostrictive material creates a uniaxial magnetoelastic anisotropy

$$K_\sigma = \frac{3}{2}\lambda_s\sigma \tag{7}$$

that modifies the permeability of the material. If the magnetostriction of the material is positive ($\lambda_s > 0$), a tensile stress ($\sigma > 0$) produces an easy axis in the direction of the applied stress. The magnetoelastic anisotropy is then perpendicular to the axis of the original anisotropy K, so the effective anisotropy in the sample is $K_{eff} = K - K_\sigma$. An increase of the applied tensile stress produces an increase of the permeability, as schematized in Fig. 4b. If the magnetostriction is negative, a ($\lambda_s < 0$), a tensile stress, ($\sigma > 0$), produces an easy axis perpendicular to the applied stress. The magnetoelastic anisotropy K_σ adds up to the original K, and the applied tensile stress produces a decrease of the permeability as depicted in Fig. 4c. If the applied stress in compressive ($\sigma < 0$), the situation reverses, according to the resulting sign of the product $\lambda_s\sigma$ as expressed in Eq. (6).

The indirect methods for measuring the magnetostriction use the effects of the magnetoelastic anisotropy created when the sample is stressed to determine the saturation magnetostriction coefficient λ_s. Some strategies are briefly described in the following.

Dependence of Permeability on Stress

The actual shape of the hysteresis loop in a magnetostrictive sample is greatly modified by the applied stress. Figure 5 clearly illustrates this fact in two amorphous ribbons with opposite sign of magnetostriction. In a more general way than the one outlined in the previous paragraph, the effect of stress on ferromagnetic hysteresis was modeled by Jiles and co-workers [46]. As a tool for measuring magnetostric-

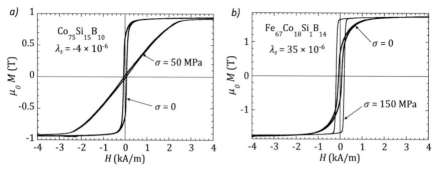

Fig. 5 Influence of the applied tensile stress on the hysteresis loop of magnetostrictive amorphous ribbons 20 μm thick and 2 mm wide. (**a**) Negative magnetostriction sample. The stress is applied by hanging a load of 200 g. (**b**) Sample with positive magnetostriction. The load for applying the stress is 600 g

tion, hysteresis loops are recorded for different values of the applied stress. λ_s is determined from the dependency on the applied stress σ of either the anisotropy field, the permeability at low field, or the area A enclosed by the magnetization curve and vertical axis ($A = 1/2 \; \mu_0 M_s H^k = K_{\text{eff}}$). To simplify the analysis, the hysteresis of the magnetization curve can be eliminated by averaging the ascending and descending branches. These methods were early used to determine the magnetostriction of Nickel wires [47] and amorphous ribbons [48]. Since then, the method has been intense and routinely used, for example, to investigate the origin of the stress dependence of magnetostriction in amorphous ribbons [49], and it is still presently in active use [50]. The analysis of the dependence of the anisotropy field with the applied stress is particularly interesting in amorphous ribbons, where a distribution of anisotropy and magnetostriction values can exists [51].

Small Angle Magnetization Rotation

This technique, developed by Narita and co-workers [52] in 1980, determines the saturation magnetostriction through the effect of the magnetoelastic anisotropy on the amplitude of the oscillation of the magnetization produced by an alternating magnetic field. The schematic procedure is represented in Fig. 6. A longitudinal *dc* magnetic field H_0 is applied to magnetically saturate the sample. A small transverse *ac* magnetic field $h = h_0 \sin \omega t$ makes the magnetization M_s oscillate about the equilibrium position. In the presence of an applied stress σ, the total energy is given by the sum of the magnetostatic energy $E_m = -\mu_0 H_0 M_s \cos \theta - \mu_0 h M_s \sin \theta$ and the magnetoelastic energy $E_{me} = -\frac{3}{2}\lambda_s \sigma \cos^2\theta$ (Narita also considered the demagnetizing field contribution, which we ignore here). The equilibrium angle is given by

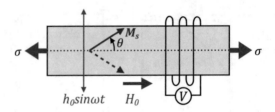

Fig. 6 Schematic setup for the small angle magnetization rotation method of measuring the saturation magnetostriction. A pickup coil receives the voltage induced by the oscillation of the magnetization produced by a transversal *ac* magnetic field, while the sample is subjected to a longitudinal *dc* magnetic field and an applied stress

$$\sin \theta = -\frac{h}{H_0 + 3\frac{\lambda_s \sigma}{\mu_0 M_s}}. \tag{8}$$

The longitudinal magnetization is $M = M_s \cos \theta$ and the voltage induced in a coil with N turns of section S

$$V = -NS\frac{dM}{dt} = NSM_s \left(\frac{h_0}{H_0 + 3\frac{\lambda_s \sigma}{\mu_0 M_s}} \right)^2 \omega \sin(2\omega t) \tag{9}$$

where $\cos \theta = 1$ has been considered since the angle is small.

Therefore, the amplitude of the second harmonic depends on the magnetostriction coefficient. The usual mode of operation consists in adjusting the magnitude of the longitudinal field in order to maintain the amplitude of the induced voltage constant under changes of the applied stress. In this case, the magnetostriction coefficient is obtained from

$$\lambda_s = \frac{\mu_0 M_s}{3\sigma} \Delta H_0. \tag{10}$$

The procedure is well adapted to magnetostriction measurements in samples in the form of strips. It has also been used in wide amorphous ribbons [53] and wires [54], where this measurement method can benefit from the modification introduced by Hernando and co-workers [55], consisting in producing the oscillation of the magnetization by the circumferential field produced by an alternating current flowing through the sample. Another variation of the method was proposed by Kraus, in which the *dc* field is applied transversally, instead of longitudinally, effectively measuring the transverse susceptibility, from which the saturation magnetostriction was derived [56].

Ferromagnetic Resonance Methods

Ferromagnetic resonance (FMR) is very sensitive to changes in the effective internal field and can therefore be used to determine the magnetostriction coefficient through the shift in the resonance spectra caused by the magnetoelastic anisotropy field

$$H_{me}^k = 3\frac{\lambda_s \sigma}{\mu_0 M_s} \tag{11}$$

as derived from Eqs. (5) and (7). A direct measurement of the magnetostriction coefficient using this method was first proposed by Smith and Jones [57], in a traditional resonance cavity in which the stress on the sample was produced by means of a quartz plunger. Alternatively, transmission lines can be used to measure the changes of FMR, where the stress is applied by bending the sample, which is particularly useful in thin films [58]. A variation of the method, which has been named *strain-modulated FMR*, consists in applying an alternating strain to the sample by means of a piezoelectric driver and evaluating the frequency shift of FMR in a resonance cavity with phase sensitive detection [59]. Related with FMR methods, novel techniques are being proposed in which the study of the dynamics of the magnetization in time-resolved experiments would allow to determine the influence of the magnetoelastic effects, for example, after the application of a step-like stress to the sample [60].

Thin Film Methods

Magnetostrictive materials in the form of thin films are especially interesting for integration in miniaturized sensors and actuators or, in general, in micro electromechanical systems or MEMS. As with bulk materials, both direct and indirect methods have been developed to determine the magnetoelastic coefficients of thin films. Some of the examples and references described in the previous paragraphs already dealt with thin film systems. Here, we want to emphasize how the presence of the substrate onto which the films are deposited conditions the magnetostriction measurements.

In the direct approach, a cantilever-style system is employed, in which the sample is usually clamped by one end. When the film strains magnetostrictively, a bending is produced, as shown in Fig. 7a. If the film is strained λ, the deflection d of the cantilever and the slope angle ϕ at a distance l from the clamping point are given by [61]:

$$d = \frac{3 t_f l^2}{t_s^2}\frac{E_f (1 - \nu_s)}{E_s (1 - \nu_f)}\lambda; \quad \phi = \frac{2d}{l} = \frac{6 t_f l}{t_s^2}\frac{E_f (1 - \nu_s)}{E_s (1 - \nu_f)}\lambda \tag{12}$$

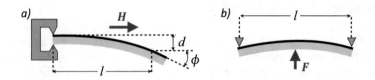

Fig. 7 Configuration for magnetostriction measurements in thin films. (**a**) Direct measurements of the magnetostrictive strain by the deflection of a cantilever. (**b**) Indirect determination of the magnetostriction coefficient through the magnetoelastic anisotropy induced by the bending

where t_f, t_s are the thickness of the film and the substrate, and E_f, E_s and v_f, v_s are Young's modulus and Poison's ratio of the film and substrate, respectively. Equation (12) evidences that long samples and thin substrates are preferred to maximize the deflection, but in any case, to determine the magnetostrictive strain of the film, the elastic constants must be known.

The deflection can be measured by several sensitive methods, either capacitive [30, 62, 63] or different optical [61, 64, 65] and interferometric [40, 66] strategies. The deflection of the cantilever has also been measured directly using a nanoindentation apparatus [67] or an atomic force microscope [68]. Nevertheless, it is also possible to measure directly the magnetostrictive strain without clamping the sample. For instance, Varghese and co-workers use laser Doppler vibrometry to determine the deflections of a free sample under an alternating magnetic field [69].

Indirect methods in thin films determine the anisotropy induced magnetoelastically when deforming the substrate film system. The changes in the anisotropy field H_{me}^k can be observed in the hysteresis loop of the bent film. In the three-point bending configuration of Fig. 7b, the stress produced by bending is given by

$$\sigma = \frac{3}{2} \frac{Fl}{bd^2} \tag{13}$$

being F the applied force, l the distance between the outer supports, and b and d the width and the thickness of the substrate, respectively. The produced strain ϵ is.

$$\epsilon = \frac{1 - v_s}{E_s} \sigma \tag{14}$$

where E_s and v_s are the Young's modulus and Poison's ratio of the substrate. Since the strain is the same in the film and the substrate, Eq. (11) allows calculating the magnetostriction coefficient from the relation:

$$\lambda_s = \frac{\mu_0 M_s}{3\sigma} \frac{E_s (1 - v_f)}{E_f (1 - v_s)} H_{me}^k \tag{15}$$

There are many examples of the application of this method in the literature [70, 71]. It has even been described as a metrology tool for MEMS using whole-wafer measurements [72]. Jiles and co-workers provide a comparison of direct and indirect methods in thin films [73].

To finalize this section, we must remark that the different techniques and procedures to determine the magnetostriction have been reviewed extensively along the time. Some of these compilations are the ones of Squire [74], O'Handley [7], Ekreem [75], and Grossinger [76].

3 Applications

The magnetostrictive effect produces a conversion of electric energy (used to magnetize the material) to mechanical energy, and in this respect magnetostrictive materials, especially the giant magnetostriction ones, have been intensively investigated for developing useful actuators. Conversely, the *Villari* or magnetoelastic effect produces an electrical output (derived from magnetization changes) when a magnetostrictive material is strained under the action of a stress. Thus, suitable materials, with a predominance of metallic glasses, have been used to develop sensing devices that take advantage of this effect. Magnetostrictive materials occupy a distinguished position in the collection of active materials for modern transducer (sensor and actuator) technologies [77]. Additionally, the electrical output produced by vibrations transmitted to magnetostrictive materials has motivated the proposal of magnetostrictive energy harvesters, an area that has gained traction recently with the develop of magnetoelectric composites constituted by piezoelectric and magnetostrictive laminates. These three are the main areas of application of the magnetostrictive effect and will be briefly surveyed in the following.

Magnetostrictive Actuators

Magnetostrictive and piezoelectric actuators share the same ground for applications. The former promises a higher energy density, increased robustness, and a potentially interesting noncontact activation. Terfenol-D is the material with the largest magnetostriction at room temperature (up to 2000 ppm), which makes it the ideal candidate for actuation. However, its poor workability and brittleness limit the size and shape of the active elements. Galfenol displays a reduced value of magnetostriction (~300 ppm) but with improved mechanical properties that makes it suitable for more flexible designs. Other materials as CoFe-based (Permedur) have also been used in actuators.

The simplest device configuration is the linear actuator that can be used to design micropositioners. These can achieve nanometric resolution [78], although the repeatability and accuracy are greatly compromised by the large hysteresis of the active materials. This base configuration is also employed for active vibration control [79]. This application can benefit from self-sensing actuation, which exemplifies the bidirectionality of the transducer, since the same magnetostrictive element can be used to sense the structural vibrations and provide the actuation to attenuate them [80]. The stroke of magnetostrictive actuators is intrinsically small, and different strategies are used to produce an amplified actuation from the linear configuration. This can be achieved mechanically by means of levers and flexible hinges [81] or hydraulically using pistons of different areas [82].

Another field of actuation for magnetostrictive materials consists on motors based on the inchworm principle [83], in which, basically, the magnetostrictive elongation and contraction is used to produce motion by fixing alternatively each extreme of the material. The clamping can be achieved by piezoelectric actuation [84], illustrating the possibilities of hybrid systems. A discussion of various configurations and actuation modes can be found in Ref. [85].

A significant number of diverse applications have been proposed for magnetostrictive actuators, including injectors [86], pumps [87], valves [88], alignment of segmented mirrors for telescopes [89], nondestructive testing [90], and loudspeakers [91] to cite some.

The mentioned hysteresis and nonlinearity of the response of magnetostrictive materials makes that useful applications require a control strategy usually implemented over a model of the material behavior. The review of Apicella and co-workers compiles such procedures, together with a very complete introduction about magnetostrictive actuators [92]. Some commercial devices (although listed as prototypes at the time of this work) can be found from Cedrat Technologies [93].

Magnetoelastic Sensors

The magnetoelastic phenomena have given rise to the development of a great number of sensing procedures. Magnetoelastic sensors are often classified as static or dynamic according to the principle of operation, the latter being related with the use of magnetoelastic waves and resonance. There are many exhaustive reviews covering the subject [94–96], so a broad summary will be provided here.

Static sensors make direct use of the changes in permeability produced by the Villari effect and are therefore conceived to measure strain, stress, or any other magnitude that can be related to them. From Eq. (6), we can derive the sensitivity of the permeability ($\mu = 1 + \chi = 1 + M/H$) to applied stresses:

$$\frac{\mathrm{d}\mu}{\mathrm{d}\sigma} = \frac{3\mu_0 M_\mathrm{s}}{(2K - 3\lambda_\mathrm{s}\sigma)^2}\lambda_\mathrm{s}, \tag{16}$$

that for small stresses becomes

$$\left(\frac{\mathrm{d}\mu}{\mathrm{d}\sigma}\right)_{\sigma\to0} = \frac{3\chi_0^2}{\mu_0 M_\mathrm{s}^2}\lambda_\mathrm{s} \tag{17}$$

with $\chi_0 = \mu_0 M_s^2/2K$ being the susceptibility of the unstressed material (Eq. 4). Therefore, ideal sensing materials should display large magnetostriction (λ_s) and low anisotropy (K).

In this range of applications, magnetoelastic sensors compete with resistive strain gauges (that were briefly introduced in Sect. 2.1.1), featuring two main advantages. First, they present greater sensitivity: the gauge factor $K = (\Delta R/R)/\epsilon$ is about 2 for metallic gauges and few hundreds for semiconductor ones, whereas the relative change in permeability in a magnetoelastic gauge made of an amorphous FeSiB alloy can produce a gauge factor $G = (\Delta\mu/\mu)/\epsilon \sim 4 \times 10^5$ [97]. Second, magnetoelastic detection is inherently contactless, since the permeability changes can be detected using a pickup coil. This second advantage becomes of paramount importance in measuring the deformation of rotating shafts, which makes torque sensors a niche application for magnetoelastic sensors. ABB commercializes different devices based on this technology [98]. The detailed principle of operation is explained in Ref. [95], but, basically, the torsion created by the torque to be measured produced magnetoelastic strains in the surface of the shaft that changes the permeability, either of the shaft material, if it is magnetic as happens in ship propulsion, or on specialized materials attached to the surface. The permeability changes are detected differentially by a system of coils, because torsion creates both compressive and tensile stress in the surface (at 45° with the axis of the shaft).

Similar procedures are used in magnetoelastic sensors to measure force and pressure, which are reviewed in Ref. [99]. Particularly, amorphous ferromagnetic materials have been used to develop a great number of applications [100, 101].

Dynamic magnetoelastic sensors make use of the propagation of magnetoelastic waves. A sudden change of the magnetic state at a given point of a magnetostrictive material generates a stress that propagates as a sound wave through the material. Conversely, the deformation at each point produced by the wave magnetoelastically changes the magnetic state, which can be measured inductively by a pickup coil. The most obvious application of this principle is distance or position sensing, and sensors under the name of magnetostrictive delay line (MDL) were developed according to this principle [102]. The compilation of Hristoforou reviews the fundamentals and applications of MDLs [103]. Position and level sensors based in this principle are commercialized by MTS [104].

As introduced Magnetoelastic Waves and Resonance, if the frequency of the excitation is suitably tuned, a magnetoelastic resonance occurs. There are two main ways of using the magnetoelastic resonance for sensing purposes. One successful application is in the field of antishoplifting labels [105]. The tag attached to the good that is to be protected contains a magnetostrictive amorphous strip whose resonance frequency can be conveniently changed due to the ΔE effect (Eq. 3). This permits an easy deactivation of the tag and to build a very robust system that avoids false alarms by detecting the ring-down oscillations of the label after a short pulse activation. The other type of promising applications based on the magnetoelastic resonance analyzes the changes in the resonance parameters (frequency and amplitude) in response to physical parameters that affect the magnetostrictive element, such as temperature, pressure or viscosity of the medium, mass of substance deposited onto it, etc. Grimes and co-workers have performed an extensive work in this regard [106]. If the surface of the magnetoelastic resonator is conveniently functionalized, the principle of operation can be extended to the detection of chemical and biological substances [107].

Energy Harvesting

Small-scale energy harvesting is acquiring increasing importance to power autonomous systems, such as sensors or IoT (internet of things) devices. Magnetostrictive materials can harvest energy from mechanical vibrations, being itself an active research field [108]. Additionally, in association with piezoelectrics, they provide the possibility of energy harvesting from magnetic fields: the magnetic field causes a deformation in the magnetostrictive material that is transmitted to the piezoelectric and transformed to electrical energy. These kinds of systems have become known as magneto-mechano-electric (MME) generators. The materials that present both magnetostrictive and piezoelectric response are quite few in single phase (multiferroics) [109], so much of the research in energy harvesting is performed on magnetoelectric composites, in the form of sandwiched laminates of piezoelectric (such as PZT or PVDF [110]) and magnetostrictive materials (amorphous or giant magnetostrictive ones). The parameter that characterizes the performance is the magnetoelectric coefficient:

$$\alpha_{ME} = \frac{1}{t}\left(\frac{dV}{dH}\right) \tag{18}$$

where dV is the measured voltage induced when applying a magnetic field of amplitude dH and t is the thickness of the piezoelectric material. The conversion is greatly enhanced if the composite is driven at resonance, with values of α_{ME} that can reach 7000 V cm^{-1} Oe^{-1} [111]. A review of the use and performance of magnetoelectrics for energy harvesting can be found in Ref. [112].

4 Summary

Magnetostriction and its related phenomenology and effects have been being studied for nearly two centuries now. During this time, many useful applications in the form of sensors and actuators have been proposed, some of them becoming successful commercial devices. Despite this long history, magnetostriction and magnetostrictive materials are still actively investigated and new applications discovered. In this chapter we have first reviewed the basic principles of magnetostriction and the magnetoelastic coupling, both as a static or stationary effect and as a dynamic one, where the coupling is propagated as magnetoelastic waves. Secondly, we have briefly described the main types of magnetostrictive materials, highlighting among them amorphous and giant magnetostriction alloys for their superior properties. The central part of the chapter is dedicated to the description of the techniques most commonly used for the characterization of the magnetostrictive properties. These techniques are classified in direct and indirect methods. The first ones measure directly the magnetostrictive strain, using resistive (strain gauges), capacitive, or optical sensors. The second ones make use of the inverse magnetostrictive effects, determining the magnetostriction coefficients from the change in the magnetic properties. A section is specially dedicated to describe the measurements in materials in the form of thin films, a very active field due to their use in microelectromechanical systems (MEMS). The chapter concludes with a short revision of the application of magnetostrictive materials in sensors and actuators, together with a final part on energy harvesting systems, where magnetostrictive materials are used in combination with piezoelectric ones.

Acknowledgments The author wishes to thank the research projects MAT2017-83632-C3 from the Spanish government and KK-2021/00082 and IT1245-19 from the Basque government.

References

1. J.P. Joule, On a new class of magnetic forces. Ann. Electr. Magn. Chem. **8**, 219–224 (1842)
2. E.W. Lee, Magnetostriction and magnetomechanical effects. Rep. Prog. Phys. **18**, 184 (1955)
3. B.D. Cullity, C.D. Graham, *Introduction to Magnetic Materials*, 2nd edn. (Wiley-IEEE Press, Piscataway NJ, 2009)
4. S. Chikazumi, *Physics of Ferromagnetism*, 2nd edn. (Oxford University Press, Oxford, 1997)
5. J.M.D. Coey, *Magnetism and Magnetic Materials* (Cambridge University Press, Cambridge, UK, 2009)
6. E. de Trémolet de Laichesserie, *Magnetostriction, Theory and Applications of Magnetoelasticity* (CRC Press, Boca Ratón, FL, 1993)
7. R. O'Handley, *Modern Magnetic Materials: Principles and Applications* (Wiley, New York, 2000), p. 225
8. R.M. Bozorth, An extensive compilation of data can be found, in *Ferromagnetism*, (IEEE Press, Piscataway NJ, 1978)
9. G.Y. Chin, J.H. Wernick, Soft magnetic metallic materials, in *Ferromagnetic Materials*, ed. by E. P. Wohlfart, vol. 2, (North Holland-Elsevier, Amsterdam, 1980), p. 166
10. R. O'Handley, *Modern Magnetic Materials: Principles and Applications* (John Wiley & Sons, New York, 2000), p. 362
11. Z. Kaczkowski, Magnetostrictive Alloy, US Patent 4033791, July, 1977
12. A.L. Greer, Metallic glasses. Science **267**, 1947 (1995)
13. A. Hernando, M. Vázquez, J.M. Barandiarán, J. Phys. Instrum. **21**, 1129 (1988)
14. D. Berlincourt, D.R. Curran, H. Jaffe, Piezoelectric and piezomagnetic materials and their function in transducers, in *Physical Acoustics*, ed. by W. E. Mason, vol. 1 Part A, (Academic Press, New York, 1964), p. 169
15. F.J. Darnell, Magnetostriction in dysprosium and terbium. Phys. Rev. **132**, 128 (1963)
16. A.E. Clark, Magnetostrictive rare earth-Fe2 compounds, in *Ferromagnetic Materials*, ed. by E. P. Wohlfarth, vol. 1, Chapter 7, (North-Holland Publishing Company, Amsterdam, 1980), pp. 531–589

17. G. Engdahl, *Handbook of Giant Magnetostrictive Materials* (Academic Press, San Diego, 2000)
18. A.E. Clark, K.B. Hathaway, M. Wun-Fogle, J.B. Restorff, T.A. Lograsso, V.M. Keppens, G. Petculescu, R.A. Taylor, Extraordinary magnetoelasticity and lattice softening in bcc Fe–Ga alloys. J. Appl. Phys. **93**, 8621 (2003)
19. J.B. Restorff, M. Wun-Fogle, K.B. Hathaway, A.E. Clark, T.A. Lograsso, G. Petculescu, Tetragonal magnetostriction and magnetoelastic coupling in Fe–Al, Fe–Ga, Fe–Ge, Fe–Si, Fe–Ga–Al, and Fe–Ga–Ge alloys. J. Appl. Phys. **111**, 023905 (2012)
20. H. Wang, R. Wu, First-principles investigation of the effect of substitution and surface adsorption on the magnetostrictive performance of Fe–Ga alloys. Phys. Rev. B **99**, 205125 (2019)
21. R. Pallás-Areny, J.G. Webster, *Sensors and Signal Conditioning* (Wiley, New York, 2001), p. 80
22. J.E. Goldman, Magneto-striction of annealed and cold worked nickel rods. Phys. Rev. **72**, 529 (1947)
23. R.M. Bozorth, R.W. Hamming, Measurement of magnetostriction in single crystals. Phys. Rev. **89**, 865 (1953)
24. R.D. Greenough, C. Underhill, Strain gauges for the measurement of magnetostriction in the range 4 K to 300 K. J. Phys. E: Sci. Instrum. **9**, 451 (1976)
25. W. Wang, H. Liu, R. Huang, Y. Zhao, C. Huang, S. Guo, Y. Shan, L. Li, Thermal expansion and magnetostriction measurements at cryogenic temperature using the strain Gauge method. Front. Chem. **6**, 72 (2018)
26. R.K. Kirby, Methods of measuring thermal expansion, in *Compendium of Thermophysical Property Measurement Methods*, ed. by K. D. Maglić, A. Cezairliyan, V. E. Peletsky, (Springer, Boston, MA, 1992)
27. N. Tsuya, K.I. Arai, K. Ohmori, Y. Shiraga, Magnetostriction measurement by three terminal capacitance method. Jpn. J. Appl. Phys. **13**, 1808 (1974)
28. R.R. Birss, G.J. Keeler, P. Pearson, R.J. Potton, A capacitive instrument for the measurement of a large range of magnetostriction at low temperatures and high magnetic fields. J. Phys. E: Sci. Instrum. **11**, 928 (1978)
29. M.S. Boley, W.C. Shin, D.K. Rigsbee, D.A. Franklin, Capacitance bridge measurements of magnetostriction. J. Appl. Phys. **91**, 8210 (2002)
30. B. Kundys, Y. Bukhantsev, S. Vasiliev, D. Kundys, M. Berkowski, V.P. Dyakonov, Three terminal capacitance technique for magnetostriction and thermal expansion measurements. Rev. Sci. Instrum. **75**, 2192 (2004)
31. R. Brizzolara, Magnetostriction measurements using a tunneling-tip strain detector. J. Magn. Magn. **88**, 343 (1990)
32. J.J. Park, E.C. Estrine, S.M. Reddy, B.J.H. Stadler, A.B. Flatau, Technique for measurement of magnetostriction in an individual nanowire using atomic force microscopy. J. Appl. Phys. **115**, 17A919 (2014)
33. K.-E. Kim, C.-H. Yang, Local magnetostriction measurement in a cobalt thin film using scanning probe microscopy. AIP Adv. **8**, 105125 (2018)
34. L.W. McKeehan, P.P. Cioffi, Magnetostriction in permalloy. Phys. Rev. **28**, 146 (1926)
35. P.T. Squire, M.R.J. Gibbs, Fibre-optic dilatometer for meauring magnetostriction in ribbon samples. J. Phys. E: Sci. Instrum. **20**, 499 (1987)
36. H. Samata, Y. Nagata, T. Uchida, S. Abe, New optical technique for bulk magnetostriction measurement. J. Magn. Magn. Mater. **212**, 355 (2000)
37. V. De Manuel, R.P. del Real, J. Alonso, H. Guerrero, Magnetostriction measuring device based on an optical fiber sensor with an annular photodiode. Rev. Sci. Instrum. **78**, 095104 (2007)
38. T. Kwaaitaal, The measurement of small magnetostrictive effects by an interferometric method. J. Magn. Magn. Mater. **6**, 290 (1977)
39. K. Kakuno, Y. Gondó, A new measuring method of magnetostrictive vibration. J. Appl. Phys. **50**, 7713 (1979)

40. G.H. Bellesis, P.S. Harllee, A. Renema, D.N. Lambe, Magnetostriction measurement by interferometry. IEEE Trans. Magn. **29**, 2989 (1993)
41. M.H. Kim, K.S. Lee, S.H. Lim, Magnetostriction measurements of metallic glass ribbon by fiber-optic Mach-Zehnder interferometry. J. Magn. Magn. Mater. **191**, 107 (1999)
42. F. Salazar, A. Bayón, J.M. Chicharro, Measurement of magnetostriction coefficient λ_s by speckle photography. Opt. Commun. **282**, 635 (2009)
43. J.P. Willson, R.E. Jones, Magnetostrictive fiber-optic sensor system for detecting dc magnetic fields. Opt. Lett. **8**, 333 (1983)
44. G.N. Smith, T. Allsop, K. Kalli, C. Koutsides, R. Neal, K. Sugden, P. Culverhouse, I. Bennion, Characterisation and performance of a Terfenol-D coated femtosecond laser inscribed optical fibre Bragg sensor with a laser ablated microslot for the detection of static magnetic fields. Opt. Express **19**, 363 (2011)
45. J. Ascorbe, J.M. Corres, Magnetic field sensors based on optical fiber, in *Fiber Optic Sensors: Current Status and Future Possibilities*, ed. by I. R. Matias, S. Ikezawa, J. Corres, (Springer, 2017), p. 269
46. M.J. Sablik, H. Kwun, G.L. Burkhardt, D.C. Jiles, Model for the effect of tensile and compressive stress on ferromagnetic hysteresis. J. Appl. Phys. **61**, 3799 (1987)
47. R. Becker, M. Kersten, Die Magnetisierung von Nickeldraht unter starkem Zug. Z. Physik **64**, 660 (1930) (in German)
48. T.H. O'Dell, Magnetostriction measurements on amorphous ribbons by the Becker-Kersten method. Phys. Status Solidi **68**, 221–226 (1981)
49. A. Hernando, C. Gómez-Polo, E. Pulido, G. Rivero, M. Vázquez, A. García-Escorial, J.M. Barandiaran, Tensile-stress dependence of magnetostriction in multilayers of amorphous ribbons. Phys. Rev. B **42**, 6471 (1990)
50. V.S. Severikov, A.M. Grishin, V.S. Ignakhin, Magnetostriction in $Fe_{80-x}Co_xP_{14}B_6$ amorphous ribbons evaluated by Becker-Kersten method. J. Phys.: Conf. Ser. **1038**, 012066 (2018)
51. J.M. Baradiarán, A. Hernando, Amorphous magnetism: the role of anisotropy and magnetostriction distributions. J. Magn. Magn. Mater. **104–107**, 73–76 (1992)
52. K. Narita, Y. Yamasaki, H. Fukunada, Measurement of saturation magnetostriction of a thin amorphous ribbon by means of small-angle magnetization rotation. IEEE Trans. Magn. **16**, 435–439 (1980)
53. F. Alves, P. Houée, M. Lécrivain, F. Mazaleyrat, New design of small-angle magnetization rotation device: evaluation of saturation magnetostriction in wide thin ribbons. J. Appl. Phys. **81**, 4322 (1997)
54. J. Yamasaki, Y. Ohkubo, F.B. Humphrey, Magnetostriction measurement of amorphous wires by means of small-angle magnetization rotation. J. Appl. Phys. **67**, 5472 (1990)
55. A. Hernando, M. Vázquez, V. Madurga, H. Kronmüller, Modification of the saturation magnetostriction constant after thermal treatments for the $Co_{58}Fe_5Ni_{10}B_{16}Si_{11}$ amorphous ribbon. J. Magn. Magn. **37**, 161–166 (1987)
56. L. Kraus, A novel method for measurement of the saturation magnetostriction of amorphous ribbons. J. Phys. E: Sci. Instrum. **22**, 943 (1989)
57. A.B. Smith, R.V. Jones, Magnetostriction constants from ferromagnetic resonance. J. Appl. Phys. **34**, 1283 (1963)
58. S.E. Bushnell, W.B. Nowak, S.A. Oliver, C. Vittoria, The measurement of magnetostriction constants of thin films using planar microwave devices and ferromagnetic resonance. Rev. Sci. Instrum. **63**, 2021 (1992)
59. J.C.M. Henning, J.H. den Boef, Magnetostriction measurement by means of strain modulated ferromagnetic resonance (SMFMR). Appl. Phys. **16**, 353 (1978)
60. J.-Y. Duquesne, C. Hepburn, P. Rovillain, M. Marangolo, Magnetocrystalline and magnetoelastic constants determined by magnetization dynamics under static strain. J. Phys. Condens. Matter **394002**, 30 (2018)
61. A.C. Tam, H. Schroeder, A new high-precision optical technique to measure magnetostriction of a thin magnetic film deposited on a substrate. IEEE Trans. Magn. **25**, 2629 (1989)

62. E. Klokholm, The measurement of magnetostriction in ferromagnetic thin films. IEEE Trans. Magn. **12**, 819 (1976)
63. M. Ciria, J.I. Arnaudas, A. del Moral, G.J. Tomka, C. de la Fuente, P.A.J. de Groot, M.R. Wells, R.C.C. Ward, Determination of magnetostrictive stresses in magnetic rare-earth superlattices by a cantilever method. Phys. Rev. Lett. **75**, 1634 (1995)
64. S.-U. Jen, C.-C. Liu, S.-T. Chen, Magnetostriction Measurement of a ferromagnetic thin film using digital holographic microscope. IEEE Trans. Magn. **50**, 6000404 (2014)
65. C. Dong, M. Li, X. Liang, H. Chen, H. Zhou, X. Wang, Y. Gao, M.E. McConney, J.G. Jones, G.J. Brown, B.M. Howe, N.X. Sun, Characterization of magnetomechanical properties in FeGaB thin films. Appl. Phys. Lett. **113**, 262401 (2018)
66. P.S. Harllee III, G.H. Bellesis, D.N. Lambeth, Anisotropy and magnetostriction measurement by interferometry. J. Appl. Phys. **75**, 6884 (1994)
67. T. Shima, H. Fujimori, An accurate measurement of magnetostriction of thin films by using nano-indentation system. IEEE Trans. Magn. **35**, 3831 (1999)
68. B.L.S. Lima, F.L. Maximino, J.C. Santos, A.D. Santos, Direct method for magnetostriction coefficient measurement based on atomic force microscope, illustrated by the example of Tb–Co film. J. Magn. Magn. Mater. **395**, 336–339 (2015)
69. R. Varghese, R. Viswan, K. Joshi, S. Seifikar, Y. Zhou, J. Schwartz, S. Priya, Magnetostriction measurement in thin films using laser Doppler vibrometry. J. Magn. Magn. Mater. **363**, 179–187 (2014)
70. M. Ali, R. Watts, Measurement of saturation magnetostriction using novel strained substrate techniques and the control of the magnetic anisotropy. J. Magn. Magn. Mater. **202**, 85–94 (1999)
71. B. Buford, P. Dhagat, A. Jander, Measuring the inverse magnetostrictive effect in a thin film using a modified vibrating sample magnetometer. J. Appl. Phys. **115**, 17E309 (2014)
72. C.B. Hill, W.R. Hendren, R.M. Bowman, P.K. McGeehin, M.A. Gubbins, V.A. Venugopal, Whole wafer magnetostriction metrology for magnetic films and multilayers. Meas. Sci. Technol. **24**, 045601 (2013)
73. A. Raghunathan, J.E. Snyder, D.C. Jiles, Comparison of alternative techniques for characterizing magnetostriction and inverse magnetostriction in magnetic thin films. IEEE Trans. Magn. **45**, 3269 (2009)
74. P.T. Squire, Magnetomechanical measurements of magnetically soft amorphous materials. Meas. Sci. Technol. **5**, 67–81 (1994)
75. N.B. Ekreem, A.G. Olabi, T. Prescott, A. Rafferty, M.S.J. Hashmi, An overview of magnetostriction, its use and methods to measure these properties. J. Mater. Process. Technol. **191**, 96–101 (2007)
76. R. Grössinger, H. Sassik, D. Holzer, N. Pillmayr, *Accurate measurement of the magnetostriction of soft magnetic materials* (Tech. Univ., Technical Report Institute of Experimental Physik, Vienna, Austria, 2006)
77. S.A. Wilson et al., (27 authors) New materials for micro-scale sensors and actuators: an engineering review. Mater. Sci. Eng.: R: Rep. **56**, 1–129 (2007)
78. B.T. Yang, M. Bonis, H. Tao, C. Prelle, F. Lamarque, A magnetostrictive mini actuator for long-stroke positioning with nanometer resolution. J. Micromech. Microeng. **16**, 1227–1232 (2006)
79. T. Zhang, C. Jiang, H. Zhang, H. Xu, Giant magnetostrictive actuators for active vibration control. Smart Mater. Struct. **13**, 473–477 (2004)
80. L. Jones and E. Garcia. Application of the self-sensing principle to a magnetostrictive structural element for vibration suppression, ASME Proceedings of 1994 International Mechanical Engineering Congress and Exposition, Chicago IL, Vol. 45, 155–165, 1994
81. M. Niu, B. Yang, Y. Yang, G. Meng, Modeling and optimization of magnetostrictive actuator amplified by compliant mechanism. Smart Mater. Struct. **26**, 095029 (2017)
82. S. Chakrabarti, M.J. Dapino, Hydraulically amplified terfenol-D actuator for adaptive powertrain mounts. J. Vib. Acoust. **133**, 611–619 (2011)

83. P.S. Pan, B.T. Yang, G. Meng, J.Q. Li, Design and simulation of a mini precision positioning magnetostrictive inchworm linear motor. Appl. Mech. Mater. **130**, 2846–2850 (2012)

84. J. Kim, J.-D. Kim, S.-B. Choi, A hybrid inchworm linear motor. Mechatronics **12**, 525–542 (2002)

85. F. Claeyssen, N. Lhermet, R. Le Letty, P. Bouchilloux, Actuators, transducers and motors based on giant magnetostrictive materials. J. Alloys Compd. **258**, 61–73 (1997)

86. G. Xue, P. Zhang, X. Li, Z. He, H. Wang, Y. Li, C. Rong, W. Zeng, B. Li, A review of giant magnetostrictive injector (GMI). Sensors Actuators A Phys. **273**, 159–181 (2018)

87. E. Quandt, A. Ludwig, Magnetostrictive actuation in microsystems. Sensors Actuators A Phys. **81**, 275–280 (2000)

88. S. Karunanidhia, M. Singaperumal, Design, analysis and simulation of magnetostrictive actuator and its application to high dynamic servo valve. Sensors Actuators A Phys. **157**, 185–197 (2010)

89. B. Yang, D. Yang, P. Xu, Y. Cao, Z. Feng, G. Meng, Large stroke and nanometer-resolution giant magnetostrictive assembled actuator for driving segmented mirrors in very large astronomical telescopes. Sensors Actuators A Phys. **179**, 193–203 (2012)

90. Y.Y. Kim, Y.E. Kwon, Review of magnetostrictive patch transducers and applications in ultrasonic nondestructive testing of waveguides. Ultrasonics **62**, 3–19 (2015)

91. J.-J. Zhou, Y.-S. Wang, X. Wang, A.-H. Meng, Y.-L. Pan, Design of a flat-panel loudspeaker with giant magnetostrictive exciters, in *Proceedings of the 2008 Symposium on Piezoelectricity, Acoustic Waves, and Device Applications, Nanjing, China, 5–8 December 2008*, pp. 528–532

92. V. Apicella, C.S. Clemente, D. Davino, D. Leone, C. Visone, Review of modeling and control of magnetostrictive actuators. Actuators **8**, 45 (2019)

93. Cedrat Technologies. Available online: https://www.cedrat-technologies.com (accessed on 13 March 2019)

94. G. Hinz, H. Voigt, Magnetoelastic sensors, in *Sensors, A Comprehensive Survey, Magnetic Sensors* (R. Boll and K. J. Overshott, Eds.), ed. by W. Gel, J. Hesse, J. N. Zemel, vol. 5, (VCH, New York, 1989), p. 98

95. A. García-Arribas, J.M. Barandiarán, J. Gutiérrez, Magnetoelastic sensors, in *Encyclopedia of Sensors*, ed. by C. A. Grimes, E. C. Dickey, vol. 5, (American Scientific Publishers, California, 2005), p. 467

96. F.T. Calkins, A.B. Flatau, M.J. Dapino, Overview of magnetostrictive sensor technology. J. Int. Mater. Syst. Struct. **18**, 1057–1066 (2007)

97. M. Wun-Fogle, H.T. Savage, M.L. Spano, Enhancement of magnetostrictive effects for sensor applications. J. Mater. Eng. **11**, 103 (1989)

98. ABB Torductor-Marine brochures: Torductor 500 Shaft Torque and Fuel Efficiency Metering. Available online: https://library.e.abb.com/public/400afcacc0304c8fc1257b2300693495/Engine%20performance%20-%20Brochure%20-%20Torductorr%20500%20Shaft%20Torque%20and%20Fuel%20Efficiency%20Metering.pdf (accessed on July 7, 2021); Torductor-S, Torque measurement with Pressductor® Technology. Available online: https://library.e.abb.com/public/eef8846525950f25c1257380003623c6/3BSE022227R0101_-001.pdf (accessed on July 7, 2021)

99. D.M. Ştefănescu, Magnetoelastic force transducers, in *Handbook of Force Transducers*, (Springer, Berlin, Heidelberg, 2011)

100. A. Hernando, M. Vázquez, J.M. Barandiaran, Metallic glasses and sensing applications. J. Phys. E: Sci. Instrum. **21**, 1129–1139 (1988)

101. T. Meydan, Application of amorphous materials to sensors. J. Magn. Magn. Mater. **133**, 525 (1994)

102. J. Tellerman, *Linear Distance Measuring Device Using a Moveable Magnet Interacting with a Sonic Waveguide*, US Patent No 389855, August, 1975

103. E. Hristoforou, Magnetostrictive delay lines: engineering theory and applications. Meas. Sci. Technol. **14**, R15 (2003)

104. MTS Systems Corporation, Sensors Division, http://www.mtssensor.de (accessed on July 7, 2021)
105. G. Herzer, Magnetic materials for electronic article surveillance. J. Magn. Magn. Mater. **254–255**, 598–602 (2003)
106. C.A. Grimes, C.S. Mungle, K. Zeng, M.K. Jain, W.R. Dreschel, M. Paulose, K.G. Ong, Wireless magnetoelastic resonance sensors: a critical review. Sensors **2**, 294–313 (2002)
107. C. Menti, J.A.P. Henriques, F.P. Missell, M. Roesch-Ely, Antibody-based magneto-elastic biosensors: potential devices for detection of pathogens and associated toxins. Appl. Microbiol. Biotechnol. **100**, 6149–6163 (2016)
108. Z. Deng, M. Dapino, Review of magnetostrictive vibration energy harvesters. Smart Mater. Struct. **26**, 103001 (2017)
109. N. Ortega, A. Kumar, J.F. Scott, R.S. Katiyar, Multifunctional magnetoelectric materials for device applications. J. Phys. Condens. Matter **27**, 504002 (2015)
110. A. Lasheras, J. Gutiérrez, S. Reis, D. Sousa, M. Silva, P. Martins, S. Lanceros-Mendez, J.M. Barandiarán, D.A. Shishkin, A.P. Potapov, Energy harvesting device based on a metallic glass/PVDF magnetoelectric laminated composite. Smart Mater. Struct. **24**, 65024 (2015)
111. Z. Chu, H. Shi, W. Shi, G. Liu, J. Wu, J. Yang, S. Dong, Enhanced resonance magnetoelectric coupling in (1-1) connectivity composites. Adv. Mater. **29**, 1606022 (2017)
112. V. Annapureddy, H. Palneedi, G.-T. Hwang, M. Peddigari, D.-Y. Jeong, W.-H. Yoon, K.-H. Kim, J. Ryu, Magnetic energy harvesting with magnetoelectrics: an emerging technology for self-powered autonomous systems. Sustain. Energy Fuels **1**, 2039 (2017)

Magnetic Properties of Granular L1₀ FePt Films for Heat-Assisted Magnetic Recording (HAMR) Applications

Cristian Papusoi, Mrugesh Desai, Sergiu Ruta, and Roy W. Chantrell

Abstract The merit of heat-assisted magnetic recording (HAMR) that enables to increase the recording density beyond the classical perpendicular magnetic recording (PMR) is a contribution to the writing field gradient additional to that generated by the magnetic writer. This contribution is of thermal origin, and it is proportional to $|\mathrm{d}H_\mathrm{K}/\mathrm{d}T|_{T=T_\mathrm{W}}$, where H_K is the anisotropy field and T_W is the writing temperature of the recording medium. In order to improve the sharpness of the recorded transitions, $|\mathrm{d}H_K/\mathrm{d}T|_{T=T_\mathrm{W}}$ should be increased and the intensity of grain interactions should be decreased. The former requires a judicious choice of the material used for the grain core alloy and also a tight control of the granular medium intrinsic distributions, namely, the grain Curie temperature T_C, the H_K, and the grain size distributions. In addition to these factors, the grain magnetization reversal mechanism at T_W plays an important role on the transition noise. This chapter describes a set of experimental methods used in the characterization of HAMR media, from setup description to experimental data analysis. The temperature dependence of the AC susceptibility measured along the film easy axis is used to evaluate the T_C distribution (of average $<T_\mathrm{C}>$ and standard deviation σ_TC). Thermal erasure of the remanent magnetization using ns duration laser pulses is used to evaluate σ_TC. The dependence of the AC susceptibility on the DC field applied along a hard-axis direction is used to evaluate the H_K distribution (of average $<H_\mathrm{K}>$ and standard deviation σ_HK) as a function of temperature. A method to evaluate the intensity of intergranular exchange coupling based on high-field polar-Kerr magnetometry is described. The grain magnetization reversal mechanism is investigated based on measurements of thermal stability and of remanence coercivity as a function of temperature. These measurements enable the validation of theoretical concepts regarding the origin and the law governing the temperature dependence of H_K in chemically ordered alloys used as HAMR media.

C. Papusoi (✉) · M. Desai
Western Digital, San Jose, CA, USA
e-mail: cristian.papusoi@wdc.com

S. Ruta · R. W. Chantrell
Department of Physics, The University of York, York, UK

© Springer Nature Switzerland AG 2021
V. Franco, B. Dodrill (eds.), *Magnetic Measurement Techniques for Materials Characterization*, https://doi.org/10.1007/978-3-030-70443-8_25

The described experimental methodologies are exemplified on a series of samples having the structure glass/seed/heat-sink/MgO/MAG/C, where the MAG layer is an $L1_0$ FePt-based granular alloy of 3.8–10.5 nm thickness.

Keywords Heat-assisted magnetic recording · $L1_0$ FePt · Coherent and incoherent magnetization reversal · Curie temperature · AC susceptibility · Pump-probe measurements

The principle of heat-assisted magnetic recording (HAMR) consists of heating a region of a thin film recording medium, including several magnetic grains, for a very short time (~1 ns) at a high enough temperature for the medium coercive field to become lower than the magnetic field applied by a writing head. Subsequent to the heat pulse, the heated region rapidly cools down in the presence of the applied field, leading to a "quenching" of the magnetization along the applied field direction. The quenching or "blocking" temperature, defined as the temperature required for a system of identical grains to reach thermal equilibrium during the application of the heating pulse, is just below the grain Curie temperature T_C. This is a consequence of both the short (ns) heat pulse duration and the abrupt temperature variation of the anisotropy field, defined as $H_K = 2\,K/M_S$ where K is the uniaxial anisotropy constant and M_S the saturation magnetization, in the vicinity of the Curie temperature. HAMR responds to the following two requirements for magnetic recording extendibility. On the one hand, it enables to decrease the grain size while preserving a high thermal stability at ambient temperature $T_{RT} = 300$ K. This is accomplished using high anisotropy alloys, such as the chemically ordered $L1_0$ FePt featuring a uniaxial magneto-crystalline anisotropy as high as $K \sim 7 \times 10^7$ erg/cm^3 at T_{RT} [1]. On the other hand, HAMR enables a higher value of effective magnetic field gradient with respect to the applied magnetic field gradient dH/dx used in classical perpendicular magnetic recording (PMR), by involving an additional thermal contribution $|\mathrm{d}H_K/\mathrm{d}T|_{T=T_C}$ (dT/dx) [2]. The tetragonal $L1_0$ FePt phase is the favorite candidate for the grain core alloy due to its high uniaxial magneto-crystalline anisotropy, moderate Curie temperature of $T_C \sim 750$ K in bulk, low susceptibility to corrosion, and high $|\mathrm{d}H_K/\mathrm{d}T|_{T=T_C}$ of ~200 Oe/K. Since the film lateral temperature gradient dT/dx can be enhanced up to values of ~10 K/nm by optimizing the optical and thermal properties of the heat-sink and thermally resistive layers located below the recording layer in the HAMR film stack [3], the thermal field gradient can be substantially enhanced (>10 times) with respect to the head field gradient (~150 Oe/nm), giving HAMR a leap to achieve higher recording densities than classical PMR.

Recording transition noise (jitter) is directly related to the thermal field gradient, which can be expressed as $|\mathrm{d}H_{CR}/\mathrm{d}T|_{T=T_W}$ (dT/dx), where T_W is the writing temperature and $H_{CR}(T)$ is the remanence coercivity on the recording time scale, i.e., the field that can reverse the remanent saturation magnetization during the application of

Fig. 1 **(a)** STEM plan-view image of a 7.6 nm thick MAG layer. **(b)** Grain size distribution, $<D>$ (•) and $\sigma_D/<D>$ (○), and grain pitch (center-to-center) distribution, $<D_{CC}>$ (■) and $\sigma_{CC}/<D_{CC}>$ (□), as a function of MAG thickness. Reproduced from Ref. [4] with permission from Elsevier

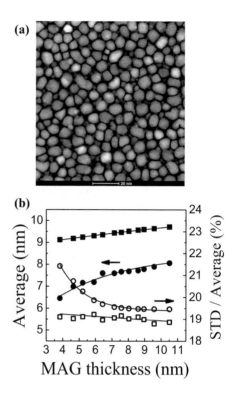

a temperature pulse of amplitude T and ~1 ns duration. Various distributions specific to a granular film, such as the H_K distribution (of average $<H_K>$ and standard deviation σ_{HK}), T_C distribution ($<T_C>$, σ_{TC}), and the grain size distribution ($<D>$, σ_D) decrease $|dH_{CR}/dT|_{T=T_W}$, leading to higher noise and preventing narrow transition widths. In addition to these factors, grain interactions and the mechanism of spin reversal in exchange isolated grains or grain clusters have also an important influence on $|dH_{CR}/dT|_{T=T_W}$.

The following paragraphs describe various experimental methodologies to evaluate the T_C and H_K distributions as well as $|dH_{CR}/dT|_{T=T_W}$. These methodologies are exemplified on a series of samples having the structure glass/seed/heat-sink/MgO/MAG/C deposited using an industrial sputtering system. The MAG layer is an FePt–X alloy, where X = C, SiO$_2$ fulfills the role of segregant. The MAG layer thickness is varied in a range of 3.8–10.5 nm. The thicknesses of seed, heat-sink and MgO layers are 120 nm, 110 nm, and 5 nm, respectively. A scanning transmission electron microscopy (STEM) plan-view image of a 7.6 nm MAG is presented in Fig. 1(a), and the dependence of the grain size and pitch distributions of the MAG thickness are presented in Fig. 1(b).

1 Grain Interactions

Since the introduction of granular-type recording media, considerable effort has been dedicated to establishing methods to characterize the magnitude of grain interactions [5–11]. The motivation is twofold. On the one hand, it enables the extraction of the intrinsic (unbiased by the effect of grain interactions) distribution of grain reversal fields, usually called switching field distribution (SFD). The latter is directly related to the recording track width as well as to the media writability. On the other hand, the characterization of grain interactions enables the evaluation of its intensity and, in particular, of the intergranular exchange coupling per unit grain surface J_{EX}. The latter impacts the recording transition noise and defines the smallest achievable recorded bit size. In the following, the dependence of intergranular interaction magnitude on the MAG thickness is investigated at T_{RT} using the method described in Refs. [4, 8, 12] which involves the following two magnetization curves: the major easy axis (EA) hysteresis loop and the minor EA hysteresis loop measured in the field range of $(-H_C, H_{SAT})$ where H_C is the coercive field and H_{SAT} is the positive saturation field of the EA loop.

The granular film is theorized as a collection of grains, each of them having a rectangular hysteresis loop of intrinsic coercive field H_{SW} distributed according to $F(H_{SW})$ ($\int_0^\infty F(H_{SW}) \, dH_{SW} = 1$), interacting via a mean field $H_{INT} = -4\pi NM$ where M is the grain system magnetization and N is a mean field parameter. Grain system parameters $F(H_{SW})$ and N can be estimated using the following magnetization states: $(H_1, +M_S/2)$ and $(H_2, -M_S/2)$ on the ascending branch of the major hysteresis loop and $(H_3, +M_S/2)$ on the minor loop (Fig. 2(a)). The intrinsic switching field distribution SFD_I, defined as the H_{SW} range corresponding to 25–75% variation of the cumulative distribution function of $F(H_{SW})$, is given by:

$$SFD_I = H_1 - H_3. \tag{1}$$

The extrinsic switching field distribution SFD_E is given by:

$$SFD_E = H_3 - H_2 = 4\pi NM_S \tag{2}$$

This denomination is justified by the observation that SFD_E represents the contribution of grain interactions to the switching field distribution, or SFD, which is a widely used technical parameter of an $M(H)$ loop, defined as:

$$SFD = H_1 - H_2. \tag{3}$$

In this sense, SFD_E is complementary to the intrinsic, grain interaction-free SFD_I in the expression of the SFD:

$$SFD = SFD_E + SFD_I. \tag{4}$$

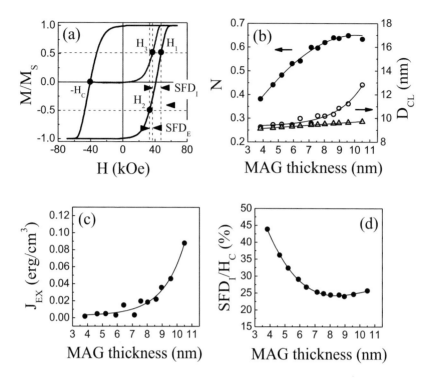

Fig. 2 (a) Principle of the method used to evaluate the grain interactions. Mean interaction field parameter N (•), cluster size D_{CL} (○), and the average grain pitch $<D_{CC}>$ (△) (b), intergranular exchange coupling constant J_{EX}, (c) and intrinsic switching field distribution SFD$_I$ normalized to the coercive field H_C(d) as a function of MAG thickness. Figures (b–d) reproduced from Ref. [4] with permission from Elsevier

Equation (2) is used to extract N using the experimental values of H_2, H_3 and M_S. Knowledge of the mean field parameter N enables the evaluation of the average intergranular exchange coupling energy per grain unit surface J_{EX} according to [13]:

$$J_{EX} = \pi M_S^2 < D_{CC} > \left(\frac{t}{\sqrt{< D_{CC} >^2 + t^2}} - N \right). \tag{5}$$

where t is the MAG thickness and $<D_{CC}>$ is the average grain pitch (center-to-center distance). The grain cluster size D_{CL} is evaluated based on the following relationship [14]:

$$D_{CL} = t \frac{\sqrt{1 - N^2}}{N} \tag{6}$$

where t is the film thickness. The meaning of D_{CL} is the in-plane diameter of a cylindrical film region of height equal to the film thickness t that experiences a magnetostatic field from the rest of the medium equal to $-4\pi NM$ where N is the experimental value of the mean field parameter and M is the sample magnetization. The dependencies of N, D_{CL}, J_{EX}, and SFD_I/H_C on the MAG thickness are presented in Fig. 2(b–d). As one can notice in Fig. 2(b), D_{CL} is close to $<D_{CC}>$ (average grain pitch) for MAG thicknesses lower than ~7 nm and noticeably increases with respect to $<D_{CC}>$ above this thickness. This behavior is interpreted to be a consequence of lateral grain growth in the MAG layer, leading to narrower grain boundaries with increasing MAG thickness [4]. As shown in Fig. 2(d), SFD_I/H_C experiences a significant decrease with increasing MAG thickness in the range of 3.8–7.5 nm. Three effects contribute to this behavior: σ_{HK} decreases, σ_D decreases, and thermal stability increases with increasing MAG thickness.

2 Thermal Stability

In order to evaluate the film thermal stability at T_{RT}, we use the procedure illustrated in Fig. 3(a–c). The EA major hysteresis loop (MHL) is measured using a 7 T polar-Kerr magnetometer. As measured, the MHL exhibits a nonlinear background caused by the fact that for applied fields H higher than the saturation field of the material constituting the magnetometer pole pieces H_{PS} ~ 18 kOe, the linearly polarized laser beam ($\lambda = 405$ nm) is exposed to a considerably higher magnetic field orientated parallel to the beam trajectory inside the superconducting magnet. As a consequence, the laser beam polarization experiences a rotation proportional to the applied field via the Faraday effect. This results in a field linear background for $H > H_{PS}$ and $H < -H_{PS}$, as illustrated in Fig. 3(a). The following approach is used to remove the Faraday background and extract the polar-Kerr loop of the magnetic film. The region on the MHL descending branch $M_{desc}(H)$ extending from $H_{MAX} = 70$ kOe down to a predefined valued H_{LIM} (typically 10 kOe) lower than H_{PS} but higher than the nucleation field of the sample H_N (corresponding to the shoulder of the MHL descending branch) is extracted. In the range of applied fields $H_{LIM} < H < H_{MAX}$, the film magnetization is saturated. Therefore, the shape of the extracted curve $M(H > H_{LIM})$ is a consequence of both the Faraday and the pole piece saturation effects. Consequently, the background for the range of applied fields (H_{LIM}, H_{MAX}) can be obtained as $M_{back}(H_{LIM} < H < H_{MAX}) = M_{desc}(H) - M(H_{LIM})$. Similarly, the background in the range of applied fields $(-H_{MAX}, -H_{LIM})$ is obtained as $M_{back}(-H_{MAX} < H < -H_{LIM}) = M_{asc}(H) + M(H_{LIM})$, where $M_{asc}(H)$ is the ascending branch of the measured MHL. In the applied field range $(-H_{LIM}, H_{LIM})$, the magnet pole pieces act as a magnetic shield for the laser beam; therefore the Faraday effect is expected to be negligible. Consequently, the background magnetization $M_{back}(-H_{LIM} < H < H_{LIM}) = 0$. The background curve for the entire range of applied field $M_{back}(H)$ can be constructed by merging the three curves $M_{back}(H) = M_{back}(-H_{MAX} < H < -H_{LIM})$ U $M_{back}(-H_{LIM} < H < H_{LIM})$ U $M_{back}(H_{LIM} < H < H_{MAX})$. Since $M_{back}(H)$ is reversible with respect to the applied field, it is interpolated and then subtracted from the measured signal. Subsequent to

Fig. 3 (a) MHL as measured using polar-Kerr (A), extracted Faraday background using $H_{LIM} = 10$ kOe (B) and MHL corrected for the Faraday background (C). **(b)** Time decay of magnetization for several applied fields H: $-26{,}475$ Oe (○), $-27{,}518$ Oe (□), $-28{,}626$ Oe (△), $-29{,}674$ Oe (▽), $-30{,}701$ Oe (◇), and $-31{,}712$ Oe (◁). **(c)** Dependence of the applied field H on the time τ_C required for the film magnetization to relax to zero (symbols) and fit using Eq. (7) (solid line)

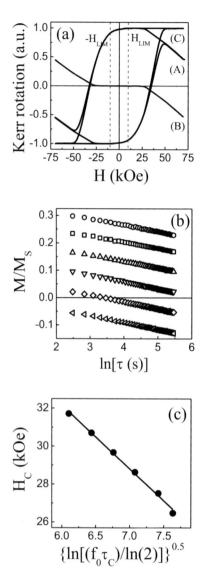

the background correction, the MHL loop is normalized to saturation as shown in Fig. 3(a).

In order to evaluate the thermal stability factor $KV/(k_B T)$, the time decay of magnetization in a constant applied field H is measured according to the following procedure. Initially, the MHL loop is measured and the background $M_{back}(H)$ is extracted. The resultant MHL loop is normalized to saturation. The film magnetization is saturated via the application of $H_{MAX} = +7$ T using the highest field sweep rate available (700 Oe/s). Then the field is ramped with the same sweep rate down to a predefined value $H > -H_C$ and subsequently kept constant. The time dependence of magnetization $M(\tau, H = \text{const})$ is measured during a time frame of 6 min. The measured $M(\tau, H)$ curve is background corrected by subtracting $M_{back}(H)$ and then normalized to saturation using the same normalization constant as that used for MHL. As evidenced in Fig. 3(b), the magnetization decay curves are linear on a logarithmic time scale. The time $\tau_C(H)$ corresponding to the intersection of the linearly interpolated $M(\ln(\tau))$ with the $M = 0$ axis is the time required for the applied field H to become the coercive field $-H_C$ (Fig. 3(c)).

According to the Stoner-Wohlfarth (SW) model [15] that assumes that magnetization reversal occurs by coherent rotation of spins in isolated grains, the relationship between H_C and τ_C is given by the Sharrock law [16]:

$$H_C(\tau_C) = H_0 \left(1 - \sqrt{\frac{1}{\beta} \ln \frac{f_0 \tau_c}{\ln 2}} \right) \qquad (7)$$

where $\beta = K^C V_{SW}/(k_B T)$ is the thermal stability factor, K^C is the grain core anisotropy constant, H_0 is the grain coercivity in the absence of thermal relaxation, and $f_0 \cong 10^{11}$ s^{-1} [17, 18] is the transition attempt frequency of thermal fluctuations. The switching volume V_{SW} represents the grain volume that nucleates the reversal by coherent rotations of spins. In the frame of the SW model, V_{SW} is close to the rms value of the grain volume distribution $\sqrt{\langle V^2 \rangle}$ [19]. Thus, the assumption of a coherent rotation reversal mechanism can be validated by comparing the experimental value of β evaluated from the fit of $H_C(\tau)$ using Eq. (7) with that theoretically predicted by the SW model $\beta = K^C \sqrt{\langle V^2 \rangle}/(k_B T)$. This comparison is presented in Fig. 4(a). One notices that for a MAG thicknesses below the threshold value $t_C \sim 8$ nm, the SW prediction matches the experimental value of β, indicative of a coherent spin reversal in isolated grains. For MAG thicknesses larger than t_C, $V_{SW} < \sqrt{\langle V^2 \rangle}$ suggesting that only a fraction of a grain switches coherently at H_C. This is consistent with a domain-wall (DW) nucleation/displacement (incoherent) reversal mechanism occurring above t_C. Figure 4(b) presents the dependence of H_0/H_K on the MAG thickness. One notices that H_0/H_K is reasonably close to the SW predicted value of ~0.82 corresponding to an EA dispersion of $\Delta\theta_{50} = 5°$ evaluated using X-ray diffraction, if the MAG thickness is lower than t_C and it decreases with respect to the SW prediction above t_C. This behavior is consistent with the non-monotonic dependence of EA loop coercivity on the MAG thickness (Fig. 4(c)) displaying a maximum at t_C.

In order to understand the significance of t_C, let us recall that the high anisotropy of the L1$_0$ Fe$_{50}$Pt$_{50}$ phase results in a DW width $\delta = \pi\sqrt{A/K^C} = 6.82$ nm ($A \cong 1.53 \times 10^{-6}$ erg/cm, $K^C \cong 3.25 \times 10^7$ erg/cm^3). If the grain size becomes larger than δ an incoherent magnetization reversal mediated by DW nucleation/displacement becomes more energetically favorable than coherent rotations of spins inside individual grains, leading to a lower H_C and a lower β than in the case of coherent reversal. However, the MAG thickness threshold for

Fig. 4 (**a**) Thermal stability factor $\beta = K^C V_{SW}/(k_B T_{RT})$, evaluated by fitting the time dependence of coercivity $H_C(\tau)$ using the Sharrock law, as a function of MAG thickness; the values of β calculated based on the SW model ($V_{SW} = \sqrt{\langle V^2 \rangle}$) are represented by line+empty symbols. (**b**) Short-time coercivity H_0, extracted using the Sharrock law fit, normalized to H_K, as a function of MAG thickness. (**c**) Coercive field of EA loop as a function of MAG thickness ($\bullet$). Reproduced from Ref. [4] with permission from Elsevier

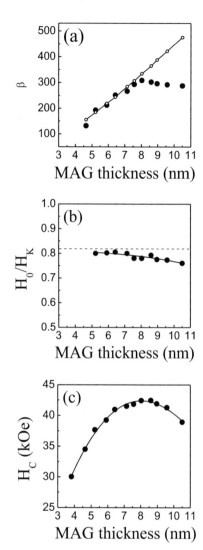

the onset of intergranular exchange interactions is also close to t_C. Therefore, the resulting increase of grain cluster size illustrated in Fig. 2(b) can be a contributing factor in addition to the reduced DW width to the incoherent switching above t_C.

3 H_K and T_C Distributions

Measurements of AC susceptibility χ_{AC} are used to evaluate the H_K and the T_C distributions as a function of MAG thickness. AC susceptibility is measured by applying an AC field, of frequency in the kHz range and amplitude in the 50 Oe range, along the MAG film EA, and probing the induced AC magnetization along the same direction. Depending on the measurement conditions, the AC susceptibility can be sensitive to different properties of the MAG layer. The H_K distribution is extracted using the imaginary component of the AC susceptibility $\chi_{AC}^i (H_{DC})$ measured as a function of a DC field H_{DC} applied along a hard-axis (HA) direction of the MAG layer (perpendicular to the AC field), at constant temperature. This type of experiment, which is essentially a HA magnetization process, is usually denominated "AC transverse susceptibility" (ACTS) [20–22]. The H_K distribution is extracted by fitting $\chi_{AC}^i (H_{DC})$ using the Monte Carlo (MC) model developed in Ref. [23]. The MC model takes into account the grain size, EA orientation and T_C distributions, and also the grain interaction (magnetostatic and exchange). The latter is evaluated using a Voronoi tessellation of the granular film structure and the experimental value for the grain core saturation magnetization $M_S^C \cong 800$ emu/cm^3 and intergranular exchange coupling J_{EX}.

The T_C distribution is extracted using the real component of the AC susceptibility measured as a function of temperature T, in the absence of a transverse field ($H_{DC} = 0$), $\chi_{AC}^r(T)$. This type of measurement, which is essentially an EA magnetization process, is usually denominated "AC susceptibility" (ACS) [3, 4, 12]. The T_C distribution is evaluated by fitting $\chi_{AC}^r(T)$ using the same MC model as that used for extracting the H_K distribution from ACTS. Since ACS is a variable temperature process, the laws describing $M_S^C(T)$ and $H_K (T)$ are directly involved in the interpretation of $\chi_{AC}^r(T)$, and they must be known before attempting to extract the T_C distribution from the fit of $\chi_{AC}^r(T)$. For this reason, these laws are evaluated in the following. In the case of ACS, the input parameters of the MC model are the grain size, EA orientation and H_K distributions, and the intensity of grain interactions.

The schematic of the AC susceptibility setup is presented in Fig. 5. The sample (S) (thin film) is exposed to the simultaneous action of two fields: (a) a DC field H_{DC} having an orientation parallel to the film plane and a value in the range of ± 80 kOe, generated by a superconducting coil (SC), and (b) an AC field h_{AC} of amplitude ~50 Oe and frequency variable in the range 100 Hz–30 kHz, having an orientation perpendicular to the film plane, generated by a coil (HAC) connected to an AC power supply (AC). In this work, the AC field frequency is fixed at 1 kHz during the ACTS measurements and 500 Hz during the ACS measurements.

The DC field is measured using a Gaussmeter (GM). The AC field h_{AC} acting along the film EA induces an AC magnetization along the same direction which is probed via polar-Kerr effect using a stabilized HeNe laser (LAS) ($\lambda = 632$ nm), a polarizer (P), a plano-convex lens (L), a half-wave plate (HWP), a Wollaston prism (WP), and a dual detector (DD). The output signal of the detector is measured using a dual-phase lock-in amplifier (LIA) triggered by the AC coil power supply. The lock-in amplifier measures both components of the output signal: real (in-phase with the AC field) which is proportional to χ_{AC}^r and imaginary (90° phase shifted with respect to the AC field) which is proportional to χ_{AC}^i. The sample holder contains

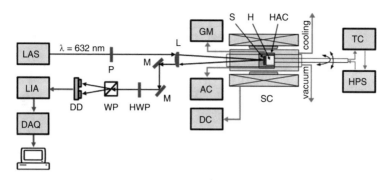

Fig. 5 Schematics of the AC susceptibility setup, containing: HeNe laser (LAS), polarizer (P), plano-convex lens (L), mirrors (M), sample (S), half-wave plate (HWP), Wollaston prism (WP), dual detector (DD), superconducting coil (SC) and its power supply (DC), Gaussmeter (GM), AC coil (HAC) and its power supply (AC), temperature controller (TC), heater (H) and heater power supply (HPS), lock-in amplifier (LIA) and data acquisition card (DAQ). Reproduced from Ref. [4] with permission from Elsevier

a filament that is used to heat the sample in the temperature range of 300–800 K. A thermocouple, embedded in the sample holder, is connected to a temperature controller (TC) that reads the sample temperature. The temperature controller is also used in a feedback loop with the heater power supply (HPS) to control the sample temperature by adjusting the voltage applied to the heater. Measurements are performed in a vacuum of 10 mTorr which is low enough to prevent sample oxidation/degradation during heating.

The SFD is intrinsically related to the H_K distribution. The latter is thought to originate in the finite grain size effect [24]. The poorer atomic coordination of the Fe and Pt atoms at the grain surface with respect to the grain core leads to a lower anisotropy of the Fe spins at the grain boundary than inside the grain. As a result, K^C and H_K decrease with decreasing grain size. The decrease is more substantial the lower the grain size due to the increasing grain surface/volume ratio. As a result, the grain size and the H_K distributions are related. Furthermore, the spin disorder at the grain surface, originating in the finite size effect, leads to a profile of the chemical ordering parameter S across the grain and thereby to a grain size dependence of the average chemical ordering [25, 26]. Since the uniaxial anisotropy of L1$_0$ FePt $K \sim S^2$ [27, 28], the dependence of S on the grain size leads to a relationship between H_K and the grain size. The H_K distribution is evaluated from the MC fit of $\chi^i_{AC}(H_{DC})$. Figure 6(a) presents the $\chi^i_{AC}(H_{DC})$ curves measured at T_{RT} and their MC fit for the 3.8, 5.2, 6.4, and 9.6 nm MAG. The resulting values of $<H_K>$, $\sigma_{HK}/<H_K>$ are presented as a function of MAG thickness in Fig. 6(b). ACTS is not a standard technique for evaluating $<H_K>$, being more suitable for systems of weekly interacting single-domain grains, such as recording media. Therefore, in order to validate the obtained values of $<H_K>$, two other methods are used to evaluate H_K at T_{RT}, such as the 45° torque [29] and the singular point detection (SPD) [30] techniques, the results being presented in Fig.

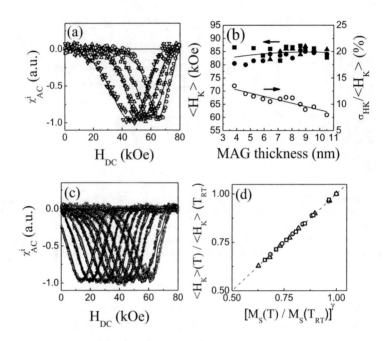

Fig. 6 (a) Ambient temperature measurements of χ^i_{AC} as a function of the transverse field H_{DC} for a 3.8 nm (○), 5.2 nm (□), 6.4 nm (△), and 9.6 nm (▽) MAG. (b) $<H_K>$ (solid symbols) and $\sigma_{HK}/<H_K>$ (empty symbols) evaluated using ACTS (circles), SPD (rectangles), 45° torque (triangles), and average over the three methods (solid line) as a function of MAG thickness. (c)χ^i_{AC} (H_{DC}) for a 7.1 nm MAG measured at: 300 K (○), 323 K (□), 373 K (△), 398 K (▽), 423 K (◇), 448 K (◁), 473 K (▷), 498 K (□), 523 K (⋆), and 548 K (□); solid lines represent the MC fit used to evaluate $<H_K>$ (T) and $\sigma_{HK}/<H_K>$ (T). (d) $<H_K>$ (T)/$<H_K>$ (T_{RT}) as a function of $[M_S(T)/M_S(T)]^\gamma$ for a 3.8 nm (○), 7.1 nm (□), and 9.6 nm (△) MAG. Reproduced from Ref. [4] with permission from Elsevier

6(b). One notices that the $<H_K>$ extracted using the three methods are in reasonable agreement. According to Fig. 6(b), $<H_K>$ increases in the range of 83–85 kOe, and the normalized standard deviation $\sigma_{HK}/<H_K>$ decreases in the range of 8–14% with increasing MAG thickness in the range of 3.8–10.5 nm. The increase of $<H_K>$ and decrease of $\sigma_{HK}/<H_K>$ with increasing MAG thickness correlate with the increase of the chemical ordering parameter S, as shown in Ref. [4].

The evaluation of the H_K distribution based on χ^i_{AC} (H_{DC}) can be performed at various temperatures up to a certain threshold where the peak of χ^i_{AC} (H_{DC}) vanishes. This enables the correlation between the temperature dependences of H_K and M_S. The dependence $M_S(T)$ is extracted using Vibrating Sample Magnetometry, and then it is used in the MC fit of χ^i_{AC} (H_{DC}) curves to evaluate $H_K(T)$. The experimental χ^i_{AC} (H_{DC}) curves for three samples of 3.8, 7.1, and 9.6 nm MAG thickness are measured at various temperatures in the range of 300–573 K and fitted using the MC model to evaluate $<H_K>$ and σ_{HK} at each temperature T. The experimental curves measured for the 7.1 nm MAG are presented in Fig. 6(c).

Fig. 7 **(a)**$\chi^r_{AC}(T)$ (solid symbols), $\chi^i_{AC}(T)$ (empty symbols), and MC fit (solid line) for a 3.8 nm (circles), 4.6 nm (squares), 5.9 nm (up-triangles), and 9 nm (down-triangles) MAG. **(b)** <T_C> (•) evaluated from the inflection $\chi^r_{AC}(T)$ of and σ_{TC}/<T_C> (○) evaluated from the MC fit of $\chi^r_{AC}(T)$ as a function of MAG thickness. Reproduced from Ref. [4] with permission from Elsevier

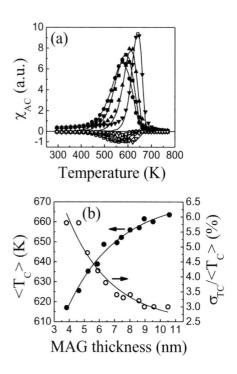

Details about the fitting procedure are presented in Ref. [4]. Based on previous experimental and theoretical results [31, 40], it is assumed that <H_K> has a similar temperature dependence as $(M_S)^\gamma$, i.e., <H_K> is proportional to $(M_S)^\gamma$ in the temperature range of 0 K–T_C. By using the values of <H_K> and M_S extracted from the MC fitting of the ACTS curves measured at various temperatures T (Fig. 6(c)), a least-squares fit is used to evaluate the exponent γ for each of the three samples, as shown in Fig. 6(d). One obtains $\gamma = 1.13$ for the 3.8 nm MAG, $\gamma = 1.19$ for the 7.1 nm MAG, and $\gamma = 1.16$ for the 9.6 nm MAG. This range of γ values is close to the value of $\gamma = 1.1$ previously evaluated experimentally [27] and theoretically [31] for bulk L1$_0$ FePt films and explained to be a consequence of the anisotropic exchange coupling between Fe spin moments intermediated by the itinerant Pt induced moments (two-ion anisotropy).

The T_C distribution originates in the grain size distribution via the finite size effect, and it is an important parameter for HAMR, being the major contributor to the recorded transition jitter. The T_C distribution for the investigated samples is evaluated using the temperature dependence of ACS. The real $\chi^r_{AC}(T)$ and imaginary $\chi^i_{AC}(T)$ components of ACS are normalized to the peak amplitude of $\chi^i_{AC}(T)$, as illustrated in Fig. 7(a). $\chi^r_{AC}(T)$ is fitted using the MC model. The fitting parameters are <T_C> and σ_{TC}. The calculated $\chi_{AC}(T)$ curves are represented by solid lines in Fig. 7(a). The σ_{TC}/<T_C> extracted from the fit are presented as a function of MAG thickness in Fig. 7(b). One notices that σ_{TC}/<T_C> decreases in the range of 2.9–6% with increasing MAG thickness in the range of 3.8–10.5 nm. For increasing accuracy, the average Curie temperature of the MAG layer $<T_C>$ is evaluated directly from experimental data as the temperature corresponding to the inflection point of $\chi^r_{AC}(T)$ in the superparamagnetic regime, the results being presented in Fig. 7(b). One notices that $<T_C>$ increases with increasing film thickness in the range of 615–665 K. The most important advantage of ACS over other methods to evaluate $<T_C>$ is the low amplitude of the AC field that minimizes the influence of the applied field on the value of M_S close to criticality, enabling an accurate evaluation of $<T_C>$.

$$|\mathbf{d}H_{CR}/\mathbf{d}T\|_{\mathbf{T=TW}}.$$

In HAMR, the thermal field gradient is proportional to $|dH_{CR}/dT|_{T=T_W}$. The latter is influenced by the T_C distribution as well as by the magnetization reversal mechanism. A wider T_C distribution causes a wider H_K distribution close to criticality. If the magnetization reversal occurs by coherent rotations of spins in isolated grains [15], the remanence coercivity H_{RC} is expected to approach H_K (due to the poor effect of thermal relaxation during the writing process) and $|dH_{CR}/dT|_{T=T_W} \cong |dH_K/dT|_{T=T_W}$. In these circumstances, a wider T_C distribution causes a lower value of $|dH_{CR}/dT|_{T=T_W}$ which is detrimental to the recording performance. Thermally assisted writing may also occur via an incoherent magnetization reversal if the DW width at T_W is smaller than the grain size. According to the atomistic simulations results of Refs. [32, 33], DW width in L1$_0$ FePt films has a T_{RT} value in the range of 4–5 nm and exhibits a poor increase with temperature from T_{RT} up to T_C where it diverges sharply. Consequently, H_{RC} is expected to be close to H_K for T_W close to T_C, where the larger DW size with respect to the grain size favors a coherent magnetization reversal mechanism and to rapidly decrease with respect to H_K for $T_W < T_C$ where the DW size becomes smaller than the grain size favoring an incoherent magnetization reversal. Accordingly, $|dH_{CR}/dT|_{T=T_W}$ is expected to decrease with respect to $|dH_K/dT|_{T=T_W}$ the more substantially the larger is the film average grain size with respect to the DW size at T_{RT}. In order to corroborate this statement, the dependence of $|dH_{CR}/dT|_{T=T_W}$ on the film thickness is evaluated below.

In order to extract the intrinsic properties of the MAG layer close to criticality, which are relevant for the thermally assisted recording process, the effect of thermal relaxation must be minimized by decreasing the duration of the film exposure to heat. The method proposed in Ref. [34] involves the measurement of remanent magnetization subsequent to the application of an ultrafast heat pulse via a ns pulsed laser. The schematic of a pump-probe setup developed for this purpose is presented in Fig. 8. The pump laser (LAS 1) is a Q-switched Nd:YVO$_4$ pulsed laser of $\lambda = 1064$ nm that delivers laser pulses of energy in the range of 0–20 μJ, of 25 ns FWHM and 50 kHz frequency, used for heating the sample. The probe laser (LAS 2) is a HeNe stabilized laser, having $\lambda = 632$ nm and $P = 1.1$ mW that is used to detect the remanent magnetization along the perpendicular to the film plane by polar-Kerr effect. An acousto-optic modulator (AOM) enables the application

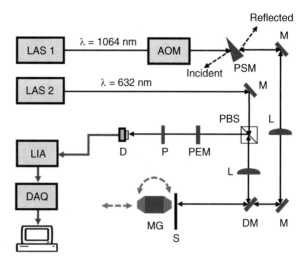

Fig. 8 Schematics of the pump-probe setup, containing: Nd:YVO4 pulsed laser (LAS1), HeNe laser (LAS2), acousto-optic modulator (AOM), wedged prism (PSM), plano-convex lens (L), mirrors (M), dichroic mirror (DM), polarizing beam splitter (PBS), photoelastic modulator (PEM), polarizer (P), sample (S), magnet (MG), lock-in amplifier (LIA) and data acquisition card (DAQ). Reproduced from Ref. [4] with permission from Elsevier

of a single pump pulse to the sample. The time delay between two consecutive measurements is of a few seconds, long enough to prevent heat accumulation in the sample. A permanent magnet (MG) is used to apply a magnetic field in the range of ± 7 kOe that is uniform over the heated region of the sample. The probe beam passes through a leaky polarizing beam splitter (PBS), and it is reflected by a dichroic mirror (DM) that combines the pump and probe beams before reaching the sample. The two beams are aligned, such that the probe beam hits the sample at the center of the pump beam and thereby probes the region exhibiting the highest temperature. The diameter of the pump beam is ~70 µm, and that of the probe beam is ~8 µm. The large ratio between the two beam diameters ensures the probed region is uniformly heated by the pump beam. After reflection from the sample, the probe beam passes through the PBS, through a photoelastic modulator (PEM) operated at 50 kHz and having the optical axis oriented at 0° with respect to the incident probe beam polarization, through an analyzer (P) oriented at 45° with respect to the incident probe beam polarization and reaches the photodetector (D). The polarization rotation of the probe beam due to the polar-Kerr effect is evaluated using the second harmonic of the detected signal, measured by a lock-in amplifier (LIA) triggered by the PEM oscillator. A wedged prism (PSM) is used to split both the incident and the reflected pump beams in order to measure their energies per pulse E_I and E_R respectively. This enables evaluation of the fraction of pump pulse energy absorbed by the sample (100% converted into heat) $E_A = (1 - R)E_I$ where $R = E_R/E_I$ is the sample reflectivity. The latter is measured during the application of the pump pulse by monitoring the signals of two fast detectors measuring the incident and reflected beams split by PSM using an oscilloscope. The relationship between the absorbed pulse energy E_A and the corresponding film temperature is given by:

$$E_A = \varepsilon \ (T - T_{RT}) \tag{8}$$

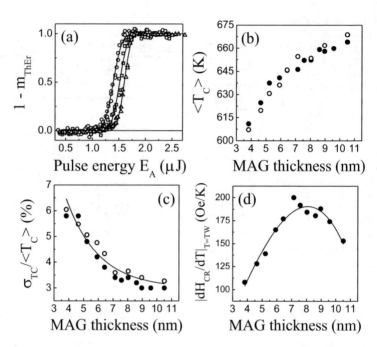

Fig. 9 (a) Experimental $1 - m_{ThE}$ as a function of pump pulse energy E_A for a 3.8 nm (○), 5.2 nm (□), and 10.5 nm (△) MAG and their fit using the CDF of a Gaussian distribution of average $<E>$ and standard deviation σ_E (solid line). (b) Dependence of $<T_C>$ on the MAG thickness evaluated using ACS (•) and that evaluated based on Eq. (6) where $\varepsilon = 0.00433$ μJ/K (○). (c) Dependence of $\sigma_{TC}/<T_C>$ on the MAG thickness evaluated using ACS (•) and ThEr (○). (d) $|dH_{CR}/dT|_{T=T_W}$ as a function of MAG thickness (•). Reproduced from Ref. [4] with permission from Elsevier

where ε is a calibration constant [34]. The latter is evaluated using measurements of thermal erasure (ThEr) of remanent magnetization as a function of the absorbed pump pulse energy E_A, $M_{ThEr}(E_A, H = 0)$. In this experiment, the sample remanent magnetization is initially saturated in the positive direction along the sample EA. The applied field H is set to zero. A laser pulse of energy E_A is applied, and the remanent magnetization M_{ThEr} is measured at T_{RT} following the suppression of the laser pulse. These steps are repeated for progressively increasing pulse energies E_A to obtain $M_{ThEr}(E_A, H = 0)$. Figure 9(a) presents the dependence of $1 - m_{ThEr}$, where $m_{ThEr} = M_{ThEr}/M_{RS}$, on the pump pulse energy E_A for a 3.8, 5.2, and 10.5 nm MAG. Equation (8) establishes the relationship between T_C distribution and the erasure pulse energy distribution. By assuming a Gaussian shape for the T_C distribution, of parameters $<T_C>$ and σ_{TC}, the corresponding distribution of erasure pulse energies is also Gaussian, of average $<E>$ and standard deviation σ_E. The experimental $1 - m_{ThEr}(E_A)$ curves are fitted with the CDF of a Gaussian distribution $CDF(E_A) = 0.5\left\{1 + \mathrm{erf}\left[(E_A - \langle E \rangle)/\left(\sigma_E \sqrt{2}\right)\right]\right\}$ in order to evaluate $<E>$ and σ_E as a function of MAG thickness.

An example of ThEr curves and their fit is represented by a solid line in Fig. 9(a). The value of $<T_C>$ was already evaluated using the temperature dependence of ACS (Fig. 7(b)). Thus, ε can be evaluated using Eq. (8) where $E_A \equiv <E>$ and $T \equiv <T_C>$. As shown in Ref. [4], ε is essentially independent of the MAG thickness. Accordingly, ε is evaluated by matching the experimental values of $<T_C>$, evaluated using the AC susceptibility, with those evaluated using Eq. (8) and the experimental values of $<E>$ for all MAG thicknesses using a least-squares approach, as illustrated in Fig. 9(b). One obtains $\varepsilon = 43.3 \times 10^{-4}$ µJ/K. The standard deviation of the T_C distribution σ_{TC} can be obtained for both series based on the relationship:

$$\sigma_E = \varepsilon \sigma_{TC} \qquad (9)$$

where σ_E is evaluated from the fit of the ThEr curves. The dependence of $\sigma_{TC}/<T_C>$ on the MAG thickness evaluated using Eq. (9) is presented in Fig. 9(c) alongside with that evaluated using the AC susceptibility. As one can notice in Fig. 9(c), $\sigma_{TC}/<T_C>$ evaluated using the two techniques are in agreement in the entire investigated MAG thickness range.

In order to evaluate $H_{CR}(T)$ and $|dH_{CR}/dT|_{T=T_W}$, a family of remanent magnetization curves $M_R(H, E_A = \text{const})$ is measured for progressively increasing values of E_A. The field $H = 0$–7 kOe is applied along the sample EA, and the remanence is saturated in the opposite direction to that of the applied field H previous to each measurement of M_R using the highest value of E_A of ~3 µJ. The MAG temperature corresponding to this energy of the pump pulse is well above the MAG T_C distribution for all the samples investigated in this study. The value of ε evaluated above is used to convert $H_{CR}(E_A)$ into $H_{CR}(T)$ based on Eq. (8). $|dH_{CR}/dT|_{T=T_W}$ is evaluated as the slope of $H_{CR}(T)$ in the temperature range where H_{CR} decreases from the maximum applied field of $H_{MAX} = 7$ kOe down to 0. This temperature range is presumably bracketing the HAMR writing temperature T_W due to the ns duration of the heating pulse and the range of applied fields which reproduce the recording conditions. The dependence of $|dH_{CR}/dT|_{T=T_W}$ on the MAG thickness is represented in Fig. 9(d). One notices that $|dH_{CR}/dT|_{T=T_W}$ exhibits a maximum close to the MAG thickness threshold $t_C \sim 8$ nm. Below t_C, $|dH_{CR}/dT|_{T=T_W}$ increases with increasing MAG thickness. This behavior is attributed to the decrease of $\sigma_{TC}/<T_C>$, as illustrated in Fig. 9(c). Above t_C, $|dH_{CR}/dT|_{T=T_W}$ decreases with increasing MAG thicknesses. This behavior is attributed to the incoherent magnetization reversal occurring in the MAG layer, which corroborates the previous conclusions based on measurements of thermal stability at T_{RT}.

4   Summary

Several parameters of chief importance for HAMR performance, such as the Curie temperature T_C and the anisotropy field H_K distributions and the temperature slope of the remanence coercivity H_{CR} at the HAMR writing temperature T_W,

$|dH_{CR}/dT|_{T=T_W}$, are investigated for a series of $L1_0$ FePt films of thickness in the range of 3.8–10.5 nm. The T_C distribution, including the average $<T_C>$ and the standard deviation $\sigma_{TC}/<T_C>$, are evaluated using the temperature dependence of the AC susceptibility (ACS). $<T_C>$ increases in the range of 615–665 K, and $\sigma_{TC}/<T_C>$ decreases in the range of 2.9–6% with decreasing film thickness. The H_K distributions, both average $<H_K>$ and standard deviation $\sigma_{HK}/<H_K>$, are evaluated using the dependence of the imaginary component of the AC transverse susceptibility (ACTS) measured along the film easy axis on the DC field applied along the film hard axis. It is shown that $<H_K>$ increases in the range of 83–85 kOe and $\sigma_{HK}/<H_K>$ decreases in the range of 8–14% with increasing film thickness. The temperature dependence of $<H_K>$ is evaluated using ACTS, and it is shown to correlate with the temperature dependence of the saturation magnetization $M_S(T)$ according to the law $<H_K>(T) \sim [M_S(T)]^\gamma$ with $\gamma \cong 1.1$. This observation supports previous experimental and theoretical results interpreting this behavior as a consequence of the two-ion anisotropy of Fe spins in a chemically ordered FePt lattice.

The remanence coercivity $H_{CR}(T)$ is defined as the positive field that has to be applied along the film easy axis to reduce the remanent magnetization, previously saturated in the negative direction at ambient temperature $T_{RT} = 300$ K, to zero subsequent to the application of a ns duration heating pulse of amplitude T. The dependence $H_{CR}(T)$ is evaluated for a range of heating temperatures T close to criticality that corresponds to an H_{CR} range of 0–7 kOe. In this temperature range, $H_{CR}(T)$ is almost linear, which enables the extraction of $|dH_{CR}/dT|_{T=T_W}$ as the slope of $H_{CR}(T)$. A pump-probe method to evaluate H_{CR} is described that uses ns duration laser pulses to heat the film. The calibration of the relationship between the laser-pulse energy and the sample temperature is enabled via the proportionality between the laser pulse energy $<E>$ required to reduce the remanent magnetization to 50% of its saturation value in the absence of an applied field (thermal erasure or ThEr) and $(<T_C> - T_{RT})$ where $<T_C>$ is evaluated using ACS. The pump-probe method enables the evaluation of $\sigma_{TC}/<T_C>$ using the temperature dependence of remanent magnetization $M_{ThEr}(T, H = 0)$ and of $H_{CR}(T)$ using measurements of remanent magnetization as a function of the applied field performed at constant temperature $M_R(H, T = \text{const})$. This enables the extraction of $H_{CR}(T)$ and of $|dH_{CR}/dT|_{T=T_W}$. The latter features a non-monotonic dependence on the film thickness, reaching a maximum at $t_C \sim 7.5$ nm. The dependence of the T_C distribution on the film thickness is consistent with a monotonic increase of $|dH_K/dT|_{T=T_W}$ with increasing film thickness, suggesting that $|dH_{CR}/dT|_{T=T_W} < |dH_K/dT|_{T=T_W}$ above t_C. This behavior is attributed to a domain-wall (DW) mechanism of magnetization reversal in the MAG layer above t_C, and it has a negative impact on the recording performance since it decreases the thermal field gradient. As a consequence, increasing the MAG thickness above t_C leads to a tradeoff between writing resolution and readback signal which is an obstacle in the development of HAMR media. However, this difficulty can be overcame in the framework of the Superparamagnetic Writing (SPW) model [35, 36]. The SPW involves a writing-assist layer (CAP), of higher $<T_C>$ than that of the recording layer (MAG), deposited on top of the MAG layer. If the CAP is ferromagnetic at the MAG writing temperature (T_W) and it preserves a perpendicular anisotropy at T_W, it is easily switched by the head magnetic field at T_W and assists the orientation/writing of the MAG layer during the refreezing process from T_W to T_{RT} via the MAG/CAP positive exchange coupling. The writing occurs with a negligible influence of the CAP T_C distribution which is well above T_W. As demonstrated in Refs. [35, 36], SWP enables the recording of sharp transitions in the MAG layer and narrows the T_W distribution of a MAG/CAP bilayer with respect to that of an isolated MAG. This suggests that the thickness of MAG does not need to be larger than t_C in order to enhance the readback signal. The later task can be fulfilled by the CAP layer. As a result, the SPW scheme complies with both requirements of large stack thickness and large $|dH_{CR}/dT|_{T=T_W}$ or large thermal field gradient. Also, since SPW can mitigate the negative impact of the MAG $\sigma_{TC}/<T_C>$ on the recording performance, it could potentially enable to decrease $<T_C>$ of MAG, e.g., by doping the MAG alloy with Ni [37], and thereby decrease T_W and enhance the reliability of the HAMR head-media system.

References

1. D. Weller, O. Mosendz, G. Parker, S. Pisana, T.S. Santos, Phys. Stat. Sol. A **210**, 1245 (2013)
2. M.H. Kryder, E.C. Gage, T.W. McDaniel, W.A. Challener, R.E. Rottmayer, G. Ju, Y.T. Hsia, M.F. Erden, Proc. IEEE **96**, 1810 (2008)
3. G. Ju, Y. Peng, E.K.C. Chang, Y. Ding, A.Q. Wu, et al., IEEE Trans. Magn. **51**, 3201709 (2015)
4. C. Papusoi, T. Le, P.O. Jubert, D. Oswald, B. Ozdol, D. Tripathy, P. Dorsey, M. Desai, J. Magn. Magn. Mater. **483**, 249 (2019)
5. R.J.M. van de Veerdonk, X. Wu and D. Weller. IEEE Trans. Magn. **39**, 590 (2003)
6. A.N. Dobrynin, T.R. Gao, N.M. Dempsey, D. Givord, Appl. Phys. Lett. **97**, 192506 (2010)
7. A. Berger, Y. Xu, B. Lengsfield, Y. Ikeda, E.E. Fullerton, IEEE Trans. Magn. **41**, 3178 (2005)
8. I. Tagawa, N. Nakamura, IEEE Trans. Magn. **27**, 4975 (1991)
9. C.R. Pike, A.P. Roberts, K.L. Verosub, J. Appl. Phys. **85**, 6660 (1999)
10. C.R. Pike, Phys. Rev. B **68**, 104424 (2003)
11. C. Papusoi, K. Srinivasan, R. Acharya, J. Appl. Phys. **110**, 083908 (2011)
12. C. Papusoi, S. Jain, H. Yuan, M. Desai, R. Acharya, J. Appl. Phys. **122**, 123906 (2017)
13. C. Papusoi, M. Desai, R. Acharya, J. Phys. D. Appl. Phys. **48**, 215005 (2015)
14. H. Nemoto, I. Takekuma, H. Nakagawa, T. Ichihara, R. Araki, Y. Hosoe, J. Magn. Magn. Mater. **320**, 3144 (2008)
15. E.C. Stoner and E.P.Wohlfarth, Phil. Trans. Roy. Soc. **A240**, 599 (1948)
16. M.P. Sharrock, J. Appl. Phys. **76**, 6413 (1994)
17. W.F. Brown, Phys. Rev. **130**, 1677 (1963)
18. W.T. Coffey, Y.P. Kalmykov, J. Appl. Phys. **112**, 121301 (2012)
19. H.N. Bertram, X. Wang, L. Safonov. IEEE Trans. Magn. **37**, 1521 (2001)
20. C. Papusoi, Phys. Lett. A **265**, 391 (2000)
21. G. Ju, H. Zhou, R.W. Chantrell, B. Lu, D. Weller, J. Appl. Phys. **99**, 083902 (2006)
22. G. Ju, B. Lu, R.J.M. van de Veerdonk, X. Wu, T.J. Klemmer, H. Zhou, R.W. Chantrell, A. Sunder, P. Asselin, Y. Kubota, D. Weller, IEEE Trans. Magn. **43**, 627 (2007)
23. C. Papusoi, S. Jain, R. Admana, B. Ozdol, C. Ophus, M. Desai, R. Acharya, J. Phys. D. Appl. Phys. **50**, 285003 (2017)
24. A. Lyberatos, D. Weller, G.J. Parker, J. Appl. Phys. **114**, 233904 (2013)

25. Y.K. Takahashi, T. Koyama, M. Ohnuma, T. Ohkubo, K. Hono, J. Appl. Phys. **95**, 2690 (2004)
26. T. Miyazaki, O. Kitakami, S. Okamoto, Y. Shimada, Z. Akase, Y. Murakami, D. Shindo, Y.K. Takahashi, K. Hono, Phys.Rev. B **144419**, 72 (2005)
27. S. Okamoto, N. Kikuchi, O. Kitakami, T. Miyazaki, Y. Shimada, K. Fukamichi, Phys. Rev. B **66**, 024413 (2002)
28. J.B. Staunton, S. Ostanin, S.S.A. Razee, B. Gyorffy, L. Szunyogh, B. Ginatempo, E. Bruno, J. Phys. Condens. Matter **16**, S5623 (2004)
29. H. Miyajima, K. Sato, T. Mizoguchi, J. Appl. Phys. **47**, 4669 (1976)
30. C. Papusoi, T. Suzuki, J. Magn. Magn. Mater. **240**, 568 (2002)
31. O.N. Mryasov, U. Nowak, K.Y. Guslienko, R.W. Chantrell, EPL **69**, 805 (2005)
32. D. Hinzke, I.N. Kazantseva, U. Nowak, O.N. Mryasov, P. Asselin, R.W. Chantrell, Phys. Rev. B **77**, 094407 (2008)
33. D. Hinzke, U. Nowak, R.W. Chantrell, O.N. Mryasov, Appl. Phys. Lett. **90**, 082507 (2007)
34. S. Pisana, S. Jain, J.W. Reiner, G.J. Parker, C.C. Poon, O. Hellwig, B.C. Stipe, Appl. Phys. Lett. **104**, 162407 (2014)
35. Z. Liu, Y. Jiao, R.H. Victora, Appl. Phys. Lett. **108**, 232402 (2016)
36. Z. Liu, R.H. Victora, AIP Adv. **7**, 056516 (2017)
37. J.-U. Thiele, K.R. Coffey, M.F. Toney, J.A. Hedstrom, A.J. Kellock, J. Appl. Phys. **91**, 6595 (2002)

Biological and Medical Applications of Magnetic Nanoparticles

María Salvador, José C. Martínez-García, M. Paz Fernández-García, M. Carmen Blanco-López, and Montserrat Rivas

Abstract This chapter aims to give insight into the successes, challenges, and opportunities of magnetic nanoparticles for bio-applications. It reviews their general requirements and how they can be met by different synthesis methods and characterization techniques. It then focuses on examples of applications such as magnetic cell separation, magnetic detection (including imaging), and magnetic particle-based therapies, such as hyperthermia, magneto-mechanical destruction of tumors, localized drug delivery, and tissue engineering. We hope to guide the reader interested in applied research of magnetic nanoparticles through an exciting collection of investigations on their application in the life sciences.

Keywords Magnetic nanoparticles · Bio-application · Biomedicine · Biosensor · Bio-detection · Magnetic resonance imaging · Magnetic particle imaging · Magnetic immuno-separation · Magnetic hyperthermia · Drug delivery · Tissue regeneration · Magnetic detoxification

1 Introduction

Today, more than ever, research is called upon to solve social problems, often those that affect public health and safety. The coronavirus that paralyzed the world in 2020 has an average size of 120 nm. The nanoscale is the dimension at which biological

M. Salvador
Department of Physics, University of Oviedo, Campus de Viesques, Gijón, Spain

Istituto di Struttura della Materia—Consiglio Nazionale delle Ricerche (CNR), Rome, Italy

J. C. Martínez-García · M. P. Fernández-García · M. Rivas (✉)
Department of Physics, University of Oviedo, Campus de Viesques, Gijón, Spain
e-mail: rivas@uniovi.es

M. C. Blanco-López
Department of Analytical and Physical Chemistry, Institute of Biotechnology of Asturias, University of Oviedo, Oviedo, Spain

© Springer Nature Switzerland AG 2021
V. Franco, B. Dodrill (eds.), *Magnetic Measurement Techniques for Materials Characterization*, https://doi.org/10.1007/978-3-030-70443-8_26

events occur at the cellular and molecular level. Nanoparticles have the ideal size to interact with biological entities, such as proteins, genes, cells, viruses, and bacteria. They can be stabilized in colloidal suspensions to mix them with biological fluids or to inject them, and they can flow through blood vessels and cross some physiological barriers. Their large surface to volume ratio and, more importantly, the fundamental properties specific to their nanosize have the potential to significantly contribute to solving major problems in biomedicine.

Additionally, magnetism adds capabilities such as remote manipulation, heat generation, and inductive detection. In the fields of life sciences (biomedicine, environment, and food safety), magnetic nanoparticles (MNPs) offer new tools to develop innovative methods for diagnosis and therapy in clinical practice, toxin detection in agrifood control, and environmental remediation.

Advances in the production of MNPs—control of their size, morphology, and crystalline quality—have stimulated their practical application. The design of new nanoarchitectures enables hybrid applications that take advantage of the combination of magnetism and optical or catalytic properties. For biological applications, the superficial coatings of the particles are crucial. They determine important properties of the chemical interaction of the MNPs with biological molecules or environments, such as biocompatibility, bioconjugation, and circulation time (Fig. 1).

2 Requirements of Magnetic Nanoparticles for Bio-Applications

An essential feature of MNPs is their tunability. It allows them to be engineered for each application. Therefore, their crystal structure, size, shape, composition, agglomeration, superficial, and magnetic properties must be optimized. Biomedical applications can be classified into two main categories: in vitro and in vivo. The latter is the most challenging as it includes biocompatibility as a requirement.

Most of the advantages of MNPs come from their size, which confers them with properties entirely different from those of their bulk counterparts. Additionally, their size is particularly interesting for biomedical applications because it is comparable to that of most biological entities of interest. The particles' size plays a vital role in their magnetic behavior, which may be superparamagnetic below a critical volume [1]. Superparamagnetism is preferable to other magnetic states because it reduces the magnetic attraction helping to stabilize colloidal suspensions required for biomedical applications [2]. For in vivo applications, the particles' size, shape, and superficial properties strongly influence their circulation time, targeting, and final biodistribution or elimination [3–7].

For in vivo use, toxicity restricts the permissible composition. Iron oxide nanoparticles are approved by the US Food and Drug Administration and the European Commission for clinical use because they can be metabolized [8]. On the contrary, Zn, Co, Ni, and Cu may be toxic [9–11]. Coating or encapsulation is an option to overcome this handicap [12, 13].

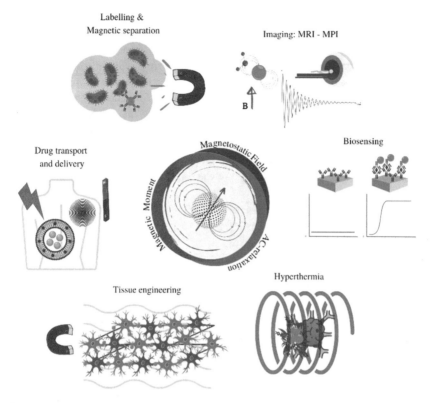

Fig. 1 Examples of applications of magnetic nanoparticles in biomedicine include magnetic cell isolation, magnetic imaging, biosensing, hyperthermia, tissue regeneration, and drug delivery

Biomedical applications require colloidal stability, but nanoparticles tend to aggregate and precipitate in aqueous suspensions. Proper surface preparation can avoid this by electrostatic forces and/or steric effects. Interestingly, aggregation can be either detrimental or advantageous, depending on the application. For in vivo processes, it increases the effective size, which may influence the circulation lifetime or heating efficiency [14, 15]. On the other hand, for in vitro purposes, as in reporter labeling of immunoassays, an increase in the number of MNPs per biomolecule enhances the detected signal [16, 17].

Surface properties are also crucial for functionalization. The adsorption of additional polymeric moieties onto the surface of the MNPs, known as electrosteric stabilization, is the most used approach to avoid agglomeration and prepare the particles for bioconjugation [18]. These coatings are usually polymers such as dextran, polyacrylic acid, polyethyleneimine, polyvinyl amine, and polyethylene glycol or small molecules such as dimercaptosuccinic acid, silica, metals, or carbon [19–21].

Due to their small size, the MNPs' magnetic behavior is strongly dominated by their surface layer spins. The broken crystalline symmetry at a particle's surface may lead to undercoordination, relaxed magnetic exchange bonds, and spin canting; all of this translates into modifications of the magnetic behavior compared to that of the bulk material. The surface magnetic anisotropy increases, and the saturation magnetization decreases, which is usually undesired for bio-applications. The action of the surfactant or polyelectrolyte can play a substantial role in improving this. Although capping organic molecules are not magnetic, they can restore the magnetic ordering of the core to the surface [22]. To explain this unexpected effect of nonmagnetic capping, Salafranca et al. [23] studied the magnetic order at the surface of Fe_3O_4 MNPs coated by oleic acid. They reported that some surface Fe ions bond to the organic acid's oxygen, so that the number of O nearest neighbors and their average distance closely resemble that of the bulk magnetite resulting in improved magnetic properties.

The specific magnetic requirements for various applications in the life sciences are treated in the following sections.

3   Synthesis Methods

Many different synthesis routes have been proposed to obtain MNPs that accomplish the characteristics detailed above (Fig. 2). The methods are commonly divided into "bottom-up" or "top-down" approaches. The latter, also known as physical

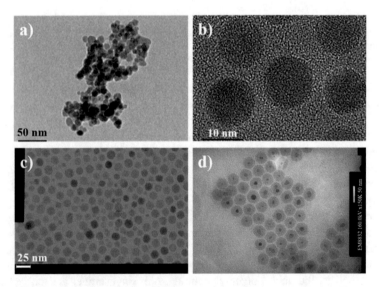

Fig. 2 TEM images of Fe_3O_4 MNPs synthesized (**a**) by coprecipitation and uncapped; (**b**) by coprecipitation and coated with polyacrylic acid; (**c**) by thermal decomposition with oleic acid coating (image courtesy of M. Puerto Morales, ICMM-CSIC); (**d**) by coprecipitation and coated by a silica shell

methods (such as milling or grinding), consist of dividing a larger material into smaller units. These methods are not usually suitable for biomedical applications because they produce particles with high polydispersity. Bottom-up processes, or chemical methods, provide more precise control over the resulting characteristics. Starting from small units of the materials, further self-assembly processes lead to the formation of the nanostructures.

Coprecipitation, developed by Massart in 1981, is one of the most popular techniques due to its simplicity and high yield [24]. The oxides are precipitated from a solution with the metal precursors as the pH is tuned to a basic range. The addition of coating polymers allows obtaining a stable colloidal suspension and to introduce reactive chemical groups necessary for subsequent functionalization.

Thermal decomposition involves breaking down metal precursors, such as acetylacetonates, carbonyls, or oleates, in organic solvents, as octyl ether, benzyl ether, or octadecene, at their boiling temperatures and in the presence of dispersants and/or hydrophobic ligands. The high control over the particle size, shape, and polydispersity is obtained by separating the nucleation and growth stages, as stated by LaMer's theory [25]. However, the hydrophobic reaction media (which is usually toxic and cannot be used for functionalization of biomolecules) must be replaced or modified for biological applications. Different approaches such as ligand exchange or extra coating layers can be used for this purpose [26, 27].

Microwave synthesis is a variation of thermal decomposition methods in which heating is achieved by microwave induction. It is attracting attention because it improves the nanoparticle quality, reaction yield, and reproducibility and reduces the synthesis duration [28]. The heating homogeneity in the reactor ensures a narrow size distribution and controlled physicochemical properties of the MNPs obtained directly in the hydrophilic medium. Although the reactors are small, their scale-up has already been proposed [29].

In **microemulsion methods**, micro- or nano-drops called micelles are formed thanks to the combination of the precursors, surfactants, and dispersing medium. The nucleation and growth of the particles occur inside them. This produces a narrow size distribution because the micelles, acting as nanoreactors, constrain the MNP growth [30].

Solvothermal synthesis is performed at high pressures above the solvent's boiling temperature (it is called hydrothermal if the solvent is water). The sophisticated equipment and operation conditions render high-quality monodisperse particles but may complicate the scale-up of this method.

The **sol-gel method** is a multistep process based on the metal precursors' hydrolysis and polycondensation near room temperature. The first step forms the sol, transformed into the gel by successive condensation, vaporization, and solidification. Further heat treatments are needed to produce crystallization. However, it is still difficult to control particle morphology and reproducibility due to the complex mechanisms involved during the steps of the reaction.

Polyol methods use polyethylene glycols of different hydrocarbon chain length that act both as reducing and stabilizing agents during the particles' formation. Their amphiphilic character allows the direct obtention of hydrophilic suspensions for bio-applications. The high temperatures enable control over the size and crystallinity while reducing the starting materials' cost and toxicity.

Further information about traditional synthesis methods and new approaches like microfluidic systems and dendrimer templates may be found in Refs. [31–33].

4 Characterization Methods

Because the performance of colloidal MNPs essentially depends on their physicochemical properties, their thorough characterization is critical for finding the best routes to optimize their biomedical applications and to obtain a meaningful interpretation of the outcome. Additionally, standardization of the analysis protocols is crucial for interlaboratory comparison and translation to the clinics and industry [34].

In this section, we review the physical characterization methods most frequently used for MNPs. Size, morphology, agglomeration, structure, and superficial and magnetic properties significantly influence the particles' biodistribution, safety, and efficacy. They are evaluated by techniques commonly used for other materials but which have peculiarities when applied to MNPs for bio-applications. In this section, we list some of the most popular characterization techniques, with a special focus on one of the most specific problems: sample preparation. We provide starting points and references for interested readers to follow.

Sample Preparation

Depending on the characterization technique and the specific application, particles are either dried, suspended in a liquid medium, or immobilized in a solid phase. Many characterization devices cannot handle liquids, so colloidal nanoparticles need to be dried. Although spontaneous evaporation of the solvent is an option, it often produces undesired oxidation of the particles' surface that would alter the characterization results. Oven drying at moderate temperatures in a protective atmosphere is a better choice. Freeze-drying (lyophilization) is an alternative that is especially recommended if the particles are conjugated to biological material. It consists of cooling the sample below the triple point of the solvent, reducing the pressure, heating it to produce sublimation (vacuum speeds it up), and heating again to produce desorption of the remaining water molecules [35, 36].

For some techniques or applications, the powder samples need to be suspended or resuspended. The nature of the particle surface needs to be considered to find a suitable solvent (organic for hydrophobic particles; water, or other polar solvents for hydrophilic coatings.) The protocols involve mixing, sonicating, and syringe filtering.

Mainly for magnetometry, the colloidal particles may need to be immobilized because some instruments cannot handle liquids or to prevent Brownian motion or physical rotation of the particles. They can be immobilized in a compatible solid matrix. For hydrophobic particles (coated with hydrophobic polymers or suspended in organic phases), paraffin wax, docosane, or hydrocarbons of high molecular weight are adequate. Hydrophilic particles (coated with hydrophilic polymers, peptides, amino acids with amino or carboxylic residual groups) can be immobilized in silica, dimethacrylate triethylene glycol, or agar. The protocols involve mixing a known mass of nanoparticles with the initiator solution, sonication to disperse them, and placing them in an oil bath at 70 °C for some hours. The initiator polymerizes, forming a solid matrix.

Morphology, Size, and Structural Characterization

The microstructure, size, and morphology of the MNPs are typically determined combining diffraction techniques, electron and atomic force microscopies, and Mössbauer spectroscopy.

The most direct way to obtain the particles' metal (or metal oxide) core size and shape is transmission electron microscopy (TEM). It uses a beam of accelerated electrons, benefiting from their short wavelength to resolve distances in the nanoscale. TEM uses a static beam transmitted through a thin sample (the thickness must be below 100 nm) to get the structural information and provide magnified bidimensional images. The sample preparation starts with a diluted suspension. A droplet is deposited on a carbon membrane and placed on a copper grid (carbon-coated copper grids are commercialized). The sample must be then dried either by natural evaporation or in flowing nitrogen. Careful sample preparation is essential to avoid problems like aggregation while drying [37] or the formation of crystalline salts or films. Specimen preparation is more difficult when it contains biological material, such as cancer cells. In this case, the material must be embedded in a pure resin mixture, dried, and cured. Then, an ultramicrotome is used to cut thin sections [38]. TEM can achieve lateral spatial resolution and magnifications up to factors of 10^7. The MNPs' shape and size distribution can be inferred from the TEM images; its drawback is that a significant sampling requires the analysis of many micrographs. Transmitted beams also provide insights into the morphology and sample's structure and reveal atomic defects or dislocations. In TEM, the diffracted electrons can be used to obtain a selected area electron diffraction pattern (SAED), which provides information about the crystal structure, and the energy-dispersive X-ray spectroscopy, to get qualitative information about the chemical composition of individual particles. High-resolution TEM uses hardware to improve the resolution to 0.05 nm, enabling imaging of the atomic structure.

In scanning electron microscopy (SEM), the electron beam scans the specimen line by line. The electrons interacting with matter are scattered and can be collected in a variety of forms. The secondary electrons (from the sample's atoms) and the backscattered electrons (from elastic collisions) provide topological images and composition maps. Although monodisperse MNPs can be too small to be seen by SEM, this technique can help analyze beads containing embedded or encapsulated particles. For SEM experiments, the sample must be in powder, fixed on an adhesive tape on which a thin layer (15–20 nm) of a conductive metal is sputtered. This is an optimum thickness for obtaining good-quality images and, at the same time, a compositional analysis of the sample by energy-dispersive X-ray spectroscopy, which is frequently included in SEM. Field-emission SEM uses an electromagnetic electron generation, which notably improves the spatial resolution and avoids conducting coatings on insulating materials. Scanning TEM is a variation of TEM in which the focus electron beam is scanned across the specimen.

Atomic force microscopy (AFM) gives information in the three spatial dimensions. It scans the specimen with a probe that mechanically interacts with the specimen and measures the force as a function of their mutual separation. It creates three-dimensional images of nanoparticles with sizes between 0.5 and 50 nm. The MNPs can be deposited on a cleaned silicon substrate after sonication, left to dry for 5–10 min, and blown by a nitrogen stream to eliminate the residual solution [39]. AFM does not require conductive coating of the sample and can work in liquid medium. Magnetic force microscopy (MFM) is a variation of AFM that uses a magnetic tip to interact with the sample and visualize magnetic domains [40]. MFM has been used to distinguish between blocked and superparamagnetic magnetite particles in aggregates [41].

For many aspects of the bio-applications, the hydrodynamic size plays a vital role. This size includes not only the inorganic core of the particle but also the coating and solvation molecules. It can be obtained by dynamic light scattering (DLS). In this technique, a laser beam irradiates the sample, and the scattered light is collected. The intensity fluctuates depending on the particles' Brownian motion, which in turn is related to their size. One can also use DLS to assess the aggregation of the particles [42, 43] and to check the success of the biofunctionalization. When a bio-receptor is attached to the particles' surface, the hydrodynamic size increases. Specimen preparation for DLS usually requires dilution to avoid multiple scattering. The hydrodynamic size depends on the diluent salinity. Nanoparticle tracking analysis (NTA) is an increasingly popular technique that combines light scattering with a microscope to analyze the individual particles' Brownian motion. It then provides particle size distributions ranging from approximately 10–20 nm to 1000–2000 nm [44]. Because this technique individually tracks and sizes each particle, a low initial concentration to avoid crowded observation fields is required to achieve accurate results (usually 10^6–10^{10} particles/mL). This dilution process and its time-consuming handling make DLS measurements more widespread for a routine size characterization. On the other hand, NTA offers a more accurate distinction between the smaller and larger particles and information about nanoparticle concentration in the liquid sample.

A comparison of experimental methods to measure metal nanoparticle dimensions is found in Ref. [45]. One can learn more about nanoparticle and biological sample preparation for electron microscopy in [46] and references therein.

X-ray diffraction (XRD) and neutron diffraction (ND) allow investigating the long-range crystal and magnetic structure, respectively, of the samples [47]. In both cases, the beam interacts with the material, and the raw data consist of the diffracted intensity as a function of the diffraction angle. The main difference is that X-rays interact with the atom's electron cloud, whereas neutrons interact with the nuclei (nuclear scattering) and with the unpaired electrons (magnetic scattering). In XRD, the diffracted intensity is related to the atomic number of the atom, revealing the symmetry and lattice parameters of the crystallographic unit cell based on the diffraction peaks' angular position. Additionally, the peaks' widths relate to the material's intrinsic properties, such as crystallinity, strain, or grain size. XRD is more commonly used for structural studies because it is versatile and accessible to most researchers. ND, on the other hand, requires a large facility's neutron source but provides the magnetic structure. In materials combining light elements, XRD may not be adequate, but ND can be an alternative because it is not directly related to the atomic number. Additionally, the atoms that are neighbors in the table of elements are not easy to distinguish in an XRD pattern, but they can be with ND.

A full-profile fit based on Rietveld refinement can yield quantitative information from XRD and ND experiments. It is remarkable that even though Cu Kα_1 radiation ($\lambda = 0.15418$ nm) is the most common wavelength for XRD experiments when the MNPs contain elements such as Cr, Mn, or Ni but especially Fe and Co, XRD with a source with a Mo anode ($\lambda = 0.07107$ nm) is more suitable because these elements fluoresce with X-rays from a Cu anode.

Both liquid and powder samples can be examined, although the latter is preferred in long experiments to prevent the change in the sample volume and particle position due to the solvent's evaporation [36]. The powder for XRD experiments is placed in quartz sample holders of small depth. The preparation of the powdered specimen for characterization is critical because its surface must be extremely homogeneous. In the case of ND, double-walled vanadium cylinder sample holders are traditionally used to reduce neutron absorption.

X-ray absorption spectroscopy (XAS) is used to study the local environment around the metallic atoms. Even though XRD requires long-range order of atoms (>10 nm), XAS is not so limited. The principle of this technique is based on the absorption of X-rays by the atoms. Depending on the absorption spectrum's energy range, XAS is split into X-ray absorption near-edge spectroscopy (XANES) and extended X-ray absorption fine structure (EXAFS) [48, 49]. The first is sensitive to the oxidation state and coordination chemistry of the absorbing atoms, whereas the second is used to determine distances and coordination numbers of the neighboring shells around the central atom. Additionally, if the MNPs contain mixtures of phases with different valence states of a given element (e.g., Fe in α-Fe, γ-Fe, Fe_3O_4, γ-Fe_2O_3), XANES analysis can also be used to estimate phase ratios [48]. The preparation of samples for XAS experiments is crucial because the thickness of the sample affects the absorption coefficient and, thus, the intensity of the oscillations. Two different procedures are commonly used. The first involves hydraulically compacted pellets that are a mixture of boron nitride and the powdered sample. In the second, a small amount of powder is spread over an adhesive Kapton tape. In both cases, one must optimize the samples' thickness and homogeneity to attain the best signal-to-noise ratio.

MNPs for bio-applications are often colloidal suspensions of mixed magnetite and maghemite. Both are spinel ferrites with almost identical lattice parameters, so their identification by XRD is not a trivial matter. However, Mössbauer spectroscopy (MS) has proved to be extremely helpful in distinguishing their composition and stoichiometry. The task is challenging in nanoparticle specimens because magnetic interparticle interactions cause broadening of the spectral lines, which hampers a profile-based distinction. A useful guide for Fe^{57} MS discrimination of maghemite and magnetite nanoparticles is in Ref. [50]. The basis of MS is recoilless γ-ray resonance absorption, and the recoil energy can be obtained only when both source and absorber are embodied in a solid lattice. The sample preparation varies with the configuration. For measurement at room temperature, several grams of powder sample must be spread on adhesive Kapton tape, whereas for low-temperature spectra, small Teflon boxes may encompass the powder. The samples' thickness and homogeneity are crucial, and the thin-powdered layer must be optimized to contain around 5 mg of Fe per square centimeter.

Surface Properties

Routine characterization of the surface condition of the nanoparticles for bio-applications involves ζ-potential and functional groups. The ζ-potential is the electrostatic potential difference between the dispersion medium and the stationary layer of fluid attached to the particle (Stern layer). It can be obtained by electrophoretic light scattering (included in some DLS analyzers). It determines the stability of the particles' suspension (absolute values of ζ-potential above 30 mV usually result in repulsive electrostatic forces strong enough to attain colloidal stability). Samples may require dilution, avoiding the addition of any electrolytes [51].

Fourier transform infrared spectroscopy (FTIR) enables investigation of the molecules attached to the particles' surface [52]. The spectrum gives information about the strength and nature of the bonds and the functional groups. The measurements are preferably on powder samples, usually grounded with potassium bromide (KBr). FTIR is one of the basic techniques to prove that a ligand exchange process has been successful [53].

Raman spectroscopy (RS) provides chemical and structural information. For quantification, some precautions regarding the laser-induced oxidation must be taken [54]. RS is suitable for all kinds of samples, even biological. It distinguishes the composition of the different phases in the core and shell [55, 56]. A thorough characterization of the surface can be done by X-ray photoelectron spectroscopy (XPS) [57]. This technique gives the elemental composition of the outer layers of the material and the chemical and electronic state of the elements at the surface. It is especially useful to get information about the oxidation state in iron oxide particles

[58], core-shell structures [59], and ligands bound to the surface [60]. Samples for XPS must be dried powder because the measurements are made in ultrahigh vacuum.

Concentration

The concentration of the inorganic magnetic cores of the particles in a solution is a parameter required for most studies and applications. One can give the total mass per unit volume, iron mass per unit volume, and number of particles per unit volume. It can be obtained by thermogravimetrical analysis (TGA), taking great care to measure the deposited volume. Additionally, TGA provides information about the organic layers, the extent of coverage, and layer-mass to core-mass ratio [61]. Some instruments include differential scanning calorimetry, which shows the endothermic or exothermic nature of the processes. TGA requires a large sample volume and is a destructive test. Micro-TGA eliminates this drawback [62].

Inductively coupled plasma mass spectrometry (ICP-MS) and optical emission spectroscopy (ICP-OES) are used to determine the concentration of elements, providing high precision, accuracy, and sensitivity even with trace amounts of material. They are compatible with any kind of samples, although digestion processes are usually required [63, 64].

NTA devices, discussed above, can also estimate the concentration as the number of particles per volume [65].

Magnetic Characterization

Classical and advanced magnetic characterization techniques are described in other chapters of this book. This section aims to focus on the most used techniques and sample preparation for MNPs that are typically used in the field of life sciences.

The most used magnetometers for MNPs are vibrating sample magnetometers VSM [66], alternating gradient force magnetometers AGFM [67], and magnetometers based on superconducting quantum interference devices SQUIDs, often including AC susceptometry based on mutual induction [68]. Fluxgate magnetometers are often used to measure magnetorelaxometry.

The magnetic properties of MNPs that more substantially affect their biomedical applications are their DC and AC hysteresis, remanent magnetization M_R, saturation magnetization M_S, magnetic effective anisotropy K_{eff} and magnetic interactions, blocking temperature T_B, DC and AC permeability, and relaxation. The methods to obtain them are with DC magnetization curves as a function of applied field or temperature (zero-field cooling and field cooling ZFC-FC; isothermal remanent magnetization IRM; demagnetization remanence curves DCD; and first-order reversal curves (FORC); and field, temperature, or frequency dependence of AC magnetic susceptibility).

ZFC curves are obtained by cooling the sample in zero applied field from room temperature to, usually, liquid-helium temperature (4.2 K) and then applying a DC field and measuring the magnetization (M_{ZFC}) while the sample is warmed. FC curves are obtained in the same way but with the field applied during cooling (M_{FC}) [69]. The blocking temperature determines the transition between the blocked and the superparamagnetic state. Some authors identify the temperature at which the M_{ZFC} has a maximum (T_{max}) with the blocking temperature [70, 71], which is adequate for monodisperse populations. In other cases, the T_B is obtained from the T_B-distribution curve (calculated as the temperature derivative of the difference $M_{FC} - M_{ZFC}$). For monodisperse populations, the FC and ZFC curves would coincide for temperatures above T_B; their separation, quantified by the difference of T_{max} and T_{irr} (temperature at which the two curves merge), is an indication of the degree of polydispersity [69]. We can calculate $K_{eff} = 25k_B T_B/V$ for average particle volume V and Boltzmann constant k_B. ZFC–FC curves are also effective to study the MNP interactions, which cause spin glass or superferromagnetic behavior [72, 73].

The remanent magnetization technique can provide a thorough insight into the MNP interaction. It consists of two curves: IRM and DCD. Both are obtained applying a field H and measuring M_R after its removal. For IRM, the sample starts demagnetized while for DCD it starts from negative saturation (The IRM is measured after the application and removal of a field with the sample initially demagnetized. The DCD is measured from the saturated state by application of increasing demagnetizing fields). In an ideal non-interacting system of monodomain particles, Wohlfarth's relation [74] holds

$$\partial M = m_r^{DCD}(H) - \left(1 - 2m_r^{IRM}(H)\right) = 0 \tag{1}$$

where $m_r^{DCD}(H)$ and $m_r^{IRM}(H)$ stand for the remanent magnetization in DCD and IRM, respectively, normalized by the saturation magnetization value at $H \to \infty$. A nonzero value of ∂M is an indication of the magnetic interactions. This is traditionally represented in a Henkel plot or ∂M plot [70, 75]. Given that these techniques are related to the remanent state, the curves need to be obtained below the particles' blocking temperature.

First-order reversal curve (FORC) analysis can provide distributions of the local interaction and switching fields [76]. To obtain a FORC, an initial saturating field (H_S) is applied and reduced to the return field (H_R); H is then increased up to H_S while measuring the magnetization $M(H_R, H)$; the procedure is repeated for equi-spaced values of H_R. For a non-interacting system of monodomain particles, the differential susceptibility

$$\chi_m(H, H_R) = \left.\frac{\partial M}{\partial H}\right|_{H_R} \tag{2}$$

is the same for all the values of H_R in the common range of H, but this equality is broken due to interactions. These interactions are then shown in the FORC diagram [77], that is, the bidimensional plot of the FORC distribution

$$\rho(H, H_R) = -\frac{1}{2}\frac{\partial^2 M}{\partial H \partial H_R} \tag{3}$$

on the H–H_R plane [78].

Dynamic magnetic susceptibility (DMS) (also called AC magnetic susceptibility) frequency spectra are essential to evaluate the MNPs' potential in many biomedical applications. For example, the DMS (10^3–10^8 Hz) of blocked nanoparticles can be used to determine their interaction with proteins in their protein corona [79]. DMS and AC magnetization curves and susceptibility are also crucial to assess the heating capacity of nanoparticles for magnetic hyperthermia or their use as nanotags in inductive biosensing.

Magnetorelaxometry consists of magnetizing the MNPs by applying a static magnetic field, removing it, and measuring the sample's decaying magnetization as a function of time. For this purpose, one or several sensitive magnetometers monitor the magnetization evolution. For strictly monodisperse particles, one expects an exponential decay with a size-dependent time constant. A MNPs suspension has a nontrivial, non-exponential decay that depends on the hydrodynamic size, size distribution, and aggregation.

Most magnetometers and susceptometers can measure either liquid or solid samples. Liquid samples are preferred, for example, when there is an interest in distinguishing effects associated with Brownian relaxation or changes in the magnetization dynamics produced by bioconjugation, chaining, or agglomeration of particles. Nevertheless, solid samples are usually easier to handle and the only possibility for measurements with a varying temperature that goes through the melting point of the solvent. One rough option is to dry the samples, encapsulate the powder in a gelatin capsule, and press it with cotton (to avoid movement during measurement). However, this may increase the interaction among the particles, which can affect the effective anisotropy constant, the distribution of blocking temperatures, or the characteristic relaxation times. In general, the particles' dispersion with a concentration smaller than 1 wt.% in a solid matrix can assure that the measured magnetic properties are intrinsic [80]. Ref. [81] has a detailed discussion of precautions needed for sample preparation for measurements of samples with low magnetic moments.

Preclinical Characterization

For biomedical applications, especially in vivo, nanomedicines must be safe for the patient. A thorough study of the transport and fate of MNPs in the human body is vital before any diagnostic or therapeutic concept is transferred to the clinical field.

Preclinical characterization involves assays for toxicity, sterility, and interactions with living biological matter, including MNP uptake by cells or macrophages, and thrombogenic or hemolytic effects. Ref. [82] and references therein contains a detailed description of these techniques for both in vitro and in vivo models.

When the nanoparticles enter a biological system, some effects can crucially affect their behavior: size growth due to protein corona, agglomeration due to interactions among proteins, the variation of their magnetic properties caused by biodegradation, or the change of Brownian relaxation due to viscosity. It is then essential to characterize the nanomaterials in biological matrices that mimic the application environment [83].

5 Magnetic Bio-Separation

Many areas of biosciences and biotechnology require the separation of specific biomolecules from a complex medium. The two main processes that involve bio-separation are (i) the isolation or purification of a target biomolecule for its study or further use (e.g., antibodies, viruses, or genes) and (ii) the enrichment or concentration of a molecule in the sample, for example, to ease its detection or quantification. Additionally, magnetic separation can be employed for environmental applications in water and soil remediation, either to remove iron-containing particles or to eliminate other contaminants adsorbed at the surface of a MNP [84, 85].

The basic principle of the magnetic bio-separation is simple: MNPs bearing a specific surface functionalization to capture the target molecule are mixed with the complex sample and incubated for a few minutes to allow the molecule-particle binding, after which the particles are trapped by a magnetic field gradient, carrying the desired molecules. The supernatant can be removed and the specific biomolecules resuspended in a fresh medium.

This kind of magnetic "fishing" relies on the affinity constant of the ligand (which determines the specificity and recovery), the magnetic moment of the particles, and the magnetic separator, as Fig. 3 exemplifies.

The surface of magnetite or other ferrite nanoparticles can be modified for separation purposes. Although superparamagnetic particles can be stable in suspension and easily resuspended, their magnetic moment is small and corresponds to their size. In this case, high magnetic gradients are needed to separate them. Small superparamagnetic particles can be encapsulated in biopolymers, porous glass, or magneto-liposomes, forming micrometer-sized beads to attain the large magnetic moments that favor efficient separation in low magnetic gradients.

Magnetism has some advantages versus alternative techniques such as column chromatography, precipitation, ultracentrifugation, or filtering. One of them is that all steps can be produced in only a tube, avoiding the sample losses associated with transfer. Magnetic separation does not involve costly reagents or equipment, and it can easily handle large-scale operations and automated processes. The magnetic technique can deal with samples containing particulate material, and, contrary to

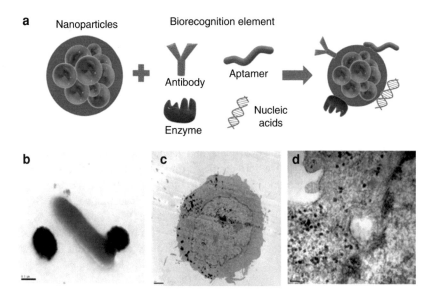

Fig. 3 (**a**) Schematic representation of the affinity ligands that can be attached to a MNP. (**b**) TEM image of MNP *Escherichia coli* recognition via anti-*E. coli* antibody attached to fluidMAG-Streptavidin (Chemicell GmbH, Germany). (**c**) MNP internalized by a HeLa cell. (**d**) Detail of the internalized MNP by a HeLa cell [86]

chromatography or filtering, it does not lead to very diluted molecule solutions. In fact, magnetic separation is useful for a preliminary concentration of bio-analytes for their detection, which can be done by the same particles used in the role of detection labels. Another excellent feature of magnetic separation compared to, for example, centrifugation, is that it involves only small shearing forces, which ensures the integrity of the molecules during the procedure.

To attract a particle of magnetic moment $\vec{m}$ and separate it from the matrix, a magnetic gradient is required to produce a force $\vec{F} = \nabla \left(\vec{m} \cdot \vec{B} \right)$. The simplest setup to create a magnetic gradient is a permanent magnet, especially if it has a pole with a sharp or pointed shape.

In the literature and market, one can find various designs with different geometries and coatings depending on the specific needs, such as laboratory tubes, sample volume, biocompatibility, or disinfection. Flow-through magnetic separators obtain a high gradient useful to capture small volume or weak magnetization particles [87]. They consist of column filters made of densely packed ferromagnetic spheres or fibers magnetized by a permanent magnet or superconducting solenoid. The sample containing MNPs flows through the magnetized column and is trapped by it. After removal of the applied field, the nanoparticles and their cargo are released from the filter by pushing water or another solvent through it. The column filters are disposable.

6 Magnetic Nanoparticles for Biosensing

Today, more than ever, biomedicine, food industry, and environmental safety demand procedures for biomolecule detection that are as fast, specific, and sensitive as possible. Depending on the precise application, other specifications may be crucial, such as low cost, ease of use, and portability.

Various research groups are responding to these demands with solutions that rely on MNPs as signal reporters. Most biosensing platforms adapt magnetic field sensors to excite the nanoparticles and detect their magnetization. Small devices are based mainly on Hall effect, magnetoresistive, or inductive sensors and microfluidics or surface biofunctionalization to bring particles into the sensing area [88–90]. Some bio-analytical systems with extremely high accuracy involve SQUID or atomic magnetometers or nuclear magnetic resonance [91, 92].

Other chapters of this book treat in detail the basics of magnetic field sensing. We focus in this section on how to adapt them for biological specimens tagged with MNPs.

Spintronic Sensors

Spintronic sensors based on either giant magnetoresistance GMR, tunnel mag-netoresistance TMR, or planar Hall effect PHE can be used for ultrasensitive detection of MNPs and, consequently, for biosensing [89, 93]. Among them, GMR devices, specifically spin valves SPV, have been the most widely used for biosensing applications. As in the case of the bio-separation, for bio-detection the MNPs will tag the target analyte by affinity ligands attached to their surface. Once this step accomplished, the next steps consist of (i) removing the non-bound nanoparticles and (ii) detecting and quantifying the nanotags. The sensor arrays are coated with a corrosion-resistant passivation layer, and then a small area is biologically activated by immobilizing the bio-recognition molecules that selectively capture the target analyte. The latter is tagged with MNPs before or after this capture. In this way, the MNPs are very close to the sensing layer, which is essential to detect the magnetic fringing fields. The geometry and size of the biologically activated area and the particles' spatial distribution are essential variables to consider for precise quantification.

The transformation of a spintronic sensor into a biosensor involves challenging tasks, such as integration with microfluidic channels to bring the fluid sample into the sensing area, passivation of the sensor surface and electronics, chemical surface modification to bind the target molecule, and washing out of the unbound particles and molecules. Ref. [94] has further details on sensors' surface modification.

Uncontrolled agglomeration of MNPs should be avoided, as it could cause obstructions in microfluidic channels. Also, uneven agglomeration can produce calibration failures, as different-sized agglomerates can attach to the biomolecule.

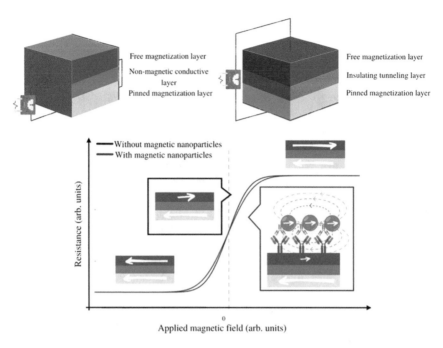

Fig. 4 Scheme of a GMR sensor (top left), a TMR sensor (top right), and the resistance response in a biosensing application (bottom)

Finally, magnetic dipolar interactions can modify the magnetic response of the particles and consequently, their quantification. For all these reasons, superparamagnetic particles are preferred for this application.

The magnetic nanotags must be excited by either a DC or an AC magnetic field to produce a detectable fringing field (Fig. 4) In addition, another biasing field may also be applied to define a proper working point for the sensing multilayer.

Using SPV sensors, some authors report the detection of only 20 Fe_3O_4 MNPs of 16 nm, equivalent to a total magnetic moment of 15×10^{-18} A m^2. After the adequate surface biofunctionalization, they can detect proteins and DNA with limits of detection below 10 pM (0.4 ng/mL) and procedure durations below 2 h [93]. Cubic nanoparticles of $Fe_{70}Co_{30}$ with saturation magnetization as large as 226 A m^2/kg of FeCo have been used for GMR detection of proteins achieving sensitivities 13 times larger than that of standard ELISA techniques [95]. Nanometric trapezoidal multilayers consisting of two ferromagnetic nanolayers magnetostatically coupled through a ruthenium spacer have been used for GMR quantification of proteins combined with magnetic tweezers. Their high saturation magnetization of 150 A m^2/kg of CoFe allowed limits of detection of 10 pM [96]. Configurations to detect influenza viruses, proteins, DNA strands, *Escherichia coli*, the liver cancer biomarker α-fetoprotein, and sentinel lymph nodes [97] are other examples of direct contact GMR and TMR biosensors. Biosensing platforms based on micro-Hall sensors have been reported for the detection of DNA [98] and circulating tumor cells in unprocessed biological samples [99].

SQUID and Atomic Magnetometers Applied to Biosensing

Magnetometers based on superconducting quantum interference devices (SQUIDs) are among the most sensitive magnetometers, capable of measuring magnetization relaxation, remanence, and susceptibility. For biosensing, the molecule to detect is tagged with MNPs in suspension. Magnetic relaxation immunoassays are based on modifying the magnetization relaxation time of the MNPs when they are bound to the molecule. The free MNPs relax fast by Brownian motion (involving physical rotation of the particles), but when they are bound, their relaxation takes place mainly by Néel relaxation (the magnetic moment of the particle rotates to align with the magnetic field), which is significantly slower. The measurements are usually done in liquid samples containing both the bound and free particles, thus avoiding the time-consuming washing steps or the microfluidic devices to separate them. For this reason, this method is usually called wash-free or liquid-phase immunoassay. They can also be done by immobilization on a surface or latex beads much larger than the MNP. In general, various measurements are necessary to interpret the relaxation rates and allow quantification, which is mainly done by comparing to reference curves for the relaxation of bound and unbound MNPs. For this application, iron oxides and spinel ferrites are the most used. The MNP solution must be very homogeneous in size and magnetic properties. Its colloidal stability and exhaustive control of their agglomeration must be assured to avoid false-positive detections.

SQUID magnetometers can also be used for detection based on AC susceptibility or remanence measurements and their change upon immuno-attachment of the MNP to the target analyte.

Atomic magnetometers are made of gases of rubidium or cesium. Their working principle is detection of the interaction between the atomic gas moments and the external magnetic field. They are highly sensitive, and unlike SQUIDs, they do not require cryogenics. For bio-detection, the magnetic response of the MNP before and after binding to the target molecule is studied.

Ref. [89] has an exhaustive description of these magnetometry bio-detection techniques and an extensive collection of applications on biomarkers, tumor cells, and DNA detection.

Faraday Induction Coil Biosensors

Faraday coil magnetometers and susceptometers can also be adapted for bio-detection mediated by MNPs. A variety of approaches differ both in the transducer coil and the electronics. In most inductive biosensors, an excitation coil produces a time-varying field that affects the MNPs, which, in turn, produces a changing magnetic flux that is picked up by a second coil. The pickup coil's response can be measured as a change of its self-inductance by an impedance analyzer, a variation of its resonance frequency, or by monitoring the induced voltage. The system can be reduced to a single coil with a dual role: excitation and detection if high frequencies are used to achieve a significant coil quality factor. In this case, superparamagnetic particles are advantageous because their initial magnetic permeability drops only at very high frequencies, above those corresponding to their Néel relaxation times.

Depending on the coil geometry, inductive magnetometers can encompass a vial with the liquid sample or an immune-active substrate on which the magnetically tagged analyte is captured [89, 100].

Magnetic Lateral Flow Immunoassays

Bringing a liquid sample close to the transducer usually involves coating the sensing surface and electronics with a corrosion-resistive passivation layer, integrating microfluidics, and washing out the remaining liquid after immobilization of the target molecule. Additionally, the regeneration of binding sensor surfaces is necessary to make them reusable and keep the cost down. Removing the biological material without damaging the sensor is complex and a research matter itself [101].

Paper-based microfluidics consists of a porous cellulose membrane that guides the liquid by capillary action and therefore without pumps. It may include filtering stages and has a high surface to volume ratio, which enables a large number of bio-recognition events. Additionally, the membranes are disposable and biodegradable.

Among the paper-based microfluidic analytical devices, probably the best known is the lateral flow immunoassay (LFIA) [102]. A popular example of LFIA is the home pregnancy test. Its main element is a strip of nitrocellulose membrane along which the liquid sample can run (Fig. 5). In the first stage, the target molecule (e.g., the human chorionic gonadotropin hCG) is selectively labeled by a signal particle (which is attached to a specific anti-hCG antibody). Two lines of selective recognition molecules have been dispensed across the membrane strip: the test line, at which the target analyte is captured together with its signal tags, and the control line that traps unbound particles and is used to assess the completion of the assay. This type of test is called a sandwich or direct assay. If it is not possible to attach two antibodies to the target molecule, an indirect or competitive assay is used.

A binary presence/absence response may be adequate for pregnancy tests, but many other biomedical, food safety, or environmental targets require quantification. In such cases, the LFIA must be combined with a sensor that does not compromise the advantages of low cost, portability, and easy use. MNPs used as tags can provide a quantifiable signal along with other benefits, such as magnetic immuno-separation or concentration of the analyte, long-lasting signal, and lack of interference and background signal from the biological matrix or the paper.

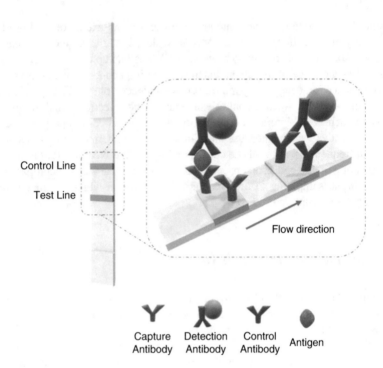

Fig. 5 Schematic representation of a LFIA in sandwich format

Besides the conditions common to other magnetic immunoassays and biosensing applications, like the homogeneity of the magnetic properties and size, LFIA requires that the MNPs are dispersed in a water solution. The particles' hydrodynamic size must be kept below a threshold that depends on the membrane pore size (typically in the order of 0.1 to 1 μm), which determines the flow rate and the sensitivity. Other specifications are related to the sensing technology, mainly GMR or Faraday inductive sensors [103, 104]. Ref. [105] has more information on magnetic LFIA.

Magnetic Nanoparticles in Nonmagnetic Detection

Nonmagnetic sensors can also use MNPs for pre-concentration, separation, and location of the target molecules.

In some nonmagnetic affinity sensors (e.g., electrochemical, piezo-immunoassays, or impedance sensors), the MNPs are used as an interesting alternative to immobilize the antibody instead of a direct attachment. Disposal and renewal of the antibody layer is easy, allowing regeneration of the sensing surface. Additionally, in some enzymatic sensors (with conductimetric, amperometric, or potentiometric transducers), MNPs simplify the enzymes' storage and handling [106, 107]. In some Bio-Bar-Code sensors [108, 109] for high sensitivity detection of protein and DNA, MNPs are used for immuno-separation.

7 Magnetic Nanoparticles for Bio-Imaging

Magnetic resonance imaging (MRI) is one of the most used techniques for noninvasive biomedical diagnosis due to its innocuousness, good spatial resolution, and penetration depth. It is based on the detection of water protons' magnetic relaxation; the contrast in soft tissues comes from the different proton density in different tissues, intra- or extracellular spaces, or lipidic environments, or different relaxation times.

MRI devices consist of a superconducting magnet, gradient magnetic fields, RF coils, and signal processing equipment. An intense magnetic field is applied to align the protons' magnetic moment in the field's direction. A magnetic gradient is superimposed on it to encode the signal spatially (because it causes the field strength to vary linearly in space). A radiofrequency field then reorients the protons' magnetic moment to align either at 90° or 180° with the applied magnetic field. When this pulse is turned off, the relaxation can be measured in longitudinal (T_1, spin-lattice relaxation) and transverse (T_2, spin-spin relaxation) modes. Protons that relax fast produce a large signal (the time variation of the magnetic moment is picked up by inductive coils), whereas those that relax slowly produce a smaller one. The different signals lead to the contrast between tissues. MNPs enormously contribute to the enhancement of this contrast. In addition, MNPs can be functionalized to attach specifically to tumor cells, facilitating early cancer diagnosis.

Paramagnetic particles containing chelated gadolinium are widely used to shorten T_1 by coordinating with the water molecules, which increases the positive contrast. But their potential toxicity has led to the expansion of research on superparamagnetic iron oxide nanoparticles as contrast agents thanks to the dipolar coupling of their large magnetic moments to the water protons' moments. (MNPs have magnetizations that are 10^8 times larger and relaxations 10^4 times faster than those of water protons.) T_1 relaxation comes from the fluctuating electron spin moment of the contrast agents coupling with neighboring water protons' moments, which quickly relax back to their equilibrium magnetization, resulting in an increased $1/T_1$ rate MRI signal (positive contrast) and reduce the acquisition times [110, 111]. On the other hand, T_2 relaxation comes from long-range magnetic dipolar fields, which locally shift the proton resonance frequency and dephase the proton spin precession. This leads to a faster T_2 relaxation and a reduced MRI signal (negative contrast) so that the affected regions appear darker [112]. The effect of particles' size and microstructure on the magnetic relaxivity are crucial [113]. The relaxation rate $1/T_2$ increases quadratically with the particle's size (the outer-sphere spins dominate it), until it reaches a limit value. On the other hand, ultrasmall superparamagnetic iron oxide particles have higher longitudinal relaxivity, leading to improved T_1-weighted images.

While toxicity concerns hamper the in vivo application of many nanoparticle-based compounds, iron oxide MNPs of several formulations have been approved by the US Food and Drug Administration and the European Medicines Agency as MRI contrast agents. The particle coating and thickness are crucial as they have a strong influence on MRI performance, pharmacokinetics, and biodistribution [114–116].

The emerging magnetic particle imaging (MPI) technique entirely relies on the MNPs used as tracers for image formation. The magnetic tracers are detected by Faraday induction with a pickup coil. The exciting field is its most peculiar feature. It consists of a time-varying magnetic field gradient (selection field) carefully designed to have a field-free point (FFP). The selection field is strong enough to align the MNPs except around the FFP. The FFP is rapidly moved in the image volume so that particles are misaligned around it, while the magnetization changes are picked up by the detecting coil. The signal is proportional to the number of particles and related to the location of the FFP, yielding a map of the tracer accumulation. This map is frequently superimposed on anatomical images such as those obtained by MRI, which provides accurate information about the tracer's distribution in the patient's body. In MPI there is no background noise coming from biological tissues, which enables high contrast. Its high spatial resolution (about 1 mm) combines with low acquisition times (on the order of or below 1 s) to allow real-time in vivo imaging [117, 118]. These characteristics make it an exciting tool for different clinical applications such as cardiovascular, pulmonary or gastrointestinal imaging, cancer diagnosis, brain injury detection, and in vivo tracking of stem cells [119, 120].

As in other in vivo applications, iron oxide nanoparticles stand out for MPI tracers. Sensitivity depends on the magnetic moment, which can be increased by increasing the particle size (staying below the superparamagnetic limit) and by agglomerating or encapsulating small particles. A large initial magnetic permeability makes differences in magnetic field amplitude translate into large variations of the magnetic moment in the measurement direction, resulting in a better spatial resolution. Therefore, the optimal size depends on the operating frequency [121]. Gleich and Weizenecker did the first experiments on MPI in 2005 [117]. They recommended a magnetic core size of 30 nm to work at 25 kHz with field amplitudes of 10–20 mT. Since then, the commercial contrast agent Resovist® (Schering AG, Berlin) [122] is considered the gold standard for MPI tracers. Although most MNPs in Resovist® are very small, they form aggregates with average sizes of 24 nm that behave like monodomains. A variety of aggregation strategies are being investigated to increase the number of magnetic nuclei at a spot of interest while providing biocompatibility and functionality: micelles, liposomes, and bacterial magnetosomes (Ref. [121] and references therein.)

8 Magnetic Hyperthermia, Magneto-mechanical Disruption, Drug Delivery, and Tissue Regeneration

MNPs can enable new tools for biomedical therapy, standing alone or in combination with plasmonic particles. Moreover, they are used for theranostics, a term derived from the contraction of "therapy" and "diagnostics."

Localized therapy that avoids collateral damage is possible thanks to the small size of the MNPs and the possibility of remotely guiding them and targeting a specific location of interest. Whether thermal, mechanical, or chemical, the therapy can be focused, thereby reducing the required dosage. The most significant innovations are in magnetic hyperthermia, magneto-mechanical tumor destruction, and drug transport and delivery.

Heating cancer cells to 40–46 °C can destroy them or, at least, make them more susceptible to chemo- and radiotherapy. For localized tumors, local heating can be more efficient and safer than heating extensive body areas. This is the motivation for magnetic hyperthermia (MH) development, firstly proposed by Gilchrist et al. in 1957 [123]. This technique uses the MNPs as mediators that transform the energy provided by an alternating magnetic field into heat delivered to the tumor cell. This transformation mechanism is the magnetization rotation, which can take place by Brownian or Néel relaxation. Inside the biological tissue, the MNP are practically immobilized, so Néel relaxation is the dominant mechanism. The delivered energy per cycle for a volume of particles V_{MNP} can be calculated as the hysteresis loop area: $E = \mu_0 V_{\mathrm{MNP}} \oint M dH$, where μ_o is the permeability of free space. Of course, E depends on the frequency and amplitude of the driving field H. Even for particles with minimal hysteresis in DC mode, the magnetization loop area increases significantly for frequencies in the order of several kilohertz because of the time lag between the induced magnetization and the applied field. This makes magnetic heating possible even for moderate frequencies. At higher frequencies, the particles behave as blocked and therefore do not respond to the magnetic field or have hysteresis. This is relevant because high frequencies and amplitudes are not safe for healthy organs (they can affect nerve synapses which produce undesired muscle stimulation and heart malfunctioning). Some studies suggest that the product of frequency and amplitude should not exceed 4.85×10^8 A m^{-1} s^{-1} [124].

The MNPs' heating capacity is characterized by the specific absorption rate (SAR, also known as specific loss power), defined as their heating power per unit mass:

$$\mathrm{SAR} = \frac{P}{m} = \frac{C}{m}\frac{dT}{dt} \tag{4}$$

where C is the specific heat capacity of the colloidal suspension, m is the mass of the magnetic core, and dT/dt is the rate of temperature change. Obtaining the SAR from magnetometry, instead of calorimetry, makes it easier to maintain adiabatic conditions and obtain its dependence on temperature [125]. In this case,

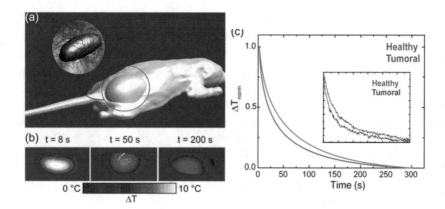

Fig. 6 (**a**) Computable mouse virtual phantom used for in silico testing. (**b**) Temperature maps of a melanoma corresponding to different times after starting thermal relaxation. (**c**) Time evolution of temperature extracted from the analysis of the thermal maps in (**b**) for an advanced-stage tumor and a healthy tissue. Copyright 2018 Wiley; used with permission from H. D. A. Santos et al. [130]

SAR can be calculated from the hysteresis loop area, the frequency f, and the particle concentration c as:

$$\text{SAR} = \frac{f}{c}\mu_0 \oint M\mathrm{d}H \qquad (5)$$

In general, the maximum SAR is achieved when the frequency of the applied field equals the inverse of the relaxation time of the MNPs, but it also depends on the field amplitude H_0 (in the linear response regime, the area of the magnetization curve is directly proportional to H_0^2). To facilitate comparison of the results of different laboratories, the heating capacity is identified by the intrinsic loss power ILP, defined as $\text{ILP} = \text{SAR}/(f H_0^2)$, with the advantage of it being independent of f and H_0 for moderate values.

For details on MH instrumentation, one can refer to Ref. [126].

The MNPs' heating power depends on extrinsic parameters, such as the exposure time, frequency, and amplitude of the applied field, and on intrinsic ones, such as their size, saturation magnetization, and anisotropy constant, which are interrelated [127]. If we compare, for example, magnetite and cobalt ferrite nanoparticles, we see that the latter have larger magnetic anisotropy and, consequently, wider hysteresis loops but at the same time, will require larger field amplitudes. Additionally, the biological environment in which the MNPs are located, their agglomeration, and distribution have an enormous influence on their mobility and, hence, on the type of magnetization relaxation permitted [128]. *Magforce*® developed a human-sized MH device working at 100 kHz and 0–18 kA/m, which obtained European Union regulatory permission to treat patients with brain tumors [129].

In silico testing is an emerging research line in the field of MH (Fig. 6) to predict heating efficiency. Modeling and simulating the temperature distribution, mass concentration, and heat transfer can significantly reduce the number of in vivo trials on animals and humans. Numerical computations must include the relevant parameters of the MNPs and the intracellular environment (i.e., the tumor cell). This technique's long-term goal is to guide clinicians in adjusting dosage and other variables to the tumor's evolution.

Magneto-mechanical actuation has appeared recently as a new application of alternating magnetic fields in cancer cell destruction (some studies use rotating or precessing magnetic fields). In this case, the alternating magnetic field's desired action is the physical rotation or oscillation of the particles, which induces the cancer cells' death. The cell damage is provoked mainly by cell membrane disruption and the trigger of programed cell death (apoptosis).

Magneto-mechanical action is mediated by anisometric particles that rotate or vibrate under rotating or alternating magnetic fields of moderate amplitudes and frequencies on the order of millitesla and tens of hertz. Magnetic microdisks that possess a spin-vortex configuration ground state are good candidates for this application, as they do not have remanent magnetization, which avoids their undesired agglomeration. When a field is applied, the magnetization vortex shifts, making the particle oscillate [131, 132]. The advantages of this technique over hyperthermia are that it avoids collateral thermal damage, and the power source requirements are much lower. On the other hand, their current microsize can be a problem for in vivo applications as they can be cleared by the immune system or do not cross biological barriers. Smaller anisometric particles [133, 134] such as nanodisks, nanorectangles, or nanorods can produce significant mechanical action if stimulated with stronger magnetic fields, easily achievable because the frequency required is low. However, one of the challenges is upscaling their fabrication.

Drug delivery is another biomedical challenge that can profit from MNPs. It is well recognized that large systemic drug doses may have adverse effects on healthy organs. Magnetic nanocarriers can significantly reduce them by local delivery of therapeutic doses of the drug, which can be sustained in time and have deeper penetration (e.g., inside of tumors).

In 1908 Paul Ehrlich received the Nobel Prize in Medicine for his advances in immunology. He popularized the concept of "magic bullet" to refer to an ideal therapeutic agent that would act locally and specifically against a pathogen without relevant undesired side effects. This is the goal pursued by investigations on MNPs as drug or gene carriers. It involves three aspects: the cargo, the transport, and the release. The drug's linkage can be obtained by covalent bonding, entrapment, adsorption, or microencapsulation in biocompatible polymers or vesicles [135]. The latter may facilitate the drug administration, favoring dosage control, and patients' compliance.

Once attached, the MNP-drug complex is injected in or close to the target and can be guided by magnetic field gradients. Targeting of the diseased site can be active or passive. Active targeting points to specific receptors that only tumors express or overexpress on their surface. The nanocarrier must then be accordingly functionalized against such a receptor. On the other hand, the passive targeting strategy relies on the selective accumulation of the MNPs at the tumor, which has

Fig. 7 TEM image of a MTB
with magnetosomes aligned
along its body (image
courtesy of M. Luisa
Fernández-Gubieda,
University of Basque
Country, Spain)

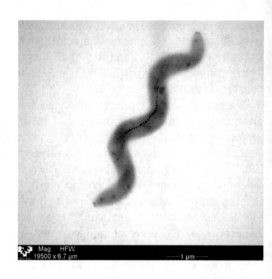

limited lymphatic drainage and is particularly permeable compared to healthy cells. Once at the target, the MNPs can act at the tissue or be internalized by the cell, either by endocytosis or phagocytosis. The drug released may be triggered upon environmental changes or external stimuli, such as the temperature rise produced by magnetic hyperthermia (with a thermosensitive porous coating on the magnetic nanocarriers).

One of the most challenging drug delivery goals is crossing the blood-brain barrier (BBB) to reach the central nervous system. Besides an adequate coating or encapsulation in liposomes, MNPs can be forced across the BBB by a magnetic force. Some studies propose the help of magnetic drilling enabled by helical dynamic gradients and multilayered Au-Fe nanorods [136].

Some researches study magnetotactic bacteria MTB (see Fig. 7) as "smart" drug or gene carriers [137]. MTB are frequently found in freshwater reservoirs. They contain magnetosomes, which are highly crystalline magnetite nanoparticles (with sizes of 30–120 nm) surrounded by a phospholipid layer. MTB usually have around 15–20 magnetosomes aligned along their body, which act as a natural compass. They are natural swimmers that follow the magnetic field lines propelled by their flagella. Besides the targeting enabled by a magnetic gradient, MTB can use other natural properties. For example, aerotactic MTB have an affinity for oxygen-depleted hypoxic regions, which are the most active in tumors [138]. MTB can bear a high cargo loading, such as nanoliposomes attached to the bacteria or gold nanoparticles that favor endocytosis [139]. To release the therapeutic agent, MTB are subjected to an alternating magnetic field to produce hyperthermia [140], which eventually dissociates the bacteria.

Tissue regeneration refers to procedures that combine scaffolds, cells, and therapeutic agents to replace or regenerate tissues to undertake normal functions. Success at regeneration requires the precise delivery of transplanted cells and growth factors and their retention and controlled release. Some keys for using MNPs in tissue regeneration have been already reviewed in this chapter: biocompatibility, magnetic targeting, surface modification with targeting moieties, magnetic heating, and ability to load therapeutic compounds.

MNPs can target cells, drugs, and nucleic acid and control their retention in diseased tissue until differentiation occurs. They are sometimes combined with magnetic scaffolds fabricated by dip-coating a conventional synthetic bone or vascular scaffold in a MNP solution. Such a scaffold can be magnetized by a magnetic field external to the body to attract the MNPs loaded with drugs. Alternatively, a porous magnetic scaffold can be loaded with MNPs. After transplantation, the particles' biological load is triggered and controlled by magnetic hyperthermia [141].

Magnetic stents can open vascular conduits and create magnetic gradients to capture MNPs. This procedure can avoid re-obstructions and reduce the drug dose [142]. Additionally, an external field can change the stent's shape and volume to deliver therapeutic agents on demand.

Some types of regeneration involve introducing genes into cells, which is known as transfection. "Magnetofection" has been developed to enhance its efficiency. It consists of magnetically guided and forced delivery of nucleic acids by MNPs.

MNPs can promote cell proliferation thanks to their coatings' biochemical properties. (For example, calcium phosphate coating significantly improves bone tissue osteoblast density [143].)

Vascularization is an essential requirement for the success of implants. Remote magneto-mechanical action based on MNPs can significantly enhance the formation of new vascularization, as this process is extremely sensitive to mechanical stimulation. More details and references on MNP applied to tissue regeneration can be found in Ref. [144].

9   Conclusion and Prospects

The growing evidence of the usefulness of MNPs for biomedical applications stimulates researchers to persist in their investigations. The worldwide 2020 coronavirus outbreak exemplifies the need for solutions to emerging biomedical and biosafety problems, like affordable rapid diagnostic tests. In this context, magnetism and, specifically, MNPs have unique properties that make them very interesting alone or combined with other materials and procedures (e.g., optical, chemical, or immunological) whose development and progress are also growing fast.

Thanks to the scientific community's intense activity in multidisciplinary research involving MNPs, one can envisage new effective and safe diagnostics and treatment procedures. Novel routes of fabrication give rise to particles with improved quality and varied morphologies, such as hollow MNPs, nanorods, or molecular imprinted MNPs. Hybrid nanostructures, combining different materials, hold promise for multiple functionalities such as simultaneous therapy and imaging. In this exciting scenario, much interest focuses on the standardization of terminology, synthesis routes (to ensure the batch-to-batch reproducibility), and characterization and metrology techniques (to enable reliable quality control) [173–175]. In 2019, ISO/TS 19807–1:2019 established the first standard on magnetic nanomaterials, specifying the magnetic nanosuspensions' characteristics to be measured and listing measurement methods for them [145].

After two decades of creative ideas on how MNP can improve health, food, and environment, there is still plenty of room for interdisciplinary laboratories to mimic the in vivo conditions for MNP applications. There are also opportunities for mathematical modeling and simulation to explain and even predict magnetic nanostructures' behavior in biological media and their effects on them.

A few technologies enabled by MNP already have been translated to the health care industry (MRI contrast agents), and some of them are currently undergoing human clinical trials (drug delivery and magnetic hyperthermia). Nevertheless, there is still some work to do for many promising inventions to be attractive enough to be adopted. One of the biggest challenges is the technology transfer to the medical industry and clinical medicine. Issues related to safety, protein corona (the layer of proteins that rapidly forms around MNPs and largely influence their performance), the noneffectiveness of magnetic gradients in deep body regions, and matters related to manufacturing and commercialization (large-scale fabrication, sustainability, packaging and storage, and standardization) are stimuli for future research and innovation on the bio-application of magnetic nanoparticles.

Acknowledgments The authors sincerely thank R. B. Goldfarb, J. A. Blanco, and P. Gorria for fruitful discussions and valuable suggestions. The authors acknowledge the financial support of the Spanish Ministry of Economy and Competitiveness with grants MAT2017-84959-C2-1-R and RTI2018-094683-B-C52, the Principality of Asturias with project IDI/2018/000185 and the University of Oviedo with PAPI-18-EMERG-8/2018/00061/008. M. Salvador thanks the Council of Gijon/IUTA, University of Oviedo, and the Principality of Asturias for their pre-doctoral grants (SV-20-GIJON-1-22 and PA-20-PF-BP19-141), the Spanish Ministry of Education, Culture and Sport and Banco Santander for grant CEI15-24.

References

1. M. Knobel et al., Superparamagnetism and other magnetic features in granular materials: a review on ideal and real systems. J. Nanosci. Nanotechnol. **8**(6), 2836–2857 (2008)
2. C. Vasilescu et al., High concentration aqueous magnetic fluids: structure, colloidal stability, magnetic and flow properties. Soft Matter **14**(32), 6648–6666 (2018)
3. N. Hoshyar et al., The effect of nanoparticle size on in vivo pharmacokinetics and cellular interaction. Nanomedicine **11**(6), 673–692 (2016)
4. A. Jurj et al., The new era of nanotechnology, an alternative to change cancer treatment. Drug Des. Devel. Ther. **11**, 2871–2890 (2017)
5. Y. Geng et al., Shape effects of filaments versus spherical particles in flow and drug delivery. Nat. Nanotechnol. **2**(4), 249–255 (2007)
6. A.B. Jindal, The effect of particle shape on cellular interaction and drug delivery applications of micro- and nanoparticles. Int. J. Pharm. **532**(1), 450–465 (2017)
7. G. Sharma et al., Polymer particle shape independently influences binding and internalization by macrophages. J. Control. Release **147**(3), 408–412 (2010)

8. A.L. Cortajarena et al., Engineering iron oxide nanoparticles for clinical settings. Nano **1**, 2 (2014)
9. J. Ahmad et al., Differential cytotoxicity of copper ferrite nanoparticles in different human cells. J. Appl. Toxicol. **36**(10), 1284–1293 (2016)
10. A. Hanini et al., Nanotoxicological study of polyol-made cobalt-zinc ferrite nanoparticles in rabbit. Environ. Toxicol. Pharmacol. **45**, 321–327 (2016)
11. S.Y. Srinivasan et al., Applications of cobalt ferrite nanoparticles in biomedical nanotechnology. Nanomedicine **13**(10), 1221–1238 (2018)
12. Q. Feng et al., Uptake, distribution, clearance, and toxicity of iron oxide nanoparticles with different sizes and coatings. Sci. Rep. **8**(1), 2082 (2018)
13. D. Bobo et al., Nanoparticle-based medicines: a review of FDA-approved materials and clinical trials to date. Pharm. Res. **33**(10), 2373–2387 (2016)
14. U. Engelmann et al., Agglomeration of magnetic nanoparticles and its effects on magnetic hyperthermia, in *Current Directions in Biomedical Engineering*, (2017), p. 457
15. B. Halamoda-Kenzaoui et al., The agglomeration state of nanoparticles can influence the mechanism of their cellular internalisation. J. Nanobiotechnol. **15**(1), 48–48 (2017)
16. Y. Wang et al., Study of superparamagnetic nanoparticles as labels in the quantitative lateral flow immunoassay. Mater. Sci. Eng. C **29**(3), 714–718 (2009)
17. M. Salvador et al., Improved magnetic lateral flow assays with optimized nanotags for point-of-use inductive biosensing. Analyst **145**(17), 5905–5914 (2020)
18. E. Amstad, M. Textor, E. Reimhult, Stabilization and functionalization of iron oxide nanoparticles for biomedical applications. Nanoscale **3**(7), 2819–2843 (2011)
19. R.A. Bohara, N.D. Thorat, S.H. Pawar, Role of functionalization: strategies to explore potential nano-bio applications of magnetic nanoparticles. RSC Adv. **6**(50), 43989–44012 (2016)
20. R. Mout et al., Surface functionalization of nanoparticles for nanomedicine. Chem. Soc. Rev. **41**(7), 2539–2544 (2012)
21. A. Moyano et al., Carbon-coated superparamagnetic nanoflowers for biosensors based on lateral flow immunoassays. Biosensors **10**(8), 80 (2020)
22. M. Vasilakaki et al., Effect of albumin mediated clustering on the magnetic behavior of $MnFe_2O_4$ nanoparticles: experimental and theoretical modeling study. Nanotechnology **31**(2), 025707 (2019)
23. J. Salafranca et al., Surfactant organic molecules restore magnetism in metal-oxide nanoparticle surfaces. Nano Lett. **12**(5), 2499–2503 (2012)
24. R. Massart, Preparation of aqueous magnetic liquids in alkaline and acidic media. IEEE Trans. Magn. **17**(2), 1247–1248 (1981)
25. V.K. LaMer, R.H. Dinegar, Theory, production and mechanism of formation of monodispersed hydrosols. J. Am. Chem. Soc. **72**(11), 4847–4854 (1950)
26. G. Cotin et al., Unravelling the thermal decomposition parameters for the synthesis of anisotropic iron oxide nanoparticles. Nano **8**(11), 881 (2018)
27. M. Unni et al., Thermal decomposition synthesis of iron oxide nanoparticles with diminished magnetic dead layer by controlled addition of oxygen. ACS Nano **11**(2), 2284–2303 (2017)
28. M.E. Brollo Fortes et al., Key parameters on the microwave assisted synthesis of magnetic nanoparticles for MRI contrast agents. Contrast Media Mol. Imaging **2017**, 13 (2017)
29. L. Gonzalez-Moragas et al., Scale-up synthesis of iron oxide nanoparticles by microwave-assisted thermal decomposition. Chem. Eng. J. **281**, 87–95 (2015)
30. A. Scano et al., Microemulsions: the renaissance of ferrite nanoparticle synthesis. J. Nanosci. Nanotechnol. **19**(8), 4824–4838 (2019)
31. S.S. Staniland et al., Chapter 3: Novel methods for the synthesis of magnetic nanoparticles, in *Frontiers of Nanoscience*, ed. by C. Binns, (Elsevier, 2014), pp. 85–128
32. M. Salvador et al., Synthesis of superparamagnetic iron oxide nanoparticles: SWOT analysis towards their conjugation to biomolecules for molecular recognition applications. J. Nanosci. Nanotechnol. **19**(8), 4839–4856 (2019)

33. A.G. Roca et al., Progress in the preparation of magnetic nanoparticles for applications in biomedicine. J. Phys. D: Appl.Phys. **42**(22), 224002 (2009)

34. S. Bogren et al., Classification of magnetic nanoparticle systems—synthesis, standardization and analysis methods in the NanoMag Project. Int. J. Mol. Sci. **16**(9), 20308–20325 (2015)

35. L. Chang et al., Facile one-pot synthesis of magnetic Prussian blue core/shell nanoparticles for radioactive cesium removal. RSC Adv. **6**(98), 96223–96228 (2016)

36. S. Upadhyay, K. Parekh, B. Pandey, Influence of crystallite size on the magnetic properties of Fe$_3$O$_4$ nanoparticles. J. Alloys Compd. **678**, 478–485 (2016)

37. B. Michen et al., TEM sample preparation of nanoparticles in suspensions—understanding the formation of drying artefacts. Imaging Microsc. (2014)

38. S. Akhtar, F.A. Khan, A. Buhaimed, Functionalized magnetic nanoparticles attenuate cancer cells proliferation: Transmission electron microscopy analysis. Microsc. Res. Tech. **82**(7), 983–992 (2019)

39. A. Rao et al., Characterization of nanoparticles using atomic force microscopy. J. Phys. Conf. Ser. **61**, 971–976 (2007)

40. A. Krivcov, T. Junkers, H. Möbius, Understanding electrostatic and magnetic forces in magnetic force microscopy: towards single superparamagnetic nanoparticle resolution. J. Phys. Commun. **2**(7), 075019 (2018)

41. C. Moya et al., Superparamagnetic versus blocked states in aggregates of Fe3−xO4 nanoparticles studied by MFM. Nanoscale **7**(42), 17764–17770 (2015)

42. C. Bantz et al., The surface properties of nanoparticles determine the agglomeration state and the size of the particles under physiological conditions. Beilstein J. Nanotechnol. **5**, 1774–1786 (2014)

43. K. Rausch et al., Evaluation of nanoparticle aggregation in human blood serum. Biomacromolecules **11**(11), 2836–2839 (2010)

44. V. Filipe, A. Hawe, W. Jiskoot, Critical evaluation of Nanoparticle Tracking Analysis (NTA) by NanoSight for the measurement of nanoparticles and protein aggregates. Pharm. Res. **27**(5), 796–810 (2010)

45. P. Eaton et al., A direct comparison of experimental methods to measure dimensions of synthetic nanoparticles. Ultramicroscopy **182**, 179–190 (2017)

46. A. Méndez-Vilas, J. Díaz, *Microscopy: Science, Technology, Applications and Education* (Formatex Research Center, Badajoz, 2010)

47. T. Chatterji, *Neutron Scattering from Magnetic Materials* (Elsevier Science, Boston, 2006)

48. M.P. Fernández-García et al., Microstructure and magnetism of nanoparticles with γ-Fe core surrounded by α-Fe and iron oxide shells. Phys. Rev. B **81**(9), 094418 (2010)

49. M.P. Fernández-García et al., Enhanced protection of carbon-encapsulated magnetic nickel nanoparticles through a sucrose-based synthetic strategy. J. Phys. Chem. C **115**(13), 5294–5300 (2011)

50. J. Fock et al., On the 'centre of gravity' method for measuring the composition of magnetite/maghemite mixtures, or the stoichiometry of magnetite-maghemite solid solutions, via57Fe Mössbauer spectroscopy. J. Phys. D: Appl. Phys. **50**(26), 265005 (2017)

51. A.V. Delgado et al., Measurement and interpretation of electrokinetic phenomena. J. Colloid Interface Sci. **309**(2), 194–224 (2007)

52. D.L. Pavia et al., *Introduction to Spectroscopy* (Cengage Learning, London, 2008)

53. O. Bixner et al., Complete exchange of the hydrophobic dispersant shell on monodisperse superparamagnetic iron oxide nanoparticles. Langmuir **31**(33), 9198–9204 (2015)

54. R.L. McCreery, *Raman Spectroscopy for Chemical Analysis*, vol 225 (Wiley, New York, 2005)

55. L. Slavov et al., Raman spectroscopy investigation of magnetite nanoparticles in ferrofluids. J. Magn. Magn. Mater. **322**(14), 1904–1911 (2010)

56. J.C. Rubim et al., Raman spectroscopy as a powerful technique in the characterization of ferrofluids. Braz. J. Phys. **31**, 402–408 (2001)

57. H. Konno, Chapter 8: X-ray photoelectron spectroscopy, in *Materials Science and Engineering of Carbon*, ed. by M. Inagaki, F. Kang, (Butterworth-Heinemann, 2016), pp. 153–171

58. T. Radu et al., X-Ray photoelectron spectroscopic characterization of iron oxide nanoparticles. Appl. Surf. Sci. **405**, 337–343 (2017)
59. A.G. Shard, A straightforward method for interpreting XPS data from core–shell nanoparticles. J. Phys. Chem. C **116**(31), 16806–16813 (2012)
60. D. Farhanian, G. De Crescenzo, J.R. Tavares, Large-scale encapsulation of magnetic iron oxide nanoparticles via syngas photo-initiated chemical vapor deposition. Sci. Rep. **8**(1), 12223 (2018)
61. P. Gabbott, *Principles and Applications of Thermal Analysis* (Wiley, Singapore, 2008)
62. E. Mansfield et al., Determination of nanoparticle surface coatings and nanoparticle purity using microscale thermogravimetric analysis. Anal. Chem. **86**(3), 1478–1484 (2014)
63. D. Pröfrock, A. Prange, Inductively Coupled Plasma–Mass Spectrometry (ICP-MS) for quantitative analysis in environmental and life sciences: a review of challenges, solutions, and trends. Appl. Spectrosc. **66**(8), 843–868 (2012)
64. R. Costo et al., Improving the reliability of the iron concentration quantification for iron oxide nanoparticle suspensions: a two-institutions study. Anal. Bioanal. Chem. **411**(9), 1895–1903 (2019)
65. C. Fornaguera, C. Solans, Characterization of polymeric nanoparticle dispersions for biomedical applications: size, surface charge and stability. Pharmaceut. Nanotechnol. **6**(3), 147–164 (2018)
66. S. Foner, Vibrating sample magnetometer. Rev. Sci. Instrum. **27**(7), 548–548 (1956)
67. P.J. Flanders, A vertical force alternating-gradient magnetometer. Rev. Sci. Instrum. **61**(2), 839–847 (1990)
68. R.B. Goldfarb, M. Lelental, C. Thompson, Alternating-field susceptometry and magnetic susceptibility of superconductors, in *Magnetic Susceptibility of Superconductors and Other Spin Systems*, ed. by R. A. Hein, T. L. Francavilla, D. H. Liebenberg, (Springer, New York, 1991), pp. 49–80
69. M.F. Hansen, S. Mørup, Estimation of blocking temperatures from ZFC/FC curves. J. Magn. Magn. Mater. **203**(1), 214–216 (1999)
70. D. Peddis et al., Chapter 4: Magnetic interactions: a tool to modify the magnetic properties of materials based on nanoparticles, in *Frontiers of Nanoscience*, ed. by C. Binns, (Elsevier, 2014), pp. 129–188
71. K.L. Livesey et al., Beyond the blocking model to fit nanoparticle ZFC/FC magnetisation curves. Sci. Rep. **8**(1), 11166 (2018)
72. O. Petracic, Superparamagnetic nanoparticle ensembles. Superlattice. Microst. **47**(5), 569–578 (2010)
73. J.A. De Toro et al., Controlled close-packing of ferrimagnetic nanoparticles: an assessment of the role of interparticle superexchange versus dipolar interactions. J. Phys. Chem. C **117**(19), 10213–10219 (2013)
74. E.P. Wohlfarth, Relations between different modes of acquisition of the remanent magnetization of ferromagnetic particles. J. Appl. Phys. **29**(3), 595–596 (1958)
75. J.L. Dormann, D. Fiorani, E. Tronc, Magnetic relaxation in fine-particle systems, in *Advances in Chemical Physics*, (Wiley, New York, 1997), pp. 283–494
76. E. De Biasi, J. Curiale, R.D. Zysler, Quantitative study of FORC diagrams in thermally corrected Stoner– Wohlfarth nanoparticles systems. J. Magn. Magn. Mater. **419**, 580–587 (2016)
77. F. Groß et al., gFORC: A graphics processing unit accelerated first-order reversal-curve calculator. J. Appl. Phys. **126**(16)), 163901 (2019)
78. C.R. Pike, A.P. Roberts, K.L. Verosub, First order reversal curve diagrams and thermal relaxation effects in magnetic particles. Geophys. J. Int. **145**(3), 721–730 (2001)
79. A.C. Bohorquez, C. Rinaldi, In situ evaluation of nanoparticle–protein interactions by dynamic magnetic susceptibility measurements. Part. Part. Syst. Charact. **31**(5), 561–570 (2014)
80. V.L.C.-D. Castillo, C. Rinaldi, Effect of sample concentration on the determination of the anisotropy constant of magnetic nanoparticles. IEEE Trans. Magn. **46**(3), 852–859 (2010)

81. M.A. Garcia et al., Sources of experimental errors in the observation of nanoscale magnetism. J. Appl. Phys. **105**(1), 013925 (2009)
82. S.E. McNeil, *Characterization of Nanoparticles Intended for Drug Delivery*, vol 697 (Springer, Humana Press, Totowa, NJ, 2011)
83. J.W. Wills et al., Characterizing nanoparticles in biological matrices: tipping points in agglomeration state and cellular delivery in vitro. ACS Nano **11**(12), 11986–12000 (2017)
84. D. Baragaño et al., Magnetite nanoparticles for the remediation of soils co-contaminated with As and PAHs. Chem. Eng. J. **399**, 125809 (2020)
85. O.V. Kharissova, H.V.R. Dias, B.I. Kharisov, Magnetic adsorbents based on micro- and nano-structured materials. RSC Adv. **5**(9), 6695–6719 (2015)
86. D. Lago-Cachón et al., HeLa cells separation using MICA antibody conjugated to magnetite nanoparticles. Phys. Status Solidi C **11**(5–6), 1043–1047 (2014)
87. S.P. Schwaminger et al., Magnetic separation in bioprocessing beyond the analytical scale: from biotechnology to the food industry. Front Bioeng Biotechnol **7**, 233 (2019)
88. N.J. Darton, A. Ionescu, Sensing magnetic nanoparticles, in *Magnetic Nanoparticles in Biosensing and Medicine*, ed. by A. Ionescu, J. Llandro, N. J. Darton, (Cambridge University Press, Cambridge, 2019), pp. 172–227
89. Y.T. Chen et al., Biosensing using magnetic particle detection techniques. Sensors (Basel) **17**(10), 2300 (2017)
90. G. Lin, D. Makarov, O.G. Schmidt, Magnetic sensing platform technologies for biomedical applications. Lab Chip **17**(11), 1884–1912 (2017)
91. D. Budker, M. Romalis, Optical magnetometry. Nat. Phys. **3**(4), 227–234 (2007)
92. D. Alcantara et al., Iron oxide nanoparticles as magnetic relaxation switching (MRSw) sensors: current applications in nanomedicine. Nanomed.: Nanotechnol. Biol. Med. **12**(5), 1253–1262 (2016)
93. S.X. Wang, G. Li, Advances in giant magnetoresistance biosensors with magnetic nanoparticle tags: review and outlook. IEEE Trans. Magn. **44**(7), 1687–1702 (2008)
94. D. Su et al., Advances in magnetoresistive biosensors. Micromachines **11**(1), 34 (2020)
95. B. Srinivasan et al., A three-layer competition-based giant magnetoresistive assay for direct quantification of Endoglin from human urine. Anal. Chem. **83**(8), 2996–3002 (2011)
96. A. Fu et al., Protein-functionalized synthetic antiferromagnetic nanoparticles for biomolecule detection and magnetic manipulation. Angew. Chem. Int. Ed. **48**(9), 1620–1624 (2009)
97. A. Cousins et al., Novel handheld magnetometer probe based on magnetic tunnelling junction sensors for intraoperative sentinel lymph node identification. Sci. Rep. **5**, 10842 (2015)
98. K. Togawa et al., Detection of magnetically labeled DNA using pseudomorphic AlGaAs/InGaAs/GaAs heterostructure micro-Hall biosensors. J. Appl. Phys. **99**(8), 08P103 (2006)
99. D. Issadore et al., Ultrasensitive clinical enumeration of rare cells ex vivo using a micro-hall detector. Sci Transl Med **4**(141), 141ra92 (2012)
100. B. Cao et al., Development of magnetic sensor technologies for point-of-care testing: fundamentals, methodologies and applications. Sensors Actuat. A: Phys. **312**, 112130 (2020)
101. J.A. Goode, J.V.H. Rushworth, P.A. Millner, Biosensor regeneration: a review of common techniques and outcomes. Langmuir **31**(23), 6267–6276 (2015)
102. K.M. Koczula, A. Gallotta, Lateral flow assays. Essays Biochem. **60**(1), 111–120 (2016)
103. A. Moyano et al., Magnetic immunochromatographic test for histamine detection in wine. Anal. Bioanal. Chem. **411**(25), 6615–6624 (2019)
104. C. Marquina et al., GMR sensors and magnetic nanoparticles for immuno-chromatographic assays. J. Magn. Magn. Mater. **324**(21), 3495–3498 (2012)
105. A. Moyano et al., Magnetic lateral flow immunoassays. Diagnostics **10**(5), 288 (2020)
106. A. Elyacoubi et al., Development of an amperometric enzymatic biosensor based on gold modified magnetic nanoporous microparticles. Electroanalysis **18**(4), 345–350 (2006)
107. M.-H. Liao, J.-C. Guo, W.-C. Chen, A disposable amperometric ethanol biosensor based on screen-printed carbon electrodes mediated with ferricyanide-magnetic nanoparticle mixture. J. Magn. Magn. Mater. **304**(1), e421–e423 (2006)

108. J.-M. Nam, C.S. Thaxton, C.A. Mirkin, Nanoparticle-based bio-bar codes for the ultrasensitive detection of proteins. Science **301**(5641), 1884–1886 (2003)
109. S.I. Stoeva et al., Multiplexed DNA detection with biobarcoded nanoparticle probes. Angew. Chem. Int. Ed. **45**(20), 3303–3306 (2006)
110. B.H. Kim et al., Large-scale synthesis of uniform and extremely small-sized iron oxide nanoparticles for high-resolution T1 magnetic resonance imaging contrast agents. J. Am. Chem. Soc. **133**(32), 12624–12631 (2011)
111. L. Wang et al., Exerting enhanced permeability and retention effect driven delivery by ultrafine iron oxide nanoparticles with T1–T2 switchable magnetic resonance imaging contrast. ACS Nano **11**(5), 4582–4592 (2017)
112. X. Yin et al., Large T1 contrast enhancement using superparamagnetic nanoparticles in ultralow field MRI. Sci. Rep. **8**(1), 11863 (2018)
113. M.P. Morales et al., Contrast agents for MRI based on iron oxide nanoparticles prepared by laser pyrolysis. J. Magn. Magn. Mater. **266**(1), 102–109 (2003)
114. J. Estelrich, M.J. Sánchez-Martín, M.A. Busquets, Nanoparticles in magnetic resonance imaging: from simple to dual contrast agents. Int. J. Nanomed. **10**, 1727–1741 (2015)
115. J.-t. Jang et al., Critical enhancements of MRI contrast and hyperthermic effects by Dopant-controlled magnetic nanoparticles. Angew. Chem. Int. Ed. **48**(7), 1234–1238 (2009)
116. E. Peng, F. Wang, J.M. Xue, Nanostructured magnetic nanocomposites as MRI contrast agents. J. Mater. Chem. B **3**(11), 2241–2276 (2015)
117. B. Gleich, J. Weizenecker, Tomographic imaging using the nonlinear response of magnetic particles. Nature **435**(7046), 1214–1217 (2005)
118. L.M. Bauer et al., Magnetic particle imaging tracers: state-of-the-art and future directions. J. Phys. Chem. Lett. **6**(13), 2509–2517 (2015)
119. S. Zanganeh et al., Chapter 5: Magnetic Particle Imaging (MPI), in *Iron Oxide Nanoparticles for Biomedical Applications*, ed. by M. Mahmoudi, S. Laurent, (Elsevier, Amsterdam, 2018), pp. 115–133
120. E.U. Saritas et al., Magnetic Particle Imaging (MPI) for NMR and MRI researchers. J. Magn. Reson. **229**, 116–126 (2013)
121. N. Panagiotopoulos et al., Magnetic particle imaging: current developments and future directions. Int. J. Nanomed. **10**, 3097–3114 (2015)
122. R. Lawaczeck et al., Magnetic iron oxide particles coated with carboxydextran for parenteral administration and liver contrasting:pre-clinical profile of SH U555A. Acta Radiol. **38**(4), 584–597 (1997)
123. R.K. Gilchrist et al., Selective inductive heating of lymph nodes. Ann. Surg. **146**(4), 596–606 (1957)
124. W.J. Atkinson, I.A. Brezovich, D.P. Chakraborty, Usable frequencies in hyperthermia with thermal seeds. IEEE Trans. Biomed. Eng. **BME-31**(1), 70–75 (1984)
125. E. Garaio et al., Specific absorption rate dependence on temperature in magnetic field hyperthermia measured by dynamic hysteresis losses (ac magnetometry). Nanotechnology **26**(1), 015704 (2014)
126. D. Cabrera et al., Instrumentation for magnetic hyperthermia, in *Nanomaterials for Magnetic and Optical Hyperthermia Applications*, (Elsevier, Amsterdam, 2019), pp. 111–138
127. L. Asín et al., Controlled cell death by magnetic hyperthermia: effects of exposure time, field amplitude, and nanoparticle concentration. Pharm. Res. **29**(5), 1319–1327 (2012)
128. N. Telling, Chapter 7: High-frequency magnetic response and hyperthermia from nanoparticles in cellular environments, in *Nanomaterials for Magnetic and Optical Hyperthermia Applications*, ed. by R. M. Fratila, J. M. De La Fuente, (Elsevier, Amsterdam, 2019), pp. 173–197
129. Magforce®: The nanomedicine company. June 2020. Available from: https://www.magforce.com/en/home/
130. H.D.A. Santos et al., In vivo early tumor detection and diagnosis by infrared luminescence transient nanothermometry. Adv. Funct. Mater. **28**(43), 1803924 (2018)

131. R. Mansell et al., Magnetic particles with perpendicular anisotropy for mechanical cancer cell destruction. Sci. Rep. **7**(1), 4257 (2017)

132. E.A. Vitol, V. Novosad, E.A. Rozhkova, Microfabricated magnetic structures for future medicine: from sensors to cell actuators. Nanomedicine **7**(10), 1611–1624 (2012)

133. H. Chiriac et al., Fe–Cr–Nb–B ferromagnetic particles with shape anisotropy for cancer cell destruction by magneto-mechanical actuation. Sci. Rep. **8**(1), 11538 (2018)

134. M. Goiriena-Goikoetxea et al., High-yield fabrication of 60 nm Permalloy nanodiscs in well-defined magnetic vortex state for biomedical applications. Nanotechnology **27**(17), 175302 (2016)

135. R. Singh, J.W. Lillard, Nanoparticle-based targeted drug delivery. Exp. Mol. Pathol. **86**(3), 215–223 (2009)

136. S. Jafari et al., Magnetic drilling enhances intra-nasal transport of particles into rodent brain. J. Magn. Magn. Mater. **469**, 302–305 (2019)

137. D. Kuzajewska et al., Magnetotactic bacteria and magnetosomes as smart drug delivery systems: a new weapon on the battlefield with cancer? Biology **9**(5), 102 (2020)

138. O. Felfoul et al., Magneto-aerotactic bacteria deliver drug-containing nanoliposomes to tumour hypoxic regions. Nat. Nanotechnol. **11**(11), 941–947 (2016)

139. S.K. Alsaiari et al., Magnetotactic bacterial cages as safe and smart gene delivery vehicles. OpenNano **1**, 36–45 (2016)

140. D. Gandia et al., Unlocking the potential of magnetotactic bacteria as magnetic hyperthermia agents. Small **15**(41), 1902626 (2019)

141. M. Bañobre-López et al., Hyperthermia induced in magnetic scaffolds for bone tissue engineering. IEEE Trans. Magn. **50**(11), 1–7 (2014)

142. T. Kobayashi et al., Augmentation of degenerated human cartilage in vitro using magnetically labeled mesenchymal stem cells and an external magnetic device. Arthrosc.: J. Arthrosc. Rel. Surg. **25**(12), 1435–1441 (2009)

143. R.A. Pareta, E. Taylor, T.J. Webster, Increased osteoblast density in the presence of novel calcium phosphate coated magnetic nanoparticles. Nanotechnology **19**(26), 265101 (2008)

144. Y. Gao et al., Emerging translational research on magnetic nanoparticles for regenerative medicine. Chem. Soc. Rev. **44**(17), 6306–6329 (2015)

145. ISO. ISO/TS 19807–1:2019 Nanotechnologies — Magnetic nanomaterials — Part 1: Specification of characteristics and measurements for magnetic nanosuspensions. June 2020]; Available from: https://www.iso.org/standard/66237.html

Index

© Springer Nature Switzerland AG 2021
V. Franco, B. Dodrill (eds.), *Magnetic Measurement Techniques for Materials
Characterization*, https://doi.org/10.1007/978-3-030-70443-8

Printed in the United States
by Baker & Taylor Publisher Services